HANDBOOK OF PHILOSOPHICAL LOGIC

VOLUME IV

SYNTHESE LIBRARY

STUDIES IN EPISTEMOLOGY,

LOGIC, METHODOLOGY, AND PHILOSOPHY OF SCIENCE

VOLUME 167

HANDBOOK OF PHILOSOPHICAL LOGIC

Volume IV: Topics in the Philosophy of Language

Edited by
D. GABBAY
Department of Computing, Imperial College, London, England

and

F. GUENTHNER
SNS, University of Tübingen, West Germany

D. REIDEL PUBLISHING COMPANY

A MEMBER OF THE KLUWER  ACADEMIC PUBLISHERS GROUP

DORDRECHT / BOSTON / LONDON

Library of Congress Cataloging-in-Publication Data

Topics in the philosophy of language / edited by D. Gabbay and F.
Guenthner.
p. cm. -- (Handbook of philosophical logic ; v. 4) (Synthese
library ; v. 167)
Includes bibliographies and indexes.
ISBN 9027716064 (Netherlands)
1. Language and logic. 2. Languages--Philosophy. I. Gabbay, Dov
M., 1945- . II. Guenthner, Franz. III. Series. IV. Series:
Synthese library ; v. 167.
BC6.H36 1983 vol. 4
[P39]
160 s--dc19
[401] 88-18404
CIP

ISBN 90-277-1542-4 (v. I)
ISBN 90-277-1604-8 (v. II)
ISBN 90-277-1605-6 (v. III)
ISBN 90-277-1606-4 (v. IV)

Published by D. Reidel Publishing Company
P.O. Box 17, 3300 AA Dordrecht, The Netherlands

Sold and distributed in the U.S.A. and Canada
by Kluwer Academic Publishers,
101 Philip Drive, Norwell, MA 02061, U.S.A.

In all other countries, sold and distributed
by Kluwer Academic Publishers Group,
P.O. Box 322, 3300 AH Dordrecht, The Netherlands

Printed in The Netherlands

CONTENTS TO VOLUME IV

ACKNOWLEDGEMENTS

The preparation of the *Handbook of Philosophical Logic* was generously supported by the Lady Davis Fund at Bar-Ilan University, Ramat-Gan, Israel and the Werner-Reimers-Stiftung, Bad Homburg, West Germany, which provided us with the chance of discussing the chapters in the *Handbook* at various workshops with the contributors. It is a great pleasure to acknowledge the assistance of these institutions during the preparation of this collection. We benefitted further from the editorial and personal advice and help from the publisher.

Most important of all, we would like to express our thanks to all the contributors to the *Handbook* for their unlimited goodwill, their professional counsel, as well as their friendly support. They have made the preparation of this collection a stimulating and gratifying enterprise.

D. GABBAY (*Imperial College*)
F. GUENTHNER (*University of Tuebingen*)

PREFACE*

There is hardly any branch of philosophical logic that is not closely connected in one way or the other to issues in the semantics of natural languages. Historically as well, the philosophy of language and philosophical logic have been intertwined from the very beginning. In the last hundred years both mathematical logic and philosophical logic have matured into major disciplines and the ties to issues in the philosophy and the formal semantics of language are – it is fair to say – stronger than ever before.

The present volume of the *Handbook* attempts to survey some of, but by no means all, the central areas of the formal philosophy of language. The emphasis of the overviews in this volume is on the whole not so much on the exposition of particular theoretical approaches in semantics as on the presentation of problem areas that need to be taken into account in any full-fledged general semantic theory of natural language.

The first chapter in Volume IV is a comprehensive description of perhaps the (to date) most successful application of logical techniques to the semantics of natural language, namely, generalized quantifier theory. This very prominent interaction between logical and linguistic semantics is significant for several reasons. First, the formal semantics of natural language was profoundly marked by Richard Montague's work on quantification in the late sixties. Second, even though it is not surprising that it is in the study of quantification that modern logic would have much to contribute to the semantic analysis of natural language, a closer look at what are called 'quantifiers' here has revealed a much richer semantic structure than one would previously have expected on the basis of working in standard predicate logic systems. Third, the detailed study of quantifiers in natural language is of interest to many disciplines and it is likely that it will give rise to new research themes in all of these.

The next three chapters deal with various issues connected to problems of predication. Chapter IV.2 reviews arguments for introducing properties as primitive entities and discusses their role in a theory of predication couched in intensional logic. Chapter IV.3 complements the previous chapter in that it provides formal reconstructions of several (nominalistic, conceptual, realist) theories of predication. Chapter IV.4 centers on an important class of expressions used for predication in connection with quantities: mass expressions. This chapter reviews the most well-known approaches to mass terms and the ontological proposals related to them.

In addition to quantification and predication, matters of reference have constituted the other overriding theme for semantic theories in both philosophical logic and the semantics of natural languages. Chapter IV.5 presents an overview of how the semantics of proper names and descriptions have been dealt with in recent theories of reference. Chapter IV.6 is concerned with the context-dependence of reference, in particular, with the semantics of indexical expressions.

The topic of Chapter IV.7 is related to predication as it surveys some of the central problems of ascribing propositional attitudes to agents. Chapter IV.8 deals with the analysis of the main temporal aspects of natural language utterances. Together these two chapters give a good indication of the intricate complexities that arise once modalities of one or the other sort enter on the semantic stage.

Chapter IV.9 deals with another well-known topic in philosophical logic: presupposition, an issue on the borderline of semantics and pragmatics. The volume closes with an extensive study of the Liar paradox and its many implications for the study of language (as for example, self-reference, truth concepts and truth definitions).

D. M. GABBAY
F. GUENTHNER

NOTE

*Due to circumstances beyond the control of the editors, this volume does not contain all the papers that were originally foreseen. We hope that we will be able to make a step towards a more complete coverage of applications of philosophical logic in the analysis of natural language in the near future.

A NOTE ON NOTATION

Writings in the field of philosophical logic abound with logical symbols and every writer seems to have avowed or non-avowed (strong) preferences for one or the other system of notation. It had at one point been our intention to adopt a completely uniform set of symbols and notational conventions for the *Handbook*. For various reasons, we have left the choice of minor bits of notation (e.g. sentential symbols and the like) to the particular authors, unifying only whenever confusion might arise. The reader is invited to translate whatever seems to go against his or her notational taste into his or her favorite variant.

CHAPTER IV.1

QUANTIFIERS IN FORMAL AND NATURAL LANGUAGES

by Dag Westerståhl

D. Gabbay and F. Guenthner (eds.), Handbook of Philosophical Logic, Volume IV, 1–131.

INTRODUCTION

For a long time, the word 'quantifier' in linguistics and philosophy simply stood for the universal and existential quantifiers of standard predicate logic. In fact, this use is still prevalent in elementary textbooks. It seems fair to say that the dominance of predicate logic in these fields has obscured the fact that the quantifier expressions form a *syntactic category*, with characteristic interpretations, and with many more members than $\forall$ and $\exists$.

Actually, when Frege discovered predicate logic, it was clear to him that the universal and existential quantifiers were but two instances of a general notion (which he called *second level concept*). That insight, however, was not preserved during the early development of modern logic. It took quite some time before the mathematical machinery behind quantification received, once more, an adequate general formulation. This time, the notion was called *generalized quantifier*; a first version of it was introduced by Mostowski in the late 1950s. Logicians gradually realized that generalized quantifiers were an extremely versatile syntactic and semantic tool – practically anything one would ever want to say in any logic can be expressed with them. The power of expression, properties and interrelations of various logics with generalized quantifiers is now a well established domain of study in mathematical logic.

This is the mathematical side of the coin. The linguistic side looks a bit different. Syntactically, there are many expressions one could place in the same category as *some* and *every*: *no, most, many, at least five, exactly seven, all but three,* .... These expressions – the *determiners* – occur in *noun phrases*, which in turn occur as subjects, objects, etc. in the *NP–VP* analysis of sentences usually preferred by linguists. Logically, however, subject-predicate form had fallen into disrepute since the breakthrough of predicate logic. So it was not obvious how to impose a semantic interpretation on these syntactic forms – except by somehow rewriting them in predicate logic. This may explain why the systematic study of quantifiers in natural language is of a much later date than the one for mathematical language.

The starting-point of this study was when Montague showed that linguistic syntax is, after all, no insurmountable obstacle to systematic and rigorous semantics. Montague did not yet have the quantifiers in a separate category. But in 1981 Barwise and Cooper united Montague's insights with the work on generalized quantifiers in mathematical logic in a study of the characteristics of natural language quantification (Barwise and Cooper [1981]). At about the same time, but independently and from a slightly different perspective, Keenan and Stavi were investigating the semantic properties of determiner interpretations (Keenan and Stavi [1986]). It became clear that, in natural language too, the quantifier category is quite rich and semantically powerful. In the few years that have passed since then, the subject has developed considerably. In particular, van Benthem has discovered an interesting logical theory behind the mechanisms of natural language quantification – often with no direct counterpart for mathematical language (van Benthem [1984a]).

My main aim in this chapter is to give a comprehensive survey of the logic and semantics of natural language quantification, concentrating on the developments in the last five years or so. The basic tools are the generalized quantifiers from mathematical logic. But it is the *questions* asked about quantifiers, not the methods used, that distinguishes our present perspective on quantifiers from that of mathematical logic.

The basic question facing anyone who studies natural language quantification from a semantic viewpoint can be formulated as follows. Logically, the category of quantifiers is extremely rich. For example, even on a universe with *two* elements, there are $2^{16} = 65536$ possible (binary) quantifiers (the reader who finds this hard to believe may wish to turn directly to Section 4.6 for the explanation). But, in natural languages, just a small portion of these are 'realized' (512, according to Keenan and Stavi). Which ones, and why? What are the *constraints* on determiner interpretations in natural language? What are the *properties* of quantifiers satisfying those constraints?

Most of this paper presents various answers to such questions. But we start, in Section 1, with a selective history of quantifiers: from Aristotle via Frege to modern generalized quantifiers. It will be seen that both Aristotle's and Frege's contributions compare interestingly to the recent developments. That section also gives a thorough introduction to generalized quantifiers, and to some logical issues pertaining to them. In particular, the *logical expressive power* of *monadic* quantifiers is discussed in some detail. Section 2 presents basic ideas of the Montague–Barwise–Cooper–Keenan–Stavi approach to natural language quantification. A number of examples of English quantifier expressions are also collected, as emprical data for later use. In Section 3, several constraints on quantifiers are formulated and discussed and various properties of quantifiers are introduced. The constraints can also be seen as potential *semantic universals*. Section 4 then presents various results in the *theory* of quantifiers satisfying certain basic constraints; results on how to classify them under various aspects, on how to represent them, on their inferential behavior and other properties. The paper ends with a brief further outlook and two appendices, one on branching quantification and the other on quantifiers as variables.

This chapter is concerned with the *semantics* of quantification. It examines certain precisely delimited classes of quantifiers that arise naturally in the context of natural language. These classes are related in various ways to the (loosely delimited) class of *natural language quantifiers*, i.e., those that are denotations of natural language determiners. I will make few definite claims about the exact nature of this relationship, but I will discuss several tentative proposals. The idea is to present the *possibilities* for determiner interpretation, and to give a framework sufficiently general for serious discussion of natural language quantifiers, yet restricted in significant ways compared with the generalized quantifier framework of mathematical logic. (I also hope to make it clear that interesting logical issues arise in the restricted framework (and sometimes only in that framework), and thus that logic can fruitfully be inspired by natural language as well as by the language of mathematics.)

So, except for a few rather straightforward things, I shall have little to say about the *syntax* of quantification here. And except for an introductory overview, I will not attempt to survey generalized quantifiers in mathematical logic. For more on quantification and linguistic theory, cf. Cooper [1983] or van Eijck [1985]. A very comprehensive survey of quantifiers in mathematical logic is given in Barwise and Feferman [1985].

The semantic framework used here is that of classical model theory. It is simple, elegant and well known. That it works so well for natural language quantification too is perhaps a bit surprising. However, there are certain things it does not pretend to handle, for example, intensional phenomena, vagueness, collectives, or mass terms. So these subjects will not be taken up here. But then, they receive ample treatment in other parts of this Handbook.

The logical techniques we need are usually quite elementary. The reader should be used to logical and set-theoretic terminology, but, except on a few occasions, there are no other specific prerequisites (Chapter I.1 of this Handbook (Volume I), by Hodges, gives a suitable background; occasionally, part of Chapter I.4 on higher-order logic, by van Benthem and Doets, will be useful). I have intended to make the exposition largely self-contained, in the sense that a) most proofs and arguments are given explicitly, and b) when they are not given, references are provided, but the reader should be able to get a feeling for what is going on without going to the references. Naturally, if these intentions turn out not to be realized, it does not follow that the fault lies with the reader.

This is a survey, and most results are from the literature, although several are new, or generalized, or proved differently here. I have tried to give reasonable credit for known results.

1. BACKGROUND: FROM ARISTOTELIAN TO GENERALIZED QUANTIFIERS

This section gives a condensed account of the development of what can be called *the relational view of quantifiers*. As a chapter in the history of logic, it seems not to be widely known, which is why I have included a subsection on Aristotle and a subsection on Frege. My main purpose, however, is to introduce a precise concept of quantifier sufficiently general to serve as a basis for what will follow. This is the notion of a *generalized quantifier* from mathematical logic. In the last subsections, I will also mention some of the things mathematical logicians do with quantifiers, as a background to what linguistically minded logicians might do with them.

1.1. *Aristotle*

Aristotle's theory of syllogisms, for ages considered the final system of logic, is now often seen as a rather pointless formal exercise, whose main achievement is to have hampered the development of logic for some two thousand years. But to understand Aristotle's contribution to logic one must distinguish his views from those of his followers. It is a fact that most of his followers and commentators were unable, for various reasons, to appreciate his logical insights (to take one simple but important example: the point of using *variables*).[1] From the standpoint of modern logic, on the other hand, these insights ought to be easily visible.

There is, however, one obscuring issue. According to widespread opinion, the breakthrough of modern logic rests upon the *rejection* of a basic Aristotelian idea, namely, that sentences have *subject-predicate form*. This was Russell's view, apparently vindicated by the absence of subject-predicate form in today's standard predicate logic. Hence, Aristotle's logic seems to be built on a fundamental mistake.

If we set aside questions concerning the historical causes of the long standstill in logic after Aristotle, there is, however, no necessary incompatibility between modern logic and subject-predicate form.[2] It is quite feasible to give an adequate account of both relations and quantification while preserving subject-predicate form, as we shall see in 2.3. Thus, although it is true that Aristotle's logic could not adequately account for these things, and thus was unable to express many common forms of reasoning, this weakness is not necessarily tied to his use of subject-predicate form.

In addition to matters of syntactic form, however, one ought to consider the *concepts* Aristotle introduced with his logic, the *questions* he raised about it, and the *methods* he used to answer them. Herein lies his greatest contribution.

Thousands of pages have been written on Aristotle's logic, most of them about irrelevant and futile matters (such as the order between the premisses in a syllogism, why he didn't mention the fourth figure, whether a valid syllogism can have a false premiss – Aristotle himself had no doubts about this –, etc.). Readable modern expositions, with references to the older literature, are Łukasiewicz [1957] and Patzig [1959]. Below I wish to point, without (serious) exegetic pretensions, to one important aspect of Aristotle's logic.

The syllogistics is basically a theory of *inference patterns among quantified sentences*. Here a quantified sentence has the form

(1) QXY,

where X, Y are *universal terms* (roughly 1-place predicates) and Q is one of the quantifiers *all, some, no, not all*. In practice, Aristotle treated these quantifiers as *relations* between the universal terms.[3]

Aristotle chose to study a particular type of inference patterns with sentences of the form (1), the syllogisms. A *syllogism* has two premisses, one conclusion, and three universal terms (variables). Each sentence has two different terms, all three terms occur in the premisses, and one term, the 'middle' one, occurs in both premisses but not in the conclusion. It follows that the syllogisms can be grouped into four different 'figures', according to the possible configurations of variables:

$$\frac{\begin{array}{c}Q_1ZY\\ Q_2XZ\end{array}}{Q_3XY} \qquad \frac{\begin{array}{c}Q_1YZ\\ Q_2XZ\end{array}}{Q_3XY} \qquad \frac{\begin{array}{c}Q_1ZY\\ Q_2ZX\end{array}}{Q_3XY} \qquad \frac{\begin{array}{c}Q_1YZ\\ Q_2ZX\end{array}}{Q_3XY}$$

Here the Q_i can be chosen among the above quantifiers, so there are $4^4 = 256$ syllogisms. As a matter of historical fact, Aristotle's specification of the syllogistic form was not quite accurate; he had problems with defining the middle term, and his systematic exposition does not mention the fourth figure (although he in practice admitted syllogisms of this form), but these are minor defects.

Now, the question Aristotle posed – and, in essence, completely answered – can be formulated as follows:

For what choices of quantifiers are the above figures valid?

For example, if we in the first figure let $Q_1 = Q_2 = Q_3 =$ *all*, a valid syllogism results ('Barbara', in the medieval mnemonic); likewise if $Q_1 = Q_3 =$ *no* and $Q_2 =$ *all* ('Celarent'). Note that Aristotle's notion of validity is essentially the modern one: a syllogism is valid if each instantiation of X,Y,Z verifying the premisses also verifies the conclusion (a slight difference is that Aristotle didn't allow the empty or the universal instantiation; this can be ignored here).

There are interesting variants of this type of question. Given some common quantifiers, we can ask for their inference patterns, and try to systematize the answer in some perspicuous way (axiomatically, for example). This is a standard procedure in logic. But we can also turn the question around and ask which quantifiers satisfy the patterns we found: only the ones we started with or others as well? If our common schemes of inference *characterize* our common quantifiers, we have one kind of explanation of the privileged status of the corresponding 'logical constants', and one goal of a theory of quantifiers has been attained.

The latter question is somewhat trivialized in Aristotle's framework, since there were only four quantifiers. For example, the question of which quantifiers satisfy the scheme

$$\begin{array}{l} QZY \\ QXZ \\ \hline QXY \end{array}$$

has the obvious answer: just *all*. But the question itself does not depend on the quantifier concept you happen to use. In 4.1 we shall return to it (and in 4.5 to the characterization of our most common quantifiers), this time with infinitely many quantifiers to choose from, and find some non-trivial answers.

Thus, not only did Aristotle introduce the relational concept of quantifiers, he also asked interesting questions about it. His methods of answering these questions were *axiomatic* (for example, he derived all valid syllogisms from the two syllogisms 'Barbara' and 'Celarent' mentioned above) as well as *model-theoretic* (non-validity was established by means of counterexamples). Even from a modern point of view, his solution leaves only some polishing of detail to be desired. Perhaps this finality of his logic was its greatest 'fault'; it did not encourage applications of the new methods to, say, other inference patterns. Instead, his followers managed to make a sterile church out of his system, forcing logic students to rehearse syllogisms far into our own century. But we can hardly blame Aristotle for that.

It should be noted that outside of logic Aristotle studied quantifiers without restriction to syllogistic form. For example, he made interesting observations on sentences combining negation and quantification (cf. Geach [1972]).

We shall not pursue the fate of the relational view of quantifiers between Aristotle and Frege. Medieval logicians spent much time analyzing quantified sentences, but they were more or less prevented from having a *concept* of quantifier by their insistence that quantifier words are *syncategorematic*, without independent meaning (this view, incidentally, is still common). Later logicians applied the mathematical theory of relations (converses, relative products, etc.) to give explicit formulations of Aristotle's relational concept, and to facilitate the proofs of his results on syllogisms (cf. de Morgan [1847] or, for a more recent account, Lorenzen [1958]). These methods were in general only applied to the quantifiers in the traditional *square of opposition* and their converses. A systematic study of quantifiers as binary relations did not appear until the 1980s (cf. Section 4.1).

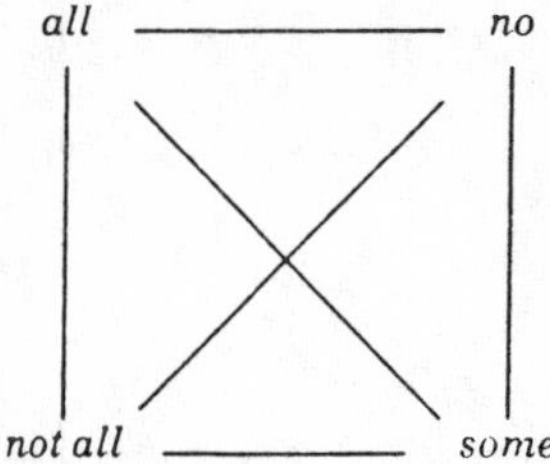

1.2. *Frege*

It is undisputed that Frege is the father of modern logic. He invented the language of predicate calculus, and the concept of a formal system with syntactic formation and inference rules. Moreover, his work was characterized by an exceptional theoretical clarity, greatly surpassing that of his contemporaries, and for a long time also his successors, in logic.

There is some difference of opinion, however, as to how 'modern' Frege's conception of logic was. According to Dummett [1973], [1981], we find in Frege, implicitly if not explicitly, just that dualism between a syntactic (proof-theoretic) and a semantic (model-theoretic) viewpoint which is characteristic of modern logic. "Frege would have had within his grasp the concepts necessary to frame the notion of completeness of a formalization of logic as well as its soundness" (Dummett [1973], p. 82). Dummett also traces the notion of an interpretation of a sentence, and thereby the semantic notion of logical consequence, to Frege's work.

This evaluation is challenged in Goldfarb [1979], a paper on the quantifier in early (modern) logic. Goldfarb holds the notion of an interpretation to be non-existent in Frege's logic: first, because there are no non-logical symbols to interpret, and, second, because the universe is fixed once and for all. The quantifiers range over this universe, and the laws of logic are *about* its objects. Furthermore, the logicism of Frege and Russell prevented them, according to Goldfarb, from raising any metalogical questions at all.

Although it takes us a bit beyond a mere presentation of Frege's notion of quantifier, it is worthwhile trying to get clear about this issue. The main point to be made is, I think, that Frege was the only one of the logicians at the time who maintained a sharp distinction between syntax and semantics, i.e., between the expressions themselves and their denotations. This fact alone puts certain metalogical questions 'within the reach' of Frege that would have been meaningless to others. Thus, one cannot treat Frege and Russell on a par here. Moreover, if one loses sight of this, one is also likely to miss the remarkable fact that, while the invention of predicate logic with the universal and existential quantifiers can also be attributed to Peano and Russell, Frege was the only one who had a mathematically precise *concept* of quantifier. This concept seems indeed to have gone largely unnoticed among logicians, at least until the last decade or so; in particular, the inventors of the modern generalized quantifiers do not seem to have been aware of it.

For this reason, Frege, but not Russell, has a prominent place in an historical overview of the relational view of quantifiers – in fact, Russell's explanations of the meaning of the quantifiers are in general quite bewildering (for example, Russell [1903], Chapter IV, Sections 59–65, or Russell [1956], pp. 64–75 and 230–231). I will present Frege's concept below, and then return briefly to the issue of how questions of soundness and completeness relate to Frege's logic.

1.2.1. *Quantifiers as second level concepts.* Let us first recall some familiar facts about Frege's theoretical framework.[4] All entities are either *objects* or *functions*. These categories are primitive and cannot be defined. Functions, however, are distinguished from objects in that they have (one or more) empty places (they are 'unsaturated'). When the empty places are 'filled' with appropriate *arguments* a *value* is obtained. The value is always an object, but the arguments can either be objects, in the case of *first level* functions, or other functions: *second level* functions take first level functions as arguments, etc. – no mixing of levels is permitted. All functions are total (defined for all arguments of the right sort). They can be called *unary, binary*, etc., depending on the number of arguments.[5]

Concepts are identified with functions whose values are among the two truth values *True* and *False*. Thus they have levels and 'arities' just as other functions.

The meaningful expressions in a logical language ('Begriffsschrift') are simple or complex *names* standing for objects or functions.[6] Names have both a sense ('Sinn') and a denotation ('Bedeutung'); only the denotation matters here. There is a strong parallelism between the syntactic and the semantic level: function names also have empty places (marked by special letters) that can (literally) be filled with appropriate object or function names. In particular, sentences are (complex) object names, denoting truth values.

Complex function names can be obtained from complex object names by deleting simple names, leaving corresponding empty places. For example, from the sentence

23 is greater than 14

we obtain the first level function (concept) names

x is greater than 14,

23 is greater than y,

x is greater than y,

and also the second level

$\Psi(23, 14)$.

Now, suppose the expression

(1) $F(x)$

is a unary first level concept name. Then the following is a sentence.[7]

(2) $\forall x F(x)$.

According to Frege, (2) is obtained by inserting the concept name (1) as argument in the *second level concept name*

(3) $\forall x \Psi(x)$.

(3) is a *simple* name in Frege's logic. It denotes a unary second level concept, namely, the function which, when applied to any unary first level function $f(x)$, gives the value *True* if $f(x)$ has the value *True* for all its arguments, *False* otherwise.[8]

This, of course, is a version of the usual truth condition for universally quantified sentences: (2) is true iff $F(x)$ is true for all objects x. But Frege's formulation makes it clear that (3) denotes just one of many possible second level concepts, for example,

(4) $\neg \forall x \neg \Psi(x)$

(5) $\forall x(\Phi(x) \rightarrow \Psi(x))$.

(4) is the existential quantifier. (5) is the *binary* second level concept of *subordination* between two unary first level concepts. Both can be defined by means of (3) in Frege's logic, and are thus denoted by complex names.

In a similar fashion, quantification over first level functions can be introduced by means of third level concepts, and so on.

Summarizing, we find that there is a welldefined *Fregean concept of quantifier*:

> *Syntactically, (simple) quantifier names can be seen as variable-binding operators (but see Note 7 on Frege's use of variables). Semantically, quantifiers are second level concepts.*

If we let, in a somewhat un-Fregean way, the *extension of an n-ary first level concept* be the class of n-tuples of objects falling under it, and the *extension of an n-ary second level concept* the class of n-tuples of extensions of first level concepts falling under it, then the extensions of the quantifiers (3)–(5) are

(6) $\forall_U = \{X \subseteq U: X = U\}$

(7) $\exists_U = \{X \subseteq U: X \neq \emptyset\}$

(8) $\mathbf{all}_U = \{\langle X, Y \rangle: X \subseteq U \,\&\, Y \subseteq U \,\&\, X \subseteq Y\}$,

where U is the class of all objects. Apart from the fact that the universe is fixed here (and too big to be an element of a class), these extensions are generalized quantifiers in the model-theoretic sense; cf. Section 1.4.

1.2.2. *Unary vs. binary quantifiers.* Frege was well aware that the usual quantifier words in natural language stand for *binary* quantifiers. For example, in 'On Concept and Object' he writes

> . . . the words 'all', 'any', 'no', 'some' are prefixed to concept-words. In universal and particular affirmative and negative sentences, we are expressing *relations between concepts*; we use the words to indicate a special kind of relation (Frege [1892], p. 48, my italics).

But he also found that these binary (Aristotelian) quantifiers could be defined by means of the unary (3) and sentential connectives. This was no trivial discovery at the time, and Frege must have been struck by the power and simplicity of the unary universal quantifier. In his logical language he always chose it as the sole primitive quantifier.

The use of unary quantifiers was to become a characteristic of predicate logic, and the success of formalizing mathematical reasoning in this logic can certainly be said to have vindicated Frege's choice. It does not follow from this, however, that the same choice is adequate for formalizing natural language reasoning. Indeed, we will see later that unary quantifiers are unsuitable as denotations of the usual quantifier words, and that, furthermore, it is simply not the case that all binary natural language quantifiers can be defined by means of unary ones and sentential connectives.

Such a preference for binary quantifiers in a natural language context is, as we can see from the foregoing, in no way inconsistent with Frege's view on quantifiers.[9]

1.2.3. *Logical truth and metalogic.* Let us return to the Dummett–Goldfarb dispute about whether metalogical issues such as completeness were in principle available to Frege. The usual notion of completeness of a logic presupposes the notion of logical truth (or consequence), i.e. truth in all models. But the latter notion was clearly not considered by Frege. As Goldfarb remarks, he had no non-logical constants whose interpretations could vary (it seems that he explicitly rejected the use of such constants; cf. Hodges [I.1], Section 17), nor did he consider the idea that the universe could be varied. One universe was enough, namely, the universe U of all objects, and only simple truth in U interested Frege.

However, the notion of truth in U is very close to the notion of logical truth. To fix ideas, consider some standard version of *higher-order logic* (say, the logic L_ω presented in van Benthem and Doets [I.4], Section 3.1). For the purposes of the present discussion we may identify *Frege's logic* with higher-order logic *without* non-logical symbols.[10] Then we can observe that Frege did not 'miss' any standard logical truths. For, each sentence ψ in L_ω has an obvious translation ψ^* in Frege's logic, obtained by 'quantifying out' the non-logical constants. For example,

$$\forall x Px \to Pa$$

translates as

$$\forall X \forall y(\forall x Xx \to Xy),$$

and similarly for higher-order sentences. It is evident that

(9) if ψ is logically true then ψ^* is true in U.

A parenthetical observation is necessary here. Logical truth is often defined as truth in all *set* models, instead of truth in *all* models, whether sets or not. The latter notion is *real* logical truth, and it is with respect to this notion that (9) is evident. As Kreisel has stressed, use of the former notion is only justified for first-order logic, since there the two notions coincide (this follows from the usual completeness proofs). For higher-order sentences, on the other hand, this is open; cf. Kreisel [1967].

For first-order logic, there is a converse to (9), provided we disregard sentences such as

$$\exists x \exists y(x \neq y),$$

which have *finite* counter-examples but are still true in the infinite U:

THEOREM. *Let M be any infinite class and ψ a first-order sentence. Then ψ is true in all infinite models iff ψ^* is true in M.*

Proof: (This proof uses some standard techniques of first-order model theory; they can be found in Chang and Keisler [1973], but will not be employed in the sequel.) From left to right this is similar to (9); if only set models are considered we employ Kreisel's observation mentioned above. For the other direction, suppose that $\neg\psi = \neg\psi(P, \ldots)$ has an infinite model $\mathbf{N} = \langle N, R, \ldots \rangle$. Again by Kreisel's observation, we can assume that N is a set. Now distinguish two cases, depending on whether M is a set or not. If M is a set, application of the Löwenheim–Skolem–Tarski theorem gives us a model $\boldsymbol{M}_0$ of $\neg\psi$ with the same cardinality as M. Via a bijection from M to M_0, $\boldsymbol{M}_0$ is isomorphic to a model $\langle M, S, \ldots\rangle$ of $\neg\psi$ with universe M. Thus, $\exists X \ldots \neg\psi(X, \ldots)$ is true in M, i.e. ψ^* is false in M, as was to be proved. Now suppose M is a proper class. Starting with $\boldsymbol{N} = \boldsymbol{N}_0$ as before, define uniformly for each ordinal α a model $\boldsymbol{N}_\alpha$ such that $\boldsymbol{N}_\alpha$ is a proper elementary extension of $\boldsymbol{N}_\beta$ when $\beta < \alpha$. The union $\boldsymbol{M}'$ of all these is then a model of $\neg\psi$ (Tarski's union lemma). Moreover, M' is a proper class, whence there is a bijection from M to M'. It follows as before that ψ^* is false in M. □

Thus, in a sense it makes no difference for first-order logic if we have, as Frege did, a fixed infinite universe (such as U) and no non-logical constants. More precisely, it follows from the above that the true Π^1_1 sentences of Frege's logic correspond exactly to the standard first-order logical truths on infinite models.

In conclusion, then, we have seen that notions such as completeness and soundness were not directly available to Frege, since they presuppose a notion of logical truth he did not have. But Dummett's position is still essentially correct, I think: Frege's work does introduce a version of the dualism between model theory and proof theory. For, Frege had the notion of *truth*, which he certainly did not confound with *provability*. Clearly he considered all theorems of his system to be true. He did not, as far as we know, raise the converse question of whether all true sentences are provable, but surely it was 'within his grasp'. And for his *logic*, this question turns out to be a version of the completeness question, as noted above. Moreover, the answer is *yes* if we restrict attention to Π^1_1 sentences (by the above result and the completeness of first-order logic), *no* otherwise (higher-order logic is not complete).

1.3. *Mostowskian Quantifiers*

As we know, Frege's work was neglected in the early phase of modern logic, and the rigor he attained, especially in semantics, was not matched for a long time. But the language of predicate logic was powerful enough to be a success even in the absence of a solid semantic basis. In the history of quantifiers, this period is mainly interesting for its discussions on the role of quantification over infinite domains for the foundation of mathematics, but that is not a subject here.

The idea of a mathematically sharp dividing line between syntax and semantics began to reappear gradually in the 1920s, but not until Tarski's truth definition in 1936 did the notion of truth (in a model) become respectable. Tarski's truth conditions for universally and existentially quantified formulas treat $\forall$ and $\exists$ syncategorematically, but it is natural to try some other quantifiers here, i.e. to consider formulas

$$Qx\psi$$

for Q other than $\forall$ and $\exists$. For example, it is clear what the truth conditions for $\exists_{\geqslant n}$ and $\exists_{=n}$ should look like. To get a general concept, however, we must treat Q non-syncategorematically, i.e. we must have a *syntactic category* 'quantifier' with a specified range of interpretations. Such a general concept appeared in Mostowski [1957].

Recall that Tarski defines the relation

$$\boldsymbol{M} \models \phi[g],$$

("g satisfies ϕ in $\boldsymbol{M}$"), where $\boldsymbol{M}$ is model, g an assignment of elements in M to the variables, and ϕ a formula. When ϕ is $\forall x\psi$ or $\exists x\psi$, this can be expressed as a condition on the *set*

$$\psi^{\boldsymbol{M},g,x} = \{a \in M: \boldsymbol{M} \models \psi[g(a/x)]\}.$$

Thus,

$$\boldsymbol{M} \models \forall x\psi[g] \Leftrightarrow \psi^{\boldsymbol{M},g,x} = M,$$

$$\boldsymbol{M} \models \exists x\psi[g] \Leftrightarrow \psi^{\boldsymbol{M},g,x} \neq \varnothing,$$

$$\boldsymbol{M} \models \exists_{\geqslant n} x\psi[g] \Leftrightarrow |\psi^{\boldsymbol{M},g,x}| \geqq n.$$

A condition on subsets of M is, extensionally, just a set of subsets of M. So Mostowski defines a (local) *quantifier on* M to be a set of subsets of M, whereas a (global) *quantifier* is a function(al) $\mathbf{Q}$ assigning to each non-empty set M a quantifier $\mathbf{Q}_M$ on M. Syntactically, a quantifier symbol Q belongs to $\mathbf{Q}$, such that $Qx\psi$ is a formula whenever x is a variable and ψ is a formula, with the truth condition

$$\boldsymbol{M} \models Qx\psi[g] \Leftrightarrow \psi^{\boldsymbol{M},g,x} \in \mathbf{Q}_M.$$

Examples of such quantifiers are

$$\forall_M = \{M\},$$

$$\exists_M = \{X \subseteq M: X \neq \varnothing\},$$

$$(\exists_{\geqslant n})_M = \{X \subseteq M: |X| \geqq n\},$$

$$(\mathbf{Q}_\alpha)_M = \{X \subseteq M: |X| \geqq \aleph_\alpha\}, \text{ (the cardinality quantifiers)}$$

$$(\mathbf{Q}_C)_M = \{X \subseteq M: |X| = |M|\}, \text{ (the Chang quantifier)}$$

$$(\mathbf{Q}_R)_M = \{X \subseteq M: |X| > |M - X|\}$$

(Rescher's 'plurality quantifier').

All of these satisfy the following condition:

ISOM *If f is a bijection from M to M′ then* $X \in \mathbf{Q}_M \Leftrightarrow f[X] \in \mathbf{Q}_{M'}$.

In fact, Mostowski included *ISOM* as a defining condition on quantifiers, expressing the requirement that "quantifiers should not allow us to distinguish between elements/of M/" ([1957], p. 13).

1.4. *Generalized Quantifiers*

Rescher, introducing the quantifier $\mathbf{Q}_R$, noted that $Q_R x\psi(x)$ expresses

(1) Most things (in the universe) are ψ,

but that the related (and more common)

(2) Most ϕs are ψ

cannot be expressed by means of Q_R (Rescher [1962]). From our discussion of Frege we recognize in (2) a *binary* quantifier, **most**, giving, on each M, a binary relation between subsets of M:

$$\mathbf{most}_M = \{\langle X, Y\rangle \in M^2: |X \cap Y| > |X - Y|\}.$$

To account for this, the construction of formulas must be generalized. This was noted by Lindström [1966], who introduced the concept of a *generalized quantifier*, defined below.

(2) can be formalized as

$$most\ x,y(\phi(x), \psi(y)).$$

Here the free occurrences of x (y) in ϕ (ψ) are bound by the quantifier symbol. In fact, the choice of variables is arbitrary; we can write

$$most\ x,x(\phi(x), \psi(x)),$$

or, more simply,

$$most\ x(\phi(x), \psi(x)).$$

In this way Mostowskian quantifiers on M are generalized to n-ary relations between subsets of M. A further generalization is to consider relations between *relations* on M. Here is an example:

$$\mathbf{W}^r_M = \{\langle X, R\rangle : X \subseteq M \ \&\ R \subseteq M^2 \ \&\ R \text{ wellorders } X\}$$

(the name of this quantifier will be explained later). The statement that (the set) ϕ is wellordered by (the relation) ψ is formalized as

$$W^r x, yz(\phi(x), \psi(y, z))$$

(note that y and z are simultaneously bound in ψ).

Quantifiers are associated with *types* (finite sequences of positive numbers: Mostowskian quantifiers have type $\langle 1\rangle$, **most** has type $\langle 1, 1\rangle$, and $\mathbf{W}^r$ has type $\langle 1, 2\rangle$; the principle should be clear. We are now prepared for the following

1.4.1. DEFINITION. A (local) *generalized quantifier of type* $\langle k_1, \ldots, k_n\rangle$ *on M* is an n-ary relation between subsets of $M^{k_1}, \ldots M^{k_n}$, respectively, i.e. a subset of $P(M^{k_1}) \times \cdots \times P(M^{k_n})$. A (global) *generalized quantifier of type* $\langle k_1, \ldots, k_n\rangle$ is a function(al) $\mathbf{Q}$ which to each set M assigns a generalized quantifier $\mathbf{Q}_M$ of type $\langle k_1, \ldots, k_n\rangle$ on M. To $\mathbf{Q}$ belongs a quantifier symbol Q (of the same type) with the following rule: If $\phi_1, \ldots, \phi_n$ are formulas and $\bar{x}_1, \ldots, \bar{x}_n$ are strings of distinct variables of length $k_1, \ldots, k_n$, respectively, then $Q\bar{x}_1 \ldots \bar{x}_n(\phi_1, \ldots, \phi_n)$ is a formula with the truth condition

$$\mathbf{M} \models Q\bar{x}_1 \ldots \bar{x}_n(\phi_1, \ldots, \phi_n)[g] \Leftrightarrow$$
$$\Leftrightarrow \langle \phi_1^{\mathbf{M},g,\bar{x}_1}, \ldots, \phi_n^{\mathbf{M},g,\bar{x}_n}\rangle \in \mathbf{Q}_M.$$

This definition expresses our final version of the relational view of quantifiers, the one we will use in the sequel. It should be clear that, apart from the relativization to an arbitrary universe M, the notion of a generalized quantifier (or a *Lindström quantifier*, as it is sometimes called) is essentially the same as Frege's notion of a second level concept.[11]

Most of the time we will restrict attention to quantifiers of type $\langle 1, 1, \ldots, 1\rangle$. These are the *monadic generalized quantifiers*; we will usually call them just *quantifiers*. We can then continue to talk about *unary, binary*, etc. quantifiers, when we mean generalized quantifiers of type $\langle 1\rangle$, $\langle 1, 1\rangle$, etc.

Like Mostowski, Lindström included *ISOM* in the definition of generalized quantifiers:

ISOM *If f is a bijection from M to M′ then* $\langle R_1, \ldots, R_n\rangle \in \mathbf{Q}_M$
$\Leftrightarrow \langle f[R_1], \ldots, f[R_n]\rangle \in \mathbf{Q}_{M'}$.

(If R is k-ary, $f[R] = \{\langle f(a_1), \ldots, f(a_k)\rangle : \langle a_1, \ldots, a_k\rangle \in R\}$.)

Here are some further examples of generalized quantifiers:

$$\mathbf{all}_M = \{\langle X, Y\rangle \in M^2 : X \subseteq Y\},$$
$$\mathbf{some}_M = \{\langle X, Y\rangle \in M^2 : X \cap Y \neq \emptyset\},$$
$$\mathbf{I}_M = \{\langle X, Y\rangle \in M^2 : |X| = |Y|\},$$
$$\mathbf{more}_M = \{\langle X, Y\rangle \in M^2 : |X| > |Y|\},$$
$$\mathbf{W}_M = \{R \subseteq M^2 : R \text{ wellorders } M\}.$$

I is the *Härtig quantifier*. **more** is sometimes called the *Rescher quantifier* (although Rescher only considered the quantifiers $\mathbf{Q}_R$ and **most** above). **W** is the *wellordering quantifier*. The generalized quantifier $\mathbf{W}^r$ given before is the *relativization* of **W**. This notion is defined as follows.

1.4.2. DEFINITION. If **Q** is of type $\langle k_1, \ldots, k_n\rangle$, the *relativization of* **Q** is the generalized quantifier $\mathbf{Q}^r$ of type $\langle 1, k_1, \ldots, k_n\rangle$ defined by

$$\langle X, R_1, \ldots, R_n\rangle \in \mathbf{Q}^r_M \Leftrightarrow \langle R_1 \cap X^{k_1}, \ldots, R_n \cap X^{k_n}\rangle \in \mathbf{Q}_X$$

(for all $X \subseteq M$ and $R_i \subseteq M^{k_i}$).

Thus, for $X \subseteq M$ we can use Q^r to express in M what Q says in X; this will be made precise in 1.6. Note that **all** $= \forall^r$, **some** $= \exists^r$, and **most** $= \mathbf{Q}^r_R$.

1.5. *Partially Ordered Prefixes*

At this point it is appropriate to mention another generalization of quantifiers, although not directly related to the relational view. In standard predicate logic each formula can be put in *prenex form*, i.e. with a *linear prefix* $Q_1x_1 \ldots Q_nx_n$, where Q_i is either $\forall$ or $\exists$, in front of a quantifier-free formula. Henkin [1961] suggested a generalization of this to *partially ordered* or *branching* prefixes, e.g. the following

$$(1) \qquad \begin{matrix} \forall x - \exists y \\ \forall z - \exists u \end{matrix} \Big> \phi(x, y, z, u)$$

The prefix in (1) is called the *Henkin prefix*. The intended meaning of (1) is that for each x there is a y and for each z there is a u such that $\phi(x, y, z, u)$, where y and u are chosen *independently* of one another. To make this precise one uses *Skolem functions*. (1) can then be written

(1′) $\exists f \exists g \forall x \forall z \phi(x, f(x), z, g(z))$.

The method of Skolem functions works for all prefixes with $\forall$ an $\exists$. For example, the first-order

(2) $\forall x \forall z \exists y \exists u \phi(x, y, z, u)$,

(3) $\forall x \exists y \forall z \exists u \phi(x, y, z, u)$

become

(2′) $\exists f \exists g \forall x \forall z \phi(z, f(x, z), z, g(x, z))$,

(3′) $\exists f \exists g \forall x \forall z \phi(x, f(x), z, g(x, z))$.

But the dependencies in (1′) cannot be expressed in ordinary predicate logic; somewhat surprisingly, the Henkin prefix greatly increases the expressive power, as we shall see in 1.6.

Although branching quantification generalizes another feature of ordinary quantification than the one we have been considering here, it can, in fact, be subsumed under the relational view of quantifiers. To the Henkin prefix, for example, corresponds the *Henkin quantifier* **H** of type $\langle 4 \rangle$, defined by

$$\mathbf{H} = \{R \subseteq M^4\text{: there are functions } f, g \text{ on } M \text{ such that for all } a, b \in M, \langle a, f(a), b, g(b) \rangle \in R\}.$$

The formula (1) is then written, in the notation of 1.4,

(1″) $Hxyzu\phi(x, y, z, u)$.

Observe that branching was only defined for $\forall$ and $\exists$. Can we let other quantifiers branch as well, and consider formulas such as

(4) $$\begin{matrix} Q'x & \searrow & \\ & & \phi(x, y)\ ? \\ Q''y & \nearrow & \end{matrix}$$

It is not immediate what this should mean. Compare the linear

(5) $Q'xQ''y\phi(x, y)$;

this is true in $\boldsymbol{M}$ iff $X = \{a \in M: \boldsymbol{M} \models Q''y\phi[a, y]\}$ is in $\mathbf{Q}'_M$, and, for each $a \in M$, $\boldsymbol{M} \models Q''y\phi[a, y]$ iff $Y_a = \{b \in M: \boldsymbol{M} \models \phi[a, b]\}$ is in $\mathbf{Q}''_M$. But the idea with (4) is to evaluate the quantifiers *independently* of each other, and then it is not clear which sets to look for in $\mathbf{Q}'_M$ and $\mathbf{Q}''_M$. Nevertheless, Barwise [1979] shows that for certain Q' and Q'' a reasonable interpretation of (4) can be given, and Westerståhl [1987] extends this to arbitrary Q' and Q''.

Branching quantification is not only of mathematical interest. It can be argued that both the Henkin prefix and the form (4) (for certain non-first-order Q' and Q'') occur essentially in natural languages. Barwise [1979] contains a good presentation of the issues involved here; a brief review will be given in Appendix A.

1.6. *Model-Theoretic Logics*

The introduction of generalized quantifiers opens up a vast area of logical study. Let *EL* (elementary logic) be standard predicate logic, and, if $\mathbf{Q}^i$ are generalized quantifiers for $i \in I$, let $L(\mathbf{Q}^i)_{i \in I}$ be the logic obtained from *EL* by adding the syntactic and semantic rules for each $\mathbf{Q}^i$ as in Definition 1.4.1. The study of such *model-theoretic logics* is sometimes called *abstract model theory*.[12] For a comprehensive survey of this field of mathematical logic the reader is referred to Barwise and Feferman [1985], in particular, the chapter Mundici [1985]. Below, just a few examples of such logics and their properties will be given.

The expressive power of a logic is most naturally measured by the classes of models its sentences can define. Define $L \leqq L'$ (L' is an *extension* of L) to mean that for each sentence of L there is an equivalent sentence (i.e. one with the same models) of L'. Clearly $\leqq$ is reflexive and transitive, and every logic $L = L(\mathbf{Q}^i)_{i \in I}$ is an extension of *EL*. We write $L \equiv L'$ when $L \leqq L'$ and $L' \leqq L$, and $L < L'$ when $L \leqq L'$ and $L' \nleqq L$.[13]

Since formulas are defined inductively, to prove that $L(\mathbf{Q}^i)_{i \in I} \leqq L'$ it suffices to show that each $\mathbf{Q}^i$ is definable in L'. For example, if $\mathbf{Q}^i$ is of type $\langle 2, 1\rangle$, it suffices to show that the sentence

$$Q^i xy,z(P_1xy, P_2z)$$

is equivalent to a sentence in L'.

The inductive characterization of formulas also gives the following result, which explains why *ISOM* is normally assumed for generalized quantifiers in mathematical logic: If each $\mathbf{Q}^i$ satisfies *ISOM*, then truth of sentences in $L(\mathbf{Q}^i)_{i\in I}$ is preserved among isomorphic models. In fact, the inductive proof of this gives slightly more:

1.6.1. PROPOSITION. *If each* $\mathbf{Q}^i$ *satisfies ISOM,* ϕ *is a formula in* $L(\mathbf{Q}^i)_{i\in I}$, *f an isomorphism from* M_1 to M_2, *and g an assignment in* M_1, *then*

$$M_1 \models \phi[g] \Leftrightarrow M_2 \models \phi[fg].$$

Here is the relative strength of some of the logics we have considered:

1.6.2. THEOREM. $EL < L(\mathbf{Q}_0) < L(\mathbf{I}) < L(\mathbf{more}) < L(\mathbf{H})$.

The easiest part of the proof of this theorem is to show that one logic is an extension of the previous one. That $L(\mathbf{Q}_0) \leqq L(\mathbf{I})$ follows from the equivalence

$$Q_0xPx \leftrightarrow \exists y(Py \wedge Ix(Px, Px \wedge x \neq y))$$

(P is infinite iff removal of one element does not change its cardinality). That $L(\mathbf{I}) \leqq L(\mathbf{more})$ is obvious, and that $L(\mathbf{more}) \leqq L(\mathbf{H})$ follows by the following trick (due to Ehrenfeucht):

$$\begin{aligned} \neg more\ x(P_1x, P_2x) &\leftrightarrow \exists f(f \text{ is a 1–1 function from } P_1 \text{ to } P_2) \\ &\leftrightarrow \exists f\forall x\forall z(x = z \leftrightarrow f(x) = f(z) \wedge \\ &\qquad \wedge P_1x \rightarrow P_2f(x)) \\ &\leftrightarrow \exists f\exists g\forall x\forall z(x = z \leftrightarrow f(x) = g(z) \wedge \\ &\qquad \wedge P_1x \rightarrow P_2f(x)) \\ &\leftrightarrow Hxyzu(x = z \leftrightarrow y = u \wedge \\ &\qquad \wedge P_1x \rightarrow P_2y). \end{aligned}$$

To prove that one logic is *not* an extension of another, one can either show directly that some sentence in the first is not equivalent to any sentence in the second, or, more indirectly, use *properties* of the two logics to distinguish them. For example, the following wellknown properties of *EL* can sometimes be used:

1. *The compactness property: If every finite subset of a set of sentences has a model, the whole set has a model.* Consider the following set of $L(\mathbf{Q}_0)$-sentences:

$$\{\neg Q_0x(x = x)\} \cup \{\exists_{\geqq n}x(x = x): n = 1, 2, 3, \ldots\}.$$

This set has no models, but each finite subset has one. So $L(\mathbf{Q}_0)$ (and all its extensions) is not compact. In particular, $L(\mathbf{Q}_0) \nleq EL$.

2. *The Tarski property: If a sentence has a denumerable model it has an uncountable model.* Let ϕ be an EL-sentence saying that $<$ is a discrete linear ordering with a first element. Then the $L(\mathbf{Q}_0)$-sentence

$$(1) \qquad \phi \wedge \forall x \neg Q_0 y(y < x)$$

characterizes the natural number ordering $\langle N, <\rangle$ (i.e. $\langle M, R\rangle$ is a model of (1) iff it is isomorphic to $\langle N, <\rangle$). All models of (1) are denumerable, so $L(\mathbf{Q}_0)$ does not have the Tarski property.

3. *The completeness property: The set of valid sentences is recursively enumerable.* Adding to (1) sentences (of EL) defining addition and multiplication, and saying that 0 is the least element and $x + 1$ the immediate successor of x, we obtain a sentence θ which characterizes the standard model of arithmetic $\boldsymbol{N} = \langle N, <, +, \times, 0, 1\rangle$. Then, for every $L(\mathbf{Q}_0)$-sentence ψ in this vocabulary,

$$\boldsymbol{N} \models \psi \Leftrightarrow \theta \rightarrow \psi \text{ is valid.}$$

Thus, since the set of true arithmetical sentences is not recursively enumerable, $L(\mathbf{Q}_0)$ is not complete. This time there is no immediate consequence for extensions of $L(\mathbf{Q}_0)$. For the extensions mentioned in Theorem 1.6.2, however, sentences characterizing $\boldsymbol{N}$ can be constructed in a similar way, so they are not complete either.

4. *The Löwenheim property: If a sentence has an infinite model it has a denumerable model.* It is not very difficult to show that $L(\mathbf{Q}_0)$ in fact has the Löwenheim property. But $L(\mathbf{I})$ (and its extensions) does not: we can write down a sentence of $L(\mathbf{I})$ saying that $<$ is a dense linear ordering without endpoints, and that there is an element which does not have as many predecessors as it has successors. In a model, the set of predecessors and the set of successors of this element are infinite and of different cardinalities, so the model must be uncountable. It follows, in particular, that $L(\mathbf{I}) \nleq L(\mathbf{Q}_0)$.

In the proof of Theorem 1.6.2, it only remains to show that $L(\mathbf{I})$ is not an extension of $L(\mathbf{more})$, and that $L(\mathbf{more})$ is not an extension of $L(\mathbf{H})$. A convenient way to prove the former will be given in 1.7. A proof of the latter can be found in Cowles [1981].

Recall the definition of relativized quantifiers in 1.4.2. We say that $L = L(\mathbf{Q}^i)_{i \in I}$ *relativizes*, if

$$L^r = L((\mathbf{Q}^i)^r)_{i \in I} \leqq L,$$

i.e., if the relativization of each $\mathbf{Q}^i$ is definable in L. EL, $L(\mathbf{Q}_\alpha)$, $L(\mathbf{I})$, $L(\mathbf{most})$, $L(\mathbf{more})$, and $L(\mathbf{H})$ all relativize. For example,

$$\forall^r x(Px, P_1x) \leftrightarrow \forall x(Px \rightarrow P_1x),$$

$$most^r x(Px, P_1x, P_2x) \leftrightarrow most\ x(Px \wedge P_1x, P_2x),$$

$$H^r v, xyzu(Pv, P_1xyzu) \leftrightarrow Hxyzu((Px \wedge Pz) \rightarrow$$

$$\rightarrow (Py \wedge Pu \wedge P_1xyzu)).$$

$L(\mathbf{Q}_R)$, $L(\mathbf{Q}_C)$, and $L(\mathbf{W})$, on the other hand, do not relativize (cf. Section 1.7).

As the above equivalences show, relativized quantifier symbols are used to make relativized statements. This extends to all L-sentences. Define, for each L-formula ϕ and each unary predicate symbol P, the *relativized formula*

$$\phi^{(P)}$$

in L^r inductively by letting $\phi^{(P)} = \phi$ if ϕ is atomic, $(\neg\psi)^{(P)} = \neg\psi^{(P)}$, $(\psi \wedge \theta)^{(P)} = \psi^{(P)} \wedge \theta^{(P)}$, and, when ϕ is quantified, beginning with Q^i of type $\langle 2, 1\rangle$, say,

$$Q^i xy, z(\psi, \theta)^{(P)} = (Q^i)^r v, xy, z(Pv, \psi^{(P)}, \theta^{(P)}).$$

$\phi^{(P)}$ expresses exactly what ϕ says about the universe restricted to (the denotation of) P. We can formulate this precisely as follows. Call a subset X of the universe of the model $\boldsymbol{M}$ *universe-like* if $X \neq \varnothing$, the denotations of all individual constants in the vocabulary for $\boldsymbol{M}$ are in X, and X is closed under the denotations of all function symbols in the vocabulary. In that case, let $\boldsymbol{M}|X$ be the model with universe X, and all the relations etc. in $\boldsymbol{M}$ restricted to X. Then it can be shown by induction that if X is universe-like and ϕ is an L-sentence,

$$(REL) \quad (\boldsymbol{M}, X) \models \phi^{(P)} \Leftrightarrow \boldsymbol{M}|X \models \phi$$

(here we assume that P does not occur in ϕ and that it denotes X in $(\boldsymbol{M}, X)$).

If L relativizes, all this can be done in L, since $\phi^{(P)}$ is then clearly equivalent to an L-sentence.

So far we have only discussed particular logics and their properties. The most exciting part of abstract model theory, however, concerns results relating various properties of logics to each other, and results *characterizing* certain logics in terms of their properties. Most famous of these characterizations is still *Lindström's theorem* (Lindström [1969]), which characterizes *EL* in terms of the four properties mentioned above (for proofs, cf. Flum [1985], van Benthem and Doets [I.4], or Hodges [I.1]).

THEOREM. *If L is compact and has the Löwenheim property, then* $L \equiv EL$. *Also, if L relativizes, then:* (a) *if L is complete and has the Löwenheim property then* $L \equiv EL$; (b) *if L has the Löwenheim and Tarski properties then* $L \equiv EL$.

1.7. *The Strength of Monadic Quantifiers*

In general, it may be quite difficult to determine whether $L \leqq L'$ or not, where L and L' are logics with generalized quantifiers. In the case of *monadic* quantifiers, however, things become much easier. Since this case is what we shall mainly be dealing with, I will devote the present subsection to developing some machinery for comparing the expressive power of logics with monadic quantifiers. The machinery will be applied in particular to the quantifiers **more** and **most**. I use these quantifiers later to illustrate some important points concerning natural language quantification, and it will then be instructive to have established their logical properties.

This subsection is a bit more technical than the previous ones; I have written out proofs of results that are new or not easily found in the literature (cf. the bibliographical note at the end). The reader can skip or glance through it now, and return to it for a definition or a result that is used later.

From now on, when **Q** is an *m*-ary monadic quantifier, we will write simply

$$\mathbf{Q}_M X_1 \ldots X_m,$$

instead of $\langle X_1, \ldots, X_m \rangle \in \mathbf{Q}_M$. Thus,

$$\mathbf{all}_M AB \Leftrightarrow A \subseteq B,$$

$$\mathbf{most}_M AB \Leftrightarrow |A \cap B| > |A - B|,$$

$$\mathbf{more}_M AB \Leftrightarrow |A| > |B|,$$

etc.

Let $\boldsymbol{M} = \langle M, A_0, \ldots, A_{k-1} \rangle$ be a *k-ary monadic structure* (i.e., the A_i are subsets of M, and the vocabulary consists of k unary predicate symbols). The following terminology will be used here and in later sections. If $X \subseteq M$, let $X^0 = X$ and $X^1 = M - X$. If s is a function from $\{0, \ldots, k-1\}$ to $\{0, 1\}$, i.e., if $s \in 2^k$, let

$$P_s^{\boldsymbol{M}} = A_0^{s(0)} \cap \ldots \cap A_{k-1}^{s(k-1)}.$$

$\{P_s^{\boldsymbol{M}}\}_{s \in 2^k}$ is a *partition* of M, and, up to isomorphism, the number of elements in these partition sets is all there is to say about $\boldsymbol{M}$. In other words, if $|P_s^{\boldsymbol{M}}| = |P_s^{\boldsymbol{M}'}|$ for all $s \in 2^k$, then $\boldsymbol{M}$ and $\boldsymbol{M}'$ are isomorphic. Finally, let

$$U_i^{\boldsymbol{M}},$$

for $1 \leqq i \leqq 2^{2k}$, be all possible *unions* of the partition sets (including $\varnothing$), in some fixed order.

If L is a logic, $\boldsymbol{M}$ a structure (not necessarily monadic), $X \subseteq M$, and $a_1, \ldots, a_n \in M$, X is said to be *L-definable in* $\boldsymbol{M}$ *with parameters* $a_1, \ldots, a_n$, if there is an L-formula ϕ in the vocabulary of $\boldsymbol{M}$ such that

$$a \in X \Leftrightarrow \boldsymbol{M} \models \phi[a, a_1, \ldots, a_n].$$

The following is an almost immediate consequence of this definition and Proposition 1.6.1:

1.7.1. LEMMA. *If L satisfies ISOM, X is L-definable in $\boldsymbol{M}$ with parameters $a_1, \ldots, a_n$, and f is an automorphism on $\boldsymbol{M}$ (i.e. an isomorphism from $\boldsymbol{M}$ to $\boldsymbol{M}$) with $f(a_i) = a_i$, then $f[X] = X$.*

If A, B are sets, $A \oplus B$, the *symmetric difference* between A and B, is $(A - B) \cup (B - A)$. We say that B is an *X-variant* of A, if $A \oplus B \subseteq X$.

1.7.2. LEMMA. *Suppose that L satisfies ISOM and that $\boldsymbol{M}$ is a monadic structure. Then the L-definable sets in $\boldsymbol{M}$ with parameters $a_1, \ldots, a_n$ are precisely the $\{a_1, \ldots, a_n\}$-variants of the unions $U_i^{\boldsymbol{M}}$.*

Proof: Clearly all these sets are so definable. Now suppose X is L-definable in $\boldsymbol{M}$ from $a_1, \ldots, a_n$. Then so is $X' = X - \{a_1, \ldots, a_n\}$. It suffices to show that X' has the desired form. Let $s_1, \ldots, s_p$ be those $s \in 2^k$ for which $X' \cap P_s^{\boldsymbol{M}} \neq \varnothing$. Thus,

$$X' \subseteq P_{s_1}^{\boldsymbol{M}} \cup \ldots \cup P_{s_p}^{\boldsymbol{M}}.$$

Suppose X' is *not* an $\{a_1, \ldots, a_n\}$-variant of $P^M_{s_1} \cup \ldots \cup P^M_{s_p}$. Then, for some i, there is $a \in P^M_{s_i} - X'$ such that $a \neq a_1, \ldots, a_n$. But, by the construction, there is $b \in P^M_{s_i} \cap X'$ such that $b \neq a_1, \ldots, a_n$. Let $f(a) = b$, $f(b) = a$, and $f(x) = x$ when $x \neq a,b$. Then f is an automorphism on $\boldsymbol{M}$ leaving $a_1, \ldots, a_n$ fixed, so $f[X'] = X'$, by Lemma 1.7.1. But this contradicts the fact that $a \in f[X'] - X'$. □

Now we restrict attention to logics with monadic quantifiers satisfying *ISOM*. For simplicity, assume that $L = L(\mathbf{Q})$, where $\mathbf{Q}$ is *binary*; the results below extend immediately to logics $L(\mathbf{Q}^i)_{i \in I}$, with monadic $\mathbf{Q}^i$.

The *quantifier rank* of L-formulas is defined inductively as follows:

$$qr(\phi) = 0, \text{ if } \phi \text{ is atomic,}$$

$$qr(\neg\phi) = qr(\phi)$$

$$qr(\phi \wedge \psi) = max(qr(\phi), qr(\psi)),$$

$$qr(\exists x\phi) = qr(\phi) + 1$$

$$qr(Qx(\phi, \psi)) = max(qr(\phi), qr(\psi)) + 1.$$

We write

$$\boldsymbol{M} \equiv_{n,\mathbf{Q}} \boldsymbol{M}'$$

to mean that the same $L(\mathbf{Q})$-sentences of quantifier rank at most n are true in $\boldsymbol{M}$ and $\boldsymbol{M}'$. $\boldsymbol{M} \equiv_{\mathbf{Q}} \boldsymbol{M}'$ ($\boldsymbol{M}$ and $\boldsymbol{M}'$ are $L(\mathbf{Q})$-*equivalent*) if, for all n, $\boldsymbol{M} \equiv_{n,\mathbf{Q}} \boldsymbol{M}'$. Our main tool will be an equivalent but more workable formulation of the relation $\equiv_{n,\mathbf{Q}}$. This is accomplished in the next definition. If $a_1, \ldots, a_n \in M$ and $b_1, \ldots, b_n \in M'$ we write $(a_1, \ldots, a_n) \simeq_p (b_1, \ldots, b_n)$ to mean that $\{\langle a_i, b_i\rangle: 1 \leqq i \leqq n\}$ is a *partial isomorphism* from $\boldsymbol{M}$ to $\boldsymbol{M}'$ (i.e., $a_i = a_j$ iff $b_i = b_j$, and $a_i \in A_m$ iff $b_i \in A'_m$).

In what follows, $\boldsymbol{M}$ and $\boldsymbol{M}'$ are k-ary monadic structures.

1.7.3. DEFINITION.

(a) $X \approx_n Y$ iff either $|X| = |Y| < n$ or $|X|, |Y| \geqq n$.

(b) $\boldsymbol{M} \approx_n \boldsymbol{M}'$ iff $P^{\boldsymbol{M}}_s \approx_n P^{\boldsymbol{M}'}_s$ for all $s \in 2^k$.

(c) $\boldsymbol{M} \approx_{n,\mathbf{Q}} \boldsymbol{M}'$ iff

(i) $\boldsymbol{M} \approx_n \boldsymbol{M}'$

(ii) If $(a_1, \ldots, a_{n-1}) \simeq_p (b_1, \ldots, b_{n-1})$, X_i, X_j are $\{a_1, \ldots, a_{n-1}\}$-variants of $U^{\boldsymbol{M}}_i$, $U^{\boldsymbol{M}}_j$, and Y_i, Y_j the *corresponding* $\{b_1, \ldots, b_{n-1}\}$-variants of $U^{\boldsymbol{M}'}_i$, $U^{\boldsymbol{M}'}_j$, then

$$\mathbf{Q}_M X_i X_j \Leftrightarrow \mathbf{Q}_{M'} Y_i Y_j.$$

1.7.4. THEOREM. $\boldsymbol{M} \equiv_{n,\mathbf{Q}} \boldsymbol{M}' \Leftrightarrow \boldsymbol{M} \approx_{n,\mathbf{Q}} \boldsymbol{M}'$.

Proof. $\Rightarrow$: It is clear that (i) holds. As for (ii), let $\psi_i(y, x_1, \ldots, x_{n-1})$, $\psi_j(y, x_1, \ldots, x_{n-1})$ be formulas which L-define X_i, X_j in $\boldsymbol{M}$ with parameters $a_1, \ldots, a_{n-1}$. Each a_p belongs to exactly one $P^{\boldsymbol{M}}_{s_p}$; let this set be defined by $\theta_p(x)$. If $\mathbf{Q}_M X_i X_j$, then

$$\boldsymbol{M} \models \exists x_1 \ldots x_{n-1}(\theta_1(x_1) \wedge \ldots \wedge \theta_{n-1}(x_{n-1}) \wedge$$
$$\wedge\ Qy(\psi_i(y, x_1, \ldots, x_{n-1}), \psi_j(y, x_1, \ldots, x_{n-1}))).$$

This sentence has quantifier rank n. Thus, it is also true in $\boldsymbol{M}'$, whence there are $b'_1, \ldots, b'_{n-1} \in M'$ such that $b'_p \in P^{\boldsymbol{M}'}_{s'_p}$ and

$$\boldsymbol{M}' \models Qy(\psi_i, \psi_j)[b'_1, \ldots, b'_{n-1}].$$

Let f map b'_p on b_p and leave everything else in M' as it is. It follows that f is an automorphism on $\boldsymbol{M}'$, so

$$\boldsymbol{M}' \models Qy(\psi_i, \psi_j)[b_1, \ldots, b_{n-1}].$$

But this means that $\mathbf{Q}_{M'} Y_i Y_j$. The converse is similar.

$\Leftarrow$: We prove by (downward) induction over $p \leqq n$ that

(*) if $(a_1, \ldots, a_p) \simeq_p (b_1, \ldots, b_p)$ and $qr(\phi) \leqq n - p$, then $\boldsymbol{M} \models \phi[a_1, \ldots, a_p] \Leftrightarrow \boldsymbol{M}' \models \phi[b_1, \ldots, b_p]$.

The case $p = 0$ gives the result. (*) is clear for $p = n$. So suppose (*) holds for p, $(a_1, \ldots, a_{p-1}) \simeq_p (b_1, \ldots, b_{p-1})$, and $qr(\phi) = n - p + 1$. We may suppose that ϕ begins with a quantifier symbol. If this symbol is $\exists$, the result follows easily from the induction hypothesis and the fact that $\boldsymbol{M} \approx_n \boldsymbol{M}'$. So suppose ϕ is $Qx(\psi_1, \psi_2)$. Let $\psi_i^{\boldsymbol{M}} = \{a \in M$: $\boldsymbol{M} \models \psi_i[a, a_1, \ldots, a_{p-1}]\}$, $i = 1,2$, and similarly for $\psi_i^{\boldsymbol{M}'}$. By Lemma 1.7.2, each $\psi_i^{\boldsymbol{M}}$ is an $\{a_1, \ldots, a_{p-1}\}$-variant of some union $U^{\boldsymbol{M}}_{j_i}$ of partition sets.

CLAIM: $\psi_i^{\boldsymbol{M}'}$ is the corresponding $\{b_1, \ldots, b_{p-1}\}$-variant of $U^{\boldsymbol{M}'}_{j_i}$.

The result follows immediately from the chain and (ii) above. The proof of the claim is straightforward, using the induction hypothesis together with the fact that $\boldsymbol{M} \approx_n \boldsymbol{M}'$. □

As noted, the theorem extends to logics with several monadic quantifiers (satisfying *ISOM*). We use this in the next corollary. A k-ary quantifier $\mathbf{Q}$ is said to be *closed under* $\approx_{n,\mathbf{Q}^1 \ldots \mathbf{Q}^m}$ if $\mathbf{Q}_M A_0 \ldots A_{k-1}$ and $\langle M, A_0, \ldots, A_{k-1}\rangle \approx_{n,\mathbf{Q}^1 \ldots \mathbf{Q}^m} \langle M', A'_0, \ldots, A'_{k-1}\rangle$ implies $\mathbf{Q}_{M'} A'_0 \ldots A'_{k-1}$.

1.7.5. COROLLARY. *A monadic quantifier* $\mathbf{Q}$ *is definable in* $L(\mathbf{Q}^1, \ldots, \mathbf{Q}^m)$ *if and only if, for some natural number n,* $\mathbf{Q}$ *is closed under* $\approx_{n,\mathbf{Q}^1 \ldots \mathbf{Q}^m}$.

Proof: (outline) If $\mathbf{Q}$ is defined by a sentence ϕ in $L(\mathbf{Q}^1, \ldots, \mathbf{Q}^m)$, i.e., if

$$\mathbf{Q}_M A_0 \ldots A_{k-1} \Leftrightarrow \langle M, A_0, \ldots, A_{k-1} \rangle \models \phi,$$

just let n be the quantifier rank of ϕ and use the theorem. Conversely, note that, with a fixed finite vocabulary there are, up to logical equivalence, only finitely many $L(\mathbf{Q}^1, \ldots, \mathbf{Q}^m)$-sentences of quantifier rank at most n. Now take the disjunction of all such sentences which are 'complete n-descriptions' of the models $\langle M, A_0, \ldots, A_{k-1} \rangle$ for which $\mathbf{Q}_M A_0 \ldots A_{k-1}$; this disjunction defines $\mathbf{Q}$. □

We will now apply these results to some particular monadic quantifiers. First, note the following special cases of Theorem 1.7.4:

(1) $\quad \mathbf{M} \equiv_n \mathbf{M}' \Leftrightarrow \mathbf{M} \approx_n \mathbf{M}'$,

(2) $\quad$ *If* $\mathbf{Q}$ *is first-order definable, then* $\mathbf{M} \equiv_{n,\mathbf{Q}} \mathbf{M}' \Leftrightarrow \mathbf{M} \approx_n \mathbf{M}'$.

Using this, one easily shows that quantifiers such as $\mathbf{Q}_\alpha$, $\mathbf{Q}_C$, $\mathbf{Q}_R$ are *not* first-order definable. Next, note that an $\{a_1, \ldots, a_{n-1}\}$-variant of U_i^M has cardinality $\geqq \aleph_\alpha$ iff U_i^M has cardinality $\geqq \aleph_\alpha$ iff one of the partition sets in U_i^M has cardinality $\geqq \aleph_\alpha$. Thus, when $\mathbf{Q} = \mathbf{Q}_\alpha$, we need only consider the partition sets (not variants of unions of them) in Definition 1.7.3(c). This makes it easy to show, for example, that if $\alpha \neq \beta$, $L(\mathbf{Q}_\alpha)$ and $L(\mathbf{Q}_\beta)$ have *incomparable* expressive power.

(3) $\quad L(\mathbf{Q}_R) \nleqq L(\mathbf{I})$.

Proof. By the theorem, it suffices to find, for each n, structures $\langle M, A \rangle$ and $\langle M', A' \rangle$ such that $\langle M, A \rangle \approx_{n,I} \langle M', A' \rangle$, $(\mathbf{Q}_R)_M A$, and $\neg(\mathbf{Q}_R)_{M'} A'$. But this is easy. For example, let $|A| = 4n$, $|M - A| = 2n$, $|A'| = 2n$, $|M' - A'| = 4n$. There are just four unions of partition sets to consider in each structure, and it is easy to verify that the conditions in Definition 1.7.3(c) are satisfied. □

(4) $\quad$ "$|A|$ *is even*" *is not expressible in* $L(\mathbf{more})$.

Proof. For each n, choose $M, M', A \subseteq M, A' \subseteq M'$ such that $|A| = 4n$, $|M - A| = |M' - A'| = n$, $|A'| = 4n + 1$. Then $\langle M, A \rangle \approx_{n,\mathbf{more}} \langle M', A' \rangle$, so $\langle M, A \rangle \equiv_{n,\mathbf{more}} \langle M', A' \rangle$ by the theorem, but $|A|$ is even and $|A'|$ is odd. □

The following result is from Barwise and Cooper [1981]:

(5) $L(\mathbf{most}) \nleq L(\mathbf{Q}_R)$, *i.e., $L(\mathbf{Q}_R)$ does not relativize.*

Proof. Given n, choose $\langle M, A_0, A_1\rangle, \langle M', A_0', A_1'\rangle$ such that $A_0 \cap A_1 = \varnothing, A_0' \cap A_1' = \varnothing, |A_0| = |A_1| = n, |M| = 6n, |A_0'| = n, |A_1'| = n + 1$, $|M'| = 6n + 2$. So $\varnothing$, A_0, A_1, $A_0 \cup A_1$ all have cardinalities less than their complements, and this continues to hold if $n - 1$ elements are 'moved around' in the model. The same holds for M', and it is then easy to see that $\mathbf{M} \approx_{n,\mathbf{Q}_R} \mathbf{M}'$. However, $\neg\mathbf{most}_M, A_0 \cup A_1\ A_1$ and $\mathbf{most}_{M'} A_0' \cup A_1'\ A_1'$. □

Similarly, we can prove that $\mathbf{Q}_C$ does not relativize. Note that only finite structures have been used so far. The next and final application involves infinite structures.

(6) $L(\mathbf{Q}_0) \nleq L(\mathbf{most})$.

Proof. This time, choose $\langle M, A\rangle$, $\langle M', A'\rangle$ such that $|M - A| = |M' - A'| = n$, $|A| = \aleph_0$, and $|A'| = 3n$. Again, it is not hard to see that $\langle M, A\rangle \approx_{n,\mathbf{most}} \langle M', A'\rangle$ (especially if we use the characterization of $\approx_{n,\mathbf{most}}$ given in Theorem 1.7.6 below), but A is infinite and A' is finite. □

Finally, we shall consider more closely the relative expressive power of **most** and **more**. Note first that the four properties of logics mentioned in 1.6 do not enable us to distinguish between these two quantifiers: we saw that $L(\mathbf{more})$ does not have any of these properties, and similar arguments establish that neither does $L(\mathbf{most})$. For example, if we replace the second conjunct in the sentence (1) in 1.6 by a sentence saying that, for each x (except the first) there is a *greatest* element $y < x$ with the property that most of the x-predecessors are not predecessors of y, then we again obtain a characterization of the natural number ordering.

The next result characterizes the relations $\equiv_{n,\mathbf{Q}}$ and $\equiv_{\mathbf{Q}}$ for monadic structures, when $\mathbf{Q}$ is **most** or **more**.

1.7.6. THEOREM.
(a) $\mathbf{M} \equiv_{n,\mathbf{more}} \mathbf{M}'$ *iff, whenever* $(a_1, \ldots, a_{n-1}) \simeq_p (b_1, \ldots, b_{n-1})$, X_i, X_j *are* $\{a_1, \ldots, a_{n-1}\}$*-variants of* $U_i^{\mathbf{M}}, U_j^{\mathbf{M}}$ *and* Y_i, Y_j *the corresponding* $\{b_1, \ldots, b_{n-1}\}$*-variants of* $U_i^{\mathbf{M}'}, U_j^{\mathbf{M}'}$, *we have* $|X_i| > |X_j| \Leftrightarrow |Y_i| > |Y_j|$.

(b) *For* $\equiv_{n,\mathbf{most}}$ *we have the same condition, except that* X_i, X_j (Y_i, Y_j) *are required to be disjoint.*

(c) $\boldsymbol{M} \equiv_{\mathbf{more}} \boldsymbol{M}'$ *iff* $\boldsymbol{M} \equiv_{\mathbf{most}} \boldsymbol{M}'$ *iff* $\boldsymbol{M} \approx_{\aleph_0} \boldsymbol{M}'$ *and, for all* $s, t \in 2^k$, $|P_s^{\boldsymbol{M}}| > |P_t^{\boldsymbol{M}}| \Leftrightarrow |P_s^{\boldsymbol{M}'}| > |P_t^{\boldsymbol{M}'}|$.

Proof. (a): This is Theorem 1.7.4, except that we must show that the condition on the right hand side of the equivalence ((ii) in Definition 1.7.3 (c)) implies that $\boldsymbol{M} \approx_n \boldsymbol{M}'$. So suppose first $|P_s^{\boldsymbol{M}}| < n$. Suppose also that $|P_s^{\boldsymbol{M}'}| \neq |P_s^{\boldsymbol{M}}|$, say $|P_s^{\boldsymbol{M}'}| < |P_s^{\boldsymbol{M}}|$ (the other case is similar). If $P_s^{\boldsymbol{M}'} = \{b_1, \ldots, b_r\}$, choose $a_1, \ldots, a_r \in P_s^{\boldsymbol{M}}$ and let $Y_i = \varnothing = P_s^{\boldsymbol{M}'} - \{b_1, \ldots, b_r\}$ and $Y_j = \varnothing$. It follows from the condition that $X_i = P_s^{\boldsymbol{M}} - \{a_1, \ldots, a_r\}$ is empty, contradicting our assumption. The case when $|P_s^{\boldsymbol{M}}| \geqq n$ is similar.

(b): From left to right, note that **most** allows us to compare the cardinalities of *disjoint* sets $X, Y \subseteq M$: then $|X| > |Y|$ iff $\mathbf{most}_M X \cup Y\ X$. In the other direction, observe first that the argument in (a) above goes through under the disjointness requirement. Moreover, the proof of Theorem 1.7.4 (⇐:) also goes through under this requirement, since the formula *most* $x(\psi_1, \psi_2)$ only 'compares' disjoint sets.

(c): Clearly $\boldsymbol{M} \equiv_{\mathbf{more}} \boldsymbol{M}'$ implies $\boldsymbol{M} \equiv_{\mathbf{most}} \boldsymbol{M}'$, which in turn implies the rightmost condition in (c). Now suppose that condition holds; we must show that, for all n, $\boldsymbol{M} \approx_{n,\mathbf{more}} \boldsymbol{M}'$. So take n, and suppose $a_1, \ldots, a_{n-1}$, $b_1, \ldots, b_{n-1}, X_i, X_j, Y_i, Y_j$ are as in (a) above. We assume $|X_i| > |X_j|$ and show that, in this case, $|Y_i| > |Y_j|$; the other direction is similar.

Case 1: X_i and X_j are both finite. Then the partition sets in $U_i^{\boldsymbol{M}}$ are finite and thus have the same cardinality as the corresponding partition sets in $U_i^{\boldsymbol{M}'}$, since $\boldsymbol{M} \approx_n \boldsymbol{M}'$ for all n. X_i differs from $U_i^{\boldsymbol{M}'}$ only by certain of the $a_1, \ldots, a_{n-1}$, and Y_i differs in the same way from $U_i^{\boldsymbol{M}'}$. Therefore, $|X_i| = |Y_i|$ and $|X_j| = |Y_j|$, and the conclusion follows.

Case 2: X_i and X_j are both infinite. Then $|X_i|$ is the *max* of the cardinalities of the partition sets making up $U_i^{\boldsymbol{M}}$; say, $|X_i| = |P_s^{\boldsymbol{M}}|$, and similarly $|X_j| = |P_t^{\boldsymbol{M}}|$. It then follows from the condition in (c) that $|Y_i| = |P_s^{\boldsymbol{M}'}|$ and $|Y_j| = |P_t^{\boldsymbol{M}'}|$. Since $|P_s^{\boldsymbol{M}}| > |P_t^{\boldsymbol{M}}|$ we have, again by the condition, $|P_s^{\boldsymbol{M}'}| > |P_t^{\boldsymbol{M}'}|$.

Case 3: X_i is infinite and X_j is finite. Arguing as in Cases 1 and 2, we see that Y_i is infinite and Y_j is finite. □

Thus, the relations $\equiv_{\mathbf{most}}$ and $\equiv_{\mathbf{more}}$ coincide on monadic structures (but *not* the relations $\equiv_{n,\mathbf{most}}$ and $\equiv_{n,\mathbf{more}}$). Nevertheless, L(**more**) is more expressive than L(**most**), even if we restrict attention to monadic structures, as the next result will show. Another instance of the same phenomenon is given by the fact that

$$\boldsymbol{M} \equiv_{\mathbf{Q}_0} \boldsymbol{M}' \Leftrightarrow \boldsymbol{M} \equiv \boldsymbol{M}'$$

(this is an easy consequence of Theorem 1.7.4), but $EL < L(\mathbf{Q}_0)$ (even on monadic structures).

The following theorem holds in general, but it is also true if only monadic structures are considered.

1.7.7. THEOREM.
(a) $L(\mathbf{most}) < L(\mathbf{more})$.
(b) $L(\mathbf{most}) \equiv L(\mathbf{more})$ *on finite structures.*
(c) $L(\mathbf{more}) \equiv L(\mathbf{most}, \mathbf{Q}_0)$.

Proof: (a): Clearly $L(\mathbf{most}) \leqq L(\mathbf{more})$. That $\mathbf{L(more)} \nleqq L(\mathbf{most})$ follows from (6) and Theorem 1.6.2.
(b): This follows from the fact that, if $A \cap B$ is finite, then $\mathbf{more}_M AB \Leftrightarrow |A| > |B| \Leftrightarrow |A - B| > |B - A| \Leftrightarrow \mathbf{most}_M A \oplus B A$.
(c): We must show that $L(\mathbf{more}) \leqq L(\mathbf{most}, \mathbf{Q}_0)$. Take any M and $A,B \subseteq M$. If $A \cap B$ is finite, $\mathbf{more}_M AB$ is expressed as in (b). If $A \cap B$ is infinite, then $|A| = max(|A - B|, |A \cap B|)$ and $|B| = max(|B - A|, |A \cap B|)$. It follows that

$$|A| > |B| \Leftrightarrow |A - B| > |B - A| \;\&\; |A - B| > |A \cap B|,$$

and the right hand side of this is again expressible with *most* (since only disjoint sets are compared). Moreover, $\mathbf{Q}_0$ allows us to distinguish the two cases, in one sentence of $L(\mathbf{most}, \mathbf{Q}_0)$. □

This theorem tells us that the difference between $L(\mathbf{more})$ and $L(\mathbf{most})$ is *precisely* that the former, but not the latter, can distinguish between infinite and finite sets.

The results of this section allow us to extend Theorem 1.6.2 to the following picture:

$L(\mathbf{Q}_0)$ — $L(\mathbf{I})$
EL < > $L(\mathbf{more})$ — $L(\mathbf{H})$
$L(\mathbf{Q}_R)$ — $L(\mathbf{most})$

Here each logic is strictly stronger than its immediate predecessor(s), and logics not on the same branch are incomparable.

Remark: The only thing in the figure above not proved with the simple methods used here is the fact that $L(\mathbf{H})$ is strictly stronger than $L(\mathbf{more})$. However, if we consider the logic L^{po}, where not only the Henkin prefix but *all* partially ordered prefixes with $\forall$ and $\exists$ are allowed, then it follows from (4) that $L^{po} \nleq L(\mathbf{more})$. For,

$$|A| \text{ is even} \Leftrightarrow \exists X \subseteq A(|X| = |A - X|),$$

which can be expressed as a Σ^1_1 sentence, and it is shown in Enderton [1970] and Walkoe [1970] that all such sentences are expressible in L^{po}.

Is "$|A|$ is even$|$" expressible in $L(\mathbf{H})$? More generally, is $L(\mathbf{H})$ strictly stronger than $L(\mathbf{more})$ if we restrict attention to monadic structures? I don't known the answer to these questions, but it may be noted that it follows from Theorem 1.7.6 and a result in Lachlan and Krynicki [1979] that $\equiv_{\mathbf{more}}$ and $\equiv_{\mathbf{H}}$ coincide for monadic structures.

Bibliographical remark: The theorems in this section have not, to my knowledge, appeared in the literature, although no doubt they belong to the folklore in some circles. Most of the applications to particular logics are known, but it should be noted that the methods used here are much more elementary than the ones that have been used in the literature (the proof of (5) in Barwise and Cooper [1981] is an exception). For example, it is proved in Hauschild [1981] and Weese [1981] that $L(\mathbf{more})$ is strictly stronger than $L(\mathbf{I})$, by establishing that these logics have different properties w.r.t. the decidability of certain theories formulated in them. The same result follows from the simple observation (3); in a sense, (3) gives more, since it concerns monadic structures, whereas the theories just mentioned use non-monadic languages.

2. NATURAL LANGUAGE QUANTIFIERS

A main objective of Montague's PTQ (Montague [1974]) was to show that intensional phenomena, such as quantification into intensional contexts, could be handled rigorously with model-theoretic methods. But even if one completely disregards the intensional aspects of PTQ, its approach to quantification was novel. Although it had no category 'quantifier' or 'determiner', a general pattern is discernible from its treatment of the three quantifier expressions (*every*, *a*, and *the*) it in fact did account for. The basic idea is that *quantifier expressions occur as determiners in noun phrases*. By the close correspondence between syntax and semantics in Montague Grammer, this also determines the interpretation of such expressions.

In this section, I will describe this idea in somewhat more detail, and its later development in Barwise and Cooper [1981] and Keenan and Stavi [1986], within the generalized quantifier framework of Section 1.

2.1. *Determiners*

Suppose that the expressions of the categories *common noun* (N) and *noun phrase* (NP) have somehow been (roughly) identified.[1] Since we are disregarding intensions, the semantic types of these expressions are such that Ns are interpreted, in a model $\boldsymbol{M} = \langle M, \| \ \| \rangle$ with universe M and interpretation function $\| \ \|$, as *subsets* of M, and NPs as *sets of subsets* of M. Here are three examples from PTQ:

$$\| \textit{every man} \| = \{X \subseteq M: \| \textit{man} \| \subseteq X\},$$

$$\| \textit{a man} \| = \{X \subseteq M: \| \textit{man} \| \cap X \neq \varnothing\},$$

$$\| \textit{the man} \| = \{X \subseteq M: | \| \textit{man} \| | = 1 \;\&\; \| \textit{man} \| \subseteq X\}.$$

Many NPs, like the above ones, are naturally regarded as the result of applying a syntactic *operator* to Ns. We introduce the syntactic category *determiner* (DET) for this sort of operator:

(DET) *DETs form NPs from Ns.*

This is a rough criterion, but, in a Montagovian framework, it is enough to fix the syntax and semantics of determiners. In particular, DETs are interpreted as *functions* from N denotations to NP denotations. For example,

$$\| \textit{every} \| (A) = \{X \subseteq M: A \subseteq X\},$$

$$\| \textit{a} \| (A) = \{X \subseteq M: A \cap X \neq \varnothing\},$$

$$\| \textit{the} \| (A) = \{X \subseteq M: |A| = 1 \;\&\; A \subseteq X\}.$$

Another thing, of course, is to apply the criterion to identify simplex and complex English DETs; we will return to this in 2.4.

2.1.1. *Three apparent problems.* As noted, the basic idea of the present Montague-style treatment of quantification is this:

(Q) *Quantifier expressions are DETs.*

This may not yet seem very exciting, but note at least that it differs, syntactically as well as semantically, from the standard predicate logic treatment of quantification. The import of (Q) will become clear as we go along. For the moment, however, let us look at a few apparent *counter-instances* to (Q) that come to mind.

I. In sentences like

(1) All cheered,

(2) Some like it hot,

(3) Few were there to meet him,

the words *all, some, few* are not applied to arguments of category *N*. Isn't the standard predicate logic analysis more plausible here? No, it is very natural to assume that the *DET*s have 'dummy' arguments in these sentences (with context-given interpretations); in this case (Q) still holds (cf. 2.4.5).

II. Words like *something, everything, nothing, nobody*, etc. look like quantifier expressions but are certainly not *DET*s. We have two options here. The first is to regard them as simplex *NP*s, denoting quantifiers of type $\langle 1 \rangle$ in the sense of 1.4. They would then correspond (roughly) to the standard logical $\forall$ and $\exists$. The other option, which we will take here, is to regard them as *complex*: *something* = *some*(*thing*), *nothing* = *no*(*thing*), etc.; i.e. obtained by applying a *DET* to a (perhaps logical) *N* like *thing*. In this way, (Q) can be maintained.

III. In 1.4 we defined the binary quantifier **more**. The word *more*, however, is not a *DET* by our criterion; compare

(4) Some boys run,

(5) Most boys run,

(6) *More boys run.[2]

Still, *more* does occur in quantified sentences, for example,

(7) There are more girls than boys,

which in generalized quantifier notation becomes

(8) *more x*(*girl*(x), *boy*(x)).

This is an objection to (Q) that must be taken seriously. It involves (i) finding a semantic distinction between the quantifiers **more** and, say, **most**, which explains why one but not the other is a *DET* denotation; (ii) the analysis of "there are"-sentences; (iii) the semantics of words like *more*. These matters will be taken up in 2.2.

2.1.2. *Determiner interpretations as generalized quantifiers.* Following Montague, we interpreted *DET*s as *functions* from subsets of the universe M to sets of such subsets. From now on, however, we return to the generalized quantifier framework of Section 1, where quantifiers on M are *relations* between subsets of M. Thus, to each n-place function $\mathbf{D}$ from $(P(M))^n$ to $P(P(M))$ we associate the following $(n + 1)$-ary quantifier on M:

$$\mathbf{Q}_M A_1 \ldots A_n B \Leftrightarrow B \in \mathbf{D}(A_1, \ldots, A_n).$$

In what follows, *DET* interpretations will be such monadic quantifiers on the universe.

The functional interpretation of *DET*s emphasizes similarity of structure between syntax and semantics, which is one of the characteristics of Montague Grammar. From the present semantic perspective, however, relations turn out to be easier to work with. But keep in mind that the relational approach increases the number of arguments by one: n-place *DET*s will denote $(n + 1)$-ary quantifiers (so far we have only seen 1-place *DET*s, but cf. 2.2). It should also be noted that for *some* semantic issues, the functional framework seems more natural; cf. Keenan and Moss [1985].

Terminological Remark: The use of the words "determiner" and "quantifier" is rather shifting in the literature. Here, the idea is to use "determiner" and "*DET*" *only* for syntactic objects, and "quantifier" *only* for semantic objects. The extension of "quantifier" was given in Section 1.4, and a criterion for *DET*-hood at the beginning of 2.1.

2.1.3. *Determiners as constants.* In a Montague-style model $\boldsymbol{M} = \langle M, \| \ \| \rangle$, *DET*s are on a par with expressions of other categories. Nothing in principle prevents, for example, that a determiner D is interpreted as **every** in one model and as **most** in another. But there is usually no point in allowing this generality. Moreover, there is a clear intuition, I think, that determiners are *constants*. We therefore lay down the following *methodological postulate*:

(*MP*) *Simplex DETs are constants: each one denotes a fixed quantifier* (modulo, of course, lexical ambiguity, vagueness, etc.; cf. 2.4).

(*MP*) allows us to dispense with the interpretation function for (simplex) *DET*s, and to resume the notation from 1.4, using boldface letter for quantifiers: Q denotes $\mathbf{Q}$, *most* denotes **most**, *some* denotes **some**, etc.

What about complex *DET*s? In case such a *DET* contains a non-constant expression, there seems to be a choice. We can either persist in treating them as constants, or let their interpretation depend on the interpretation of the non-constant expressions occurring in them. To take a simple example, consider *some red*. This expression *can* be construed as giving an *NP* when applied to an *N*, thus *can* be classified as a *DET* by our criterion. As a constant, it would denote the quantifier defined by

$$\mathbf{some\ red}_M AB \Leftrightarrow A \cap B \cap \{a \in M\colon a \text{ is (in fact) red}\} \neq \varnothing,$$

for each universe M. As an expression consisting of a constant and a non-constant symbol, i.e., of the form *some P*, it is interpreted in a model $\boldsymbol{M}$ as

$$\| some\ P \| AB \Leftrightarrow A \cap B \cap \| \mathrm{P} \| \neq \varnothing.$$

Given $\boldsymbol{M}$, this is a quantifier on M, but the expression does not denote a fixed quantifier on each universe.

No doubt many readers will find the latter option more natural, but we need not take a stand on this methodological issue here. Our model-theoretic machinery provides adequate semantic objects for both cases, quantifiers, and quantifiers *on* universes, respectively.

Note, however, that our decision to treat simplex *DET*s as constants does not necessarily imply that they are *logical* constants. It can be argued that logicality requires a lot more; this theme will be resumed 3.4 and 4.4 (cf. also Westerståhl [1985a]). For example, the quantifier **some red** defined above is not logical, the reason being that it violates the condition *ISOM* from 1.4.

In Appendix B we will indicate what happens if the postulate (*MP*) is dropped.

2.1.4. *Global vs. local perspective.* To study quantifiers from a *global* perspective means to concentrate on properties which are *uniform* over universes. A typical example is first-order definability: $\mathbf{Q}$ is first-order definable if there is some first-order sentence which defines it on *every* universe. Sometimes, however, it is natural to take a *local* viewpoint: Fix a universe M and restrict attention to quantifiers on M. Then other definability notions become interesting as well, involving parameters from M in an essential way.

Our perspective here will be predominantly global. The main reason for this is that global definitions and results are more general: they usually have an immediate 'local version'. The converse, however, does not hold. Quantifiers from a local perspective are studied extensively in Keenan and Stavi [1986]. Some of their results will be reviewed in 4.6.

2.2. *The interpretation of determiners*

The basic quantifier postulate (*Q*) from 2.1.1 can be split into a syntactic and a semantic part as follows:

(Q_{syn}) *Quantifier expressions are DETs.*

(Q_{sem}) *DETs denote* $(n + 1)$*-ary quantifiers,* $n \geqq 1$.

In contrast with standard predicate logic, there are no unary quantifiers on this approach. And although some binary *DET* denotations (e.g. Montague's **every**, **a**, **the**) are definable in standard predicate logic, others are not: we saw in 1.7 that **most** is an example. Consequently, *EL* is inadequate for formalizing even the pure quantificational part of natural languages.

However, (*Q*) is not yet quite satisfactory. In particular, we need to account for the apparent counter-examples mentioned in 2.1.1, III. Nothing so far precludes **more** from being a *DET* denotation.

The starting-point of a systematic study of natural language quantification was the isolation, in Barwise and Cooper [1981], and independently in Keenan and Stavi [1986] (although the latter paper was published much later, they were written at about the same time), of a purely model-theoretic property characteristic of those quantifiers that are *DET* denotations. This is the property of *conservativity*, defined below (Barwise and Cooper used a different terminology, in terms of an *NP* denotation *living on* a given set). Actually, the property (and the term) first appeared in Keenan [1981], but in the two first-mentioned papers it was proposed as a significant semantic universal for determiners (although with rather different motivations; cf. below).

2.2.1. *Conservativity.* A binary quantifier **Q** is called *conservative* if the following holds:

CONSERV *For all M and all* $A, B \subseteq M$, $\mathbf{Q}_M AB \Leftrightarrow \mathbf{Q}_M A\ A \cap B$.

It is easily checked that **most** is conservative, but **more** is not. As we will see in 2.4, practically all English *DET*s denote conservative quantifiers (a few possibly doubtful cases will be noted).

CONSERV gives the *first* argument of **Q** a privileged role: only the part of B which is common to A matters for whether $\mathbf{Q}_M AB$ holds or not. This semantic difference between the arguments A and B matches the syntactic difference between the corresponding expressions:

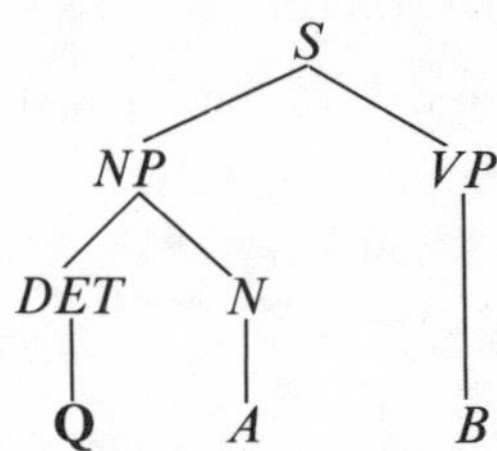

Conservativity is a very fruitful postulate, as will be seen in Sections 3 and 4. Still, one may ask what, if any, is the idea or intuition behind it. As for Barwise and Cooper, they seem to regard it mainly as a successful empirical generalization. Keenan and Stavi, on the other hand, give an interesting theoretical motivation: they prove that, on a given (finite) universe M, the conservative quantifiers on M are precisely those which can be generated from certain initial quantifiers by means of a few natural closure operations; an exact statement (and proof) will be given in 4.6. Yet another motivation, discussed in Westerståhl [1985a], is that *CONSERV* is related to the notion of *restricted domains of quantification*: an *NP* 'restricts' the universe to the denotation of the *N*; this will be formulated in 3.2.

CONSERV resolves the first doubt concerning (*Q*) expressed in 2.1.1, III. We still have to deal with "there are"-sentences and with the semantics of *more*.

2.2.2. *"there are"-sentences.* Consider sentences such as

(1) There are no flowers,

(2) There are many patients waiting outside,

(3) There are some philosophers who like logic,

(4) There are a few errors in the text.

Without commitment to their syntactic form, let us write such sentences

(5) *There are* $\mathbf{Q}_M A$,

where **Q** is the quantifier denoted by the *DET* and *A* is the set contributed, in a model ***M***, by the expression following the *DET*.[3] There are in fact two questions here. The first is to interpret quantified sentences of the form (5) in a way consonant with the basic postulate (*Q*). The second concerns the fact that certain *DETs* do not fit in (5): *all, most, not all*, for example. Is there a semantic explanation for this phenomenon?

We shall review the answers proposed by Barwise and Cooper to both of these questions, first, because they show a way to handle "there are"-sentences, and second, because this case can serve as a model of the kind of linguistic explanation one may expect from the present theory of quantifiers.

The first proposal is simple: interpret (5) as

(6) $\mathbf{Q}_M AM$.

This interpretation *works* in the sense that it gives (1)–(3) the right truth conditions. Moreover, one can argue that it accounts for the idea that the phrase "there are" serves to ascribe *existence*, i.e. the property that everything in the universe has, to the rest of the sentence.

But why are some choices of *Q* apparently forbidden in (5)? First, a definition. Call a *DET strong*, if its denotation, as a binary relation, is either reflexive or irreflexive; otherwise the *DET* is *weak*. Now observe that the *DETs* that fit in (5) are weak, whereas the exceptions are strong.[4] This is still no explanation, but it is a *fact* which may point to one. The next move is theoretical: we *prove* in our theory that (6) is equivalent to

(7) $\mathbf{Q}_M AA$;

this is actually an immediate consequence of *CONSERV*. It follows that

If Q is strong, (5) *is either trivially true or trivially false.*

Thus, the connection between the strong/weak distinction and our problem has not merely been *described*; it has been *explained*, given the plausible assumption that it is in general 'strange' to utter trivial truths or falsities.

This simple but instructive model of explanation shows the typical interplay between linguistic facts, theoretical concepts, and results in the theory. Here the results used were quite trivial, but this may not always be the case.

Let me hasten to add that the above by no means exhausts the many interesting problems connected with "there are"-sentences. Moreover,

Keenan and Stavi [1986] argue against the explanation in terms of the strong/weak distinction; they propose another semantic characterization of the relevant class of determiners (a detailed discussion of these matters can be found in Keenan [1984]). But it is the *type* of explanation that I have tried to illustrate here.

2.2.3. *(n + 1)-ary conservative quantifiers.* Now, what about *more*? We noted that

(8) There are more P than Q

is the natural rendering of the logical

(9) *more* $x(Px, Qx)$.

Observe further that *more P than Q* is very naturally considered as an *NP*, obtained by applying the 2-place *DET more . . . than* to two *N*s, and typically occurring in sentences such as

(10) More men than women voted for Smith.

(10) means that the number of men who voted for Smith is greater than the number of women who voted for Smith. So *more . . . than* denotes a *ternary* quantifier:

$$\textbf{more . . . than}_M A_1 A_2 B \Leftrightarrow |A_1 \cap B| > |A_2 \cap B|.$$

Other examples of such ternary quantifiers are

$$\textbf{fewer . . . than}_M A_1 A_2 B \Leftrightarrow |A_1 \cap B| < |A_2 \cap B|,$$

$$\textbf{as many . . . as}_M A_1 A_2 B \Leftrightarrow |A_1 \cap B| = |A_2 \cap B|.$$

We now see that (8) can be written as a generalization of (5) to ternary quantifiers:

(11) *There are* $\mathbf{Q}_M A_1 A_2$.

Furthermore, (11) can be interpreted on exactly the same principle as (5), namely, as

(12) $\mathbf{Q}_M A_1 A_2 M$.

For example, if P denotes A and Q denotes B, the interpretation of (8) is

$$\textbf{more . . . than}_M ABM,$$

which is equivalent to

$$|A| > |B|,$$

i.e., to

$$\textbf{more}_M AB,$$

as predicted. So the previous analysis of "there are"-sentences with binary quantifiers extends naturally to ternary (in fact, $(n + 1)$-ary) quantifiers. (The reader might wish to ponder whether the characterization in terms of the strong/weak distinction also generalizes; cf. Keenan [1984]).

Finally, the notion of conservativity also extends to $(n + 1)$-ary quantifiers: the set to which the *VP* denotation can be restricted is then the *union* of the *N* denotations. We get the following general version of *CONSERV* for $(n + 1)$-ary quantifiers:

CONSERV *For all M and all* $A_1, \ldots, A_n, B \subseteq M$,
$\mathbf{Q}_M A_1 \ldots A_n B \Leftrightarrow \mathbf{Q}_M A_1 \ldots A_n (A_1 \cup \ldots \cup A_n) \cap B.$

It is easily verified that **more . . . than**, **fewer . . . than**, **as many . . . as** are all conservative, in contrast with the binary **more**.

In conclusion, our findings about the use of *more* do not contradict the basic idea (*Q*), on the contrary, they support it. A final formulation of this idea goes as follows (cf. the beginning of 2.2):

(Q_{syn}) *Quantifier expressions are DETs.*

(Q_{sem}) *n-place DETs denote* $(n + 1)$*-ary conservative quantifiers,* $n \geqq 1$.

We should perhaps note that there are other uses of *more* in determiners, for example, *more than ten* or *six or more*. These are ordinary (complex) 1-place *DET*s, and denote binary conservative quantifiers, just as (Q_{sem}) predicts (cf. also 2.4.7).

2.3. *Subject-predicate Logic*

As in Montague Grammar, Barwise and Cooper use an intermediate logical language, called *L*(*GQ*), into which a fragment of English is translated. *L*(*GQ*) has two unusual features:

(i) Quantified sentences have *NP–VP* form (subject-predicate form).
(ii) Quantifier symbols are not used as variable-binding operators.

The wellformed expressions in *L*(*GQ*) are of two kinds: *formulas* and *set terms*. A set term is either a unary predicate symbol or an expression of the form

$$\hat{x}[\psi],$$

where x is a variable and ψ a formula; in models, set terms denote subsets of the universe. Variable-binding is done with the *abstraction operator* $\hat{}$. Quantifier symbols are (certain) 1-place *DET*s, and quantified formulas are of the form

(*) $D(\eta)(\delta)$,

where D is a *DET* and η, δ are set terms. There are the usual atomic formulas, plus formulas of the form $\eta(t)$, where η is a set term and t an individual term, and the formulas are closed under sentential connectives. *DET*s are interpreted as binary conservative quantifiers; the truth condition for (*) in a model is then obvious.

The result is that logical form in $L(GQ)$ corresponds more closely to syntactic form in the fragment than usual. (*) can be said to have *NP–VP* form with $D(\eta)$ as the *NP* and δ as the *VP* (the formation rules actually give (*) this structure). Another pleasant feature is that some unnecessary uses of bound variables are avoided. For example,

(1) Some boys run

is translated

(1′) $some(boy)(run)$

instead of the usual

(1″) $\exists x(boy(x) \wedge run(x))$.

The example also shows that certain unnecessary sentential connectives in the standard formalization are avoided. In more complex cases, e.g., with transitive verbs or relative clauses, $L(GQ)$ must introduce variables and connectives (though English often can avoid them): consider

(2) Most women who love Harry have a cat,

(2′) $most\ (\hat{x}[woman(x) \wedge love(x, Harry)])(\hat{x}[a(cat)(\hat{y}[have(x,y)])])$,

(2″) $most\ x(woman(x) \wedge love(x, Harry), \exists y(cat(y) \wedge have(x,y)))$.

These examples should make it plausible that there is no deep difference between $L(GQ)$ and the standard language for generalized quantifiers as in 1.4. In fact, they are even syntactically intertranslatable in a rather obvious way. Still, quantified formulas in $L(GQ)$ have subject-predicate form. It is hard to avoid the conclusion that the importance of the issue of whether subject-predicate form occurs in logic has been greatly overestimated, from Russell and onwards.

2.4. *Some Natural Language Quantifiers*

A quantifier **Q** will be called a (*simple*) *natural language quantifier*, if it is denoted by some (simplex) natural language *DET*.

This notion is somewhat loose, but it serves our purposes. A more exact specification would presuppose, among other things, (i) that the class of *DET*s has been more precisely delimited; (ii) that it has been decided how to treat complex non-logical *DET*s (2.1.3); (iii) that a global or a local perspective has been chosen (2.1.4). We may think of the notion of a natural language quantifier as having various *parameters*, which can be set at different values. It turns out that, for many of the things we shall have to say about natural language quantifiers, the value of these parameters is immaterial. This is why the above 'loose' notion is useful. And in other cases, we will indicate how a particular observation on natural language would depend on different parameter settings.

To take a first and crude example, consider the assertion that *not all binary quantifiers are natural language quantifiers*. From a global perspective, or from a local perspective with a given infinite universe *M*, this is true for cardinality reasons: there are uncountably many binary quantifiers (on *M*), but at most countably many natural language quantifiers. But, even from a finite local perspective, the assertion is true for another reason, namely, the conservativity universal (e.g. **more** or $\mathbf{more}_M$ is not a natural language quantifier). The other parameter settings are clearly irrelevant here, so the assertion is true however the parameters are set. An example of an assertion whose truth does depend on the parameters is this: *All natural language quantifiers satisfy ISOM*. We will see in 3.3 that this is in fact a candidate for a quantifier universal, but only under a certain delimitation of the class of *DET*s.

In the remainder of this section, I will present a list of examples of natural language quantifiers. Some of them will be used later on, but the list is also intended to give the reader a feeling for the perhaps surprising richness of the class of natural language quantifiers.

The method is simply to list various English *DET*s, together with their semantic interpretations (when these are not obvious). The *DET*s are selected by using the criterion for *DET*-hood in 2.1 as liberally as possible, but with some 'common sense' (standard co-occurrence criteria for constituenthood, etc.). Thus I will be listing *possible DET*s – there may be syntactic, semantic, or methodological reasons for discarding several of them from a more definitive list. In fact, some such reasons will be discussed in what follows.

The main sources for the list that follows are Keenan and Stavi [1986] and Keenan and Moss [1985]. The reader is referred to these works for further examples, and for detailed arguments that most of the expressions listed really belong to the category *DET*.

2.4.1. *Some simplex DETs.*

(1) *all, every, each, some, a, no, zero, most*

(2) *both, neither*

(3) *one, two, three, . . .*

(4) *many, few, several, a few*

(5) *the*

(6) *this, that, these, those*

(7) *more . . . than, fewer . . . than, as many . . . as*

Here are some interpretations, a few of which have already been given

$$\mathbf{all}_M AB \Leftrightarrow \mathbf{every}_M AB \Leftrightarrow \mathbf{each}_M AB \Leftrightarrow A \subseteq B,$$

$$\mathbf{some}_M AB \Leftrightarrow \mathbf{a}_M AB \Leftrightarrow A \cap B \neq \varnothing,$$

$$\mathbf{no}_M AB \Leftrightarrow \mathbf{zero}_M AB \Leftrightarrow A \cap B = \varnothing,$$

$$\mathbf{most}_M AB \Leftrightarrow |A \cap B| > |A - B|,$$

$$\mathbf{both}_M AB \Leftrightarrow \mathbf{all}_M AB \;\&\; |A| = 2,$$

$$\mathbf{neither}_M AB \Leftrightarrow \mathbf{no}_M AB \;\&\; |A| = 2,$$

$$\mathbf{one} = \mathbf{some},$$

$$\mathbf{two}_M AB \Leftrightarrow |A \cap B| \geqq 2,$$

$$\mathbf{three}_M AB \Leftrightarrow |A \cap B| \geqq 3, \ldots$$

So **n** is interpreted as **at last n** here, although it can be argued that it sometimes means **exactly n**. As for (4)–(6), cf. 2.4.2–6 below. The denotation of the 2-place *DET*s in (7) were given in 2.2.3.

2.4.2. *Vague DETs.* Vagueness in the sense of the occurrence of *borderline cases* (in some suitable sense) pertains to *DET*s as well as to other expressions. We do not incorporate a theory of vagueness here, but choose idealized precise versions instead.

Two examples of vague *DETs* are *several* and *a few*. Here one may, following Keenan and Stavi, stipulate that

several = **three**,

a few = **some**.

2.4.2. *Context-dependent DETs*. The *DETs* *many* and *few* are not only vague but also context-dependent in the sense that the 'standard of comparison' may vary with the context. For example, in

(8) Many boys in the class are right-handed,

(9) Lisa is dating many boys in the class,

some 'normal' standard for the least number considered to be many is used, but probably different standards in the two cases. Even within one sentence different standards may occur, as in the following example (due to Barbara Partee):

(10) Many boys date many girls.

Other, complex, *DETs* with a similar behavior are, for example,

a large number of, unexpectedly few, unusually many.

Westerståhl [1985a] discusses various interpretations of *many*. Basically, there are two possible strategies. Either one excludes this type of *DETs* from extensional treatments such as the present one (this is what Keenan and Stavi do), or one tries to capture what *many* might mean in a *fixed* context (this is the approach of Barwise and Cooper). Here are some suggestions for the second strategy:

$$\mathbf{many}^1_M AB \Leftrightarrow |A \cap B| \geqq k|M| \qquad (0 < k < 1),$$

$$\mathbf{many}^2_M AB \Leftrightarrow |A \cap B| \geqq k|A| \qquad (0 < k < 1),$$

$$\mathbf{many}^3_M AB \Leftrightarrow |A \cap B| \geqq (|B|/|M|)|A|.$$

$\mathbf{many}^1$ relates the standard to the size of the universe: in a universe of 10, 5 may be many, but not in a universe of 1000. $\mathbf{many}^2$ is a *frequency* interpretation: the number of *A*s that are *B*, compared to the total number of *A*s, is at least as great as a 'normal' frequency of *B*s, given by *k*. In both cases, *k* has to be supplied by the context. But in $\mathbf{many}^3$, the 'normal' frequency of *B*s is just the actual frequency of *B*s in the universe.

Notice that $\mathbf{many}^1$ and $\mathbf{many}^3$ make essential reference to the *universe* of the model. As we shall see, this is in contrast with most other natural language quantifiers. Also notice that $\mathbf{many}^3$ is *not conservative*. Since the conservativity universal is so central, this observation gives a (methodological) argument for discarding $\mathbf{many}^3$ as an interpretation of *many*.

As for *few*, we may simply interpret it as *not many*.

2.4.4. *Ambiguous DETs.* Ambiguity in the sense of a small number of clearly distinguishable meanings of a *DET* is another phenomenon than context-dependence. We have already noted that the *DET*s *one, two, three, . . .* may be ambiguous with respect to the quantifiers **at least n** and **exactly n**. Another possibly ambiguous *DET* is *most*: it can be argued that aside from the interpretation we have given, *most* can also mean something like *almost all*; cf. Westerståhl [1985a].

The fact that certain *DET*s may be ambiguous is not a problem in the present context, as long as we make sure to include each of their interpretations among the natural language quantifiers.

2.4.5. *Pronominal DETs.* Most 1-place *DET*s can occur without their *N* arguments, as was noted in 2.1.1. Such *DET*s may be called *pronominal.* The natural analysis of sentences with pronominally occurring *DET*s is that the argument (or the set it denotes) is given by the context. So

All cheered

is interpreted as

$$\mathbf{all}_M X \,\|\, \textit{cheered} \,\|,$$

where the set X is provided by the context. The use of such *context sets* is studied further in Westerståhl [1985b].

The only non-pronominal 1-place *DET*s encountered so far are, as the reader can check,

a, every, no, the.

Moreover, *DET*s taking two or more arguments are *never* pronominal, it seems.

Note that the pronominal *all* and the non-pronominal *every* denote the same quantifier. So pronominality is not a semantic property of *DET*s in the present framework.

2.4.6. *Definites.* By the *simple definites* we shall understand here

(i) the definite article *the*,

(ii) the *simple possessives*, like *John's, Susan's, my, his, their,*

(iii) the demonstratives: *this, that, these, those.*

We have already given an interpretation for *the*:

$$\mathbf{the}_M AB \Leftrightarrow \mathbf{all}_M AB \;\&\; |A| \;=\; 1.$$

This is the *singular the*, as in

(11) The boy is running.

For a sentence like

(12) The boys are running

we must use instead

$$\mathbf{the}^{\mathrm{pl}}_M AB \Leftrightarrow \mathbf{all}_M AB \;\&\; |A| > 1.$$

Thus *the* is ambiguous on this analysis. Demonstratives can be interpreted similarly; there we have singular and plural forms and thus no ambiguity. But the simple possessives exhibit the same ambiguity as *the*:

(13) John's car is clean,

(14) John's cars are clean

can be interpreted, respectively, with the quantifiers

$$\mathbf{John's}_M AB \Leftrightarrow \mathbf{all}_M P_{\mathrm{John}} \cap AB \;\;\&\;\; |P_{\mathrm{John}} \cap A| \;=\; 1,$$

$$\mathbf{John's}^{\mathrm{pl}}_M AB \Leftrightarrow \mathbf{all}_M P_{\mathrm{John}} \cap AB \;\;\&\;\; |P_{\mathrm{John}} \cap A| > 1,$$

where P_{John} is the set of things possessed by John; a possession relation is then supposed to be given in the model. There are also *relational* uses of possessives, where the relation is given explicitly, as in

(15) John's friends are nice

Here it is doubtful whether *John's* applies to an *N*, and thus whether it is a *DET* in our sense. (In any case, the truth condition for sentences like (15) can be given by

$$\mathbf{John's}^{\mathrm{pl}}_M RB \Leftrightarrow \mathbf{all}_M R_{\mathrm{John}} B \;\&\; |R_{\mathrm{John}}| > 1,$$

where R is a binary relation on M and $R_a = \{b \in M: Rab\}$ – here we have a generalized quantifier of type $\langle 2, 1 \rangle$.)

We see that the definites come with a *number condition*, concerning the number of elements in a certain set. It is also possible to let sentences with definites *presuppose* that the number condition is satisfied, instead of making them false when it isn't, as we did above. This could be effected by extending the model-theoretic framework to allow *partial* quantifiers: **the**, $\mathbf{the}^{pl}$, **John's**, $\mathbf{John's}^{pl}$ would then be *undefined* when the number condition is not met. We return to this in 3.7.

2.4.7. *Complex DETs with definites.* There are several ways to construct complex *DET*s with definites in English, in particular with partitive constructions. I will present a rather uniform way of interpreting such *DET*s. The starting-point is the observation that one function of the simple definites is to indicate the occurrence of *context sets* (cf. 2.4.5). For simple possessives, this is usually the set of things possessed by the individual (it may also be a subset of this set). But also *the* and the demonstratives need context sets to make the interpretation come out right. For example, in (11) or (12) we are usually *not* talking about the set of all boys in the universe M (as the interpretations given in 2.4.6 would have us believe), but a context-given subset of it (in the singular case, this set has one element).

Consider sentences (with *DET*s as indicated) like

(16) *Some of the seven* men survived,

(17) *Most of John's few* books were stolen.

We interpret these on the following scheme:

(18) $(\mathbf{Q}_1 \textbf{ of Def } \mathbf{Q}_2)BC \Leftrightarrow \mathbf{Q}_1 X \cap B\ C\ \ \&\ \ \mathbf{Q}_2 X \cap B\ M,$

where $\mathbf{Q}_1$, $\mathbf{Q}_2$ are quantifiers and *Def* is a simple definite with X as associated context set (the subscript 'M' is omitted for readability). Note that the second conjunct in (18) can be written, as in 2.2.2,

There are $\mathbf{Q}_2 X \cap B,$

expressing the condition that, in (17), John's books were few, and, in (16), that the set of men under consideration has (exactly?) seven elements.

Some other constructions with definites can be obtained as special cases of (18). We define

(19) $(\mathbf{Def\ Q_2})BC \Leftrightarrow (\mathbf{all\ of\ Def\ Q_2})BC$,

(20) $(\mathbf{Q_1\ of\ Def})BC \Leftrightarrow (\mathbf{Q_1\ of\ Def\ all})BC$

$\Leftrightarrow \mathbf{Q_1} X \cap B\ C$ (by (18) with $\mathbf{Q_2} = \mathbf{all}$),

(21) $\mathbf{Def}\ BC \Leftrightarrow (\mathbf{all\ of\ Def})BC$

$\Leftrightarrow \mathbf{all}\ X \cap B\ C$ (by (20) with $\mathbf{Q_1} = \mathbf{all}$).

(19) takes care of complex *DET*s such as

the five, these few, John's several, etc.

(20) deals with partitives such as

some of Susan's, many of these, at least five of the, etc.

And (21) returns to the simple definites: the truth conditions are essentially the same as in 2.4.6, except that context sets are mentioned.

(18)–(21) can be seen to give the right truth conditions for sentences of these forms, *except* that we have, for readability, omitted the number conditions belonging to these interpretations: in (18) and (20) a *plural condition*, i.e. that $|X \cap B| > 1$, should be added, and in (19) and (21) the cases with singular and plural conditions should be distinguished (syntactically they are distinguished by the singular or plural form of the *N* denoting *B*).

More complicated *DET*s with definites can be treated along similar lines. For example, there are *DET*s which quantify over the *possessor a* in a simple possessive

$a\text{'}\mathbf{s}BC \Leftrightarrow \mathbf{all}\ P_a \cap B\ C$

(we continue to leave out the number condition, and assume for simplicity, in the rest of this subsection, that everything is in the plural). One example is with *DET*s like

some students', most boys', several girls', etc.,

as in

(22) Some students' books were stolen.

The interpretation of these *DET*s is given by

(23) $(\mathbf{Q_1} A\text{'}\mathbf{s})BC \Leftrightarrow \mathbf{Q_1} A\ \{a \in M\text{: } a\text{'}\mathbf{s}\ BC\}$.

Another example is with *iterated* definites. Here is one scheme, which generalizes (20):

(24) $(\mathbf{Q}_1 \textbf{ of Def } A\textbf{'s})BC \Leftrightarrow \mathbf{Q}_1 X \cap A\ \{a \in M\text{: } a\textbf{'s } BC\}$

(we could have generalized (18) similarly, but examples of this form seem rare). This covers *DET*s like

most of the students', some of these boys', three of John's cars', etc.

It could be argued that a sentence like

(25) Most of the students' books were stolen

is ambiguous; then (24) gives the sense where *most* takes *students* as argument, whereas the sense where it takes *books* as arguments is given by

(26) $(\mathbf{Q}_1 \textbf{ of Def } A\textbf{'s})BC \Leftrightarrow \textbf{all } X \cap A\ \{a \in M\text{: } (\mathbf{Q}_1 \textbf{ of } a\textbf{'s})BC\}$.

As before, if *the* is replaced by *John's* in (25), $X = P_{\text{John}}$ (or a subset of it) in (24) and (26). Also as before, we get *DET*s like

the students', those boys', Susan's cars', etc.

as a special case of (24):

(27) $(\textbf{Def } A\textbf{'s})BC \Leftrightarrow (\textbf{all of Def } A\textbf{'s})BC$,

and similarly for *DET*s like

the five students', those few boys', Susan's two cars', etc.

We have given uniform truth conditions for a number of sentences with complex *DET*s by proposing a semantics for the *DET constructions* involved there. This is one task of a theory of natural language quantification. Another is to describe and if possible explain the *restrictions* that often belong to such constructions (cf. 2.2.2).

Consider, for example, the construction in (18). One can see that only *pronominal DET*s can be in the $\mathbf{Q}_1$ position here. As for the **Def** position, the definites, and no others, will work. And there are restrictions on $\mathbf{Q}_2$ too: e.g. *most, all, every, no, some* sound strange here. This last restriction can actually be explained by combining the Barwise and Cooper explanation of the restrictions on "there are"-sentences (2.2.2) with the *plural condition* holding for (18): the exceptions will then once more be those quantifiers making the truth condition trivial. This and other restrictions at work here are discussed further in Westerståhl [1985b].

There is one notable feature of the constructions with definites given here: although the analysis is compositional, it does not use the quantifiers taken to interpret the simple definites in 2.4.6. The function of simple definites was merely to provide context sets. If our analysis is viable, it opens the possibility to leave out the definites from the class of *DET*s, i.e. to treat them as not denoting quantifiers. This move has in fact been viewed desirable for independent reasons which I will not discuss here. My point is merely that such a move can be accomodated in the present quantifier framework.

Likewise, it is not strictly necessary to regard the constructions in this subsection as giving new *DET*s and thereby new natural language quantifiers. Instead, the definitions (18)–(21), (23)–(25), (27) *could* be seen as uniform truth conditions for *sentences* involving (among other things) quantifiers $\mathbf{Q}_1$ and $\mathbf{Q}_2$, but not as defining new quantifiers. The class of natural language quantifiers will then become correspondingly smaller.

If, on the other hand, these constructions are regarded as quantifier definitions, it should be noted that they always yield *conservative* quantifiers, provided $\mathbf{Q}_1$ and $\mathbf{Q}_2$ are conservative.

Clearly we have merely scratched the surface of the many problems pertaining to the analysis of definites, possessives, partitives, etc. It seems, however, that the present quantifier framework can be applied quite fruitfully to these wellknown linguistic questions; cf., for example, Keenan and Stavi [1986], Partee [1984a,b], Thijsse [1983].

2.4.8. *Numerical DETs*. There are many variations of the simplex numerical *DET*s *one, two, three, . . .* , e.g.,

> *at least five, at most five, exactly five, five or more, between five and ten, more than five, fewer than five, infinitely many, at most finitely many, an even number of, an infinite number of, every other, every third, around ten, almost ten, nearly ten, approximately ten, . . . ;*

the interpretations are more or less obvious. A particular group of numerical expressions is

> *half, more than half, less than half, at least half, not more than half, two thirds, at least two thirds, ...*

These are not really *DET*s by our criterion (they don't apply to *N*s), but if a phrase of the form *of Def* is appended to them (after *half*, the *of* is optional), the resulting expressions are quite similar to those in (20): *more than half of the, two thirds of John's, not more than half of these, ...* The interpretation given in (20) fits well here, but to use it we must have suitable quantifiers $\mathbf{Q}_1$ available. Thus, it seems reasonable, even if the above expressions are not *DET*s, to include the quantifiers

$$\textbf{at least m/n}\ AB \Leftrightarrow |A \cap B| \geqq m/n|A|$$

$(n > m > 0)$ among the natural language quantifiers (Boolean combinations of these will then give the other quantifiers needed here).

2.4.9. *Comparative DETs.* The words *more, less, fewer, . . .* can be used in *DET*s for comparison with a fixed number or proportion, as in 2.4.8. We also have the 2-place simplex *DET*s *more . . . than, fewer . . . than,* etc. Some complex variants of these are

> *more than twice as many . . . as, less than half as many . . . as, etc.*

Keenan and Stavi discuss other comparative *DET*s, e.g., those in

(28) *More male than female* students stayed home,

(29) *More* students attended *than stayed home,*

(30) *More* students attended *than teachers who stayed home;*

the respective 1-place *DET*s are italicized. That they are putative *DET*s follows by our criterion (nothing prevents a 1-place *DET* from being syntactically discontinuous!). However, it is also possible to analyze (28)–(30) with the 2-place *more . . . than*: rewrite them as

(28′) *More* male students *than* female students stayed home,

(29′) There are *more* students who attended *than* students who stayed home,

(30′) There are *more* students who attended *than* teachers who stayed home.

The last two "there are"-sentences are then treated as in 2.2.3.

These examples illustrate nicely that more than one structural analysis of an *NP* is often possible. Since no semantic ambiguity is involved here, one would like to make a choice. For a further illustration, consider

(31) *More* men *than* women voted for Smith,

(31′) *More* men *than women* voted for Smith.

(31) uses *more . . . than,* whereas (31′) uses the 1-place *more than women.* But this latter *DET* is not conservative, as one easily sees, so we have good reason to prefer (31). The *DETs* in (28)–(30), on the other hand, are all conservative. For example,

$$\textbf{more than stayed home}_M AB \Leftrightarrow$$

$$\Leftrightarrow |A \cap B| > |A \cap \| \textit{stayed home} \| |$$

$$\Leftrightarrow |A \cap (A \cap B)| > |A \cap \| \textit{stayed home} \| |$$

$$\Leftrightarrow \textbf{more than stayed home}_M A\ A \cap B.$$

Still, there are reasons to prefer (28′)–(30′). One is that they are simpler and more uniform. Another will be given in Section 3.3.

Keenan and Stavi also consider comparatives with definites, such as

> *more of John's than of Susan's, fewer of the male than of the female,* etc.

These can be dealt with, if one wishes, by combining the simplex 2-place comparatives with the treatment of definites in 2.4.6. and 2.4.7; we omit details.

2.4.10. *"Only".* Consider the sentence

(32) Only women voted for Smith.

If *only* is a *DET* here, its interpretation is

$$\textbf{only}_M AB \Leftrightarrow B \subseteq A.$$ [5]

This is *not* a conservative quantifier (indeed, $\textbf{only}_M A\ A \cap B$ is trivially true for all A,B). So let us look for alternatives. Now, *only* can modify many other things besides *N*s, e.g. *NP*s:

(33) Only Susan voted for Smith.

An alternative analysis is then to treat *women* in (32) as a full *NP* (a 'bare plural'); then *only* is not a *DET* at all.

There are also complex *DETs* with *only*. Consider the following example (essentially from Keenan and Stavi):

(34) Only liberal students voted for Smith.

This sentence is three ways ambiguous: (i) as an answer to "Which students voted for Smith?"; (ii) as an answer to "Which liberals voted for Smith?"; and (iii) as an answer to "Who voted for Smith?". Writing (34) in the form *only ABC*, we can represent its three meanings as

(i) $\textit{only } ABC \Leftrightarrow B \cap C \subseteq A,$

(ii) $\textit{only } ABC \Leftrightarrow A \cap C \subseteq B,$

(iii) $\textit{only } ABC \Leftrightarrow C \subseteq A \cap B.$

There are various possibilities here. One is to treat *only* as a 2-place *DET* with three possible interpretations, as in (i)–(iii). One readily verifies that (i) and (ii), but not (iii), are conservative. Or, if one wants to analyze (34) with a 1-place *DET*, we have, in case (i),

$$\mathbf{only\ liberal}_M AB \Leftrightarrow A \cap B \subseteq \|\textit{liberal}\|;$$

in case (ii),

$$\mathbf{only\ \ldots\ students}_M AB \Leftrightarrow A \cap B \subseteq \|\textit{student}\|$$

(but *only . . . students* isn't really a *DET* since it applies to an adjective); and in case (iii) the ordinary *only*, as in (32). Again, the first two are conservative, but not the third.

Only can also combine with numerical expressions, as in

(35) Only five students voted for Smith.

This time, there is no analysis with a 2-place *DET*, and there are just two possible meanings: (i) as an answer to "How many students voted?"; and (ii) as an answer to "How many voted?". So, writing (35) as *only five AB*, we get

(i) $\textit{only five } AB \Leftrightarrow \mathbf{exactly\ five}_M AB,$[6]

(ii) $\textit{only five } AB \Leftrightarrow \mathbf{exactly\ five}_M AB\ \&\ B \subseteq A.$

In case (ii), *only five* would be a non-conservative *DET*, but it is more natural to treat *only* as an *NP*-modifier here. In case (i), on the other hand, *only five* works fine as a conservative *DET*. Here one would like to see a uniform treatment of *DET*s of the form

(36) *only Q;*

we have already seen that *only* 'transforms' *n* into *exactly n*, but when *Q* is a definite, things get more complicated, as the reader can check by considering the example

(37) Only John's students voted for Smith

(three possible readings). Also, one would like to explain the restrictions on *Q* in (36). For example, *a few, between five and ten, around ten* are fine, but not *several, all, most.*

These are just a few hints about some phrases with *only*, and nothing like a uniform semantic analysis. For further discussion, cf. Keenan and Stavi [1986], Rooth [1984], [1985].

2.4.11. *Exception DETs.* This term is used by Keenan and Stavi for *DET*s like

all but three, all but at most five, all but finitely many, ...

As for interpretations, we have

$$\textbf{all but three}_M AB \Leftrightarrow |A - B| = 3,$$

$$\textbf{all but at most five}_M AB \Leftrightarrow |A - B| \leqq 5,$$

$$\textbf{all but finitely many}_M AB \Leftrightarrow A - B \text{ is finite.}$$

The construction *all but Q* apparently obeys certain restrictions – we will return to these in 3.4. It can create ambiguities similar to the ones discussed for *only* in 2.4.10; cf.

(38) All but five liberal students voted for Smith.

There are also exception *DET*s with proper names and with definites:

every but John, no but John, every but John's, all but the liberal, ...

Some of these are discontinuous, as in

(39) *Every* student *but John* voted for Smith,

(40) *Every* car *but John's* was stolen,

(41) *Every* book *but this* (one) was returned.

If we were to treat proper names as definites in the sense of 2.4.7, i.e., as providing suitable sets (in this case: the unit set of the denoted individual), we could interpret these on the uniform scheme

(42) $\textbf{every but Def}_M AB \Leftrightarrow |X \cap A| = 1\ \&$

$\&\ \textbf{every}_M A - X\ B\ \&\ \textbf{no}_M A \cap X\ B,$

where, in (39), $X = \{\text{John}\}$, and, in (40), $X = P_{\text{John}}$; note that e.g. (39) says that John is a student, that he didn't vote for Smith, but that all other students voted for Smith. Note also that (42) gives conservative quantifiers.

2.4.12. *Boolean combinations.* First, *negation*, as in

not every, not all, not many, not more than five, not fewer than three, not more than half (*of the*), ...

The semantics of negated quantifiers is obvious,

$$(\textbf{not Q})_M AB \Leftrightarrow \neg \textbf{Q}_M AB,$$

but *not* cannot stand in front of all *DET*s: e.g. *not some, not most, not at most five* are not well-formed. It is not clear that there is a semantic explanation for this. An interesting question, however, is whether the class of natural language quantifiers is *closed under negation*. For example, even though *not most* is not a *DET*, we can express the intended quantifier with another *DET*:

$$\neg \textbf{most}_M AB \Leftrightarrow |A \cap B| \leqq |A - B|$$

$$\Leftrightarrow |A \cap B| \leqq 1/2|A| \quad \text{(on finite sets, of course)}$$

$$\Leftrightarrow \textbf{not more than half (of the)}_M AB.$$

Likewise, we have $\neg$**(at most five)** = **more than five**. But there are other cases which seem more doubtful, for example, the exception *DET*s: what *DET* would express the negation of *all but three* or *every but John*? We return to this question in 3.4.

As for conjunction and disjunction, we have

some but not all, some but not many, most but not all, at least five and at most ten, either exactly five or more than ten, neither less than five nor more than ten, John's but not Susan's, neither John's nor Susan's, both John's and Susan's, ...

Again the semantics is clear. It is tempting to claim that *any* two 1-place *DET*s can in principle be conjoined with *and* or *or* (another matter is that many such conjunctions and disjunctions would be long and cumbersome, express trivial or otherwise 'strange' quantifiers, etc.). n-place *DETs* for $n > 1$ are discontinuous, which makes the claim less plausible in this case.[7] But the class of binary natural language quantifiers would, if the claim is correct, be closed under conjunction and disjunction.

Boolean operators can also be used to create n-place *DETs* for $n > 1$, e.g., the 2-place

every . . . and, some . . . or,

as in

(43) Every businessman and lawyer knows this,

(44) Some mother or father will react.

Note that (43) is ambiguous. In general, there are two possible readings of sentences of the form *QA and/or B C*:

(45) $Q^1 A \text{ and } B\ C \Leftrightarrow QA \cap B\ C,$

$Q^2 A \text{ and } B\ C \Leftrightarrow QAC \ \&\ QBC,$

(46) $Q^1 A \text{ or } B\ C \Leftrightarrow QA \cup B\ C,$

$Q^2 A \text{ or } B\ C \Leftrightarrow QAC \vee QBC.$

In the one sense of (43) we have the ordinary *every* applied to the complex *N businessman and lawyer*, and in the other we have *every*^{2} applied to the two *N*s *businessman* and *lawyer*. Of course, it is not absolutely necessary to use 2-place *DETs* here, since the interpretations are definable with 1-place *DETs*. For several arguments that 2-place *DETs* are in fact the natural choice, and for more examples, we refer to Keenan and Moss [1985].

We may note that

(47) $\textit{every}^2 A \text{ and } B\ C \Leftrightarrow \textit{every}^1 A \text{ or } B\ C,$

(48) $\textit{some}^1 A \text{ or } B\ C \Leftrightarrow \textit{some}^2 A \text{ or } B\ C.$

(47) explains why the second reading of (43) can also be expressed by

(49) Every businessman or lawyer knows this.

(48) explains why (44) isn't in fact ambiguous.

The same method as above can be used to create n-place *DET*s for all $n > 1$; cf.

(50) Every professor and assistant and secretary and student has a key.

This 4-place *DET* would be interpreted by a 5-ary quantifier similarly to (45) (the second reading seems to be preferred here, which again is manifested in the fact that *and* can be replaced by *or* in (50)).

3. QUANTIFIER CONSTRAINTS AND SEMANTIC UNIVERSALS

A natural way to approach the class of natural language quantifiers is to study the effect of linguistically motivated *constraints*, such as conservativity, on the class of all quantifiers. These constraints are related to *semantic universals*, i.e. general statements about semantic interpretation true for all natural languages. In this section we discuss some such constraints; a number of *possible* semantic universals will be noted along the way.

3.1. *The Restriction to Monadic Quantifiers*

In Section 2 we tacitly assumed that natural language quantifiers are monadic, i.e. of type $\langle 1, 1, \ldots, 1 \rangle$. Is there some reason natural language should not employ non-monadic generalized quantifiers like those used in mathematical logic?

Towards an answer to this, recall first that generalized quantifiers are *second-order* properties or relations (cf. 1.2.1 and 1.4). Thus, *any* sentence which attributes, say, a (second-order) property to a (first-order) property can in principle be formalized as a quantified sentence. For example, consider

(1) Red is a color.

Even in our extensional framework we *could* define a quantifier C of type $\langle 1 \rangle$ by

$$\mathrm{C}_M = \{X \subseteq M: X \text{ is the extension in } M \text{ of some color}\}.$$

So C_M would contain the set of all blue things in M, the set of all red things in M, etc. Then (1) can be formalized as

(2) $Cx\ red(x)$,

which is true in a model *M* iff the set which *red* denotes in *M* is (the extension of) a color. This quantifier is monadic, but a similar story could be told for properties of binary relations, i.e. generalized quantifiers of type $\langle 2 \rangle$.

But from our perspective, (2) is clearly an *unreasonable* formalization of (1). It is useful to understand why. Compare (2) with

(3) $\exists x$ *red*(x),

which formalizes

(4) Something is red.

There is a match in *logical form* between (3) and (4),[1] which is lacking between (1) and (2). Roughly, the difference is that *some* and *color* are of completely different syntactic categories (*some* is an operator and *color* is a predicate). In a natural language context, such matching appears to be essential. It is not always essential in mathematical contexts; cf. the quantifier **W**, where

$WxyPxy$

expresses that

P is a wellordering.

These remarks are really just another way of putting our basic idea that, in natural language, quantifier expressions are *DET*s. So the question is this: Are there *DET*s denoting non-monadic quantifiers? Put differently, are there *DET*s whose corresponding quantifier symbols bind more than one variable in the succeeding formula(s)?

The following example was suggested by Hans Kamp:

(5) Most lovers will eventually hate each other.

This sentence makes good sense,[2] and, looking closely, one sees that it does not talk about the *set* of people who love and are loved by someone, but instead about *pairs*[3] of people who love each other: most such pairs will end up as pairs whose members hate each other. In other words, (5) is *not* equivalent to

(6) Most people who love and are loved by someone will eventually hate and be hated by everyone (or someone) they love.

This follows from the observation that one person may belong to different 'loving pairs'; using this it is easy to construct models where (5) and (6) (in either version) differ in truth value.[4]

In the terminology of Section 1.4 we would formalize (5) as

(7) $most^{(2)}xy(love^*(x,y), will\ eventually\ hate^*(x,y))$,

where $R^*(x,y)$ means $R(x,y) \wedge R(y,x)$ and

$$\mathbf{most}^{(2)}_M = \{\langle R_1,R_2\rangle: R_1, R_2 \subseteq M^2\ \&$$
$$\&\ |R_1 \cap R_2| > |R_1 - R_2|\},$$

a generalized quantifier of type $\langle 2, 2\rangle$.

Another suggestion to use quantification over pairs instead of individuals appears in Fenstad, Halvorsen, Langholm and van Benthem [1987]. They consider sentences like

(8) Every boy who owns a dog kicks it.

There is a question as to the meaning of this, but the preferred reading appears to be that every boy who owns a dog kicks every dog he owns; in other words, using the binary *every* and *some*,

(9) $every\ x(boy(x) \wedge some\ y(dog(y), owns(x,y)),$
$every\ y(dog(y) \wedge owns(x,y), beats(x,y)))$.

The traditional problem here has been to get (9) (or something equivalent to it) from a compositional analysis of (8); note that *it* refers back to *a dog*, but does not correspond to a bound variable in (9)! Fenstad et. al. propose a way to do this; their analysis (whose details need not concern us here) leads, essentially, to the formalization

(10) $every^{(2)}xy(boy(x) \wedge dog(y) \wedge owns(x,y), beats(x,y))$,

where $every^{(2)}$ denotes the type $\langle 2,2\rangle$ generalized quantifier

(11) $\mathbf{every}^{(2)}_M = \{\langle R_1,R_2\rangle: R_1,R_2 \subseteq M^2\ \&\ R_1 \subseteq R_2\}$.

Note that (10) and (9) are equivalent. (Note also, however, that, as Johan van Benthem has pointed out, this analysis does not seem to work for all quantifiers: consider

(12) Most boys who own a dog kick it.

Here, the sentence obtained from (9) by replacing the first occurrence of *every* with *most* is *not* equivalent to the sentence obtained from (10) by replacing $every^{(2)}$ with $most^{(2)}$. Moreover, the former sentence appears to give the preferred reading.[5])

A third and final example that *could* be construed as quantification over pairs in natural language is branching quantification as discussed in section 1.5. To take an example from Barwise [1979], consider

(13) Most boys in my class and most girls in your class know each other.

The preferred reading of this sentence (which has a *conjoined NP*) can be formalized as

$$(14)\qquad \begin{matrix} \textit{most } x \textit{ boy-in-my-class}(x) \\ \\ \textit{most } y \textit{ girl-in-your-class}(y) \end{matrix} \Big> \textit{know}^*(x, y),$$

where this is taken to mean that there is a subset X of the boys in my class, containing most of these boys, and a subset Y of the girls in your class, containing most of those girls, such that if $a \in X$ and $b \in Y$ then a knows* b (cf. Appendix A).

(14) involves branching of the ordinary monadic **most**. But, as noted in 1.5, it is *possible* to 'simulate' branching of two (or more) quantifiers by means of one generalized quantifier. That generalized quantifier will be non-monadic – in the present case, it has type $\langle 1, 1, 2\rangle$, since it relates two sets (the set of the boys and the set of the girls) and one binary relation (know*).

What can be concluded from these examples? Two things should be noted. The first is that *the logical power of expression increases* if the constructions in the examples are included. Consider the logic $L(\mathbf{most}^{(2)})$. It is easy to see that **most** is expressible in this logic, so $L(\mathbf{most}) \leqq L(\mathbf{most}^{(2)})$. But the converse does not hold; the following result was pointed out by Per Lindström:

3.1.1. THEOREM. $L(\mathbf{most}) < L(\mathbf{most}^{(2)})$ (*even on finite models*).

Proof. (Cf. Section 1.7.) Given a natural number d, choose two finite models $\boldsymbol{M} = \langle M, A_0, A_1, A_2\rangle$ and $\boldsymbol{M}' = \langle M', A_0', A_1', A_2'\rangle$ such that the A_i (A_i') are pairwise disjoint sets whose union is M (M'), and, if $|A_0| = k, |A_1| = m, |A_2| = n$, then $|A_0'| = k - 1, |A_1'| = m, |A_2'| = n$, and

(a) $(k - 1)m \leqq n < km$,

(b) $k < m < n$ and $k, m - k, n - m > 2d$.

Now, consider the sentence

$$most^{(2)}xy((P_0x \wedge P_1y) \vee (P_2x \wedge x = y), P_0x \wedge P_1y).$$

In $\boldsymbol{M}$, this expresses that

$$km > n$$

(note that $P_2x \wedge x = y$ denotes $\{(a, a) \in M^2 : a \in A_2\}$, whose cardinal is n). Likewise, it expresses in $\boldsymbol{M}'$ that $(k - 1)m > n$, so, by (a), it is true in $\boldsymbol{M}$ but false in $\boldsymbol{M}'$. On the other hand, using (b) and Theorem 1.7.4 it is easily seen that $\boldsymbol{M} \equiv_{d,\mathbf{most}} \boldsymbol{M}'$. Thus, since d was arbitrary, $L(\mathbf{most}^{(2)}) \not\leq L(\mathbf{most})$. □

The same holds for the branching of **most**. Let $L_b(\mathbf{most})$ be the logic which extends $L(\mathbf{most})$ by allowing formulas of the form (14), interpreted as indicated for that example. It can be shown that "$|A|$ is even" is expressible in $L_b(\mathbf{most})$. Thus, by (4) in Section 1.7, we get the

3.1.2. THEOREM: $L(\mathbf{most}) < L_b(\mathbf{most})$ (*even on finite models*).

The second observation to make, however, is that there are clear senses in which the non-monadic quantification considered here is *reducible* to monadic quantification. Thus, branching may be seen as a linguistic construction on its own, taking monadic quantifiers as *arguments*. And as for the first two examples, $\mathbf{most}^{(2)}$ is really just the old **most** applied to the new universe M^2:

$$\mathbf{most}^{(2)}_M = \mathbf{most}_{M^2},$$

and similarly for $\mathbf{every}^{(2)}$. Here we have *lifted* a relation on sets to a relation on binary relations. In general, any k-ary monadic quantifier $\mathbf{Q}$ can be lifted to any $n > 1$: define $\mathbf{Q}^{(n)}$, of type $\langle n, n, \ldots, n\rangle$ by letting, for all $R_1, \ldots, R_k \subseteq M^n$,

$$\langle R_1, \ldots, R_k\rangle \in \mathbf{Q}^{(n)}_M \Leftrightarrow \mathbf{Q}_{M^n} R_1 \ldots R_k.$$

In view of the foregoing discussion we have a possible semantic universal of the form

(*U1*) *Natural language quantifiers are either monadic or reducible to monadic quantifiers*,

where 'reducible' may be specified along the lines suggested above.

NB. This universal has been challenged recently, however, in Keenan [1987]. He considers sentences like

(15) Every boy read a different book

and shows that, although this may seem as simple *iteration* of two monadic quantifiers, the truth conditions for (15) cannot be so obtained, nor can they be obtained by branching or lifting monadic quantifiers. For further discussion of this matter, cf. also van Benthem [1987b]. In what follows, however, we will restrict attention to monadic quantifiers.

3.2. *The Universe of Quantification*

Recall the definition of conservativity for an $(n + 1)$-ary quantifier $\mathbf{Q}$:

$$\textit{CONSERV} \quad \mathbf{Q}_M A_1 \ldots A_n, B \Leftrightarrow \mathbf{Q}_M A_1 \ldots A_n, (A_1 \cup \ldots \cup A_n) \cap B$$

(for all M and all $A_1, \ldots, A_n, B \subseteq M$; we will usually omit this). We have put a comma before 'B' here to indicate that '$\mathbf{Q}_M A_1 \ldots A_n$' corresponds to the *NP* and 'B' to the *VP*. *CONSERV* says that the *VP* denotation can be restricted to (the union of) the *N* denotation(s). Another way to put this is

(*) *If B and C have the same intersections with all the A_i, then* $\mathbf{Q}_M A_1 \ldots A_n, B \Leftrightarrow \mathbf{Q}_M A_1 \ldots A_n, C$.

It is easily checked that *CONSERV* and (*) are equivalent conditions.

It is *almost* true that *CONSERV* restricts the universe of quantification to (the union of) the first (n) argument(s); cf. the discussion in 2.2.1. But not quite: the *DET* denotation may depend essentially on the universe M. The following condition, which we formulate for arbitrary n-ary quantifiers, expresses the requirement of 'universe-independence' for quantifiers ('*EXT*' for 'extension'):

$$\textit{EXT} \quad \textit{If } A_1, \ldots, A_n \subseteq M \subseteq M' \quad \textit{then } \mathbf{Q}_M A_1 \ldots A_n \Leftrightarrow \mathbf{Q}_{M'} A_1 \ldots A_n.$$

This has nothing directly to do with *CONSERV*; rather, it is a strengthening of the postulate, discussed in 2.1.3, that quantifier expressions are *constants*. For example, *EXT* excludes a quantifier which is $\mathbf{all}_M$ when M has less than 10 elements and $\mathbf{some}_M$ otherwise. But *together* with *CONSERV*, *EXT* gives the exact sense in which *DET*s can be said to restrict the universe of quantification:

$$\textit{UNIV} \quad \mathbf{Q}_M A_1 \ldots A_n, B \Leftrightarrow \mathbf{Q}_{A_1 \cup \ldots \cup A_n} A_1 \ldots A_n, (A_1 \cup \ldots \cup A_n) \cap B.$$

It is an easy exercise to show

3.2.1. PROPOSITION. *UNIV is equivalent to CONSERV + EXT.*

Some further discussion of universe-restriction can be found in Westerståhl [1985a], [1983].

CONSERV and *EXT* are related to the logician's notion of *relativization* (Sections 1.4. and 1.6). Let us first note

3.2.2. PROPOSITION. *If* $\mathbf{Q}^i$ *satisfies EXT for* $i \in I$, *then* $L(\mathbf{Q}^i)_{i \in I}$ *relativizes.*

Proof. Since *EXT* implies that

$$(Q^i)^r x(Px, P_1 x, \ldots, P_n x) \leftrightarrow$$

$$\leftrightarrow Q^i x(Px \wedge P_1 x, \ldots, Px \wedge P_n x)$$

is valid. □

If in addition *CONSERV* holds we can say more: the *binary* quantifiers satisfying *CONSERV* and *EXT* are precisely the relativized ones. Moreover, the *sentences* (in any logic) with two unary predicate symbols which satisfy *CONSERV* and *EXT* (in the obvious sense) are precisely the ones equivalent to relativized sentences. This is the content of the next result:

3.2.3. THEOREM.

(a) *A binary quantifier* $\mathbf{Q}$ *satisfies CONSERV and EXT iff* $\mathbf{Q} = (\mathbf{Q}')^r$, *for some unary* $\mathbf{Q}'$.

(b) *A sentence* $\phi(P_1, P_2)$ *with two unary predicate symbols in a logic L satisfies CONSERV and EXT iff it is equivalent to* $\psi^{(P_1)}$, *for some L-sentence* ψ.

Proof. We prove (b); (a) then follows (it is also easily proved directly). Recall the basic property of relativized sentences from 1.6, in this case, with $\boldsymbol{M} = \langle M, A, B\rangle$,

$$(REL) \quad \langle M, A, B\rangle \vDash \psi^{(P_1)} \Leftrightarrow \langle A, A \cap B\rangle \vDash \psi.$$

From this it is immediate that $\psi^{(P_1)}$ satisfies *CONSERV* and *EXT*. Conversely, if $\phi(P_1, P_2)$ satisfies *CONSERV* and *EXT*, let $\psi = \phi(x = x, P_2)$.

Then

$$\langle M, A, B\rangle \models \psi^{(P_1)} \Leftrightarrow \langle A, A \cap B\rangle \models \psi \qquad (REL)$$

$$\Leftrightarrow \langle A, A, A \cap B\rangle \models \phi(P_1, P_2)$$

$$\text{(by def. of } \psi)$$

$$\Leftrightarrow \langle M, A, B\rangle \models \phi(P_1, P_2)$$

$$\text{(by } UNIV). \quad \square$$

The interest of (b) is that it relates a semantic notion (*CONSERV* and *EXT*) to a syntactic property of sentences – a typical sort of logical result.

Notice that, for *unary* quantifiers, *CONSERV* makes no sense, and *EXT*, although it can be formulated, is *not true* for e.g. the standard universal quantifier $\forall$. This is another aspect of the advantage of binary quantifiers. Any unary quantifier can be replaced by a binary one (its relativization) which does (at least) the same work and has the additional property of restricting the universe of quantification to the first argument. As Theorem 3.2.3 shows, this move gives us *all* the binary quantifiers with that property, in particular, it gives us all the binary natural language quantifiers (provided ($U2$) and ($U3$) below hold).

For n-ary quantifiers with $n > 1$, it is also possible to secure *CONSERV* and *EXT* by raising the number of arguments, though not quite as simply as when $n = 1$. The next proposition surveys the possibilities.

3.2.4. PROPOSITION. *Let* $\mathbf{Q}$ *be an n-ary quantifier. Then*
(i) *there is an* $(n + 1)$*-ary quantifier* $\mathbf{Q}'$ *satisfying CONSERV such that* $\mathbf{Q}_M A_1 \ldots A_n \Leftrightarrow \mathbf{Q}'_M A_1 \ldots A_n, M$;
(ii) *there is an* $(n + 1)$*-ary quantifier* $\mathbf{Q}''$ *satisfying EXT such that* $\mathbf{Q}_M A_1 \ldots A_n \Leftrightarrow \mathbf{Q}''_M A_1 \ldots A_n, M$;
(iii) *there is an* $(n + 2)$*-ary quantifier* $\mathbf{Q}^+$ *satisfying both CONSERV and EXT such that* $\mathbf{Q}_M A_1 \ldots A_n \Leftrightarrow \mathbf{Q}^+_M A_1 \ldots A_n M, M$.

Proof. (i): Define $\mathbf{Q}'_M A_1 \ldots A_n, B \Leftrightarrow \mathbf{Q}_M A_1 \cap B \ldots A_n \cap B$. The verification of *CONSERV* is immediate. (ii): Let $\mathbf{Q}''_M A_1 \ldots A_n, B \Leftrightarrow \mathbf{Q}_B A_1 \cap B \ldots A_n \cap B$; again *EXT* is immediate. (iii): Define $\mathbf{Q}''$ as in (ii), and then form $\mathbf{Q}^+$ from $\mathbf{Q}''$ as in (i); the result follows from (i) and (ii). □

For the record, we formulate the semantic universals corresponding to *CONSERV* and *EXT*:

($U2$) *Natural language quantifiers are conservative.*
($U3$) *Natural language quantifiers satisfy EXT.*

We saw in 2.4 that the few apparent exceptions to ($U2$) could be accounted for by reasonable methodological decisions (2.4.3, 2.4.9–10). As for ($U3$), the only exceptions found were certain interpretations of context-dependent *DET*s like *many*. For example, if

$$\mathbf{Q}_M AB \Leftrightarrow |A \cap B| \geqq 1/3|M|,$$

Q violates *EXT*. Again, it is mainly a methodological question whether one wants to allow this kind of context-dependence or not.

3.3. *Quantity*

The condition *ISOM*, repeated below, was formulated for generalized quantifiers of any type $\langle k_1, \ldots, k_n \rangle$:

ISOM *If f is a bijection from M to M′ then* $\mathbf{Q}_M R_1 \ldots R_n \Leftrightarrow \mathbf{Q}_{M'} f[R_1] \ldots f[R_n]$.

The idea is that **Q** does not distinguish between different elements of the universe, or even across two universes. This requirement, which is a version of what is sometimes called *topic-neutrality*, can be formulated for arbitrary syntactic categories (cf. van Benthem [1983b]). It is a general requirement on *logical constants*.

For monadic quantifiers, *ISOM* has a particularly conspicuous formulation. Roughly, it says that quantifiers deal only with *quantities*. The latter assertion can be made precise with the terminology from Section 1.7 as follows:

QUANT *If* $\mathbf{M} = \langle M, A_0, \ldots A_{k-1} \rangle$, $\mathbf{M}' = \langle M', A'_0, \ldots, A'_{k-1} \rangle$, *and* $|P_s^{\mathbf{M}}| = |P_s^{\mathbf{M}'}|$ *for all* $s \in 2^k$, *then* $\mathbf{Q}_M A_0 \ldots A_{k-1} \Leftrightarrow \mathbf{Q}_{M'} A'_0 \ldots A'_{k-1}$.

This means that the truth value of $\mathbf{Q}_M A_0 \ldots A_{k-1}$ depends only on 2^k quantities, namely, the number of elements in the partition sets.

A bijection from M to M' splits into bijections of the respective partition sets, and, conversely, bijections between these sets can be joined to one from M to M'. Thus we have the

3.3.1. PROPOSITION. *ISOM and QUANT are equivalent* (*for a monadic* **Q**).

If we consider only one universe M in *ISOM* (letting $M' = M$), and thus *permutations* on M, we get a slightly weaker version, called *PERM*.[6] From a local perspective on quantifiers (2.1.4), *PERM* is the natural notion. Our global condition *EXT*, however, says that the choice of universe is unimportant. Indeed, it is straight-forward to prove

3.3.2. PROPOSITION. *Under EXT, ISOM and PERM are equivalent.*

All the simplex *DET*s from 2.4.1–3 denote quantitative quantifiers. To see this, it is sufficient to check that the defining conditions can be expressed as conditions on the cardinalities of the relevant sets. For example, $\mathbf{all}_M AB \Leftrightarrow |A - B| = 0$, $\mathbf{some}_M AB \Leftrightarrow |A \cap B| \neq 0$, $\mathbf{most}_M AB \Leftrightarrow |A \cap B| > |A - B|$, $\mathbf{both}_M AB \Leftrightarrow |A - B| = 0 \; \& \; |A \cap B| = 2$, $\mathbf{many}^2_M AB \Leftrightarrow |A \cap B| \geqq k(|A \cap B| + |A - B|)$, etc.

As for complex *DET*s, there are just a few of the constructions in 2.4.6–12 which yield non-quantitative quantifiers. One example is *DET*s with fixed adjective phrases, or similar expressions, such as *more male than female, some red, only liberal.* We saw, however, that sentences with such expressions can also be interpreted using only quantitative quantifiers (2.4.9–10). Another major example are the possessives, either simple ones such as *John's*, or complex constructions with possessives. The quantifier **John's** from 2.4.6 violates *ISOM* since the ownership relation need not be preserved under permutations of the objects in the universe. For example, John may own two white shirts but no red tie, even though it is possible to permute the shirts and the ties, and the white things and the red things, in a one–one fashion. Then

John's shirts are white

is true, but not

John's ties are red,

as *ISOM* would require.

In 2.4.7, we mentioned an alternative analysis of definites, and thus in particular of possessives. Under this analysis, one can dispense with quantifiers denoted by simple possessives, also in various complex constructions. Quantitative quantifiers would suffice, it seems, for all of these constructions (the same holds for *every but John* (2.4.11), another counter-instance to *ISOM*). It would then be possible to propose the following rather appealing universal:

($U4$) *Natural language quantifiers are quantitative.*

If one does not want to take this methodological step, on the other hand, one will settle for the more modest

($U4'$) *Simple natural language quantifiers are quantitative.*

3.4. *Logical Quantifiers, Negations and Duals*

Whichever version of the last universal one prefers, the following class of quantifiers is a natural object of study:

3.4.1. DEFINITION. An n-ary quantifier ($n > 1$) is *logical* if it satisfies *CONSERV*, *EXT*, and *QUANT*.

The terminology is meant to suggest that these three requirements are *necessary* for logicality; further conditions will be discussed in 4.4.

For *binary* quantifiers, logicality means that the truth value of $\mathbf{Q}_M AB$ depends only on the two numbers $|A - B|$ and $|A \cap B|$:

3.4.2. PROPOSITION. *A binary quantifier* $\mathbf{Q}$ *is logical iff, for all* M, M' *and all* $A, B \subseteq M$ *and* $A', B' \subseteq M'$, $|A - B| = |A' - B'|$ *and* $|A \cap B| = |A' \cap B'|$ *implies that* $\mathbf{Q}_M AB \Leftrightarrow \mathbf{Q}_{M'} A'B'$.

Proof. If $\mathbf{Q}$ is logical and $|A - B| = |A' - B'|$ and $|A \cap B| = |A' \cap B'|$, then, by *QUANT*, $\mathbf{Q}_A A A \cap B \Leftrightarrow \mathbf{Q}_{A'} A'A' \cap B'$, and so, by *UNIV* (Proposition 3.2.1), $\mathbf{Q}_M AB \Leftrightarrow \mathbf{Q}_{M'} A'B'$. Conversely, if the right hand side of the equivalence holds, *QUANT* is immediate. Take M and $A, B \subseteq M$ and let $M' = A' = A$ and $B' = A \cap B$. Thus, $\mathbf{Q}_M AB \Leftrightarrow \mathbf{Q}_A AA \cap B$, i.e., *UNIV* holds. □

This means that a logical binary relation between *sets* can be replaced by a binary relation between *cardinal numbers*; we exploit this in 4.2. Proposition 3.4.2 can be generalized to n-ary logical quantifiers: *QUANT* transforms an n-ary $\mathbf{Q}$ to a relation between 2^n cardinal numbers, and *CONSERV* + *EXT* eliminate the dependence of *two* of these.

The class of logical quantifiers has some nice closure properties. It is straightforward to verify that if $\mathbf{Q}_1$ and $\mathbf{Q}_2$ are *CONSERV and EXT* (*QUANT*), then so are $\mathbf{Q}_1 \wedge \mathbf{Q}_2$, $\mathbf{Q}_1 \vee \mathbf{Q}_2$, and $\neg\mathbf{Q}_1$. Thus,

3.4.3. PROPOSITION. *For each* $n > 1$, *the class of n-ary logical quantifiers is closed under the usual Boolean operations.*

In a natural language context, there are also *inner* Boolean operations. We noted in 2.4.12 that from a binary $\mathbf{Q}$ one can construct two $(n + 1)$-ary inner conjunctions:

$$\mathbf{Q}_M^{\wedge 1} A_1 \ldots A_n, B \Leftrightarrow \mathbf{Q}_M A_1 \cap \ldots \cap A_n\ B,$$

$$\mathbf{Q}_M^{\wedge 2} A_1 \ldots A_n, B \Leftrightarrow \mathbf{Q}_M A_1 B \ \& \ldots \& \ \mathbf{Q}_M A_n B.$$

Inner disjunctions $\mathbf{Q}^{\vee 1}$ and $\mathbf{Q}^{\vee 2}$ are defined similarly. As for negation, we make the

3.4.4. DEFINITION. If $\mathbf{Q}$ is $(n + 1)$-ary, the *inner negation* of $\mathbf{Q}$ is the quantifier $\mathbf{Q}\neg$, defined by

$$(\mathbf{Q}\neg)_M A_1 \ldots A_n, B \Leftrightarrow \mathbf{Q}_M A_1 \ldots A_n, M - B.$$

Also, the *dual* of $\mathbf{Q}$, $\check{\mathbf{Q}}$, is the quantifier $\neg(\mathbf{Q}\neg)$ $(= (\neg\mathbf{Q})\neg)$.

Outer and inner negation correspond to sentence negation and *VP* negation, respectively; cf.

Not many boys are lazy,

Many boys are not lazy,

with the respective truth conditions

$$(\neg\mathbf{many})_M \ \|boy\| \ \|lazy\|,$$

$$(\mathbf{many}\neg)_M \ \|boy\| \ \|lazy\|.$$

3.4.5. PROPOSITION. *The class of logical quantifiers is closed under inner conjunctions and disjunctions (both kinds), and inner negation (hence also duals).*

Proof. This is again a routine check; let us take one case and verify that $\mathbf{Q}\neg$ satisfies *EXT* if $\mathbf{Q}$ satisfies *CONSERV* and *EXT*. Suppose $A_1, \ldots, A_n, B \subseteq M \subseteq M'$. Then

$$\begin{aligned}(\mathbf{Q}\neg)_M A_1 \ldots A_n, B &\Leftrightarrow \mathbf{Q}_M A_1 \ldots A_n, M - B \\ &\Leftrightarrow \mathbf{Q}_M A_1 \ldots A_n, (A_1 \cup \ldots \cup A_n) - B \quad (CONSERV) \\ &\Leftrightarrow \mathbf{Q}_{M'} A_1 \ldots A_n, (A_1 \cup \ldots \cup A_n) - B \quad (EXT) \\ &\Leftrightarrow \mathbf{Q}_{M'} A_1 \ldots A_n, M' - B \quad (CONSERV) \\ &\Leftrightarrow (\mathbf{Q}\neg)_{M'} A_1 \ldots A_n, B. \end{aligned}$$

□

It should be noted that other inner negations than *VP* negation do *not* preserve logicality. For example, if we define, for a binary $\mathbf{Q}$,

$$\mathbf{Q}^*_M AB \Leftrightarrow \mathbf{Q}_M M - A\ B,$$

then *CONSERV* will not be preserved.

The following propositions lists some de Morgan-like laws for inner Boolean operations on quantifiers:

3.4.6. PROPOSITION.

(a) $(\neg\mathbf{Q})^{\wedge 1} = \neg(\mathbf{Q}^{\wedge 1}),\ (\neg\mathbf{Q})^{\wedge 2} = \neg(\mathbf{Q}^{\vee 2}),$

(b) $(\neg\mathbf{Q})^{\vee 1} = \neg(\mathbf{Q}^{\vee 1}),\ (\neg\mathbf{Q})^{\vee 2} = \neg(\mathbf{Q}^{\wedge 2}),$

(c) $(\mathbf{Q}_1 \wedge \mathbf{Q}_2)\neg = \mathbf{Q}_1\neg \wedge \mathbf{Q}_2\neg,$

(d) $(\mathbf{Q}_1 \vee \mathbf{Q}_2)\neg = \mathbf{Q}_1\neg \vee \mathbf{Q}_2\neg,$

(e) $(\mathbf{Q}^{\wedge i})\neg = (\mathbf{Q}\neg)^{\wedge i}\quad (i = 1, 2),$

(f) $(\mathbf{Q}^{\vee i})\neg = (\mathbf{Q}\neg)^{\vee i}\quad (i = 1, 2).$

In 2.4.12 we considered the suggestion that the class of binary natural language quantifiers is closed under (outer) conjunction and disjunction, i.e., that the following universal holds:

($U5$) *If* $\mathbf{Q}_1$ *and* $\mathbf{Q}_2$ *are binary natural language quantifiers then so are* $\mathbf{Q}_1 \wedge \mathbf{Q}_2$ *and* $\mathbf{Q}_1 \vee \mathbf{Q}_2$.

The case of negation was more doubtful. In the table opposite, some examples of *DET*s for negations and duals in English are given. '–' means that it seems hard to find a *DET*, simplex or complex, denoting the negation or dual in question. Of course these quantifiers are always expressible by some suitable paraphrase, but the question here is whether there are *determiners* denoting them.

This table suggests certain questions. When is the (inner or outer) negation of a simple quantifier again simple? Barwise and Cooper have several proposals here, e.g., that the negations of the cardinal quantifiers **at least n** and **exactly n** are never simple, and that if a language has a pair of simple duals, that pair consists of **every** and **some**; cf. also 3.6.

Here we shall look a bit closer at the '–' signs for the binary quantifiers in the table. Note that if these signs are correct, the class of binary natural language quantifiers is not closed under inner or outer negation. Discussing

TABLE I

Q	¬Q	Q¬	Q̆
some	*no*	*not every*	*every*
every	*not every*	*no*	*some*
no	*some*	*every*	*not every*
most	*at most half*	*less than half*	*at least half*
many	*few*	–	*all but a few*
infinitely many	*at most finitely many*	–	*all but finitely many*
(at least) n	*less than n*	–	*all but less than n*
at most n	*more than n*	*all but at most n*	–
(exactly) n	*not exactly n*	*all but n*	–
more . . . than	*at most as many . . . as*	–	–
fewer . . . than	*at least as many . . . as*	–	–

this question will give us an occasion to look at some typical issues, and to introduce a few useful notions. The purpose, as usual, is to illustrate problems and ideas, rather than making definite empirical claims.

Note first that part of what Table I claims is that certain expressions of the form *all but Q* are anomalous. Thus, while *all but five, all but at most five, all but finitely many* are fine, *all but at least five, all but not exactly five, all but* (*infinitely*) *many* are not. It might be claimed that the anomaly in the latter cases is pragmatic rather than semantic. I will not argue about this directly, but instead try to see if there are in fact significant semantic differences between the normal and the anomalous cases.

Exception *DET*s of the form *all but Q* (cf. 2.4.11) are interpreted on the scheme

(1) **all but Q** $=$ **Q¬**.

When is **Q¬** a natural language quantifier? Before trying to give some answers to this, we need to introduce a new concept.

3.4.7. DEFINITION. A binary quantifier **Q** is *VP-positive* (*VP-negative*) if, for all M, M' and all A, $B \subseteq M$, $A'B' \subseteq M'$ such that $A \cap B = A' \cap B' (A - B = A' - B')$, $\mathbf{Q}_M AB \Leftrightarrow \mathbf{Q}_{M'} A'B'$.[7]

As the terminology indicates, *VP*-positivity means that **Q** amounts solely to a condition on the *VP* denotation (intersected with the *N* denotation, since we assume *CONSERV*), whereas a *VP*-negative quantifier reduces to a condition on the *complement* of the *VP* denotation. For example, **some**, **no**, **many**,[8] **few**, **infinitely many**, **at least n**, **at most n**, **exactly n** are

VP-positive, whereas **every**, **not every**, **all but n**, **all but at most n** are *VP*-negative. **most**, **at least half**, and other 'proportional' quantifiers are neither *VP*-positive nor *VP*-negative, and the same holds for the interpretations of the definites (because of the *number condition* on the *N* denotation; cf. 2.4.6–7).

For a *conservative* **Q**, *VP*-positivity (-negativity) is related to inner and outer negation as follows:

(2) $\mathbf{Q}$ is *VP*-positive (-negative) $\Leftrightarrow \neg\mathbf{Q}$ is *VP*-positive (-negative) $\Leftrightarrow \mathbf{Q}\neg$ is *VP*-negative (-positive).

The next result, essentially due to Barwise and Cooper, shows that *VP*-positivity is in fact a simple relational property of quantifiers. A binary quantifier is *symmetric* if it is symmetric as a relation, i.e., if for all M and all $A,B \subseteq M$,

$$\mathbf{Q}_M AB \Rightarrow \mathbf{Q}_M BA.$$

3.4.8. PROPOSITION. *If* **Q** *satisfies CONSERV and EXT the following are equivalent:*

(a) **Q** *is VP-positive.*

(b) **Q** *is symmetric.*

(c) $\mathbf{Q}_M AB \Leftrightarrow \mathbf{Q}_M A \cap B A \cap B$ (*for all M and all* $A,B \subseteq M$).

Proof. (a) $\Rightarrow$ (b): Suppose $\mathbf{Q}_M AB$. Let $A' = B$ and $B' = A$. Thus $A \cap B = A' \cap B'$, so, by *VP*-positivity, $\mathbf{Q}_M A'B'$, i.e., $\mathbf{Q}_M BA$.
(b) $\Rightarrow$ (c): Suppose **Q** is symmetric. Then $\mathbf{Q}_M AB \Leftrightarrow \mathbf{Q}_M A\, A \cap B$ (*CONSERV*) $\Leftrightarrow \mathbf{Q}_M A \cap B A$ (symmetry) $\Leftrightarrow \mathbf{Q}_M A \cap B A \cap B$ (*CONSERV*).
(c) $\Rightarrow$ (a): If (c) holds and $A \cap B = A' \cap B'$, where A, $B \subseteq M$ and A', $B' \subseteq M'$, then $\mathbf{Q}_M AB \Leftrightarrow \mathbf{Q}_M A \cap B A \cap B \Leftrightarrow \mathbf{Q}_M A' \cap B' A' \cap B' \Leftrightarrow \mathbf{Q}_{M'} A' \cap B' A' \cap B'$ (by *EXT*) $\Leftrightarrow \mathbf{Q}_{M'} A'B'$ (by (c)). □

The following corollary is easy using (2):

3.4.9. COROLLARY. *Under CONSERV and EXT the following are equivalent:*

(a) **Q** *is VP-negative.*

(b) $\mathbf{Q}\neg$ *is symmetric.*

(c) $\mathbf{Q}_M AB \Leftrightarrow \mathbf{Q}_M A - B\, \varnothing$.

From our list of English *DETs* in 2.4, it appears much easier to find *VP*-positive quantifiers than *VP*-negative ones. Moreover, it seems that for each *DET* giving a condition on the complement of the *VP* denotation, there is another *DET* giving the *same* condition on the *VP* denotation itself. For example, if the first *DET* is of the form *all but Q*, the corresponding positive condition is given by *Q*, and if the first *DET* is *every* or *not every*, the second is *no* or *some*, respectively. This lets us propose the following universal:

(*U*6) *If* **Q** *is a VP-negative natural language quantifier, then* **Q**$\neg$ *is also a natural language quantifier.*

A related observation is that when *Q* denotes a *VP*-negative quantifier, the form *all but Q* is not allowed: *all but every, all but not every, all but all but five*, etc. are ruled out. The reason, one imagines, is that this would be a very cumbersome way of expressing a 'double *VP* negation', which in any case is equivalent to the more easily expressed positive condition.

(*U*6) gives one (partial) answer to our question about when **Q**$\neg$ is a natural language quantifier. But, to return to Table I, the most interesting case concerns *VP*-positive quantifiers: all the '–' signs (for binary quantifiers) are examples of failure of **Q**$\neg$ to be a natural language quantifier for *VP*-positive **Q**. What, then, is wrong with a *DET* such as *all but at least five*?

Here is one suggestion: Sentences of the form *all but Q A B* imply the *existence* of *A*s that are *B* (in contrast with *all A B*). More precisely, let us say that a quantifier **Q** has *existential import*, if

(3) for sufficiently large A (and M), $\mathbf{Q}_M AB \Rightarrow \mathbf{some}_M AB$.

(3) holds for **all but five, all but at most five, all but finitely many**, etc., but fails for **(at least five)**$\neg$, **(not exactly five)**$\neg$, **(infinitely many)**$\neg$, etc. E.g.,

$$\textbf{(at least five)}\neg_M AB \Leftrightarrow |A - B| \geqq 5,$$

so for each A with at least five elements, we have **(at least five)**$\neg_M A\varnothing$ but not $\mathbf{some}_M A\varnothing$. Note that the qualification "for sufficiently large A" is necessary: $\mathbf{all\ but\ at\ most\ five}_M AB$ implies $\mathbf{some}_M AB$ only when $|A| > 5$, and $\mathbf{all\ but\ finitely\ many}_M AB$ implies $\mathbf{some}_M AB$ only when A is infinite.

What condition on **Q** corresponds to the fact that **Q**$\neg$ has existential import? For *VP*-positive quantifiers, the answer is as follows. Call **Q** *bounded*, if

(4) there is an n such that for all M and all $A, B \subseteq M$, $\mathbf{Q}_M AB \Rightarrow |A \cap B| \leqq n$.

3.4.10. PROPOSITION. *Suppose* **Q** *is VP-positive and satisfies CONSERV and EXT. Then* $\mathbf{Q}\neg$ *has existential import iff* **Q** *is bounded.*

Proof. If **Q** is bounded by n, then $|A| > n$ & $\mathbf{Q}\neg_M AB \Rightarrow |A| > n$ & $|A - B| \leqq n \Rightarrow A \cap B \neq \varnothing$, so (3) holds for $\mathbf{Q}\neg$. On the other hand, if **Q** is not bounded, it follows from Proposition 3.4.8 that there are arbitrarily large A (and M) such that $\mathbf{Q}_M AA$. But this means that $\mathbf{Q}\neg_M A\varnothing$, so (6) fails for $\mathbf{Q}\neg$. □

From these observations it is tempting to suggest the universal: For *VP*-positive **Q**, $\mathbf{Q}\neg$ is a natural language quantifier only if **Q** is bounded. But this would be premature. The universal concerns arbitrary quantifiers $\mathbf{Q}\neg$, whereas the above discussion concerned the interpretations of *DET*s of the form *all but Q′*. In fact, there is a simple counter-example to this universal: **some** is *VP*-positive, **some**$\neg$ = **not every** is a natural language quantifier, but **some** is not bounded!

Of course we cannot require in the universal that $\mathbf{Q}\neg$ be the interpretation of a *DET all but Q′*; that would make $\mathbf{Q}\neg$ *trivially* a natural language quantifier! But all is not lost: it seems that if we require **Q** to be *non-simple*, the universal holds; possibly, the simple **some** was the *only* counter-example.

What about the converse statement, i.e., if **Q** is bounded, does it follow that $\mathbf{Q}\neg$ is a natural language quantifier? Here we can say something more definite:

3.4.11. PROPOSITION. *If* **Q** *is logical, VP-positive, and bounded, then* **Q** *is a finite disjunction of quantifiers of the form* **exactly n**.

(The proof is best postponed until Section 4.2.) Thus, if **Q** is as in this proposition, **Q** is clearly a natural language quantifier, and so is $\mathbf{Q}\neg$, which by Proposition 3.4.6 is a finite disjunction of quantifiers of the form **all but n**.

Some of the last observations are collected in the following tentative universal:

(*U7*) *If* **Q** *is a VP-positive, non-simple, logical quantifier, then* $\mathbf{Q}\neg$ *is a natural language quantifier iff* **Q** *is bounded.*

This universal, then, would be an explanation of the empty spaces (for the binary quantifiers) in Table I.

3.5. *Non-triviality*

Call an n-ary quantifier $\mathbf{Q}$ *trivial on* M, if $\mathbf{Q}_M$ is either the empty or the universal n-ary relation on $P(M)$. Consider the condition

NONTRIV $\mathbf{Q}$ *is non-trivial on some universe.*

Quantifiers violating *NONTRIV* are not very interesting: Either any sentence beginning with a *DET* denoting such a quantifier (satisfying *EXT*) is true in each model, or any such sentence is false in each model. Nevertheless, natural language permits the construction of such *DET*s, for example, *at least zero, fewer than zero, at least ten and at most nine, more than infinitely many*, as pointed out in Keenan and Stavi [1986]. But the following universal seems true:

(*U*8) *Simple natural language quantifiers satisfy NONTRIV.*

Note that the *NONTRIV* quantifiers are *not* closed under Boolean operations: for any $\mathbf{Q}$, the quantifier $\mathbf{Q} \vee \neg\mathbf{Q}$ is trivial on every universe.

NONTRIV requires a very modest amount of 'activity' of $\mathbf{Q}$; a stronger variant is

ACT $\mathbf{Q}$ *is non-trivial on each universe.*

ACT holds for many natural language quantifiers, but there are exceptions even among the simple ones, e.g. **both**, **two**, **three**, **four**, . . . (if M has less than 4 elements, $\mathbf{four}_M AB$ is always false).

van Benthem [1984a] considers an even stronger requirement of activity, called 'variety', for binary quantifiers. Here is a generalization to $(n + 1)$-ary quantifiers:

VAR *For all* M *and all* $A_1, \ldots, A_n \subseteq M$ *such that* $A_1 \cap \ldots \cap A_n \neq \varnothing$, *there are* $B_1, B_2 \subseteq M$ *such that* $\mathbf{Q}_M A_1 \ldots A_n, B_1$ *and* $\neg\mathbf{Q}_M A_1 \ldots A_n, B_2$.

In the binary case, we could say that *VAR* transfers the requirement of activity to each non-empty first argument. For quantifiers satisfying *CONSERV* and *EXT*, this seems a reasonable strengthening of *ACT*.

Clearly,

$$VAR \Rightarrow ACT \Rightarrow NONTRIV;$$

the implications cannot be reversed: an example of a (logical) quantifier satisfying *ACT* but not *VAR* is

$$\mathbf{Q}_M AB \Leftrightarrow |A| = 1.$$

Note that this does not seem to be a natural language quantifier. In fact, inspection of the *DET*s in 2.4 shows that the *ACT* ones – e.g., **some, no, all, not all, most, more . . . than, fewer . . . than, every . . . and/or, some . . . and/or** (both interpretations) – also satisfy *VAR*. So one may propose

(*U*9) *Natural language quantifiers satisfying ACT also satisfy VAR.*

3.6. *Monotonicity*

The monotonicity behavior of a quantifier **Q** concerns the preservation of the truth value of $\mathbf{Q}_M A_1 \ldots A_n$ when the arguments are decreased or increased. For simplicity, we shall only consider *binary* quantifiers here, although many of the definitions and results below can easily be extended to $(n + 1)$-ary quantifiers.

3.6.1. DEFINITION. A binary quantifier **Q** is

$$MON\uparrow, \text{ if } \mathbf{Q}_M AB \ \& \ B \subseteq B' \Rightarrow \mathbf{Q}_M AB',$$

$$MON\downarrow, \text{ if } \mathbf{Q}_M AB \ \& \ B' \subseteq B \Rightarrow \mathbf{Q}_M AB',$$

$$\uparrow MON, \text{ if } \mathbf{Q}_M AB \ \& \ A \subseteq A' \Rightarrow \mathbf{Q}_M A'B,$$

$$\downarrow MON, \text{ if } \mathbf{Q}_M AB \ \& \ A' \subseteq A \Rightarrow \mathbf{Q}_M A'B.$$

Also, **Q** is *RIGHT MON* (*LEFT MON*) if it is *MON*↑ or *MON*↓ (↑*MON* or ↓*MON*), and **Q** is ↑*MON*↑ if it is both *MON*↑ and ↑*MON*; similarly for ↑*MON*↓, ↓*MON*↑, and ↓*MON*↓.

Barwise and Cooper call *RIGHT MON* monotonicity, ↑*MON persistence*, and ↓*MON anti-persistence*.

Many natural language quantifiers have simple monotonicity properties. The four types of *double monotonicity* are exemplified by the square of opposition:

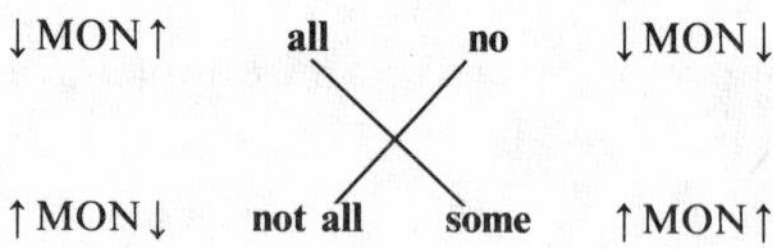

Other doubly monotone quantifiers are **at least n**, **infinitely many**, which are $\uparrow MON\uparrow$, and **at most n**, **at most finitely many**, **only liberal** (cf. 2.4.10), which are $\downarrow MON\downarrow$. **most** is $MON\uparrow$ but not *LEFT MON*, as is easily seen, and the same holds for simple definites like **the** and **John's** (as defined in 2.4.6). Of the interpretations of *many* from 2.4.3, **many**[1] is $\uparrow MON\uparrow$, **many**[2] is $MON\uparrow$ but not *LEFT MON*, and **many**[3] is neither *LEFT* nor *RIGHT MON*. Other examples of neither *LEFT* nor *RIGHT MON* quantifiers are **exactly n**, **all but n**, **between five and ten**.

The monotonicity behavior of **Q** determines that of its negations and dual:

3.6.2. PROPOSITION. (a) *Outer negation reverses the direction of both RIGHT and LEFT MON.* (b) *Inner negation reverses RIGHT MON but preserves LEFT MON.* (c) *Dual-formation preserves RIGHT MON but reverses LEFT MON.*

For example, from the monotonicity behavior of one column of Table I, we can infer that of all the other columns (for the binary quantifiers).

For doubly monotone quantifiers, we have the following pleasing result from van Benthem [1983c]. The proof is a nice demonstration of the strength and flexibility of the quantifier constraints we are using.

3.6.3. THEOREM (van Benthem). *Under CONSERV and VAR, the only doubly monotone quantifiers are those in the square of opposition.*

Proof. Suppose **Q** is $\downarrow MON\downarrow$. We prove that $\mathbf{Q} = \mathbf{no}$; the theorem then follows from Proposition 3.6.2. Take a universe M and $A, B \subseteq M$. First assume that $A \cap B = \varnothing$. We claim that there is C such that $\mathbf{Q}_M AC$. This is immediate from *VAR* if $A \neq \varnothing$; otherwise, note that $\mathbf{Q}_M \varnothing\varnothing$ holds by $\downarrow MON\downarrow$ and the fact that **Q** is non-trivial on M. By $MON\downarrow$ it then follows that $\mathbf{Q}_M A\varnothing$, i.e., $\mathbf{Q}_M A A \cap B$. Thus, by *CONSERV*, $\mathbf{Q}_M AB$. Conversely, suppose that $\mathbf{Q}_M AB$ holds. By $\downarrow MON\downarrow$, $\mathbf{Q}_M A \cap B A \cap B$. But then $\mathbf{Q}_M A \cap BC$ holds for *all* $C \subseteq M$, since, for any such C, it suffices (by *CONSERV*) to show $\mathbf{Q}_M A \cap B\ A \cap B \cap C$, and this holds by $MON\downarrow$. Hence, *VAR* tells us that $A \cap B = \varnothing$, and the proof is finished. □

For logical quantifiers, we can replace double monotonicity by *LEFT MON*:

3.6.4. THEOREM (van Benthem). *The only logical and LEFT MON quantifiers satisfying VAR are the ones in the square of opposition.*

A convenient method to prove this for *finite* universes (the case van Benthem considers) will be given in 4.2; actually, the result holds for all universes. Note the use of *VAR* here; without it, room is left for many other *LEFT MON* quantifiers, as is clear from the examples above.

Barwise and Cooper propose several universals involving monotonicity. One of them is the following:

(U10) *Simple binary natural language quantifiers are either RIGHT MON or conjunctions of RIGHT MON quantifiers.*

Note that **exactly n** (which probably is simple) is the conjunction of the *RIGHT MON* **at least n** and **at most n**. This and other examples of neither *LEFT* nor *RIGHT MON* quantifiers suggest a weaker notion of monotonicity, which will be called *continuity*:

3.6.5. DEFINITION. A binary quantifier $\mathbf{Q}$ is

$$\begin{aligned}&RIGHT\ CONT, \text{ if } \mathbf{Q}_M AB \ \&\ \mathbf{Q}_M AB'' \ \&\\ &\quad \&\ B \subseteq B' \subseteq B'' \Rightarrow \mathbf{Q}_M AB',\\ &LEFT\ CONT, \text{ if } \mathbf{Q}_M AB \ \&\ \mathbf{Q}_M A''B \ \&\\ &\quad \&\ A \subseteq A' \subseteq A'' \Rightarrow \mathbf{Q}_M A'B.\end{aligned}$$

Let us further call a quantifier *STRONG RIGHT* (*LEFT*) *CONT* if both it and its outer negation are *RIGHT* (*LEFT*) *CONT*. We have

$$\begin{aligned}&RIGHT\ (LEFT)\ MON \Rightarrow\\ &\quad \Rightarrow STRONG\ RIGHT\ (LEFT)\ CONT \Rightarrow\\ &\quad \Rightarrow RIGHT\ (LEFT)\ CONT.\end{aligned}$$

None of the implications can be reversed: for example, **exactly n** is *RIGHT* (and *LEFT*) *CONT*, but not *STRONG RIGHT* (or *LEFT*) *CONT*.

Thijsse [1983] observes that the property of quantifiers identified in (U10) is in fact *RIGHT CONT*:

3.6.6. PROPOSITION. *A binary quantifier is RIGHT CONT iff it is the conjunction of a MON*↑ *and a MON*↓ *quantifier.*

The proof is similar to the proof of Proposition 3.6.8 (b) below.

Our use of the conservativity constraint on binary quantifiers gives the right and the left arguments quite different roles, so it is not surprising that right monotonicity and left monotonicity are very different properties. This is clear from Theorem 3.6.4, and will become even more apparent in Section 4.3. A further illustration of the difference is afforded by the following model-theoretic characterization of the left monotonicity properties. Note first that any quantifier **Q** can be identified with a *class of structures*: in the binary case,

$$\mathbf{Q} = \{\langle M, A, B\rangle : \mathbf{Q}_M AB\}.$$

Call such a class *sub-closed* (*ext-closed*) if it is closed under substructures (extensions), and *inter-closed* if, whenever two structures, one a substructure of the other, are in **Q**, then so is every structure 'between' these two. It is straightforward to verify that

3.6.7. PROPOSITION. *Under CONSERV and EXT, a binary quantifier is sub-closed (ext-closed, inter-closed) iff it is* $\downarrow$*MON* ($\uparrow$*MON, LEFT CONT*).

For *first-order definable* quantifiers, the semantic property of being sub-closed has a wellknown syntactic counterpart, namely, definability by a *universal* sentence (cf. Chang and Keisler [1973], p. 128). Thus, among first-order definable quantifiers satisfying *CONSERV* and *EXT*, the $\downarrow$*MON* ones are precisely those definable by universal sentences. Corresponding results for $\uparrow$*MON* and *LEFT CONT* quantifiers follow from the previous proposition and

3.6.8. PROPOSITION. *For any binary quantifier* **Q**,

(a) $\mathbf{Q}$ *is ext-closed* $\Leftrightarrow \neg\mathbf{Q}$ *is sub-closed,*

(b) $\mathbf{Q}$ *is inter-closed* $\Leftrightarrow \mathbf{Q} = \mathbf{Q}' \wedge \mathbf{Q}''$, *for some sub-closed* $\mathbf{Q}'$ *and some ext-closed* $\mathbf{Q}''$.

Proof: (a) is obvious. As for (b), a conjunction of the sort indicated is clearly inter-closed. Conversely if **Q** is inter-closed, define

$\mathbf{Q}'_M AB \Leftrightarrow \mathbf{Q}_{M'} A'B'$, for some extension $\langle M', A', B'\rangle$ of $\langle M, A, B\rangle$,

$\mathbf{Q}''_M AB \Leftrightarrow \mathbf{Q}_{M'} A'B'$, for some substructure $\langle M', A', B'\rangle$ of $\langle M, A, B\rangle$;

then $\mathbf{Q}'$ and $\mathbf{Q}''$ are as desired. □

Another syntactic characterization of monotonicity from first-order logic is the following. Call a sentence $\phi(P)$, containing the unary P among its non-logical symbols, *upward monotone* (*in P*), if

$$\phi(P) \wedge \forall x(Px \to P'x) \to \phi(P')$$

is valid, and similarly for downward monotonicity. For example, sentences defining *LEFT* or *RIGHT MON* quantifiers will be monotone in certain predicate symbols. An occurrence of P in ϕ is said to be *positive* (*negative*), if it is within the scope of an even (odd) number of negations, when $\to$ and $\leftrightarrow$ have been eliminated. The next result is wellknown from first-order model theory (the proof is an application of Lyndon's interpolation theorem; cf. Chang and Keisler [1973], p. 90).

3.6.9. PROPOSITION. *A first-order sentence $\phi(P)$ (which may contain other predicate symbols but no function or constant symbols) is upward (downward) monotone iff it is equivalent to a sentence where P occurs only positively (negatively).*

Monotonicity properties have been quite useful in describing and explaining linguistic phenomena; cf. Barwise and Cooper [1981], Keenan and Stavi [1986], and, in connection with so-called polarity items, Ladusaw [1979], Zwarts [1986]. We will have several further uses of monotonicity in Section 4. In mathematical logic, monotone quantifiers have been studied in model theory and recursion theory. The beginnings of the model theory for monotone quantifiers will be given in Appendix B; further information can be found in Barwise and Feferman [1985]. On the more recursion-theoretic side, cf., for example, Aczel [1975] and Barwise [1978], and the references therein.

3.7. *Partial and Definite Quantifiers*

In 2.4.6 we mentioned that the number conditions belonging to the definites have been taken to indicate that the corresponding quantifiers are *partial*. This is the approach of Barwise and Cooper, who furthermore identify a semantic property of partial quantifiers, called *definiteness*, characteristic of the interpretation of the definites.[9]

Consider (in this subsection) binary quantifiers which are *partial in the first argument* (i.e. for certain A, $\mathbf{Q}_M AB$ may be *undefined* for all B). For example, the partial quantifier **the** coincides with the total **the** when $|A| = 1$, but is undefined when $|A| \neq 1$.

3.7.1. DEFINITION. **Q** is *definite*, if, for all M and all $A \subseteq M$ for which **Q** is defined, there is a non-empty set B_A such that, for all $B \subseteq M$, $\mathbf{Q}_M AB \Leftrightarrow B_A \subseteq B$.

The simple definites of 2.4.6 all have this property, when treated as partial quantifiers: e.g., for **the**, $B_A = A$ (or $B_A = X \cap A$ for some context set X), and for **John's**, $B_A = P_{\text{John}} \cap A$. That the use of partial quantifiers is necessary here follows from

3.7.2. PROPOSITION. *Under CONSERV, no definite quantifier is total.*

Proof. This follows from the fact that a definite and conservative quantifier must be undefined for $A = \varnothing$: Suppose **Q** is defined for $\varnothing$ and consider $B_\varnothing$ that exists by definiteness. Since $B_\varnothing \subseteq B_\varnothing$ we have $\mathbf{Q}_M \varnothing B_\varnothing$ and thus, by *CONSERV*, $\mathbf{Q}_M \varnothing\varnothing$. But then $B_\varnothing \subseteq \varnothing$, by definiteness, contradicting the stipulation that $B_\varnothing$ is non-empty. □

In view of this proof it is natural to weaken the requirements in Definition 3.7.1 slightly. Call **Q** *universal*, if it is as in 3.7.1, *except* that $B_\varnothing$ is allowed to be empty (i.e. that B_A is required to be non-empty only when A is). All definite quantifiers are universal, but not conversely, since **all** is universal. This is indeed the prime example of a universal quantifier, as the next result shows.

3.7.3. THEOREM. *Suppose* **Q** *is logical. Then* **Q** *is universal iff* **Q** = **all** *whenever defined.*

Proof. If **Q** coincides with **all** whenever defined it is clearly universal (with $B_A = A$). Conversely, suppose **Q** is universal and defined for A. We need to show that $B_A = A$. If $A = \varnothing$ we get $B_A = \varnothing$ just as in the proof above. Suppose, then, that $A \neq \varnothing$. Then $B_A \neq \varnothing$ by universality. Also, $B_A \subseteq A$; this follows from *CONSERV*, since $\mathbf{Q}_M AB_A$, whence $\mathbf{Q}_M A$ $A \cap B_A$, and thus $B_A \subseteq A \cap B_A$ by universality. Now assume that $B_A \neq A$. Take $a \in B_A$ and $a' \in A - B_A$. Let f be a function which permutes a and a' but leaves everything else in M as it is. By *ISOM*, $\mathbf{Q}_M f[A] f[B_A]$, i.e., $\mathbf{Q}_M A\ (B - \{a\}) \cup \{a'\}$. Thus, by universality, $B_A \subseteq (B_A - \{a\}) \cup \{a'\}$. But this contradicts $a \in B_A$. □

Thus the logical universal quantifiers, and in particular the definite ones, are just partial versions of **all**. This is one reason to restrict attention to total quantifiers, as we have done in preceding sections and shall continue to do in what follows. Another reason is that partial quantifiers make the model theory more cumbersome, and that many results for total quantifiers can rather easily be extended to the partial case by inserting phrases of the form "whenever . . . is defined" in suitable places.

Note finally that even if partial quantifiers are admitted in principle, the alternative treatment of definites suggested in 2.4.7 makes it possible to propose the universal

(*U*11) *Natural language quantifiers are total,*

while still preserving the intuition that statements involving definites lack truth value when the corresponding number conditions are not met.

3.8. *Finite Universes*

Many *DET*s more or less presuppose that the *N* and *VP* denotations under consideration are finite sets. Examples are *more than half, 30 percent of, many,* but also *DET*s like *most, more . . . than, fewer . . . than,* where the interpretations we gave actually work for infinite sets as well. It seems that in many natural language contexts we can make the blanket assumption

FIN *Only finite universes are considered.*

For *DET*s like *infinitely many* or *all but finitely many*, on the other hand, infinite models seem to be needed. So our strategy will be to keep track of those results that need *FIN* and those that don't. Interestingly, it turns out that *FIN* is a very natural constraint for the quantifier theory in the next section, in the sense that it *simplifies results and proofs*. Most of the results have generalizations to the case when *FIN* is dropped, but the added information does not appear to be very exciting from a natural language point of view.

This should be contrasted with the situation in mathematical logic. There infinite sets are crucial, and finite models are often just a nuisance. Consider the effect *FIN* would have in classical model theory. Most standard methods of constructing models (compactness, ultraproducts, etc.) would become ineffective, and many of the usual logical questions would become pointless. For example, the four properties of logics mentioned in section 1.6 lose their interest. This is clear for the Tarski and the Löwenheim property, and for compactness and completeness it follows from

3.8.1. PROPOSITION. *Under FIN, no logic is compact or complete.*

Proof. Under *FIN*, the set $\{\exists_{\geqslant n}x(x = x): n = 1, 2, \ldots\}$ has no model, so *EL* (and hence all its extensions) fail to be compact. The statement about completeness follows from a result by Trakhtenbrot, by which the set of all finitely valid *EL*-sentences (i.e. the set of valid sentences under *FIN*) is not recursively enumerable. For any logic $L = L(\mathbf{Q}^i)_{i \in I}$, this set is the intersection of the set of finitely valid *L*-sentences with the (recursive) set of *EL*-sentences. It follows that the set of finitely valid *L*-sentences is not recursively enumerable. □

Some standard logical questions remain, though. For example, we may still compare the *power of expression* of various logics under *FIN*, though some of the facts may change: we showed in 1.7 that *L*(**most**) < *L*(**more**) in general, but that *L*(**most**) ≡ *L*(**more**) under *FIN*. Likewise, *definability* issues are affected by *FIN*; for example, **all but finitely many** is not first-order definable in general, although it is trivially first-order definable under *FIN*.

It should be noted, however, that the main definability results in Section 1.7 (Theorem 1.7.4 and Corollary 1.7.5) continue to hold in the presence of *FIN*.

4. THEORY OF BINARY QUANTIFIERS

Binary quantifiers are the most common ones in natural language; they are also the most manageable relations, and we restrict attention to them from now on. A similar study of $(n + 1)$-ary quantifiers appears quite feasible, cf. Keenan and Moss [1985]. The important step is abandoning unary quantifiers: most of the results in this section have no counterpart for the unary case.

If nothing else is said, we assume in what follows that *all quantifiers involved are logical and satisfy NONTRIV*. Other constraints, such as *ACT*, *VAR*, and *FIN*, will be stated explicitly.

As a consequence of the assumption that *EXT* holds, we can often skip reference to the universe *M*, and write

$$\mathbf{Q}AB$$

instead of $\mathbf{Q}_M AB$. More precisely, let $\mathbf{Q}AB$ mean that, for *some M* such that $A, B \subseteq M$, $\mathbf{Q}_M AB$. *EXT* then guarantees that this is well defined.

Most of the results in 4.1–5 below originate from van Benthem [1984a], [1983c].

4.1. *Relational Behavior*

We have already encountered standard properties of binary relations, such as (ir)reflexivity (2.2.2) and symmetry (3.4), in the context of natural language quantification. A first start in quantifier theory is to exploit this perspective systematically. As we shall see, this turns out to be both rewarding in itself and useful for other purposes. Here are a few common properties of relations, and some quantifiers exemplifying them:

TABLE II

Property	*Definition*	*Examples*
symmetry	$\mathbf{Q}AB \Rightarrow \mathbf{Q}BA$	**some, no, at least n, at most n, exactly n, between n and m**
antisymmetry	$\mathbf{Q}AB \;\&\; \mathbf{Q}BA \Rightarrow A = B$	**all**
asymmetry	$\mathbf{Q}AB \Rightarrow \neg\mathbf{Q}BA$	–
reflexivity	$\mathbf{Q}AA$	**all, at least half, all but finitely many**
quasireflexivity	$\mathbf{Q}AB \Rightarrow \mathbf{Q}AA$	**some, most, at least n**
weak reflexivity	$\mathbf{Q}AB \Rightarrow \mathbf{Q}BB$	**some, most, at least n**
quasiuniversality	$\mathbf{Q}AA \Rightarrow \mathbf{Q}AB$	**no, not all, all but n**
irreflexivity	$\neg\mathbf{Q}AA$	**not all, all but n**
linearity	$\mathbf{Q}AB \vee \mathbf{Q}BA \vee A = B$	**not all**
transitivity	$\mathbf{Q}AB \;\&\; \mathbf{Q}BC \Rightarrow \mathbf{Q}AC$	**all, all but finitely many**
circularity	$\mathbf{Q}AB \;\&\; \mathbf{Q}BC \Rightarrow \mathbf{Q}CA$	–
euclidity	$\mathbf{Q}AB \;\&\; \mathbf{Q}AC \Rightarrow \mathbf{Q}BC$	–
antieuclidity	$\mathbf{Q}AB \;\&\; \mathbf{Q}CB \Rightarrow \mathbf{Q}AC$	?

One project is to find informative characterizations of (logical) quantifiers having such properties. As for symmetry, two useful equivalent formulations were given in Proposition 3.4.8. To deal with the other properties, we first state a

4.1.1. LEMMA. *If* $\mathbf{Q}AB$ *holds, there exists* B' *such that* $\mathbf{Q}AB'$ *and* $\mathbf{Q}B'A$.

Proof. Choose B' such that $A \cap B = B' \cap A$ and $|A - B| = |B' - A|$ (this may require extending the original universe, which is permitted by *EXT*). Since $\mathbf{Q}AB$, we get $\mathbf{Q}AB'$ by *CONSERV*, and then $\mathbf{Q}B'A$ by *QUANT*. □

Note the use of logicality here; the lemma fails if any of *CONSERV*, *EXT*, or *QUANT* are dropped. The following corollary is immediate (since we are assuming *NONTRIV*):

4.1.2. COROLLARY (van Benthem). *There are no asymmetric quantifiers.*

A characterization of antisymmetry is also forthcoming:

4.1.3. COROLLARY. **Q** *is antisymmetric iff* $\mathbf{Q}AB \Rightarrow A \subseteq B$.

Proof. If the condition holds, **Q** is clearly antisymmetric. Conversely, if **Q** is antisymmetric and $\mathbf{Q}AB$ holds, take B' as in the proof of Lemma 4.1.1. Thus $A = B'$ by antisymmetry, and $|A - B| = |B' - A| = 0$, i.e., $A \subseteq B$. □

This also gives a characterization of linearity, since **Q** is linear iff $\neg\mathbf{Q}$ is antisymmetric. As to the reflexivity properties and quasiuniversality, their main interest is in combination with other properties, as we shall see. The following consequences of Lemma 4.1.1 may nevertheless be noted:

4.1.4. COROLLARY. *Weak reflexivity implies quasireflexivity.*

This leaves only the properties in Table II involving *three* set variables. The '–' signs here are explained by the following results from van Benthem [1984a]:

4.1.5. THEOREM (van Benthem). *There are no euclidean quantifiers.*

We omit the proof, but show how to obtain the following corollary with the aid of Lemma 4.1.1:

4.1.6. COROLLARY (van Benthem). *There are no circular quantifiers.*

Proof. Suppose **Q** is circular. If $\mathbf{Q}AB$, take B' as in Lemma 4.1.1. By circularity, $\mathbf{Q}AA$. Thus, $\mathbf{Q}AB \Rightarrow \mathbf{Q}AA$ & $\mathbf{Q}AB \Rightarrow \mathbf{Q}BA$ (again by circularity), i.e., **Q** is symmetric. But it is easy to see that a circular and symmetric quantifier is euclidean, contradicting the theorem. □

Actually, some of these results, e.g. Corollary 4.1.2 and Theorem 4.1.5, were first proposed as semantic universals, based on empirical evidence (Frans Zwarts). Only later was it realized that they are consequences of more fundamental properties of quantifiers. This provided a first illustration of the potential usefulness of quantifier theory for linguistic explanation.

We left a question mark for antieuclidity in Table II. Here is an example, though: $\mathbf{Q}AB \Leftrightarrow |A| = n$. The following result from Westerståhl [1984] explains the situation:

4.1.7. THEOREM. **Q** *is antieuclidean iff* $\mathbf{Q}AB \Rightarrow \mathbf{Q}AC$ (*for all* A, B, C).

Two corollaries follow easily:

4.1.8. COROLLARY. **Q** *is antieuclidean iff there is a class* X *of cardinal numbers such that* $\mathbf{Q}AB \Leftrightarrow |A| \in X$.

4.1.9. COROLLARY (Zwarts). *Under VAR there are no antieuclidean quantifiers.*

Thus antieuclidean quantifiers put no condition at all on the *second* argument, i.e., the *VP* denotation. It seems safe to conclude that there are no antieuclidean natural language quantifiers.

Finally, consider transitivity. Here are some examples of transitive quantifiers:

(a) **all, all but finitely many,**

(b) $\mathbf{all}_e AB \Leftrightarrow \varnothing \neq A \subseteq B$ (**all** with existential import; cf. 3.4)

(c) $\mathbf{all}_n AB \Leftrightarrow A \subseteq B \vee |A| < n$ ($n \geqq 1$; note that $\mathbf{all}_1 = \mathbf{all}$)

(d) any antieuclidean quantifier (by Theorem 4.1.7)

(e) $\mathbf{Q}AB \Leftrightarrow (A \subseteq B \,\&\, |A| \geqq 5) \vee |A| = 3$.

Let us check (e): Suppose $\mathbf{Q}AB$ and $\mathbf{Q}BC$. In case $|A| = 3$ we get $\mathbf{Q}AC$ automatically, so suppose $A \subseteq B \,\&\, |A| \geqq 5$. But then $|B| \neq 3$, so we must have $B \subseteq C \,\&\, |B| \geqq 5$, whence $A \subseteq C \,\&\, |A| \geqq 5$, i.e., $\mathbf{Q}AC$. □

The following characterization of transitivity from Westerståhl [1984] depends essentially on *FIN*. It shows that (e) above is in a sense the typical case. If X, Y are sets of natural numbers, let $X < Y$ mean that every number in X is smaller than every number in Y; this is taken to hold trivially if X or Y are empty.

4.1.10. THEOREM. (*FIN*) **Q** *is transitive iff there are sets* X, Y *of natural numbers such that* $X < Y$ *and* $\mathbf{Q}AB \Leftrightarrow |A| \in X \vee (A \subseteq B \,\&\, |A| \in Y)$.

The proof combines a result from van Benthem [1984a] with techniques that will be introduced in 4.2 below. Note that the transitive **all but finitely many** fails to satisfy the condition in the theorem, if infinite universes are allowed. The next corollary shows that *VAR* has drastic effects on transitivity:

4.1.11. COROLLARY. (*FIN*) *Under VAR, the only transitive quantifiers are* **all** *and* $\mathbf{all}_e$.

Proof. This follows from the observation that *VAR* implies that either $X = \varnothing$ and $Y = N$, or $X = \{0\}$ and $Y = N - \{0\}$ in the theorem. □

Having thus looked at single properties of quantifiers, we can go on to *combinations* of such properties. For example, using Theorem 4.1.5 and Proposition 3.4.8 we obtain the

4.1.12. COROLLARY. *No quantifiers are both*

(a) *symmetric and transitive,*
(b) *symmetric and antieuclidean,*
(c) *symmetric and (ir)reflexive,*
(d) *quasiuniversal and reflexive.*

Reflexivity often has strong effects in combination with other properties. Note that, if **Q** is reflexive, $A \subseteq B \Rightarrow \mathbf{Q}AB$ (by *CONSERV*). From this and Corollary 4.1.3 we immediately get

4.1.13. COROLLARY. *The only reflexive and antisymmetric quantifier is* **all**.[1]

Furthermore, it is not hard to see that reflexivity together with the condition in Theorem 4.1.10 implies that, for some $n \geqq 1$, $X = \{0, \ldots, n - 1\}$ and $Y = \{k\colon k \geqq n\}$. This gives

4.1.14. COROLLARY (van Benthem). (*FIN*) *The only reflexive and transitive quantifiers are* $\mathbf{all}_n$, *for* $n \geqq 1$.

Again, **all but finitely many** is a counterexample if *FIN* is dropped.

4.1.15. COROLLARY. (*FIN*) *Under ACT, the only reflexive and transitive quantifier is* **all**.

Proof. Suppose that $\mathbf{Q} = \mathbf{all}_n$, for some $n \geqq 2$. Let M be a universe with exactly one element. It follows that **Q** is trivial on M, contradicting *ACT*. □

The next result connects our simple properties of relations with the monotonicity properties of Section 3.6:

4.1.16. THEOREM (Zwarts).
(a) *If* **Q** *is reflexive and transitive, then* **Q** *is* $\downarrow MON \uparrow$.
(b) *If* **Q** *is symmetric, then*
(i) **Q** *is quasireflexive iff* **Q** *is* $MON\uparrow$,
(ii) **Q** *is quasiuniversal iff* **Q** *is* $MON\downarrow$.

Proof. We prove (a); (b) is similar. If $\mathbf{Q}AB$ and $A' \subseteq A$, then $\mathbf{Q}A'A$, by reflexivity and CONSERV, and hence $\mathbf{Q}A'B$ by transitivity. Similarly, if $\mathbf{Q}AB$ and $B \subseteq B'$, then $\mathbf{Q}BB'$ and hence $\mathbf{Q}AB'$. □

From this and Theorem 3.6.3 we get the following variant of Corollary 4.1.15:

4.1.17. COROLLARY. *Suppose that* **Q** *satisfies CONSERV and VAR (but not necessarily EXT or QUANT), and is reflexive and transitive. Then* **Q** = **all**.

Proof. It suffices to note that neither Theorem 3.6.3 nor Theorem 4.1.16 uses *EXT* or *QUANT*. □

Instead of characterizing properties in terms of which quantifiers satisfy them, one may turn the question around and ask for characterizations of our most common quantifiers in terms of their properties. For the quantifier **all** and its variants, such characterizations were in fact obtained in 4.1.11, 4.1.13–15, 4.1.17. We end by giving a corresponding result for **some**. Let, for each cardinal κ, $\mathbf{some}_\kappa$ be the quantifier **at least** κ, i.e.,

$$\mathbf{some}_\kappa AB \Leftrightarrow |A \cap B| \geqq \kappa$$

(so $\mathbf{some}_1$ = **some**).

4.1.18. THEOREM (van Benthem). **Q** *is symmetric and quasireflexive iff* $\mathbf{Q} = \mathbf{some}_\kappa$, *for some* $\kappa \geqq 1$.

A proof will be given in Section 4.2. The following corollary is obtained similarly to Corollary 4.1.15:

4.1.19. COROLLARY. *Under ACT, the only symmetric and quasireflexive quantifier is* **some**.

4.2. *Quantifiers in the Number Tree*

By Proposition 3.4.2, each binary logical quantifier **Q** can be identified with a binary relation between cardinal numbers. We use the same notation for this relation, which is thus defined by

(1) $\mathbf{Q}xy \Leftrightarrow$ for some A,B with $|A - B| = x$ and $|A \cap B| = y$, $\mathbf{Q}AB$.

Inversely, given any binary relation **Q** between cardinal numbers, we get the corresponding logical quantifier by

(2) $\mathbf{Q}AB \Leftrightarrow \mathbf{Q}|A - B||A \cap B|$.

With (1) and (2) we can switch back and forth between a *set-theoretic* and a *number-theoretic perspective* on quantifiers. The latter perspective is the subject of the present subsection.

Here are the number-theoretic versions of a few well known quantifiers:

$$\mathbf{all}\,xy \Leftrightarrow x = 0,$$

$$\mathbf{no}\,xy \Leftrightarrow y = 0,$$

$$\mathbf{some}_n xy \Leftrightarrow y \geqq n,$$

$$\mathbf{all}_n xy \Leftrightarrow y = 0 \vee x + y < n,$$

$$\mathbf{most}\,xy \Leftrightarrow y > x,$$

infinitely many $xy \Leftrightarrow y$ is infinite.

all but finitely many $xy \Leftrightarrow x$ is finite.

Properties of quantifiers also have their number-theoretic versions. In the case of universal properties, such as those in Table II, there is a simple translation from the set-theoretic to the number-theoretic framework. Details can be found in Westerståhl [1984]; here we just consider a few examples. If two sets A,B are involved, let x correspond to $|A - B|$, y to $|A \cap B|$, and z to $|B - A|$. Then, for example,

(3) *quasireflexivity* is the property: $\mathbf{Q}xy \Rightarrow \mathbf{Q}0\ x + y$ (for all x, y),

(4) *symmetry* is the property: $\mathbf{Q}xy \Rightarrow \mathbf{Q}zy$ (for all x, y, z), or, equivalently, $\mathbf{Q}xy \Leftrightarrow \mathbf{Q}0\,y$ (for all x, y);

the last equivalence follows from Proposition 3.4.8 (it is also easy to see directly).

Sometimes proofs are simpler to carry out in the number-theoretic framework. This holds for several of the results in 4.1, in particular Theorems 4.1.7 and 4.1.10. As an illustration, we give the following

Proof of Theorem 4.1.18. Let κ be the least cardinal x such that $\mathbf{Q}0x$. κ exists, by *NONTRIV* and (4). Also, $\kappa > 0$; otherwise, for any x, y, we get $\mathbf{Q}y0$ (from $\mathbf{Q}00$ by (4)), whence $\mathbf{Q}0y$ (by (3)), and so $\mathbf{Q}xy$ (by (4)), contradicting *NONTRIV*. We claim that $\mathbf{Q} = \mathbf{some}_\kappa$. Clearly, $\mathbf{Q}xy$ implies $y \geqq \kappa$, by (4). Conversely, given x, y such that $y \geqq \kappa$, take x' such that $\kappa + x' = y$. By (4) and the definition of κ, $\mathbf{Q}x'\kappa$. Thus, by (3), $\mathbf{Q}0\,x' + \kappa$, i.e., $\mathbf{Q}0y$, so $\mathbf{Q}xy$ by (4). □

An operation that becomes nicely represented in the number-theoretic framework is *inner negation*, since we have

4.2.1. PROPOSITION. $(\mathbf{Q}\neg)xy \Leftrightarrow \mathbf{Q}yx$.

The number-theoretic perspective becomes particularly attractive if *FIN* is assumed. Quantifiers are then subsets of N^2. N^2 can be represented as a *number tree*, where each point (x, y) has two *immediate successors* $(x + 1, y)$ and $x, y + 1)$, which in turn are the *immediate predecessors* of the point $(x + 1, y + 1)$:[2]

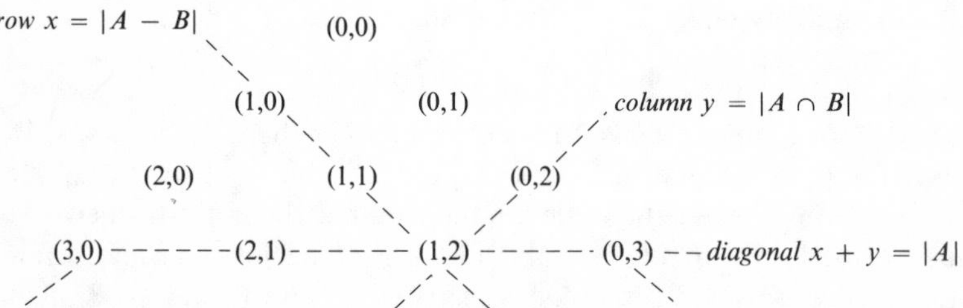

Quantifiers and their properties can be visualized in the number tree, and proofs can often be carried out directly in it. For an illustrative example, the reader is invited to carry out the above proof of Theorem 4.1.18 in the number tree (assuming *FIN*). Note that symmetry (quasi-reflexivity) means that if a point is in $\mathbf{Q}$ then so are all the points on the column (so is the rightmost point on the diagonal) through it. Another illustration, also left to the reader, is the proof of Proposition 3.4.11 in the number tree.

When representing a quantifier **Q** in the tree it is often practical to write a '+' on the points in **Q** and a '−' on the other points. For example,

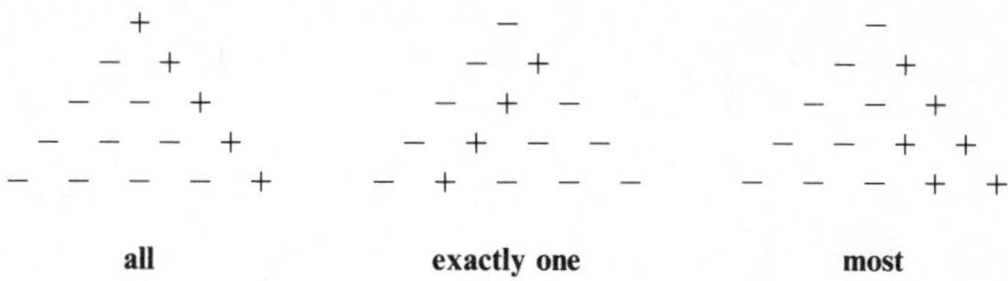

With this technique we can give our *non-triviality* conditions the following perspicuous formulations (we assume *FIN* for the rest of this subsection):

(5) *NONTRIV* ⇔ there is at least one + and one − in the tree,

(6) *ACT* ⇔ there is at least one + and one − *in the top triangle* (0,0), (1,0), (0,1),

(7) *VAR* ⇔ there is at least one + and one − *on each diagonal* (except (0,0)).

This illustrates that *VAR* is a much stronger assumption than *ACT*, i.e., that the universal (*U*9) in 3.5 really has content.

Monotonicity properties turn out to be particularly suited to number tree representation. Beginning with the *RIGHT* monotonicity properties, we can easily verify that

(8) *MON*↑ ⇔ each + fills the diagonal to its right with +'s,

(9) *MON*↓ ⇔ each + fills the diagonal to its left with +'s,

(10) *RIGHT CONT* ⇔ between two +'s on a diagonal there are only +'s.

Also observe that *STRONG RIGHT CONT*, i.e. *RIGHT CONT* for both **Q** and ¬**Q**, amounts to (10) together with the same condition with '+' replaced by '−'. It follows that

(11) *STRONG RIGHT CONT* ⇔ on each diagonal there is at most one change of sign.

The *LEFT* monotonicity properties can be illustrated as follows:

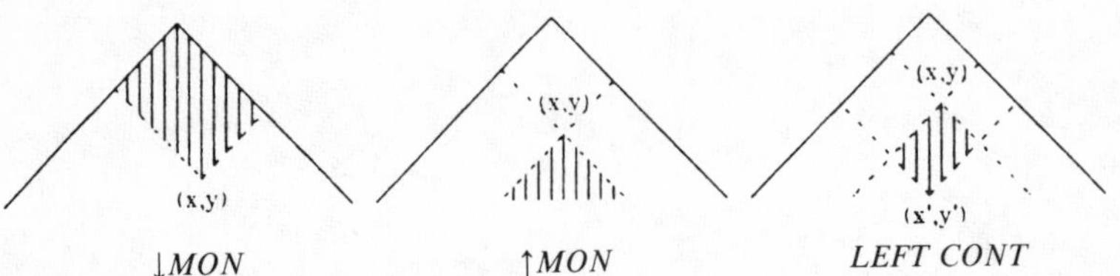

I.e., if (x, y) (and (x', y')) is in **Q** then so are all the points in the shaded area.

Working in the number tree, we can introduce several variants of the above monotonicity properties. Define $\uparrow_c MON$, $\downarrow_c MON$, $LEFT_c\ CONT$, and $STRONG\ LEFT_c\ CONT$ by replacing, in (8)–(11), respectively, "diagonal" with "column", and do the same for $\uparrow_r MON$, $\downarrow_r MON$, $LEFT_r\ CONT$, and $STRONG\ LEFT_r\ CONT$, replacing "diagonal" with "row". The terminology is motivated by the fact that

(12) $\uparrow_c MON \Leftrightarrow (\mathbf{Q}AB\ \&\ A' \subseteq A\ \&\ A \cap B = A' \cap B \Rightarrow \mathbf{Q}A'B)$,

and similarly for the other properties; in other words, they are as the previous *LEFT* properties, only we keep $A \cap B$ fixed in the '*c*' case, and $A-B$ fixed in the '*r*' case. To make the intuitive picture clear, here is yet another way to illustrate the downward monotone properties we have so far encountered:

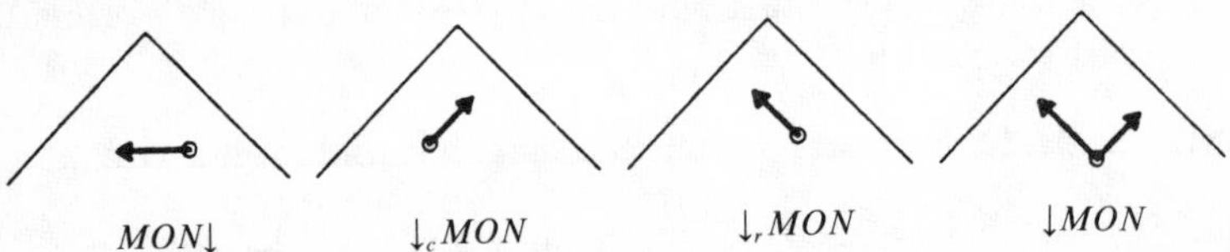

In the tree it is easy to check whether particular quantifiers have such properties. For example, it is clear from the above illustrations that **most** is $MON\uparrow$ and $\downarrow_c MON$, but not $\downarrow_r MON$. It is also clear that

(13) $\uparrow MON \Leftrightarrow \uparrow_c MON\ \&\ \uparrow_r MON$,

(14) $\downarrow MON \Leftrightarrow \downarrow_c MON\ \&\ \downarrow_r MON$.

The corresponding statement for *CONT* fails, however (as can also be seen from the tree).

An interesting application of 'tree techniques' is given in van Benthem [1983c] to an idea in Barwise and Cooper [1981] concerning how hard it is (psychologically) to 'process' (verify or falsify) quantified statements. Barwise and Cooper speculated that quantifiers with monotonicity properties were easier to process and would therefore be preferred in natural language. Now, verifying a sentence of the form *allAB* in a universe with n elements takes n observations, and falsifying it takes at least 1 observation. If we look at *mostAB* instead (and suppose that n is even for simplicity), the least possible number of observations it takes to verify it is $n/2 + 1$, and the corresponding number for falsification is $n/2$. In both cases the sum is $n + 1$. This holds for many basic quantifiers, but not all: e.g. *exactly one AB* requires n observations for verification and 2 for falsification.

van Benthem defines, with reference to the number tree, **Q** to be of *minimal count complexity* if, on each universe with n elements (this corresponds to the finite top triangle of the tree with the diagonal $x + y = n$ as base), there is a minimal confirmation pair (x_1, y_1) and a minimal refutation pair (x_2, y_2) $(x_i + y_i \leqq n)$ such that every pair (x, y) on the diagonal $x + y = n$ is determined by them:

$$x \geqq x_1 \;\&\; y \geqq y_1 \Rightarrow \mathbf{Q}xy,$$

$$x \geqq x_2 \;\&\; y \geqq y_2 \Rightarrow \neg\mathbf{Q}xy.$$

One can verify that $x_1 + y_1 + x_2 + y_2 = n + 1$, and thus that **all** and **most** are of minimal count complexity, but not **exactly one**.

Now consider the following very strong continuity property:

$$\textit{SUPER CONT} =_{\mathrm{df}} \textit{STRONG RIGHT CONT} \;\&$$
$$\&\; \textit{STRONG LEFT}_c\ \textit{CONT} \;\&$$
$$\&\; \textit{STRONG LEFT}_r\ \textit{CONT}.$$

In other words, *SUPER CONT* means that there are no changes of sign in any of the three main directions in the number tree. It can be seen that the *SUPER CONT* quantifiers are precisely those determined by a *branch* in the tree (which can start anywhere on the edges; not necessarily at the top) with the property that, going downward, it always contains one of the immediate successors of each point on it:

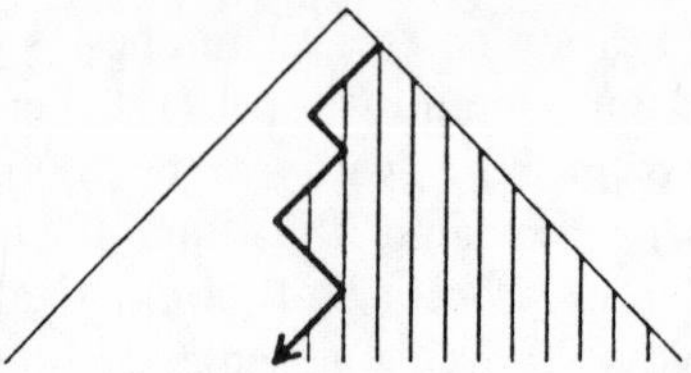

The connection with count complexity is now the following:

4.2.2. THEOREM (van Benthem): (*FIN*) *Under ACT*, **Q** *is of minimal count complexity iff it is SUPER CONT.*

The proof of this consists simply in showing that the two combinatorial descriptions give the same tree pattern.

From the above description of *SUPER CONT* one also obtains the following results:

4.2.3. PROPOSITION. *SUPER CONT* ⇒ *RIGHT MON.*[3]

4.2.4. PROPOSITION. *There are uncountably many SUPER CONT logical quantifiers* (*even under FIN*).

What is the relation between *SUPER CONT* and *LEFT CONT*? Using the tree it is easy to see that neither property implies the other. In the next subsection we shall find, moreover, that there are only countably many *LEFT CONT* logical quantifiers, under *FIN*.

4.3. *First-order Definability and Monotonicity*

We shall prove a theorem characterizing the first-order definable quantifiers in terms of monotonicity, under *FIN*. The most general form of the result has nothing directly to do with logicality, so we begin by assuming that **Q** is an arbitrary k-ary quantifier ($k \geqq 1$). We noted in 3.6 that **Q** can be identified with the class of structures $\langle M, A_0, \ldots, A_{k-1} \rangle$ such that $\mathbf{Q}_M A_0 \ldots A_{k-1}$, and we defined the properties of being sub-closed, ext-closed, and inter-closed for classes of structures.

The key to the result is the following lemma from van Benthem [1984a]:

4.3.1. LEMMA (van Benthem). (*FIN*) *Suppose* **K** *is a class of* (*finite*) *structures which is definable in EL by a set of monadic universal sentences. Then* **K** *is definable already by one such sentence.*

4.3.2. THEOREM. (*FIN*) **Q** *is first-order definable iff there are interclosed quantifiers* $\mathbf{Q}_1, \ldots, \mathbf{Q}_m$ *satisfying ISOM such that* $\mathbf{Q} = \mathbf{Q}_1 \vee \ldots \vee \mathbf{Q}_m$.

Proof. Suppose first **Q** is a disjunction of this kind. By Proposition 3.6.8, each $\mathbf{Q}_i$ can be written $\neg\mathbf{Q}_i' \wedge \mathbf{Q}_i''$, where $\mathbf{Q}_i'$ and $\mathbf{Q}_i''$ are sub-closed. Moreover, it easily follows from the proof of that proposition that both $\mathbf{Q}_i'$ and $\mathbf{Q}_i''$ satisfy *ISOM* if $\mathbf{Q}_i$ does. Thus, it will suffice to show that every sub-closed quantifier satisfying *ISOM* is first-order definable. Assume, then, that **Q** has these properties. Under *FIN, any* class of structures closed under isomorphism is definable by a *set* of *EL*-sentences, by a standard argument: a finite structure can be completely described (up to isomorphism) by *one EL*-sentence, and the relevant set consists of all negated descriptions of models *not* in the class. If the class is in addition sub-closed, a variant of this argument shows that the sentences can be taken universal (one takes the negations of the existentially quantified diagrams of structures not in the class).[4] Since in our case the class is also monadic, **Q** is first-order definable by Lemma 4.3.1.

Now suppose **Q** is definable by an *EL*-sentence $\psi = \psi(P_0, \ldots, P_{k-1})$. By Corollary 1.7.5 (with $L = EL$), there is a natural number n such that **Q** is closed under the relation $\approx_n$ (cf. Section 1.7). Consider sentences expressing conditions

$$|P_s^M| = i,$$

for some $i < n$, or

$$|P_s^M| \geqq n.$$

It follows that any conjunction of such sentences where s runs through all the functions from k to 2, is a complete description of a model $\langle M, A_0, \ldots, A_{k-1}\rangle$, as far as **Q** is concerned. There are finitely many such descriptions, and ψ must be equivalent to the disjunction of all complete descriptions of structures in **Q**. Moreover, each disjunct defines a quantifier, which, by the form of the definition, is easily seen to be inter-closed. Since any *EL*-definable quantifier satisfies *ISOM*, the theorem is proved. □

Returning now to the case of binary logical quantifiers, we get from the theorem and Proposition 3.6.7 the

4.3.3. COROLLARY. (*FIN*) *If* **Q** *is binary and logical, then* **Q** *is first-order definable iff* **Q** *is a finite disjunction of LEFT CONT* (*binary and logical*) *quantifiers.*

There is a simpler direct proof of the corollary. This is because we can work in the number tree. In one direction, it suffices to show that $\uparrow$*MON* quantifiers are first-order definable. If **Q** is $\uparrow$*MON*, each point in **Q** generates an infinite downward triangle. From a given triangle within **Q**, only finitely many steps can be taken towards the edges of the tree. It follows that **Q** is a finite union of such triangles,

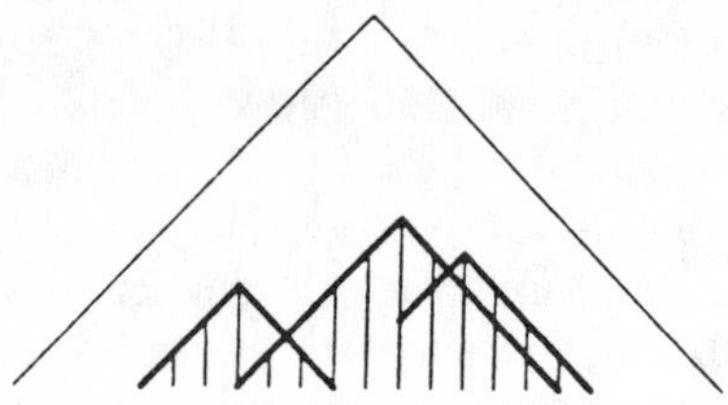

and therefore clearly first-order definable. The proof in the converse direction, using Corollary 1.7.5, also becomes simpler in the number tree.

Corollary 3.4.3 shows, once more, that the *LEFT* monotonicity properties are much stronger than the *RIGHT* ones, due to the special role *CONSERV* gives to the left argument of a quantifier. In particular, there are only denumberably many *LEFT CONT* logical quantifiers (under *FIN*); this should be contrasted with Proposition 4.2.4.

Note that *FIN* is essential here. For example, **at most finitely many** is $\downarrow$*MON* but not definable by any first-order sentence (or set of such sentences).[5]

Definability results such as these have not only logical interest: they also tell us something about the extent to which a certain logic – first order logic, in this case – is adequate for natural language semantics. Of course, we knew already that first-order logic is not adequate, e.g. by the non-definability of **most**, but Corollary 4.3.3 places such isolated facts in a wider perspective.

The results here concern definability in the set-theoretic framework for quantifiers. What about *number-theoretic definability* (for logical quantifiers, under *FIN*)? Here we should consider formulas $\phi(x, y)$ in some suitable arithmetical language, containing at least the individual constant 0 and the unary successor function symbol S (and hence the *numerals* $\mathbf{0} = 0$, $\mathbf{1} = S0$, $\mathbf{2} = SS0$, etc.). Then ϕ *defines* **Q** iff, for all m, n,

$$\mathbf{Q}mn \Leftrightarrow \langle N, 0, S, \ldots \rangle \vDash \phi(\mathbf{m}, \mathbf{n}).$$

Examples of definable quantifiers, some of which in languages with the relation $<$ or the operation $+$, were given at the beginning of Section 4.2. Now which arithmetical definability notion corresponds to first-order definability in the set-theoretic sense? Notice first that even the simple formula

$$x = y$$

defines a non-first-order definable quantifier, namely, **exactly half**. However, let the *pure number formulas* be those formulas in the language $\{0, S\}$ obtained from atomic formulas of the form

$$x = \mathbf{n}$$

by closing under Boolean connectives. Clearly every pure number formula with variables among x, y defines a first-order definable quantifier. But also conversely, for it can be seen by inspecting more closely the proofs of Theorem 4.3.2 and Corollary 4.3.3 that every first-order definable quantifier is in fact a Boolean combination of quantifiers of the form **at most n** and **all but at most n**, and the former, for example, is defined by the pure number formula

$$y = \mathbf{0} \vee \ldots \vee y = \mathbf{n}.$$

Thus, we have the

4.3.4. COROLLARY. (*FIN*) **Q** *is first-order definable iff* **Q** *is arithmetically defined by some pure number formula.*

This of course raises new definability questions. Which quantifiers are defined by arbitrary formulas in $\{0, S\}$? Which are defined by formulas in $\{0, S, +\}$? It can be seen that **most** belongs to the second, but not the first, class. These questions are studied in connection with *computational complexity* in van Benthem [1985], [1987a]. He shows, among other things, that the second class of quantifiers mentioned above consists precisely of those computable by a push-down automaton (under *FIN*). He also characterizes the first-order definable quantifiers computationally, namely, as those computable by a certain type of finite-state machine. This illustrates another aspect of the interest of definability questions: *classification* of quantifiers w.r.t. various notions of complexity. For the relevant definitions, and for several other interesting results along the same lines, we must refer to the two papers by van Benthem mentioned above.

4.4. *Logical Constants*

Clearly not all the $2^{\aleph_0}$ logical quantifiers deserve the title *logical constant*. We have already presented conditions that severely restrict the range of quantifiers. For example, *LEFT MON* plus *VAR* leaves only the quantifiers in the square of opposition (3.6). But there is no immediate reason why these two constraints should apply to logical constants. In this subsection, we look at some conditions which can be taken to have an independent connection with logical constanthood.

One idea seems natural enough, namely, that quantifiers that are logical constants should be *simple* natural language quantifiers (Section 2.4). Thus, the semantic universals holding for simple quantifiers apply to them. It follows that they should be logical (i.e. obey *CONSERV, EXT*, and *QUANT*) and satisfy *NONTRIV* and *RIGHT CONT* (by (*U*10) and Proposition 3.6.6).

As for constraints specifically related to logical constanthood, we will concentrate on one rather strong property often claimed to be characteristic of logical constants, namely, that *they do not distinguish cardinal numbers*. The idea is that such distinctions belong to mathematics, not logic. We will consider two rather different ways of making this idea precise.

FIN is used in what follows, so that we can argue in the number tree. It is possible, however, to generalize the results (with suitable changes) to infinite universes.

The first version of the above idea goes back to Mostowski [1957], although he only applied it to infinite cardinalities. Given **Q***AB*, the relevant cardinality here is that of the universe, or in our case, by *CONSERV* and *EXT*, that of *A*. We must of course separate 0 from the other cardinalities, since distinguishing non-zero numbers from 0 is precisely what basic quantifiers such as **some** and **all** do. With these observations, we can transplant Mostowski's idea to the finite case as follows:

4.4.1. DEFINITION. Suppose $m, n > 0$. **Q** *does not distinguish m and n*, if

(a) $\mathbf{Q}m0 \Leftrightarrow \mathbf{Q}n0$,

(b) $\mathbf{Q}0m \Leftrightarrow \mathbf{Q}0n$,

(c) if $x_1 + y_1 = m$ and $x_2 + y_2 = n$, where $x_i, y_i > 0$, then

$$\mathbf{Q}x_1 y_1 \Leftrightarrow \mathbf{Q}x_2 y_2.$$

For example, **at least k** does not distinguish any $m,n < k$, but distinguishes all m,n for which at least one is $\geqq k$.

Note that no restriction at all is put on the point (0,0). To avoid trivial complications in the next result, we shall restrict attention to the number tree *minus* (0,0) (we write '-0' to indicate this). Also, we replace, in this subsection, *NONTRIV* by the slightly stronger condition that in the tree *minus* (0,0), there is at least one + and one −.

It is not surprising that the present logicality constraint has rather drastic effects on the range of quantifiers:

4.4.2. THEOREM. (*FIN*, -0) *Suppose that* **Q** *does not distinguish any pair of non-zero natural numbers and satisfies RIGHT CONT. Then* **Q** *is one of the quantifiers* **some**, **no**, **all**, **not all**, *and* **some but not all**.

Proof. There are four possible patterns for the top triangle minus (0,0). *Case* 1: + +. By the cardinality property and *RIGHT CONT*, this puts a + everywhere, contradicting (our present version of) *NONTRIV*. *Case* 2: + −. Then the left edge will have only +, and the right edge only −. For the remaining interior triangle, there are two possibilities, giving either **no** or **not all**. *Case* 3: − +. This is symmetric to Case 2 and gives **some** and **all**. *Case* 4: − −. Besides the trivial case with only −, there is the case with + in the interior triangle and − on both edges, i.e., **some but not all**. □

We may note that *VAR*, or *STRONG RIGHT CONT*, will exclude **some but not all**, but it is not clear that we have the right to assume any of these. (On the other hand, it could be argued that one interpretation of the *DET some*, especially when focussed or stressed, *is* **some but not all**.)

The second version of the requirement that logical constants do not distinguish cardinal numbers is from van Benthem [1984a]. Here the idea is that no point in the tree is special: you always *proceed downward in the same way*. Proceeding one step downward can be regarded as a thought experiment, whereby one, given A and B, adds one element to $A - B$ or $A \cap B$. The condition is then that the outcome is *uniform* in the tree, i.e., that it does not depend on the number of elements in these sets (the point (0,0) need not be excluded here, although it can be):

UNIF *The sign of any point in the tree determines the sign of its two immediate successors.*

4.4.3. THEOREM (van Benthem). (*FIN*) *The UNIF and RIGHT CONT quantifiers are precisely* **some**, **no**, **all**, **not all**, *and the quantifiers* $|A|$ **is even** *and* $|A|$ **is odd**.

Proof. Again a simple tree argument suffices. There are eight top triangles to consider; let us look at two. First, consider $\begin{smallmatrix} & - & \\ - & & + \end{smallmatrix}$. Here the − successors are determined, but for the right + successor there is a choice, and we get the two patterns

$$\begin{array}{ccccccc} & & & - & & & \\ & & - & & + & & \\ & - & & + & & - & \\ - & & + & & - & & + \end{array} \qquad\qquad \begin{array}{ccccccc} & & & - & & & \\ & & - & & + & & \\ & - & & + & & + & \\ - & & + & & + & & + \end{array}$$

The first of these is excluded by *RIGHT CONT*, and the second is **some**. Now consider the top triangle $\begin{smallmatrix} & + & \\ - & & - \end{smallmatrix}$. Here the − successors are either both − or both +. In the first case we get only − in the rest of the tree, contradicting *NONTRIV*. In the second case we get $|A|$ **is even**. The other cases are similar. □

The last two quantifiers in the theorem are not natural language quantifiers and should be excluded somehow. The following slight strengthening of *NONTRIV* would suffice:

*NONTRIV** *One some diagonal in the tree, there is at least one + and at least one −.*

In fact, it seems that we may safely replace *NONTRIV* by *NONTRIV** in the universal (*U*8) in 3.5.

It is interesting that these two quite different implementations of the idea that logical constants are insensitive to changes in cardinal number give so similar results. There are of course other ideas than cardinal insensitivity on which one can base constraints for logical constanthood. Further ideas and results in this direction can be found in van Benthem [1984a], [1983c]. For example, he shows that by slightly weakening *UNIF* one can obtain, in addition to the quantifiers in the square of opposition, **most**, **not most**, **least** (i.e. **least** $AB \Leftrightarrow |A \cap B| < |A - B|$), **not least**, and no others, as logical constants. The number tree is an excellent testing ground for experiments in this area.

4.5. *Inference Patterns*

The universal properties of quantifiers we have considered can be seen as *inference schemes for quantified sentences*:

$$\frac{QAB}{QBA} \qquad \frac{QAB \quad QBC}{QAC} \qquad \frac{QAB \quad QAC}{QBC} \qquad \text{etc.}$$

(symmetry) (transitivity) (euclidity)

There are also schemes with fixed quantifiers, such as

$$\frac{QAB \quad allBC}{QAC}$$

($MON\uparrow$).

In 4.1 we answered some questions of the type: Which quantifiers satisfy inference scheme *S*? This is familiar from Aristotle's study of syllogisms, cf. Section 1.1. Aristotle aimed at systematic survey, and he answered the question for *all schemes of a certain form*.

EXAMPLE. Consider schemes with 2 premisses, 1 conclusion (all of the form QXY with distinct X, Y), at most 3 variables, and 1 quantifier symbol. There are 6 possibilities for each formula in a scheme, and hence, up to notational variants (permutations of the variables), $6^3/3! = 36$ possible schemes. Identifying schemes that differ only by the order of the premisses, and deleting the trivially valid schemes whose conclusion is among the premisses, 15 schemes will remain. Then, it can be shown, using Lemma 4.1.1 and Theorems 4.1.5–7, that for *logical* quantifiers these reduce to *symmetry, transitivity, anti-euclidity,* and the following property, which we may call *weak symmetry*:

$$\frac{QAB \quad QBC}{QBA}$$

(ignoring unsatisfiable schemes, such as euclidity). Weak symmetry is strictly weaker than symmetry; a number-theoretic characterization of it can be found in Westerståhl [1984].

Thus, there are no other schemes than these (of the present form), and the results of 4.1 (e.g. Corollary 4.1.8 and Theorem 4.1.10) give us a pretty good idea of which quantifiers satisfy them.

EXAMPLE. *Aristotelian syllogisms.* The schemes are as in the first example, except that there are 3 quantifier symbols, and that one variable (the 'middle term') is required to occur in both premisses but not in the conclusion. Aristotle solved this problem in the special case that quantifiers are taken among **some**, **all**, **no**, **not all**. In the general case of logical quantifiers the solution is of course much more complicated.

The last example indicates that systematic survey of all possible cases is not necessarily an interesting task. In this subsection we shall consider a more specific problem: Given the wellknown inference schemes for basic quantifiers such as **some** and **all**, are these quantifiers *determined* by the schemes, or are the schemes, as it were inadvertently, satisfied by other quantifiers as well?

The logical interest of such questions should be clear. They concern the extent to which the syntactic behavior of logical constants determine their semantic behavior. Negative results will tell us that inference rules of a certain type underdetermine semantic interpretation – a familiar situation in logic. Positive results, on the other hand, can be viewed as a kind of *completeness* or *characterization* theorems.[6]

These questions are also related to deeper issues in the philosophy of language, namely, whether the 'concrete manifestations' of linguistic expressions determine their meaning; cf. post-Wittgensteinian discussions of meaning and use, or Quine's idea on the indeterminacy of translation, or the debate on whether the meaning of logical constants are given by their introduction rules, and more generally on the relation between meaning and proofs (in the context of classical vs. intuitionistic logic; cf. Prawitz [1971], [1977], Dummett [1975]).

Clearly, inference patterns concern the 'concrete' side of language, whereas model theory deals with abstract entities. It would seem that results which relate these two perspectives may be of interest regardless of one's position on the deeper philosophical issues.

A first observation is that the content of our question depends crucially on which kind of inference scheme one allows, i.e. on the choice of *inferential language*. We will look at two such languages here, with quite different properties. But the point is illustrated even more clearly by the following

EXAMPLE. Let the inferential language be predicate logic with the (binary) quantifiers **some** and **all** (this is not essential; we could use $\forall$ and $\exists$ instead). The standard rules for **some**, but with an arbitrary quantifier symbol Q in place of *some*, can be formulated as follows:

(1) $$\frac{\phi(t)\ \psi(t)}{Qx(\phi(x),\ \psi(x))} \qquad \frac{\phi(x) \wedge \psi(x) \rightarrow \theta}{Qx(\phi(x),\ \psi(x)) \rightarrow \theta} \quad (x \text{ not free in } \theta).$$

Q *satisfies* a rule of this type if, for each model $\boldsymbol{M}$ and each sequence $\bar{a}$ of individuals from M, if the premisses are true in $(\boldsymbol{M}, \bar{a})$ (with Q interpreted as $\mathbf{Q}_M$), then so is the conclusion. But then it is practically *trivial* that

(2) **Q** satisfies the rules (1) iff $\mathbf{Q} = \mathbf{some}$.

For suppose **Q** satisfies (1). Take any M. We must show that $\mathbf{Q}_M AB \Leftrightarrow A \cap B \neq \varnothing$. If $a \in A \cap B$ then P_1x and P_2x are true in $\langle M, A, B, a\rangle$, and hence, by the first rule, so is $Qx(P_1x, P_2x)$, i.e. $\mathbf{Q}_M AB$ holds. If, on the other hand, $A \cap B = \varnothing$, let θ be a logically false sentence and b any element of M. Then $P_1x \wedge P_2x \rightarrow \theta$ is true in $\langle M, A, B, b\rangle$, and thus also $Qx(P_1x, P_2x) \rightarrow \theta$, by the second rule. So $Qx(P_1x, P_2x)$ is false in the model, i.e., $\mathbf{Q}_M AB$ does not hold. (Similar remarks apply to **all**.) □

Why does the inferential language of this example trivialize the question of whether the rules characterize the quantifiers? One suggestion might be that rules like (1) are *circular* (in some sense to be specified) as explanations of meaning. In any case, we shall now define two other inferential languages, IL_{syll} and IL_{boole}, for which the problem has non-trivial solutions. These languages have no individual variables, only *set* variables. Most of the inference schemes we have seen so far can be expressed in them. The idea to pose the present characterization problem for quantifiers was introduced in van Benthem [1984a], and the results on IL_{syll} below are from van Benthem [1983c].

DEFINITION of IL_{syll}. (a) *Syntax: Elementary schemes* in IL_{syll} are of the form QAB or $\neg Q'AB$, where $A,B, \ldots$ are set variables and $Q,Q', \ldots$

quantifier symbols. A *scheme* in IL_{syll} is either an elementary scheme or has the form

(3) $\phi_1 \wedge \ldots \wedge \phi_n \rightarrow \theta_1 \vee \ldots \vee \theta_k$,

where ϕ_i and θ_j are elementary schemes.
(b) *Semantics*: Suppose ψ is a scheme in IL_{syll} with quantifier symbols among $Q^1, \ldots, Q^m$, and with p set variables. For any quantifiers $\mathbf{Q}^1, \ldots, \mathbf{Q}^m$, a $(\mathbf{Q}^1, \ldots, \mathbf{Q}^m)$*-model* (for ψ) is a model $\boldsymbol{M} = \langle M, A_0, \ldots, A_{p-1} \rangle$, where Q^i is interpreted as $\mathbf{Q}^i_M$. We say that

$$(\mathbf{Q}^1, \ldots, \mathbf{Q}^m) \text{ } \textit{satisfies} \text{ the scheme } \psi,$$

if ψ is true (in the obvious sense) in all $(\mathbf{Q}^1, \ldots, \mathbf{Q}^m)$-models. Similarly, $(\mathbf{Q}^1, \ldots, \mathbf{Q}^m)$ satisfies a *set* Ψ of IL_{syll}-schemes if it satisfies each element of Ψ. Finally, the *syllogistic theory* of $(\mathbf{Q}^1, \ldots, \mathbf{Q}^m)$ is

$$Th_{\text{syll}}(\mathbf{Q}^1, \ldots, \mathbf{Q}^m) = \{\psi\colon (\mathbf{Q}^1, \ldots, \mathbf{Q}^m) \text{ satisfies } \psi\}.$$

This definition just gives more formal versions of notions we have been using all along. For example, all the properties in Table II (4.1), *except* antisymmetry and linearity, can be expressed in IL_{syll} (these two would be expressible if we had allowed quantifier *constants* above). That a quantifier $\mathbf{Q}$ satisfies a scheme just means that the scheme expresses a valid inference rule for $\mathbf{Q}$. For example, $\mathbf{Q}$ satisfies

$$QAB \rightarrow QBA$$

just in case $\mathbf{Q}$ is symmetric. Note that more than one quantifier symbol may occur in a scheme. For instance, the scheme

$$Q^1AB \wedge Q^2CA \rightarrow Q^1CB$$

is satisfied by the pair (**no**, **all**) (this is the syllogistic inference 'Celarent'; cf. Section 1.1).

DEFINITION of IL_{boole}. (a) *Syntax*: As for IL_{syll}, except that elementary schemes now have the form QXY or $\neg Q'XY$, where X, Y are (Boolean) combinations of set variables with the symbols $\cap$, $\cup$, and $-$.
(b) *Semantics*: As before, where the Boolean symbols have their usual meaning.

Examples of schemes in IL_{boole} but not in IL_{syll} are

$$QAB \rightarrow QA\ A \cap B,$$

$$QAA \cap B \rightarrow QAB,$$

$$QAB \rightarrow QA \cap B\, A \cap B,$$

$$QA \cap B\, A \cap B \rightarrow QAB;$$

the first two together express *CONSERV*, and the other two are (together) equivalent to symmetry.

There is one last

4.5.1. DEFINITION. Let Ψ be a set of schemes in IL_{syll} (or IL_{boole}), in the quantifier symbols $Q^1, \ldots, Q^m$. Let $\mathbf{Q}^1, \ldots, \mathbf{Q}^m$ be quantifiers. We say that

$$\Psi \textit{ determines } (\mathbf{Q}^1, \ldots, \mathbf{Q}^m),$$

if (a) $(\mathbf{Q}^1, \ldots, \mathbf{Q}^m)$ satisfies Ψ, and (b) no other sequence of m quantifiers satisfies Ψ. Also, $(\mathbf{Q}^1, \ldots, \mathbf{Q}^m)$ is *determined in* IL_{syll} (IL_{boole}), if some set of schemes in IL_{syll} (IL_{boole}) determines $(\mathbf{Q}^1, \ldots, \mathbf{Q}^m)$.

Note that if $(\mathbf{Q}^1, \ldots, \mathbf{Q}^m)$ is determined in IL_{syll} (IL_{boole}), it is determined by the set $Th_{\text{syll}}(\mathbf{Q}^1, \ldots, \mathbf{Q}^m)$ ($Th_{\text{boole}}(\mathbf{Q}^1, \ldots, \mathbf{Q}^m)$).

As an example, consider the set consisting of the two IL_{syll}-schemes expressing *symmetry* and *quasireflexivity*. **some** satisfies this set, but, by Theorem 4.1.18, the set does *not* determine **some**. The obvious question is then whether some larger set determines **some**, i.e., whether **some** is determined in IL_{syll}. A negative answer follows from the next theorem.

We assume *FIN* from now on (but see the comments at the end). The quantifiers $\mathbf{some}_n$ and $\mathbf{all}_n$ were defined in Section 4.1.

4.5.2. THEOREM (van Benthem): $Th_{\text{syll}}(\mathbf{some}, \mathbf{all})$ *is satisfied precisely by the pairs* $(\mathbf{some}_n, \mathbf{all}_n)$, *for* $n \geqq 1$.

Thus not even (**some**, **all**) is determined in IL_{syll}. That **some** (or **all**) is not determined follows immediately, since $Th_{\text{syll}}(\mathbf{some}) \subseteq Th_{\text{syll}}(\mathbf{some}, \mathbf{all})$.

This theorem is an immediate consequence of the next two theorems, which give additional information about the pair (**some**, **all**).

4.5.3. THEOREM (van Benthem). $Th_{\text{syll}}(\mathbf{some}, \mathbf{all}) = Th_{\text{syll}}(\mathbf{some}_n, \mathbf{all}_n)$, *for* $n \geqq 1$.

For the next result, let Φ consist of the IL_{syll}-schemes saying that Q^1 is symmetric and quasireflexive and that Q^2 is reflexive and transitive, plus the following schemes:

(4) $\quad Q^1AB \wedge Q^2AC \rightarrow Q^1BC,$

(5) $\quad \neg Q^1AA \rightarrow Q^2AB.$

4.5.4. THEOREM (van Benthem). *If* $(\mathbf{Q}^1, \mathbf{Q}^2)$ *satisfies* Φ, *then, for some* $n \geqq 1$, $\mathbf{Q}^1 = \mathbf{some}_n$ *and* $\mathbf{Q}^2 = \mathbf{all}_n$.

The proof uses Theorems 4.1.18 and 4.1.14, which tell us that $\mathbf{Q}^1 = \mathbf{some}_m$ and $\mathbf{Q}^2 = \mathbf{all}_k$, for some m,k. It can then be seen that (4) implies that $k \leqq m$, and (5) that $m \leqq k$.

As to the proof of Theorem 4.5.3, we shall indicate the basic technique that is used. The first step is reformulation. Note that the negation of a scheme of the form (3) is equivalent to

$$\phi_1 \wedge \ldots \wedge \phi_n \wedge \neg\theta_1 \wedge \ldots \wedge \neg\theta_k,$$

i.e., that *negated schemes* are (equivalent to) conjunctions of elementary schemes. Since

$$\psi \in Th_{\text{syll}}(\mathbf{Q}^1, \ldots, \mathbf{Q}^m) \Leftrightarrow \neg\psi \text{ has no } (\mathbf{Q}^1, \ldots, \mathbf{Q}^m)\text{-model},$$

we are done if any (**some**, **all**)-model for a negated scheme can be transformed into a ($\mathbf{some}_n$, $\mathbf{all}_n$)-model for the scheme, and *vice versa*.

Now let $\boldsymbol{M} = \langle M, A_0, \ldots, A_{p-1}\rangle$ be a (**some**, **all**)-model for $\neg\psi$. Each conjunct in $\neg\psi$ expresses either that a set of the form $A_i \cap A_j$ or $A_i - A_j$ is empty, or that it is non-empty. Each $A_i \cap A_j$ or $A_i - A_j$ can be written uniformly as a union of partition sets of the form P_s^M (cf. Section 1.7). The two types of condition expressed are thus

(a) $\quad x = x_1 + x_2 + \ldots > 0,$

(b) $\quad x = x_1 + x_2 + \ldots = 0,$

where x is the cardinal of $A_i \cap A_j$ (or $A_i - A_j$) and the x_k are the cardinals of the relevant partition sets. Now add $n - 1$ new elements to each *non-empty* partition set. This gives a model $\boldsymbol{M}^+ \langle M^+, A_0^+, \ldots, A_{p-1}^+\rangle$, where the conditions (a) and (b) are transformed into

(a)$^+$ $\quad x^+ = x_1^+ + x_2^+ + \ldots \geqq n,$

(b)$^+$ $\quad x^+ = x_1^+ + x_2^+ + \ldots = 0.$

But then it is easy to check that $\boldsymbol{M}^+$ is a ($\mathbf{some}_n$, $\mathbf{all}_n$)-model of $\neg\psi$.

Note that this method does not work if we start with a ($\mathbf{some}_n$, $\mathbf{all}_n$,)-model and want to get a ($\mathbf{some}_{n+1}$, $\mathbf{all}_{n+1}$)-model, say. For example, with $n = 3$, we may have

$$x = x_1 + x_2 < 3$$

with $x_1 = x_2 = 1$; then adding 1 gives

$$x^+ = x_1^+ + x_2^+ \geqq 4,$$

which means that schemes of the form $\neg Q^1 A_i A_j$ will not be preserved.

Nevertheless, by an ingenious elaboration of this technique, van Benthem shows that a ($\mathbf{some}_{n+1}$, $\mathbf{all}_{n+1}$)-model can in fact always be obtained, and, combining this with yet another construction, he also shows how to obtain a (**some**, **all**)-model from a ($\mathbf{some}_n$, $\mathbf{all}_n$)-model.

In view of these negative results about IL_{syll}, it is natural to ask if there is a stronger inferential language where the basic logical constants are determined. Indeed, IL_{boole} is such a language. First observe that in IL_{boole} it is sufficient to look at *one* of the quantifiers **some** and **all**. This follows from the next, easily verified, proposition.

4.5.5. PROPOSITION. (a) **Q** *is determined in* IL_{syll} *iff* $\neg$**Q** *is determined in* IL_{syll}.
(b) **Q** *is determined in* IL_{boole} *iff* **Q**$\neg$ *is determined in* IL_{boole} *iff* (**Q**, $\check{\mathbf{Q}}$) *is determined in* IL_{boole}.

We therefore concentrate on **some**. Let Φ_0 consist of schemes saying that Q is symmetric and quasireflexive, plus the following IL_{boole}-scheme:

$$(6) \qquad \neg QAA \wedge \neg QBB \rightarrow \neg QA \cup B\, A \cup B.$$

4.5.6. THEOREM. Φ_0 *determines* **some**.

Proof. Clearly **some** satisfies these schemes. Now suppose **Q** is any (logical) quantifier satisfying Φ_0. As before, the first two schemes imply that $\mathbf{Q} = \mathbf{some}_n$, for some $n \geqq 1$. Since **Q** satisfies (6) we also have

$$|A| < n \;\&\; |B| < n \Rightarrow |A \cup B| < n$$

(for all *sets* A, B). But this means that $n = 1$. □

Now let us look at the other $\mathbf{some}_n$ in IL_{boole}. From the last result, $Th_{\text{boole}}(\mathbf{some}) \neq Th_{\text{boole}}(\mathbf{some}_n)$ when $n > 1$. The proof technique for IL_{syll} works for IL_{boole} as well – indeed, it works better since conditions on (the cardinal number of) *any* Boolean combinations of $A_0, \ldots, A_{p-1}$ can be expressed there. We thus get a **some**$_n$-model from a **some**-model as before. In fact, even from a **some**$_2$-model we get a **some**$_n$-model with this method: adding $n - 2$ to each non-empty partition set transforms

(a) $\quad x = x_1 + x_2 + \cdots \geqq 2,$

(b) $\quad x = x_1 + x_2 + \cdots < 2$

into

(a)$^+$ $\quad x^+ = x_1^+ + x_2^+ + \cdots \geqq n,$

(b)$^+$ $\quad x^+ = x_1^+ + x_2^+ + \cdots < n,$

since at most one x_i in (b) is non-zero. This gives us

4.5.7. THEOREM. *Th_{boole}(**some**$_n$) $\subseteq$ Th_{boole}(**some**$_2$) $\subseteq$ Th_{boole}(**some**), for $n > 2$.*

No such method works if we start with a **some**$_m$-model with $m > 2$, however. This was pointed out by Per Lindström: in fact, we have the

4.5.8. THEOREM. (a) *Th_{boole}(**some**$_{n+1}$) $\nsubseteq$ Th_{boole}(**some**$_n$), for $n \geqq 2$.*
(b) *On the other hand, if $m \geqq n^2$ then Th_{boole}(**some**$_m$) $\subseteq$ Th_{boole}(**some**$_n$).*

Proof. (a): The case $n = 3$ will give the general idea. Let $\neg\psi$ be a negated scheme in IL_{boole} expressing the conditions

$$\begin{array}{llll} x_1 + x_2 + x_3 \geqq k, & x_1 + x_4 < k, & x_2 + x_4 < k, & x_3 + x_4 < k, \\ x_4 + x_5 + x_6 \geqq k, & x_1 + x_5 < k, & x_2 + x_5 < k, & x_3 + x_5 < k, \\ & x_1 + x_6 < k, & x_2 + x_6 < k, & x_3 + x_6 < k, \end{array} \tag{7}$$

when **Q** is interpreted as **some**$_k$ (6 partition sets are needed, so a negated scheme with 3 set variables suffices). First note that for $k = 3$, (7) is satisfied when all the x_i are 1. Thus $\neg\psi$ has a **some**$_3$-model. But (7) cannot be true when $k = 4$. For, the first two conditions would give an x_i $(1 \leqq i \leqq 3)$ and an x_j $(4 \leqq j \leqq 6)$ which both are $\geqq 2$, and this contradicts one of the remaining conditions. So $\neg\psi$ has no **some**$_4$-model.

(b): Suppose $m \geqq n^2$, and take k such that $(k - 1)n \leqq m < kn$. It follows that $n \leqq k$, and hence that $k(n - 1) \leqq (k - 1)n < m$. Now, given conditions

(a) $\quad x = x_1 + x_2 + \cdots \geqq n,$

(b) $\quad x = x_1 + x_2 + \cdots \leqq n - 1,$

multiply all the x_i by k. Then, $x^+ \geqq m$ in (a)$^+$ and $x^+ < m$ in (b)$^+$; this gives the desired $\mathbf{some}_m$-model. □

As to the converse inclusions, we have the

4.5.9. THEOREM. *$Th_{\text{boole}}(\mathbf{some}_n) \nsubseteq Th_{\text{boole}}(\mathbf{some}_m)$, for $1 \leqq n < m$.*

Proof. Generalizing (6), we can write a scheme ψ in IL_{boole} with $n + 1$ set variables which expresses

$$\bigwedge |A_{i_1} \cup \ldots \cup A_{i_n}| < k \Rightarrow |A_1 \cup \ldots \cup A_{n+1}| < k$$

(here the conjunction is taken over all subsets of $\{1, \ldots, n + 1\}$ with exactly n elements), when $\mathbf{Q}$ is interpreted as $\mathbf{some}_k$. Then $\mathbf{some}_n$ satisfies ψ. For otherwise, there are sets $A_1, \ldots, A_{n+1}$ such that $|A_1 \cup \ldots \cup A_{n+1}| \geqq n$ and $|A_{i_1} \cup \ldots \cup A_{i_n}| < n$ for $1 \leqq i_1, \ldots, i_n \leqq n + 1$. It follows that, for all i,

$$A_i \nsubseteq \bigcup_{j \neq i} A_j$$

So in every A_i there is an element not in the other A_j. But this means that $|A_1 \cup \ldots \cup A_n| \geqq n$, a contradiction.

Now let $m > n$. Choose pairwise disjoint $A_1, \ldots, A_{n+1}$ such that $|A_1| = m - n$ and $|A_i| = 1$ for $1 < i \leqq n + 1$. Then, if $1 \leqq i_1, \ldots, i_n \leqq n + 1$, the cardinal of $A_{i_1} \cup \ldots \cup A_{i_n}$ is either n or $m - 1$, i.e., in both cases $< m$, whereas $|A_1 \cup \ldots \cup A_{n+1}| = m$. So $\mathbf{some}_m$ does not satisfy ψ. □

Summarizing, we find once more than **some** behaves in a significantly different way than $\mathbf{some}_n$ for $n > 1$ (and similarly for **all**):

4.5.10. COROLLARY. *Of the quantifiers* $\mathbf{some}_n$, *only* **some** *is determined in* IL_{boole}.

Proof. **some** is determined, by Theorem 4.5.6. Further, if Ψ determines $\mathbf{some}_n$, then, by Theorem 4.5.7,

$$\Psi \subseteq Th_{\text{boole}}(\mathbf{some}_n) \subseteq Th_{\text{boole}}(\mathbf{some}).$$

Thus **some** satisfies Ψ, and it follows that $n = 1$. □

As for the quantifiers satisfying $Th_{\text{boole}}(\mathbf{some}_n)$, it follows from our results here that they are all of the form $\mathbf{some}_k$ with $k \leqq n$, that **some**, $\mathbf{some}_2$, and $\mathbf{some}_n$ are always among them, but that $\mathbf{some}_{n-1}$ never is if $n > 3$.

The results in this subsection depend on *FIN*. For IL_{boole}, the proof technique works without *FIN*, but the facts are different. More precisely, with the previous methods one easily proves

4.5.11. THEOREM. *For each infinite cardinal* κ, $Th_{\text{boole}}(\textbf{some}) = Th_{\text{boole}}$ $(\textbf{infinitely many}) = Th_{\text{boole}}(\textbf{some}_\kappa)$.

Thus, as one would expect, **some** is not determined in IL_{boole} without *FIN*.

4.6. *Local Perspective*

Let M be a fixed finite universe, with n elements. We can then study local quantifiers *on* M, with much the same aim as before: Of all these quantifiers, which ones are 'realized' in natural language?

Most of our global constraints have local versions. *CONSERV* is the same as before (with M fixed), and so are the monotonicity properties of 3.6 and the relational properties of 4.1. *ISOM* reduces to the local *PERM* (3.3). But one constraint which lacks a local version is *EXT*. As a consequence, results not depending on *EXT* have more or less immediate local versions, but when *EXT* is used, such versions may be harder to get. For example, Theorem 3.6.3 on double monotonicity holds locally as well, whereas Corollary 4.1.2 on the non-existence of asymmetric quantifiers, which uses *EXT*, fails: $\mathbf{Q}_M AB \Leftrightarrow A = M \;\&\; B = \varnothing$ is an asymmetric quantifier on M, satisfying *CONSERV* and *PERM*. Suitably modified versions of Corollary 4.1.2 and similar results do exist, however, cf. Westerståhl [1983].

One advantage of a local and finite perspective is that the effects of constraints such as *CONSERV* and *PERM* can be assessed in a rather perspicuous way, namely, by the *number* of quantifiers they allow. Here are some examples for binary quantifiers on M:

TABLE III

number of quantifiers on M under	no constraints	*CONSERV*	*CONSERV* & *VP*-positivity	*CONSERV* & *MON*↑
no constraints	2^{4^n}	2^{3^n}	2^{2^n}	?
when $n = 2$:	65536	512	16	108
PERM	$2^{\binom{n+3}{3}}$	$2^{\binom{n+2}{2}}$	$2^{\binom{n+1}{1}}$	$(n + 2)!$
when $n = 2$:	1024	64	8	24

There is a simple uniform calculation for the first three entries in both rows of this table (these and other calculations have appeared in Higginbotham and May [1981], Keenan and Stavi [1986], Keenan and Moss [1985], van Benthem [1984a], Thijsse [1983]). Consider a pair (A,B), with $A,B \subseteq M$, as a function f from M to $\{0,1\}^2$: $f(x) = (1,1)$ if $x \in A \cap B$, $f(x) = (0,1)$ if $x \in B - A$, etc. There are 4^n such functions and hence 2^{4^n} quantifiers on M. *CONSERV* means that $B - A$ can be assumed to be empty, removing the value (0,1), and reducing the number of functions to 3^n. By Proposition 3.4.8, *CONSERV* + *VP*-positivity means that only the pairs (A,A) need be considered, reducing the number of functions to 2^n.

Under *PERM*, $\mathbf{Q}_M$ is a relation between 4 numbers whose sum is n. To choose such numbers is essentially to put n indistinguishable objects in 4 (distinguished) boxes; there are $\binom{n+3}{3}$ ways to do this, by standard combinatorics. As before, addition of *CONSERV* or *CONSERV* + *VP*-positivity reduces the number of boxes to 3 and 2, respectively.

PERM and *CONSERV* are defined for k-ary quantifiers on M ($k \geqq 2$), and the above calculations extend straightforwardly to this case: just replace '4' by '2^k' (= the number of partition sets induced by $(A_0, \ldots, A_{k-1})$), '3' by '$2^k - 1$', and '2' by '$2^k - 2$' (in the exponent) in the first two columns of Table III.

The value $(n + 2)!$ for *PERM* + *CONSERV* + *MON*$\uparrow$ can be obtained by looking in the *number tree for M*, i.e., the number tree restricted to pairs (x, y) such that $x + y \leqq n$. But the corresponding value without *PERM* is unknown: Thijsse [1983] shows that a calculation of this appears to require an explicit calculation of the number of *anti-chains* in $P(M)$; the latter is an unsolved mathematical problem. Thijsse's paper contains several further counting results for quantifiers under various constraints (e.g. the number 108 for the case $|M| = 2$), and so does the paper by Keenan and Moss.

It is rather amazing at first sight that there are 65536 possible quantifiers on a universe with only two elements. The strength of the conservativity universal appears clearly from Table III, which indicates that counting quantifiers is not just pleasant combinatorics – see the papers by Keenan and Stavi and by Keenan and Moss for linguistic applications of such counting results.

Another distinguishing feature of the local perspective on quantifiers is that new *definability* issues arise here. Suppose certain *DET* denotations are *given* in M, and likewise denotations of other expressions: proper names, common nouns, transitive and intransitive verbs, etc. (we may think of a *model* $\boldsymbol{M}$ being given, not just a universe). Suppose further that we have identified certain constructions in natural language which can be interpreted as operations producing new quantifiers from given denotations. We can then ask which quantifiers can be *generated* from the given denotations by means of these operations. Such generated quantifiers are 'realized' in a definite sense; in fact, if the operations and the starting-point were chosen wisely, one may expect each generated quantifier to be *denoted* by some complex *DET* expression (relative to $\boldsymbol{M}$).

This approach is pursued in Keenan and Stavi [1986]. We will present one of their main results, which shows that *conservativity* is a crucial invariant here. Let $CONSERV_M$ be the class of binary conservative quantifiers on M. Also, if K is any class of binary quantifiers on M, let $B(K)$ be the smallest class containing K which is *closed* under conjunction, disjunction, and inner and outer negation. Finally, for each $a \in M$, define the quantifiers $\mathbf{S}_a$ on M by

$$\mathbf{S}_a AB \Leftrightarrow a \in A \cap B.$$

Keenan and Stavi argue that each $\mathbf{S}_a$ can be taken as a basic, initially given quantifier. For, if b is an individual in M who *owns* a and nothing else, i.e., if $P_b = \{a\}$ (cf. Section 2.4.6), then

$$\begin{aligned}\textbf{b's one or more}_M AB &\Leftrightarrow P_b \cap A \subseteq B \;\&\; |P_b \cap A| \geqq 1 \\ &\Leftrightarrow \mathbf{S}_a AB.\end{aligned}$$

Note that the $\mathbf{S}_a$ are conservative (but *PERM* fails), and that, to regard them as *given*, we also need each element of M to be given (by proper names or other means), and enough ownership relations to guarantee that for each a in M there is a b in M such that $P_b = \{a\}$. These are not implausible assumptions, and the Boolean operations are natural enough.[7]

4.6.1. THEOREM (Keenan and Stavi). *Suppose* $K \subseteq CONSERV_M$ *and that* $\mathbf{S}_a \in K$ *for* $a \in M$. *Then* $B(K) = CONSERV_M$.

Proof. We know from 3.4 that Boolean operations preserve conservativity, so $B(K) \subseteq CONSERV_M$. Now let $\mathbf{Q}$ be any element of $CONSERV_M$. We then have

$$\begin{aligned}\mathbf{Q}AB &\Leftrightarrow \mathbf{Q}A\ A \cap B \\ &\Leftrightarrow \exists X \exists Y \subseteq X(\mathbf{Q}XY \wedge X = A \wedge Y = A \cap B) \\ &\Leftrightarrow \bigvee_{\substack{X \subseteq Y \subseteq M \\ \& \mathbf{Q}XY}} (X = A \wedge Y = A \cap B).\end{aligned}$$

Note that the last disjunction is finite. It only remains to show that each disjunct can be generated from the $\mathbf{S}_a$ by Boolean operations. We claim that each disjunct is equivalent to the conjunction of

(1) $\bigwedge_{a \in Y} \mathbf{S}_a AB,$

(2) $\bigwedge_{a \in X-Y} (\mathbf{S}_a \neg) AB,$

(3) $\bigwedge_{a \in M-X} (\neg \mathbf{S}_a AB \wedge \check{\mathbf{S}}_a AB).$

For, (1) expresses that $Y \subseteq A \cap B$, (2) that $X - Y \subseteq A - B$, and (3) that $A \cap B \subseteq X$ and $A - B \subseteq X$, and it easily verified that the conjunction of these expresses that $X = A \wedge Y = A \cap B$. □

By this theorem, *precisely* the conservative quantifiers on M are generated from certain basic ones by Boolean operations. This lends new significance to the conservativity universal ($U2$). By ($U2$) and the theorem, precisely these quantifiers on M are 'realized', in the sense of being denoted by *DET*s (relative to a model; cf. also Note 8).

Note that the complex *DET* expression resulting from the proof of the theorem depends crucially on M. That is, conservative quantifiers, such as **most**, will get *different* 'definitions' on different universes, and there is in general no way of giving a global definition working for all universes. Keenan and Stavi prove a theorem (the 'Ineffability Theorem') to the effect that no fixed *DET* expression, containing symbols for simplex *DET*s, k-place predicates, adjectives, *NP*s, and prepositions, can be made to denote, by varying the interpretation of these symbols, an arbitrary conservative quantifier on an arbitrary universe. The reason is that the number of possible denotations of such expressions grows slower with $|M| = n$ than 2^{3^n}.[8]

5. PROBLEMS AND DIRECTIONS

A basic theme of this paper has been to point to natural language as a source for logical investigation. This theme is by no means limited to quantifiers. Thus, one main direction for further study is *extension to other categories*. Some of the constraints we have studied can be transferred to other categories, and new constraints emerge. A typical trans-categorical constraint is *ISOM*, which has significant effects in most categories. For instance, Westerståhl [1985a] shows that, for *relations between individuals*, *ISOM* leaves essentially just Boolean combinations with the *identity relation*, and van Benthem [1983b] proves that, for arbitrary *operations on subsets* of the universe, *ISOM* leaves precisely the operations whose values are Boolean combinations of the arguments. For further results in this area, and for a broad assessment of the present approach to logical semantics, the reader is referred to van Benthem [1986], which also lists several topics for further research, both in the quantifier area and beyond, complementing the brief suggestions given below.

Within the area of quantifiers there is, to begin with, the whole field of the *syntax* of various constructions with *DET*s, and of how to treat them semantically. We have mentioned (Section 2) constructions with *only*, the treatment of definites, of partitives, and of "there are"-sentences, to take just a few examples. The papers Keenan and Stavi [1986] and Keenan and Moss [1985] provide ample evidence that these linguistic questions may be fruitfully pursued from the present model-theoretic perspective.

Another linguistic concern is the search for *universals*. As we have seen, universals can be used as basic theoretical postulates, or they can appear as empirical generalizations, sometimes amenable to explanation by means of other principles. The list of universals in Section 3 was not meant to be complete, and some of the formulations were quite tentative. Further proposals can be found in the papers by Barwise and Cooper and by Keenan and Stavi.

The use of semantic theory to explain linguistic facts, such as the privileged status of certain constants, the restrictions on various syntactic constructions, or the discrepancies between possible and actual interpretations of expressions of a certain category, can most likely be carried a lot further. Recall, for example, the discussion after Table I in 3.4. Other similar questions are easily found. Why are there so few simple *VP*-negative quantifiers? Why so few simple *MON*$\downarrow$ ones? Why isn't **not every** a *simple* natural language quantifier (like the other quantifiers in the square of opposition)? Such questions may warrant psychological considerations, but van Benthem's analysis of 'count complexity' in 4.2 shows that simple model theory may be useful even in this context.

In connection with the last remark, it should be mentioned that van Benthem [1985], [1987a] carries the study of *computational complexity* in semantics much further. He shows (cf. the end of Section 4.3) that the well known complexity hierarchies of automata theory are eminently suitable for classification of quantifiers. Moreover, these investigations carry the promise of a new field of *computational semantics*, which, in addition to questions of logical and mathematical interest, has applications to *language learning* and to mental *processing* of natural language.

On the *logical* side of quantifier theory, many further questions suggest themselves. One natural direction is *generalization* by weakening the assumptions. For example:

(a) *Drop EXT*. This allows for 'universe-dependent' quantifiers, such as some of the interpretations of *many* in 2.4.3. Some hints on how this admission affects the theory can be found in Westerståhl [1983].

(b) *Drop QUANT*. If possessives are allowed, this is a natural move. One can then replace *QUANT* (*ISOM*) by postulates of *quality*, requiring closure under 'structure-preserving' bijections. Other new constraints can also be formulated for this case, which is studied in van Benthem [1983b].

(c) Allow *ternary* quantifiers, or arbitrary *n*-ary ones ($n \geqq 2$). We did this in Section 3 for the basic concepts, but the corresponding generalization of the theory in Section 4 is by no means straight-forward; cf. Keenan and Moss [1985].

Dropping *CONSERV*, on the other hand, does not seem fruitful (except for purely logical issues such as definability; cf. Section 4.3). (a)–(c) are not (only) generalizations for their own sake, but linguistically motivated. The next generalization is more mathematical:

(d) *Drop FIN*. Many of the results using *FIN* can in fact be generalized, as we have noted from time to time. Two apparent exceptions were the results on transitivity, Theorems 4.1.10 and 4.1.14 (without *VAR*; cf. Corollary 4.1.17). Are there generalizations of these to infinite universes? But perhaps these generalizations lead in the wrong direction. It could be that *FIN*, or some similar constraint, is an essential characteristic of natural language quantification (cf. Section 3.8). In any case, the assessment of some minimal model-theoretic means for handling 'natural language infinity' appears to be an interesting task. Some results in this direction can be found in van Deemter [1985].

But, even without generalizing, the type of logical study conducted in Section 4 can be pursued further. The properties in 4.1 were chosen in a rather conventional way; there may be more interesting *properties of relations* to study. *Definability* questions need not be confined to first-order definability – as we saw in 4.3, *arithmetical definability* is a natural concept in the realm of (logical) quantifiers.

A particularly interesting aspect of definability concerns the *expressive power of natural language*. Various global notions of definability may be used here, e.g. definability *from* given quantifiers. There is also the local definability question mentioned in 4.6: Of the possible denotations of expressions of a certain category, which ones a 'generated' in a given model? The conservativity theorem of Keenan and Stavi gives one answer, for *DET* denotations. Perhaps *NP* denotations are even more interesting; this aspect of expressive power is studied in Keenan and Moss [1985], where several results on which *NP* denotations are obtainable from quantifiers with certain properties (conservative, logical, *VP*-positive, etc.) are proved.

The study of *inferential languages* from Section 4.5 gives rise to a number of logical questions. This appears to be a new field, though related to well-known questions on the correlation between a proof-theoretic and a model-theoretic perspective on logic.[1] Note that the results of 4.5 depend crucially on our use of *binary* quantifiers instead of unary ones. As for particular questions, one would like to know which quantifiers are determined in these languages. Are *any* (non-trivial) quantifiers determined in IL_{syll}? Are any quantifiers *besides* those in the square of opposition determined in IL_{boole}? One can also pose 'finiteness' (compactness) questions, e.g.: If $\mathbf{Q}$ is determined by Ψ, is $\mathbf{Q}$ by necessity determined by a finite subset of Ψ? This may of course be a trivial question, depending on the answer to the first two. Another compactness question is: If every finite subset of Ψ is satisfied by some quantifier (or sequence of quantifiers), must Ψ itself be satisfiable? Actually, this question can be seen to have a negative answer for IL_{boole}, but the case of IL_{syll} seems open. Other inferential languages could also be considered. In general, one would like to have a better understanding of what is required of a good inferential language. An obvious extension of IL_{syll} and IL_{boole}, however, is to add **some** and **all** as constants. This allows, e.g., monotonicity properties to be expressed in IL_{syll}, and the logical questions are reopened.

In this connection we should also mention an application of the present theory outside the domain of quantifiers: van Benthem [1984b] analyzes *conditional* sentences *If X then Y* as relations between *sets* $\| X \|$ and $\| Y \|$ (of possible worlds, situations, etc.), i.e., as binary quantifiers, and obtains several interesting results for the logic of conditionals.

Finally, all of the logical questions mentioned so far presuppose the classical model-theoretic framework we have used in this paper. If one wants to treat such linguistically interesting phenomena as *plurals, collective quantification* (as opposed to the *distributive* quantification we have studied; cf. sentences such as *five boys lifted the piano*), or *mass terms* (with new determiners such as *much* or *a little*), this framework has to be extended.

From a natural language point of view, such extension seems imperative. For some steps taken in these directions, cf., e.g., van Benthem [1983b], Hoeksema [1983], Link [1987], Lønning [1987a,b]. An even more radical change would be the switch from the traditional 'static' model theory to a *dynamic* view on interpretation, e.g., along the lines suggested in Kamp [1981] or Barwise and Perry [1983]. It would be pleasant if the insights gained from the present quantifier perspective were preserved in such a transition. But, however that may be, standard model-theoretic semantics has already, I think, proved unexpectedly useful for a rich theory of quantifiers, and this theory is in turn a fair illustration of the possibilities of a logical study which starts not from mathematical but from natural language.

APPENDIX A: BRANCHING QUANTIFIERS AND NATURAL LANGUAGE

This appendix presents a brief summary of the main issues related to the occurrence of branching quantification (Section 1.5) in natural language. A more detailed presentation is given in Barwise [1979].

Let us say, somewhat loosely, that a sentence exhibits *proper branching*, if its formalization requires a partially ordered quantifier prefix which is not equivalent to a linear one. There has been some debate over the following question:

(I) *Does proper branching occur in natural languages*?

The debate started with the claim in Hintikka [1973] that proper branching occurs in English. Here is the most well-known of his examples:

(1) Some relative of each villager and some relative of each townsman hate each other.

The idea is that (1) should be analyzed with the Henkin prefix. Arguing that the branching reading of (1) is preferred over linear versions requires a detailed and quite complicated analysis of what we actually mean when using such a sentence, and not all linguists agreed with Hintikka. In Barwise [1979], where the main arguments are summarized, it is argued that the most natural logical form of (1) does involve a branching reading, but one which is equivalent to a linear one, so that this branching is not proper. But the answer to (I) does not necessarily depend on sentences like (1). Barwise, who was sympathetic to Hintikka's general claim, argued that with other quantifiers than ∀ and ∃ one can find clearer examples of proper branching. One of his examples was

(2) Most boys in your class and quite a few girls in my class have all dated each other.

It seems that (2) does *not* mean the same as

(3) Most boys in your class have dated quite a few girls in my class

or

(4) Quite a few girls in my class have dated most boys in your class.

The preferred reading of (2) is *stronger* than both of these: it says that there is a set X containing most boys in your class, and a set Y containing quite a few girls in my class, such that *any* boy in X and *any* girl in Y have dated each other. Note that X and Y are *independent* of each other. This is a branching reading, which is (provably) not equivalent to any linear sentence in L(**most**, **quite a few**). We could formalize (2) as

(5) *most x boy*(x) / *quite a few y girl*(y) > *have dated each other*(x, y).

Barwise pointed out that the above truth definition for such sentences gives the desired reading when, as in the present case, both quantifiers are $MON\uparrow$, and gave a similar (but different) truth condition for the case when both are $MON\downarrow$. He also noted that sentences of this form with one $MON\uparrow$ and one $MON\downarrow$ quantifier are anomalous:

(6) Few of the boys in my class and most girls in your class have dated each other.

Even though it seems perfectly grammatical, (6) makes no sense, and this may be explained by means of the monotonicity behavior of the quantifiers involved. Further discussion of the circumstances under which it makes sense to branch two quantifiers can be found in Westerståhl [1987].

For another example, van Benthem has noted that we can have proper branching with certain first-order definable quantifiers that are *not* monotone. Consider

(7) Exactly one boy in your class and exactly one girl in my class have dated each other.

The meaning of (7) is clear and unambiguous, and it is easily seen that (7) is not equivalent to any of its 'linear versions' (or to their conjunction). (Note that we are talking about prefixes with *exactly one* here; it is in this sense the branching is proper, even though (7) is clearly equivalent to a (linear) first-order sentence.)

In conclusion, it seems that there are good arguments for an affirmative answer to (I). Then, one may ask:

(II) *What are the consequences for the 'logic of natural language' of the occurrence of proper branching?*

One of the aims of Hintikka's original paper was to use the occurrence of proper branching to give lower bounds of the complexity of this logic. From 1.5 and 1.6 it should be clear that logic with the Henkin quantifier has many affinities with *second-order logic*. In fact, it can be shown that the set of valid sentences with the Henkin quantifier, or with arbitrary partially ordered prefixes with $\forall$ and $\exists$, is recursion-theoretically just as complex as the set of valid second-order sentences, and this is an extremely complicated set. It is tempting to conclude that natural language is at least as complicated. This last inference, however, is not unproblematic. The result about second-order logic depends crucially on the fact that second-order variables vary over *all* subsets (relations) of the universe. In a natural language context, on the other hand, it may be reasonable to *restrict* the range of these variables, and thus to alter the strength of the resulting logic. More on these issues can be found in van Benthem and Doets [I.4]. Some other types of consequences of the occurrence of proper branching are discussed in Barwise [1979].

In addition to the principled questions (I) and (II), there is also the more pragmatical:

(III) *Should branching quantification be used more extensively in the analysis of logical and linguistic form?*

Both Hintikka and Barwise suggest that in many cases a branching reading may be preferable regardless of whether the branching is proper or not: the actual *order* between two (or more) quantifier expressions in a sentence sometimes seems irrelevant, syntactically *and* semantically, and a logical form where these expressions are unordered is then natural. Certain syntactic constructions appear to trigger such branching readings, in particular, conjoined noun phrases with a reciprocal object (*each other*). An even more extensive use of branching is proposed in van Benthem

[1983a]: he suggests using branching instead of 'substitution' to explain certain well-known scope ambiguities with $\forall$ and $\exists$; cf. also van Eijck [1982]. There seem to be a lot of interesting possibilities in this field.

APPENDIX B: LOGIC WITH FREE QUANTIFIER VARIABLES

Quantifier symbols have been *constants* in this paper (cf. Section 2.1.3). What happens if they are treated as free variables instead, or, more precisely, as symbols whose interpretation varies with models? From a logical perspective at least, this is a natural question. Some answers are reviewed in this appendix.

To fix ideas, consider a language, L_Q, of standard first-order logic with one binary quantifier symbol Q added (for simplicity; we could have added several monadic quantifier symbols), and a fixed (countable) vocabulary of other non-logical symbols. L_Q is a language for logics like L(**most**), except that this time Q does not denote a fixed quantifier. Instead, a *model* is now a pair $(\boldsymbol{M}, \mathbf{q})$, where $\boldsymbol{M}$ is as before and $\mathbf{q}$ is a binary quantifier on M. Such models are often called *weak models* (since nothing in particular is required of $\mathbf{q}$). *Truth* (satisfaction) in $(\boldsymbol{M}, \mathbf{q})$ is defined in the obvious way, with Q interpreted as $\mathbf{q}$. A *valid* sentence is thus true regardless of the interpretation of Q (and other non-logical symbols). Here is a trivial example:

$$Qx(x \neq x, \psi) \rightarrow (\exists x\phi \vee Qx(\phi, \psi))$$

(where ϕ, ψ only have x free). Are there non-trivially valid sentences in L_Q? This is answered below.

B.1. *The weak logic.* Add to a standard axiomatization of first-order logic the axioms

(1) $$Qx(\phi(x), \psi(x)) \leftrightarrow Qy(\phi(y), \psi(y))$$

(y free for x in $\phi(x)$, $\psi(x)$)

(2) $$\forall x(\phi_1 \leftrightarrow \phi_2) \rightarrow (Qx(\phi_1, \psi) \rightarrow Qx(\phi_2, \psi))$$

(3) $$\forall x(\phi_1 \leftrightarrow \phi_2) \rightarrow (Qx(\psi, \phi_1) \rightarrow Qx(\psi, \phi_2))$$

(the last two are extensionality axioms for Q). Call this the *weak logic*. Provability (from assumptions) is defined as usual, the deduction theorem holds, and the axiomatization is obviously sound. The following completeness theorem goes back to Keisler [1970]:

B.1.1. THEOREM. *If Σ is a consistent set of sentences in the weak logic, then Σ has a weak model.*

Proof (outline). A slight extension of the usual Henkin-style proof suffices. Extend Σ to Σ' by witnessing existentially quantified sentences, and then to a maximally consistent Γ. Let M consist of the usual equivalence classes $[c]$ of new individual constants, and interpret relation and constant symbols as usual. For each $\psi(x)$ with at most x free, let $\psi(x)^{\Gamma} = \{[c] \in M: \Gamma \vdash \psi(c)\}$. Then define $\mathbf{q}$ as follows:

$$\mathbf{q}AB \Leftrightarrow \text{there are } \phi, \psi \text{ such that } \phi^{\Gamma} = A, \psi^{\Gamma} = B, \text{ and } \Gamma \vdash Qx(\phi, \psi).$$

One then shows that, for all sentences θ,

$$(M, \mathbf{q}) \models \theta \Leftrightarrow \Gamma \vdash \theta$$

by a straight-forward inductive argument, using (1)–(3) and properties of Γ when θ is of the form $Qx(\phi, \psi)$. □

B.1.2. COROLLARY. *The weak logic is complete, compact, and satisfies the downward Löwenheim–Skolem theorem.*

B.2. *Axiomatizable properties of quantifiers.* By the last results, if *all* weak models are allowed, no 'unexpected' new valid sentences appear. However, it may be natural to *restrict* the interpretation of Q to, say, *conservative* quantifiers, or *transitive and reflexive* ones, or $MON\uparrow$ ones. Such properties are *second-order,* and hence in general not directly expressible in L_Q. Nevertheless, in many cases the resulting logic is still axiomatizable, by adding the obvious axioms to the weak logic.

Let P be a property of $\mathbf{q}$ expressible by a universal second-order sentence

(4) $\forall X_1 \ldots \forall X_n \Psi(X_1, \ldots, X_n)$,

where the X_i are unary set variables and Ψ is in L_Q (with the X_i acting as predicate symbols). Let the *corresponding set of L_Q-sentences*, Σ_P, consist of the universal closures of all formulas obtained by replacing all occurrences of $X_1, \ldots, X_n$ in Ψ by L_Q-formulas $\phi_1, \ldots, \phi_n$. For example, $\Sigma_{CONSERV}$ and $\Sigma_{MON\uparrow}$ consist, respectively, of universal closures of formulas of the form

$$Qx(\phi, \psi) \leftrightarrow Qx(\phi, \phi \wedge \psi),$$

$$Qx(\phi, \psi) \wedge \forall x(\psi \rightarrow \theta) \rightarrow Qx(\phi, \theta).$$

Let $\mathbf{K}_P$ be the class of models $(\boldsymbol{M}, \mathbf{q})$ such that $\mathbf{q}$ satisfies P. Clearly,

(5) $\qquad (\boldsymbol{M}, \mathbf{q}) \in \mathbf{K}_P \Rightarrow (\boldsymbol{M}, \mathbf{q}) \models \Sigma_P,$

but the converse fails in general. To $\mathbf{K}_P$ corresponds a logic, which we write $L(\mathbf{K}_P)$, where truth and validity is as for the weak logic, except that models are restricted to $\mathbf{K}_P$. When is $L(\mathbf{K}_P)$ axiomatized by Σ_P? A sufficient condition is given below.

A subset A of M is called $(\mathbf{M}, \mathbf{q})$-*definable*, if, for some L_Q-formula ψ and some finite sequence $\bar{b}$ of elements of M, $a \in A \Leftrightarrow (\boldsymbol{M}, \mathbf{q}) \models \psi[a, \bar{b}]$. Consider the following property of P:

(*) *If* $(\boldsymbol{M}, \mathbf{q}) \models \Sigma_P$ *then there is a* $\mathbf{q}'$ *satisfying* P *which agrees with* $\mathbf{q}$ *on the* $(\boldsymbol{M}, \mathbf{q})$*-definable sets.*

We need one more definition: $(\boldsymbol{M}', \mathbf{q}')$ is an *elementary extension* of $(\boldsymbol{M}, \mathbf{q})$, in symbols, $(\boldsymbol{M}, \mathbf{q}) < (\boldsymbol{M}', \mathbf{q}')$, if $\boldsymbol{M}'$ is an extension of $\boldsymbol{M}$ and, for all L_Q-formulas ψ and all finite sequences $\bar{b}$ of elements of M, $(\boldsymbol{M}, \mathbf{q}) \models \psi[\bar{b}] \Leftrightarrow (\boldsymbol{M}', \mathbf{q}') \models \psi[\bar{b}]$. Now a straight-forward induction proves the

B.2.1. LEMMA. *If* $\mathbf{q}$ *and* $\mathbf{q}'$ *agree on the* $(\boldsymbol{M}, \mathbf{q})$*-definable sets, then* $(\boldsymbol{M}, \mathbf{q}) < (\boldsymbol{M}, \mathbf{q}')$.

From this lemma and Theorem B.1.1 we immediately obtain the

B.2.2. THEOREM. *If* (*) *holds for* P *then each set of* L_Q*-sentences consistent with* Σ_P *in the weak logic has a model in* $\mathbf{K}_P$. *Hence,* $L(\mathbf{K}_P)$ *is complete, compact, and satisfies the Löwenheim–Skolem theorem.*

Instances of this result appear, for example, in Keisler [1970], Broesterhuizen [1975], Sgro [1977], Makowsky and Tulipani [1977], and Barwise [1978]. To see its utility we consider some examples.

EXAMPLE. Given $(\boldsymbol{M}, \mathbf{q})$, let M^d be the set of $(\boldsymbol{M}, \mathbf{q})$-definable subsets of M, and let $\mathbf{q}^d = \mathbf{q} \cap (M^d)^2$. If $(\boldsymbol{M}, \mathbf{q}) \models \Sigma_P$ then, since P is universal, $\mathbf{q}^d$ satisfies P *on* M^d. In some cases, $\mathbf{q}^d$ actually satisfies P on the whole of $P(M)$, i.e. (*) holds with $\mathbf{q}' = \mathbf{q}^d$. This is true for all the properties of quantifiers in Table II (Section 4.1), *except* reflexivity, quasiuniversality, and linearity, as is easily checked. So, for example. the logic $L(\mathbf{K}_P)$, where P is the property of being a *strict partial order* (irreflexive and transitive), is axiomatizable.

EXAMPLE. P = *strict linear order*. If $(\mathbf{M}, \mathbf{q}) \models \Sigma_P$, let $\mathbf{q}^*$ be any strict linear order on $P(M) - M^d$, and let $\mathbf{q}' = \mathbf{q}^d + \mathbf{q}^*$ (order type addition). Then $\mathbf{q}'$ is a strict linear order coinciding with $\mathbf{q}$ on M^d, so $L(\mathbf{K}_P)$ is axiomatizable. A similar construction can be used to show that each of the three properties left over in the preceding example is axiomatizable.

EXAMPLE. $P = MON\uparrow$. If $(\mathbf{M}, \mathbf{q}) \models \Sigma_{MON\uparrow}$, define $\mathbf{q}'$ by: $\mathbf{q}'AB \Leftrightarrow$ for some $C \in M^d$, $C \subseteq B$ and $\mathbf{q}AC$. Since $\mathbf{q}$ is $MON\uparrow$, $\mathbf{q}'$ agrees with $\mathbf{q}$ on M^d. Also, $\mathbf{q}'$ is $MON\uparrow$ (on all subsets of M). Other monotonicity (or continuity) properties can be treated similarly.

EXAMPLE. $P = CONSERV$. If $(\mathbf{M}, \mathbf{q}) \models \Sigma_{CONSERV}$, let $\mathbf{q}'AB \Leftrightarrow \mathbf{q}A\ A \cap B$. Again, the verification that (*) holds is immediate.

EXAMPLE. In the following mathematical example, $\mathbf{q}$ is *unary*, and satisfies P iff $\mathbf{q}^- = P(M) - \mathbf{q}$ is a *proper, non-principal ideal* in $P(M)$, i.e., iff, for all $A, B \subseteq M$, (i) $A, B \in \mathbf{q}^- \Rightarrow A \cup B \subseteq \mathbf{q}^-$; (ii) $A \in \mathbf{q}^-$ & $B \subseteq A \Rightarrow B \in \mathbf{q}^-$; (iii) $M \notin \mathbf{q}^-$; (iv) $\{a\} \in \mathbf{q}^-$ for all $a \in M$. In $L(\mathbf{K}_P)$, $Qx\psi$ can be read "for many x in the (infinite) universe, ψ". Now suppose $(\mathbf{M}, \mathbf{q}) \models \Sigma_P$. Then $\mathbf{q}^{d-} = M^d - \mathbf{q}^d$ is a proper, non-principal ideal in M^d. Also, $\mathbf{q}^{d-}$ generates a proper, non-principal ideal $\mathbf{q}'^-$ in $P(M)$: let $A \in \mathbf{q}'^- \Leftrightarrow A \subseteq B_1 \cup \ldots \cup B_n$, for some $B_1, \ldots, B_n \in \mathbf{q}^{d-}$. Then (*) holds for $\mathbf{q}' = P(M) - \mathbf{q}'^-$, so $L(\mathbf{K}_P)$ is axiomatizable. $L(\mathbf{K}_P)$ is studied in Bruce [1978], mainly as a means for obtaining results about the logic $L(\mathbf{Q}_1)$, where $\mathbf{Q}_1$ is the quantifier "for uncountably many".

Note that even though axiomatizability comes rather easy in these examples, other properties, such as interpolation, unions of chains, etc., may be much harder and require new methods (cf. Bruce [1978]).

B.3. *A non-axiomatizable property*. In view of the above examples one may ask if the property of *quantity* is also axiomatizable. After all, *PERM* is a universal second-order property (with a binary relation variable in addition to the unary set variables), and a corresponding Σ_{PERM} can be found much as before. However, $L(\mathbf{K}_{PERM})$ is a rather strong logic, and *not* axiomatizable. The reason is, roughly, that it can express that two sets have different cardinalities. For example, if $(\mathbf{M}, \mathbf{q}) \in \mathbf{K}_{PERM}$, and $\mathbf{q}MA$ is *not* equivalent to $\mathbf{q}MB$, it follows that either $|A| \neq |B|$ or $|M - A| \neq |M - B|$. This is used in the following result, which is due to Yasuhara [1969].

B.3.1. THEOREM. *The natural number ordering, $\langle N, <\rangle$, is characterizable in $L(\mathbf{K}_{PERM})$ in the sense that there is an L_Q-sentence θ such that $\langle M, R\rangle$ is isomorphic to $\langle N, <\rangle$ iff, for some* $\mathbf{q}$ *satisfying PERM*, $(\langle M, R\rangle, \mathbf{q}) \models \theta$.

Proof. Let θ be the conjunction of a sentence saying that $<$ is a linear ordering with immediate successors and a first but no last element, and the sentence

$$\forall x \forall y(\text{“}y \text{ is the successor of } x\text{”} \rightarrow$$

$$\neg(Qz(z = z, z < x) \leftrightarrow Qz(z = z, z < y))).$$

If $(\langle M, R\rangle, \mathbf{q}) \models \theta$, where $\mathbf{q}$ satisfies *PERM*, it is easy to see that, for each $a \in M$, $|M_a| < |M_{a+1}|$ (where M_a is the set of predecessors to a), and thus that $\langle M, R\rangle$ is isomorphic to $\langle N, <\rangle$. Conversely, if the quantifier $\mathbf{q}$ on N is defined by $\mathbf{q}AB \Leftrightarrow A = N\ \&\ |B|$ is even, then *PERM* holds and $(\langle N, <\rangle, \mathbf{q}) \models \theta$. □

As in Section 1.6, we obtain the

B.3.2. COROLLARY. *$L(\mathbf{K}_{PERM})$ is neither complete nor compact.*

Väänänen [1978] extends these results to show that, in terms of *implicit definability* (definability with extra non-logical symbols), $L(\mathbf{K}_{PERM})$ is equivalent to the logic $L(\mathbf{I})$ (cf. 1.6), and that its set of valid sentences is very complicated: it is neither Π^1_1 nor Σ^1_1 in the analytical hierarchy.

The above theorem and corollary extend, with the same proof, to the logic $L(\mathbf{K}_{PERM+CONSERV})$. They also extend to *logical* quantifiers. To see this, note that in this appendix we have used *local* quantifiers in our models, for which *ISOM* or *EXT* have no immediate meaning. An alternative procedure would be to consider models of the form $(M, \mathbf{Q})$, where $\mathbf{Q}$ is a global quantifier, and interpret Q as $\mathbf{Q}_M$ on such a model. It is then easy to check that, for each model $(M, \mathbf{q})$ in $\mathbf{K}_{PERM+CONSERV}$, there is a *logical* quantifier $\mathbf{Q}$ such that $\mathbf{Q}_M = \mathbf{q}$. From this it follows that a sentence is valid in $L(\mathbf{K}_{PERM+CONSERV})$ iff it is valid when Q varies over arbitrary logical quantifiers.

Let us remark, finally, that the results of this appendix depend on the fact that the usual universal and existential quantifier constants occur in L_Q. Anapolitanos and Väänänen [1981] show that, if we drop these, and also drop identity, then $L(\mathbf{K}_{PERM})$ becomes axiomatizable; actually, it becomes *decidable*.

ACKNOWLEDGEMENT

I am grateful for many helpful comments and suggestions made by several people at various stages of the preparation of this paper, among them, Jens Allwood, Jan van Eijck, Mats Furberg, Franz Guenthner, Björn Haglund, David Israel, Hans Kamp, Ed Keenan, Per Lindström, Barbara Partee, and, in particular, Johan van Benthem.

University of Gothenburg, Göteborg.

NOTES

Notes to Section 1:

[1] Actually, contemporaries of Aristotle, like Theophrastus, seem to have understood him rather well. But the medieval reintroduction of Aristotle's logic lost track of many important points. Even 19th century commentators continue in the medieval vein; cf. Łukasiewicz [1957].

[2] About the historical causes Russell may well be right. Note that we are also setting aside here the metaphysical claims of Russell's logical atomism, according to which the logical form of sentences mirror the structure of reality.

[3] He sometimes comes very close to an explicit statement; cf. the last pages of Patzig [1959].

[4] More precisely, the system of *Grundgesetze* [1893]. The English translation of the first part of this work by M. Furth is prefaced with an excellent introduction, where more details about Frege's conceptual framework can be found.

[5] This notion of 'arity' does not tell us the number of arguments of the arguments, etc., for levels greater than one; we will not need that here.

[6] Actually, Frege did not use "name" for expressions referring to functions. Instead, he used "incomplete expression" and the like.

[7] Here I depart from Frege by (i) using modern quantifier notation, and (ii) using the same letter 'x' in (1) and (2). According to Frege, the variable in (1) just marks a place and does not really belong to the concept name, whereas in (2) it is an inseparable part of a function name (cf. below). These distinctions, while interesting, are not essential in the present context.

[8] Note that the quantifier (3) must be defined for all unary first level functions (not only for concepts), since functions are total. As we can see, $\forall x\Psi(x)$ is *false* for arguments that are not concepts.

[9] There may be deeper reasons for preferring binary quantifiers. For example, Dummett [1981] regards Frege's decision to use a unary quantifier as *the* fatal step which eventually led to paradox in his system. This is because in the unary case we quantify over all objects, whereas binary quantifiers can restrict the domain to that part of the universe denoted by the first argument (as we will see in Section 2), thereby avoiding the need to consider a total universe (Dummett [1981], p. 227).

This argument may point to one cause of Frege's actual choice of an inconsistent system, but it is not by itself conclusive against unary quantifiers. The lesson of the paradoxes is not necessarily that one must not quantify over all objects. Indeed, the Tarskian account of the truth conditions for universally quantified sentences is quite independent of the size of the universe, and logicians often quantify over total domains, e.g., the domain of all sets in

Zermelo–Fraenkel set theory, without fearing paradox. (It is another matter that they, for 'practical' reasons, often prefer set domains when this is possible.) So the above argument can only have force, I think, when combined with a general theory of meaning of the type that Dummett advocates (and which in some sense rejects the Tarskian account). These deeper issues in the theory of meaning will not be discussed here.

[10] Frege's logic, that is, not his whole system with its (inconsistent) principles of set existence (abstraction). The proposed identification slurs over some details, but is consistent with Frege's idea that logic is about a domain of *objects* (U), upon which a structure of functions of different levels is built, with no mixing between functions and objects, or between functions of different levels.

[11] Neither Mostowski nor Lindström seem to have been aware of Frege's concept. There is, however, a tradition within type theory which builds on Frege's work, starting with Church's logic of sense and denotation (cf. Church [1951]). More recent works are e.g. Cocchiarella [1975] and Daniels and Freeman [1978].

[12] There are more general concepts of logic, used in abstract model theory. A comparison of various abstract notions of a logic is given in Westerståhl [1976].

[13] This partial order concerns *explicit* power of expression, by *single* sentences. One can also consider *implicit* strength (cf. Appendix B.3), or expressibility by *sets* of sentences.

Notes to Section 2:

[1] We don't need to assume that proper *definitions* of these categories exist, only that there is agreement about them in a large number of cases.

[2] Even if there are contexts where (6) might be uttered, it is unreasonable to interpret *more* as an independent *DET*: the standard of comparison is missing, and has to be supplied to get at the meaning. So *more* in (6) would then stand for something like *more than 10, more . . . than the number of girls*, etc. These are *DET*s by our criterion, but not the single *more*.

[3] The 'hybrid' form (5) is used in order to avoid discussion of the syntactic structure of "there are"-sentences. This structure is quite varied, as already (1)–(4) indicate, and there may be divergent opinions about it, but it still seems that (5) is *semantically* adequate in a large number of cases.

[4] Actually, *most*, as we have interpreted it, is not reflexive, since $\mathbf{most}_M AA$ is false when $A = \varnothing$. One remedy is to redefine it for this argument.

[5] One may argue that (32) also says that *some* women voted for Smith. We ignore the possible existence implications of *only* here, but they could easily be added without affecting the discussion.

[6] There is also the idea that five is unexpectedly few here. It would be possible to add $\mathbf{few}_M AB$ as a further condition.

[7] We had a few examples of discontinuous 1-place *DET*s too, e.g., *every but John*, and here the claim is more dubious. But note that in all these cases, an alternative analysis was proposed, which eliminates the need for the *DET*s in question.

Notes to Section 3:

[1] The match would be even better if we had used the binary **some** instead of the usual existential quantifier.

[2] Other similar sentences are harder to make sense of, for example,

Most schoolboys tease each other.

Is this about pairs of schoolboys, or does it mean that most schoolboys tease some other schoolboy, or most other schoolboys, . . . ? The problem seems to be that *schoolboy* denotes a set but *each other* indicates a relation.

[3] I take the pairs to be ordered, but this doesn't really matter.

[4] In other cases equivalence would obtain. Consider, for example,

Most twins like each other.

Since everyone is the twin of at most one other person, there are as many individual twins as there are ordered twin pairs, and thus the same proportion of 'liking' twin pairs as that of twins who like their other twin.

[5] Consider a situation with two boys, one of which owns and kicks two dogs, the other owning, but not kicking, one dog. The formalization with $\mathbf{most}^{(2)}$ would be true in this case, which seems counter-intuitive.

[6] To get a 'quantity version' of *PERM*, let $M' = M$ in *QUANT*.

[7] *VP*-positivity is related to the notions of *existential* and *cardinal* quantifiers in Keenan and Stavi [1986]. In fact, under *CONSERV*, *VP*-positivity is equivalent to existentiality, and cardinality is equivalent to *VP*-positivity + *QUANT*.

[8] This is for $\mathbf{many}^1$ (2.4.3); $\mathbf{many}^2$ is neither *VP*-positive nor *VP*-negative.

[9] They consider (the singular) *the, both*, and *DET*s of the form *the n*, but not possessives.

Notes to Section 4:

[1] Actually, only *CONSERV* is needed for this result (van Benthem [1984a]).

[2] Without *FIN* one may represent logical quantifiers as subsets of $Card^2$ ($Card$ = the class of cardinal numbers). This is not as easy to visualize as N^2. For example, diagonals and columns get mixed up: $(0, \aleph_0)$, $(1, \aleph_0)$, ... are in the column given by $\aleph_0$, but also in the diagonal $\{(x, y): x + y = \aleph_0\}$.

[3] This does not need *FIN*.

[4] This observation is also from van Benthem [1984a]. For the notion of a diagram, cf. Chang and Keisler [1973], p. 68.

[5] Michał Krynicki has observed (private communication) that, without *FIN*, *LEFT CONT* quantifiers are definable in logic with the cardinality quantifiers $\mathbf{Q}_\alpha$ (Section 1.3).

[6] One analogy is with the usual completeness theorems in logic, relating provability to truth in models. Or, one may think of the extent to which axiomatic characterizations of a relation (say) determine an intended interpretation (e.g., questions of categoricity). Since the relations in the present case are basic logical constants, a third analogy suggests itself: characterizations of *EL*, such as Lindström's theorem (Section 1.6).

[7] Cf. Keenan and Stavi [1986] for the plausibility of the assumptions. Unlike Keenan and Stavi, I have included inner negation in the closure operations, but this can be avoided at the cost of adding a variant of $\mathbf{S}_a$ (namely, **b's zero or more**, when $P_b = \{a\}$) to the initial quantifiers. In 3.4 I expressed some doubts as to the closure of natural language quantifiers under inner or outer negation. These doubts do not affect Theorem 4.6.1, however, for, in the proof, we only apply inner and outer negation to the quantifiers $\mathbf{S}_a$, and, as Keenan and Stavi show, $\neg\mathbf{S}_a$ and $\mathbf{S}_a\neg$ are expressible with familiar *DET*s.

[8] This makes heavy use of the universal (*U4′*) that simplex *DET*s denote *PERM* quantifiers: $2^{(n+1)(n+2)/2}$ grows slower than 2^{3^n}. Without (*U4′*), a simplex *DET* symbol could denote any conservative quantifier on any M.

Note to Section 5:

[1] Zucker [1978] adopts a point of view similar to the present one. There seems to be a connection between his notion of a quantifier being *implicitly definable* and our notion of it being *determined*, even though the settings are different.

REFERENCES

Aczel, P.: 1975, 'Quantifiers, games, and inductive definitions', in *Proc. of the 3rd Scandinavian Logic Symposium*, North-Holland, Amsterdam, pp. 1–14.

Anapolitanos, D. A. and Väänänen, J.: 1981, 'Decidability of some logics with free quantifier variables', *Zeit. Math. Logik und Grundl. der Math.* **27**, 17–22.

Barwise, J.: 1978, 'Monotone quantifiers and admissible sets', in J. E. Fenstad *et al.* (eds.), *Generalized Recursion Theory II*, North-Holland, Amsterdam, pp. 1–38.

Barwise, J.: 1979, 'On branching quantifiers in English', *J. Phil. Logic* **8**, 47–80.

Barwise, J. and Cooper, R.: 1981, 'Generalized quantifiers and natural language', *Linguistics and Philosophy* **4**, 159–219.

Barwise, J. and Feferman, S. (eds.): 1985, *Model-Theoretic Logics*, Springer-Verlag, Berlin.

Barwise, J. and Perry, J.: 1983, *Situations and Attitudes*, MIT Press/Bradford, Cambridge, U.S.A.

Broesterhuizen, G.: 1975, 'Structures for a logic with additional generalized quantifier', *Colloquium Mathematicum* **33**, 1–12.

Bruce, K.: 1978, 'Ideal models and some not so ideal problems in the model theory of $L(Q)$', *J. Symbolic Logic* **43**, 304–321.

Chang, C. C. and Keisler, K. J.: 1973, *Model Theory*, North-Holland, Amsterdam.

Church, A.: 1951, 'A formulation of the logic of sense and denotation', in E. Henle *et al.* (eds.), *Structure, Method, and Meaning. Essays in Honor of Henry M. Sheffer*, New York, pp. 3–24.

Cocchiarella, N.: 1975, 'A second-order logic of variable-binding operators', *Rep. Math. Logic* **5** (Univ. of Cracow), 3–18.

Cooper, R.: 1983, *Quantification and Syntactic Theory*, D. Reidel, Dordrecht.

Cowles, J. R.: 1981, 'The Henkin quantifier and real closed fields', *Zeit. Math. Logik und Grundl. der Math.* **27**, 549–555.

Daniels, C. B. and Freeman, J. B.: 1978, 'A logic of generalized quantification', *Rep. Math. Logic* **10** (Univ. of Cracow), 9–42.

de Morgan, A.: 1847, *Formal Logic*, London; repr. (ed. A. E. Taylor) London, 1926.

Dummett, M.: 1973, *Frege: Philosophy of Language*, Duckworth, London.

Dummett, M.: 1975, 'The philosophical basis of intuitionistic logic', in H. E. Rose and J. C. Shepherdson (eds.), *Logic Colloquium* **73**, North-Holland, Amsterdam, pp. 5–40.

Dummett, M.: 1981, *The Interpretation of Frege's Philosophy*, Duckworth, London.

Enderton, H. B.: 1970, 'Finite partially-ordered quantifiers', *Zeit. Math. Logik und Grundl. der Math.* **16**, 393–397.

Fenstad, J. E., Halvorsen, P.-K., Langholm, T., and van Benthem, J.: 1987, *Situations, Language and Logic*, D. Reidel, Dordrecht.

Flum, J.: 1985, 'Characterizing logics', ch. III in Barwise and Feferman [1985], pp. 77–120.

Frege, G.: 1892, 'On concept and object', in P. Geach and M. Black (eds.), *Translations from the Philosophical Writings of Gottlob Frege*, Blackwell, Oxford, 1952.

Frege, G.: 1893, *Grundgesetze der Arithmetik I*, Jena; partial transl. and introduction by M. Furth: *The Basic Laws of Arithmetic*, Univ. Calif. Press, Berkeley, 1964.

Geach, P.: 1972, 'A program for syntax', in D. Davidson and G. Harman (eds.), *Semantics of Natural Language*, D. Reidel, Dordrecht, pp. 483–497.

Goldfarb, W.: 1979, 'Logic in the twenties: the nature of the quantifier', *J. Symbolic Logic* **44**, 351–368.

Hauschild, K.: 1981, 'Zum Vergleich von Härtigquantor und Rescherquantor', *Zeit. Math. Logik und Grundl. der Math.* **27**, 255–264.

Henkin, L.: 1961, 'Some remarks on infinitely long formulas', in *Infinitistic Methods*, Pergamon Press, Oxford, pp. 167–183.

Higginbotham, J. and May, R.: 1981, 'Questions, quantifiers and crossing', *The Linguistic Review* **1**, 41–79.

Hintikka, J.: 1973, 'Quantifiers vs. quantification theory', *Dialectica* **27**, 329–358.

Hodges, W.: 1983, [I.1], 'Elementary predicate logic', *Handbook*, Vol. I, ch. 1, pp. 1–131.

Hoeksema, J.: 1983, 'Plurality and conjunction', in A. ter Meulen (ed.), *Studies in Model-theoretic Semantics*, Foris, Dordrecht, pp. 63–83.

Kamp, H.: 1981, 'A theory of truth and semantic representation', in J. Groenendijk *et al.* (eds.), *Formal Methods in the Study of Language*, Math. Centre, Amsterdam, pp. 277–322 (also in *Truth, Interpretation and Information*, Foris, Dordrecht, pp. 1–41).

Keenan, E. L.: 1981, 'A Boolean approach to semantics', *ibid.*, pp. 343–379.

Keenan, E. L.: 1984, 'A semantic definition of "indefinite *NP*"', to appear.

Keenan, E. L.: 1987, 'Unreducible *n*-ary quantification in natural language', in P. Gärdenfors (ed.), *Generalized Quantifiers: Linguistic and Logical Approaches*, D. Reidel, Dordrecht, pp. 109–150.

Keenan, E. L. and Moss, L. S.: 1985, 'Generalized quantifiers and the expressive power of natural language', in J. van Benthem and A. ter Meulen (eds.), *Generalized Quantifiers in Natural Language*, Foris, Dordrecht, pp. 73–124.

Keenan, E. L. and Stavi, J.: 1986, 'A semantic characterization of natural language determiners', *Linguistics and Philosophy* **9**, 253–326.

Keisler, H. J.: 'Logic with the quantifier "there exists uncountably many"', *Annals of Math. Logic* **1**, 1–93.

Kreisel, G.: 1967, 'Informal rigour and completeness proofs', in I. Lakatos (ed.), *Problems in the Philosophy of Mathematics*, North-Holland, Amsterdam, pp. 138–157.

Lachlan, A. H. and Krynicki, M.: 1979, "On the semantics of the Henkin quantifier", *J. Symbolic Logic* **44**, 184–200.

Ladusaw, W.: 1979, *Polarity Sensitivity and Inherent Scope Relations*, diss., Univ. of Texas, Austin.

Lindström, P.: 1966, "First-order predicate logic with generalized quantifiers", *Theoria* **32**, 186–195.

Lindström, P.: 1969, "On extensions of elementary logic", *Theoria* **35**, 1–11.

Link, G.: 1987, "Generalized quantifiers and plurals", in P. Gärdenfors (ed.), *Generalized Quantifiers: Linguistic and Logical Approaches*, D. Reidel, Dordrecht, pp. 151–180.

Lorenzen, P.: 1958, *Formale Logik*, W. de Gruyter, Berlin.

Lønning, J. T.: 1987a, "Mass terms and quantification", *Linguistics and Philosophy* **10**, 1–52.

Lønning, J. T.: 1987b, "Collective readings of definite and indefinite noun phrases", in P. Gärdenfors (ed.), *Generalized Quantifiers: Linguistic and Logical Approaches*, D. Reidel, Dordrecht, pp. 203–235.

Łukasiewicz, J.: 1957, *Aristotle's Syllogistic*, Clarendon Press, Oxford.

Makowsky, J. A. and Tulipani, S.: 1977, "Some model theory for monotone quantifiers", *Archiv f. Math. Logik* **18**, 115–134.

Montague, R.: 1974, *Formal Philosophy*, ed. R. M. Thomason, Yale U.P., New Haven.

Mostowski, A.: 1957, "On a generalization of quantifiers", *Fund Math.* **44**, 12–36.

Mundici, D.: 1985, "Other quantifiers: an overview", ch. VI in Barwise and Feferman [1985], pp. 211–233.

Partee, B.: 1984a, "Compositionality", in F. Landman and F. Veltman (eds.), *Varieties of Formal Semantics*, Foris, Dordrecht.

Partee, B.: 1984b, "Genitives and 'have'", abstract, UMass., Amherst.

Patzig, G.: 1959, *Die Aristotelische Syllogistik*, van den Hoeck and Ruprecht, Göttingen; transl. *Aristotle's Theory of the Syllogism*, D. Reidel, Dordrecht, 1968.

Prawitz, D.: 1971, "Ideas and results in proof theory", in J. E. Fenstad (ed.), *Proceedings of the Second Scandinavian Logic Symposium*, North-Holland, Amsterdam, pp. 235–307.

Prawitz, D.: 1977, "Meaning and proofs", *Theoria* **43**, 2–40.

Rescher, N.: 1962, "Plurality-quantification", abstract, *J. Symbolic Logic* **27**, 373–374.

Rooth, M.: 1984, "How to get even with domain selection", in *Proceedings of NELS 14*, UMass., Amherst, pp. 377–401.

Rooth, M.: 1985, *Association with Focus*, diss., UMass., Amherst.

Russell, B.: 1903, *The Principles of Mathematics*, Allen and Unwin, London.

Russell, B.: 1956, *Logic and Knowledge. Essays 1901–1950*, ed. R. C. Marsh, Allen and Unwin (reference to "Mathematical logic as based on the theory of types", 1908, and "The philosophy of logical atomism", 1918).

Sgro, J.: 1977, "Completeness theorems for topological models", *Annals of Math. Logic* **11**, 173–193.

Thijsse, E.: 1983, *Laws of Language*, thesis, Rijksuniversiteit Groningen.

van Benthem, J.: 1983a, "Five easy pieces", in A. ter Meulen (ed.), *Studies in Modeltheoretic Semantics*, Foris, Dordrecht, pp. 1–17.

van Benthem, J.: 1983b, "Determiners and logic", *Linguistics and Philosophy* **6**, 447–478.

van Benthem, J.: 1983c, "A linguistic turn: New directions in logic" in *Proceedings of the 7th International Congress of Logic, Methodology and Philosophy of Science*, Salzburg, 1983, eds. R. Barcan Marcus *et al.*, North-Holland, Amstersam, 1986.

van Benthem, J.: 1984a, "Questions about quantifiers", *J. Symbolic Logic* **49**, 443–466.

van Benthem, J.: 1984b, "Foundations of conditional logic", *J. Phil. Logic* **13**, 303–349.

van Benthem, J.: 1985, "Semantic automata", Report No. CSLI-85-27, Stanford.

van Benthem, J.: 1986, *Essays in Logical Semantics*, D. Reidel, Dordrecht (*Studies in Linguistics and Philosophy*). Contains, among other things, revised versions of van Benthem [1983b,c], [1984a,b], [1985].

van Benthem, J.: 1987a, "Towards a computational semantics", in P. Gärdenfors (ed.), *Generalized Quantifiers: Linguistic and Logical Approaches*, D. Reidel, Dordrecht, pp. 31–71.

van Benthem, J.: 1987b, 'Polyadic quantifiers', to appear in *Linguistics and Philosophy*.

van Benthem, J. and Doets, K.: 1983, [I.4], "Higher-order logic", Handbook, vol. I, ch. 4, 275–329.

van Deemter, K.: 1985, "Generalized quantifiers: finite versus infinite", in J. van Benthem and A. ter Meulen (eds.), *Generalized Quantifiers in Natural Language*, Foris, Dordrecht, pp. 147–159.

van Eijck, J.: 1982, "Discourse representation, anaphora and scope", in F. Landman and F. Veltman (eds.), *Varieties of Formal Semantics*, Foris, Dordrecht.

van Eijck, J.: 1985, *Aspects of quantification in natural language*, diss., Rijksuniversiteit Groningen.

Väänänen, J.: 1979, "Remarks on free quantifier variables", in J. Hintikka *et al.* (eds.), *Essays on Mathematical and Philosophical Logic*, D. Reidel, Dordrecht, pp. 267–272.

Walkoe, W. Jr.: 1970, Finite partially ordered quantification", *J. Symbolic Logic* **35**, 535–550.

Weese, M.: 1981, "Decidability with respect to the Härtig and Rescher quantifiers", *Zeitschrift f. Math. Logik und Grundl. der Math.* **27**, 569–576.

Westerståhl, D.: 1976, *Some Philosophical Aspects of Abstract Model Theory*, diss., Dept. of Philosophy, Univ. of Göteborg.

Westerståhl, D.: 1983, "On determiners", in *Abstracts from the 7th Int. Congress of Logic, Methodology and Phil. of Science*, vol. 2, Salzburg, pp. 223–226.

Westerståhl, D.: 1984, "Some results on quantifiers", *Notre Dame J. of Formal Logic* **25**, 152–170.

Westerståhl, D.: 1985a, "Logical constants in quantifier languages", *Linguistics and Philosophy* **8**, 387–413.

Westerståhl, D.: 1985b, "Determiners and context sets", in J. van Benthem and A. ter Meulen (eds.), *Generalized Quantifiers in Natural Language*, Foris, Dordrecht, pp. 45–71.

Westerståhl, D.: 1987, "Branching generalized quantifiers and natural language", in P. Gärdenfors (ed.), *Generalized Quantifiers: Linguistic and Logical Approaches*, D. Reidel, Dordrecht, pp. 269–298.

Yasuhara, M.: 1969, "The incompleteness of L_p languages", *Fund. Math.* **66**, 147–152.

Zucker, J. I.: 1978, "The adequacy problem for classical logic", *J. Phil. Logic* **7**, 517–535.

Zwarts, F.: 1983, "Determiners: a relational perspective", in A. ter Meulen (ed.), *Studies in Modeltheoretic Semantics*, Foris, Dordrecht, pp. 37–62.

Zwarts, F.: 1986, *Modeltheoretic Semantics and Natural Language: the case of modern Dutch*, diss., Nederlands inst., Rijksuniversiteit Groningen.

CHAPTER IV.2

PROPERTY THEORIES

by George Bealer and Uwe Mönnich

INTRODUCTION: WHAT IS A PROPERTY THEORY?

We begin with a truism.[1] A property theory is a theory that deals with properties. More precisely, it is a theory that formulates general, non-contingent laws that deal with properties. There are two salient ways of talking of properties. First, they can be talked about as *predicables* (i.e., as *instantiables*). Accordingly, one sort of property theory would be a

D. Gabbay and F. Guenthner (eds.), Handbook of Philosophical Logic, Volume IV, 133–251.

theory that provides general, noncontingent laws for the behavior of the predication relation (instantiation relation). Nothing prevents the logical framework of such a theory from being extensional; that is, it could be formulated in a logical framework in which equivalent formulas are intersubstitutable *salva veritate*. For example, this sort of property theory could be formulated in a first-order extensional language with identity and a distinguished binary logical predicate for the predication relation. The major challenge facing this sort of property theory is to resolve various paradoxes that result from naive predication principles such as the following analogue of Russell's paradox: $(\exists x)(\forall y)$ (x is predicable of $y \leftrightarrow y$ is not predicable of itself). The second salient way of talking about properties is by means of *property abstracts* such as 'the property of being a man'. Property abstracts belong to a family of complex singular terms known as *intensional abstracts*. These include gerundive phrases, infinitive phrases, and 'that'-clauses. These singular terms are intensional in the sense that expressions occurring within them do not obey the substitutivity principles of extensional logic. Accordingly, another sort of property theory would be a theory that provides general, noncontingent laws for the behavior of intensional abstracts. The major challenge facing this sort of property theory is to systematize various subtle nonextensional substitutivity phenomena such as the nonsubstitutivity of necessarily equivalent formulas, the nonsubstitutivity of co-denoting names and indexicals, the paradox of analysis, and Mates' puzzle. These two types of property theory can be developed independently. Once this is done, one would then want to combine them to arrive at a single theory that treats both predication and intensional abstraction.

Although both types of property theory are important, the second type has an epistemological primacy, which we will now explain. Evidently, the best argument for the existence of properties – and for intensional entities, generally – is the following *argument from intensional logic*.[2] (Intensional logic is that part of logic in which the principle of the substitutivity of equivalent formulas fails.) The argument has three premises, which of course must be established. First, the best way to formulate intensional logic is to adjoin an intensional abstraction operation to standard extensional logic with identity and then to formulate laws governing the substitutivity conditions on expressions occurring within intensional abstracts. Second, on any acceptable interpretation of this intensional logic, intensional abstracts must be interpreted as being semantically correlated with real intensional entities, specifically, entities whose identity conditions match the substitutivity conditions of the intensional abstracts with which they are semantically correlated. From this premise it follows that intensional logic is committed to the existence of intensional entities. Third, intensional logic is an indispensable part of any acceptable comprehensive theory of the world. (It is understood here that any acceptable comprehensive theory of the world should include an account of its own acceptability.)[3] From these three premises it follows that intensional entities – and properties, in particular – are indispensable to any acceptable comprehensive theory of the world. Every acceptable comprehensive theory of the world is committed to the existence of intensional entities.[4]

However, suppose *per impossibile* that intensional logic could be omitted from an acceptable comprehensive theory. In that event, it is plausible that intensional entities – and properties, in particular – could be dispensed with. True enough, the sort of property theory that codifies laws for the predication relation might have nice theoretical payoffs; for example, it might provide an elegant construction of the foundations of mathematics or of the extensional semantics of extensional language. However, these payoffs on their own do not appear to justify an ontology of properties. The reason is that, for all we can tell, each of these theoretical payoffs could be provided by a theory of extensional entities (such as sets, or mathematical categories, or perhaps some new type of extensional theoretical posit). Thus, in the absence of the argument from intensional logic, intensional entities would, for all we can tell, be dispensable in favor of such extensional entities. It is for this reason that property theory construed as an intensional logic is epistemologically more primary than property theory construed as a theory of predication. It is needed to show that properties really exist.

Once one has established the need for the former sort of property theory, one would be justified in going on to develop the latter sort of property theory. There are two reasons. First, once one has established that properties exist, Ockham's razor directs one to attempt to dispense with the more complex ontology of both sets and properties in favor of the simpler ontology of just properties. However, one can accomplish this ontological simplification only if one has a property theory that has all the theoretical payoffs that set theory has. This is what a satisfactory theory of predication promises to do. Second, when a theory of intensional abstraction is combined with a satisfactory theory of predication, the resulting theory promises to yield several additional theoretical payoffs (for example, a definition of truth for propositions, a definition of necessity for propositions, a definition of logical validity and analyticity for propositions, and so on).[5]

Although there are good reasons to look forward to a unified theory of intensional abstraction and predication, there are nevertheless good methodological reasons for proceeding in separate stages. For example, the immediate prospects of finding a truly satisfactory (as opposed to merely workable) resolution of the paradoxes of predication are much less bright than those for a satisfactory intensional logic. This and other methodological reasons for keeping the two projects separate at this stage of research will be elaborated upon below.

Our plan in this paper is the following. In Part I, we will spell out the argument from intensional logic. In Part II, we will show in detail how to construct a property theory that is suited to serve as an intensional logic. In Part III, we will close with a discussion of a few somewhat more sophisticated issues in property theory, namely, the propositional-function thesis, type-free predication theories, and a proof of nonextensionality within predication theories with unrestricted abstraction principles.

The propositional-function thesis is the thesis that there is a strong correlation between properties and propositional functions. A theory that takes properties as primitive entities can capture the extremely fine-grained substitutivity conditions that hold in propositional-attitude contexts. The question we will address in connection with the propositional-function thesis is whether a propositional-functional theory can capture these extremely fine-grained substitutivity conditions as well.

When we speak of type-free predication theories, we have in mind theories in the style of Gilmore [1974], Feferman [1975], Scott [1975], Aczel [1980], Feferman [1984], Reinhardt [1985], Flagg and Myhill [1987], Turner [1987], and others. As we indicated above, we do not believe that any of the existing theories can be singled out as embodying a final resolution of the paradoxes. Nevertheless, these recent type-free predication theories have an attractive feature from our point of view: one can actually prove within them that the most general form of abstraction principle for the predication relation entails a principle of nonextensionality. We will present an outline of the argument in Part III.

Despite the bewildering diversity of the diagnoses and cures for the paradoxes offered within a type-free setting, all these theories share one common theme, namely, that the mathematics that can be derived within them is rather weak. Because of the missing link between our favored theory of intensional abstracts and an ideal resolution of the paradoxes and because of the limited success of existing predication theories in providing a non-set-theoretic foundation of mathematics, we have forgone a critical review of these theories. Nevertheless, Part III does contain some discussion of the ubiquitous technique in this area, namely, the use of fixed-point constructions to establish the consistency of systems that admit unlimited self-reference in the presence of a principle of full abstraction.

Finally, a word about other research on property theory. Given the purpose of the present volume, we thought it would be valuable to defend and to formulate in detail one particular version of a property theory rather than to attempt a comprehensive overview. The reader will nevertheless find references to competing property theories at various points in our discussion and also in the bibliography.

I. THE ARGUMENT FROM INTENSIONAL LOGIC

The hallmark of extensional logic is the principle that equivalent expressions can be substituted for one another *salva veritate*. That is, whenever two expressions of the same syntactic type apply to exactly the same things – or, in the case of sentences, are alike in truth value – they can be substituted for one another without altering the truth values of the whole sentences in which they occur. The safest general characterization of intensional logic is that it is the part of logic in which there are exceptions, or at least apparent exceptions, to this substitutivity principle.

Sometimes, however, a second criterion is used to characterize intensional logic. According to this criterion, intensional logic is the part of logic in which the rule of existential generalization fails or at least appears to fail. For example, the inference from 'Pythagoras was looking for the rational $\sqrt{2}$' to 'There exists something such that Pythagoras was looking for it' is intuitively invalid; therefore, the occurrence of 'the rational $\sqrt{2}$' in the first sentence would qualify as intensional according to this second criterion. And this is as it should be. However, this criterion is not quite right as it is usually stated, for existential generalization appears to fail in some cases that would not standardly be counted as intensional. For example, the inference from 'The rational $\sqrt{2}$ does not exist' to 'There exists something such that it does not exist' is intuitively invalid, but the occurrence of 'the rational $\sqrt{2}$' in the first sentence would not standardly be counted as intensional.[6]

The first criterion – substitutivity failure – avoids this sort of difficulty. Since 'the rational $\sqrt{2}$' is a vacuous term, the only other terms that apply to the same (real) things must themselves be vacuous; and whenever another vacuous term is substituted for 'the rational $\sqrt{2}$' in 'The rational $\sqrt{2}$ does not exist', the resulting sentence has the same truth value as the original. So according to the substitutivity criterion, 'the rational $\sqrt{2}$' does not occur intensionally in this sentence. And this is the desired result. At the same time, in sentences like 'Pythagoras was looking for the rational $\sqrt{2}$', this vacuous term does occur intensionally according to the substitutivity criterion, for when we put in some other vacuous term, the resulting sentence will often have a different truth value. (For example, 'Pythagoras was looking for the largest integer' is false.) So once again this criterion fits our standard notion of intensional occurrence. It appears, therefore, that failure of substitutivity is indeed the best criterion to use in characterizing intensional logic. This, at least, is what we will assume in the remainder of this paper.

There are a number of interconnected logical phenomena that any adequate formulation of intensional logic ought to accommodate. Although some of them are widely recognized, others are not (for example, the existence of transcendental and self-embeddable predicates). Taken together, these phenomena more or less force one to formulate intensional logic as a certain sort of first-order theory of properties. Our argument for this thesis will be divided into the following sections: (1) generality, (2) 'that'-clauses, gerundive phrases, and infinitive phrases, (3) quantifying-in, (4) learnability, (5) referential semantics for intensional language, (6) what intensional abstracts denote, (7) nominalism, (8) conceptualism, (9) realism, (10) transcendental predicates and type-free languages, (11) self-embeddable properties, relations, and propositions, (12) the first-order vs. higher-order language controversy, (13) names and indexicals, (14) Mates' puzzle, the paradox of analysis, and the need for fine-grained intensional distinctions.[7]

1. GENERALITY

Substitutivity failures typically arise in connection with talk about such matters as intentionality (assertion, belief, desire, intention, perception, etc.), the logical modalities (necessity, possibility, contingency), definition, analyticity, meaning, strict implication, relevant implication, moral obligation, purpose, probability, causation, explanation, epistemic justification, evidence, counterfactuality, etc. However, many philosophers, logicians, and linguists have failed to notice that, when we talk about these matters in a *general* way, our discourse is typically *extensional*. For example, the sentence 'Whatever is necessary is possible' says something general about necessity and possibility, and it is a fully extensional sentence. Any adequate formulation of intensional logic should be able to accommodate, not just intensional talk about intentionality, modality, etc., but also this sort of general extensional talk. In particular, an adequate formulation of intensional logic ought to be able to represent intuitively valid extensional arguments like the following:

(I) Whatever x believes is necessary.
Whatever is necessary is possible.
∴ Whatever x believes is possible.

Suppose that 'is necessary' and 'is possible' are treated as 1-place predicate expressions and 'believes', as a 2-place predicate expression. Then this argument can be represented as valid in a standard quantifier logic:

(I′)
$$(\forall y)(B^2xy \rightarrow N^1y)$$
$$(\forall y)(N^1y \rightarrow P^1y)$$
$$\therefore (\forall y)(B^2xy \rightarrow P^1y).$$

Now in theoretical matters, if a currently accepted theory can be easily and naturally employed to account for new phenomena, then other things being equal it is desirable to do so. In logical theory, the currently accepted theory includes quantifier logic. By treating 'is necessary' and 'is possible' as 1-place predicate expressions and 'believes' as a 2-place predicate expression, we can easily and naturally account for the validity of (I) within this currently accepted theory. Therefore, other things being equal it would seem desirable to do so. Indeed, none of the alternatives appears to be satisfactory.

For example, one alternative is the sentential-operator approach, which posits an open-ended list of special sentential operators (that is, operators that can be applied to sentences to yield new sentences). On this approach, there is a separate operator for each of the topics mentioned earlier – assertion, belief, desire, intention, necessity, possibility, contingency, definition, analyticity, meaning, strict implication, relevant implication, moral obligation, purpose, probability, causation, explanation, epistemic justification, counterfactuality, etc. A major problem with this approach is that it does not provide a *unified* theory of intensionality; it is eclectic and incomplete at best. Furthermore, on the first-order version of this approach, elementary arguments like (I) cannot even be *expressed*. The reason is that in a first-order language sentential operators like 'B_x', '$\Box$', and '$\Diamond$' may only be applied to *specific* first-order sentences. (By 'first-order' we mean syntactically first-order. In a language that is syntactically first-order there are no sentential variables.)[8] But arguments like (I) are *general*; in such arguments expressions like '*x* believes', 'is necessary', and 'is possible' are not applied to any specific sentences at all. Plainly, a variable-*cum*-quantifier apparatus – or something comparable – is needed.

Of course, such an apparatus is available on a higher-order sentential-operator approach that contains sentential variables, that is, variables that are themselves counted as sentences and that take other sentences as substituends. (By 'higher-order' we mean syntactically higher-order.)[9] On such an approach (I) would be represented along the following lines:

$$\begin{array}{l} (\forall p)(B_x p \rightarrow \Box p) \\ (\forall p)(\Box p \rightarrow \Diamond p) \\ \hline \therefore\ (\forall p)(B_x p \rightarrow \Diamond p). \end{array}$$

On this approach, however, there is no clear distinction between sentential operators, on the one hand, and predicates that take sentences as argument expressions, on the other hand. Consequently, this approach may be viewed as a variant of – rather than a genuine alternative to – our official logical syntax which treats 'believes', 'is necessary', 'is possible', etc. as predicate expressions.

Should the higher-order sentential-operator approach be adopted? In the next paragraph we will give some surface grammatical evidence against this approach. We wish to emphasize that our larger line of argument does not depend on this. So if this grammatical evidence should strike our readers as unconvincing, they should not for this reason be doubtful about our other conclusions. In particular, the argument in the next section that 'that'-clauses should be treated as singular terms will hold with at most minor alterations: on the higher-order sentential-operator approach, sentences are already treated as singular terms, and expressions like '*x* believes that *p*' and '*x* believes *p*' are in effect counted as mere notational variants of one another.

Now although the higher-order sentential-operator approach succeeds in representing arguments like (I), it has *prima facie* implausible side-effects. In particular, it is forced to treat sentences as full-fleged singular terms. Consider the sentence 'There is something that x believes that is different from something that y believes'. On the higher-order approach this sentence would be represented as follows:

$$(\exists p)(B_x p \ \& \ (\exists q)(\neg p = q \ \& \ B_y q)).$$

This shows that these special sentential variables are able to flank the identity symbol as in the (open) sentence '$p = q$'. But since specific (closed) sentences are supposed to be substituends for these variables, specific (closed) sentences may themselves flank the identity symbol, for example, '$5 + 7 = 12 = 7 < 9$', '$2 + 2 = 4 = 3 + 5 = 8$', etc. This, however, is ungrammatical nonsense. The problem seems plain: sentences are not genuine singular terms. But if they are not, then sentential variables – that is, variables for which specific closed sentences are supposed to be substituends – must themselves be illegitimate. It follows that, if a sentential-operator approach makes use of such variables, it too is illegitimate. This, of course, is not to say that the use of a special sort of variable (say, 'p', 'q', 'r', . . .) whose *values* are supposed to be sentences is illegitimate; what is illegitimate is the use of a variable whose *substituends* are supposed to be sentences. But the latter sort of variable is what is needed to enable a sentential-operator approach to represent general sentences and general arguments.

Whether or not this grammatical evidence against the higher-order sentential-operator approach is convincing, this approach does not in any event provide a genuine alternative to our favored approach, which treats 'believes', 'asserts', etc. as 2-place predicate expressions and 'is necessary', 'is possible', etc. as 1-place predicate expressions.[10] This conclusion is all that matters for the rest of our argument.

There are two further alternatives to our favored approach that deserve to be mentioned – the adverbial approach and the adjectival approach. According to the adverbial approach 'x believes that $7 < 9$' would be represented by '((that $7 < 9$)-ly B^1)x', where 'B^1' is a 1-place predicate that expresses the property of believing, '(that $7 < 9$)-ly' is a complex adverbial phrase that expresses a certain "mode of believing", and '(that $7 < 9$)-ly B^1' is a complex 1-place predicate that expresses the property of believing under that mode. According to the adjectival approach, 'x believes that $7 < 9$' would be represented by '($\exists y$)(x is-in-state y & $B^1 y$ & ((that $7 < 9$)-ish)y)', where 'B^1' is a 1-place predicate that expresses the property of being a belief state and '(that $7 < 9$)-ish' is a complex 1-place predicate that expresses the property of being a state with a "$7 < 9$-ish content."

But how do these approaches represent *general* sentences, for example, the sentence 'x believes something', which we would represent with '$(\exists y)B^2xy$'? Evidently, the adverbial approach must use something like '$(\exists\alpha)((\alpha\text{-ly } B^1)x)$', and the adjectival approach, something like '$(\exists\alpha)(\exists y)$ (x is-in-state y & B^1y & (α-ish)y)'. But notice that on both approaches the everyday use of the verb 'believes' is in effect represented by a complex 2-place predicate expression: either '(②-ly B^1)①' or '$(\exists y)$(① is-in-state y & B_1y & (②-ish)y)'. Now our thesis is that the verb 'believes' should be treated as a 2-place predicate expression. So, evidently, the adverbial and adjectival approaches just turn out to be complex variants of our logical syntax and, as such, are not genuine alternatives to it. At the same time, these two approaches involve significant additional complexity, and with no apparent gain. Let us take a moment to show that this is really so.

First, the issue of complexity. The adverbial approach requires that we complicate our logical syntax by adding a new syntactic category, namely, the category of predicate adverb and by adjoining an adverb-forming operator '-ly'.[11] In turn, it requires that we devise a new semantical method to deal with these new syntactic structures, a semantical method that inevitably will lead to ontological and ideological complications of its own. And the adjectival approach requires that we complicate our logical syntax by adjoining special vocabulary for dealing with states and by adjoining a predicate-forming operator '-ish'. In turn, it requires that we develop an associated semantical method, which also will carry with it new ontological and ideological complications. [Of course, the operators '-ly' and '-ish' might be contextually defined. For example, (α-ish)y iff_{df} $(\exists z)$ (z is-a-propositional-content-fixing-property-that-corresponds-to α and y is an instance of z). However, such contextual definitions would themselves invoke new logical machinery, for example, the new primitive predicate 'is-a-propositional-context-fixing-property-that-corresponds-to' and a device like 'is an instance of' for attributing properties to their instances, and such machinery would lead to corresponding complications in the semantics. Later on, we will give an argument that devices like 'is an instance of' have no place in intensional logic *per se*. In our introduction, we outlined our reason for thinking that intensional logic is epistemologically more primary than the logic for the instantiation (predication) relation.

Now these complications in syntax and semantics are considerable. Yet they are gratuitous inasmuch as they do nothing whatsoever to advance the formulation of a comprehensive intensional logic, as we will now show in detail. (We will focus on the adverbial approach, but the argument applies *mutatis mutandis* to the adjectival approach.)

Recall that the adverbialist's rendering of 'x believes that $7 < 9$' is '((that $7 < 9$)-ly B^1)x'. In this formula, '$7 < 9$' occurs intensionally, for we cannot replace it with an equivalent sentence – for example, '$7 <$ the number of planets' – without risking an alteration in the truth value of the whole formula. This might lead one to think that the adverb-forming operator '-ly' is what generates the intensionality in sentences concerning intentionality, modality, etc. But this would be an error. We have already seen that sentences like 'x believes something' force the adverbialist to apply the adverb-forming operator '-ly' to a straightforward externally quantifiable free variable like 'α': $(\exists\alpha)(\alpha$-ly $B^1)x$. Moreover, kindred sentences show that the adverbialist is forced to use variables like α as terms in elementary identity statements. For example, the adverbialist has no choice but to represent 'x believes something that is different from everything u believes' along the following lines: $(\exists\alpha)((\alpha$-ly $B^1)x$ & $(\forall\beta)$ $((\beta$-ly $B^1)u \rightarrow \alpha \neq \beta))$. Given that variables like 'α' and 'β' may occur as terms in elementary identity statements and given that 'α' and 'β' occur in '(α-ly B^1)x' and '(β-ly B^1)x' as straightforward externally quantifiable free variables, the adverbialist has no choice but to accept the following sentence as well-formed and logically true: $(\forall\alpha)(\forall\beta)(\alpha = \beta \rightarrow ((\alpha$-ly $B^1)x \equiv$ $(\beta$-ly $B^1)x$. Now, we may assume that 'that'-clauses are permissible substituends for the variables 'α' and 'β'. (If the adverbialist were perversely to require instead that the substituends of these variables be in some other syntactic category – say, the category of sentence – the remainder of our argument still would go through *mutatis mutandis*. It is true that intuitively ill-formed expressions – for example, 'The cat is on the mat $= 7 < 9$' or '$7 < 9 = 7 <$ the number of planets' – would result from the adverbialist's requirement, but that would be the adverbialist's responsibility, not ours.) Accordingly, the adverbialist must accept the following sort of instantiation of the above logically true sentence:

$$\text{that } 7 < 9 \;=\; \text{that } 7 < \text{the number of planets} \rightarrow$$
$$(((\text{that } 7 < 9)\text{-ly } B^1)x \;\equiv$$
$$((\text{that } 7 < \text{the number of planets})\text{-ly } B^1)x).$$

However, the original intention of the adverbial theory is that '((that $7 < 9$)-ly B^1)x' and '((that $7 <$ the number of planets)-ly B^1)x)' should be able to differ in truth value. It follows that the adverbialist is forced to accept the truth of the sentence 'that $7 < 9 \neq 7 <$ the number of planets'. (Other things being equal, this outcome should be welcome, for intuitively this sentence is true!) But if this sentence is true, it follows that '$7 <$ the number of planets' occurs intensionally in it. After all, '$7 < 9$' has the same truth values as '$7 <$ the number of planets'. But when we substitute '$7 < 9$' in the true sentence 'that $7 < 9 \neq$ that $7 <$ the number of planets', we obtain a logically false sentence 'that $7 < 9 \neq$ that $7 < 9$'. A clear case of intensionality, a case that does not involve adverbial constructions even implicitly. Now how are such cases of intensionality to be handled logically? All the complicated apparatus of the adverbial theory is of no help whatsoever. Moreover, once we have a theory able to handle complex singular-term cases of intensionality like this, we can easily and economically extend it to handle *all* standard cases of intensionality, and we can do so without recourse to any of the complications of the adverbial theory. For this reason, then, the complications of adverbial theory are gratuitous, having nothing special to contribute to the formulation of a comprehensive intensional logic.

In view of this, the only reasonable decision is to reject the adverbial theory in favor of the essentially simpler theory that treats 'believes', 'asserts', etc. at face value as ordinary 2-place predicates. Furthermore, as we have already indicated, a fully analogous argument can be given against the adjectival theory. And so our original conclusion stands: 'believes', etc. should be treated as ordinary 2-place predicates.

2. 'THAT'-CLAUSES, GERUNDIVE PHRASES, AND INFINITIVE PHRASES

Our conclusion that 'believes', 'asserts', 'is necessary', 'is possible', 'is true', etc. are predicative expressions has an important consequence. Consider the following intuitively valid argument, where A is any formula:

(II) Whatever x believes is possible.
x believes that A.
――――――――――――――――
∴ It is possible that A.

Suppose, as we have concluded, that one should treat 'is possible' as a 1-place predicate and 'believes' as a 2-place predicate. In this case, we seem to be left with no alternative but to parse the second and third lines of (II) as follows:

$\underline{x}$ $\underline{\text{believes}}$ $\underline{\text{that } A}$

$\underline{\text{It is possible}}$ $\underline{\text{that } A}$

where ⌜that A⌝ is counted as a singular term syntactically. As a notational convenience, let us represent the singular term ⌜that A⌝ by means of ⌜$[A]$⌝.[12] When this bracket notation is adopted, (II) can be naturally represented as follows:

$$\text{(II}')\qquad \frac{\begin{array}{l}(\forall y)(B^2x, y \rightarrow P^1y)\\ B^2x, [A]\end{array}}{\therefore\ P^1[A].}$$

The conclusion of (II′) is straightforwardly derivable from the two premises by an application of universal instantiation (UI) and *modus ponens* (MP), two rules of inference valid in standard quantifier logic. Thus, one can bring arguments like (II) within the scope of standard quantifier logic simply by adopting the hypothesis that 'that'-clauses are singular terms representable with the bracket notation. To represent such arguments successfully, one needs no new logical principles, and one needs no knowledge about the logic of expressions occurring within the singular term ⌜$[A]$⌝. It would seem, therefore, this is the simplest way to represent such arguments. Thus, on the assumption that the logic for the new singular terms ⌜$[A]$⌝ can be satisfactorily worked out, we conclude that it is desirable to treat 'that'-clauses as singular terms that may represented by means of the bracket notation.[13]

Now analogous considerations show that certain other complex nominal expressions – for example, gerundive phrases ⌜being something that is F⌝ and infinitive phrases ⌜to be something that is F⌝ – are also best treated as singular terms. An easy extension of the bracket notation provides a natural way to represent these complex singular terms. Accordingly, let A be a formula and $v_1, \ldots, v_m$ be distinct variables where $m \geqslant 1$. Then ⌜$[A]_{v_1 \ldots v_m}$⌝ will be our canonical singular term corresponding to the gerundive phrase ⌜being $v_1, \ldots, v_m$ such that A⌝ and to the infinitive phrase ⌜to be $v_1, \ldots, v_m$ such that A⌝.

We shall see that what is logically distinctive about these singular terms ⌜$[A]$⌝ and ⌜$[A]_\alpha$⌝ is that expressions occurring within them do not obey the substitutivity principles characteristic of extensional logic. That is, when a formula A is enclosed within square brackets (followed by appropriate subscripts), an intensional context is generated. This bracketing operation may therefore be viewed as a generalized *intensional abstraction operation.* Now most types of substitutivity failures result from the fact that the offending expressions explicitly occur within intensional abstracts.

This suggests the general working hypothesis that *all* substitutivity failures can be traced to explicit or implicit occurrences of intensional abstracts. (Consider an example of intensionality that does not involve an explicit occurrence of an intensional abstract, say, 'Pythagoras was seeking the rational $\sqrt{2}$'. The idea would be that such a sentence can be treated as a transform of an underlying sentence that explicitly contains an appropriate intensional abstract, for example, the sentence 'Pythagoras was seeking to find the rational $\sqrt{2}$'. The fact that 'the rational $\sqrt{2}$' occurs intensionally in the transform would then be explained by the fact that it occurs (with narrow scope) in the intensional abstract 'to find the rational $\sqrt{2}$' in the underlying sentence.) Although we need not commit ourselves to this hypothesis, its attractiveness is striking: if true, it would have considerable explanatory power, and it would serve to simplify and unify the entire subject of intensional logic. Indeed, intensional logic could be identified with the logic for intensional abstraction. So we urge it as a working hypothesis.

3. QUANTIFYING-IN

Consider the following argument:

(III) x believes that he believes something.
∴ There is someone v such that x believes that v believes something.

There is a reading according to which (III) is intuitively valid. This reading provides an example of the logical phenomenon of quantifying-in. It is desirable that all valid cases of quantifying-in should be representable in a comprehensive intensional logic. In the previous two sections we reached these conclusions: 'believes' should be treated as a 2-place predicate; 'is possible', as a 1-place predicate, and 'that'-clauses, as singular terms. These conclusions all but entail an answer to our problem. Consider the following instance of argument (II), which we considered in the previous section:

(IV) Whatever x believes is possible.
x believes that v believes something.
∴ It is possible that v believes something.

Given our previous conclusions, we must represent (IV) as follows:

(IV′) $(\forall y)(B^2xy \rightarrow P^1y)$
$B^2x[(\exists u)B^2vu]$
∴ $P^1[(\exists u)B^2vu]$.

And by analogy we must represent (III) as follows:

$$\text{(III}'\text{)} \qquad \frac{B^2x[(\exists u)B^2xu]}{\therefore\ (\exists v)B^2x[(\exists u)Bv^2u].}$$

What is important about this is that the occurrence of 'v' in the singular term '$[(\exists u)B^2vu]$' is an *externally quantifiable* occurrence of a variable.[14] We are thus led to conclude that 'that'-clauses ought to be treated as singular terms which may contain externally quantifiable occurrences of variables.

Now there are several alternate treatments of quantifying-in, but we find none of them acceptable. Before proceeding, let us give a critical survey of these alternatives. Perhaps the most popular one involves multiplying the senses of 'believe' so that, e.g., 'x believes that v believes something' would be represented as '$B^3x,v,[(\exists u)B^2w,u]_w$' (roughly, x believes of v that it has the property of being something w such that w believes something). But on this approach one cannot even begin to represent *mixed* arguments like (IV) – that is, arguments that "mix" the intentional verbs like 'believe' and modals like 'possible' – unless one also multiplies the senses of modals as well. And this is only the tip of the iceberg; senses of all expressions that take 'that'-clauses as arguments must similarly be multiplied – all intentional verbs, modals, 'imply', 'explain', 'justify', 'probable', etc., and even 'true' and '='. Furthermore, this multiple-sense approach is unable to represent formulas that intuitively involve only *one* sense of 'believe' but *two* 'that'-clauses, one containing an externally quantifiable variable and the other containing none: for example, 'x believes *both* that v believes something and that everything is self-identical'. On our approach we would use 'B^2x, $[(\exists u)B^2v, u]$ & B^2x, $[(\forall u)u = u]$' to represent this sentence. Evidently, the best someone can do using B^3 is 'B^3x, ,$[(\exists u)(\exists z)B^3w,z,u]_w$ & $B^3x,\phi,[(\forall u)u = u]$' (where ϕ represents the null sequence), thus abandoning altogether the familiar 2-place sense of 'believe'. Moreover, similar examples would then seem to force proponents of the 'B^3'-approach to abandon the familiar 2-place senses of '=', 'assert', 'explain', 'justify', etc., and the familiar 1-place senses of 'true', 'necessary', 'possible', 'contingent', 'probable', etc. For example, whereas we would represent 'x asserted two things; one was that v asserted something and the other was that everything is self-identical' with

$$(\exists y)(\exists z)(A^2x,y \ \&\ A^2x,z \ \&\ y = [(\exists u)A^2v,u] \ \&\ z = [(\forall w)w = w],$$

the opposing approach is forced to use something like

$$(\exists y)(\exists y')(\exists z)(\exists z')(A^3 x,y,y' \ \& \ A^3 x,z,z' \ \&$$
$$y,y' =^4 \ v, \ [(\exists u)(\exists u')A^3 v,u,u']_v \ \&$$
$$z,z' =^4 \ \phi, \ [(\forall w)(\forall w')w,w' \ =^4 \ w,w'],$$

where '$=^4$' is a new 4-place "identity" predicate. Not only are these consequences extremely unintuitive in themselves, but also they evidently make it impossible to represent important "cross-referential" sentences like the following: 'x asserted exactly *one* thing and we have a name for *it*'. Furthermore, given that the familiar 2-place sense of '=' must be abandoned on the present approach in favor of '$=^4$', the prospect of any coherent identity theory is seriously threatened. A final problem with these approaches arises in connection with multiple embedding. For example, suppose that someone u is consciously and explicitly thinking that u is consciously and explicitly thinking something; that is, $T^2 u, [(\exists y)T^2 u,y]$. On the 3-place approach, this must be represented along the following lines: $T^3 u,u,[(\exists x)(\exists y)T^3 v,x,y]_v$. However, this is implausible; in the example, u is not consciously and explicitly thinking of u that he has the property of being someone v who is consciously and explicitly thinking of something that it has some property. This thought – and any other one of its ilk – is intuitively different from the thought that u is having; u's thought is simply that u is consciously and explicit thinking something. This subtle difference is just lost on the 3-place approach.

Another approach to 'that'-clauses containing externally quantifiable variables is to associate them with certain *sequences* and to treat 'believe' as a 2-place predicate: for example, 'u believes that Fu' might be represented by '$B^2 u, \langle u, [Fu]_u \rangle$'. However, this seemingly simple idea seems impossible to formulate in a satisfactory general way. Here are some of the problems that confront it.

First, certainly some identities of the following form hold: that Fu = that Gu,v. (For example, is it not true that the following identity holds:

$$\text{That } u \text{ is a Moonie} \ = \ \text{that } u \text{ is a follower of } v.$$

where v is Reverend Sun Yen Moon?) However, such identities would be impossible if we were to represent 'that Fu' with '$\langle u, [Fu]_u \rangle$' and 'that Gu, v with '$\langle\langle u, v\rangle, [Gu, v]_{uv}\rangle$' as our sequence theorists would do. For plainly $\langle u, [Fu]_u\rangle \neq \langle\langle u, v\rangle, [Gu, v]_{uv}\rangle$.

Second, we find it extremely counterintuitive that what a person perceives, believes, asserts, hopes, decides, etc. is ever *really* a sequence. How can sequence theorists accept such an implausible theory? (See Section 7 for further discussion of this sort of intuitive objection.)

Third, the sequence approach runs into difficulties in connection with multiple embeddings. For example, it is intuitively possible for someone u to believe that, for every v, that Fv = that Fv and, nevertheless, not believe that, for every v, that $Fv = \langle\langle v\rangle, [Fv]_v\rangle$. In our bracket notation:

$$B^2[(\forall v)[Fv] \;=\; [Fv]] \;\&\; \neg B^2[(\forall v)[Fv] \;=\; \langle\langle v\rangle, [Fv]_v\rangle].$$

However, on the sequence approach, this would presumably be equivalent to the following contradiction:

$$B^2[(\forall v)\langle\langle v\rangle, [Fv]_v\rangle \;=\; \langle\langle v\rangle, [Fv]_v\rangle] \;\&$$
$$\neg B^2[(\forall v)\langle\langle v\rangle, [Fv]_v\rangle \;=\; \langle\langle v\rangle, [Fv]_v\rangle].$$

You might try to mitigate this problem by invoking your favorite resolution of the paradox of analysis. However, we believe that such ploys will not succeed. (For more on the paradox of analysis, see Section 13.)

Fourth, by attempting their reduction, the sequence theorists prejudge certain questions concerning the identity conditions for the items denoted by intensional abstracts. For example, on one important traditional conception (dubbed "Alternative (2)" by Alonzo Church)[15] the following principle of identity is valid for any formulas A and B:

> If it is necessary that A and it is necessary that B, then that A = that B.

(Analogous principles hold for intensional abstracts that are gerundive and infinitive in form.) However, for all u and v, it is necessary that u is self-identical and it is necessary that v is self-identical. This fact and the above principle of identity imply:

> That u is self-identical = that v is self-identical.

for all u and v. However, if $u \neq v$, the sequences that would be associated with these 'that'-clauses on the sequence approach would plainly not be identical: that is, $\langle u, [x \text{ is self-identical}]_x\rangle \neq \langle v, [x \text{ is self-identical}]_x\rangle$. In this way, therefore, the sequence approach is incompatible with the above traditional conception. A prudent approach to intensional logic would not prejudge such questions, for it is quite conceivable that our best overall theory of intensionality will invoke this coarse-grained conception perhaps in tandem with various fine-grained conceptions. Indeed, it is quite plausible that, unlike the fine-grained objects of the propositional attitudes, conditions (or states of affairs) in the world conform to the coarse-grained conception. For example, intuitively, the glass's being half empty is the same condition (state of affairs) in the world as the glass's being half full. And this is so despite the fact that someone could be thinking that the glass is half empty without thinking that the glass is half full.

Fifth, the sequence approach runs into difficulty concerning a kind of type-freedom that is called "self-constituency". This kind of type freedom appears to arise in connection with the problem of mutual knowledge.[16] For example, suppose that two enemy soldiers x and y suddenly spot each other in the bush. Conceivably, a full description of this situation should include the following: x sees z and y sees z', where $z = [y \text{ sees } z']$ and $z' = [x \text{ sees } z]$. But what are z and z'? On the sequence approach, $z = \langle\langle y, z'\rangle, [u \text{ sees } v]_{uv}\rangle$ and $z' = \langle\langle x, z\rangle, [u \text{ sees } v]\rangle_{uv}$. Now suppose that sequences are identified with ordered sets. (If instead sequences are ordered properties, one runs into the regress discussed below.) In this case, z and z' would be non-well-founded sets (i.e., $z \in \ldots \in z' \in \ldots \in z$). However, according to the standard conception of sets, non-well-founded sets are impossible. The upshot is this. To represent z and z' on the set theoretical version of the sequence approach, one must revolutionize the standard conception of set. However, such a radical move seems quite unjustified merely to deal with the simple task at hand. No such radical move is required on the more cautious treatment of quantifying-in that we are advocating. (We will return to this topic in Section 11.)

Our sixth objection to the sequence approach is ontological. In Sections (5)–(9) we will argue that intensional abstracts that do not contain externally quantifiable variables denote properties, relations, and propositions. This conclusion, together with the premise that intensional logic is indispensable to any acceptable comprehensive theory of the world, leads to the conclusion that properties, relations, and propositions are indispensable. In view of this, ontological economy demands that we try to replace the more complex ontology consisting of both sets and properties with the simpler ontology consisting of just properties. In fact, this ontological simplification can be easily accomplished within our intensional framework. So on grounds of ontological economy, the set-theoretical verison of the sequence approach to quantifying-in should not be adopted. A related problem with the set-theoretical version of the sequence approach is that intensional abstracts containing externally quantifiable variables are not semantically correlated with properties, relations, or propositions; instead, they are semantically correlated with items from an entirely distinct ontological category, namely, sets. Thus, although intensional abstracts that do not contain externally quantifiable variables (e.g., 'that something is red') are semantically correlated with one kind of thing (namely, propositions), intensional abstracts that do contain externally quantifiable variables (e.g., 'that x is red') are semantically correlated with an entirely different kind of thing (namely, sets). This sort of *ad hoc* theoretical disunity is quite unjustified. Our approach to quantifying-in avoids both the ontological excess and the *ad hoc* theoretical disunity of the set-theoretical version of the sequence approach. The upshot is that the set-theoretical version of the sequence approach should not be adopted.

Our fifth and sixth objections to the sequence approach depended on the identification of sequences with extensional entities (namely, ordered sets). What would happen if the sequence theorist tried instead to identify sequences with intensional entities (namely, ordered properties)? In set theory, sequences are identified with ordered sets; for example, $\langle u, v\rangle$ might be identified with $\{\{u\}, \{u, v\}\}$. (That is, $\{x: x = \{y: y = u\} \vee x = \{y: y = u \vee y = v\}\}$.) By analogy, in a property theory, sequences are identified with ordered properties; for example $\langle u, v\rangle$ might be identified with $[x = [y = u]_y \vee x = [y = u \vee y = v]_y]_x$. So far so good. But notice that this property is, on the face of it, a *de re* property; the intensional abstract that is semantically correlated with it contains externally quantifiable variables (namely, 'u' and 'v'). So advocates of the property-theoretic version of the sequence approach to quantifying-in would be committed to identifying this property with a sequence such as:

$$\langle\langle u, v\rangle, [x = [y = u]_y \vee x = [y = u \vee y = v]_y]_{xuv}\rangle.$$

But this outcome is plainly unacceptable because:

$$\langle u, v\rangle \neq \langle\langle u, v\rangle, [x = [y = u]_y \vee x = [y = u \vee y = v]_y]_{xuv}\rangle.$$

And this is only the beginning. With which property would the latter sequence be identified? On the property-theoretic version of the sequence approach, this sequence would have to be identified with still another property that, on the face of it, is *de re*. In this way, the property-theoretic version of the sequence approach leads to a regress. You cannot eliminate *de re* properties in favor of sequences, for those very sequences must, in turn, be identified with further *de re* properties; and so on *ad infinitum*. To put the point linguistically, suppose that you have a language fitted out with just *de dicto* intensional abstracts and an apparatus for expressing the predication (instantiation) relation. In this language you could never, even in principle, identify the sequences (=properties) with which *de re* properties are supposed to be identical.[17]

Summing up, the overall verdict on the sequence approach to quantifying-in is that it is fatally flawed.

There is a final approach to quantifying-in that should be mentioned, namely, the *self-ascription theory* of externally quantified belief sentences. Such an approach has been advocated independently by David Lewis [1979] and by Roderick Chisholm [1981]. (We will confine our discussion to Chisholm; however, our comments will by and large apply to Lewis as well.) Chisholm's approach is not intended to be a general treatment of quantifying-in; rather, it was developed primarily to help solve certain recalcitrant substitutivity puzzles involving indexicals in propositional-attitude sentences. Chisholm imposes a special ontological constraint on his solution to these and other substitutivity puzzles, namely, that a solution should avoid ontological commitment to *de re* properties, relations, and propositions; for Chisholm, a solution should be ontologically committed to what may be called "pure Platonic" properties, relations, and propositions. For this reason, if his theory is adequate, Chisholm ought to be able to extend it to cover *all* 'that'-clause sentences that on the surface seem to contain externally quantifiable variables. We will give two criticisms of the self-ascription theory – one aimed directly against its treatment of *de re* belief sentences and the other intended to show that it cannot be generalized to yield a uniform treatment of 'that'-clause sentences containing externally quantifiable variables.

According to Chisholm, a person x directly believes that Fx if and only if x self-ascribes the property of being F.[18] (In symbols. $A^2x, [Fv]_v$). And x indirectly believes that Fy if and only if, for some relation R, y is the unique item bearing R to x and x self-ascribes the property of being something v such that there is a unique item u bearing R to v and Fu. (In symbols, $(\exists R)(R!y,x$ & $A^2x, [(\exists!u)(Ru, v$ & $Fu)]_v$.) Finally, x believes that Fy if and only if x directly believes that Fy or x indirectly believes that Fy. The first problem is that these three biconditionals do not yield the right results for the belief sentences they are intended to cover. Specifically, the second biconditional (and, in turn, the third biconditional) is far too weak. For example, suppose that 'Ry,x' is 'y = the tallest man & $x = x$'. Suppose x self-ascribes the property of being something v such that there is a unique item u bearing R to v and Fu. That is, x self-ascribes $[(\exists!u)(Au, v \;\&\; Fu)]_v$. The problem is that in a typical situation this would not be sufficient for believing *of* the tallest man that *he* is F. On the contrary, this self-ascription would constitute a run-of-the-mill *de dicto* belief to the effect that the tallest man is F. Without stricter constraints on R, the proposal virtually erases the distinction between genuine *de re* beliefs and run-of-the-mill *de dicto* beliefs. Now there might be ways to impose suitably strict constraints on R *on a case-by-case basis*. However, there seems to be no *systematic* way to impose suitable constraints on R.

The second problem is that the self-ascription approach does not mesh well with a *general* treatment of 'that'-clause sentences. To illustrate the problem, consider the following intuitively valid argument form:

For all y, if x believes y, then . . . y . . . .
x believes that Fx.

∴ . . . that Fx . . . .

In our notation:

$(\forall y)(B^2x,y \rightarrow \ldots y \ldots)$
$B^2x, [Fx]$

∴ . . . $[Fx]$ . . . .

Suppose that ⌜. . . y . . .⌝ is syntactically simple. For example, suppose it is 'y is true', 'y is necessary', 'y is logically true', 'y is probable', 'y is explainable', etc. Then, presumably, self-ascription theorists would then be led to adopt the following representations, respectively: 'x has $[Fy]_y$', 'x necessarily-has $[Fy]_y$', 'x logically-has $[Fy]_y$', 'x probably-has $[Fy]_y$', 'x explainably-has $[Fy]_y$', etc. The idea here is to introduce special new primitive predicates ('has', 'necessarily-has', 'logically-has', etc.) to represent syntactically simple cases of ⌜. . . y . . .⌝. However, this pattern of representation breaks down when ⌜. . . y . . .⌝ is more complex. For example, suppose ⌜. . . y . . .⌝ is 'y implies that Hxz', 'That Hxz explains y', 'y = that Hxz', 'Given the premise that if Hxz then Fx and given the premise that Hxz, then y follows immediately by *modus ponens*, where y=that Fx', etc. There appears to be no systematic way to treat all such cases by straightforward extension of the technique used to represent the syntactically simple cases. It appears, therefore, that the self-ascription theorist has no alternative but to adopt the sequence approach that we discussed a few paragraphs above. (This assessment is fortified by the following consideration. Notice that the three biconditionals that comprise the self-ascription theory do not constitute a *general* analysis (definition) of belief in terms of self-ascription. That is, for an *arbitrary* w, we are not told the conditions under which a person could be said to believe w; we are told merely the conditions under which a person could be said to believe w for certain special w. It appears that the only way to arrive at a *general* analysis would be to utilize the sequence approach.) However, if the sequence approach were adopted, the self-ascription theory would then inherit *all* of the problems inherent in that approach.

Our conclusion, therefore, is that the self-ascription theory is seriously flawed. But this does not mean that the problems that the self-ascription theory was designed to handle (i.e., the explanation of substitutivity failures involving co-denoting indexicals) cannot be solved in some other way. Indeed, they can be solved within the framework for treating quantifying-in that we have advocated. (See Sections 12 and 13 below.) In view of this, it would be a mistake to abandon our treatment of quantifying-in in favor of an alternate treatment inspired by the self-ascription theory.[19]

Our primary conclusion, then, is this: 'that'-clauses are best represented by means of an intensional abstraction operation (such as the bracket notation $\ulcorner[A]\urcorner$) and these intensional abstracts may contain externally quantifiable variables. Now fully analogous considerations lead to the conclusion that gerundive phrases and infinitive phrases likewise are best represented by means of an appropriate intensional abstraction operation (such as our generalized bracket notation $\ulcorner[A]_\alpha\urcorner$) and these abstracts may also contain externally quantifiable variables. Thus, in the formal language for our intensional logic, $\ulcorner[A]_{v_1 \ldots v_m}\urcorner$ will be a well-formed singular term, for any $m \geqslant 0$, even if the formula A contains free variables that are not among the variables $v_1, \ldots, v_m$; such free variables are externally quantifiable.

4. LEARNABILITY

Donald Davidson has argued persuasively that human beings can learn a language only if it contains a finite number of semantical primitives and, hence, that a formal language can serve as an idealized representation of (a fragment of) a human natural language only if it too contains a finite number of semantical primitives.[20]

There has been some confusion about what Davidson's learnability requirement comes to. It does not imply that all learnable languages – and all idealized representations of them – must have a finite number of syntactically primitive constants. This would be too strong. For we humans are able to learn certain specialized languages that have an infinite number of syntactically primitive constants; for example, we could learn a language for arithmetic in which all the numerals are syntactically primitive. But what makes this possible? The explanation, of course, is that each numeral following '0' can be *defined* (e.g., in terms of '0' and '+ 1') and, hence, is not *semantically* primitive. When we generalize on examples of this sort, we arrive at the following highly plausible principle: for any infinitary language L that a human could learn, there must be a finitary language L′ in which all the constants in L (besides those that are already in L′) either could be defined or could be introduced in some comparable manner (for example, by means of Kripkean reference-fixing descriptions). Given this principle, we may infer that, if an infinitary language L is to qualify as an idealized representation of the logical syntax of natural language, there must be an associated finitary language L′ that satisfies the condition just stated. In view of this, the safest and most direct way to insure that a candidate infinitary language L qualifies as an idealized representation of the logical syntax of natural language is just to produce the requisite finitary language L′. Notice, however, that in place of L this finitary language L′ should itself be able to serve as an idealized representation of the logical syntax of natural language. So , in practice, we are entitled to demand from people seeking to construct an idealized representation of the logical syntax of natural language. So, in practice, bill directly with a finitary language like L′. (See Section 11 for further elaboration of this argument.)

The foregoing – or something quite like it – is what Davidson's learnability requirement comes to. And it seems basically right to us. Now the formal language in which intensional logic is formulated should be able to serve as an idealized representation of the logical syntax of the intensional fragment of natural language. Given this fact, together with Davidson's learnability requirement, we conclude that this formal language should have a finite number of primitive constants.

5. REFERENTIAL SEMANTICS FOR INTENSIONAL LANGUAGE

We come now to the question of the semantics for sentences containing 'that'-clauses, gerundive phrases, or infinitive phrases. What, if anything, corresponds semantically to these abstract singular terms? In seeking the answer to this question, we may assume that any adequate semantics either includes an explicit specification of the truth conditions for the sentences of the language or is set up so that these truth conditions can be derived. For simplicity, therefore, let us examine what an explicit specification of truth conditions would have to be like. The crucial issue for us arises in connection with atomic sentences. For example, under what conditions is an atomic sentence of the form $\ulcorner F[A] \urcorner$ true? In a referential semantics, an ordinary atomic sentence $\ulcorner Fa \urcorner$ is true on an interpretation $\mathscr{I}$ iff, according to $\mathscr{I}$, the singular term $\ulcorner a \urcorner$ denotes an item of which the predicate $\ulcorner F \urcorner$ is true. The thesis we will defend is that the truth conditions for atomic sentences containing intensional abstracts must be specified in the analogous way: $\ulcorner F[A] \urcorner$ is true on interpretation $\mathscr{I}$ iff, according to $\mathscr{I}$, the singular term $\ulcorner [A] \urcorner$ denotes an item of which the predicate $\ulcorner F \urcorner$ is true. (In our discussion we will refer to these atomic sentences as *atomic intensional sentences*.)

Our argument will proceed in two steps. First, we will argue that everyone who aspires to an acceptable comprehensive theory of the world must acknowledge the truth of an infinite variety of atomic intensional sentences. Second, we will argue that, besides a referential semantics, there is no viable alternative semantics that will account systematically for the truth of these atomic intensional sentences.

To expedite the first step, let us consider the radical theory that *no* atomic intensional sentences are, strictly speaking, true. This theory is wildly implausible, for true sentences like the following would have to be deemed false: 'It is true that someone has a hand', 'It is true that 7 < 9', 'It is logically valid that everything is self-identical', 'It is probable that the sun will rise', 'That I have sense experiences has an explanation', and so forth. How much more plausible it would be to accept the truth of these sentences and to try to devise a semantics for them. At this stage, one need not make any assumptions about whether it would have to be a referential semantics.

However, suppose that supporters of the radical theory persist. How could they justify their position? Since it is *locally* so implausible, they have no choice but to try to justify it *globally*, that is, by showing that their best comprehensive theory deems all atomic intensional sentences to be false. But if their theory is truly comprehensive, it must among other things be able to account for its own acceptability (justification). We will argue that the radical theorists cannot show this without resorting to intensional idioms and so their position is essentially self-defeating. As a result, it is not acceptable.

How might the radical theorists try to show that their position is acceptable? The *standard* idiom for discussing acceptability (justification) is intensional. For example, it is standard to say: ⌜It is acceptable that *A*⌝, ⌜The theory that *A* is justified⌝, and so forth. (There is also a metalinguistic idiom for discussing acceptability. This will be considered in a moment.) So if the radical theorists are to defend the acceptability of their theory by this standard means, they will have to make various positive assertions with intensional sentences, sentences which they deem to be not true. Specifically, the conclusion of their argument would be (the proposition expressed by) 'It is acceptable that no atomic intensional sentence is true'. But this sentence is itself an atomic intensional sentence. So the radical thesis (i.e., that no atomic intensional sentence is true) implies that this conclusion is not true. But if this conclusion is not true, then it is not acceptable that no atomic intensional sentence is true. Thus, the radical thesis implies that the radical thesis is not acceptable. The radical thesis is, in this sense, self-defeating.[21]

To avoid this self-defeat, the radical theorists might try to invoke some new, *nonstandard* idiom with which to show that their comprehensive theory is acceptable (justified). However, to succeed at this strategy, they must in addition be able to show that this new, nonstandard idiom is *relevantly like* the standard idiom, for otherwise there would be no reason to think that their argument, which uses a new idiom, has any bearing on *acceptability*. After all, *acceptability*, or something relevantly like it, is what is at issue. There can be many similarities between a standard idiom and a new idiom (e.g., length or sound of constituent expressions, etc.); only some of them are *relevant*. Therefore, it is incumbent on the radical theorists to show that their new idiom is relevantly like the standard one.[22]

(As we indicated above, there is also a metalinguistic idiom for discussing acceptability. For example, someone might say, ⌜The sentence ⌜*A*⌝ is acceptable⌝. However, to the extent that it is standard, this metalinguistic idiom bears the following systematic relation to the standard intensional idiom: the sentence ⌜*A*⌝ is acceptable if and only if it is acceptable that *A*. Suppose that this systematic relationship is affirmed by the radical theorists. In this case, they are led to the same sort of self-defeat described above. On the other hand, suppose that this systematic relationship is not affirmed by the radical theorists. In this case, their use of the metalinguistic idiom would, for all anyone could tell, be nonstandard. That is, for all anyone could tell, it might be just some new idiom. Therefore, if the standard systematic relationship is not affirmed by the radical theorists, they would be obliged to show either that their metalinguistic idiom is, despite this, still the standard idiom or that, if it is not the standard idiom, it nevetheless has bearing on acceptability. In either case, they would need to show that this idiom is *relevantly like* the standard idiom. So no real progress has yet been made.)

The conclusion so far is this. To avoid self-defeat, the radical theorists have no choice but to use an idiom that either appears to be or is nonstandard and then to show that this idiom is relevantly like the standard intensional idiom for talking about acceptability. How might the radical theorists try to show that their idiom is relevantly like the standard intensional idiom? There are two ways. One would be to show that the *meanings* of expressions in the new idiom are relevantly like the meanings of expressions in the standard idiom.[23] The other way would be to show that the *purpose* or *function* of the new idiom is relevantly like that of the standard idiom. (Or radical theorists might try to show that the two idioms share something that is relevantly like meaning, or they might try to show that they share something that is relevantly like purpose or function.) But both ways inevitably fail.

Stated briefly, the problem is this. The standard idioms for talking about meaning, purpose, and function are intensional: ⌜⌜*A*⌝ means that *A*⌝; ⌜The purpose of *F*-ing is to *G*⌝, ⌜The function of *F*-ing is to *G*⌝, and so on. So if they use these idioms, the radical theorists once again end up in self-defeat.[24] Moreover, although there are standard extensional idioms for talking about meaning, purpose, and function, they bear systematic relations to the standard intensional idioms for talking about meaning, purpose, and function. (For example, the standard extensional idiom for talking about synonymy bears the following systematic relationship to the standard intensional idiom for talking about meaning: ⌜*A*⌝ is synonymous to ⌜*B*⌝ iff ⌜*A*⌝ and ⌜*B*⌝ mean the same iff that *A* = that *B*.) If the radical theorists affirm these systematic relations, they again end up in self-defeat. If they do not affirm these systematic relationships, then they are obliged to show either that their use of the extensional idiom is standard or, if it is not standard, that it has bearing on meaning, purpose, or function. In either case, they must be able to show that their idiom is relevantly like the standard intensional idiom. Alternatively, the radical theorists could invoke some new, nonstandard idiom for (allegedly) talking about meaning, purpose, or function (or for talking about something that is relevantly like meaning, purpose, or function). But, then, they would once again be obliged to show that their idiom is relevantly like the standard intensional idiom for talking about meaning, purpose, or function. Now how is the required relevant similarity to be shown? Well, by demonstrating that the meaning, purpose, or function of the questionable idiom is relevantly like the meaning, purpose, or function of the standard intensional idiom for talking about meaning, purpose, or function. However, if this demonstration is conducted in the standard intensional idiom, self-defeat results once again. On the other hand, if the demonstration is conducted in the questionable idiom (i.e., an idiom whose relevance to the standard idiom is the very question at issue), this demonstration simply begs the question. For at no stage will it have been established that any conclusion stated in the questionable idiom has any bearing on meaning, purpose, or function (or anything that is relevantly like meaning, purpose, or function).

The overall pattern, then, is this. In the effort to establish the acceptability of their anti-intensionalist theory, the radical theorists get caught either in self-defeat or in begging-the-question. The epistemic situations, if you will, hermeneutical: the standard idioms are intensional, and to show the relevance of a nonstandard idiom, one must use a standard intensional idiom or one must beg the question by using a nonstandard idiom whose relevance is equally in question. There is no epistemically acceptable way to go from where we are to the radical anti-intensionalist theory. And more generally, there is no epistemically acceptable way to make out the possibility of beings who have an acceptable comprehensive theory (or something relevantly like an acceptable comprehensive theory) that includes the radical anti-intensionalist theory.[25]

We have established that every acceptable comprehensive theory of the world must admit a wide variety of atomic intensional sentences as true. In this connection, it would be unacceptable to exclude any part of the standard network of atomic intensional sentences bearing systematic relations to one another; specifically, atomic intensional sentences dealing with acceptability, truth, meaning, purpose, function, definition, intention, belief, causation, explanation, probability, evidence, necessity, and so forth. Given that such a variety of atomic intensional sentences must be counted as true, what semantical theory will account for their truth? As we have indicated, a standard referential semantics provides the most straightforward answer: an atomic intensional sentence $\ulcorner F[A] \urcorner$ is true on an interpretation $\mathscr{I}$ if and only if, according to $\mathscr{I}$, the singular term $\ulcorner [A] \urcorner$ denotes an item of which the predicate $\ulcorner F \urcorner$ is true.

What alternative is there to a standard referential semantics? Evidently, there is only one alternative that is even faintly promising. Namely, the sort of non-referential semantics that *anti*-Meinongian realists often envisage for positive sentences containing ordinary vacuous names, sentences like 'Apollo is a Greek god' and 'Pegasus is a mythical flying horse'. On this theory, such sentences are deemed to be literally true. In this respect, the theory is like Meinong's. However, contrary to Meinong's theory, this theory treats terms such as 'Apollo' and 'Pegasus' as genuinely vacuous. Since these terms refer to nothing, the truth of sentences containing them needs to be explained in some new, nonreferential way. The idea is that, in the semantical description of the truth conditions of these everyday vacuous-name sentences, all purported references to nonactual objects (Apollo, Pegasus, etc.) is to be replaced by references to actual human beings in relevant actual mental states. For example, the truth conditions for 'Apollo is a Greek god' would on this nonreferential approach be characterized in terms of actual religious beliefs (and other mental states) of the ancient Greeks. Now concerning atomic sentences like $\ulcorner F[A] \urcorner$, the proposal would be to characterize their truth conditions along this sort of nonreferential lines.

It is true that ordinary vacuous-name sentences are standardly "backed" by an identifiable body of actual myths, legends, rumors, works of fiction, and so forth. Let us suppose for the sake of argument that the truth conditions for ordinary vacuous-name sentences might be specified in terms of these. (We need not take any stand on whether this sort of nonreferential semantics is really feasible. If it is not feasible even for ordinary vacuous-name sentences, it certainly is not feasible for atomic intensional sentences.) For example, as a first approximation, a nonreferentialist might hold that an ordinary vacuous-name sentence $\ulcorner Fa \urcorner$ is true if and only if it is derivable from a maximal consistent set of sentences extracted from a standard linguistic statement of a community's myths, legends, rumors, and works of fiction. But there is no comparable proposal for atomic intensional sentences. There are at least two decisive reasons.

First, unlike the true atomic vacuous-name sentences, the true atomic intensional sentences are not even recursively enumerable. (Consider sentences of the form $\ulcorner$It is true that $A \urcorner$. $\ulcorner$It is probable that $A \urcorner$, $\ulcorner$It is explainable that $A \urcorner$, $\ulcorner$It is possible that $A \urcorner$, etc.) So there is no body of actual beliefs (and other actual mental states) that could play a role comparable to a community's body of actual myths, legends, rumors, and works of fiction. If there is nothing mental to play this role, the semanticist has no alternative but to posit actual reference to real things.

Second, let us suppose *per impossible* that the true atomic intensional sentences can be fixed on the basis of some relevant body of our beliefs. Which beliefs would these be? They would not be myths, legends, rumors, or works of fiction; rather, they would be straightforward acceptable (justified) theoretical beliefs. For example, as the above argument concerning acceptability indicates, they would include acceptable (justified) beliefs about the acceptability of our overall theory. Such beliefs – unlike myths, legends, etc. – would therefore need to be counted as true in our best overall theory. Now because certain beliefs (myths, legends, etc.) are not true, the nonreferentialist holds that they can "back" the truth of associated vacuous-name sentences without implying thereby that these names are semantically associated with any relevant entity. By contrast, the beliefs that would presumably "back" the truth of our atomic intensional sentences are true. Accordingly, they do imply that there are relevant entities semantically associated with the intensional abstracts occurring in these atomic sentences. After all, the way in which intensional abstracts are used in our acceptable (justified) theorizing is entirely like the way in which standard nonvacuous referring expressions are used in such theorizing. So the nonreferentialists' strategy of likening intensional abstracts to names whose use is sustained by mere myths, legends, rumor, and fiction breaks down. To single out intensional abstracts as vacuous is then nothing but an arbitrary attack. If this were acceptable, it would be equally acceptable to single out any other family of singular terms (e.g., place names, names of people, etc.) as vacuous. And if this were acceptable, it would lead to an absurd form of skepticism that not even our nonreferentialists could tolerate.

Now we submit that, when one surveys alternate ways of characterizing the truth conditions for atomic sentences of the form $\ulcorner F[A] \urcorner$, one will find that they all run into these difficulties or variants of them. If we are right about this, there is no reasonable choice but to give a referential semantics for such sentences.

Given this conclusion, how are we to specify the truth conditions for atomic sentences containing other intensional abstracts, namely, abstracts of the form $\ulcorner [A]_{v_1 \ldots v_n} \urcorner$, for $n \geqslant 1$? Given the conclusions we have just reached, considerations of uniformity support the conclusion that sentences of the form $\ulcorner F[A]_{v_1 \ldots v_n} \urcorner$ are true on an interpretation $\mathscr{I}$ iff, according to $\mathscr{I}$, the singular term $\ulcorner [A]_{v_1 \ldots v_n} \urcorner$ denotes an item of which the predicate $\ulcorner F \urcorner$ is true. Attempts to avoid this conclusion by means of a nonreferential semantics like that considered above only lead to variants of the problems that already undermined that style of semantics. It seems best, therefore, to accept the conclusion that sentences of the form $\ulcorner F[A]_{v_1 \ldots v_n} \urcorner$ have a standard referential semantics.

6. THE DENOTATIONS OF INTENSIONAL ABSTRACTS

Let us now turn to the question of what sorts of things are denoted by (or are semantically correlated with) the singular terms $\ulcorner [A] \urcorner$ and $\ulcorner [A]_{v_1 \ldots v_n} \urcorner$. As we have already indicated, the logically distinctive feature of these terms – and their counterparts in natural language – is that various expressions occurring within them do not obey the substitutivity principle that characterizes extensional logic. For example, neither of the following argument forms is valid:

(V)
$$\frac{\begin{array}{l} G[B] \\ B \leftrightarrow C \end{array}}{G[C]}$$

(VI)
$$\frac{\begin{array}{l} G[B(v_1, \ldots, v_n)]_{v_1 \ldots v_n} \\ (\forall v_1, \ldots, v_n)(B(v_1, \ldots, v_n) \leftrightarrow C(v_1, \ldots, v_n)) \end{array}}{G[C(v_1, \ldots, v_n)]_{v_1 \ldots v_n}}$$

That is, in many arguments having form (V) or (VI), the first two lines are true and the third line is false. Given the conclusion we have reached about the truth conditions for sentences of the form $\ulcorner F[A] \urcorner$ and $\ulcorner F[A]_{v_1 \ldots v_n} \urcorner$ there is only one way in which this pattern of truth values is possible. Consider arguments of form (V). The truth of $\ulcorner G[B] \urcorner$ implies that $\ulcorner [B] \urcorner$ denotes an item of which the predicate $\ulcorner G \urcorner$ is true, and the falsity of $\ulcorner G[C] \urcorner$ implies that $\ulcorner [C] \urcorner$ denotes an item of which the predicate $\ulcorner G \urcorner$ is not true. From these two conclusions it follows that the item denoted by $\ulcorner [B] \urcorner$ and the item denoted by $\ulcorner [C] \urcorner$ must be different. This is so despite the fact that, given the truth of the second line $\ulcorner B \leftrightarrow C \urcorner$, the formulas $\ulcorner B \urcorner$ and $\ulcorner C \urcorner$ are equivalent (in truth value) and, in turn, the items denoted by the terms $\ulcorner [B] \urcorner$ and $\ulcorner [C] \urcorner$ are equivalent (in truth value). Or consider arguments of form (VI). The truth of $\ulcorner G[B(v_1, \ldots, v_n)]_{v_1 \ldots v_n} \urcorner$ implies that $\ulcorner [B(v_1, \ldots, v_n)]_{v_1 \ldots v_n} \urcorner$ denotes an item of which $\ulcorner G \urcorner$ is true, and the falsity of $\ulcorner G[C(v_1, \ldots, v_n)]_{v_1 \ldots v_n} \urcorner$ implies that $\ulcorner [C(v_1, \ldots, v_n)]_{v_1 \ldots v_n} \urcorner$ denotes an item of which $\ulcorner G \urcorner$ is not true. From this it follows that the item denoted by $\ulcorner [B(v_1, \ldots, v_n)]_{v_1 \ldots v_n} \urcorner$ must be different from the item denoted by $\ulcorner [C(v_1, \ldots, v_n)]_{v_1 \ldots v_n} \urcorner$. This is so despite the fact that, given the truth of the second line $\ulcorner(\forall v_1, \ldots, v_n)(B(v_1, \ldots, {}_n) \leftrightarrow C(v_1, \ldots, v_n))\urcorner$, the formulas $\ulcorner B(v_1, \ldots, v_n)\urcorner$ and $\ulcorner C(v_1, \ldots, v_n)\urcorner$ are equivalent (in what they are true of) and, in turn, the items denoted by the terms $\ulcorner[B(v_1, \ldots, v_n)]_{v_1 \ldots v_n}\urcorner$ and $\ulcorner[C(v_1, \ldots, v_n)]_{v_1 \ldots v_n}\urcorner$ are equivalent (in what they are true of).

Thus, to do the semantics for the singular terms $\ulcorner[B]\urcorner$ we need a special category of objects with the following feature: they can be distinct from one another even though in some cases they are equivalent (in truth value). And to do the semantics for the singular terms $\ulcorner[B(v_1, \ldots, v_n)]_{v_1 \ldots v_n}\urcorner$, $n \geqslant 1$, we need a special category of objects with the corresponding feature: they can be distinct from one another even though in some cases they are equivalent (in what they are true of).

Now linguistic entities – sentences and n-ary predicates, respectively – have these special features. And so do certain extralinguistic entities – propositions and n-ary relations (properties if $n = 1$). Should we identify the denotations of the singular terms $\ulcorner[B]\urcorner$ and $\ulcorner[B(v_1, \ldots, v_n)]_{v_1 \ldots v_n}\urcorner$ with linguistic entities or with extralinguistic entities? Nominalists favor the former; conceptualists and realists, the latter. Let us see which theory is better.

7. NOMINALISM

According to the most straightforward version of nominalistic semantics, intensional abstracts denote linguistic expressions. Specifically, 'that'-clauses denote sentences, and infinitive and gerundive phrases denote predicates or open-sentences. The first problem with this sort of theory is that it is extremely counterintuitive. If I see that it is daytime, am I really seeing a sentence? If a prelinguistic child or lower animal knows directly that he is in pain, does he or she know directly a sentence? (If so, how is it possible that he or she should be entirely unfamiliar with the sentence?) If I have an experience of being in pain, do I have an experience of a linguistic predicate? If my dog likes swimming, does he like a predicate? Of course not. Nominalists might reply that this intuitive argument is an instance of the so-called fallacy of incomplete analysis. However, this reply is theoretically weak, for it forces nominalists to hold that the present intuitions cannot be taken at face value. But other things being equal, a theory is superior if it can take relevant intuitions at face value. The traditional realist theory that we advocate permits one to do just this. So, other things being equal, it comes out ahead of the nominalist theory.

Of course, nominalists believe that other things are not equal; specifically, they believe that their ontology is simpler than the traditional realist ontology of properties, relations, and propositions. This belief might be defensible when the debate is restricted to philosophy of mathematics. (An advantage of the argument from intensional logic is that it does not oblige one to take a stand on this issue.) However, the nominalist's belief about the relative simplicity of their ontology is not defensible when the debate is over the semantics for intensional abstracts. The problem here concerns the ontological status of linguistic expressions themselves. Let us explain.

Suppose that our nominalists try to identify linguistic expressions, not with types (e.g., shapes or sound types), but with linguistic tokens or set-theoretical constructs whose ultimate elements are linguistic tokens. Consider sentences of the following form:

It is possible that $F^1 t$.

where ⌜t⌝ is a singular term. In symbols: $P^1[F^1 t]$. Now either ⌜t⌝ has wide scope, or it does not. If it has wide scope, ⌜$P^1[F^1 t]$⌝ is true if and only if there is something z that is identical to t and it is possible that there is something y such that y is identical to z and $F^1 y$. On the other hand, if ⌜t⌝ does not have wide scope, ⌜$P^1[F^1 t]$⌝ is true if and only if it is possible that there is something y such that y is identical to t and $F^1 y$. Therefore, whether or not ⌜t⌝ has wide scope, ⌜$P^1[F^1 t]$⌝ is true only if it is possible that there exists an appropriate item y such that $F^1 y$.[26] Now for the problem. Recall that linguistic tokens are contingent particulars. Indeed, it is possible that there are *no* linguistic tokens at all. (Or it is possible that there are no *relevant* linguistic tokens. The following argument would go through *mutatis mutandis* using this weaker premise.) Accordingly, the following sentence is true:

It is possible that there are no linguistic tokens.

But this sentence is equivalent to the following intuitively true sentence:

It is possible that it is true that there are no linguistic tokens.

This sentence has the form ⌜$P^1[F^1 t]$⌝, where ⌜F^1⌝ is the predicate 'is true' and ⌜t⌝ is the singular term '[$\neg(\exists x)$Token(x)]'. So it follows by the above considerations that, whether or not this singular term has wide scope, this sentence is true only if it is possible that there is an appropriate item y such that y is true. But what could this true item y be? According to the nominalist semantics for intensional abstracts, this true item *y* would be a linguistic expression (or a set built up somehow from linguistic expressions). However, in the envisaged possible circumstance in which there is such an item that is true, there would be *no* linguistic tokens. So if linguistic expressions were identified with linguistic tokens or sets built up somehow from linguistic tokens, then in the envisaged circumstance there would not be any linguistic expression *y*. Therefore, given a nominalist semantics for intensional abstracts, the sentence 'It is possible that it is true that there are no linguistic tokens' would be false. But it is true. Therefore, given the nominalist semantics, it follows that linguistic expressions cannot be identified with linguistic tokens or sets built up somehow from linguistic tokens.[27] (In what follows, we will call this *the problem of necessary existence*. There is of course an analogous problem of eternal existence.)

The only way for our nominalists to get out of this problem of necessary existence is to refrain from identifying linguistic expressions with (items that ontologically depend on contingent) linguistic tokens and, instead, to identify them with shapes or sound types, which are entities that exist necessarily. But shapes and sound types are properties *par excellence*. So the problem of necessary existence (and the analogous problem of eternal existence) is avoided only by invoking the ontology of properties. However, if the ontology of properties is admitted to solve this problem, it would be uneconomical not to make full, systematic use of this ontology in giving the semantics for intensional abstracts. Doing so would lead one simply to drop the nominalistic semantics for intensional abstracts and to adopt instead a straightforward realist semantics.

Indeed, the perversity of the nominalistic semantics can now be brought out with special poignancy. For nominalists who accept the ontology of shapes presumably would hold that the gerund 'being square' denotes, say, the complex shape consisting in order of the shapes 's', 'q', 'u', 'a', 'r', 'e' (or some set-theoretical construct built up from the shapes 's', 'q', 'u', 'a', 'r', 'e') as opposed to simply the shape square. There is no ontological gain in this position, and it is, on its face, incredible.

Now our nominalists might reject the above argument by denying the correctness of the intuitions upon which it is based. However, to press such a counterintuitive position is to press a mere bias. Basing one's theories on a mere bias cannot be acceptable even to the nominalist, for anyone who adopts this way of proceeding loses the ability to refute opponents whose biases favor some other arbitrary (perhaps anti-nominialistic) theory. The only way out of this difficulty is to honor our intuitions as evidence in such controversies. But if intuitions are honored here, consistency demands that they be honored elsewhere. When they are, the nominalist semantics is seen to be inferior. For, as we have seen, intuitions support the argument from necessary existence (or external existence). That argument shows that nominalist semantics is no more economical than a traditional realist semantics. However, the latter semantics, unlike the nominalist semantics, permits us to take at face value our intuitions about the identity of the primary objects of perception, belief, and so on. So, by comparison with the traditional realist theory, the nominalist theory is not acceptable.

We believe that this argument is decisive. However, our positive view can be made more convincing by laying bare the defects in the various specific versions of the nominalist semantics. This is the purpose of the remainder of this section.

According to the most common version of nominalist semantics for intensional abstracts, a 'that'-clause is taken to denote the complement sentence contained within the 'that'-clause itself: for example, the intensional abstract 'that man is a rational animal' is taken to denote its complement sentence 'man is a rational animal'. This nominalist theory has the greatest intuitive appeal in connection with indirect discourse. On the simplest version of the theory, the verb 'say' of indirect discourse is just identified with the verb 'say' of direct discourse. Thus,

(1) Seneca said that man is a rational animal.

is taken to be equivalent to

(2) Seneca said 'man is a rational animal'.

However, this clearly is wrong. Whereas (1) is true, (2) is false: Seneca never spoke English.

This difficulty can be overcome by giving the 'say' of indirect discourse a more sophisticated analysis. For example, Carnap[28] would have analyzed (1) as follows:

(3) There is a language such that Seneca wrote as a sentence of L words whose translation from L into English is 'Man is a rational animal'.

However, sophisticated analyses like this are beset with fatal flaws of their own. Consider, first, Alonzo Church's famous criticism:

> For it is not even possible to infer (1) as a consequence of (3), on logical grounds alone – but only by making use of the item of factual information, not contained in (3), that 'Man is a rational animal' means in English that man is a rational animal.
>
> Following a suggestion of Langford we may bring out more sharply the inadequacy of (3) as an analysis of (1) by translating into another language, say German, and observing that the two translated statements would obviously convey different meanings to a German (whom we may suppose to have no knowledge of English). The German translation of (1) is (1′) *Seneca hat gesagt, dass der Mensch ein vernünftiges Tier sei.* In translating (3), of course 'English' must be translated as 'Englisch' (not as 'Deutsch') and '"Man is a rational animal"' must be translated as '"Man is a rational animal"' (not as '"*Der Mensch ist ein vernünftiges Tier*"').[29]

Another difficulty with the more sophisticated nominalist analysis is that it does not carry over to belief sentences in the way Carnap hoped. Carnap proposed to analyze belief behaviorally in terms of dispositions to assent. However, standard criticisms of behaviorism show that this kind of analysis is mistaken. Dispositions to assent are not correlated with beliefs taken singly; instead they are correlated with the body of a person's beliefs and desires. As a result, they cannot be used to analyze any single belief.

A related problem with the nominalist analysis is that it fails to mesh with a general theory in which 'that'-clauses and other intensional abstracts are treated as singular terms and in which 'says', 'believes', 'perceives', etc. are treated as standard two-place predicates that take 'that'-clauses as arguments and 'is necessary', 'is possible', 'is true', etc. are treated as standard one-place predicates that take 'that'-clause as arguments. This problem is dramatized by the fact that the analysis provides no clue about how to identify what it is that 'that'-clauses actually denote.

A final difficulty with this nominalist analysis is hidden in its use of the phrase 'as a sentence of L'. This restriction is needed; for, without it, the speaker (Seneca in the present example) could utter the sentence without any of the relevant linquistic intentions. For example, the speaker might utter the sentence merely *as* a pleasant sound; in this case, the speaker would not even have made a statement. Or the speaker might utter the sentence *as* a sentence of some phonologically equivalent but semantically different language; in this case, the speaker would not have made the relevant statement (i.e., that man is a rational animal). So the qualifying phrase 'as a sentence of L' is needed. The problem for our nominalist is that this phrase is a covert intensional qualifer with something like the following force: x utters A as a sentence of L iff x utters A and x intends to speak L when *x* utters *A*. However, as we have seen, an infinitive phrase such as ⌜to speak L when *x* utters *A*⌝ is an intensional abstract. So, by employing the phrase 'as a sentence of L', the sophisticated nominalist analysis only sweeps this inherent intensional aspect of indirect discourse under the rug.

Specific difficulties like these spell defeat for all natural versions of the nominalist theory. Nevertheless, there are some quite unnatural versions of nominalism that avoid these difficulties. But they all run into special new difficulties of their own.

According to one of these unnatural versions of nominalism, the denotation of a 'that'-clause is identified with the equivalence class of all sentences synonymous to the complement sentence contained within the 'that'-clause, and the denotation of a gerundive or infinitive phrase is identified with the class of all predicates (or open-sentences) synonymous to the predicate (open-sentence) that generates the gerundive or infinitive phrase. (So, for example, 'that man is a rational animal' would denote the class {*S*: for some actual language L, the sentence *S* in L is synonymous to 'man is a rational animal' in English}, and the gerund 'swimming' would denote the equivalence class {*F*: for some actual language L, predicate *F* in L is synonymous to 'swim' in English}.) On this approach, such equivalence classes of synonyms are then identified as the primary bearers of truth, necessity, logical truth, probability, etc. and as primary objects of perception, belief, desire, moral obligation, explanation, etc.

The first problem with this sort of nominalist theory is that it too is extremely counterintuitive. If I see that it is daytime, am I really seeing a set of sentences? If a prelinguistic child or lower animal knows directly that he is in pain, does he know directly a set of sentences? (If so, how is it possible that he or she should be entirely unfamiliar with every single sentence in the set?) If I have an experience of being in pain, do I have an experience of a set of predicates? If my dog likes swimming, does he like a set of predicates?

Another problem with the present nominalist theory is that it does not mesh with a satisfactory general explanation of how cognitive states succeed in *representing* things, in being *about* things. To dramatize this point, let us consider a hypothetical situation in which no one ever speaks any of the languages that, as a matter of fact, we actually speak. In such a situation, however, people still would be able to have a wide range of cognitive states, states whose objects in many cases would be the same as objects of our cognitive states. For example, in the indicated situation someone could believe *that someone feels pain*. But if the present nominalist theory were correct, the object of such a person's belief (namely, the object denoted by the 'that'-clause 'that someone feels pain') would be a class of sentences belonging to languages *we* happen actually to speak. Accordingly, the object of such a person's belief would be a class of shapes and/or sounds having nothing to do with the person (or anyone else in the hypothetical situation) and having no relevant relation to what the person's belief is about, namely, pain. On the present nominalist theory, therefore, it would be competely fortuitous that in the hypothetical situation the person's belief is *about* pain rather than some other arbitrary item.

To avoid this outcome, why not allow 'that'-clauses to denote sets of synonyms belonging to *possible*, as well as actual, languages? That is, why not identify the denotation of ⌜that A⌝ with the class $\{S$: for some possible language L, the sentence S in L is synonymous to A in English$\}$? The answer is that *all* 'that'-clauses would, wrongly, turn out to be co-denoting. After all, for *every* sentence S, there is some possible language L such that S in L is synonymous to A in English. A similar problem confronts the slightly more sophisticated nominalist semantics in which the denotation of ⌜that A⌝ is identified with the class of all possible synonym/language pairs, i.e., the class $\{S$, L: S in L is synonymous with the sentence A in English, where S is some sentence in some possible language L$\}$. For if one follows the standard extensionalist practice of identifying a language L with an ordered-pair consisting of a set of well-formed expressions and a function that assigns extensional semantical values to those expressions, then on the present more sophisticated nominalist semantics, the extensional semantical value of various intuitively non-codenoting 'that'-clauses would turn out to be the *same* set of possible synonym/language pairs.[30] Another alternative would be to identify the denotation of ⌜that A⌝ with a function that assigns to each possible world w a class $\{S$: S in L is synonymous to A in English, where L is some language that is spoken in $w\}$. However, this theory would not be acceptable to the nominalists inasmuch as it relies on an ontology of possible worlds. (For a critique of possible-worlds semantics, see Section 9 which deals with nontraditional realist semantics.)

Another problem with the equivalence-class approach is that it employs the predicate 'is synonymous' in the metalanguage. But what is the status of this predicate? According to our best theory, synonomy is to be defined in a broadly Gricean way in terms of certain complex conventional intentions of speakers. In the specification of these intentions, however, we would use 'that'-clauses. Thus, in the statement of our metatheory, we would identify the nominalists' equivalence classes in terms of certain speaker intentions that are identified by means of 'that'-clauses. So far, then, one does not end up with a purely nominalist specification of the denotation of intensional abstracts: the specification of the denotation of intensional abstracts in the object language involves the use of intensional abstracts in the metalanguage. But given their view of things, one would think that nominalists would be able to state their position without this recourse to explicit intensionality in the metalanguage. For the point of the nominalist semantics is to have syntactic entities take the place of traditional intensional entities (properties, relations, and propositions), and it should be possible to give a purely extensional description of these syntactic entities and of the key relations (e.g., synonymy) holding among them. Of course, nominalists might try to achieve such a description by insisting that 'is synonymous' is undefinable. But this claim would contradict our best theory of synonymy.

A way of trying to circumvent this difficulty is to try to define synonymy within the framework of a "language-of-thought" treatment of the propositional attitudes. The idea would be to define synonymy in a broadly Gricean way in terms of the propositional attitudes and then to identify the objects of the propositional attitudes with sentences in an ideal language-of-thought. But if one adopts this approach to synonymy, one is forced to give up the equivalence-class-of-synonyms semantics for intensional abstracts. For, as we have seen, the objects of the propositional attitudes are paradigmatic examples of the sort of items denoted by intensional abstracts, and on the language-of-thought theory these items are sentences in an ideal language-of-thought, not equivalence classes of synonyms in natural languages. So even if the language-of-thought theory were successful, it would be of no help to the equivalence-class theory.

Let us now examine this sort of language-of-thought semantics for intensional abstracts. According to the most straightforward formulation of this theory, there is a single universal ideal language that underlies all possible cognition and all possible natural languages, and intensional abstracts denote expressions in this ideal language. We have seen that the denotata of intensional abstracts are the primary bearers of truth, necessity, possibility, definition, probability, etc. and are the primary objects of belief, perception, hope, moral obligation, explanation, causation, etc. On the language-of-thought theory, therefore, it follows that expressions in this ideal language are the primary bearers of truth, necessity, possibility, definition, probability, etc. and are the primary objects of belief, perception, hope, moral obligation, explanation, causation, etc.

Like the previous nominalist semantics, this one is extremely counter-intuitive. When I see that it is daytime, do I really see a sentence in some ideal language? When I have an awareness of being in pain, do I have an experience of some hypothetical linguistic shape or sound? Certaintly not. As with the previous nominalist theories, the present one appears to be driven by a mere bias for nominalism. Moreover, as with all nominalist approaches, the interface between the sensation of phenomenal qualities (e.g., the quality of being in pain) and the cognition of phenomenal qualities is a "representationalist mystery" on the language-of-thought theory. The straightforward way to solve this mystery is with a full-fledged realist theory in which the objects of the propositional attitudes (i.e., propositions) are built up, by means of fundamental logical operations, from basic properties (including, in particular, phenomenal qualities), basic relations, and perhaps subjects of singular predications.[31]

A further difficulty with this sort of nominalist semantics is that, like the previous one, it too fails to mesh with a satisfactory general explanation of how cognitive states succeed in representing things, in being about things. The problem here is that the radical language-of-thought hypothesis suffers from the following form of radical arbitrariness: for any candidate universal ideal language L, there are an infinite number of alternative languages L′ that could serve the same theoretical roles attributed to L. Which is *the* right one? Which one provides *the* genuine primary bearers of truth, necessity, etc. and the genuine objects of belief, perception, hope, etc.? The choice is in principle utterly arbitrary. Accordingly, there is no general philosophical explanation of why some linguistic shape or sound S should be taken to represent – or to be about – one thing rather than another. No satisfactory theory can tolerate this degree of arbitrariness.

To avoid this problem of arbitrariness, the language-of-thought theorist could adopt an analysis reminiscent of Carnap's: x believes that A iff x believes a sentence that plays a causal role for x that is analogous to the causal role that the English sentence $\ulcorner A \urcorner$ plays for English speakers. However, this analysis falls prey to difficulties rather like those confronting Carnap's. First, it fails to pass the Langford–Church translation test. Second, because of the phenomenon of fine-grained intensionality, belief cannot be analyzed functionally in terms of the notion of causal role; causal role is simply too coarse a criterion for the identity of belief. (See Bealer [1984].) Finally, this analysis fails to mesh with a general theory in which 'that'-clauses and other intensional abstracts are treated as singular terms and in which 'says', 'believes', 'perceives', etc. are treated as standard two-place predicates that take 'that'-clauses as arguments and 'is necessary', 'is possible', 'is true', etc. are treated as standard one-place predicates that take 'that'-clauses as arguments. This problem is dramatized by the fact that the analysis provides no clue about how to say systematically what it is that 'that'-clauses actually denote. As a result, this analysis suggests no general treatment of sentences in which *non-psychological* predicates take 'that'-clauses as arguments, for example: ⌜It is true that A⌝, ⌜It is possible that A⌝, ⌜It is probable that A⌝, ⌜It is explainable that A⌝, and so forth.

Indeed, a common failing of language-of-thought theories is that they usually disregard the role of intensional abstracts in non-psychological contexts. (For example, many language-of-thought theorists believe that *fine-grained* intensionality arises only in connection with the propositional attitudes. But in fact this phenomenon also arises in connection with familiar logical relations such as following-by-*modus-ponens*. Witness the sentence 'Given the premise that if A then B and given the premise that A, the conclusion that B follows by *modus ponens*'.) What is needed is a unified theory of intensional language, not just a theory that treats intensionality in propositional-attitude sentences. Some language-of-thought theorists might try to respond by claiming that all intensionality (truth, necessity, logical truth, probability, counterfactuality, explainability, etc.) that is not psychological is somehow derivative. However, this kind of metaphysical idealism is a well-known dead-end. (For some of the problems with this kind of metaphysical idealism, see the next section, which deals with conceptualism.)

Another way language-of-thought theorists might try to avoid the problem of arbitrariness is to adopt a theory that is reminiscent of the equivalence-class-of-synonyms theory. Specifically, the intensional abstract ⌜$[A]$⌝ would denote the class $\{S, x$: S is a sentence that has a causal role for an actual cognitive agent x that is analogous to the causal role the sentence ⌜A⌝ has for us$\}$. But this theory runs into the same sort of difficulty that originally plagued the equivalence-class-of-synonyms theory. Namely, it makes it a mystery how, in hypothetical circumstances with altogether different cognitive agents, the mental states of those cognitive agents succeed in representing, or in being about, anything. For if these cognitive agents believe, say, that someone feels pain, the object of the belief would be a set of items having no relevant relation to these cognitive agents. Moreover, modifications of the present proposal that invoke possibility in one way or another fall prey to difficulties quite like those that beset analogous modifications of the equivalence-class-of-synonyms approach.

Before we proceed to a final version of nominalistic semantics, a remark about the merits of the language-of-thought hypothesis in cognitive science is in order. We have seen that a language-of-thought semantics for intensional abstracts is not viable, and we shall soon see that only a traditional realist semantics is defensible. Moreover, relative to the algebraic style of realist semantics that we will present, intensional logic can be given a highly fine-grained formulation, a formulation in which propositions may be treated as entities upon which computations are performed directly, without any linguistic mediation. At the same time, the problem of representationalism, which causes so much trouble for the language-of-thought hypothesis, is solved automatically by our traditional realist theory.[32] In view of these results, why invoke any form of the language-of-thought hypothesis in cognitive science? (For example, some cognitive scientists now advocate treating 'believes' as a three-place relation holding among a cognitive agent, a proposition, and a sentence.) The answer is that there is no good reason whatsoever. On a suitable formulation of fine-grained intensional logic, this residual nominalistic element is entirely extraneous, a mere throw-back to a defunct nominalistic semantics.

There is a final kind of nominalist approach to intensional abstracts that we should mention, namely, the approach of Israel Scheffler [1954]. According to this approach, a singular term $\ulcorner[A]\urcorner$ would be contextually analyzed as follows:

$$\ldots [A] \ldots \text{ iff}_{\text{df}}\ (\exists v_k)(v_k \text{ is-an-}A\text{-inscription} \ \&\ \ldots v_k \ldots)$$

where $\ulcorner$is-an-A-inscription$\urcorner$ is an undefined primitive predicate. On the intended interpretation, this predicate is satisfied by all and only inscriptions synonymous to $\ulcorner A \urcorner$. However, given that, for an infinite number of sentences $\ulcorner A \urcorner$, the sentence $\ulcorner[A] = [A]\urcorner$ is logically valid, this theory implies the actual existence of inscriptions (tokens) of every sentence. But, in fact, there are infinitely many sentences $\ulcorner A \urcorner$ of which there are no actual inscriptions (tokens). Moreover, since there are an infinite number of distinct 'that'-clauses in natural language, Scheffler's approach requires an infinite number of undefined primitive predicates $\ulcorner$is-an-A-inscription$\urcorner$. This fact amounts to a violation of Davidson's learnability requirement. Furthermore, it seems to block the systematization of the internal logic of 'that'-clauses. Finally, as we explained earlier, the need to use the predicate 'is synonymous to' in the metalanguage is inconsistent with the nominalistic point of view. For these reasons, Scheffler's approach does not help to save nominalism.

The above considerations, together with a number of others, lead us to conclude that linguistic expressions, whether types or tokens, are not the sort of entity denoted by intensional abstracts $\ulcorner[A]\urcorner$ and $\ulcorner[A]_{v_1 \ldots v_n}\urcorner$. And the same conclusion goes for sequences or sets of linguistic entities, or indeed any other kind of object that is linguistic in character.

So what sort of entities are denoted by $\ulcorner[A]\urcorner$ and $\ulcorner[A]_{v_1 \ldots v_n}\urcorner$? Given the failure of nominalism, we are left with realism and conceptualism.

8. CONCEPTUALISM

According to both realism and conceptualism, when we use 'that'-clauses and gerundive and infinitive phrases, we denote extralinguistic intensional entities. The difference between realism and conceptualism concerns the ontological character of these entities. Realists hold that they are mind-independent entities whereas conceptualists hold that they are mind-dependent. Mind-dependent in the sense that they depend for their existence on minds or mental activity; they would not exist if there were no minds or mental activity. Contemporary realists tend to call these intensional entities 'properties', 'relations', and 'propositions' (depending on their degree). By contrast, conceptualists usually call them 'concepts' and 'thoughts' (depending on their degree). But this difference is largely terminological. The real difference between conceptualism and realism lies in the alleged ontological status of these intensional entities. Are they mind-dependent or mind-independent?

In our discussion of conceptualism we will confine ourselves to the version that ascribes to intensions an ontological dependence on contingent, finite minds like ours. There is another version of conceptualism, however. On this version, even though intensions are ontologically dependent on mind, they nonetheless exist necessarily; for they exist necessarily in the infinite, necessary mind of God. We will take no stand on this version of conceptualism. Our reason is that, like realism, it implies that intensional entities exist necessarily, and this is what matters most to contemporary realists.

On the most plausible version of conceptualism, there are certain basic intensions (much like Locke's simple ideas) that are simply "given" in ordinary mental activity, and all other intensions are somehow "formed" or "constructed" out of these nonconstructed intensions. At relevant points in our critical assessment, we will focus on this version of conceptualism.

The first difficulty with conceptualism is this. Evidently, there are intensions that have never been "given" in anyone's mental activity and that could not, even in principle, be "formed" or "constructed" from intensions that have been "given". For example, various fundamental physical properties (e.g., quark-theoretic properties such as the property denoted by the intensional abstract 'having spin up') seem to be like this: they appear not to be "constructible" in any way from "given" intensions; rather, they appear to be mere theoretical posits that we can at best describe.[33] Indeed, many physicists believe that there still exist fundamental physical properties and relations (e.g., sub-quark properties, sub-sub-quark properties, etc.) that remain to be described theoretically.

In a related vein, there are no doubt primitive phenomenal qualities that no one has ever experienced (e.g., new shades, fragrances, or tastes) and that, in principle, could not be "constructed" from intensions that have already been "given". Indeed, the taste of pineapple (i.e., the familiar phenomenal quality we know in sensation) once had this ontological status, for there was a time when no one had ever tasted it.

This last example gives rise to a general defect in conceptualism that should have been predictable; namely, conceptualism falls prey to the argument from necessary existence (and also to the analogous argument from eternal existence). (This style of argument was used in the previous section to refute nominalist semantics for intensional abstracts.) Consider sentences of the following form:

It is possible that $F^1 t$.

where $\ulcorner t \urcorner$ is a singular term. In symbols: $P^1[F^1 t]$. We saw earlier that, whether or not $\ulcorner t \urcorner$ has wide scope, $\ulcorner P^1[F^1 t] \urcorner$ is true only if it is possible that there is an appropriate item y such that $F^1 y$. Now, intuitively, the following sentence is true:

It is possible that it is true that there are no finite minds.

This sentence has the form $\ulcorner P^1[F^1 t] \urcorner$, where $\ulcorner F^1 \urcorner$ is the predicate 'is true' and $\ulcorner t \urcorner$ is the singular term '$[\neg(\exists x)$ Finite-mind $(x)]$'. It follows that, whether or not this singular term has wide scope, this sentence is true only if it is possible that there is an appropriate item y such that y is true. But what could this true item y be? According to conceptualists, y would be an intensional entity that is ontologically dependent on finite minds. However, in the envisaged circumstance in which there is an intension y that is true, there would be *no* finite minds. Therefore, in the envisaged circumstance there would not be *any* intensions, and so the sentence 'It is possible that it is true that there are no finite minds' would be false. But it is true. So conceptualism must be mistaken: it cannot overcome the problem of necessary existence. (Some conceptualists might try to escape this conclusion by a "modalizing strategy." We will consider this strategy in a moment.)

We believe that the foregoing intuitive considerations tell decisively against conceptualism. However, to remove lingering doubts, we move on to a more theoretical line of argument. The problem concerns the infinite. Intuitively, there are infinitely many distinct fine-grained intensions. For example, there are infinitely many nontrivial logical truths: that A_1, that A_2, that A_3, ... (To see this, suppose that the sentence $\ulcorner A_i \urcorner$ expresses the nontrivial logical truth that A_i. Suppose that $\ulcorner A_i \urcorner$ is not in prenex normal form, and suppose that $\ulcorner A_j \urcorner$ is the result of converting $\ulcorner A_i \urcorner$ to prenex normal form. Let x = that A_i, and y = that A_j. Intuitively, it is possible that someone is consciously and explicitly thinking x and not consciously and explicitly thinking y. If so, that $A_i \neq$ that A_j.) The problem facing conceptualists is to explain why there seem to be infinitely many intensions. They are not "given" in anyone's actual mental activity, and we do not actually "construct" them. For doing so would require infinitely many acts of "construction," and our finiteness excludes this. Conceptualists have two ways of trying to solve this problem. The first is to grant that there actually exist infinitely many intensions and to identify intensions that are not "given" in actual mental activity with a certain kind of "extensional complex" (e.g., finite sequences, ordered sets, or abstract trees) whose ultimate elements are intensions that are "given" in actual mental activity. The other strategy is to deny that there actually exist infinitely many intensions and to explain why there seem to be by exploiting the distinction between intensions that have actually been "constructed" and possible acts of "construction". The latter strategy is the *modalizing* strategy.

The first strategy is defeated by considerations of ontological economy. For, on this treatment of intensional entities, conceptualists would have to posit *two* fundamentally dissimilar ontological categories – extensional complexes (finite sequences, ordered sets, abstract trees) and primitive intensional entities (namely those "given" in actual mental activity). Realists, by contrast, need only one ontological category, namely, that of intensions. One ontological category suffices for realists because the theoretical work that can be accomplished with the conceptualists' extensional complexes can be accomplished by appropriate realist intensions. For example, the theory of finite sequences, finite ordered sets, and finite abstract trees can be represented within first-order logic with identity and intensional abstraction. (E.g., the jobs done by the finite sequence $\langle v_1, \ldots, v_n \rangle$ can be done by the realists' intension $[u_1 = v_1 \& \ldots \& u_n = v_n]_{u_1 \ldots u_n}$.) And if the conceptualists' theory of extensional complexes is supplemented with a set-membership relation, the realists' theory of intensions may be supplemented with a predication (instantiation) relation. Any theoretical task that can be performed by the conceptualists' theory of membership can then be performed by this realist theory of predication. The upshot is that the conceptualist theory can perform no theoretical task that the realist theory cannot perform. At the same time, the conceptualist theory is in principle ontologically more complex than the realist theory, for it requires two fundamentally dissimilar ontological categories whereas the realist theory requires only one. So the conceptualist theory should be rejected on ontological grounds.

In response, someone might wonder whether the conceptualists' two categories (extensional complexes and actually "given" primitive mind-dependent intensions) really are fundamentally dissimilar. To dramatize the fact that they are, recall that on such a theory there would need to be extensional complexes whose elements would not even be intensions (for example, ordered sets whose elements are physical objects).

Another response to our argument would be to hold that the present version of conceptualism is not really ontologically excessive, for the posited extensional complexes can be eliminated in favor of items that everybody (including realists) already accepts. But which items could they possibly be? In debates about foundations of mathematics the usual candidates put forward at this juncture are linguistic entities. But here we encounter the power of the argument from intensional logic. Perhaps linguistic entities can play the role of sets or other extensional complexes in certain formulations of the foundations of mathematics. However, in the present context, the conceptualists' purpose for introducing extensional complexes is to provide a realm of entities to serve as the denotata of intensional abstracts. But in our critique of nominalism we saw that linguistic entities are wholly inadequate for this purpose. So this escape route is not available to conceptualists.

We will mention three other defects in the present version of conceptualism. First, it is highly unintuitive that ordered sets, sequences, or abstract trees are really the sort of thing that are perceived, believed, etc.

or that are true, necessary, probable, explainable, etc. Advocates of this theory certainly have lost the "naive eye". Second, it is *prima facie* implausible that some intensional abstracts should denote one category of entity (i.e., primitive intensions "given" in our actual mental activity) and that other intensional abstracts should denote ontologically very different sorts of entities (i.e., ordered sets or sequences). Third, by identifying the denotata of infinitely many intensional abstracts with extensional complexes, conceptualists might run into a potentially fatal difficulty in connection with "self-embeddable" intensions. (This general issue shall be discussed three sections below.)

Our overall conclusion, then, is that the first strategy is of no help to conceptualists. This outcome leaves them with no alternative but to try the "modalizing" strategy. The idea behind this strategy is to deny that there are really an infinite number of actual intensions (e.g., the nontrivial logical truths that A_1, that A_2, that A_3, . . .) and to hold instead that there are merely an infinite number of possible ways of thinking (which, if actualized, would generate the intensions that A_1, that A_2, that A_3, . . .). Our reply to the modalizing strategy will be that it does not really avoid ontological commitment to an infinity of actual intensions. To explain this reply, we must spell out the modalizing strategy more fully.

Consider intuitively true sentences of the form ⌜It is logically true that A_i⌝. We have argued that each such sentence is ontologically committed to an intensional entity (i.e., the intension that A_i). According to the modalizing strategy, however, the sentence ⌜It is logically true that A_i⌝ is in most cases not strictly speaking true; what is true is an associated modal sentence something like the following: ⌜It is possible that someone should form the thought that A_i and the resulting thought would be logically true⌝. (On a somewhat related form of the modalizing strategy, although ⌜It is logically true that A_i⌝ would be counted as true, it would be treated as a mere abbreviation of the modal sentence ⌜It is possible that someone should form the thought that A_i and the resulting thought would be logically true⌝.)

The fundamental shortcoming of the modalizing strategy is that it does not really address the problem it was supposed to solve. The problem was to find a way to avoid commitment to an actual infinity of intensional entities. However, the proposal still leaves us with an actual infinitude of such entities, namely, those denoted by the 'that'-clause occurring in the proposed modal sentence ⌜It is possible *that someone forms the thought that A_i and this thought is logically true*⌝. (In symbols, ⌜$P^1[(\exists x)(F^2x, [A_i]$ & $LT^1[A_i])]$⌝.) After all, as we showed in earlier sections, the best systematic treatment of intensionality is by means of intensional abstracts and accompanying auxiliary predicates. Just as ⌜*x* believes that *B*⌝ is represented as ⌜B^2x, [*B*]⌝, ⌜It is possible that *B*⌝ is represented as ⌜P^1[*B*]⌝. The modalizing strategy would require a systematic way of capturing the relevant possibilities of forming thoughts. The way to do this is by means of *further* intensional abstracts, ones that generate their own commitment to an actual infinitude of intensional entities. So the modalizing strategy does not work.

One way of trying to avoid this conclusion is by resorting to primitive operators that are designed to avoid use of the offending intensional abstracts. For example, instead of putting forward the above intensional-abstract sentence, modalizers would put forward the following primitive-operator sentence: ⌜Possibly someone forms the thought that A_i and this thought is logically true⌝. (In symbols, ⌜$\Diamond(\exists x)(F^2x, [A_i] \ \& \ LT^1[A_i])$⌝.) The alleged gain is that the offending intensional abstract does not explicitly occur in this new primitive-operator sentence. But the argument from intensional logic undercuts this move.[34]

There are two sorts of reasons. First, we have already established that ⌜It is possible that *B*⌝ is to be represented as ⌜P^1[*B*]⌝. However, it is intuitively obvious that it is possible that *B* iff possibly *B*. It would be entirely *ad hoc* to deny this obvious equivalence just to save conceptualism. So, on intuitive grounds, the primitive-operator move cannot be used to side-step the ontological commitment to the implicit intensional entity (i.e., that *B*). Second, we have seen that great theoretical economy can be gained by treating commonplace operator sentences as derived forms of intensional-abstract sentences. For example, by treating ⌜$\Diamond B$⌝ as a derived form of ⌜P^1[*B*]⌝. Since the latter form is already required by an acceptable *general* formulation of intensional logic (e.g., one that can represent general relations between belief and possibility), it would be highly unjustified theoretically to insist on representing ⌜$\Diamond B$⌝ as a primitive-operator sentence, rather than to bring it within a unified, general theory of intensional logic.[35]

Now conceptualists might try to avoid our critique of the modalizing strategy by resorting to other primitive operators (e.g., constructibility operators, constructibility quantifiers, etc.). However, these alternate linguistic forms create the same problems for conceptualism. First, they are intuitively equivalent to linguistic forms involving intensional abstracts and accompanying predicates. Second, these primitive-operator sentences typically generate intensional contexts. Therefore, the canonical representation of them is provided by means of intensional abstracts and accompanying predicates. This way these intensional contexts can be dealt with within a unified, general theory. It is inevitable, therefore, that the modalizing strategy does not successfully avoid the commitment to an actual infinity of intensional entities.

Notice that the foregoing critique of conceptualism did not get us involved in many of the usual worries that characterize contemporary discussions of conceptualism and realism, for example, worries about the law of the excluded middle, impredicativity, and so forth. This is a significant advantage of the argument from intensional logic. Considerations in intensional logic (notably, the treatment of generality and intensional abstraction) just on their own force one to posit an actual infinitude of intensions. When conceptualists try to give an alternate explanation, either they end up positing two fundamentally dissimilar ontological categories (mind-independent extensional complexes and mind-dependent intensions) whereas one ontological category (mind-independent intensions) suffices for realism. Or they offer a modalizing strategy that, upon closer examination, implies the very sort of infinite intensional ontology that it is designed to avoid.

For these and the other reasons we listed, our overall conclusion is that, in comparison with realism, conceptualism is not acceptable.

9. REALISM

Given the failure of nominalism and conceptualism, we are left with realism. According to traditional realism, $\ulcorner[A]\urcorner$ would denote the proposition that A; $\ulcorner[A]_v\urcorner$ would denote the property of being something v_1 such that A; and $\ulcorner[A]_{v_1 \ldots v_m}\urcorner$ would denote the relation holding among $v_1 \ldots v_m$ such that A. There are, however, non-traditional forms of realism according to which PRPs are replaced by or reduced to other sorts of mind-independent extralinguistic entities. For example, according to the possible-worlds approach, propositions are reduced to functions from possible worlds to truth values; properties are reduced to functions from possible worlds to sets of possible individuals; and m-ary relations are reduced to functions from possible worlds to sets of ordered m-tuples of possible individuals. And according to the original version of the Perry–Barwise situation semantics, although properties and relations are taken as primitive traditional realist entities, propositions (or situations) are reduced to ordered sets of primitive properties, primitive relations, and actual individuals, or they are reduced to set-theoretical compounds of such ordered-sets. [On another version of situation semantics, propositions (situations) are not reduced to such set-theoretical constructs. Instead, these constructs are used only for *model* propositions (situations). In the final analysis, propositions (situations) are to be taken as primitive, irreducible entities. Int his vein, Barwise and Perry now seem attracted to a traditional realist theory of properties, relations, and propositions much like that we have defended here and in previous work. At this stage in the history of the subject, calling such a theory 'situation theory' risks terminological confusion; it is so similar to traditional PRP theory. In the ensuing remarks, we address ourselves only to the earlier, reductionistic version of situation semantics. We wish to emphasize that Perry and Barwise no longer hold this theory.] Scott Soames and Nathan Salmon have recently advocated reducing propositions to ordered sets of primitive properties, primitive relations, and real individuals.[36] Finally, the theory developed by Max Cresswell in *Structured Meanings* is a special hybrid reductionistic approach that combines a sequence treatment of propositions and possible-worlds reductionism regarding properties and relations.

These non-traditional forms of realism, however, have several defects not found in our traditional realism. For example, some of them often identify the denotations of intensional abstracts with items whose identity conditions are not right. Possible-worlds semantics provides the most notorious case of this, for on this approach necessarily equivalent PRPs turn out to be identical. The original formulation of situation semantics has a number of equally damaging consequences.[37] Although consequences like this can be tolerated in some parts of intensional logic such as modal logic, they are quite unacceptable in other parts, notably, those dealing with intentional matters. To compensate for this defect, some people (e.g., David Lewis and Max Cresswell) propose to reduce only "coarse-grained" PRPs to sets of possibilia and, then, to reduce "fine-grained" PRPs to sequences of these reduced coarse-grained PRPs. But this revised semantics turns out to be quite flawed (see below); moreover, it turns out to be more complicated technically than our realist semantics which treats coarse-grained and fine-grained PRPs as irreducible primitive entities.

Another difficulty with the reductionistic approaches concerns "self-embeddable" PRPs. For example, both the original possible-worlds semantics and Cresswell's hybrid theory are inconsistent with the plain fact that a person can contemplate the contemplating relation, and they are inconsistent with the plain fact that the relation of being distinct is distinct from the relation of being identical. Other difficulties concerning "self-constituency" threaten the other non-traditional realist semantics, including the reductionistic version of situation semantics and the theories articulated by Soames and by Salmon. (See the section after next for a detailed discussion of self-embeddability and self-constituency.) A traditional PRP semantics, by contrast, can easily deal with these phenomena, as we will show later on.

Another critical point is that many of the non-traditional realist approaches are extremely counterintuitive. For example, it is incredible, intuitively, that sets or sequences can ever strictly and literally be the sort of thing that are perceived, believed, and so forth or that are true, necessary, valid, probable, and so forth. People who hold otherwise have lost their "semantic innocence"; they are under the spell of set-theoretical reductionism. Although set-theoretical constructs might in the short run have heuristic value in the model theory for intensional abstracts, in the long run we should like a semantical theory that provides a natural and intuitive semantics for these important expressions. A semantics that takes PRPs at face value as primitive entities does this; possible-worlds semantics, the original version of situation semantics, and the theories of Cresswell, of Soames, and of Salmon plainly do not.

In a related vein, it is doubtful that possible-worlds semantics (and Cresswell's semantics) can be made to mesh with a plausible epistemology. For example, in sense experience we can be directly aware of phenomenal properties – say, the aroma of coffee. But in sense experience can we be directly aware of the function that assigns to each possible world the set of possible individuals that smell like coffee in that world? This is hardly plausible. Here are some related questions. (1) Suppose that the taste of pineapple is a function from possible worlds to sets of possible individuals. Could a person have a sense experience of a function that is identical to this one except for the presence (or absence) of a few possible individuals in one of the sets in the range of this function? Presumably not, for there is nothing such a sense experience could *be like*. But how are we to explain this? (2) Consider two visibly similar but distinct shades of blue b_1 and b_2. Suppose that b_1 is the function from possible worlds to sets of things that are shaded b_1 in those worlds and, likewise, that b_2 is the function from possible worlds to sets of things that are shaded b_2 in those worlds. Given that functions are sets of ordered pairs, b_1 and b_2 would then be sets that have *no* members in common. What makes b_1 and b_2 look so similar? (3) Let the arguments and values of a possible-worlds function b_1' differ from those of the shade b_1 at no points except for the presence (or absence) of one non-actual individual in one of the values. On the possible-worlds theory, b_1' is a property. Does the shade b_1 resemble this property as much as the two shades b_1 and b_2 resemble each other? Presumably, we would answer no. Why? (4) Why does the shade of blue b_1 resemble the other shade of blue b_2 more than it resembles a shade of red r? Perhaps the possible-worlds answer to this question is that the individuals in the ranges of b_1 and b_2 resemble each other more than the individuals in the ranges of b_1 and of r resemble each other. But if this is the answer, a vicious regress results. What is it about the individuals in the ranges of b_1 and b_2 that makes them resemble one another more than the individuals in the ranges of b_1 and r? The answer, of course, is that the shades (b_1 and b_2) of the former individuals resemble each other more than the shades (b_1 and r) of the latter individuals resemble each other. But why? This is the question with which we started. Now, all these questions can be answered satisfactorily, but only with a traditional realist theory that takes properties and relations as primitive, irreducible entities.[38]

Moreover, possible-worlds theories (and Cresswell's theory) are beset with insurmountable epistemological problems concerning the individuation of "nonactual individuals". For example, suppose that I form a thought that is (allegedly) about a particular "nonactual individual". (If one cannot form such a thought about *any* item in the category of "nonactual individual", that is itself a count against the ontology; for no other ontological category is like this.) Suppose that years later, after forgetting all about this earlier episode, I form a thought that is qualitatively indistinguishable from my earlier one. Is the "nonactual individual" I first thought about identical to the one I thought about on the second occasion, or are they nonidentical items that are only qualitatively alike? There is in principle no way to tell! For another example, suppose that two causally separated people form thoughts that are (allegedly) about "nonactual individuals", and suppose that their thoughts are qualitatively indistinguishable. Are they thinking about the same "nonactual individual", or are they thinking about distinct "nonactual individuals" that are only qualitatively alike? Again, there is in principle no way to tell. A theory with this kind of epistemological indeterminacy is, other things being equal, unacceptable.

And then there is the problem of ontological economy. According to the reductionistic version of situation semantics and the positions advocated by Soames and by Salmon, propositions (situations) are to be reduced to set-theoretical constructs whose ultimate elements are actual individuals, primitive properties, and primitive relations. However, these reductionistic theories are guilty of an ontological excess. For they must posit, in addition to individuals, two fundamentally dissimilar ontological categories – intensions and sets. (Relatedly, they must hold that some intensional abstracts denote intensions whereas others denote sets! This kind of *ad hoc* disunity is unacceptable.) On a traditional, nonreductionistic realist theory, by contrast, there is instead only one corresponding ontological category, namely, that of intensions. Sets (i.e., extensions) are just dropped in favor of intensions. Now the main reason these reductionists have resorted to their more complex ontology of both sets and intensions is that it permits them to treat propositions (situations) and other "complex" intensions. However, we will show that this can be accomplished far more simply without resorting to set-theoretical constructs but rather by treating such intensions simply as the result of applying fundamental logical operations (e.g., conjunction, negation, existential generalization, predication, etc.) to other intensions. The result is that the identification of propositions (and other "complex" intensions) with sets is ontologically superfluous.

It might be replied that these reductionistic theories are not ontologically excessive because sets are needed for independent reasons. But this is simply false. As we have already seen in our discussion of conceptualism, the theoretical work done by finite sets and finite sequences can be accomplished within the first-order logic for identity and intensional abstraction. And the theoretical work done by a set theory with a membership relation can be done by a property theory with the predication (instantiation) relation. In foundational matters such as those we are concerned with here, there can be no justification for positing the two fundamentally dissimilar ontological categories of intensions and sets (extensions). Intensions alone suffice: sets are ontologically superfluous, mere hold-overs from the days when it was unknown how to formulate a theory of intensions. It is high time that property theorists acknowledge that they have a thoroughgoing alternative to set theory. From this perspective, it is plain that reductionistic theories that posit both intensions and sets are ontologically unjustified.

The question also arises whether the possible-worlds reduction is guilty of a similar ontological excessiveness. In addition to actual individuals, it posits both sets and non-actual possible individuals. Do the latter constitue a new fundamental ontological category? Many possible-worlds theorists would answer in the negative on the grounds that, ontologically, "nonactual individuals" are just like ordinary actual individuals except that they are nonactual. But this is a very odd statement. For it seems that there could not be a greater difference between two sorts of items, one actual and the other "nonactual". Indeed, a sign that "nonactual individuals" are fundamentally unlike actual individuals is that the former, unlike the latter, present insurmountable epistemological problems of individuation not presented by actual things (whether actual individuals or actual intensions). We have in mind the problems of individuation mentioned a moment ago: there is no way in principle to tell whether, from occasion to occasion or from person to person, you are thinking about the same "nonactual individual". By contrast, actual things (actual individuals and actual intensions) are not by nature like this. (Alternatively, if items in the category of "nonactual individual" cannot be objects of your thought, that would be grounds for deeming them to have a different ontological status than that of actual individual. For you can typically think of particular actual individuals.)

Suppose that this, and other considerations, show that "nonactual individuals" constitute a fundamentally new ontological category. And suppose that the traditional realist theory has no need to posit nonactual things. In this case, the possible-worlds theory would be guilty of ontological excess. For, in addition to actual individuals, it would posit two fundamentally dissimilar categories, namely, sets and "nonactual individuals". By contrast, the traditional realist theory would posit, in addition to actual individuals, only one further ontological category, namely, intensions. In this case, the latter theory would be ontologically more economical than the possible-worlds theory.

On the other hand, suppose that nonactual individuals do not constitute a new ontological category above and beyond actual individuals. Then, ontologically, the two theories would be on a par: the possible-worlds theory would posit individuals and sets; the realist theory would posit individuals and intensions. But in this case, the possible-worlds theory would still be confronted with the insurmountable epistemological problems of identifying nonactual individuals. Moreover, it would be confronted with all the logical and intuitive problems cited earlier. So, even if the two theories were ontologically on a par, the possible-worlds theory would have to be counted as deficient.

The possible-worlds theory is deficient on one further count. There is a compelling list of reasons for thinking that only certain properties are genuine *qualities*. (These reasons are spelled out in *Quality and Concept* [1982] and again by David Lewis, who is perhaps the leading possible-worlds theorist, in "New Work for a Theory of Universals" [1983].) On the traditional realist picture, genuine qualities can be combined, by means of fine-grained logical operations, to form properties that are not genuine qualities (e.g., the property of being grue); but properties that are not genuine qualities cannot be combined, by means of fine-grained logical operations, to form genuine qualities. So, on the traditional realist picture, qualities are *logically* distinctive. Indeed, this logically distinctive feature can be used as the basis of a definition of the notion of a genuine quality. By contrast, qualities (or "natural properties", as David Lewis calls them) are not logically distinctive on the possible-worlds theory even though they are ontologically distinctive. To deal with them, therefore, Lewis is forced simply to introduce a new undefined primitive predicate 'natural property'. Accordingly, what it is to be a quality (natural property) remains an unanswerable mystery on the possible-worlds theory. So on this score, too, the traditional realist theory comes out ahead of the possible-worlds theory.

Our overall conclusion is that the various nontraditional (reductionist) versions of realism do not compare with traditional realism. The best semantics for intensional abstracts is that based on the traditional realism.

10. TRANSCENDENTAL PREDICATES AND TYPE-FREE LANGUAGES

There are in natural language a great many "transcendental" predicates, that is, predicates that we apply freely across ontological categories. The 2-place predicate 'contemplate' is an example, for items from any ontological category can be contemplated by someone or other. The 2-place predicate 'distinct' (or '$\neq$') is another example because items belonging to any two distinct ontological categories are themselves distinct. (For example, for any universal x and particular y, $x \neq y$.) It turns out that the existence of transcendental predicates provides compelling evidence for the thesis that, syntactically, the formal language for intensional logic should *not* be a type-restricted (or categorial) language.

Consider first the matter of predicates. Suppose that all the predicates in this formal language were syntactically type-restricted. Then for every transcendental predicate in natural language – let us take 'contemplate' as our example – there would in the formal language need to be infinitely many distinct primitives $\ulcorner \text{contemplate}_{(o\iota)\alpha} \urcorner$, one for each distinct syntactic type (or category) α in the formal language. But this outcome conflicts with our previous conclusion concerning Davidson's learnability requirement. Therefore, there is no choice but to allow the transcendental predicates to be syntactically unrestricted in this formal language.

Consider next the issue of variables. If all the variables in the formal language were syntactically restricted according to ontological type, then the formal language would not be equipped to express various *general propositions* that are expressible in the natural language. (For example, the proposition that, for any item, it is possible that someone contemplates it; the proposition that, for any item, there is something that is distinct from it; etc. In our notation:

$$[(\forall x)\text{Possible}[(\exists y)y \text{ contemplates } x]];$$

$$[(\forall x)(\exists y)y \neq x]$$

etc.) This implies that the formal language should contain a sort of variable that ranges over all items regardless of ontological type.

Now suppose that certain type-restricted variables or certain type-restricted predicates (i.e., predicates whose argument expressions must belong to some preferred syntactic type) are needed for some purpose or other in an idealized representation of natural language. We have determined that transcendental predicates – and general statements we can make with them – force us to include in our formal language various type-free predicates and a syntactically unrestricted sort of variable. However, once we have adopted a syntax with such devices, the simplest and most economical way to deal with type-restricted variables and type-restricted predicates is to define them contextually by standard techniques using the unrestricted sort of variable and appropriate syntactically type-free predicates. In view of this, it would seem that the simplest and most economical construction of intensional logic would be formulated in a one-sorted, syntactically type-free language, that is, a language in which there is only one sort of variable, which ranges freely over all ontological categories, and in which all predicates are free of syntactical type-restrictions inasmuch as they all may take this single sort of variable as arguments.

11. SELF-EMBEDDING AND SELF-CONSTITUENCY

As we have seen, transcendental predicates express properties and relations that apply freely across ontological categories. An important special case of this occurs in connection with *self-embeddable predicates*, that is, predicates that take as arguments intensional abstracts in which the very same predicate occurs. Consider an example: x contemplates contemplating. (In our notation, $C^2x, [C^2u, v]_{uv}$.) Does the occurrence of the verb 'contemplates' express the very relation of contemplating that is denoted by the gerund 'contemplating'? Intuitively, the answer is that it does: one of the things you can contemplate is the relation of contemplating.

Self-embedding arises in connection, not just with transcendental predicates, but with many others as well. For example, in a sentence like 'It is necessary that something is necessary', 'necessary' hardly seems to occur ambiguously. Intuitively, its two occurrences express the very same property, necessity. Thus, when we assert this sentence, we are ascribing this property to a proposition that "involves" this very property. Similarly, in the sentence 'Someone believes that someone believes something', the verb 'believes' certainly does seem not to occur ambiguously; intuitively, both occurrences of 'believe' express the same relation, namely, believing. The sentence is true if and only if someone stands in this relation to the proposition that someone believes something, a proposition that "involves" the very same relation of believing.

With these observations in mind, let us extend our use of 'self-embeddable' from predicates – e.g., 'contemplate', 'necessary', 'believe', 'identical', 'distinct', etc. – to the corresponding properties and relations that they express. Accordingly, we will say that a property or relation is *self-embeddable* if and only if it applies either to itself or to a PRP that "involves" it. (We should emphasize that this talk of PRP's "involving" one another is heuristic only; on the algebraic semantics we advocate this heuristic talk gives way to fully literal talk of fundamental logical operations such as conjunction, negation, existential generalization, and predication.)

Of course, ramified type theorists such as Russell, Whitehead, and Church would hold that there really are no self-embeddable properties or relations and that our ordinary uses of self-embeddable predicates are instead to be explained in terms of "typical ambiguity". However, there are two sorts of considerations that count decisively against these type theorists.

First, since in ordinary English a predicate like 'believe' can be embedded *any* finite number of times within its own scope – e.g., 'Someone believes that someone believes something', 'Someone believes that someone believes that someone believes something', etc. – ramified type theorists like Russell, Whitehead, and Church must hold that 'believe' and kindred predicates actually are *infinitely* ambiguous. But such infinite ambiguity entails a violation of Davidson's learnability requirement. The natural, intuitive way around this difficulty is to admit that in belief sentences, like those above, 'believe' does not occur ambiguously and, hence, that the familiar belief relation is in fact self-embeddable. Of course, type theorists could posit a relation R^2 that holds between the infinitely many alleged belief relations B_0, B_1, B_2, . . . ; that is, they could hold that $R^2(B_\alpha, B_{\alpha+1})$, for ordinals α. The problem is that R^2 would fall outside the hierarchies with which type theory deals. As a result, type theory would not provide a theory for this logically fundamental relation that presumably holds among the very entities type theory is designed to treat, a relation that is needed simply to explain how ordinary people learn to use the predicate 'believe'. If, to deal with this problem, type theorists try to explicitly incorporate 'R^2' into their logical theory, they face a fatal dilemma. Either they subject 'R^2' itself to infinite typical ambiguity and non-self-embeddability, thereby violating the learnability requirement once again. Or they treat 'R^2' as a type-free, self-embeddable predicate. In the latter case, however, they have just come around to our way of doing things except that they do so in an *ad hoc*, disunified way; for they would still treat familiar predicates like 'believe', 'necessary', 'identical', etc. as infinitely ambiguous and non-self-embeddable. On either horn of the dilemma, therefore, the 'R^2'-approach is unacceptable.

Second, ramified type theorists seem unable to explain satisfactorily little dialogues like the following:

> A: I believe many things.
>
> B: So do I; in fact, what you have just asserted is one of them.

In this dialogue, A asserts a proposition "involving" the relation of believing, namely, the proposition that he believes many things. Then B affirms the corresponding proposition about himself, namely, that he [i.e., B] believes many things, too. And then B goes on to provide an example of one of the things to which he stands in this relation of believing, namely, the original proposition A asserted, which, as we saw, is a proposition "involving" this very relation of believing. Or consider the following little dialogue:

> A: Some things are necessary.
>
> B: I agree; in fact, what you have just asserted is an example of one of them.

Here A asserts a proposition "involving" the property of being necessary, namely, the proposition that some things are necessary. B affirms the proposition and then goes on to indicate a proposition having this property, namely, A's original proposition, which "involves" this very property. Now these little examples are not at all exceptional; they are entirely typical of our everyday thought and discourse about belief, necessity, possibility, epistemic justification, etc. Yet they would make no sense if these properties and relations were not self-embeddable. To rule out the self-embeddability of these central properties and relations would be to undermine one of the primary functions of intensional language. Indeed, we submit that it is impossible to formulate an acceptable comprehensive theory of the world unless one makes liberal use of self-embeddable properties and relations. Such a theory must include, among other things, an epistemology (and a methodology and philosophy of science) that can account for its own acceptability. Here self-embeddability is inevitable.[39] Any intensional logic that does not deal with this phenomenon cannot be deemed acceptable, even provisionally.[40]

We have seen that self-embeddability causes trouble for type-theoretical intensional logics like those of Russell and Whitehead and of Church. But it causes equally serious trouble for many other approaches to intensionality. The possible-worlds approach is a case in point. (Related problems involving propositions that are "constituents" of themselves confront the original version of the Perry–Barwise situation semantics and also the theories of Cresswell, Soames, and Salmon; see below.) According to the possible-worlds approach, all PRPs are identified (at least in the semantical model) with sets constituted or formed ultimately from actual individuals and "nonactual individuals" (and the real world and "nonactual possible worlds"). For example, a property is identified with a function (i.e., a set of ordered pairs) from possible worlds to sets of possible individuals (intuitively, the possible individuals that have the property in that possible world). And an m-ary relation is identified with a function from possible worlds to sets of ordered m-tuples of possible individuals (intuitively, the possible individuals that stand in the relation to one another in that possible world).

Most possible-worlds theorists seem unaware that this theory implies a rigid type theory. To see why, consider the transcendental predicate 'is self-identical'. (Any other transcendental predicate would do, e.g., 'contemplate', 'think of', 'identical', 'distinct', etc.) On the possible-worlds theory, the property of being self-identical is the function from possible worlds to the domain of *individuals* existing in those worlds. As a result, the intuitively true sentence 'Being self-identical is self-identical' could be counted as true, but only at the price of treating 'self-identical' as typically ambiguous. All the fatal difficulties associated with this kind of type theory follow in train.

Now possible-worlds theorists might try to escape these fatal problems by admitting properties and relations right into the domains of things existing in possible worlds. So, for example, the property of being self-identical would then be the function from possible worlds to the set of things, including the property of being self-identical itself, that exist in the possible world. But since on the possible-worlds theory a function is a set of ordered pairs, this move would imply the following: being self-identical $\in \ldots \in$ being self-identical. An ill-founded set, which is an impossibility on the standard conception of set. In a moment, we will elaborate reasons why possible-worlds theorists should be unwilling to posit ill-founded sets. However, before doing so, it would be good to show that the problem of self-embeddability in possible-worlds semantics is even more pervasive than one would initially think.

On the original formulations of possible-worlds theory, propositions were identified with functions from possible worlds to truth values (intuitively, the truth value that the proposition has in that possible world). However, this treatment of propositions has the disastrous consequence of making all necessarily equivalent propositions identical. This is plainly wrong. (For example, most uneducated people believe that $2 + 2 = 4$ and fail to believe that arithmetic is essentially incomplete, even though it is necessary that $2 + 2 = 4$ iff arithmetic is essentially incomplete.) It follows that the objects of belief – the semantical correlates of the kind of 'that'-clauses occurring in ordinary belief sentences – cannot be identified with functions from possible worlds to truth values; they must be identified with another sort of 0-ary intension, which we call *fine-grained propositions*.

If you were a possible-worlds theorist, with what possible-worlds set-theoretical construct would you identify fine-grained propositions? The most popular answer among sophisticated possible-worlds theorists (Max Cresswell, David Lewis, etc.) is that fine-grained propositions should be identified with certain ordered sets or abstract trees whose elements or nodes are either possible individuals or sets constructed ultimately from possible individuals. So, for example, when someone believes that F^2x, y, the fine-grained proposition believed would be the ordered set $\langle f, x, y \rangle$, where f is a function from possible worlds to sets of ordered pairs of possible individuals (intuitively, the possible individuals that stand in the relation F^2 to one another in that world). Or when someone believes that $(\exists y)F^2x, y$, the fine-grained proposition would be $\langle EG, \langle f, x\rangle\rangle$, where EG is some set-theoretical item selected to play the role of the operation of existential generalization. (Which item should play this role? The choice seems utterly arbitrary unless the real logical operation of existential generalization is chosen. But in this case the possible-worlds theory would have drifted very close to the traditional realist algebraic semantics we advocate. The same problem of chosing the fundamental logical operations confronts the theories of Soames and of Salmon: the choice is arbitrary unless it buys into our traditional realist algebraic semantics. But, then, why not just adopt the traditional realist algebraic semantics?)

Now, as we have already noted, it is hardly plausible that, strictly and literally, people believe, hope, or perceive ordered sets such as $\langle f, x, y\rangle$ or $\langle EG, \langle f, x\rangle\rangle$. (See Section 9 above.) But this problem, although severe, is not the one we are concerned with here. The problem, rather, concerns self-embeddability. With which set-theoretical construct is the ordinary belief relation to be identified according to the present sophisticated possible-worlds theory? The answer is that it must be some function b (i.e., a set of ordered pairs) from possible worlds to sets of ordered pairs each of which consists of (i) a possible believer and (ii) a fine-grained proposition (intuitively, a fine-grained proposition that the believer believes in that world). However, on the standard conception of set, there exists no such function b that behaves in anything like the way that the ordinary belief relation behaves.

To see why, recall that, on the standard conception of set, all sets are constituted (or formed) ultimately from ontologically primitive entities that are not sets. At the lowest ontological level, there are just the primitive non-sets (both actual individuals and non-actual individuals, if the possible worlds theory is right). At the next level come sets of these non-sets. (The null set is the degenerate case of the set of non-sets that are not self-identical.) Following that, there are sets whose elements are non-sets and/or sets of non-sets. And so on. That is, at any given level, we find sets whose elements are either sets constituted at some lower level or the non-sets given at the lowest level. Thus, on the standard conception of set, every set "has its being in" ontologically prior entities – either entities constituted at some lower level or primitive entities given at the lowest level. Consequently, nowhere in the hierarchy of sets is there a set that contains itself as an element; nor is there a set that contains a second set that contains the first set as an element, and so on. That is, the following pattern never holds: $u \in \ldots \in u$.

It is now easy to see why, on the standard conception of set, the ordinary belief relation cannot be identified with any of the set-theoretical constructs postulated in the sophisticated possible-worlds theory. Consider the little dialogue discussed earlier: A asserts a fine-grained proposition "involving" the belief relation, namely, the fine-grained proposition that he [i.e., A] believes something, and then B affirms that he [i.e., B] believes this very proposition. Given B's remark, B stands in the belief relation to a fine-grained proposition that "involves" the belief relation itself. Belief is a self-embeddable relation. On the sophisticated possible-worlds theory we are discussing, this fine-grained proposition is identified with an ordered set, say, $\langle EG, \langle b, A\rangle\rangle$, where b is the possible-worlds function with which the belief relation is to be identified. But this function b is itself only a set of ordered pairs, one of whose elements would have to be the ordered pair $\langle$the actual world, $\langle B, \langle EG, \langle b, A\rangle\rangle\rangle$. This would imply that $b \in \ldots \in b$. An impossibility on the standard conception of set. Therefore, on this conception – indeed, on any conception according to which sets have their being in ontologically prior elements – the possible-worlds theory is incompatible with the existence of self-embeddable PRPs.

The problem stems from the fact that the possible-worlds theory is *reductionistic*. It tries to reduce all PRPs to sets constituted (or formed) ultimately from actual indivuals and "nonactual individuals". Although this reductionism is formally feasible for some cases, it is not for self-embeddable PRPs, which are so central to thought and speech. At this juncture, unrelenting possible-worlds theorists have two choices. Either they can assert that self-embeddable PRPs are not sets at all but rather that they belong to an entirely new fundamental ontological category. Or they can abandon the standard conception of set and advocate instead a nonstandard conception that permits non-well-founded "sets".

One problem with the first alternative is that it is disunified. How odd that some PRPs (non-self-embeddable PRPs) should be sets and that others (self-embeddable PRPs) should belong to an entirely different, irreducible ontological category. To avoid this problem of disunity, possible-worlds theorists could identify all PRPs with items in this new ontological category. But the resulting theory would have a problem of ontological economy, for it would posit two new categories of entities, namely, the new, irreducible ones that have replaced sets and the highly questionable category of "nonactual individuals". In view of this, why not just drop these questionable new categories and take PRPs at face value as primitive, irreducible entities? The resulting theory would be more unified, more economical, and more intuitive. Furthermore. it would be free of the insurmountable epistemological problems confronting the possible-worlds theory. (How, from occasion to occasion or from person to person, can one ever tell whether one is thinking about the same "nonactual individual" as opposed to a numerically distinct one that is qualitatively like it?).

The second alternative available to the unrelenting possible-worlds theorist is to abandon the standard conception of set and to adopt instead a nonstandard conception that permits non-well-founded "sets", that is "sets" displaying the pattern $u \in \ldots \in u$. However, there are three considerations that weigh heavily against this alternative.

First, the original possible-worlds program sought to reduce PRPs to items that are constituted (or formed) ultimately from actual individuals and "nonactual individuals". A primary goal of this reduction was to provide a metaphysical explanation of PRPs by showing that they "had their being in" ontologically prior entities (namely, actual individuals and "nonactual individuals"). However, the new possible-worlds proposal undermines the prospect of this sort of metaphysical explanation. For on the new proposal self-embeddable PRPs would be identified with non-well-founded "sets", but such "sets" are not constituted (or formed) ultimately from actual individuals and "nonactual individuals". Unlike standard sets, non-well-founded "sets" do not have their being in ontologically prior entities; on the contrary, inasmuch as they have their being "in themselves", they are virtually on a par with individuals. Because such "sets" are ontologically primitive in this way, identifying PRPs with them cannot yield the kind of metaphysical explanation of the being of PRPs that was originally promised by the possible-worlds program. So on this score, we are just as well off taking PRPs at face value rather than as a queer kind of "set".

Second, it is not clear that talk of non-well-founded *sets* is really coherent to begin with. Many people believe that, as a conceptual or metaphysical necessity, all sets must have their being in ontologically prior entities; this is just the kind of thing sets are. According to these people, non-well-founded "sets" are not strictly and literally sets at all; rather, they belong to an entirely new primitive ontological category above and beyond sets. If this is right, people who are favorably inclined toward set theory would have no reason – either ontological or epistemological – to prefer this new "set" theory to a theory that takes PRPs at face value as a basic category of entities. First, there would be no *ontological* gain, for both alternatives must posit a new primitive ontological category. And let us not forget that the possible-worlds theorist has already posited the additional primitive ontological category of "nonactual individuals". Second, there would be no *epistemological* gain. The usual Quinean argument is that sets are epistemologically superior to properties and relations (-in-intension) because sets can be individuated simply by considering their elements. Let u be one of the new non-well-founded "sets" that is an element of itself. If, following Quine's procedure, we try to individuate u by considering its elements, we only get caught in a vicious circle: to individuate u by these means requires that we must already have individuated u. And must the same sort of epistemological difficulty infects all the new "sets" displaying the pattern $u \in \ldots \in u$. (Of course, one could adopt Peter Aczel's bold decision to permit exactly one "set" displaying the pattern $u \in u$ and exactly one "set" displaying the pattern $u \in v \in u \neq v$ and so on. But how could one *know* that there is exactly one "set" displaying the pattern $u \in u$? This is just Quine's worry all over again.) To overcome this sort of difficulty, some other epistemological procedure will be needed. But it would seem that, whatever this further procedure is (for example, systematization of one's first-person introspective reports and/or systematization of one's *a priori* intuitions), it would work at least as successfully on PRPs as it would on the new non-well-founded "sets". Indeed, if this further procedure is available (as of course it is), PRPs are fully as respectable epistemically as ordinary well-founded sets are commonly thought to be. And finally, let us not forget that the possible-worlds theory is beset with a number of absolutely intractable epistemological problems produced by its peculiar ontology of "nonactual individuals".

Methodology provides the third reason not to accept the possible-worlds reduction of PRPs to non-well-founded "sets". The standard view of sets – according to which, sets have their being in their instances – provides an intuitive diagnosis and resolution of the set-theoretical paradoxes. Advocates of set theory should demand a very good reason to give up this secure position. However, Russell-style antinomies are derivable in the naive version of non-well-founded "set" theory. How should these antinomies be resolved? This becomes an absolutely urgent question if PRPs are identified with non-well-founded "sets", for in that case the semantics of intensional logic cannot even be stated without first formulating a non-well-founded "set" theory and thereby taking a strong stand on how to resolve this new family of antinomies. But this situation is methodologically very unsatisfactory. If at all possible, one should find a way to do the semantics for intensional logic without taking a stand on these highly problematic issues. However, we can do just that if we drop the attempt to reduce PRPs to "sets" and instead take PRPs at face value as unreduced entities. By anyone's standard, this is a far wiser way to proceed. To do otherwise is just asking for trouble, and it is doing so with no gain whatsoever.

Our conclusion, therefore, is that one should not invoke a non-well-founded "set" theory to save possible-worlds semantics from the difficulties posed by self-embeddable PRPs. Rather, one should just abandon possible-worlds semantics and, instead, develop a traditional nonreductionist PRP semantics. On this alternative, the above problems simply do not arise.

We have shown how self-embeddable properties and relations produce grave difficulties for possible-worlds semantics (and for Cresswell's semantics). These difficulties can be avoided if one adopts the traditional realist theory that properties and relations are primitive, irreducible entities. Like our own nonreductionistic semantics, situation semantics and the proposals of Soames and of Salmon follow this path.[41] However, the original version of situation semantics and the proposals of Soames and of Salmon are still reductionistic in character, for like possible-worlds semantics, they attempt to reduce 0-ary intensions (what we call propositions and Perry and Barwise call situations) to certain kinds of sets. Predictably, then, propositions (situations) that are "constituents" of themselves produce grave problems for these theories. (It goes without saying that these problems of self-constituency also beset possible-worlds semantics. We should also emphasize again that our remarks apply primarily to the original theory of Perry and Barwise. In their more recent theory, they attempt to deal with self-constituency by adopting a non-well-founded "set" theory.)

The phenomenon of self-constituency seems to arise in connection with such matters as public information, mutual knowledge, reflexive perception, and so on. For example, suppose two opposing soldiers x and y are tracking one another down, and suppose that simultaneously each spots the other and each perceives fully what has happened. There are reasons to think that a complete specification of what has gone on may include the following (and Barwise now seems to agree):[42] x perceives s_1 and y perceives s_2, where s_1 is that y perceives s_2 and s_2 is that x perceives s_1. If this is correct, s_1 is a proposition (situation) that is a "constituent" of s_2, which, in turn, is a proposition (situation) that is a "constituent" of s_1. So s_1 is a proposition (situation) that is a "constituent" of itself. The question is how to develop a semantics to deal with such apparent self-constituency.

(One response is to deny that there really is such a phenomenon as self-constituency. This response might be correct. However, because it is controversial, one would be much better off having a theory that is equipped to handle self-constituency in case it turns out to be a genuine phenomenon. Our traditional realist semantics is like this. The original version of situation semantics and the proposals of Soames and Salmon are not. This is the point.)

The central idea of the original version of situation semantics and the proposals of Soames and Salmon is to reduce propositions (situations) to sets – for example, to ordered sets of properties, relations, and other items.[43] Therefore, in these theories one had no choice but to try to identify s_1 with an ordered set such as $\langle y$, perceiving, $s_2 \rangle$ and s_2 with an ordered set such $\langle x$, perceiving, $s_1\rangle$. But this implies that $s_1 \in \ldots \in s_1$, and this contradicts classical set theory.

Our traditional realist semantics does not fall into the trap of trying to reduce propositions (situations) to sets. Instead, it just takes them at face value. Consequently, it is able to deal with self-constituency in a direct and intuitive fashion without having to contradict classical set theory. Rather than following this natural course, Barwise now advocates abandoning the standard conception of set and adopting instead a nonstandard conception that permits non-well-founded "sets". However, in the preceding discussion of possible-worlds semantics, we found convincing reasons for not following this kind of radical course, reasons that apply equally to Barwise's proposal. A traditional realist theory is plainly superior. Indeed, it would be unreasonable to decide to revolutionize classical set theory just to save a certain style of semantics (possible-worlds semantics, situation semantics, etc.) when there is a simple and natural alternative that requires no such revolution and that, for the purpose of modeling intensional logic, makes use of a relatively weak, uncontroversial standard set theory.

Our overall conclusion, therefore, is that the phenomena of self-embedding and self-constituency cause serious difficulties for all reductionistic semantics and, hence, that a traditional nonreductionistic semantics is the best one to adopt.

12. THE FIRST-ORDER/HIGHER-ORDER CONTROVERSY

From a linguistic point of view, the upshot of the previous two sections was that the formal language for intensional logic should be a one-sorted, type-free language that may contain unambigous transcendental and self-embeddable predicates. Even though this formal language should be one-sorted – that is, even though it should contain only one sort of variable – that does not tell us whether, syntactically, this language should be first-order or higher-order. For it is possible – though quite unusual – for a one-sorted language to be syntactically higher-order. In such a language strings like '$x(x)$' would be counted as well-formed. (In a syntactically higher-order language, predicates – and perhaps sentences, as well – are counted as singular terms for which quantifiable variables may be substituted. So in a one-sorted higher-order language strings like '$x(x)$' would be well-formed. In a syntactically first-order language, by contrast, neither predicates nor sentences are counted as singular terms, and variables may not be substituted for them. Accordingly, strings like '$x(x)$' would not be well-formed.)

In this section we turn to the question of whether we should adopt this style of higher-order syntax or whether we should instead adopt a standard first-order syntax. We believe that the considerations favoring the first-order syntax decisively outweigh those favoring the higher-order syntax. The arguments are too lengthy to give in full detail here. However, we will touch on two issues, one methodological and one grammatical.

First, the methodological issue. In Part II, we shall see that first-order intensional logic – that is, first-order logic with identity and intensional abstraction – is complete: there is a recursive axiomatization of the logically valid sentences of the language.[44] Nevertheless, by a straightforward adaptation of the proof of Gödel's incompleteness theorem, we can show that first-order logic with identity and a copula is essentially incomplete, and this is so whether this logic is intensional or extensional, that is, whether or not the operation of intensional abstraction is adjoined to the language. What explains these results? From a semantical point of view, a copula ('is', 'has', 'stand in', etc.) is a distinguished logical predicate that permits one to talk in a general way about what items have what properties and about what items stand in what relations. That is, a copula is a distinguished logical predicate that expresses a predication (or instantiation) relation. This suggests the following explanation. Intensionality – the failure of substitutivity – is not responsible for incompleteness in logic. Rather, the responsibility lies with those devices that permit us to talk in a general way about what has what properties or about what stands in what relations, that is, with devices that permit us to talk generally about predication. This explanation is borne out by the fact that both higher-order intensional logic and higher-order extensional logic are essentially incomplete, and each is equipped with devices for talking generally about these matters. (Specifically, each is equipped with linguistic forms like '$u(x)$', '$u(x, y)$', '$u(x, y, z)$', . . . , where 'u' is a variable.).

For terminological convenience, let us call any logic (whether first-order or higher-order) that treats such matters a *logic of predication*. Our goal is to formalize intensional logic, which is the logic for contexts in which the substitutivity principles of extensional logic do not hold. Our goal is *not* to develop the logic of predication. In view of the fact that the logic of predication is essentially incomplete, our goal of formalizing intensional logic is best served if, initially, we separate it from the formalization of the logic of predication. In a first-order setting, this separation is possible. For in a first-order setting intensional logic is just the logic for intensional abstracts, and these terms may be adjoined to first-order quantifier logic with identity prior to singling out the copula as a distinguished logical predicate. When we do this, the result is a complete intensional logic. However, this sort of separation of goals is not feasible in a higher-order setting. For on the intended interpretation of a higher-order language, devices for dealing with the predication relation are present in the syntactic forms '$u(x)$', '$u(x, y)$', '$u(x, y, z)$', etc. right from the start. For this methodological reason, then, it is desirable to develop intensional logic in a first-order setting. (This completeness issue will be discussed further in Section 4 of Part II.)

There is a closely related, but much more important, methodological point that we touched on in our introduction. Not only do devices for dealing with the relation of predication produce incompleteness in logic, but also they invite logical paradoxes. For example, naive predication (or comprehension) principles – both higher-order principles like $(\exists u)(\forall x)(u(x) \leftrightarrow A(x))$ and first-order principles like $(\exists z)(\forall y)(y \Delta z \leftrightarrow A(y))$ – lead directly to Russell-style paradoxes. ('Δ' is our symbol for the copula.) On the first-order approach to intensional logic, however, the device that generates intensional contexts (namely, intensional abstraction) and the device for dealing with the predication relation (namely, the copula) are independent of one another. Therefore, on the first-order approach, these paradoxes can be avoided simply by not singling out a distinguished logical predicate (e.g., 'Δ') for the predication relation. On the higher-order approach, by contrast, a device for dealing with the predication relation is built into the very syntax of the language on its intended interpretation. As a result, the paradoxes must be confronted from the very start.

Faced with this demand to resolve the paradoxes, higher-order theorists usually adopt a type-theoretical resolution, and often they actually encode this resolution into the syntax of their language by dividing the variables into sorts, one sort for each distinct ontological type (e.g., one sort of variable for individuals, another for properties of individuals, a third for properties of properties of individuals, etc.). However, these type theories have extremely counterintuitive features. For example, they rule out the possibility of transcendental properties and relations – e.g., contemplating, non-identity, etc. – which we discussed earlier. Moveover, a many-sorted syntax rules out the associated possibility of a sort of variable that ranges freely across all ontological types. We conclude, therefore, that this way of responding to the paradoxes is unsatisfactory.

It is safe to say that, as yet, no one really understands the paradoxes. Despite the elegance and ingenuity of the known resolutions, every one of them is unsatisfactory in one crucial way or another, and it seems unlikely that this situation will change anytime soon. In view of this, the breakdown in type-theoretical higher-order intensional logic should not be viewed as an isolated phenomenon. Since every system of higher-order intensional logic is forced to include a resolution of the paradoxes, it is highly likely that every higher-order system of intensional logic developed in the forseeable future will be unsatisfactory in one crucial way or another. The only realistic strategy for developing a satisfactory system of intensional logic is to use a framework that does not force us to include a resolution of the paradoxes. First-order logic is the only framework like this. For in first-order logic, unlike higher-order logic, we can include a device for representing intensionality – namely, intensional abstraction – without also including a device or devices that threaten to generate the paradoxes. At the same time, the first-order strategy of treating intensionality independently of the paradoxes is not at all *ad hoc*. For, as we shall see in a moment, there are independent grammatical considerations that support a first-order logical syntax. Without any hidden costs, the first-order strategy allows us to keep our options open with respect to the paradoxes: we have a strong chance of being able to incorporate an ideal resolution if ever one is discovered, and, in the meantime, we have a wide variety of interim resolutions to choose from depending on the theoretical task at hand. From a methodological point of view, therefore, the first-order approach to intensional logic is vastly superior to the higher-order approach. There is no good reason not to adopt it.

Now for the grammatical considerations. We wish to emphasize at the outset that these considerations do not carry the same weight as the foregoing methodological considerations we have been discussing. Until one has a satisfactory general syntax of natural language, surface syntactical considerations like those we will discuss are only provisional. Nevertheless, they do suffice to show that the first-order approach is not *ad hoc*. This is all that is needed for our overall argument.

The first-order approach honors the traditional linguistic distinction between subject and predicate, between noun phrase and verb phrase; the higher-order approach does not. That is, on the first-order approach an absolute distinction is made between linguistic subjects and linguistic predicates such that a linguistic subject (noun phrase) cannot, except in cases of equivocation, be used as a linguistic predicate (verb) and conversely. The higher-order approach does not recognize this distinction.[45] On the contrary, it treats linguistic predicates (verb phrases) as substituends for variables and, hence, as a sort of subject expression. Accordingly, strings like $\ulcorner(\exists u)\, u = R\urcorner$, where $\ulcorner R\urcorner$ is a linguistic predicate, are treated as well-formed and valid. But these linguistic forms do not match the surface syntax of anything in natural language, for in natural language linguistic predicates may not (without equivocation) be used as linguistic subjects. For example, 'There is something such that it is identical to runs' makes no sense at all. Of course, we can say 'There exists something such that it is identical to running'. But here the linguistic predicate 'runs' is replaced by the linguistic subject 'running', which is a nominialization of the linguistic predicate, namely, a gerund. Gerundive phrases have exact couterparts in a first-order language with intensional abstraction. Accordingly, the above sentence would be represented by '$(\exists u)\, u = [Rx]_x$'. In this way, the surface syntax of the above natural-language sentence is directly and faithfully represented in a first-order language.

Many higher-order languages also treat sentences as linguistic subjects. For example, strings like $\ulcorner(\exists s)\, s = A\urcorner$, where $\ulcorner A\urcorner$ is a sentence (open or closed), are often treated as well-formed and valid. But these linguistic forms do not match anything in natural language: in natural language, sentences do not qualify as linguistic subjects. Strings like 'There is something such that it is identical to everyone loves someone' make no sense at all. Of course, *nominalizations* of sentences may be used as linguistic subjects. For example, 'There is something such that it is identical to the proposition that everyone loves someone' is meaningful. But here the sentence 'Everyone loves someone' is replaced by a legitimate linguistic subject 'the proposition that everyone loves someone'. This linguistic subject has an exact counterpart in the sort of first-order language we advocate, namely, the intensional abstract '$[(\forall x)(\exists y)Lxy]$'. Accordingly, the above sentence would be represented in our first-order language by '$(\exists s)\, s = [(\forall x)(\exists y)Lxy]$'. So once again the surface syntax of the natural language is directly and faithfully represented in first-order language but not in higher-order language. And this is the general pattern: in natural language no linguistic predicate or (open or closed) sentence is used (without equivocation) as a linguistic subject. Instead, an appropriate nominalization of the linguistic predicate or sentence plays this role, and such nominalizations are none other than the intensional abstracts we have been discussing in earlier sections.

The next issue concerns higher-order uses of names. First, consider names of propositions. For example, let 'c' name some proposition, say Church's thesis. In a higher-order language, such a name may just on its own be used as a sentence: c. But nothing in natural language corresponds to this. The closest we can come is 'Church's thesis holds' which can be represented in a first-order language by 'Hc'. So the naked higher-order use of a name as a sentence gives way to the use of the name as a subject together with the predicative use of an appropriate verb.

A rather similar pattern emerges for names of properties and relations. For example, let 'b' and 'g' name the colors blue and green, respectively. In a higher-order language, such names may be used both as linguistic subjects and as linguistic predicates. Accordingly, a string like '$g(a)$ & $g \neq b$' is counted as well-formed. However, as in the previous examples, there is no natural language sentence corresponding directly to this higher-order string. The closest we can come is '*a is* green and green ≠ blue. But here we have an occurrence of a verb, namely, the copula 'is'.[46] Now the most direct way to represent the copula is by means of a corresponding primitive 2-place predicate, say, 'Δ' or simply 'is' itself. With this predicate available, '*a* is green and green ≠ blue' would be represented by 'a is g & $g \neq b$' rather than by '$g(a)$ & $g = b$', and hence the intuitively ungrammatical predicative use of the linguistic subject 'g' drops out of the picture. And this pattern generalizes: once a primitive copula is available, all predicative uses of property and relation names drop out; such names are used exclusively as linguistic subjects, as in first-order formulas like 'a is g' and '$g \neq b$'.

Much the same verdict holds for higher-order uses of variables. In a higher-order language, a variable, say, 'u' may be used as a linguistic predicate; however, a higher-order string like '$(\exists u)u(x)$' corresponds directly to no natural-language sentence. The closest we can come is 'There is something that x *is*'. But here again we have an occurrence of the copula 'is', a verb which we may represent by a primitive 2-place linguistic predicate. With this primitive predicate available, the natural-language sentence would be represented by '$(\exists u)x$ is u' rather than by '$(\exists u)u(x)$', and, thus, the predicative use of the variable 'u' drops out of the picture. (Indeed, when people introduce novices to higher-order languages, they usually say that '$f(x)$' is to be read as 'x *is* f'. Presumably, then, this is what people actually understand when they grasp a higher-order formula. Is there any reason to think otherwise?) As before, the pattern generalizes: once a primitive copula is available, all predicative uses of variables become gratuitous; the use of variables is confined to their use as linguistic subjects, as in first-order formulas like 'x is u'. Finally, consider the higher-order use of variables as sentences. For example, a string like '$(\forall s)(s \to s)$' is well-formed in higher-order languages. But, as in previous examples, this string corresponds to no sentence in natural language. The closest we can come in English is something like 'whatever holds holds' or 'whatever is true is true', which can easily be represented in a first-order language by '$(\forall x)(Hx \to Hx)$' and '$(\forall x)(x$ is $t \to x$ is $t)$', respectively. So the higher-order use of a variable as a sentence gives way to the use of the variable as a linguistic subject in tandem with a predicative use of an appropriate predicative expression.

Our overall conclusion is this. First, there are decisive methodological grounds for favoring a syntactically first-order approach to intensional logic. Second, there is grammatical evidence that a first-order language with intensional abstraction and the copula (and perhaps other auxilary logical predicates) more directly and faithfully represents the syntax of natural language. In view of these considerations, there seems to be no reasonable choice but to take the first-order option.

13. NAMES AND INDEXICALS

There are many varieties of substitutivity failures. Not only are there the standard substitutivity failures involving materially equivalent formulas, but also there are ones involving necessarily equivalent formulas. There even seem to be substitutivity failures involving synonymous formulas; we have in mind those associated with the paradox of analysis and Mates' puzzle.[47] (See the next section for a discussion of these puzzles.) Finally, there are puzzles involving, not only co-referring definite descriptions, but also co-referring proper names and co-referring indexicals – expressions that may well be lacking in descriptive content, at least if the "direct reference" theory is right.

At this stage of research it is desirable to have a general technique for constructing a spectrum of intensional logics ranging from systems that treat PRPs as relatively coarse-grained to systems that treat them as extremely fine-grained. After all, it is plausible that different kinds of PRPs are responsible for different kinds of intensional phenomena. (In Part II we will develop this sort of general technique and then use it to construct in detail both a coarse-grained intensional logic and a fine-grained intensional logic.) At the same time, we should not *commit* ourselves to the strategy of always explaining new substitutivity puzzles in terms of ever more fine-grained distinctions among PRPs. In this connection, we should not rule out the possibility that some of these puzzles (perhaps Mates' puzzle or puzzles involving co-referring proper names or indexicals) are a special kind of pragmatic phenomenon to be explained, not in terms of ultra-fine-grained distinctions among PRPs, but rather in terms of subtle shifts of interest in the conversational context.

For a case in point, consider the substitutivity puzzles involving co-referring proper names. (Co-referring indexicals would be handled analogously.) There are at bottom two theories about the content of ordinary proper names, Frege's theory and Mill's theory. According to Frege's theory, each name has associated with it a descriptive content that determines the name's nominatum; according to Mill's theory, names lack such a content.

Let us suppose that the Fregean theory is right. In this case, we would treat each ordinary proper name '*a*' as an abbreviation for a definite description: $a =_{\mathrm{df}} (\iota v)F(v)$. Here $\ulcorner F \urcorner$ is a new predicate interpreted so as to capture the descriptive property Fregeans would associate with the name $\ulcorner a \urcorner$. Substitutivity failures involving co-referring definite descriptions can be explained by the fact that the associated descriptive properties are distinct. Therefore, given that ordinary proper names can be treated as abbreviated definite descriptions, substitutivity failures involving ordinary proper names can be explained by the fact that the underlying descriptive properties are distinct.

But how are definite descriptions to be treated? One way would be to treat them as contextually defined expressions much as Russell does. A standard objection to this treatment is that there are several candidate Russellian analyses, and there seems to be no way to decide which one is "the" correct one. This problem brings us close to the paradox of analysis. Consider an analogy. There are several candidate definitions of circularity (e.g., locus of coplanar points equidistant from a common point, closed plane figure every segment of which is equally curved, etc.), and there seems to be no way to decide which one is "the" correct one. However, it seems mistaken to say that circularity is simply not definable because of this. A more reasonable response would be to say that there are several correct definitions. Indeed it is easy to develop an algebraic semantics that accommodates this view and that, at the same time, provides the sort of highly fine-grained distinctions needed for the treatment of the propositional attitudes. (This solution is sketched at the end of Chapter 3 of *Quality and Concept*.) It turns out that this kind of semantics can also be adapted to solve the multiple-analyses problem that arises in connection with definite descriptions. Thus, a Russellian approach to definite descriptions can be saved after all.

Another way to deal with the multiple-analyses problem is just to treat 'ι' as a primitive, undefinable operator. This can be done in various ways. One way is to treat it as a primitive binary quantifier. Evans [1977 and 1982] gives persuasive linguistic evidence for this treatment. Moreover, this approach is extremely easy to implement within the algebraic semantic method we will present in Part II. (We simply add to our model structures a logical operation **the** that corresponds to the primitive binary operator 'ι'. The action of **the** is just what one would expect:

$$\mathbf{the}([Fu]_u, [Gu]_u) \;=\; [G(\iota u)Fu]$$

and

$$H(\mathbf{the}(x, y)) \;=\; T \textit{ iff } (\exists!u)(u \in H(x)) \;\&\; (\forall u)(u \in H(x) \rightarrow u \in H(y))$$

for all $H \in K$.)[48] Other syntactical treatments of definite descriptions can also be accommodated by the algebraic semantic method. For example, a treatment that counts 'the *F*' as a restricted unary quantifier (on a par with 'an *F*', 'no *F*', 'every *F*', etc.), and also a treatment, like Frege's, that counts 'the *F*' as an ordinary singular term.

For the present purposes, we need not make a choice about which of these treatments of definite descriptions is best. The point is that one of them is bound to be acceptable. Given this and given the Fregean supposition that ordinary proper names have descriptive contents, substitutivity problems involving co-denoting names can be solved within our general framework.

On the other hand, suppose with Mill that ordinary proper names do not have descriptive contents. In this case, we would treat them as primitive singular terms whose semantical behavior (namely, "rigid designation") is like that of free variables with fixed assignments. So, if the Mill theory is right, names can easily be incorporated as long as we have a theory that permits free variables with fixed assignments to occur in any context, including contexts that are otherwise intensional. However, on our approach to intensional logic, this condition will be met automatically, for our intensional language is expressly designed to permit this kind of unrestricted quantifying-in.

But how on this Millian approach do we explain *prima facie* substitutivity failures involving co-referring proper names? On this approach, names behave semantically like free-variables with fixed assignments. Therefore, strictly and literally, co-referring names may always be substituted for one another *salva veritate*. Consequently, *prima facie* substitutivity failures involving these expressions cannot be semantic phenomena. They must, therefore, be pragmatic phenomena. That is, in actual contexts of conversation, what one means by uttering sentences that arise from one another by replacement of co-referring names can be quite different things, and *prima facie* substitutivity failures may be traced to such differences in pragmatic meaning. More specifically, in certain conversational contexts the use of one name will (by Gricean mechanisms) implicate a descriptive content not implicated by the use of a co-referring name, and *prima facie* substitutivity failures in such contexts may be traced to these differences in implicated descriptive content.[49] (For further discussion of pragmatic solutions to substitutivity puzzles, see the next section.) So on Mill's theory we are also able to explain *prima facie* substitutivity failures involving ordinary proper names.

Until we have a final resolution of the Mill/Frege controversy, the best strategy for us is to set up an intensional logic that is neutral with respect to the two theories and yet that can be easily extended to accommodate either theory. The way to do this is as follows. First, we should construct a neutral language to which names (and indexicals) can be adjoined either as abbreviated definite descriptions or as primitive rigid designators lacking descriptive content. Second, within this language we should construct a general intensional logic that can accommodate both the sort of intensional entities posited in a Fregean semantics and the sort posited in a Millian semantics (and the pragmatics that accompanies it). Our algebraic semantic technique permits this two-step approach.

14. MATES' PUZZLE, PARADOX OF ANALYSIS, AND THE NEED FOR FINE-GRAINED INTENSIONAL DISTINCTIONS

In Part II we will present a general technique for constructing a spectrum of intensional logics ranging from coarse-grained to extremely fine-grained, and we will illustrate this technique by presenting in detail a representative coarse-grained theory and a representative fine-grained theory. However, there are certain outstanding substitutivity puzzles that seem initially to call for intensional distinctions that are even more fine-grained than those treated in this fine-grained theory.

The original formulation of Mates' puzzle is a case in point.[50] Mates holds that, for any distinct sentences D and D',

(1) Nobody doubts that whoever believes that D believes that D.

and

(2) Nobody doubts that whoever believes that D believes that D'.

can always diverge in truth value. However, let D be 'Somebody chews' and D' be 'Somebody masticates'. Then, given that the property of chewing and the property of masticating are identical, it will follow in our fine-grained theory that (1) and (2) must be equivalent, contradicting Mates.

There are two reasonable responses to this outcome. The first is to accept this outcome and to explain Mates' intuition *pragmatically*. Accordingly, sentences (1) and (2) would be deemed semantically equivalent. Nevertheless, utterances of (1) and (2) in an appropriate conversational context could express non-equivalent propositions. To determine exactly which propositions, we would appeal, not only to the semantics of the language, but also to Gricean pragmatic rules of conversation.[51]

The second response to the problem would be to construct a new theory that admits even more fine-grained intensional distinctions than our fine-grained theory. In particular, even though the propositions denoted by '$[(\exists x)Cx]$' and '$[(\exists x)Mx]$' would still be identical according to the new theory, the propositions denoted by the following more complex intensional abstracts would not:

$$[(\forall u)(B^2u, [(\exists x)Cx] \rightarrow B^2u, [(\exists x)Cx])]$$

and

$$[(\forall u)(B^2u, [(\exists x)Cx] \rightarrow B^2u, [(\exists x)Mx])].$$

How can this be? The idea, (once suggested by Putnam)[52] is to exploit the differences in *syntactic form* between these two complex abstracts. Specifically, the predicate 'C' is *repeated* in the former abstract but not in the latter; so the former has the form ⌜[(∀u)(. . . 1 . . . → . . . 1 . . .)]⌝ whereas the latter has the form ⌜[(∀u)(. . . 1 . . . → . . . 2 . . .)]⌝. The new theory, then, is built around the following general principle: two abstracts are to be codenoting only if they have exactly the same syntactic form. It turns out that such a theory is easy to formulate within our general algebraic approach.[53]

How are we to choose between these two responses to Mates? The second response is initially very appealing because it is systematic. But there is reason to question it. True, this response solves the original puzzle given by Mates. But, ironically, there are *simpler* versions of the puzzle that cannot be solved no matter how fine-grained we allow PRPs to be. Consider any two predicates that express the same property, for example, 'chew' and 'masticate'. (Or choose some predicate 'C' and then just stipulate that a new predicate 'M' expresses the same property as the one expressed by 'C'.) Consider someone "halfway" along in the process of picking up the use of 'masticate' by hearing others use it. There are conversational contexts in which such a person could assert something true by saying, "I am sure that whatever masticates chews, but I am not sure that whatever chews masticates." In this example, the two intensional abstracts have the same syntactic form: '$[(\forall x)(Mx \rightarrow Cx)]$' and '$[(\forall x)(Cx \rightarrow Mx)]$' are perfectly isomorphic syntactically. Thus, the second response to Mates will not allow us to hold that these two abstracts denote distinct propositions, and, therefore, it cannot be used to solve this instance of the puzzle. Consequently, there is no choice but to invoke the first response, that is, to solve the puzzle pragmatically. (See below for details.) However, if we must resort to a pragmatic solution of this simple version of Mates' puzzle, why not use it to solve the original, more complex version? If the pragmatic solution is adequate, the second response, which involves positing ultra-fine-grained intensional distinctions, would then appear to be extraneous.

This prospect raises a general methodological question. How are we to decide which types of fine-grained distinctions to admit in intensional logic? Principles of ontological economy would seem to suggest that we should admit those distinctions that are needed to explain substitutivity failures. But we have just seen that a pragmatic solution to at least some substitutivity puzzles is inevitable and that, once this style of explanation is available, ultra-fine-grained intensional distinctions might not be needed to explain the versions of Mates' puzzle for which they were designed. If so, ontological economy would lead us to reject such distinctions.

Let us suppose for a moment that this is right. One wonders how far this sort of elimination can go. What types of fine-grained intensional distinctions, if any, survive a systematic attempt to explain substitutivity failures pragmatically? A transcendental argument yields a partial answer to this question: the fine-grained distinctions that survive must include at least those that are needed to spell out satisfactory pragmatic explanations; it turns out that very fine-grained intensional distinctions are needed for this purpose. Here is the argument.

Consider the person "half-way" along in the process of learning to use the predicate 'masticate'; he says. "I am sure that whatever masticates chews, but I am not sure that whatever chews masticates". In an actual conversational context the person certainly would assert something true. The problem is to identify what true proposition it is that the person would assert. The ultra-fine-grained theory, developed in response to Mates' original puzzle, is forced to identify the proposition that whatever masticates chews and the proposition that whatever chews masticates, for the two abstracts 'that whatever masticates chews' and 'that whatever chews masticates' have exactly the same syntactic form: $[(\forall x)(1(x) \rightarrow 2(x)]$. So this theory implies that the sentence uttered by our person expresses something that is strictly and literally false (indeed, something that is formally contradictory). Since on the ultra-fine-grained theory the sentence expresses something that is strictly and literally false and since our person has asserted something that is true, what he has asserted must be something other than what the sentence strictly and literally expresses. Therefore, the problem of identifying what he has asserted cannot be solved semantically; it must be solved pragmatically. In pragmatics, we may take into account, not only the syntactic and semantic features, but also features of the conversational context and the Gricean rules of conversation. Given all this information, one might identify the person's true assertion with something like the following:

> I am sure that whatever satisfies the predicate 'masticate' also chews, but I am not sure that whatever chews also satisfies the predicate 'masticate'.

This pragmatic solution is a good first try, but there is a problem with it.[54]

Suppose that the person who utters the sentence is a child (or a slow-learning adult) who appears to have no command of the metalinguistic concepts we take for granted. In particular, he appears to be unfamiliar with any device (e.g., quotation names) for naming expressions, and he appears to have no articulated concepts from linguistic theory such as the syntactic concept of a linguistic predicate or the semantical concept of satisfaction(-in-English). Furthermore, when we try to teach him these bits of linguistic theory, he has great difficulty learning them. (He learns the new predicate 'masticate' much more readily.) However, a few years later when we try again to teach him these things, he learns them quickly. This shows, so the worry goes, that the above pragmatic analysis of his assertion represents him as having reached a stage of conceptual development beyond what we can plausibly attribute to him. If so, the pragmatic analysis is mistaken; the fellow's assertion could not have involved the specific metalinguistic concepts attributed to him by this analysis.

There appears to be only one successful way out of this problem, and that is to treat our fellow's apparent ignorance of metalinguistic concepts as a species of the kind of ignorance involved in the paradox of analysis. Consider two analogies. First, suppose a child can sort variously shaped objects so well that it becomes plain that he recognizes, say, the circular objects *as* circular and, therefore, that he has command of the concept of circle. However, suppose that the child displays no particular behavior to indicate that he has command of the concept of a (mathematical or physical) point, the concept of a locus of points, the concept of a (mathematical or physical) plane, the concept of degree of curvature, etc. When we try to teach him geometric theory – with its definition of circle as a locus of points in the same plane equidistant from a common point – we get nowhere. (If he were a few years older, he would be able to learn this readily.) In this situation it is natural to characterize the child as follows: he has an unanalyzed concept of circle (i.e., an unanalyzed concept of being-a-locus-of-points-in-the-same-plane-equidistant-from-a-common-point); however, he lacks the theoretical concepts (points, locus, plane, etc.) that someone might use to analyze this unanalyzed concept.

Second, consider someone who can reliably tell whether a middle-sized object *comes to a halt smoothly*. But this person seems to have no grasp of the sophisticated concepts of calculus that would be used to spell out what it is for an object to come to a halt smoothly; indeed, if the person has limited mathematical appitude, he might never be able to grasp these theoretical concepts. It would be natural to say of this person that he has an unanalyzed concept of coming-to-a-halt-smoothly but that he lacks the specific theoretical concepts needed to unpack this unanalyzed concept.[55]

With these geometry and calculus examples in mind, let us consider a linguistics example. Suppose that a child is not yet in command of various theoretical concepts from linguistics, concepts such as satisfaction-in-English, linguistic predicate, and quotation. Despite this, it should still be possible for the child to have an unanalyzed concept whose analysis involves such theoretical concepts from linguistics. This would be quite analogous to the child's having an unanalyzed concept of circle (i.e., an unanalyzed concept of being-a-locus-of-points-in-the-same-plane-equidistant-from-a-common-point) and yet not being in command of the theoretical concepts from geometry (point, locus, plane, etc.) that someone would use to analyze this concept. And it would also be analogous to the child's having an unanalyzed concept of coming-to-a-halt-smoothly and yet not being in command of the theoretical concepts from calculus that someone would use to analyze this concept. Surely nothing can prevent this sort of thing from happening in linguistics examples, too. We submit that this is exactly what is going on in the case of the child who is "half-way" along in the process of learning to use the predicate 'masticate': he has an unanalyzed concept of satisfying-the-predicate-'masticate', but he is not in command of the theoretical concepts (satisfaction, predicate, quotation) that someone would use to analyze this unanalyzed concept. If this is right, then we have the makings of a solution to the problem confronting the pragmatic analysis of what the child asserted when he said, "I am sure that whatever masticates chews, but I am not sure that whatever chews masticates". The child's assertion comes to something like this:

> I am sure that whatever satisfies-the-predicate-'masticate' also chews, but I am not sure that whatever chews also satisfies-the-predicate-'masticate'.

Since this analysis attributes to the child the unanalyzed concept of satisfying-the-predicate-'masticate', it avoids the problem of mistakenly attributing to the child theoretical concepts that he will acquire only at a more advanced developmental stage. And this is what we needed.

Notice, however, that this way of salvaging the pragmatic solution of our 'chew'/'masticate' substitutivity puzzle is based on a very fine-grained intensional distinction, namely, the distinction between the unanalyzed concept of satisfying-the-predicate-'masticate' and the analyzed concept of satisfying the predicate 'masticate'.

Now, as far as we can tell, this outcome is unavoidable. That is, there is no satisfactory way of salvaging the pragmatic solution of the original substitutivity puzzle that does not somehow invoke antecedently given intensional distinctions that are very fine-grained. Here, then, is a place where very fine-grained intensional distinctions cannot, even in principle, be eliminated by the technique of pragmatic explanation. If this is right, a very fine-grained intensional logic is inevitable.[56]

II: THE FORMULATION OF INTENSIONAL LOGIC

Using the above guidelines, we are finally ready to present our formal intensional logic. We proceed in three standard stages: (1) syntax, (2) semantics, (3) axiomatic theory. Following this, we will close with some remarks about the significance of completeness results in first-order intensional logic.

1. SYNTAX

We now construct our first-order intensional language L_ω.
Primitive symbols:

Logical operators: &, $\neg$, $\exists$,

Predicate letters: $F_1^1, F_2^1, \ldots, F_s^r, \ldots$ (for $r, s \geqslant 1$),

Variables: $x, y, z, x_1, y_1, z_1, \ldots,$

Punctuation: (,), [,].

Simultaneous inductive definition of *term* and *formula* of L_ω:

(1) All variables are terms.

(2) If $t_1, \ldots, t_j$ are terms, then $F_i^j(t_1, \ldots, t_j)$ is a formula.

(3) If A and B are formulas and v_k is a variable, then $(A \& B)$, $\neg A$, and $(\exists v_k)A$ are formulas.

(4) If A is a formula and $v_1, \ldots, v_m$ (for $m \geqslant 0$) are distinct variables, then $[A]_{v_1 \ldots v_m}$ is a term.

In the limiting case where $m = 0$, $[A]$ is a term. On the intended informal interpretation of L_ω, $[A]_{v_1 \ldots v_m}$ denotes a proposition if $m = 0$, a property if $m = 1$, and an m-ary relation-in-intension if $m \geqslant 2$.

The following are auxiliary syntactic notions. Formulas and terms are *well-formed expressions*. An occurrence of a variable v_i in a well-formed expression is *bound* (*free*) if and only if the expression is (is not) a formula of the form $(\exists v_i)A$ or a term of the form $[A]_{v_1 \ldots v_i \ldots v_m}$. A term t is said to be *free for* v_i *in* A if and only if, for all v_k, if v_k is free in t, then no free occurrence of v_i in A occurs either in a subcontext of the form $(\exists v_k)(\ldots)$ or in a subcontext of the form $[\ldots]_{\alpha v_k \beta}$, where α and β are sequences of variables. If v_i has a free occurrence in A and is not one of the variables in the sequence of variables α, then v_i is an *externally quantifiable variable* in the term $[A]_\alpha$. Let δ be the sequence of externally quantifiable variables in $[A]_\alpha$ displayed in order of their first free occurrence; $[A]_\alpha$ will sometimes be rewritten as $[A]^\delta_\alpha$. Let $A(v_1, \ldots, v_p)$ be any formula; $v_1, \ldots, v_p$ may or may not occur free in A. Then we write $A(t_1, \ldots, t_p)$ to indicate the formula that results when, for each k, $1 \leqslant k \leqslant p$, the term t_k replaces each free occurrence of v_k in A. Terms $[A(u_1, \ldots, u_p)]^\delta_{u_1 \ldots u_p}$ and $[A(v_1, \ldots, v_p)]^\delta_{v_1 \ldots v_p}$ are said to be *alphabetic variants* if and only if, for each k, $1 \leqslant k \leqslant p$, u_k is free for v_k in A and conversely. Terms of the form $[F^j_i(v_1, \ldots, v_j)]_{v_1 \ldots v_j}$ are called *elementary*. A term $[A]_\alpha$ is called *normalized* if and only if all variables in α occur free in A exactly once and α displays the order in which these variables occur free in A. The logical operators $\forall$, $\supset$, $\equiv$, $\equiv_{v_1 \ldots v_j}$ are defined in terms of $\exists$, &, and $\neg$ in the usual way. Finally, F^2_1 is singled out as a distinguished logical predicate, and formulas of the form $F^2_1(t_1, t_2)$ are rewritten as $t_1 = t_2$.

Notice that L_ω is just like a standard first-order language except for its singular terms $[A]_{v_1 \ldots v_m}$, which are intended to be intensional abstracts that denote propositions, properties, or relations, depending on the value of m.

2. SEMANTICS

Since the aim is to simply characterize the logically valid formulas of L_ω, it will suffice to construct a Tarski-style definition of validity for L_ω. Such definition will be built on Tarski-style definitions of the truth for L_ω. These definitions will in turn depend in part on specifications of the *denotations* of the singular terms in L_ω. As already indicated, every formula for L_ω is just like a formula in a standard first-order extensional language except perhaps for the singular terms occurring in it. Therefore, once we have found a method for specifying the denotations of the singular terms of L_ω, the Tarski-style definitions of truth and validity for L_ω may be given in the customary way. What we are looking for specifically is a method for characterizing the denotations of the singular terms of L_ω in such a way that a given singular term $[A]_{v_1 \ldots v_m}$ will denote an appropriate property, relation, or proposition, depending on the value of m.

Since L_ω has infinitely many singular terms $[A]_\alpha$, what is called for is a recursive specification of the denotation relation for L_ω. To do this, we will arrange these singular terms into an order according to their syntactic kind and complexity. So, for example, just as the complex formula $((\exists x)Fx\ \&\ (\exists y)Gy)$ is the conjunction of the simpler formulas $(\exists x)Fx$ and $(\exists x)Gx$, we will say that the complex term $[((\exists x)Fx\ \&\ (\exists y)Gy)]$ is the *conjunction* of the simpler terms $[(\exists x)Fx]$ and $[(\exists y)Gy]$. Similarly, just as the complex formula $\neg(\exists x)Fx$ is the negation of the simpler formula $(\exists x)Fx$, we will say that the complex term $[\neg(\exists x)Fx]$ is the *negation* of the simpler term $[(\exists x)Fx]$. The following are other examples: $[(\exists x)Fx]$ is the *existential generalization* of $[Fx]_x$; $[Fy]^y$ is the *predication* of $[Fx]_x$ of y; $[F[Gy]_y]$ is the *predication* of $[Fx]_x$ of $[Gy]_y$; $[F[Gy]^y]_y$ is the *relativized predication* of $[Fx]_x$ and $[Gy]_y$; $[Rxy]_{yx}$ is the *conversion* of $[Rxy]_{xy}$; $[Sxyz]_{xzy}$ is the *inversion* of $[Sxyz]_{xyz}$; $[Rxx]_x$ is the *reflexivization* of $[Rxy]_{xy}$; $[Fx]_{xy}$ is the *expansion* of $[Fx]_x$.

The complex singular terms of L_ω that are syntactically simpler than all other complex singular terms are those whose form is $[F^m(v_1, \ldots, v_m)]_{v_1 \ldots v_m}$. These are called *elementary*. The denotation of such an elementary complex term is just the property or relation expressed by the primitive predicate F^m. The denotation of a more complex term $[A]_\alpha$ is determined by the denotation(s) of the relevant syntactically simpler term(s). But how in detail does this work? The answer is that the new denotation is determined *algebraically*. That is, the new denotation is determined by the application of the relevant *fundamental logical operation* to the denotation(s) of the relevant syntactically simpler term(s). Let us explain.

Consider the following three propositions: $[(\exists x)Fx]$, $[(\exists y)Gy]$, $[(\exists x)Fx\ \&\ (\exists y)Gy]$. (Note: in this paragraph we will be *using* – not mentioning – terms from L_ω.) What is the most obvious relation holding among these propositions? Answer: the third proposition is the *conjunction* of the first two. And what is the most obvious relation holding between the propositions $[(\exists x)Fx]$ and $[\neg(\exists x)Fx]$? Answer: the second is the *negation* of the first. Similarly, what is the most obvious relation holding between the properties $[Fx]_x$ and $[\neg Fx]_x$? As before, the second is the *negation* of the first. In a similar manner we arrive at the following similar relationships: $[(\exists x)Fx]$ is the *existential generalization* of $[Fx]_x$; $[Fy]^y$ is the result of *predicating* $[Fx]_x$ of y; $[F[Gy]_y]$ is the result of *predicating* $[Fx]_x$ of $[Gy]_y$; $[F[Gy]^y]_y$ is the result of a *relativized predication* of $[Fx]_x$ of $[Gy]_y$; $[Rxy]_{yx}$ is the *converse* of $[Rxy]_{xy}$; $[Sxyz]_{xzy}$ is the *inverse* of $[Sxyz]_{xyz}$; $[Rxx]_x$ is the *reflexivization* of $[Rxy]_{xy}$; and $[Fx]_{xy}$ is the *expansion* of $[Fx]_x$. These fundamental logical operations, of course, correspond to the syntactic operations listed earlier.

Now choose any complex term t in L_ω that is not elementary. If t is obtained from s via the syntactic operation of negation (conversion, inversion, reflexivization, expansion, existential generalization), then the denotation of t is the result of applying the logical operation of negation (conversion, inversion, reflexivization, expansion, existential generalization) to the denotation of s. The same thing holds *mutatis mutandis* for complex terms that, syntactically, are conjunctions, predications, or relativized predications. In this way, therefore, these fundamental logical operations make it possible to define recursively the denotation relation for all of the complex intensional terms t in L_ω.

The algebraic semantics for L_ω is thus to be specified in stages. (1) An algebra of properties, relations, and propositions – or an *algebraic model structure* is posited. (2) Relative to this, an intensional *interpretation* of the primitive predicates is given. (3) Relative to this, the *denotation* relation for the terms of L_ω is recursively defined. (4) Relative to this, the notion of truth for formulas is defined. (5) In the customary Tarski fashion, the notion of *logical validity* for formulas of L_ω is defined in terms of the notion of truth.

Omitting certain details for heuristic purposes, we may characterize an algebraic model structure as a structure containing (i) a domain $\mathscr{D}$ comprised of (items playing the role of) individuals, propositions, properties, and relations, (ii) a set $\mathscr{K}$ of functions that tell us the actual and possible extensions of the items in $\mathscr{D}$, and (iii) various fundamental logical operations on the items in $\mathscr{D}$. (All items in $\mathscr{D}$ are treated on a par as primitive entities; none is constructed from the others by means of set-theoretical operations.) Once the general notion of an algebraic model structure is precisely defined, we may then go on to define a spectrum of different *types* of algebraic model structure; these types are distinguished one from another by the strictness of the identity conditions imposed on the PRPs in the domain of the various model structures. It is in this way that the algebraic method is able to provide a *general* technique for modeling any type of PRP, from coarse-grained to very fine-grained.

Our algebraic method also allows us to model transcendental and self-embeddable PRPs. To see what makes this possible let us consider the differences between algebraic model structures, on the one hand, and the usual possible-worlds model structures, on the other. Algebraic model structures contain (i) a domain consisting of individuals, properties, relations, and propositions, (ii) a set of functions that tell us the actual and possible extensions of items in the domain, and (iii) various fundamental logical operations on the items in the domain. In a possible-worlds model structure, on the other hand, (i) is typically replaced by a domain consisting of actual individuals and "nonactual individuals"; then PRP-surrogates are constructed from these items by means of set-theoretical operations; and (ii) and (iii) are omitted. The reason that (ii) can be omitted in a possible-worlds model structure is that each PRP-surrogate is a set that, in effect, encodes its own actual and possible extensions. The reason (iii) can be omitted is that the sets that play the role of "complex" PRPs are formed from other PRP-surrogates by wholly standard set-theoretical operations (like intersection, complementation, etc.), so there is no need to build these operations into the model structure itself. But notice that, if the set-theoretical construction of these PRP-surrogates is done in a standard set theory, these PRP-surrogates must form an hierarchy of well-founded sets; consequently, there are no sets in the construction that can serve as surrogates for transcendental or self-embeddable properties and relations.

In an algebraic model structure, by contrast, there can be such PRPs. The reason there can be transcendental properties and relations is that properties and relations are included in the domain as primitive entities that do not encode their own extensions; their extensions are instead specified by independent extension functions. Consequently, these functions can map items in the domain to any subset of the domain, even subsets that happen to contain the original items. For example, let s be (the element in the domain $\mathscr{D}$ that plays the role of) the transcendental property self-identity. Then each extension function H just maps s to the domain itself; that is, $H(s) = \{x \in \mathscr{D}: x = x\} = \mathscr{D}$. After all, everything in the domain, including s, is self-identical. No ill-founded set theory is involved here: s is just a primitive entity in $\mathscr{D}$ on a par with an individual, and H is just a standard well-founded function that maps s to a set that contains s.

The reason that algebraic model structures can model self-embedded PRPs is that the (items in the domain that play the role of) PRPs are not set-theoretical constructs but rather are primitive entities. So if a given PRP is self-embedded, this will be exhibited exclusively through its behavior with respect to the fundamental logical operations, in particular, the predication operations. Consider, for example, the proposition $[F[Fx]_x]$. This proposition is the result of applying the 2-place logical operation of singular predication to the property $[Fx]_x$ taken as *both* first and second argument. That is, $[F[Fx]_x] = \text{Pred}_0([Fx]_x, [Fx]_x)$. No ill-founded set theory is involved here; on the contrary, this pattern is fully analogous to an application of a standard (set-theoretically well-founded) substitution operation from formal syntax: $\ulcorner F(\ulcorner Fx \urcorner)\urcorner = \text{sub}(\ulcorner Fx \urcorner, \ulcorner Fx \urcorner)$.[57]

Having made these heuristic points, we are ready to state the semantics for L_ω. We begin by defining the general notion of an algebraic model structure. An *algebraic model structure* is any structure $\langle \mathscr{D}, \mathscr{P}, \mathscr{K}, \mathscr{G}, \text{Id}, \mathscr{T}, \text{Conj}, \text{Neg}, \text{Exist}, \text{Pred}_0, \text{Pred}_1, \text{Pred}_2, \ldots, \text{Pred}_k, \ldots \rangle$ whose elements satisfy the following conditions. $\mathscr{D}$ is a nonempty domain. $\mathscr{P}$ is a prelinear ordering on $\mathscr{D}$ that induces a partition of $\mathscr{D}$ into the subdomains $\mathscr{D}_{-1}, \mathscr{D}_0, \mathscr{D}_1, \mathscr{D}_2, \mathscr{D}_3, \ldots$. The elements of $\mathscr{D}_{-1}$ are to be thought of as particulars; the elements of $\mathscr{D}_0$ as propositions; the elements of $\mathscr{D}_1$ as properties, and the elements of $\mathscr{D}_i$, for $i \geqslant 2$, as i-ary relations-in-intension. Although $\mathscr{D}_i$, $i \geqslant 0$, must not be empty, we do permit $\mathscr{D}_{-1}$ to be empty. $\mathscr{K}$ is a set of functions on $\mathscr{D}$. For all $H \in \mathscr{K}$, if $x \in \mathscr{D}_{-1}$, then $H(x) = x$; if $x \in \mathscr{D}_0$, then $H(x) = T$ or $H(x) = F$; if $x \in \mathscr{D}_1$, then $H(x) \subseteq \mathscr{D}$; if, for $i > 1$, $x \in \mathscr{D}_i$, then $H(x) \subseteq \mathscr{D}^i$. These functions $H \in \mathscr{K}$ are to be thought of as telling us the alternate or possible extensions of the elements of $\mathscr{D}$. $\mathscr{G}$ is a distinguished element of $\mathscr{K}$ and is to be thought of as the function that determines the *actual* extensions of the elements of $\mathscr{D}$. Id is a distinguished element of $\mathscr{D}_2$ and is thought of as the fundamental logical relation-in-intension *identity*. Id must satisfy the following condition: $(\forall H \in \mathscr{K})(H(\text{Id}) = \{xy \in \mathscr{D}: x = y\})$. In order to characterize the next element $\mathscr{T}$, consider the following partial functions on $\mathscr{D}$: Exp_i, defined on $\mathscr{D}_i$, $i \geqslant 0$; Ref_i, defined on $\mathscr{D}_i$, $i \geqslant 2$; Conv_i, defined on $\mathscr{D}_i$, $i \geqslant 2$; Inv_i, defined on $\mathscr{D}_i$, $i \geqslant 3$.[58] For all $H \in \mathscr{K}$ and all $x_1, \ldots, x_{i+1} \in \mathscr{D}$, these functions satisfy the following conditions:

a. $x_1 \in H(\text{Exp}_1(u))$ *iff* $H(u) = T$ (for $u \in \mathscr{D}_0$).
$\langle x_1, \ldots, x_i, x_{i+1}\rangle \in H(\text{Exp}_i(u))$ *iff* $\langle x_1, \ldots, x_i\rangle \in H(u)$ (for $u \in \mathscr{D}_i$, $i \geqslant 1$).

b. $\langle x_1, \ldots, x_{i-2}, x_{i-1}\rangle \in H(\text{Ref}_i(u))$ *iff* $\langle x_1, \ldots, x_{i-2}, x_{i-1}, x_{i-1}\rangle \in H(u)$ (for $u \in \mathscr{D}_i$, $i \geqslant 2$).

c. $\langle x_i, x_1, \ldots, x_{i-1}\rangle \in H(\text{Conv}_i(u))$ *iff*

$\langle x_1, \ldots, x_{i-1}, x_i\rangle \in H(u)$ (for $u \in \mathscr{D}_i$, $i \geqslant 2$).

d. $\langle x_1, \ldots, x_{i-2}, x_i, x_{i-1}\rangle \in H(\text{Inv}_i(u))$ *iff*

$\langle x_1, \ldots, x_{i-2}, x_{i-1}, x_i\rangle \in H(u)$ (for $u \in \mathscr{D}_i$, $i \geqslant 3$).

A proto-transformation is defined to be a function that arises from composing a finite number of these functions in some order (repetitions permitted). A proto-transformation τ is said to be degenerate if and only if $\tau(x) = x$ for all $x \in \mathscr{D}$ for which τ is defined. A function τ is said to be *equivalent* to a proto-transformation τ' if and only if, for all $H \in \mathscr{K}$ and for all $x \in \mathscr{D}$ for which τ' is defined, $H(\tau(x)) = H(\tau'(x))$. Now $\mathscr{T}$ is a set of partial functions on $\mathscr{D}$: for every nondegenerate proto-transformation, there is exactly one equivalent function in $\mathscr{T}$, and nothing but such functions are in $\mathscr{T}$. The functions in $\mathscr{T}$ are called *transformations*. The remaining elements in a model structure are partial functions on $\mathscr{D}$. Conj is defined on each $\mathscr{D}_i \times \mathscr{D}_i$, $i \geqslant 0$; Neg, on each $\mathscr{D}_i$, $i \geqslant 0$; Exist, on each $\mathscr{D}_i$, $i \geqslant 1$; Pred_0, on each $\mathscr{D}_i \times \mathscr{D}$, $i \geqslant 1$; Pred_k, on each $\mathscr{D}_i \times \mathscr{D}_j$, $i \geqslant 1$ and $j \geqslant k \geqslant 1$. These functions satisfy the following, for all $H \in \mathscr{K}$ and all $x_1, \ldots, x_i, y_1, \ldots, y_k \in \mathscr{D}$:

1. $H(\text{Conj}(u, v)) = T$ *iff* $(H(u) = T$ & $H(v) = T)$

(for $u, v \in \mathscr{D}_0$).

$\langle x_1, \ldots, x_i\rangle \in H(\text{Conj}(u, v))$ *iff* $(\langle x_1, \ldots, x_i\rangle \in H(u)$ &

& $\langle x_1, \ldots, x_i\rangle \in H(v))$ (for $u, v \in \mathscr{D}_i$, $i \geqslant 1$).

2. $H(\text{Neg}(u)) = T$ *iff* $H(u) = F$ (for $u \in \mathscr{D}_0$).

$\langle x_1, \ldots, x_i\rangle \in H(\text{Neg}(u))$ *iff* $\langle x_1, \ldots, x_i\rangle \notin H(u)$

(for $u \in \mathscr{D}_i$, $i \geqslant 1$).

3. $H(\text{Exist}(u)) = T$ *iff* $(\exists x_1)(x_1 \in H(u))$ (for $u \in \mathscr{D}_1$).

$\langle x_1, \ldots, x_{i-1}\rangle \in H(\text{Exist}(u))$ *iff*

$(\exists x_i)(\langle x_1, \ldots, x_{i-1}, x_i\rangle \in H(u))$ (for $u \in \mathscr{D}_i$, $i \geqslant 2$).

4.0 $H(\text{Pred}_0(u, y_1)) = T$ *iff* $y_1 \in H(u)$ (for $u \in \mathscr{D}_1$).

$\langle x_1, \ldots, x_{i-1}\rangle \in H(\text{Pred}_0(u, y_1))$ *iff*

$\langle x_1, \ldots, x_{i-1}, y_1\rangle \in H(u)$ (for $u \in \mathscr{D}_i$, $i \geqslant 2$).

4.1. $\langle x_1, \ldots, x_{i-1}, y_1 \rangle \in H(\text{Pred}_1(u, v))$ *iff*

$\langle x_1, \ldots, x_{i-1}, \text{Pred}_0(v, y_1) \rangle \in H(u)$

(for $u \in \mathscr{D}_i$, $i \geqslant 1$, and $v \in \mathscr{D}_j$, $j \geqslant 1$).

4.2. $\langle x_1, \ldots, x_{i-1}, y_1, y_2 \rangle \in H(\text{Pred}_2(u, v))$ *iff*

$\langle x_1, \ldots, x_{i-1}, \text{Pred}_0(\text{Pred}_0(v, y_2), y_1) \rangle \in H(u)$

(for $u \in \mathscr{D}_i$, $i \geqslant 1$, and $v \in \mathscr{D}_j$, $j \geqslant 2$).

. . . .[59]

These functions, together with the transformations in $\mathscr{T}$, are to be thought of as fundamental logical operations on intensional entities. This completes the characterization of what a model structure is.

Now in the history of logic and philosophy there have been two competing conceptions of intensional entities, which we call conception 1 and conception 2. Conception 1 is suited to the logic for modal matters (necessity, possibility, etc.), and conception 2 appears to be relevant to the logic for psychological matters (belief, desire, decision, etc.).[60] According to conception 1, (i-ary) intensions are taken to be identical if they are *necessarily equivalent*. This leads to the following definition. A model structure is *type* 1 *iff*$_{\text{df}}$ it satisfies the following auxiliary condition: $(\forall x, y \in \mathscr{D}_i)((\forall H \in \mathscr{K})(H(x) = H(y)) \rightarrow x = y)$, for all $i \geqslant -1$. This auxiliary condition provides a precise characterization of conception 1. In contrast to conception 1, conception 2 places far stricter conditions on the identity of intensional entities. According to conception 2, when an intension is defined *completely*, it has a *unique, noncircular definition*. (The possibility that such complete definitions might in some or even all cases be infinite need not be ruled out.) This leads to the following definition. A model structure is *type* 2 *iff*$_{\text{df}}$ the transformations in $\mathscr{T}$ and the functions Conj, Neg, Exist, Pred_0, Pred_1, Pred_2, . . . are all (i) one-one, (ii) disjoint in their ranges, and (iii) noncycling. Auxiliary conditions (i)–(iii) provide us with a precise formulation of conception 2.[61]

In order to state the semantics for L_ω, we must define some preliminary syntactic notions. First, we define certain syntactic operations on complex terms of L_ω. These operations have a natural correspondence to the logical operations Conj, Neg, Exist, Pred_0, . . . in a model structure. If $[(A \;\&\; B)]_\alpha$ is normalized, it is the *conjunction* of $[A]_\alpha$ and $[B]_\alpha$. If $[\neg A]_\alpha$ is normalized, it is the *negation* of $[A]_\alpha$. If $[(\exists v_k)A]_\alpha$ is normalized, it is the *existential generalization* of $[A]_{\alpha v_k}$. Suppose that $[F_i^j(v_1, \ldots, v_{m-1}, t_m, t_{m+1}, \ldots, t_j)]_\alpha$

is normalized and that no variable occurring free in t_m occurs in α. Then this normalized term is the *predication*$_0$ of

$$[F_i^j(v_1, \ldots, v_{m-1}, v_m, t_{m+1}, \ldots, t_j)]_{\alpha v_m}$$

of t_m. (v_m is the alphabetically earliest variable not occurring in the normalized term.) Finally, suppose that, for $k \geqslant 1$,

$$[F_i^j(v_1, \ldots, v_{m-1}, [B]_\gamma^\delta, t_{m+1}, \ldots, t_j)]_{v_1 \ldots v_{m-1} u_1 \ldots u_k \alpha}$$

is normalized, that $u_1, \ldots, u_k$ occur in δ, and that no variable in δ occurs in α. Then

$$[F_i^j(v_1, \ldots, v_{m-1}, [B]_\gamma^\delta, t_{m+1}, \ldots, t_j]_{v_1 \ldots v_{m-1} \alpha u_1 \ldots u_k}$$

is the *predication*$_k$ of

$$[F_i^j(v_1, \ldots, v_{m-1}, u_1, t_{m+1}, \ldots, t_j)]_{v_1 \ldots v_{m-1} \alpha u_1}$$

of $[B]_{\gamma u_1 \ldots u_m}^{\delta'}$. ($\delta'$ is the result of deleting $u_1, \ldots, u_m$ from δ.)

Consider the following auxiliary operations on complex terms:

(a) $\quad \text{exp}_i([A]_{v_1 \ldots v_i}) =_{\text{df}} [A]_{v_1 \ldots v_i v_{i+1}}$

(where $i \geqslant 0$ and v_{i+1} is the alphabetically earliest variable not occurring in $[A]_{v_1 \ldots v_i}$).

(b) $\quad \text{ref}_i([A(v_1, \ldots, v_{i-1}, v_i)]_{v_1 \ldots v_{i-1} v_i}) =_{\text{df}}$

$$[A(v_1, \ldots, v_{i-1}, v_{i-1})]_{v_1 \ldots v_{i-1}}$$

(where $i \geqslant 2$ and v_{i-1} is free for v_i in A).

(c) $\quad \text{conv}_i([A]_{v_1 \ldots v_{i-1} v_i}) =_{\text{df}} [A]_{v_i v_1 \ldots v_{i-1}}$

(where $i \geqslant 2$).

(d) $\quad \text{inv}_i([A]_{v_1 \ldots v_{i-2} v_{i-1} v_i}) =_{\text{df}} [A]_{v_1 \ldots v_{i-2} v_i v_{i-1}}$

(where $i \geqslant 3$).

Consider the operations σ that arise from composing a finite number of these operations in some order (repetitions permitted). A relation R_σ is a *term-transforming* relation if it is associated with one of these operations σ as follows: $R_\sigma(r, s)$ *iff* $\sigma(r') = s'$, where r' is an alphabetic variant of r, s' is an alphabetic variant of s, r is either an elementary complex term, a negation, a conjunction, an existential generalization, or a predication$_k$,

$k \geqslant 0$, and s is none of these. Now for any model structure, a term-transforming relation R_σ is *associated* with a transformation τ in the set $\mathcal{T}$ in the model structure *iff*$_{\text{df}}$ (a) for some $\sigma_1, \ldots, \sigma_m$ selected from exp_i, ref_i, conv_i, inv_i, σ is the composition of $\sigma_1, \ldots, \sigma_m$; (b) for some $\tau_1, \ldots, \tau_m$ selected from Exp_i, Ref_i, Conv_i, Inv_i, τ is the transformation in $\mathcal{T}$ equivalent to the composition of $\tau_1, \ldots, \tau_m$; (c) for all k, $1 \leqslant k \leqslant m$, $\sigma_k = \text{exp}_i$ iff $\tau_k = \text{Exp}_i$; $\sigma_k = \text{ref}_i$ iff $\tau_k = \text{Ref}_i$; $\sigma_k = \text{conv}_i$ iff $\tau_k = \text{Conv}_i$; $\sigma_k = \text{inv}_i$ iff $\tau_k = \text{Inv}_i$. With these preliminary notions in hand we are finally ready to state the semantics for L_ω.

Denotation, truth, and validity. An *interpretation* $\mathcal{I}$ *for* L_ω *relative to model structure* $\mathcal{M}$ is a function that assigns to the predicate letter F_1^2 (i.e., $=$) the element $\text{Id} \in \mathcal{M}$ and, for each predicate letter F_i^j in L_ω, assigns to F_i^j some element of the subdomain $\mathcal{D}_j \subset \mathcal{D} \in \mathcal{M}$. An *assignment* $\mathcal{A}$ *for* L_ω *relative to model structure* $\mathcal{M}$ is a function that maps the variables of L_ω into the domain $\mathcal{D} \in \mathcal{M}$. Relative to interpretation $\mathcal{I}$, assignment $\mathcal{A}$, and model structure $\mathcal{M}$, the *denotation* relation for terms of L_ω is inductively defined as follows:

Variables. v_i denotes $\mathcal{A}(v_i)$.

Elementary complex terms. $[F_i^j(v_1, \ldots, v_j)]_{v_1 \ldots v_j}$ denotes $\mathcal{I}(F_i^j)$.

Nonelementary complex terms. If t is the conjunction – or predication$_k$ – of r and s, and r denotes u, and s denotes v, then t denotes $\text{Conj}(u, v)$ – or $\text{Pred}_k(u, v)$. If t is the negation – or existential generalization – of r, and r denotes u, then t denotes $\text{Neg}(u)$ – or $\text{Exist}(u)$. If R_σ is a term-transforming relation associated with a transformation $\tau \in \mathcal{T}$ and $R_\alpha(r, t)$ and r denotes u, then t denotes $\tau(u)$.

The denotation relation is clearly a function. We henceforth represent it with $D_{\mathcal{I}\mathcal{A}\mathcal{M}}$. *Truth* is then defined in terms of $D_{\mathcal{I}\mathcal{A}\mathcal{M}}$ as follows: $T_{\mathcal{I}\mathcal{A}\mathcal{M}}(A)$ *iff*$_{\text{df}}$ $\mathcal{G}(D_{\mathcal{I}\mathcal{A}\mathcal{M}}([A])) = T$.[62] And finally two notions of validity are defined. A formula A is *valid*$_1$ *iff*$_{\text{df}}$ for every type 1 model structure $\mathcal{M}$ and for every interpretation $\mathcal{I}$ and every assignment $\mathcal{A}$ relative to $\mathcal{M}$, $T_{\mathcal{I}\mathcal{A}\mathcal{M}}(A)$. A formula A is *valid*$_2$ *iff*$_{\text{df}}$ for every type 2 model structure $\mathcal{M}$ and for every interpretation $\mathcal{I}$ and every assignment $\mathcal{A}$ relative to $\mathcal{M}$, $T_{\mathcal{I}\mathcal{A}\mathcal{M}}(A)$. This completes the semantics for L_ω.

3. AXIOMATIC THEORY

The logic for PRPs on conception 1. On conception 1 intensional entities are identical if and only if necessarily equivalent. Thus, on conception 1 the following abbreviation captures the properties usually attributed to the modal operator $\Box$: $\Box A$ *iff*$_{\text{df}}$ $[A] = [[A] = [A]]$. (That is, necessarily A *iff* the proposition that A is identical to any trivial necessary truth.) The modal operator $\Diamond$ is then defined in terms of $\Box$ in the usual way: $\Diamond A$ *iff*$_{\text{df}}$ $\neg\Box\neg A$.[63]

The logic $T1$ for L_ω on conception 1 consists of the axiom schemas and rules for the modal logic S5 with quantifiers and identity and three additional axiom schemas for intensional abstracts.

Axiom schemas and rules of $T1$.

A1. Truth-functional tautologies.

A2. $(\forall v_i)A(v_i) \supset A(t)$ (where t is free for v_i in A).

A3. $(\forall v_i)(A \supset B) \supset (A \supset (\forall v_i)B)$ (where v_i is not free in A).

A4. $v_i = v_i$.

A5. $v_i = v_j \supset (A(v_i, v_i) \equiv A(v_i, v_j))$ (where $A(v_i, v_j)$ is a formula that arises from $A(v_i, v_i)$ by replacing some (but not necessarily all) free occurrences of v_i by v_j, and v_j is free for the occurrences of v_i that it replaces).

A6. $[A]_{u_1 \ldots u_p} \neq [B]_{v_i \ldots v_q}$ (where $p \neq q$).

A7. $[A(u_1, \ldots, u_p)]_{u_1 \ldots u_p} = [A(v_1, \ldots, v_p)]_{v_1 \ldots v_p}$ (where these terms are alphabetic variants).

A8. $[A]_\alpha = [B]_\alpha \equiv \Box(A \equiv_\alpha B)$.

A9. $\Box A \supset A$.

A10. $\Box(A \supset B) \supset (\Box A \supset \Box B)$.

A11. $\Diamond A \supset \Box\Diamond A$.

R1. If $\vdash A$ and $\vdash A \supset B$, then $\vdash B$.

R2. If $\vdash A$, then $\vdash (\forall v_i)A$.

R3. If $\vdash A$, then $\vdash \Box A$.

THEOREM (Soundness and Completeness). *For all formulas A in L_ω, A is valid$_1$ if and only if A is a theorem of $T1$ (i.e., $\models_1 A$ iff $\vdash_{T1} A$).*[64]

The logic for PRPs on conception 2. On conception 2 each definable intensional entity is such that, when it is defined completely, it has a unique, noncircular definition. The logic $T2$ for L_ω on conception 2 consists of (a) axioms A1–A7 and rules R1–R2 from $T1$, (b) five additional axiom schemas for intensional abstracts, and (c) one additional rule. In stating the additional principles, I will write $t(F_p^q)$ to indicate that t is a complex term of L_ω in which the primitive predicate F_p^q occurs.

Additional axiom schemas and rules for T2.

$\mathscr{A}8$. $[A]_\alpha = [B]_\alpha \supset (A \equiv B)$.

$\mathscr{A}9$. $t \neq r$ (where t and r are nonelementary complex terms of different syntactic kinds.)[65]

$\mathscr{A}10$. $t = r \equiv t' = r'$ (where $R(t', t)$ and $R(r', r)$ for some term-transforming relation R, *or* t is the negation of t' and r is the negation of r', *or* t is the existential generalization of t' and r is the existential generalization of r').

$\mathscr{A}11$. $t = r \equiv (t' = r' \;\&\; t'' = r'')$ (where t is the conjunction of t' and t'' and r is the conjunction of r' and r'' *or* t is the predication$_k$ of t' of t'' and r is the predication$_k$ of r' of r'' for some $k \geqslant 0$).

$\mathscr{A}12$. $t(F_i^j) = r(F_h^k) \supset q(F_i^j) \neq s(F_h^k)$ (where t and s are elementary and r and q are not).

$\mathscr{R}3$. Let F_k^p be a nonlogical predicate that does not occur in $A(v_i)$; let $t(F_k^p)$ be an elementary complex term, and let t' be any complex term of degree p that is free for v_i in $A(v_i)$. If $\vdash A(t)$, then $\vdash A(t')$.[66]

THEOREM (Soundness and Completeness). *For all formulas A in L_ω, A is valid$_2$ if and only if A is a theorem of $T2$ (i.e., $\models_2 A$ iff $\vdash_{T2} A$).*

The logic for PRPs and necessary equivalence on conception 2. Let the 2-place logical predicate $\approx_N$ be adjoined to L_ω. $\approx_N$ is intended to express the logical relation of necessary equivalence. A *type* 2′ model structure is defined to be just like a type 2 model structure except that it contains an additional constituent Eq_N which is a distinguished element of $\mathscr{D}_2$ satisfying the following condition:

$$(\forall H \in \mathscr{K})(H(\mathrm{Eq}_N) = \{xy\colon (\exists i \geqslant -1)(x, y \in \mathscr{D}_i) \;\&\; (\forall H' \in \mathscr{K})(H'(x) = H'(y))\}).$$

Thus, Eq_N is to be thought of as the distinguished logical relation-in-intension *necessary equivalence*. Now an interpretation $\mathscr{I}$ relative to a type 2′ model structure is just like an interpretation relative to a type 1 or type 2 model structure except that we require $\mathscr{I}(\approx_N) = \mathrm{Eq}_N$. Then type 2′ denotation, truth, and validity are defined *mutatis mutandis* as before. The following abbreviations are introduced for notational convenience;

$$\Box A \text{ } \mathit{iff}_{\mathrm{df}} \text{ } [A] \approx_N [[A] \approx_N [A]]$$

$$\Diamond A \text{ } \mathit{iff}_{\mathrm{df}} \neg \Box \neg A.$$

The intensional logic *T*2′ consists of the axioms and rules for *T*2 plus the following additional axioms and rules for $\approx_N$:

$\mathscr{A}$13. $x \approx_N x.$

$\mathscr{A}$14. $x \approx_N y \supset y \approx_N x.$

$\mathscr{A}$15. $x \approx_N y \supset (y \approx_N z \supset x \approx_N z).$

$\mathscr{A}$16. $x \approx_N y \supset \Box x \approx_N y.$

$\mathscr{A}$17. $\Box(A \equiv_\alpha B) \equiv [A]_\alpha \approx_N [B]_\alpha.$

$\mathscr{A}$18. $\Box A \supset A.$

$\mathscr{A}$19. $\Box(A \supset B) \supset (\Box A \supset \Box B).$

$\mathscr{A}$20. $\Diamond A \supset \Box \Diamond A.$

*R*4. If $\vdash A$, then $\vdash \Box A$.

Notice that these axioms and rules for $\approx_N$ are just analogues of the special *T*1 axioms and rules for $=$. Finally, the soundness and completeness of *T*2′ can be shown by applying the methods of proof used for *T*1 and *T*2.

4. THE COMPLETENESS OF FIRST-ORDER INTENSIONAL LOGIC

We have indicated that first-order logic with identity and intensional abstraction is complete relative to certain technical notions of validity that are defined by means of intensional algebraic semantics. Consider the philosophical thesis that these technical notions of validity are in fact the standard notions of validity (or at least they resemble the standard notions in all respects relevant to genuine completeness results). From the technical result and the philosophical thesis it follows that first-order intensional logic is genuinely complete. This argument is parallel to that used to show that elementary first-order logic with identity is genuinely complete: the logic is proved complete relative to a certain defined notion of validity, and this technical result is then combined with the philosophical thesis that this defined notion is the same as (or resembles in philosophically relevant respects) the standard notion. In the case of elementary first-order logic with identity, the philosophical thesis has been subjected to much critical scrutiny, and something like a consensus has emerged in support of it. In the case of first-order intensional logic, the philosophical thesis strikes many people as highly intuitive. Nevertheless, some commentators (for example, Nino Cocchiarella [1985] and I. G. McFetridge [1984]) have expressed doubts. According to such doubts, the technical completeness result might be a mere artifact of a mistakenly narrow definition of validity that results from using an overly liberal definition of a model (in much the same way that Henkin's quasi-completeness result for higher-order extensional logic is a consequence of the liberal notion of a general model). Specifically, if certain plausible auxiliary closure conditions were imposed on the models, perhaps completeness would no longer follow; indeed, perhaps incompleteness could be derived. So go the doubts. However, these doubts are unfounded as we will now explain.

Consider two plausible closure conditions on models (described by Cocchiarella [1985] and McFetridge [1984], respectively). First, the set $\mathscr{K}$ of alternate extension functions must always be maximal, that is, it should not be possible to add further extension functions H to $\mathscr{K}$ without contradicting one of the original conditions in the definition of model. Second, for every subset s of $\mathscr{D}$ and every extension function H in $\mathscr{K}$, there must be an item x in the subdomain $\mathscr{D}_1$ of properties such that $H(x) = s$. (Notice that, if a model satisfies this closure condition, $\mathscr{D}$ must be a proper class. For the closure condition implies that there are as many properties in $\mathscr{D}$ as there are sub*sets* of $\mathscr{D}$. So if $\mathscr{D}$ were a set, the closure condition would contradict Cantor's power-set theorem.)

THEOREM. *First-order intensional logic is complete even if the strong closure conditions are imposed on models.*

In broad outline the proof goes as follows. We follow the Henkin-style proof given in *Quality and Concept* except that a proper class of individual constants are adjoined to the language, and for all distinct individual constants c and d, the sentence $c \neq d$ is adjoined to the theory. In the

Henkin model that results, these individual constants will comprise the subdomain of individuals. To obtain a model meeting the second closure condition, we massage this model in appropriate ways. First, partition this subdomain into denumerably many proper classes $d_{-1}, d_0, d_1, d_2, \ldots$. The first of these proper classes d_{-1} will be the subdomain of individuals in our new model. Then the i-th proper class ($i \geqslant 0$) will be adjoined to the old subdomain $\mathscr{D}_i$ to form the new subdomain of i-ary intensions, and the functions $H \in \mathscr{K}$ and the fundamental logical operations will be adjusted accordingly. When done properly the result is a model of the theory that meets the second closure condition. What makes this construction possible is the fact that our models have a single, unified domain $\mathscr{D}$ in which individuals, propositions, properties, and relations are taken as primitive entities. Finally, concerning the first closure condition, it is straightforward to show that every $\mathscr{K}$-maximal extension of the new model is also a model of the theory. The key to the proof is the fact that, for any algebraic intensional model, the values of all identity and necessity sentences must be the same in any $\mathscr{K}$-extension of a model as they are in the model itself.

This and similar results (given in Bealer [1987]) provide strong evidence that doubts about the genuine completeness of first-order intensional logic are unfounded and that the two-stage methodology (according to which intensional logic is treated prior to treating the logic for the predication relation) is vindicated.

Incidently, Cocchiarella [1985] claims that *incompleteness* can be proven when the first closure condition is imposed. He begins with the premise that every first-order necessity sentence $\ulcorner N[A(F_1, \ldots, F_n)]\urcorner$ – where $\ulcorner A \urcorner$ is a standard first-order extensional sentence and $\ulcorner F_1 \urcorner, \ldots, \ulcorner F_n \urcorner$ are the predicates occurring in $\ulcorner A \urcorner$ – is true in a model if and only if the second-order sentence $\ulcorner(\forall F_1 \ldots F_n)A\urcorner$ is true in the model. Then he claims that this implicit second-order element in first-order intensional logic is enough to prove incompleteness. But his premise rests on a straightforward error. To see why, choose any model in which interpretation $\mathscr{I}$ assigns the property of being self-identical to the primitive predicate 'F'. That is, $\mathscr{I}$('F') $= [x = x]_x$. On this interpretation, the intensional abstract '$[(\exists x)Fx]$' would denote $[(\exists x)x = x]$, i.e., the necessary proposition that something is self-identical. So on this interpretation the first-order intensional sentence '$N[(\exists x)Fx]$' is *true*. However, according to Cocchiarella's premise, this sentence is true on an interpretation if and only if the second-order sentence '$(\forall F)(\exists x)Fx$' is true on the interpretation. But this second-order sentence is *false*, indeed, it is *logically false*. So Cocchiarella's premise does not hold, and his alleged incompleteness proof fails.

The source of Cocchiarella's error is something like this. In PRP semantics, propositions are the primary bearers of necessity; the mere fact that a syntactically simple sentence like '$(\exists x)Fx$' is not true by virtue of its syntactic form tells us nothing about whether the proposition expressed by the sentence (i.e., the proposition that $(\exists x)Fx$) is necessary and, in turn, whether the sentence '$N[(\exists x)Fx]$' is true or false in a given model. Cocchiarella's error seems to arise from a kind of generalized use/mention confusion, that the syntactic form of a linguistic expression of a proposition and the modal status of the proposition should match up. But this only happens in special cases; it does not typically happen.

By the way, incompleteness of first-order intensional logic would follow if one adopted the following premise (which is adapted from an argument that Cocchiarella gives elsewhere about completeness results in first-order modal logic): for any first-order extensional sentence $\ulcorner A \urcorner$, the first-order intensional sentence $\ulcorner N[A] \urcorner$ is true in a model if and only if $\ulcorner A \urcorner$ is a logically valid first-order extensional sentence, i.e., a theorem of first-order extensional logic. From this premise it follows that, for every such sentence $\ulcorner A \urcorner$, the intensional sentence $\ulcorner \neg N[A] \urcorner$ is valid if and only if $\ulcorner A \urcorner$ is not a logically valid first-order extensional sentence. But the sentences $\ulcorner A \urcorner$ that are not logically valid first-order extensional sentences are not recursively enumerable, so the valid intensional sentences $\ulcorner \neg N[A] \urcorner$ are not recursively enumerable. Consequently, first-order intensional logic is incomplete. However, the premise would be based on an error. To see this, let $\ulcorner A \urcorner$ be the invalid first-order extensional sentence '$(\exists x)Fx$', and consider any model like that discussed a moment ago, wherein interpretation $\mathscr{I}$ assigns to 'F' the property of being self-identical, i.e., $[x = x]_x$. In such a model the intensional abstract '$[(\exists x)Fx]$' denotes $[(\exists x)x = x]$, i.e., the necessarily true proposition that something is self-identical. Therefore, in this model the first-order intensional sentence '$N[\exists x)Fx]$' would be *true*. However, according to the premise, this sentence should be *false* because the first-order extensional sentence '$(\exists x)Fx$' is not a logically valid sentence. So the premise does not hold, and thus the alleged incompleteness proof fails. Like the earlier erroneous premise, this erroneous premise seems to be based on a kind of generalized use/mention confusion, that the syntactic form of a linguistic expression of a proposition and the modal status of the proposition should always match up. And of course this is not so.

III: THE PROPOSITIONAL-FUNCTION THESIS, TYPE-FREE PREDICATION THEORIES, AND NONEXTENSIONALITY

In this final part we will take up a few sophisticated issues in property theory, specifically, the propositional-function thesis, type-free theories of the predication relation, and the proof of nonextensionality within such theories.

We have been defending the thesis that properties, relations, and propositions are required by our best comprehensive theory of the world. Part of the defense of the thesis consisted in the rejection of reductionist approaches which would have us believe that intensional entities can somehow be constructed from other primitive material. In compliance with this position we took PRPs at face value as irreducible entities and showed how they could be modeled within the framework of certain intensional algebras.

Formalizations of PRP theory within this framework run into some technical complexities in connection with the treatment of free and bound terms occurring in intensional abstracts. Many of these complexities seem initially to be avoidable by adopting the thesis that properties and propositional functions are structurally indiscernible. We do not mean by this that properties are reducible after all to a different realm of entities. It would be a grave mistake to construe the thesis of the structural indiscernibility as a strong identity thesis. Properties and propositional functions are not the same. But according to the thesis of structural indiscernibility, propositional functions can serve as an external model that displays the structural conditions that are imposed by the characteristic axioms of $T2$.

One aim of this section is to sketch a model for the consistent implementation of the structural propositional function thesis. The model we give in outline is a structural version of one variant of a predication theory of properties. Actually it constitutes an extension of Aczel's Frege Structures.[67] We thus depart from the policy we have adhered to in the main bulk of the paper, namely, to concentrate on problems that arise within the framework of theories of intensional abstracts. The simple reason for this departure is that we do not know how to construct a fine-grained functional model without relying on the expressive power provided by (some analogue of) the predication relation.

Given the many competing versions of predication theories of properties, we thought it appropriate not to present a formal syntax and an axiom system. We have even left open the question whether the model we indicate should be constructed using a classical or an intuitionistic metalanguage. That both approaches are possible is shown by Flagg and Myhill [1987] and by Aczel [1980], respectively. With the addition of the relevant stipulations our proposal could thus be made consistent with Feferman's [1984] observation that the arguments used to derive the paradoxes are already valid constructively.

To give the reader an idea of the reduction of complexity that can initially be achieved by working with a functional structure, let us briefly consider an illustrative example. The property of being believed by John to be a spy would be denoted by the following term of the formal language introduced above:

$$[B^2 j, [S^1 x]]_x.$$

The corresponding polynomial of the intensional algebra is as follows:

$$\mathrm{Pred}_1(\mathrm{Pred}_0([B^2 z, y]_{yz}, j), [S^1 x]_x).$$

In a suitable functional setting, by contrast, B^2 and S^1 would stand for a binary and a unary propositional function, respectively. If the functional setting satisfies a closure condition to the effect that every expression constructed in the expected way from variables, constants, and propositional functions stands for a propositional function, we are assured that the following is a complex propositional function: $B^2(j, S^1(x))$. This function sends an object b into that proposition that is the value of the binary function B^2 on the pair of arguments whose first component is the individual John and whose second component is the proposition that the function S^1 connects with the object b. The last step of the interpretation procedure within a functional framework would depend on the availability of a functional that maps every unary propositional function into a corresponding object. In a Frege Structure, lambda is a map that establishes the required association. Scott aptly summarizes the general idea behind the functional approach ". . . a formula with a free variable is a *mapping* from constants to the corresponding substitution instance. In this way we eliminate all fuss with variables in the formalization – the use of **lambda** does all the work behind the scenes".[68]

As we have noted in Part I, the iterative conception of set is ultimately analysed in terms of membership. This conception, which was introduced by Cantor and formalized by Zermelo, has to be distinguished from the Fregean logical notion of a class that is ultimately to be analyzed in terms of predication. This logical notion of class is usually thought to be incoherent because of the derivability of Russell's paradox within a system that contains an unrestricted comprehension principle. The logical notion of class can be explicated somewhat more precisely as follows. To any formula $\ulcorner A(x) \urcorner$ that has 'x' as its only free variable there corresponds a class $\{x \mid A(x)\}$ such that for any object b the proposition that $b \in \{x \mid A(x)\}$ and the proposition that $A(b)$ are equivalent.[69] If we assume $\ulcorner A(x) \urcorner$ to be the Russell formula '$x \notin x$' and let 'R' denote the class $\{x \mid x \notin x\}$, then the proposition that $R \in R$ is equivalent to the proposition that $R \notin R$.

There have been many attempts to formulate type-free theories that allow the sort of unlimited self-reference implied by a general abstraction principle without suffering from the devastating effect of Russell's paradox. The story is a long and intricate one, and we are not going to recount it here. Feferman [1984] contains some very useful historical remarks. According to these remarks and other written records, it is Fitch [1948] who is to be credited for the renewed interest in type-free theories.

Our idea is to formulate a type-free theory of properties in which the notion of property corresponds closely to the Fregean logical notion of class, except of course that the identity conditions on properties will be intensional. The particular version of a type-free theory of properties on which we will base our discussion in this section introduces a primitive notion of proposition. Intuitively, propositions constitute the category of objects that can be true or false. If such a propositional formulation of the logical class concept does not contain among its theorems the assertion that $R \quad R$ is a proposition, no Russell paradox cna be derived from the axiom schema of full abstraction. This is the style of resolution we will pursue here.

It is well known that Frege espoused an extensionalist position according to which concepts differ only in so far as their extensions are different. This seems to be a basic mistake. We commented above on the intuitive justification of a theory of intensional entities that rejects the principle of extensionality. What we attempt now is to back this intuitive justification by something approximating a proof. Given full abstraction, the principle of extensionality assumes the following form:

$$(\forall x)(A(x) \leftrightarrow B(x)) \rightarrow \{x \mid A(x)\} \quad = \quad \{x \mid B(x)\}$$

(We use $\ulcorner\{x \mid A(x)\}\urcorner$ as a neutral abstract; our purpose is to show that, within a predication theory with an unrestricted abstraction principle, these abstracts *must* denote intensional entities, contrary to what Frege thought.) Let Q be the object $\{x \mid x \in x = \text{FALSE}\}$. $Q \in Q$ certainly should be a proposition because, given full abstraction, it is equivalent to $Q \in Q$ = FALSE, and we have seen above why it is highly commendable to regard identity as a transcategorial relation that can be predicated of any two objects, always yielding a proposition. Now given full abstraction and extensionality, $Q \in Q$ would be identical to the proposition $Q \in Q$ = FALSE. In this case, however, $Q \in Q$ can be neither a true proposition nor a false proposition. $Q \in Q$ cannot be a true proposition because, by this last identity, it follows that $Q \in Q$ = FALSE would be a true proposition. But, in this case, it follows by the principle of full abstraction that $Q \in Q$ is identical to FALSE, and so $Q \in Q$ is a false proposition. On the other hand, $Q \in Q$ cannot be a false proposition, for the principle of extensionality implies that there is only one false proposition, namely, FALSE. So if $Q \in Q$ were a false proposition, it would be identical to FALSE. But, in this case, the principle of full abstraction implies that $Q \in Q$ = FALSE would be a true proposition. But since this proposition is identical to $Q \in Q$, it follows that $Q \in Q$ would be a true proposition.

This argument shows that the principles of full abstraction and of extensionality, combined with the assumption that identity is a transcategorial relation that always yields a proposition when predicated of any pair of objects, leads to contradiction.[70] Since the unlimited self-reference embodied in the principle of full abstraction is part and parcel of a theory of properties, relations, and propositions and since the transcategorial characteristic of identity cannot be given up in a type-free framework, it is the extensionality principle that is at fault here. Our conclusion is that intensionality is a necessary feature of a type-free system that allows for unlimited self-reference in the form of an unrestricted abstraction principle. (Notice, incidentally, that our original derivation of Russell's paradox did not invoke any extensionalist claim. That contradiction cannot be blocked by relying on some high degree of fine-grainedness for the particular proposition presumably denoted by '$R \in R$'.)

If a predication theory of properties is committed to an intensionalist stance, we were well advised to assume the same attitude towards the meaning of intensional abstracts within the weaker framework constructed earlier. Can we provide a similar justification for our decision to regard PRPs as particular types of objects as against, e.g., certain kinds of propositional functions (where propositions would then be treated as constant propositional functions)? The debate around the strict propositional-function thesis – that is, the thesis that properties and propositional functions are identical – has been obscured by a number of confusions. We address some of them in the following discussion in order to illustrate our point of view and to forestall certain objections that might be engendered by these issues.

Frege is even better known for his ideosyncratic ontology than for his extensionalist position. He divides the world into objects and functions and attempts to impute an understanding of the latter notion by resorting to metaphors of unsaturatedness and completability. We think it is perfectly legitimate to speak of "incomplete objects" and other entities of such ilk if one wants to impart an understanding of a category otherwise proved to be ineliminable. One such well known proof runs as follows. Suppose that Fregean concepts are objects and can be designated by proper names. That an object a falls under (or is an instance of) the concept denoted by 'P' is to be indicated by '$P(a)$'. We have formed a new predicate by appending a pair of parentheses to the name of the object that is by supposition identical with the original concept. In sentences such as '$\neg P(P)$', '$\neg Q(Q)$', etc., we can discern a common pattern of predication. Assuming this pattern stands for a concept, we denote it with the proper name 'S' and construct a new predicate '$S(\ )$'. Does the object referred to by 'S' fall under the concept expressed by '$S(\ )$'? According to its definition, '$S(S)$' is true if and only if '$\neg S(S)$' is true, contradiction. A Fregean would conclude from this: "To escape this absurdity, we must deny that any concept is an object or can have a proper name; and the two sorts of quantification that answer to proper names and to predicates must be strictly distinguished".[71] Is there no other way out?

This question is connected with the problem of empty proper names and the acceptability of non-definite concepts, where a concept P is non-definite if the following assertion fails:

$$(\forall x)(P(x) \vee \neg P(x)).$$

We leave open the question whether Frege changed his mind on these issues under the weight of Russell's paradox. At one time he seems to have entertained the idea of allowing denotationless nominalizations into his system. Realists will sense a tension between their basic world-view and the notion of a non-definite concept. The same tension exists between a realist metaphysics and the primitive notion of proposition we mentioned above. But since, to repeat, it is highly unlikely that in the foreseeable future we are really going to understand the pathology of the paradoxes or that we are really going to know what form an ideal resolution of them should take, we feel justified in pursuing the way out indicated by the primitive notion of proposition.

If at the moment there is no proof that both objects and functions have to be counted among the primitive ontological categories, is it then at least technically feasible to work within a framework that models properties as propositional functions? It is possible, and we will show in outline how this can be done. But first we have to explain what we understand by 'technically feasible'.

The coarse-grained and fine-grained algebras above were presented as certain sets. We took for granted the necessary amount of set-theoretical machinery that enabled us to state the definitions and carry through the completeness proof. Nobody will want to identify properties, relations, or propositions with sets. For the same reason it does not make any sense to identify the operation Pred_0 with a particular single-valued set of ordered pairs with respect to every algebraic model. The set-theoretic definition of validity fulfills a useful role, nevertheless. Set-theoretic semantics can serve as a guide in our investigation of the realm of PRPs. This was true in the construction of the intensional algebras in Part II, and it is true in the investigation of type-free theories of predication. In the end, however, the set-theoretic ladder that we climbed has to be kicked away and supplanted by an intrinsic semantics stated wholly in terms of an applied theory of properties. According to the same line of reasoning, it would be completely implausible, for example, to equate sensible properties with functions. Propositional functions are to be invoked for no other purpose than to serve as an external criterion for such formal characteristics as soundness and completeness of a theory of properties. The two types of external modelings cannot claim any advantage over the other from this perspective.

The relevant literature contains an argument that purports to show that propositional functions do not have the right structure for being of much use as a reliable external guide in the realm of PRPs.[72] Stripped of all irrelevant details, the argument boils down to the following steps. The logic of propositional attitudes demands that very fine-grained distinctions be made among propositions. These distinctions are thought to be analogous to the syntactic pattern of expressions that convey propositional content. Let 'R' be a two-place predicate and a and b two different objects. Comprehension allows us to introduce by stipulation two properties $P = \{x \mid R(x, b)\}$ and $Q = \{y \mid R(a, y)\}$. (As before, we are using $\ulcorner\{x \mid A(x)\}\urcorner$ as a neutral kind of abstract.) If properties are indeed identical to propositional functions and if the general characteristics of functions find an adequate expression in the lambda-calculus, then the following equations are instances of the rule of β-conversion and of the intersubstitutivity of definitional identities:

$$P(a) = \{x \mid R(x, b)\}(a) = R(a, b) =$$
$$\{y \mid R(a, y)\}(b) = Q(b).$$

To illustrate the fact that these equations model a far too coarse-grained notion of proposition, let R express the relation of following and a and b stand for Jane Fonda and Rajneesh, respectively. Then we can stipulate that

> Being an x such that x rajneeshes = being an x such that x follows Rajneesh. $(= P)$

and

> Being an x such that x fondalees = being an x such that Jane Fonda follows x. $(= Q)$

From these stipulations and the propositional-function thesis we have by β-conversion:

> The proposition that Jane Fonda rajneeshes = the proposition that Rajneesh fondalees.

But this seems wrong. When a person consciously and explicitly thinks that Jane Fonda rajneeshes must that person be consciously and explicitly thinking that Rajneesh fondalees? It certainly does not seem so. The order of predication seems to be a relevant factor of propositional identity.

The same point can be made perhaps even more palatable if we restate the example within a quasi-categorial framework where every element can be required to be either a projection or a constant function of arity $n \geqslant 0$ or a definable function in some suitably restricted sense. Definability is to be understood as closure under application, functional composition of the basic set of constant and projection functions, and an additional set of functionals which counts lambda among its members and where lambda is defined as a map that sends elements of the space of unary functions into the universe of objects. When these closure conditions are made precise, it is not very difficult to check that such a structure of functions is nothing else but a model of the lambda-calculus.[73]

If we interpret the relation R of our example as a two-place function, a and b as zero-place constant functions, and the variables as one-place projections p, we obtain the equations as follows:

$$P \circ a \;=\; R \circ (p, b) \circ a \;=\; R \circ (a, b) \;=$$
$$R \circ (a, p) \circ b \;=\; Q \circ a$$

where $\circ$ indicates functional composition[74] such that

$$f \circ (g_1, \ldots, g_n)(a_1, \ldots, a_m) \;=$$
$$f(g_1(a_1, \ldots, a_m), \ldots, g_n(a_1, \ldots, a_m))$$

for f an n-place and g_i m-place functions. Something like functional composition cannot be separated from the very notion of a function. Is it logically possible for a function to behave in a way that violates the principle of functional composition we just stated?

It is, of course, possible to construct "intensional" models in which the order of execution of a complex operation determines its meaning. In such models the "subject-first" and the "object-first" R could be distinguished in the sense that, even though for all pairs of objects both operations would have the same value, they are nevertheless not the same:

$$\neg[(\forall z)(\text{“Subject-first” } R(z) \;=\; \text{“Object-first” } R(z)) \rightarrow$$
$$\{z \mid \text{“Subject-first” } R(z)\} \;=\; \{z \mid \text{“Object-first” } R(z)\}].$$

This intensional notion of a function that incorporates aspects of sequential behavior is of no help with respect to our problem of the internal structure of propositions in propositional attitude contexts. If $P(a)$ and $Q(b)$ are the same proposition, then the fact that the dynamic behavior of P is to be analyzed in terms of "Object-first" R and the dynamic behavior of Q is to be analyzed in terms of "Subject-first" R gives us no handle on the resulting proposition unless these sequential stages could be discerned as parts of the final value. But this is denied by the presupposition that the input-output behavior of the two sequential ingredients is the same.

Proponents of the propositional-function thesis seem to be caught in a dilemma. On the one hand, they have an elegant solution to the worries that have plagued those scholars who tried to reconstruct intensional entities out of the primitive elements provided by the possible-worlds approach. They have the option of relying upon a principle of individuation for intensions that is finer than necessary equivalence, the latter notion being reduced to identity of truth value on all possible worlds in the case of propositions. It is certainly true that anything that is triangular in a possible world is trilateral in that same world. This does not imply that for any object *b* the proposition that *b* is triangular is the same proposition as that *b* is trilateral. When a person consciously and explicitly thinks that *b* is triangular *must* that person be conciously and explicitly thinking that *b* is trilateral? It certainly does not seem so. The propositional-function thesis does not entail the unwanted identifications implicit in the possible-worlds approach. Thus, on this score, there is no objection against an external semantics that models properties as propositional functions and interprets the operation of predication as functional application.

On the other hand, the propositional-function thesis still seems to imply a sort of unwanted identification, namely, the unwanted identification illustrated in the rajneeshe/fondalee example. Moreover, some of these unwanted identifications that involve an inverted sequence of predications do not even depend upon the availability of a functional that permits one to assert the existence of an object that is specified (or projected or comprehended) by a complex predicate $\ulcorner A(x) \urcorner$. As we have seen, the same problem arises within a functional structure that treats the predication operation as functional composition instead of first forming a new object corresponding to the original one-place propositional function and then combining this new object and its argument with the help of the binary application operation. Both lambda-abstraction and functional composition have a specific intended meaning that is enshrined in the principle of β-conversion and the principle of functional composition that we have stated above. Giving up these basic principles or distorting the intended meaning of either lambda-abstraction or functional composition are but two sides of the same coin. And these two principles are operative in the semantic conflation of syntactically distinct expressions. Since we see no way of tampering with either lambda-abstraction or functional composition and their attending principles without violating in essential respects the basic intuition that guides our understanding of the notion of function, we have to admit that the propositional-function thesis is mistaken.

Or is it? As we have put it, the thesis is that properties can be externally modeled by propositional functions. In our discussion we have made the assumption that this "model-theoretic" decision would entail the identification – again in the sense of an external criterion – of the operation of predication with functional application. The two identificatory decisions together lead to the counterintuitive conclusion that Jane Fonda's rajneeshing and Rajneesh's fondaleeing stand for the same thought. But would it be possible to split the propositional-function thesis into two separate claims and retain the first half of the thesis of external modeling while rejecting the second half? The answer, as we have indicated already, is that a manoeuvre like this can be carried out technically. We will show one way to do this in a moment. There are two purposes in doing this. The first is to construct a "predication structure" over certain functional models; such a structure can serve as a model of PRPs. The second is to provide an illustration of how there could indeed be a distinction between functional application and predication.

To motivate the construction, it is perhaps helpful to look at the problem from a syntactic point of view. If one tries to repeat the story about fondaleeling and rajneeshing within the confines of the expressive power provided by the first-order language L_{ω} for intensional abstraction, one stumbles upon the problem of how to indicate the predication relation that is to hold between the abstracts $[Rx, b]_x$ and $[Ra, y]_y$ on the one hand and their respective arguments on the other. Abstracts are names, and the result of juxtaposing an abstract and an individual constant does not form a well-formed expression. Would a more liberal syntax that imposed no type restrictions on the concatenation of expressions bring granularity troubles on the semantic level in its wake? Not necessarily. As long as such liberal theories contain no principles to the effect that schemata of the form $\ulcorner[A(x)]_x(t)\urcorner$ have the same meaning as $\ulcorner[A(t)]\urcorner$, we still steer clear of the troubled waters of intensional individuation problems. What this amounts to semantically is that predication and functional application should be different functionals. Our construction will show that there can indeed be a functional model with distinct functionals like this.

Specifically, we are looking for a model that contains a functional Pred that sends pairs of propositional functions and objects into propositions and that satisfies the following principles:

$$\text{Pred}(P, a) \;=\; \text{Pred}(Q, b) \rightarrow (P = Q \;\&\; a = b).$$ [75]

$$\text{Pred}(P, a) \leftrightarrow P(a).$$

As we have seen, the justification for a functional satisfying these principles derives from the requirement of fine-grainedness that is generated by propositional-attitude contexts.[76]

The construction of Pred makes use of a standard inductive definability approach. The technique goes back to Fitch as we mentioned above. Closely related ideas have been used by Feferman [1975], Scott [1975], and Aczel [1980].

Let M be a model for the lambda calculus. For definiteness we assume that it belongs to the class of lambda-*systems* or lambda-*families*. Within the model we can define elements like $=$, N, $\neg$, &, $\vee$, $\rightarrow$, $\forall$, $\exists$ *and* Pred in the following way:

$$= \;=_{\text{df}}\; \langle 0, x, y\rangle \qquad N \;=_{\text{df}}\; \langle 1, x\rangle$$
$$\neg \;=_{\text{df}}\; \langle 2, x\rangle \qquad \& \;=_{\text{df}}\; \langle 3, x, y\rangle$$
$$\vee \;=_{\text{df}}\; \langle 4, x, y\rangle \qquad \rightarrow \;=_{\text{df}}\; \langle 5, x, y\rangle$$
$$\forall \;=_{\text{df}}\; \langle 6, \lambda x(fx)\rangle \qquad \exists \;=_{\text{df}}\; \langle 7, \lambda x(fx)\rangle$$
$$\text{Pred} \;=_{\text{df}}\; \langle 8, \lambda x(fx), y\rangle$$

Since we are working within a lambda-system, the closure conditions ensure that every element actually denotes a function of the model. Furthermore, these functions enjoy a certain independence property. In the special case of Pred this means that the following holds:

$$\text{Pred}(P, a) \;=\; \text{Pred}(Q, b) \rightarrow (P = Q \;\&\; a = b).$$

Based on the above definitions two subsets $\mathcal{T}$, the set of truths, and $\mathcal{P}$, the set of propositions, can be obtained as the least fixed point of a monotone operator on the lambda-system. The clauses of the monotone operator are constituted by a family of $\mathcal{T}$-positive and $\mathcal{P}$-positive formulas that express the expected characterics of the defined elements. Except for Pred these clauses (with minor variations) can be found in the cited papers by Scott, Feferman, Aczel, or Flagg-Myhill. The clause for Pred can informally be stated as follows:

PREDICATION. *If f is a propositional function (i.e., if f is a function all of whose values are propositions), then Pred(f, a) is a proposition such that: Pred(f, a) is true iff $f(a)$ is true.*

From the predication schema it becomes clear that, semantically, the new functional is nothing else but "proper-name" quantification.[77]

To emphassize the main goal of our discussion of propositional functions. It has not been our objective to provide an argument for the propositional-function thesis in its strong form. We wanted rather to make clear that the functional approach, as an external modeling technique, is flexible enough to accommodate any degree of structural discrimination that is deemed necessary.

Reed College, Portland
University of Tübingen

NOTES

[1] For financial assistance during the preparation of this work, the first author is grateful to the National Endowment for the Humanities and the second author is grateful to the Deutsche Forschungsgemeinschaft. Both authors thank the Seminar für natürlich-sprachliche Systeme at the University of Tübingen for facilitating their joint work on this project.

[2] We do not rule out the possibility of an argument to the effect that intensional entities are required for acceptable definitions of such notions as evidence (data), explanation, necessity, causation, law of nature, simplicity, and so forth. (See Section 1 of Bealer [1986].) The worry is that opponents of intensional entities might maintain that we should simply take these notions as primitives rather than trying to define them at the cost of adding intensional entities to our ontology. In the text we are seeking an argument that cannot be rebutted in this fashion. The idea is to show that any acceptable comprehensive theory would be self-defeating unless it makes use of a background intensional logic; but such use of intensional logic generates an ontological commitment to intensional entities. The conclusion is that our opponents cannot consistently deny the existence of intensional entities unless their comprehensive theory is unacceptable by their own standards. This style of argument derives from George Myro [1981]. It is developed further in Section 1 of Bealer [1986] and in Section 7 of Bealer [1988].

[3] We believe that any acceptable comprehensive theory must include an intensional semantics for natural language. But in a debate with a diehard opponent of intensional entities, this thesis must be established; it may not be assumed. The way to trap diehard opponents of intensional entities is to catch them in a self-defeat. In their own comprehensive theories opponents of intensional entities must have an account of the acceptability of their own theories. Such account, we will argue, must make use of a background intensional logic.

[4] Since one is justified in believing that one's acceptable comprehensive theory is true, it follows that one is justified in believing that intensional entities exist. Anyone who denies that intensional entities exist does so at the price of not having an acceptable comprehensive theory.

[5] This theme is developed in Chapter 9 of *Quality and Concept*.

[6] Proponents of free logic believe that existential-generalization failures in extensional logic are not confined to extraordinary cases like this one; they hold that even the most ordinary extensional instances of this rule are, strictly speaking, not valid. So free logicians would agree that failure of existential generalization cannot be used as a criterion for distinguishing intensional logic from extensional logic.

[7] Some of these arguments are fuller presentations of arguments already given in Chapter 1 of *Quality and Concept*. It is hoped that these fuller presentations will help to answer certain questions that critics have raised.

[8] In a syntactically first-order language, neither sentences nor predicates are allowed as singular terms; accordingly, one is not allowed to replace them with quantifiable variables.

[9] In a syntactically higher-order language, in contrast to a first-order language, there are variables whose substitutends are predicates and/or sentences.

[10] For other counts against the higher-order approach, see Section 12 below.

[11] We prefer treating predicate adverbs right within our sort of standard logical syntax. For example, 'x is running quickly' can be represented (very roughly) along the following lines: '$(\exists y)(y$ is a running & y is quick & x is doing $y)$'. This approach requires adding no new syntactic categories to our logical syntax; and, semantically, it requires no special ontological

items beyond PRPs, which as we will see are required to deal with 'that'-clauses and gerundive and infinitive phrases. Incidentally, once we are able to represent 'that'-clauses, sentential adverbs are easy to treat. For example, 'Necessarily, $7 < 9$' can be treated as a transformation from '$N^1[7 < 9]$', where '$[7 < 9]$' is a singular term corresponding to the 'that'-clause 'that $7 < 9$'.

[12] This notation is introduced by Quine for somewhat similar purposes (Quine [1960], §35), and it is used throughout *Quality and Concept*. For the moment we leave open what semantical significance the bracket notation shall have, and the possibility of indirectly defining the bracket notion shall also be left open here.

[13] Advocates of free logic might claim that argument (II) is not strictly speaking valid unless it is supplemented with the premise ⌜That A is something⌝ or ⌜There is something that is identical to that A⌝. We need not suppose otherwise. To accommodate the free logician, we would supplement (II′) with the premise ⌜$(\exists x)x = [A]$⌝. The philosophical point is this. We are, at the present stage, arguing merely that 'that'-clauses should be treated as singular terms. This treatment is compatible with free logic. The question of whether 'that'-clauses actually refer to anything and, accordingly, whether they have ontological significance is a separate question. Our argument that they do is given in Section 5.

So far we have established only that 'that'-clauses are singular terms. This thesis implies, for example, that 'that'-clauses cannot be treated in the way explitives are treated, that is, in the way 'it' in 'it is raining' is treated.

We should also make it clear that, strictly speaking, our treatment of 'that'-clauses in this section is consistent with the higher-order theory that entire sentences can occur as singular terms, for nothing we have said prevents a higher-order theorist from treating, say, ⌜$[A] = [B]$⌝ as a notational variant of ⌜$A = B$⌝. However, as we have already indicated, this higher-order theory has the unacceptable consequence that, for example, the grammatical nonsense 'The cat is on the mat $= 7 < 9$' is a well-formed sentence. For this and many other reasons (see Section 12), we do not advocate the higher-order theory.

[14] The possibility of externally quantifiable occurrences of variables is not allowed in Quine's original bracket notation.

[15] See Alonzo Church [1951].

[16] See Sections 1 and 6 of Bealer [1988] for further discussion of self-constituency.

[17] This regress could be avoided if a primitive notation ⌜$\langle v_i, v_j\rangle$⌝ for sequences were adjoined to the language. The problem with this approach is that one should nevertheless be able to say, specifically, what sort of entities are semantically correlated with these singular terms ⌜$\langle v_i, v_j\rangle$⌝. On the one hand, perhaps these entities are sets. But if so, this leads one right back to the ontological excess and theoretical disunity we found in the set-theoretical version of the sequence theory of quantifying-in. On the other hand, perhaps the singular terms ⌜$\langle v_i, v_j\rangle$⌝ are semantically correlated with properties. But if so, with what sort of properties? It would be unacceptably mysterious if you could not say which. However, if you confine yourself to a language fitted out with just *de dicto* intensional abstracts and an apparatus for expressing the predication relation, you get caught in the regress mentioned in the text; you can never complete your answer to the question. The only alternative is to admit into your language intensional abstracts that contain externally quantifiable variables (or some other apparatus with comparable expressive power). In this framework, it is then easy to specify the sort of properties with which the terms ⌜$\langle v_i, v_j\rangle$⌝ are semantically correlated. They are *de re* properties like $[x = [y = v_i]_y \vee x = [y = v_i \vee y = v_j]_y]_x$.

Incidentally, you might favor treating $\ulcorner\langle v_i, v_j\rangle\urcorner$ as an *indefinite description*, perhaps introduced by contextual definition in terms of the predication relation. (See p. 83, *Quality and Concept*, for an illustration.) However, this move would not entitle you to avoid the question of the general sort of entity with which these expressions are semantically correlated. For property theorists, there seems to be no satisfactory answer to this question that does not invoke *de re* properties as unreducible entities.

[18] If $\ulcorner B\urcorner$ has no externally quantifiable individual variables, then x directly believes that B if and only if x self-ascribes the property of being such that B. In symbols: A^2x, $[Fv]_v$.

[19] In a fully developed solution of the substitutivity problems involving indexicals, we would need to mark a distinction between *convictions in acquaintance* and *cognitive commitments*. (See Section 29 of *Quality and Concept* for discussion of this distinction.) We do not rule out the possibility that the notion of conviction in acquaintance might be elucidated in terms of the notion of self-ascription. The point is that such elucidation does not help to win the *ontological* point Chisholm favors, nor does it support a *logical* theory that eliminates externally quantifiable variables from 'that'-clauses, gerundive phrases, and infinitive phrases. Although the self-ascription theory might be able to make a contribution to epistemology and philosophical psychology, it appears to have little to contribute to metaphysics and logical theory.

[20] D. Davidson [1964].

[21] Indeed, given the radical theory that no atomic intensional sentence is true, it is impossible for there to be any *sound* argument for the radical theorists' conclusion about the acceptability of their theory. To see why, consider any argument whose conclusion is (the proposition expressed by) 'It is acceptable that no atomic sentence in the standard intensional idiom is true'. Since this sentence is itself an atomic sentence in the standard intensional idiom, it is not true. Therefore, (the proposition expressed by) this sentence cannot follow validly from true premises.

Another observation is in order. If (a proposition expressed by) a sentence implies so directly that it is itself not true, then (the proposition expressed by) such a sentence is not acceptable either. It follows that the radical theorists' conclusion about the acceptability of their theory is not acceptable.

[22] To be relevantly like the standard idiom, the new idiom must have systematic relations to such matters as: the simplest explanation of the evidence, the simplest coherent systematization of one's beliefs, a reliably caused body of beliefs, and so forth. However, the standard idioms for discussing these matters (explanation, evidence, belief, causation, etc.) are intensional. For example, 'It is evident to me that I am having sense experiences', 'That I have sense experiences is explained by physiology and psychology', 'I believe that so and so', 'That my brain is in such and such state causes me to have these sense experiences', 'It is causally necessary that, if my brain is in such and such state, I have these sense experiences', etc. (Of course, there are other standard ways of talking about explanation, evidence, belief, and causation. But they have systematic relations to these standard intensional idioms. If they did not, they could, for all we know, be just some new idiom whose relevance to explanation, evidence, belief, and causation would be in question.) Because the standard idioms are intensional, the radical theorist must deem our ordinary uses of it as, strictly speaking, false. For this reason, the radical theorists have no choice but to introduce some new, nonintensional idioms for discussing explanation, evidence, belief, and causation, and they must be able to show that these new idioms are relevantly like the standard intensional idioms. But

this can be done only by showing that the new idioms are relevantly like the standard ones in meaning, purpose, or function (or something relevantly like meaning, purpose, or function).

[23] To show that the meaning of an expression in a new idiom is *relevantly like* the meaning of an expression in a standard idiom, one has four options. First, one can show an actual meaning identity. But we have seen that statements of meaning identities have systematic relations to statements of intensional identities: the meaning of ⌜A⌝ = the meaning of ⌜B⌝ if and only if that A = that B. So intensionality enters in here. Second, one can show that the two expressions are definitionally related. However, the standard devices for indicating definitional relationships are intensional, for example: 'iff_{df}', '$=_{\text{df}}$', 'It is definitionally true that', and so forth. So this option does not lead to the elimination of intensionality. Third, one can show that the purpose or function served by the meanings of the two expresions is the same. However, our standard idiom for discussing purpose and function is also intensional. For example. ⌜The purpose of F-ing is to G⌝, ⌜The function of F-ing is to G⌝, etc. contain gerundive and infinitive phrases, which are intensional abstracts. Fourth, one can show that the two meanings are inherently similar. However, to show that two items are inherently similar, one must show that they share fundamental qualities and relations. But a general theory of fundamental qualities and relations is already a property theory; indeed such a theory is, on its own, sufficient for the construction of intensional logic. (See Chapter 8 of *Quality and Concept* for an elaboration of this argument. See also David Lewis [1983].) On all four options, therefore, intensionality – or a framework that implies it – plays a central role.

[24] To avoid this self-defeat, the radical theorists might try to define meaning (or something relevantly like it) in terms of Gricean intentions, or they might try to define purpose or function (or something relevantly like them) in terms of intention or causation. However, our standard idioms for giving definitions are intensional, for example: 'iff_{df}', '$=_{\text{df}}$', 'It is definitionally true that', and so forth. Moreover, our standard idioms for talking about intention and causation are also intensional: ⌜x intends to F⌝, ⌜x's F-ing caused y to G⌝, It is causally necessary that if Fx then Gy⌝, and so forth. So intensionality is not avoided. (Of course, there is a standard extensional idiom for talking about causation. However, it bears systemantic relationships to the standard intensional idiom. If the radical theorists do not affirm these systematic relationships, they are obliged to show either that their extensional idiom is still the standard one or that, if it is nonstandard, it is relevantly like the standard one.)

If, to avoid this intensionality, the radical theorists dropped these standard intensional idioms for definition, intention, and causation and if they put some new, nonstandard idioms in their place, they would then be obliged to show that the new idioms are relevantly like the standard intensional ones. But this can be done only by showing that they are relevantly like the standard ones in meaning, purpose, or function (or something relevantly like meaning, purpose, or function). To do this, the radical theorists are pushed right back into the problem.

[25] Of course, a radical theorist might simply present us with some novel scheme for determining acceptability. (This is pretty much what is done in Paul Churchland [1979] and [1981] and in Patrica Churchland [1986]. The problem is that there is no reason to pay any attention to claims made within the new scheme unless this scheme can be shown to be relevantly similar to the standard scheme for determining acceptability. We have seen that, to do this, either the proponents of the new scheme must use some standard intensional idiom, or they

must use an alternate idiom that can be shown to be relevantly similar to a standard one. And we have seen that latter option leads to a vicious regress unless at some stage the proponents of the new scheme invoke some standard intensional idiom to stop it. If they do not do this and if they persist in holding their position, they end up in groundless dogmatism. They would be like people with a magic device that purports to tell them when a candidate comprehensive theory is acceptable; the device does this by flashing the word 'acceptable' or 'not acceptable' when (the linguistic expression of) various candidate theories are typed in. Regarding the acceptability of their own comprehensive theory (which includes a theory of acceptability based on the device), they declare that their theory is acceptable on the grounds that the device has flashed the word 'acceptable' when they type in (a linguistic expression of) their theory. Plainly, their theory is not acceptable, nor is it relevantly like an acceptable theory.

[26] Meinongians might try to avoid this conclusion by invoking their (alleged) distinction between being and existence. However, to develop their views formally, Meinongians already admit the ontology of PRPs and would therefore have no good reason not to accept the natural style of PRP semantics for intensional abstracts that we are defending in the text.

[27] There is another problem with identifying linguistic expressions with linguistic tokens, namely, that it does not provide enough items for a general theory of language. A general theory of language must hold for the infinitely many expressions in a language, not just for the finitely many expressions that happen actually to be uttered or written by speakers. Because there are only finitely many (actual) linguistic tokens, tokens cannot play the role of linguistic expressions in a general theory of language. One way of trying to overcome this cardinality problem is to identify linguistic expressions with regions of physical space. Another way to overcome the problem is to identify linguistic expressions with certain set-theoretical constructs whose ultimate elements are (actual) linguistic tokens. (Quine, for example, identifies a primitive linguistic expression '*p*' with the set of actual tokens of '*p*', the primitive linguistic expression '$-$' with the set of actual tokens of '$-$', and the complex expression '$-p$' with the ordered set consisting of the set of actual tokens of '$-$' and the set of actual tokens of '*p*'.) Three observations are in order. First, both the regions-of-physical-space treatment and the set-theoretical treatment run into the problem of contingent existence, which we are discussing in the text. For neither regions of physical space nor sets that depend on linguistic tokens *necessarily* exist. Second, regions of physical space are particularly implausible candidates for being the primary bearers of truth, necessity, logical truth, etc. and the primary objects of mental representation, explanation, etc. How, for example, do regions of space succeed in *representing* things in the world? (A kindred problem besets the Quinean alternative. For more on this problem of explaining representation, see the discussion that follows shortly in the text.) Third, the set-theoretical treatment requires positing two distinct ontological categories – particulars and sets. From the point of view of pure ontological economy, this is no better than positing the two categories of particulars and properties. At the same time, the latter ontology has a clear intuitive advantage: it enables one to adopt the intuitive theory that linguistic expressions are just shapes or sound types. For shapes and sound types are properties *par excellence*.

[28] R. Carnap [1947].

[29] A. Church [1950]. For expository convenience, we have renumbered the sentences mentioned in Church's argument.

[30] These "sets" would also be non-well-founded. See section 11 for a critique of treatments of intensional logic that posit non-well-founded "sets".
[31] This theme is developed in Section 42, "Realism and Representationalism", in *Quality and Concept*.
[32] For more on this, see Section 42, *Quality and Concept*.
[33] On the Kripke-Putnam view, names of natural kinds are introduced by means of "reference-fixing descriptions" that apply only contingently to their bearers. Such descriptions therefore do not qualify as definitions and, hence, cannot double as "constructions" of these fundamental properties from intensions that are "given" in our earlier mental activity.
[34] Another problem with this move concerns quantifying-in. Suppose, for the sake of argument, that ⌜Possibly B⌝ is not equivalent to ⌜It is possible that B⌝. Nevertheless, ⌜$\Diamond(\exists x)(F^2x, [A_i] \,\&\, LT^1[A_i])$⌝ intuitively entails ⌜$(\exists u)(u = [A_i] \,\&\, \Diamond(\exists x)(F^2x, u \,\&\, LT^2u))$⌝. If this is right, the primitive-operator modalizing strategy does not even begin to avoid the ontological commitment to the intensional entity that A_i.
[35] This sort of unified intensional logic is needed to formulate a general epistemological account of why we are justified in our modal beliefs. For example, an account of what would make a person justified in believing that possibly A would go by way of an account of what "akes a person justified in believing that a proposition is possible, and then it would show that the proposition that A has these features.
[36] Soames [1985], Salmon [1986].
[37] Soames, *ibid.*, discusses some of these.
[38] Specifically, one needs the notions of quality and connection (i.e., the notions of "natural property" and "natural relation"). See Chapter 8, *Quality and Concept*, and David Lewis [1983]. More will be said about this topic at the close of this section.
[39] This argument is developed in Section 7 of Bealer [1988].
[40] Further support for the existence of self-embeddable properties and relations can be extrapolated from the arguments Kripke gives against Tarski's infinite hierarchy of distinct truth concepts for English (Sections 1 and 2, Kripke [1975]).
[41] This is not quite accurate. In situation semantics a distinction is made between *basic* properties and relations, on the one hand, and *complex* properties and relations, on the other hand. Basic properties and relations are treated as unreduced entities. However, complex properties and relations are reduced to certain kinds of sets (dubbed "event types"). As a result, self-embeddable complex properties and relations create a difficulty for situation semantics that is fully analogous to the self-embeddability problem in possible-worlds semantics that we have been discussing. Furthermore, isn't it odd that in situation semantics some properties and relations are supposed to be sets whereas others belong to an entirely different ontological category? This sort of disunity is undesirable even in reductionistic theories.
[42] See Jon Barwise [1985].
[43] Strictly speaking, the reduction is more complicated in situation semantics. But the complications do not affect the philosophical issue we are discussing, so it is convenient to suppress them.
[44] Nino Cocchiarella [1985] has claimed that this kind of first-order intensional logic can be shown to be incomplete if the semantics is modified only slightly. However, his argument is based on an elementary technical error. When the semantics is modified in the way Cocchiarella suggests, completeness still can be shown. For a discussion of this issue, see Part II, Section 4.

[45] We use 'linguistic subject' and 'linguistic predicate' to contrast with 'ontological subject' and 'ontological predicate'. Our use comes close to Strawson's use of 'logical subject' and 'logical predicate'; see Chapter 8 of his *Individuals*, for example.

[46] Linguists have assembled independent syntactic evidence for the existence of a primitive copula in English. See, for example, Williams [1980] and Higginbotham [1984].

[47] See p. 215 in Mates [1950].

[48] In the general case,

$$\mathbf{the}([F^m(u_1, \ldots, u_m)]_{u_1 \ldots u_m}, [G^n(v_1, \ldots, v_n)]_{v_1 \ldots v_n}) =$$

$$= [G^n(v_1, \ldots, v_{n-1}, (\iota u_1)(F^m(u_1, \ldots, u_m))]_{v_1 \ldots v_{n} u_2 \ldots u_m}.$$

and

$$\langle v_1, \ldots, v_{n-1}, u_2, \ldots, u_m \rangle \in H(\mathbf{the}(x, y)) \textit{ iff}$$

$$(\exists! u_1)(\langle u_1, u_2, \ldots, u_m \rangle \in H(x) \;\&$$

$$(\forall u_1)(\langle u_1, u_2, \ldots, u_m \rangle \in H(x) \rightarrow \langle v_1, \ldots, v_{n-1}, u_1 \rangle \in H(y)).$$

[49] This general approach to substitutivity failures is discussed in §39 'Pragmatics' in *Quality and Concept*, and a concrete example of the conversational pragmatics is traced out on pp. 172–4.

[50] P. 215, Mates [1950].

[51] See Section 39 in *Quality and Concept* for further discussion of this sort of pragmatic explanation.

[52] Putnam [1954].

[53] In the semantics, for example, we need only define a new type of model structure in which there is a primitive logical operation for each different syntactic form. These operations will be 1–1; their ranges will be disjoint, and their behavior with respect to the extension functions H in $\mathscr{K}$ will be just what one would expect.

[54] Tyler Burge expresses a closely related worry; see pp. 127ff., Burge [1978]; p. 97, Burge [1979], and Burge [1975]. The issue here dramatizes the fact that any adequate theory of language learning must incorporate a resolution of the paradox of analysis.

[55] A formal semantics that deals with fine-grained distinctions like this may be developed along the lines suggested in *Quality and Concept*, p. 257, n. 17.

[56] The only other known way of trying to solve this sort of substitutivity puzzle is by treating the standard propositional attitudes as three-place relations holding among a person, a proposition, and a "mode of presentation". However, there are a host of problems with this proposal. For example, it runs into trouble with iterated propositional attitudes, quantifying-in, and general sentences that mix intentional and nonintentional predicates. We plan to spell out these difficulties in a future publication.

[57] The self-embedded propositions s_1 and s_2 involving the two soldiers we discussed at the close of Section 11 are dealt with as follows:

$$s_1 = \text{Pred}_0(\text{Pred}_0(\text{perceiving}, s_2), y)$$

$$s_2 = \text{Pred}_0(\text{Pred}_0(\text{perceiving}, s_1), x).$$

Once again, no ill-founded sets are involved; indeed, this pattern is comparable to:

$$2 = (0 + 3) + -1$$

$$3 = (0 + 2) + 1$$

where + corresponds to Pred_0; 2 to s_1; 0 to perceiving; 3 to s_2; -1 to y; 1 to x.

[58] These functions – along with Conj, Neg, and Exist – are closely related to the operations Quine introduces in Quine [1966]. See also Quine [1981].

[59] In general,

$$\langle x_1, \ldots, x_{i-1}, y_1, \ldots, y_k \rangle \in H(\text{Pred}_k(u, v)) \textit{ iff}$$

$$\langle x_1, \ldots, x_{i-1}, \text{Pred}_0(\ldots \text{Pred}_0(\text{Pred}_0(v, y_k), y_{k-1}), \ldots, y_1)\rangle \in H(u)$$

where $u \in \mathscr{D}_i$, $i \geqslant 1$, and $v \in \mathscr{D}_j$, $j \geqslant k \geqslant 1$. The following examples help to explain the predication functions Pred_0, Pred_1, Pred_2, Pred_3, . . . :

$$\text{Pred}_0([Fxyz]_{xyz}, [Guvw]_{uvw}) = [Fxy[Guvw]_{uvw}]_{xy},$$

$$\text{Pred}_0([Fx]_x, [Guvw]_{uvw}) = [F[Guvw]_{uvw}],$$

$$\text{Pred}_1([Fx]_x, [Guvw]_{uvw}) = [F[Guvw]^{w}_{uv}]_w,$$

$$\text{Pred}_2([Fx]_x, [Guvw]_{uvw}) = [F[Guvw]^{vw}_{u}]_{vw},$$

$$\text{Pred}_3([Fx]_x, [Guvw]_{uvw}) = [F[Guvw]^{uvw}]_{uvw},$$

$$\text{Pred}_k([Fx]_x, [A]_{v_1 \ldots v_m u_1 \ldots u_k}) = [F[A]^{u_1 \ldots u_k}_{v_1 \ldots v_m}]_{u_1 \ldots u_k}.$$

(Note that we have just *used*, not mentioned, intensional abstracts from L_ω.) For further clarification of these predication functions Pred_0, . . . , see the definition of the associated syntactic operations given on page 220.

[60] On conception 1, PRPs are thought of as the actual qualities, connections, and conditions of things; on conception 2, PRPs are thought of as concepts and thoughts. (See §2 in Bealer [1979] and §§40–41 in *Quality and Concept* for discussion of these distinctions.) Conception 1 and conception 2 correspond very closely to what Alonzo Church calls, respectively, Alternative 2 and Alternative 0 (pp. 4ff. in Church [1951] and pp. 143ff. in Church [1973 & 1974]). Church states that he ". . . attaches greater importance to Alternative 0 because it would seem that it is in this direction that a satisfactory analysis is to be sought of statements regarding assertion and belief". (P. 7n., Church [1951]) A fuller defense of his approach to the logic for psychological matters is given in Church [1954], where he develops the criterion of strict synonymy upon which he bases Alternative 0. The importance of conception 2 is discussed at length in *Quality and Concept*, §§2, 4, 6–11, 18–20, 39.

For the present purposes, we advocate developing *both* conception 1 and conception 2 side by side without attaching greater importance to one over the other. An advantage of such a dual approach is that, once those two conceptions are well developed, it is relatively straightforward to adapt our methods to handle intermediate conceptions in the event that they should prove relevant. Consider two examples. First, according to the construction of conception 2 presented in the text, the proposition $\text{Pred}_0(\text{Pred}_0([Lxy]_{xy}, a), b)$ is treated as distinct from the proposition $\text{Pred}_0(\text{Pred}_0([Lxy]_{yx}, b), a)$. If this distinction seems artificial, then along the lines of p. 54, *Quality and Concept* one can relax the identity conditions on PRPs within a type 2 model structure so that these two propositions are treated as identical.

Secondly, there are instances of the paradox of analysis involving analyses of the logical operations themselves. (E.g., despite the usual definition of conditionalization in terms of negation and conjunction, someone might doubt that $(A \rightarrow B) \equiv \neg(A \,\&\, \neg B)$ and yet not doubt that $(A \rightarrow B) \equiv (A \rightarrow B)$.) Such puzzles can be easily resolved along the lines of Chapter 3 in *Quality and Concept* once one enriches model structures with appropriate additional logical operations (including a primitive operation for conditionalization): e.g., for each nondegenerate finite composition of the present logical operations, one might add a primitive operation that is equivalent to it in H-values. The broader philosophical point is that, if there is artificiality in the construction given in the text, it appears not to be inherent in the general algebraic approach; evidently it can be removed by some combination of the above methods. It does not follow, of course, that these methods can be used to rid other approaches to intensional logic of their forms of artificiality.

[61] Taken together, (i) and (ii) guarantee that the action of the inverses of the $\mathcal{T}$-transformations and Conj, Neg, . . . in a type 2 model structure is to decompose each element of $\mathcal{D}$ into a unique (possibly infinite) complete tree. (A decomposition tree is *complete* if it contains no terminal node that could be decomposed further under the inverses of the $\mathcal{T}$-transformations and Conj, Neg, . . .). Notice that without condition (iii) unwanted identities such as $[Fx]_x = [A \,\&\, Fx]_x$ could arise. For, as far as conditions (i) and (ii) are concerned, the property $[Fx]_x$ can have a unique complete decomposition tree in which $[Fx]_x$ occurs (denumerably many successive times) on a path descending from $[Fx]_x$. Condition (iii) rules out such a tree.

Examples of type 1 and 2 model structures are easily constructed. E.g., a type 1 model structure can be constructed relative to a model for first-order logic with identity and extensional abstraction, and a type 2 model structure can be constructed relative to a model for first-order logic with identity, extensional abstraction, and Quine's device of corner quotation.

[62] Meaning may also be defined: $M_{\mathcal{I}\mathcal{A}\mathcal{M}}(A) =_{\text{df}} D_{\mathcal{I}\mathcal{A}\mathcal{M}}([A])$.

[63] These notational conventions are adopted for convenience only. We are not reversing our earlier position on the correct parsing of natural language sentences such as ⌜It is necessary that A⌝. We would represent this sentence as ⌜$N^1[A]$⌝. The 1-place predicate 'N' may on conception 1 be defined as follows: $N^1x \text{ iff}_{\text{df}}\, x = [x = x]$.

[64] Proofs of this and the succeeding theorems are given in Bealer [1983]. A corollary of the present theorem is that first-order logic with identity and extensional abstraction (i.e., class abstraction) is complete. Notice also that, in view of the defintitions of □ and ◇ in terms of identity and intensional abstraction, modal logic may be thought of as part of the identity theory for intensional abstracts.

[65] That is, t and r are not in the range of the same term-transforming relation, nor are they in the range of the same syntactic operation – conjunction, negation, existential generalization, predication$_0$, predication$_1$, ...

[66] $\mathcal{A}8$ affirms the equivalence of identical intensional entities. Schemas $\mathcal{A}9$–$\mathcal{A}11$ capture the principle that a complete definition of an intensional entity is unique. And schema $\mathcal{A}12$ captures the principle that a definition of an intensional entity must be noncircular. The following two instances of $\mathcal{A}12$ should help to illustrate how it works:

$$[Fxy]_{xy} = [Gxy]_{yx} \rightarrow [Fxy]_{yx} \neq [Gxy]_{xy}$$

$$[Fx]_x = [\neg Gx]_x \rightarrow [\neg Fx]_x \neq [Gx]_x.$$

Finally, $\mathscr{R}3$ says roughly that, if $A(t)$ is valid$_2$ for any arbitrary elementary p-ary term t, then $A(t')$ is valid$_2$ for any p-ary term t'.

[67] The idea was presented in a talk by the second author at Augsburg (February 1987). In the meantime E. Klein informed us that a similar construction is carried through in the Edinburgh dissertation by F. D. Kamareddine. Scott's treatment of class abstraction using lambda-abstraction (Scott [1975]) must be mentioned as well. In the inductive truth definition of that paper he represents $a \in b$ as a defined formula and identifies it with $b(a)$. He comments on his choice of the particular representation and the loss of fine-grainedness caused by it: ". . . as long as we do not use the lambda-notation, the representation of formulas by elements is *unique*, because the whole system is based on tuples. (If we wanted full uniqueness, we could make the $\in$-combination a primitive . . . and save the $b(a)$-part for the truth-definition)". (Scott [1975], p. 7.) It is exactly this strategy that we will use below. We will introduce a primitive operation Pred. This allows us to abide by the requirements on the internal structure of propositions that characterize propositional-attitude contexts.

[68] Scott [1975], p. 8. Lest the reader accuse us of confusing use and mention we would perhaps observe that the cited passage appears in a context where the inductive truth definition is still to come.

[69] We want to be neutral at this stage of the discussion with respect to the proper reading of the schema $\ulcorner u \in \{x| \ldots x \ldots\}\urcorner$. '$\in$' may indicate functional application and the abstract $\ulcorner\{x| \ldots x \ldots\}\urcorner$ may accordingly stand for a propositional function. Another construal would interpret '$\in$' as a symbol for the binary application operation of a combinatory structure. Finally, '$\in$' could be read as a binary predication relation.

[70] The argument is due to Aczel [1980]. The special form in which we have presented it is taken from Flagg and Myhill [1987].

[71] Geach [1972], p. 229.

[72] Bealer [1988], Section 4.

[73] Spelled out in more detail, the functional structure adumbrated above corresponds to what Barendregt calls a *lambda-family* (Barendregt [1981], p. 110) and Aczel [1980] a *lambda-system*. The notion seems due to H. Volken.

[74] We have been "cheating" in stating the equations. R as a two-place function cannot be composed with a pair of functions one of whose components is a one-place (p) and the other component is a zero-place (a or b) function. Therefore, the constants in the argument-positions of 'R' have to be read as actually standing for the composition of the unique function which maps the universe of objects into the terminal element with a zero-place function.

[75] This is a particular case of the axiom schema 11 of $T2$.

[76] It may be questioned whether this requirement alone can provide a firm basis on which to erect a defensible, non-*ad hoc* theory of a notion of predication within a functional approach. We do not rule out there being good reasons of a different sort that would finally vindicate the introduction of a fine-grained doppelgänger of functional application.

It should be mentioned that Aczel [1985] introduces a related distinction between functional application and predication. Since he retains both properties and propositional functions within the framework he sketches, he needs a second operation pred′ and a corresponding principle which allows him to derive a bijective correspondence between properties and propositional functions. On philosophical grounds a system that encompasses both properties and propositional functions may have certain advantages. For our limited objective, though, the defense of propositional functions as external modeling objects, we thought it appropriate not to complicate the technical issue by admitting properties and propositional functions side by side into our model.

[77] We do not want to defend our choice of the particular propositional part of the schema. It would have been possible to simply stipulate that Pred(f, a) be a proposition if $f(a)$ is a proposition. Irrespective of the propositional condition, the extensional equivalence between predication and functional application provides a vivid illustration of our misgivings concerning the ultimate appropriateness of this version of a fine-grained functional structure.

REFERENCES

Aczel, P.: 1980, 'Frege structures and the notions of proposition, truth and set', in J. Barwise, H. J. Keister and K. Kunen (eds.), *The Kleene Symposium*, Amsterdam, pp. 31–59.

Aczel, P.: 1985, 'Properties and propositional functions', Ms. Manchester (To appear as an appendix to Aczel [1988]).

Aczel, P.: 1988, 'Algebraic semantics for intensional logics I', in G. Chierchia, B. Partee, R. Turner (eds.), *Property Theories, Type Theories, and Semantics*, Dordrecht.

Anderson, C. A.: 1986, 'Some difficulties concerning Russellian intensional logic', *Noûs* **20**, 35–43.

Barendregt, H. P.: 1981, *The Lambda Calculus*, Amsterdam.

Barwise, J.: 1985, 'On modeling shared understanding', unpublished manuscript.

Bealer, G.: 1979, 'Theories of properties, relations, and propositions', *The Journal of Philosophy* **76**, 643–648.

Bealer, G.: 1982, *Quality and Concept*, Oxford.

Bealer, G.: 1983, 'Completeness in the theory of properties, relations, and propositions', *The Journal of Symbolic Logic* **48**, 415–426.

Bealer, G.: 1984, 'Mind and anti-mind: why thinking has no functional definition', *Midwest Studies in Philosophy* **9**, 283–328.

Bealer, G.: 1986, 'The logical status of mind', *Midwest Studies in Philosophy* **10**, 231–274.

Bealer, G.: 1987, 'The significance of completeness results in intensional logic', unpublished manuscript.

Bealer, G.: 1988, 'Type-free fine-grained intensionality', in G. Chierchia, B. Partee, R. Turner (eds.), *Property Theory, Type Theory, and Semantics*, Dordrecht.

Burge, T.: 1975, 'Knowledge and convention', *The Philosophical Review* **84**, 249–55.

Burge, T.: 1978, 'Belief and synonymy', *The Journal of Philosophy* **75**, 119–38.

Burge, T.: 1979, 'Individualism and the mental', *Midwest Studies in Philosophy* **4**, 73–122.

Carnap, R.: 1947, *Meaning and Necessity*, Chicago.

Chierchia, G. and Turner, R.: 1988, 'Semantics and property theory', *Linguistics and Philosophy*, in press.

Chisholm, R.: 1981, *The First Person: An Essay on Reference and Intentionality*, Minneapolis.

Church, A.: 1950, 'On Carnap's analysis of statements of assertions and belief', *Analysis* **10**, 97–9.

Church, A.: 1951, 'A formulation of the logic of sense and denotation', P. Henle, H. H. Kallen, S. K. Langer (eds.), *Structure, Method, and Meaning: Essays in Honor of Henry M. Scheffer*, Liberal Arts Press, New York, pp. 3–24.

Church, A.: 1954, 'Intensional isomorphism and identity of belief', *Philosophical Studies* **5**, 65–73.

Church, A.: 1973 & 1974, 'Outline of a revised formulation of the logic sense and denotation', *Noûs* **7**, 23–33, *Noûs* **8**, 135–56.

Churchland, Patricia: 1986, *Neurophilosophy*, Cambridge, Mass.

Churchland, Paul: 1979, *Scientific Realism and the Plasticity of Mind*, Cambridge.

Churchland, Paul: 1981, 'Eliminative materialism and the propositional attitudes', *The Journal of Philosophy* **78**, 67–90.

Cocchiarella, N. B.: 1985, 'Review of Bealer's *Quality and Concept*', *The Journal of Symbolic Logic* **50**, 254–56.

Cocchiarella, N. B.: 1986, *Logical Investigations of Predication Theory and the Problem of Universals*, Naples.

Davidson, D.: 1964, 'Theories of meaning and learnable languages', in Y. Bar-Hillel (ed.), *Proceedings of the 1964 Congress for Logic, Methodology and Philosophy of Science*, pp. 383–94.

Evans, G.: 1977, 'Pronouns, quantifiers, and relative clauses (I)', *Canadian Journal of Philosophy* **7**, 467–536.

Evans, G.: 1982, *The Varieties of Reference*, J. McDowell (ed.), Oxford.

Feferman, S.: 1975, 'A language and axioms for explicit mathematics', *Algebra and Logic*, *LNM* **450**, 87–139.

Feferman, S.: 1984, 'Toward useful type-free theories', *The Journal of Symbolic* **49**, 75–111.

Fitch, F. B.: 1948, 'An extension of basic logic', *The Journal of Symbolic Logic* **13**, 95–106.

Flagg, R. C. and Myhill, J.: 1987, 'An extension of Frege structures', in D. W. Kueker, E. G. K. Lopez-Escobar and C. H. Smith (eds.), *Mathematical Logic and Computer Science*, New York, pp. 197–217.

Flagg, R. C. and Myhill, J.: 1987, 'Implication and analysis in classical Frege structures', *Annals of Pure and Applied Logic* **34**, 33–85.

Geach, P. T.: 1972, 'Class and concept', in P. T. Geach, *Logic Matters*, Oxford, pp. 226–34.

Gilmore, P. C.: 1974, 'The consistency of partial set theory without extensionality', *Proceedings of Symposia in Pure Mathematics* **8**, Part II, pp. 147–53.

Higginbotham, J.: 1985, 'Indefiniteness and predication', unpublished manuscript.

Kripke, S.: 1975, 'Outline of a theory of truth', *The Journal of Philosophy* **72**, 690–716.

Lewis, D.: 1979, 'Attitudes *De Dicto* and *De Re*', *The Philosophical Review* **87**, 513–43.

Lewis, D.: 1983, 'New work for a theory of universals', *Australasian Journal of Philosophy* **61**, 343–77.

Mates, B.: 1950, 'Synonymity', *University of California Publications in Philosophy* **25**, 201–26, reprinted in L. Linsky (ed.), *Semantics and the Philosophy of Language*, Urbana, 1952.

McFetridge, I. G.: 1984, 'Review of Bealer's *Quality and Concept*', *Mind* **93**, 455–58.

Menzel, C.: 1986, 'A complete type-free "second-order" logic and its philosophical foundations', *CSLI-Report* 86–40, Stanford.

Mönnich, U.: 1983, 'Toward a calculus of concepts as a semantical metalanguage', in R. Bäuerle, C. Schwarze, and A. von Stechow (eds.), *Meaning, Use, and Interpretation of Language*, Berlin, pp. 342–60.

Myro, G.: 1981, 'Aspects of acceptability', *Pacific Philosophical Quaterly* **62**, 107–17.

Putnam, H.: 1954, 'Synonymity and the analysis of belief sentences', *Analysis* **14**, 114–22.
Quine, W. V. O.: 1960, *Word and Object*, Cambridge, Mass.
Quine, W. V. O.: 1966, *Selected Logic Papers*, New York.
Quine, W. V. O.: 1981, 'Predicate functors revisited', *The Journal of Symbolic Logic* **46**, 649–52.
Reinhardt, W.: 1985, 'An axiomatic theory of properties', unpublished manuscript.
Salmon, N.: 1986, *Frege's Puzzle*, Cambridge, Mass.
Scheffler, I.: 1954, 'An inscriptional approach to indirect discourse', *Analysis* **14**, 83–90.
Scott, D. S.: 1975, 'Combinators and classes', *Lambda-Calculus and Computer Science Theory*, *LNCS* **37**, 1–26.
Soames, S.: 1985, 'Direct reference and propositional attitudes', Lecture at the American Philosophical Association Pacific Division Meeting, March 1985.
Strawson, P. F.: 1959, *Individuals*, London.
Strawson, P. F.: 1974, *Subject and Predicate in Logic and Grammar*, London.
Turner, R.: 1987, 'A theory of properties', *The Journal of Symbolic Logic* **52**, 455–62.
Williams, E.: 1983, 'Predication', *Linguistic Inquiry* **11**, 203–238.

CHAPTER IV.3

PHILOSOPHICAL PERSPECTIVES ON FORMAL THEORIES OF PREDICATION

by Nino B. Cocchiarella

Predication has been a central, if not the central, issue in philosophy since at least the time of Plato and Aristotle. Different theories of predication have in fact been the basis of a number of philosophical controversies in both metaphysics and epistemology, not the least of which is the problem of universals. In what follows we shall be concerned with what traditionally have been the three most important types of theories of universals, namely, nominalism, conceptualism, and realism, and with the theories of predication which these theories might be said to determine or characterize.

Though each of these three types of theories of universals may be said to have many variants, we shall ignore their differences here to the extent that they do not characterize different theories of predication. This will apply especially to nominalism where but one formal theory of predication is involved. In both conceptualism and realism, however, the different variants of each type do not all agree and form two distinct subtypes each with its own theory of predication. For this reason we shall distinguish between a constructive and a holistic form of conceptualism on the one

D. Gabbay and F. Guenthner (eds.), Handbook of Philosophical Logic, Volume IV, 254–326.

hand, and a logical and a natural realism on the other. Constructive conceptualism, as we shall see, has affinities with nominalism with which it is sometimes confused, and holistic conceptualism has affinities with logical realism with which it is also sometimes confused. Both forms of conceptualism may assume some form of natural realism as their causal ground; and natural realism in turn must presuppose some form of conceptualism as its background theory of predication. Both forms of realism may be further divided into their essentialist and non-essentialist variants (and in logical realism even a form of anti-essentialism), and though an essentialist logical realism is sometimes confused with Aristotelian essentialism, the latter is really a form of natural realism with natural kinds as the only essential properties objects can have.

1. PREDICATION AND THE PROBLEM OF UNIVERSALS

The problem of universals is sometimes all too simply put as the problem of whether there are universals or not. This is misleading, however, since even nominalism does not deny that there are universals but only that there are no universals other than predicate expressions (which we might also call nominalistic universals). The original use of the term 'universal', it should be noted, goes back to Aristotle, according to whom a universal is that which can be predicated of things (*De Interpretatione* 17^{a39}). We shall retain the core of this notion throughout this essay and assume that whatever else it may be a universal has a predicable nature and that it is this predicable nature which is what constitutes its universality.

Nothing follows from this assumption, however, regarding whether a universal is (1) merely a predicate expression (nominalism) of some language or other; (2) a concept (conceptualism) in the sense of a sociobiologically based cognitive capacity to identify (in a classificatory sense), characterize or relate things in various ways; or (3) a real property or relation (realism) existing independently of both language and the natural capacity humans have for thought and representation. We propose to take each of these interpretations or theories seriously in what follows at least to the extent that we are able to associate each with a formal theory of predication. Our particular concern in this regard, moreover, will be with the explanation each provides of the predicable nature of universals, i.e., of that in which the universality of universals consists.

Our discussion and comparison of nominalism, conceptualism and realism, accordingly, will not deal with the variety of arguments that have been given for or against each of them, but with how each as a theory of universals may be semantically associated with a formal theory of predication. Our assumption here is that insofar as such an associated formal theory of predication provides a logically perspicuous medium for the articulation of the theory of universals in question, then to that extent the formal theory may itself be identified with the explanation which that theory of universals provides of the predicable nature of universals. It is in the sense of this assumption, moreover, that we understand a philosophical theory of predication to be a formal theory of predication together with its semantically associated theory of universals.

A brief comment is in order, however, regarding how we shall represent the semantical component of a philosophical theory of predication. In particular, whereas we shall rely on our intuition regarding how logical truth and logical consequence are to be explicated in accordance with a given theory, we shall also formulate such an explication in strictly set-theoretic terms. In this regard, our semantical formulations will belong only to what might be called an *external logical semantics* for the philosophical theory of predication in question. The alternative to such an external construction would be an *internal logical semantics* which utilized an applied form of the same logically perspicuous syntax of the formal component of that theory. Such an alternative would obviate any question of distortion (such as the use of saturated sets to represent unsaturated concepts) or the use of constructions that cannot be accounted for within the theory itself; but then it would also assume that the theory is well understood in its logical structure, which is what an external semantics attempts to clarify. The problem, in other words, is the question of an independent or external criterion of the conditions under which the formal component of a philosophical theory of predication might be said to be both sound and complete, or, where appropriate, sound but essentially incomplete (as is the case with logical realism). Such an independent criterion should also avail us of the means by which proofs of these results might be obtained. Set theory, of course, provides an ideal framework for just such an external logical semantics.

Finally, it should perhaps be noted that because of limitations on length we shall provide no account of the logic of nominalized predicates for any of the theories considered here. Such an account is available and can be found in Cocchiarella [10], Chapters 4–6.

2. NOMINALISM

Of the three major types of theories of universals, nominalism is clearly the most restrictive. Some forms of nominalism, moreover, are even more restrictive in this regard than others. We shall not concern ourselves with these variations here, however, but shall simply identify nominalism with two general semantical theses.

The first general thesis of nominalism is that universals have only a formal mode of existence, i.e., that beyond the predicate expressions that occur or can occur in language there are no universals. Predicate expressions, in other words, do not stand for any universals beyond themselves; and therefore predicate expressions are the only entities according to nominalism that have a predicable nature. As already indicated, we shall occasionally refer to predicate expressions as *nominalistic universals*.

The second general thesis of nominalism is the thesis of extensionality; i.e., the thesis that, semantically, predicate expressions may make no finer distinction of content (in the sense of the individuals such expressions can be true of) than can be generated by co-extensive predicate expressions. This is the thesis, in other words, that co-extensive predicate expressions are to be interchangeable *salva veritate* in any applied formal theory of predication suitable for nominalism. This means that nominalism is committed to an extensional logic, and in particular to a non-modal form of the thesis of anti-essentialism; specifically, the thesis that no nominalistic universal is necessarily true of some of the things of which it can be truly predicated without being necessarily true of all. There is also a modal thesis of anti-essentialism, it will be noted, which is sometimes said to be compatible with nominalism. Of course if this were really the case, then nominalism would not be committed to the thesis of extensionality after all, for such a modal thesis would be formulated in a non-extensional logic. In Section 4 we shall evaluate and reject this claim.

The formal theory of predication which is commonly associated with nominalism, it is well-known, is standard first order predicate logic with identity. It is this logic, in other words, which is taken as explaining the predicable nature of nominalistic universals (i.e., of the possible predicate expressions of language). The explanation in fact is simply the identification of that nature with the logico-grammatical roles which predicatics are assigned in the logical forms of standard first order predicate logic. Predicate constants are of course assigned the paradigmatic roles in this explanation, but this is not to say that predicate constants are the only predicate expressions which must be accounted for in nominalism. In particular, any open first order formula, relative to the free individual variables occurring therein as argument indicators, must also be understood as representing a predicate expression of the (formal) language of which it is a formula; and in fact in an applied first order theory based upon that language such a formula would constitute the definiens of a possible definition for a predicate constant not already in that language. Potentially, of course there are infinitely many such predicate constants that might be introduced into an applied theory in this way, and some account must be given in nominalism of their role as predicates as well as of those taken primitively as such in the (formal) language in question.

Such an account is forthcoming, it turns out, by extending standard first order predicate logic to a second order predicate logic in which predicate quantifiers are interpreted substitutionally. That is, we can account for all of the nominalistic universals of an applied first order theory without actually introducing predicate constants for these universals by simply turning to a second order predicate logic in which predicate quantifiers are interpreted substitutionally and where predicate variables have only first order formulas as their values. There are constraints which such an interpretation imposes regarding the specification of nominalistic universals, needless to say, and such constraints are in fact precisely those involved in the comprehension principle (CP!) of standard predicative second order logic. This is not a matter of stipulation, moreover, but is in fact a consequence of the nominalistic semantics we give in Section 3 for predicate quantifiers. Our semantics for nominalism does conform, in other words, with the already established view that standard predicative second order logic is an appropriate medium for the representation of a nominalist theory of predication (cf. Parsons [20]). (The earliest proof of the type of result described here for nominalism occurs in Novak [19]. Another proof is given in Doets [12], and finitary proofs are given in Shoenfield [29] and Flanagen [13]. A detailed proof for the formulation given here is given in Cocchiarella [7].)

The formal component of nominalism, accordingly, may be described as follows. As primitive *logical constants* we need take only $\rightarrow$ (the material conditional sign), $\sim$ (the negation sign), $\forall$ (the universal quantifier sign), and $=$ (the identity sign). The remaining logical constants are assumed defined (as abbreviations of the metalanguage) in the usual way. We shall use 'x', 'y', and 'z', with or without numerical subscripts, to refer (in the metalanguage) to individual variables; and similarly we shall use 'F^n', 'G^n', and 'H^n, to refer to n-place predicate variables. (We shall usually delete the superscripts whenever the context makes clear the degree of the predicate variable.) By a (*formal*) *language*, moreover, we shall mean a finite set of predicate constants each of a specified (finite) degree. Individual constants, definite descriptions and singular terms other than variables need never be introduced into a nominalistic theory, incidentally, and in fact their role in language, according to nominalism, is said to be best explained in terms of the role of predicate constants (cf. Quine [24], Section 38). (For convenience, we shall follow this nominalist strategy and ignore singular terms other than variables throughout this essay.)

Where L is a language, the *atomic formulas* of L are all the expressions which are either of the form $(x = y)$, $F(x_1, \ldots, x_n)$, or $P(x_1, \ldots, x_n)$, where n is a positive integer, P is an n-place predicate constant in L, and F is an n-place predicater variable. The *well-formed formulas* (*wffs*) of L are then the members of the smallest class K containing all of the atomic formulas of L and such that $\sim\phi$ $(\phi \rightarrow \psi)$, $(\forall x)\phi$ and $(\forall F)\phi$ are all in K whenever ϕ, ψ are in K, x is an individual variable, and F is a predicate variable. We shall assume the usual definitions of bondage and freedom of occurrences of variables, incidentally, and we take $\phi(x_1/y_1, \ldots, x_n/y_n)$ to be the wff which results by simultaneously replacing each free occurrence of $y_1, \ldots, y_n$ in ϕ by free occurrences of $x_1, \ldots, x_n$, respectively, if such a wff exists; and otherwise we take it to be just ϕ itself. Similarly, we take $\phi[\psi/F(x_1, \ldots, x_n)]$ to be the result of properly substituting ψ (relative to the free occurrences of $x_1, \ldots, x_n$ in ψ) for F^n in ϕ, if such a wff exists; and otherwise we take it to be ϕ itself. For convenience, we shall sometimes use $\phi(x_1, \ldots, x_n)$ to indicate that $x_1, \ldots, x_n$ are among the distinct individual variables occurring free in ϕ. We shall also say that $\phi(x_1, \ldots, x_n)$ is a *fully applied* wff if no predicate variable occurs free in ϕ. By a *first order wff* we shall mean a wff in which no predicate variable occurs at all (whether bound or free). Though not every fully applied wff need be a first order wff, of course every first order wff is a fully applied wff.

Now the important axiom schema in standard predicative second order logic is the following restricted comprehension principle:

(CP!) $\quad (\exists F^n)(\forall x_1) \ldots (\forall x_n)[F(x_1, \ldots, x_n) \leftrightarrow \psi]$,

where ψ *is a wff* in which F^n does not occur and *in which no predicate variable has a bound occurrence*, and where $x_1, \ldots, x_n$ are among the distinct individual variables occurring free in ψ. The important constraint here of course is the requirement that no predicate variable should have a bound occurrence in ψ; i.e., that ψ should have only first order wffs as fully applied instances. Besides (CP!), the remaining *axioms* include all wffs which are either tautologous or of one of the following forms (where u is an individual or a predicate variable):

(A1) $(\forall u)[\phi \to \psi] \to [(\forall u)\phi \to (\forall u)\psi]$,

(A2) $\phi \to (\forall u)\phi$, where u is not free in ϕ,

(A3) $(\exists x)(y = x)$,

(LL) $x = y \to (\phi \leftrightarrow \psi)$, where ψ comes from ϕ by replacing one or more free occurrences of y by free occurrences of x.

As *inference rules* we take *modus ponens*:

(MP) if $\vdash\phi$ and $\vdash(\phi \to \psi)$, then $\vdash\psi$;

and the universal generalization of a wff with respect to an individual or a predicate variable:

(UG) if $\vdash\phi$, then $\vdash(\forall u)\phi$.

We say that a wff ϕ is a *theorem of nominalism*, in symbols $\vdash_{\overline{\text{nom}}}\phi$, if ϕ is the terminal wff a finite sequence every constituent of which is either an axiom or is obtained from preceding wffs by one of the inference rules. We also say that ϕ is *derivable* in *nominalism* from a set Σ of wffs, in symbols $\Sigma\vdash_{\overline{\text{nom}}}\phi$, if for some wffs $\psi_1, \ldots, \psi_n$ in Σ, $\vdash_{\overline{\text{nom}}}(\psi_1 \& \cdots \& \psi_n \to \phi)$. (We take this "conditional" to be just ϕ itself if $n = 0$. Also, because (UG) is an inference rule in all of the formal theories considered here, we shall speak of any universal generalization of an axiom as being itself an axiom of the same form). Occasionally, incidentally, we shall speak of a set of wffs Σ of a language L as an *applied theory* (in this case of nominalism) whose language is L. The members of such a set are then regarded as premises (or postulates) for some presumed context of application. Although an applied theory is usually defined as being closed under the derivability relation, we shall avoid that usage here, since the closure of the premise set for the same context will be different depending on the formal theory of predication being applied.

Note that by (LL), (UG), (A1)–(A2) and tautologous transformations,

($\exists$/UI$_1$) $\vdash_{\overline{\text{nom}}}(\exists x)(y = x) \to [(\forall x)\phi \to \phi(y/x)]$, and therefore by (A3),

$$(\mathrm{UI}_1) \quad \vdash_{\overline{\mathrm{nom}}} (\forall x)\phi \to \phi(y/x).$$

A second order analogue of (LL), namely,

$$(\mathrm{LL}_2) \quad \vdash_{\overline{\mathrm{nom}}} (\forall x_1) \ldots (\forall x_n)[F(x_1, \ldots, x_n) \leftrightarrow \psi] \to$$
$$\to (\phi \leftrightarrow \phi[\psi/F(x_1, \ldots, x_n)]),$$

is provable by a simple inductive argument on the structure of ϕ; and therefore if F^n is not free in ψ, then by (UG), (A1) and tautologous transformations,

$$(\exists/\mathrm{UI}_2) \quad \vdash_{\overline{\mathrm{nom}}} (\exists F^n)(\forall x_1) \ldots (\forall x_n)[F(x_1, \ldots, x_n) \leftrightarrow \psi] \to$$
$$\to ((\forall F^n)\phi \to \phi[\psi/F(x_1, \ldots, x_n)]).$$

Accordingly, *if no predicate variable has a bound occurrence in* ψ, then by (CP!),

$$(\mathrm{UI!}_2) \quad \vdash_{\overline{\mathrm{nom}}} (\forall F^n)\phi \to \phi[\psi/F(x_1, \ldots, x_n)].$$

In both (CP!) and (UI!$_2$), it should be noted, we have what amounts to the form of an assertion of the existence of a universal corresponding to the wff ψ (relative to $x_1, \ldots, x_n$ as subject position indicators). The important constraint in both cases is that no predicate variable has a bound occurrence in ψ; i.e., that ψ should have only first order wffs as fully applied instances. Note that where ψ is itself restricted to the first order wffs of an applied first order theory Σ, this notion of (second order) existence is essentially that of the different possible predicate constants which can be introduced into Σ by explicit definitions. E.g., where L is the language of Σ, P^n is an n-place predicate constant not in L, and ψ is a first order wff of L in which $x_1, \ldots, x_n$ are all the distinct individual variables occurring free, then

$$(\forall x_1) \ldots (\forall x_n)[P(x_1, \ldots, x_n) \leftrightarrow \psi]$$

is said to be a *possible definition of* P^n *in terms of* L; and it is precisely such a possible definition, according to nominalism, which is what is posited in (CP!). In other words as the definiens of such a possible definition the first order wff ψ is to be construed in effect as a complex predicate expression of the theory in question. In this sense, accordingly, all instances of (CP!) where ψ is a first order wff (and $x_1, \ldots, x_n$ are *all* of the distinct individual variables occurring therein) are to be considered as nominalistically valid.

Of course not all wffs which contain no bound predicate variables are first order wffs, and in (CP!) the individual variables $x_1, \ldots, x_n$ need not be *all* of the distinct individual variables occurring free in ψ. Before claiming that all instances of (CP!) are nominalistically valid, accordingly, we must be assured that (CP!) does not commit us to a mode of second order existence which goes beyond nominalistic universals. That is, we need a semantical account by which to show that standard predicative second order logic, as formulated above, can indeed be taken as nominalism's formal theory of predication.

3. A NOMINALISTIC SEMANTICS FOR PREDICATIVE SECOND ORDER LOGIC

In turning now to our semantics for nominalism, it must again be emphasized that we are constructing only an external logical semantics regarding how logical truth and logical consequence are to be explicated in nominalism. Of course, once a completeness theorem is established with respect to this external semantics, then we may, as it were, kick away the set-theoretic semantic ladder we have climbed and redefine logical truth and logical consequence in terms of derivability in the formal theory described above. That is, we may then switch from our external set-theoretic semantics to a semantics which is internal to nominalism, and one in particular which commits us only to nominalistic universals (cf. Quine [27], p. 297).

The models involved in our semantics, incidentally, are precisely the same models as are already involved in standard first order logic. Assignments of values (drawn from the domain of a given model) to the individual variables are extended, however, to what, relative to a language L, we shall call *nominalistic assignments* to the n-place predicate variables (for each positive integer n). These will in fact assign first order wffs of L, relative to n distinct individual variables occurring therein, to each n-place predicate variable. Given the notion of satisfaction of a first order wff in a model, the satisfaction of a second order wff will then be defined by a double recursion on the logical structure of the wff and the number of nested predicate quantifiers occurring therein.

By a *model for* a language L, or an L-*model*, accordingly, we shall understand a structure $\mathfrak{A} = \langle D, f \rangle$, where D, the domain of discourse, is a non-empty set, and f is a function with L as domain such that for all positive integers n and all $P^n \in$ L, $f(P^n) \subseteq D^n$ (where D^n is the set of all n-tuples drawn from D). Naturally, we identify all of the

(n-tuples of) individuals in D that a predicate constant P^n of L is *true of* (in the model) by means of their membership in $f(P^n)$. By a *nominalistic assignment in* $\mathfrak{A}$ we understand a function A with the set of individual and predicate variables as domain and such that (1) for each individual variable x, $A(x) \in D$, and (2) for each positive integer n and each n-place predicate variable F^n, $A(F^n) = \langle \phi, x_1, \ldots, x_n \rangle$, for some first order wff ϕ of L and pairwise distinct individual variables $x_1, \ldots, x_n$ occurring free in ϕ. It is in terms of such nominalistic assignments, it might be observed, that we are to understand how free predicate variables can be taken as dummy schema letters representing arbitrary predicate expressions of L (whether simple or complex). Bound predicate variables will be interpreted similarly in the doubly recursive definition of satisfaction we give below for nominalism.

Before defining satisfaction for arbitrary wffs, however, we must first define the standard notion of satisfaction for first order wffs, since, as already noted, the latter notion is in fact presupposed by the former. Accordingly, where L is a language, $\mathfrak{A}$ is an L-model, and A is a nominalistic assignment in $\mathfrak{A}$, then the *satisfaction by A in* $\mathfrak{A}$ of an arbitrary *first order wff* of L is defined recursively as follows:

(1) *A satisfies* $(x = y)$ *in* $\mathfrak{A}$ iff $A(x) = A(y)$;
(2) *A satisfies* $P^n(x_1, \ldots, x_n)$ *in* $\mathfrak{A}$ iff $\langle A(x_1), \ldots, A(x_n)\rangle \in f(P^n)$;
(3) *A satisfies* $\sim\phi$ *in* $\mathfrak{A}$ iff A does not satisfy ϕ in $\mathfrak{A}$;
(4) *A satisfies* $(\phi \rightarrow \psi)$ *in* $\mathfrak{A}$ iff either A does not satisfy ϕ in $\mathfrak{A}$ or A satisfies ψ in $\mathfrak{A}$;
(5) *A satisfies* $(\forall x)\phi$ *in* $\mathfrak{A}$ iff for all $d \in D$, $A(d/x)$ satisfies ϕ in $\mathfrak{A}$.

The fact that A is a nominalistic assignment in the above definition, incidentally, has no bearing on the referential interpretation which such an assignment gives to individual variables, i.e., that such variables are assigned objects in the domain of discourse. We could assume A to be referential even in what it assigns to predicate variables, in other words, without affecting the above definition (and in later sections we will in fact assume this to be the case). Thus, instead of a nominalistic assignment we could have taken A in the above definition to be a *completely referential assignment in D*, i.e., an assignment such that $A(x) \in D$, for each individual variable x, and $A(F^n) \subseteq D^n$, for each positive integer n and each n-place predicate variable F^n. A completely referential assignment differs from a nominalistic assignment, accordingly, not in how each applies to first order wffs but in how each interprets predicate variables, and in particular how each interprets bound (universally quantified) predicate variables. Thus, where A is a completely referential assignment in D, then the so-called "standard" satisfaction clauses for predicate variables and predicate quantifiers which would be added to the above five clauses are the following:

(6) *A satisfies* $F^n(x_1, \ldots, x_n)$ *in* $\mathfrak{A}$ iff $\langle A(x_1), \ldots, A(x_n)\rangle \in A(F^n)$; and
(7) *A satisfies* $(\forall F^n)\phi$ *in* $\mathfrak{A}$ iff for all $X \subseteq D^n$, $A(X/F^n)$ satisfies ϕ in $\mathfrak{A}$.

The definition of satisfaction now applies to all wffs, first order or otherwise, it is clear, but note that as now defined it is based upon a realist and not a nominalist theory of universals (or what we shall call *logical realism* as distinct from natural realism). That is, the resulting semantics is incompatible with the first general thesis of nominalism that there are no universals other than predicate expressions. This is why the corresponding satisfaction clause in our nominalistic semantics must eschew completely referential assignments in favor of nominalistic assignments.

In defining now the satisfaction by a nominalistic assignment A in $\mathfrak{A}$ of an arbitrary wff of L, first order or otherwise, we require the notion of the *rank* of a wff, in symbols $rk(\phi)$, which we take to be the maximum number of nested predicate quantifiers occurring in ϕ. That is, $rk(\phi) = 0$, if ϕ is atomic; $rk(\sim\phi) = rk(\phi)$; $rk(\phi \rightarrow \psi) = \max[rk(\phi), rk(\psi)]$; $rk((\forall x)\phi) = rk(\phi)$; and $rk((\forall F)\phi) = rk(\phi) + 1$. (Of course, if no predicate quantifiers occur in ϕ at all, then $rk(\phi) = 0$.) Note that the definition of satisfaction which follows is by a double recursion, first on the rank of a wff, and then on its structure. Accordingly, if L, $\mathfrak{A}$ and A are as above, and k is a natural number, then the *satisfaction in* $\mathfrak{A}$ *by* A *of an arbitrary wff* ϕ *of* L *of rank* $\leqslant k$, in symbols $\mathfrak{A}, A \overset{k}{\models} \phi$, is defined as follows:

(I) if $rk(\phi) \leqslant 0$, then:
  (i) if ϕ is $(x = y)$, then $\mathfrak{A}, A \overset{0}{\models} \phi$ iff $A_0(x) = A(y)$;
  (ii) If ϕ is $P^n(x_1, \ldots, x_n)$, then $\mathfrak{A}, A \overset{0}{\models} \phi$ iff $\langle A(x_1), \ldots, A(x_n)\rangle \in f(P^n)$;
  (iii) if ϕ is $F^n(x_1, \ldots, x_n)$, then $\mathfrak{A}, A \overset{0}{\models} \phi$ iff A satisfies $\psi(x_1/y_1, \ldots, x_n/y_n)$ in $\mathfrak{A}$, where $A(F^n) =_0 \langle\psi, y_1, \ldots, y_n\rangle$;
  (iv) if ϕ is $\sim\chi$, then $\mathfrak{A}, A \overset{0}{\models} \phi$ iff not $(\mathfrak{A}, A \overset{0}{\models} \chi)$;
  (v) if ϕ is $(\psi \rightarrow \chi)$, then $\mathfrak{A}, A \overset{0}{\models} \phi$ iff either not $(\mathfrak{A}, A \overset{0}{\models} \psi)$ or $\mathfrak{A}, A \overset{0}{\models} \chi$;
  (vi) if ϕ is $(\forall x)\chi$, then $\mathfrak{A}, A \overset{0}{\models} \phi$ iff for all $d \in D$, $\mathfrak{A}, A(d/x) \overset{0}{\models} \chi$; and

(II) if $rk(\phi) \leqslant k + 1$, then: repeat clauses (i)–(vi) above except for replacing "$\overset{0}{\models}$" throughout by "$\overset{k+1}{\models}$"; and

(vii) if ϕ is $(\forall F^n)\psi$, then $\mathfrak{A}, A \overset{k+1}{\models} \phi$ iff for all first order wffs χ of L and all pairwise distinct individual variables $x_1, \ldots, x_n$ occurring free in χ, $\mathfrak{A}, A \overset{k}{\models} \psi[\chi/F^n(x_1, \ldots, x_n)]$.

It should be emphasized that the notion of satisfaction involved in the definiens of clause (iii) of both (I) and (II) is that which is already defined for first order wffs. Moreover, since every first order wff is of rank 0, it is clear that the above definition for wffs of rank $\leqslant 0$ coincides with our original definition of satisfaction when it is restricted to first order wffs. That is, we have the following lemma as an immediate consequence.

LEMMA. *If* L *is a language,* $\mathfrak{A}$ *is an* L*-model,* A *is a nominalistic assignment in* $\mathfrak{A}$*, and* ϕ *is a first order wff of* L*, then* $\mathfrak{A}, A \overset{0}{\models} \phi$ *iff* A *satisfies* ϕ *in* $\mathfrak{A}$.

Another immediate consequence of the definition is the fact that the satisfaction of a wff ϕ of rank $\leqslant k$ reduces to the satisfaction of ϕ at its own rank.

LEMMA. *If* L *is a language,* $\mathfrak{A}$ *is an* L*-model,* A *is a nominalistic assignment in* $\mathfrak{A}$*, and* ϕ *is a wff of* L *of rank* $\leqslant k$*, then* $\mathfrak{A}, A \overset{k}{\models} \phi$ *iff* $\mathfrak{A}, A \overset{rk(\phi)}{\models} \phi$.

Now we shall say that a wff ϕ is a *nominalistic logical consequence* of a set of wffs Σ, in symbols $\Sigma \models_{\text{nom}} \phi$, if for every language L such that $\Sigma \cup \{\phi\}$ is a set of wffs of L and for every L-model $\mathfrak{A}$ and nominalistic assignment A in $\mathfrak{A}$, if $\mathfrak{A}, A \overset{rk(\psi)}{\models} \psi$, for all $\psi \in \Sigma$, then $\mathfrak{A}, A \overset{rk(\phi)}{\models} \phi$. We observe that where Σ is infinite it might be the case that for each natural number k there is a wff ψ in Σ such that $rk(\psi) = k$. Nevertheless, since no single wff can be of infinite rank, the above lemma allows us to ignore the question of infinite ranks altogether. Accordingly, we shall say that ϕ is *nominalistically valid* if and only if for any language L of which ϕ is a wff and for any L-model $\mathfrak{A}$ and any nominalistic assignment A in $\mathfrak{A}$, $\mathfrak{A}, A \overset{rk(\phi)}{\models} \phi$. The following is an immediate consequence of these definitions and the above lemmas (where $\vDash$ is defined in the usual way in terms of satisfaction for first order wffs).

LEMMA. *If* $\Sigma \cup \{\phi\}$ *is a set of first order wffs, then:*

(1) $\Sigma \models_{\text{nom}} \phi$ *iff* $\Sigma \vDash \phi$*; and*

(2) $\models_{\text{nom}} \phi$ *iff* $\vDash \phi$.

In regard now to the nominalistic validation of (CP!), let $F_1^{n_1}, \ldots, F_k^{n_k}$ be all the predicate variables occurring free in a wff ψ; and where A is a nominalistic assignment such that

$$\begin{aligned} A(F_1^{n_1}) &= \langle \phi_1, x_1, \ldots, x_{n_1} \rangle \\ &\vdots \\ A(F_k^{n_k}) &= \langle \phi_k, x_1, \ldots, x_{n_k} \rangle, \end{aligned}$$

let us define $A[\psi]$ as follows:

$$A[\psi] = \psi[\phi_1/F_1^{n_1}(x_1, \ldots, x_{n_1})] \ldots [\phi_k/F_k^{n_k}(x_1, \ldots, x_{n_k}].$$

The following lemma is easily seen to hold by a simple inductive argument on the rank of ψ.

LEMMA: *If* L *is a language,* $\mathfrak{A}$ *is an* L-*model,* A *is an assignment in* $\mathfrak{A}$*, and* ψ *is a wff of* L *of rank* $\leqslant k$*, then* $\mathfrak{A}, A \overset{k}{\models} \psi$ *if and only if* $\mathfrak{A}, A \overset{k}{\models} A[\psi]$.

Consider now any instance of the comprehension principle (CP!),

$$(\exists F^n)(\forall x_1) \ldots (\forall x_n)(F(x_1, \ldots, x_n) \leftrightarrow \psi),$$

where ψ is a wff in which F^n does not occur and in which no predicate variables have any bound occurrences. Then, where $\mathfrak{A}$ is an L-model for a language L of which ψ is a wff, and A is a nominalistic assignment in $\mathfrak{A}$,

$$A[(\exists F^n)(\forall x_1) \ldots (\forall x_n)(F(x_1, \ldots, x_n) \leftrightarrow \psi)]$$

reduces by definition to:

$$(\exists F^n)(\forall x_1) \ldots (\forall x_n)(F(x_1, \ldots, x_n) \leftrightarrow A[\psi]).$$

Note, however, that by definition $A[\psi]$ is a first order wff of L since $rk(\psi) = 0$; and therefore by the above lemma and the already noted validation of (CP!) when restricted to first order wffs for ψ, it follows that (CP!) is nominalistically valid for any wff ψ such that $rk(\psi) = 0$, regardless whether ψ is a first order wff or not. We conclude, in other words, that (CP!) does not commit us to a mode of second order existence which goes beyond nominalistic universals. For this reason, we shall hereafter say that a wff ψ is *predicative in nominalism's purely grammatical sense* if $rk(\psi) = 0$, i.e., if no predicate variable has a bound occurrence in ψ.

It is clear of course that not only are all instances of (CP!) nominalistically valid but so are all instances of the remaining axiom schemas described above in Section 2. It follows, accordingly, that all of the theorems of standard predicative second order logic (with identity) are nominalistically valid. Besides being nominalistically sound, however, these axioms and rules also yield the following completeness theorem whose proof we omit since it follows the well-known Henkin style proof of completeness (but for details see Cocchiarella [7]).

COMPLETENESS THEOREM: *If* $\Sigma \cup \{\phi\}$ *is a set of wffs, then:*
(1) $\Sigma \vdash_{\text{nom}} \phi$ *if and only if* $\Sigma \models_{\text{nom}} \phi$; *and*
(2) $\vdash_{\text{nom}} \phi$ *if and only if* $\models_{\text{nom}} \phi$.

4. NOMINALISM AND MODAL LOGIC

The second general thesis of nominalsim, we have said, is the thesis of extensionality, a corollary of which is the thesis of anti-essentialism, i.e., the thesis that no nominalistic universal can be necessarily true of some individuals of a given domain of discourse without being necessarily true of all of the individuals in that domain. Given $\Box$ as a new primitive logical constant such that $\Box\phi$ is a wff whenever ϕ is a wff, this thesis of anti-essentialism has the following modal counterpart:

$$\text{(A–E)} \quad (\exists x_1) \ldots (\exists x_n)[ID_j(x_1, \ldots, x_n) \;\&\; \Box\phi] \rightarrow (\forall x_1) \cdots$$
$$\ldots (\forall x_n)[ID_j(x_1, \ldots, x_n) \rightarrow \Box\phi].$$

Here $x_1, \ldots, x_n$ are to be all of the distinct individual variables occurring free in ϕ, and $ID_j(x_1, \ldots, x_n)$ is a conjunction of wffs of the form $(x_i = x_k)$ or $(x_i \neq x_k)$, for all i, k such that $1 \leqslant i < k \leqslant n$, but not also containing both $(x_i = x_k)$ and $(x_i \neq x_k)$ as conjuncts. (Where $n = 0$, the principle (A–E) is understood to be just $(\Box\phi \rightarrow \Box\phi)$; and where $n = 1$, it is understood to be $[(\exists x)\Box\phi \rightarrow (\forall x)\Box\phi]$.) For $n \geqslant 2$, the conjunction $ID_j(x_1, \ldots, x_n)$ specifies a complete identity-difference condition with respect to the variables $x_1, \ldots, x_n$, i.e., which of these variables are to have the same values and which are to have different values. Since there are only a finite number of non-equilvalent such conjuctions, we take $ID_j(x_1, \ldots, x_n)$, relative to an assumed ordering, to be the jth conjunction in the ordering. It is clear of course that relative to such a complete identity – difference condition, and where ϕ is a first order wff, the schema (A–E) is the modal counterpart of the nominalist thesis that any predicate expression which is necessarily true of some of the individuals in a given domain of discourse is necessarily true of all of the individuals in that domain. (Cf. Parsons [21] for an explication of (A–E).)

Now since the only kind of necessity which could be invoked as validating the above modal thesis of anti-essentialism is a logical or formal necessity and not a metaphysical or material necessity (which clearly cannot be accounted for in nominalism), the question naturally arises as to whether nominalism is really committed to the thesis of extensionality after all, i.e., whether in fact it need only be committed to the modal thesis of anti-essentialism. The answer, as we shall see, is that just as a metaphysical necessity goes beyond the bounds of nominalism, so too does logical necessity.

Restricting ourselves to first order wffs, we may state the semantics for $\Box$ as a logical or formal necessity in terms of the standard definition of satisfaction given for the first order wffs of a language L in Section 3. That is, we need only supplement clauses (1)–(5) of that definition with the following clause for $\Box$:

(6) *A satisfies* $\Box\phi$ *in* $\mathfrak{A} = \langle D, f \rangle$ iff for all g, if $\langle D, g \rangle$ is an L-model, then A satisfies ϕ in $\langle D, g \rangle$.

In terms of truth, in other words, a wff $\Box\phi$ is true in a given domain of discourse if and only if ϕ is true in that domain no matter which objects the predicate constants occurring in ϕ are true of in that domain. Assuming logical truth to be defined as usual, we might note that one or another version of this semantics for $\Box$ is given in Carnap [3], Kanger [16], Beth [1], Montague [18], Chapter 1, and Cocchiarella [6]. We shall refer to this semantics hereafter as *the primary semantics for* $\Box$ *as logical necessity*. That every instance of the modal thesis of anti-essentialism, (A–E), is logically true (as presently defined) is easily seen to be a consequence of the following lemma which is itself easily proved by an inductive argument on the structure of ϕ.

LEMMA. *If* L *is a language,* $\mathfrak{A}$, $\mathfrak{B}$ *are* L*-models, and h is an isomorphism of* $\mathfrak{A}$ *with* $\mathfrak{B}$*, then for all first order modal wffs* ϕ *of* L *and all assignments A in* $\mathfrak{A}$*, A satisfies* ϕ *in* $\mathfrak{A}$ *if and only if* A/h *satisfies* ϕ *in* $\mathfrak{B}$.

An appropriate consequence of the modal thesis of anti-essentialism in this semantics, it might be noted, is the reduction of all *de re* wffs to *de dicto* wffs. (A *de re* wff is one in which some individual variable has a free occurrence in a subwff of the form $\Box\psi$. A *de dicto* wff is a wff that is not *de re*.)

LEMMA. *For each de re wff* ϕ *there is a de dicto wff* ψ *such that* $(\phi \leftrightarrow \psi)$ *is logically true.*

Such a reduction, needless to say, would seem to be a necessary condition of any modal logic that might plausibly be associated with nominalism. Whatever its initial plausibility, however, such an association cannot be sustained, or at least so we shall argue, since the present semantics is in fact essentially incomplete. The incompleteness is easily seen to follow from the following lemma and the well-known fact that the modal free non-logical truths of a language containing at least one relational predicate is not recursively enumerable (cf. [6], Section 3).

LEMMA. *If* ψ *is a sentence which is satisfiable, but only in an infinite model, and* ϕ *is a modal and identity free sentence, then* $(\psi \rightarrow \sim\Box\phi)$ *is logically true if and only if* ϕ *is not logically true.*

THEOREM. *If* L *is a language containing at least one relational predicate, then the set of first order (modal) wffs of* L *that are logically true is not recursively enumerable.*

Now it is noteworthy that the above incompleteness theorem is equivalent to the incompleteness of (a proper subsystem of) second order predicate logic with respect to its "standard" or logical realist semantics (as previously described). This can be seen by means of the following transformation of first order modal wffs into second order modal free wffs. For convenience we retain the same predicate constants already occurring in the first order wffs but reconstrue them now as new predicate variables (of the same degree or number of subject positions as the constant). The transformation is defined recursively: (1) $t(x = y) = (x = y)$; (2) $t(P^n(x_1, \ldots, x_n)) = P^n(x_1, \ldots, x_n)$; (3) $t(\sim\phi) = \sim t(\phi)$; (4) $t(\phi \rightarrow \psi) = [t(\phi) \rightarrow t(\psi)]$; (5) $t((\forall x)\phi) = (\forall x)t(\phi)$; and (6) $t(\Box\phi) = (\forall P_1) \ldots (\forall P_n)t(\phi)$, where $P_1, \ldots, P_n$ are all the predicate constants (now reconstrued as predicate variables of the same degree) occurring in ϕ. (Note: because the universal quantifier commutes with itself, we ignore specifying in what order the predicates must occur in the quantifier prefix. Also, when $n = 0$, we take the quantifier prefix to be the null sequence.)

The modal operator for logical necessity is interpreted according to the above transformation as a string of universal quantifiers binding all of the predicate constants (now construed as variables) occurring in a given first order wff. The null effect of iterated occurrences of the modal operator is hereby explained as the null effect of iterated quantifiers binding variables already bound. It should be noted, incidentally, that the transformation involves only a proper subset of the (pure) second order wffs. Nevertheless, the logically true wffs in this subset, as we shall see, do not form a recursively enumerable set; i.e., the essential incompleteness of standard second order predicate logic, when interpreted in accordance with the semantics of logical realism, applies to this proper subset of (pure) second order wffs no less so than to the set of all second order wffs.

We have already described the "standard" (realist) satisfaction clause for predicate quantifiers in Section 3, relative of course to a completely referential assignment, and we may assume that definition to now be in effect. Assuming that logical truth is also defined in the usual way, it is now easily seen that a first order modal wff ϕ is logically true if and only if $t(\phi)$ is logically true (in the "standard" semantics for predicate quantifiers). Indeed, this is an immediate consequence of the following lemma which is proved by a simple induction on the structure of ϕ (cf. [6], Section 4).

LEMMA. *If ϕ is a first order modal wff of a language* L, $\mathfrak{A} = \langle D, f \rangle$ *is an* L*-model, and A is a completely referential assignment in D such that $A(P^n) = f(P^n)$, for all $P^n \in$* L, *then A satisfies ϕ in $\mathfrak{A}$ if and only A satisfies $t(\phi)$ in $\mathfrak{A}$.*

THEOREM. *A first order modal wff ϕ is logically true (in the primary semantics for logical necessity) if and only if $t(\phi)$ is logically true (in the "standard" or logical realist semantics for predicate quantifiers).*

In terms of the above connection between first order modal wffs and their second order modal free transforms we see that the supporting lemma of the incompleteness theorem cited above is really the converse transformation of the supporting lemma for the incompleteness of second order predicate logic with respect to its "standard" realist semantics.

LEMMA. *If ψ is a sentence which is satisfiable, but only in an infinite model, and ϕ is a modal and identity free first order sentence all the distinct predicate constants of which are $P_1, \ldots, P_n$, then $[\psi \rightarrow \sim(\forall P_1) \ldots (\forall P_n)\phi]$ is logically true (in the "standard" logical realist semantics for predicate quantifiers) if and only if ϕ is not logically true (in standard first order logic).*

THEOREM. *The set of (second order) wffs which are logically true (in the "standard" logical realist semantics for predicate quantifiers) is not recursively enumerable.*

We conclude then that the incompleteness of modal logic under ts primary semantics is really just a variant of the incompleteness of second order predicate logic under its "standard" or logical realist semantics. That is, under its primary semantics, the necessity operator amounts in effect to a kind of realist quantifier regarding all of the logically possible extensions drawn from a given domain of discourse that might be assigned to the predicate constants of a given language. That this semantics is realist and not nominalist is clear from the fact that if the domain is infinite then the number of such extensions which might be assigned to predicate constants is nondenumerable, and in nominalism a language with nondenumerably many predicate constants is simply out of the question. Predicate constants are human artifacts, after all, and therefore in principle there can be only a potential infinity of such (which is why languages are assumed to have a finite, even if otherwise unlimited, number of predicate constants). Thus, insofar as the modal thesis of anti-essentialism can be based only upon the primary semantics for logical necessity, then the first general thesis of nominalism (that there are no universals beyond predicate expressions) implies the second (the thesis of extensionality). There are two assumptions involved in this implication, however; namely (1) that nominalism does not exclude the possibility of there being an infinite domain of discourse and (2) that nominalism does not exclude the possibility of there being languages with relational predicates (which can be used in the formulation of a wff which is satisfiable only in an infinite domain). These are extremely plausible assumptions, needless to say, since a nominalism without them is an extremely limited philosophical theory of predication. For convenience, however, we avoid their statement in our characterization of nominalism and retain our original description instead, i.e., that nominalism is committed to the thesis of extensionality.

5. CONCEPTUALISM VS. NOMINALISM

Conceptualism, as we understand it here, is a socio-biologically based theory of the human capacity for systematic concept-formation. Concepts, on this view, are neither predicate expressions nor independently real properties or relations but are rather cognitive capacities or cognitive structures otherwise based upon such capacities whose realization in thought is what informs our mental acts with a predicable or referential nature. As capacities or cognitive structures based upon such, moreover, concepts are also neither mental images nor ideas in the sense of particular mental occurrences. That is, concepts are not individuals but are rather *unsaturated* cognitive structures or dispositional abilities. Of course, the exercise (or saturation) of a concept does result in a mental event, i.e., a mental act, and if overtly expressed, a speech act as well; but, nevertheless, the concept is not itself either the mental or speech act as an event but rather what accounts for that act's predicable or referential nature.

Referential concepts, it should be noted, are cognitive structures which are complementary to predicable concepts; i.e., each saturates the other in a given mental act. This is why in conceptualism referential concepts are what quantifiers stand for, including the predicate quantifiers that refer to concepts as unsaturated cognitive structures as well as the sortal and common noun quantifiers that refer to different kinds of individuals. That is, both predicate and individual quantifiers stand for referential concepts, and just as sortal and common noun quantifiers stand for referential concepts that refer to different kinds of individuals, so too different kinds or predicate quantifiers may be used to refer to different kinds of concepts. (For a discussion of sortal and common noun quantifiers in conceptualism, see Cocchiarella [9], Sections 11–12.) Since our primary concern in this essay is with predication, however, we shall not develop the logic of referential concepts in any detail here but shall restrict ourselves to conceptualism's account of the logic of predicable concepts. It is the predicable concepts, after all, and not the referential concepts, which are to be compared with the universals of both nominalism and realism.

Now in regard to predicable concepts, it should be noted, not all variants of conceptualism agree on what is to be admitted as such. There are in particular two types or forms of conceptualism each with its own formal theory of predication. The first type, which we shall call *constructive conceptualism*, is not only more restrictive than the second, which we shall call *holistic conceptualism*, but it is also both presupposed and contained in the latter. The restrictions imposed in the first type, moreover, are similar to those of nominalism, and for this reason it is sometimes confused with the latter, though the logics of the two, as we shall see, are really quite distinct. Holistic conceptualism, on the other hand, transcends constructive conceptualism by positing a stage in the development of concept-formation (usually occurring only in post-adolescence) at which so-called impredicative concept-formation becomes possible; and in this regard holistic conceptualism is similar to logical realism. That is, just as the full impredicative comprehension principle is logically true in logical realism, so too this principle is validated in holistic conceptualism as one of its laws of compositionality for concept-formation.

It is clear of course that both constructive and holistic conceptualism reject the first general thesis of nominalism, i.e., the thesis that there are no universals other than predicate expressions. Indeed, although the two forms of conceptualism will not agree in all cases on what is to count as a genuine predicate expression, nevertheless both agree that a genuine predicate expression will be such only through having an intelligible universal, i.e., a concept, as the semantic ground for its correct or incorrect application. This is not so say that the possession of a concept will not come to be behaviorally equivalent to the possession of a linguistic ability to correctly apply a predicate expression having that concept as its semantic grounds. Nor is it to deny, moreover, that the possession of such a linguistic ability will naturally come to serve as a criterion for the possession of the concept. Nevertheless, all questions of behavioral equivalence aside, being a criterion for the possession of a concept, according to conceptualism, is not the same as being that concept.

Conceptualism does agree, however, that concepts and predicate expressions do have analogous roles in their respective mediums, and to some extent the development of concepts in the medium of thought and cognition is determined by the development of predicate expressions in the medium of language – just as the development of language is determined in part by the development of thought and cognition. In holistic conceptualism especially, as we shall see, impredicative concept-formation is a generalized capacity which is achieved only through the capacity for language, and in particular through the capacity to use language for the expression of constructive or predicative concepts. In this regard, impredicative concept-formation is a mediated process, and language and the linguistic ability to use predicate expressions to express predicative concepts is the means used to master and direct such a process. Nevertheless, aside from this mediation, the laws of compositionality for concept-formation in holistic conceptualism greatly exceed the constraints imposed in nominalism on the restricted comprehension principle (CP!). In this regard, in other words, there can be no question of reducing or otherwise explaining only in terms of nominalistic universals the formal theory of predication we shall associate with holistic conceptualism.

Conceptualism in general also goes beyond nominalism in not being committed to the latter's second general thesis, i.e., the thesis of extensionality. Conceptualism cannot allow for the primary semantics of logical necessity, to be sure, since as a socio-biologically based theory conceptualism in general can allow only for a potential infinity of concepts; that is, conceptualism cannot account for the nondenumerably many universals involved in the primary semantics for logical necessity when the latter is applied to an infinite domain of discourse. Conceptualism can account for other modal concepts, however, including especially the temporal and causal modalities implicit in our use of natural language, and we shall later turn to such an account. On the other hand, conceptualism does not exclude an extensional formulation, and for reasons of comparison with nominalism we shall turn to such a formulation first.

6. CONSTRUCTIVE CONCEPTUALISM

The natural counterpart to nominalism in conceptualism, we have said, is constructive conceptualism. This is because, as a description of the laws of compositionality for concept-formation, the comprehension principle in constructive conceptualism's formal theory of predication agrees with nominalism in excluding wffs which are not predicative in nominalism's purely grammatical sense (i.e., wffs in which no predicate variable has a bound occurrence). But this is not all that is excluded, and in fact the comprehension principle for constructive conceptualism is even more restrictive than (CP!).

Assuming all of the same syntactical notions already defined in Section 2 for nominalism, we may describe an extensional version of constructive conceptualism in terms of the same axioms and inference rules already given for nominalism, except of course for (CP!). Instead of (CP!), the comprehension principle of constructive conceptualism is given as follows:

$$\text{(CCP!)} \quad (\forall G_1) \ldots (\forall G_k)(\exists F^n)(\forall x_1) \ldots (\forall x_n)[F(x_1, \ldots, x_n) \leftrightarrow \phi],$$

where (1) ϕ is a *pure* second order wff (i.e., one in which no non-logical constants occur), (2) neither F^n nor the identity sign occur in ϕ, (3) ϕ is predicative in nominalism's purely grammatical sense (i.e., no predicate variable has a bound occurrence in ϕ), and (4) $G_1, \ldots, G_k$ are all of the distinct predicate variables occurring (free) in ϕ and $x_1, \ldots, x_n$ are among the distinct individual variables occurring free in ϕ (though they need necessarily be all of the individual variables occurring free in ϕ). Hereafter, we shall take $\Sigma \vdash_c \phi$ to mean that ϕ is *derivable from* Σ *in constructive conceptualism* in the sense of being derivable from Σ on the basis of these axioms and rules. We still speak of Σ as an *applied theory*, only now of constructive conceptualism; i.e., the closure of Σ is now understood to be with respect to $\vdash_c$ rather than $\vdash_{\text{nom}}$.

Now because of clause (3) it is clear that every instance of (CCP!) is also an instance of (CP!) (since universal generalizations of instances of axiom schemas are also instances of those schemas). Not every instance of (CP!), however, is also an instance of (CCP!). The important differences are that in (CP!) the initial quantifiers regarding all of the predicate variables occurring free in the comprehending wff ϕ are not required as a necessary condition and that ϕ need not be a *pure* second order wff in which the identify sign does not occur; i.e., in (CP!), but not in (CCP!), ϕ may contain occurrences of the identity sign as well as occurrences of any of the predicate constants of the language or applied theory in question. That is, in nominalism every n-place predicate constant and every first order wff $\phi(x_1, \ldots, x_n)$ of the language in question is posited in (CP!) as being a "value" or legitimate substituend of the bound n-place predicate variables, and in that sense as being itself a genuine predicate expression. Of course this is because in nominalism a wff is predicative if and only if it has only first order wffs as fully applied instances; and, as indicated in Section 3, it is precisely this purely grammatical notion of predicativity which is the basis of the substitutional interpretation of predicate quantifiers in standard predicative second order logic.

In constructive conceptualism, however, predicate quantifiers are to be interpreted referentially and not substitutionally, and the only *formal criterion* as to whether a wff $\phi(x_1, \ldots, x_n)$ of an applied theory Σ is predicative or not is whether or not

$$\Sigma \vdash_c (\exists F^n)(\forall x_1) \ldots (\forall x_n)[F(x_1, \ldots, x_n) \leftrightarrow \phi].$$

It is predicativity in this logico-grammatical sense, in other words, and not predicativity in nominalism's purely grammatical sense, which is what is relevant in constructive conceptualism's formal theory of predication; and it is only with respect to this notion of predicativity that we are to understand the restrictions imposed on (CCP!). Thus, for example, the principle of universal instantiation of a wff $\phi(x_1, \ldots, x_n)$ for an n-place predicate variable is provable in constructive conceptualism only in the qualified form ($\exists/UI_2$) described in Section 2 and not also in the form ($UI!_2$). That is, as stipulated in ($\exists/UI_2$), in order to prove that a wff $\phi(x_1, \ldots, x_n)$ of an applied theory Σ is a legitimate substituend of the bound n-place predicate variables (and in that sense that $\phi(x_1, \ldots, x_n)$ represents a constructive or predicative concept of the applied context in question), we need first to prove that $\phi(x_1, \ldots, x_n)$ is predicative in the above logico-grammatical sense.

No wff can be shown to be absolutely predicative in constructive conceptualism's logico-grammatical sense, it should be noted. That is, unlike the situation in nominalism, no wff $\phi(x_1, \ldots, x_n)$ is such that for every applied theory Σ of (the language of) which ϕ is a wff:

$$\Sigma \vdash_c (\exists F^n)(\forall x_1) \ldots (\forall_{x_n})[F(x_1, \ldots, x_n) \leftrightarrow \phi].$$

Indeed, it is even possible for there to be a consistent applied theory Σ in which no wff at all of Σ is predicative in constructive conceptualism's logico-grammatical sense. Of course in nominalism such a theory would be inconsistent, since a wff which is predicative in nominalism's purely grammatical sense is therefore, by (CP!), also predicative in nominalism's logico-grammatical sense; i.e., the grammatically predicative wffs of nominalism are absolutely predicative in nominalism's logico-grammatical sense. In constructive conceptualism, on the other hand, predicativity in the logico-grammatical sense must always be relativized to an applied theory or context.

A given applied theoretical context Σ, for example, might well warrant the assertibility (for some or all values of the variables) of a wff $\phi(x_1, \ldots, x_n)$ which in fact is *impredicative* in that context; i.e., where

$$\Sigma \vdash_c \sim(\exists F^n)(\forall x_1) \ldots (\forall x_n)[F)x_1, \ldots, x_n) \leftrightarrow \phi].$$

Such a wff, moreover, might even be predicative in nominalism's purely grammatical sense. Thus, for example, Σ might be a second order set theory whose language consists of the 2-place predicate constant '$\in$' and in one of whose axioms it is claimed that there is a set corresponding to any predicative concept:

$$(\forall F)(\exists y)(\forall x)[x \in y \leftrightarrow F(x)].$$

In nominalism, it should be noted, such an axiom leads directly to Russell's paradox, because in nominalism $(x \notin x)$ is a predicative wff, and therefore it is a legitimate substituend of the 1-place predicate variables. In constructive conceptualism, on the other hand, the above axiom is both consistent and plausible; and all that follows by Russell's argument is that '$\in$' cannot stand for a constructive or predicative concept, i.e., that

$$\Sigma \vdash_c \sim(\exists R)(\forall x)(\forall y)[R(x, y) \leftrightarrow x \in y].$$

Constructive conceptualism, in other words, is "free of existential presuppositions" regarding predicate constants (and variables), and in that regard it is free of the presupposition that any wff which is predicative in nominalism's purely grammatical sense must for that reason stand for a value of the bound predicate variables.

Of course if $\phi(x_1, \ldots, x_n)$ is a first order wff of (the language of) an applied theory Σ such that (1) the identity sign does not occur in ϕ and (2) for each predicate constant P^k which occurs in ϕ,

$$\Sigma \vdash_c (\exists G^k)(\forall y_1) \ldots (\forall y_k)[G(y_1, \ldots, y_k) \leftrightarrow P(y_1, \ldots, y_k)],$$

then, by ($\exists$/UI$_2$) and (CCP!), $\phi(x_1, \ldots, x_n)$ is predicative in the context represented by Σ in the logico-grammatical sense of constructive conceptualism; i.e.,

$$\Sigma \vdash_c (\exists F^n)(\forall x_1) \ldots (\forall x_n)[F(x_1, \ldots, x_n) \leftrightarrow \phi].$$

In other words, constructive or predicative concepts regarding the individuals of an applied theoretical context can always be constructed from other predicative concepts regarding those individuals by any of the operations of first order logic, including quantificational references over the totality of individuals in question. What cannot be constructed in such a context, however, or at least not in the sense of constructive conceptualism, are concepts whose analyses involve one or another quantificational reference over the totality of predicative concepts regarding the individuals of that context. There can be no such concepts, according to constructive conceptualism, since their construction would involve a totality to which they belong; and it is for this reason that any wff of the applied theory in question which contains a predicate quantifier is to be excluded from the comprehension principle (CCP!). (Strictly speaking, of course, a predicate quantifier binding a k-place predicate variable ranges directly only over the totality of k-ary predicative concepts of the context in question, and not over the totality of all predicative concepts regarding that context. Nevertheless, as is clear from (CCP!), a quantificational reference over the one totality is indirectly a reference over the larger totality as well.)

Another difference between nominalism and constructive conceptualism, incidentally, is in the role of free predicate variables. Thus, whereas the following substitution rule,

(S!) if $\vdash_{\overline{\text{nom}}} \phi$, then $\vdash_{\overline{\text{nom}}} \phi[\psi/F(x_1, \ldots, x_n)]$

is not derivable in nominalism's standard predicative second order logic, essentially the same rule,

(S_c!) if $\vdash_{\overline{c}} \phi$, then $\vdash_{\overline{c}} \phi[\psi/F(x_1, \ldots, x_n)]$,

is derivable in constructive conceptualism (by a simple inductive argument on the proof for ϕ). (Note that by ($UI!_2$),

$$\vdash_{\overline{\text{nom}}} (\forall F^n)\phi \rightarrow \phi,$$

and therefore if (S!) were derivable in nominalism, then so would the full unrestricted principle of universal instantiation,

(UI_2) $(\forall F^n)\phi \rightarrow \phi[\psi/F(x_1, \ldots, x_n)]$;

and therefore so would the full unrestricted comprehension principle of standard impredicative second order logic. The same argument does not apply in constructive conceptualism, however, because ($UI!_2$) is not derivable in constructive conceptualism to begin with.)

Now it is noteworthy that (S_c!) validates a broader notion of a possible (explicit) definition of a predicate constant in constructive conceptualism than is allowed in nominalism, where, as already indicated, only first order wffs are allowed as definiens. Thus, where L is a language and P^n is an n-place predicate constant not in L, then in constructive conceptualism's nonstandard predicative second order logic, a *possible definition of* P^n *in terms of* L is of the form

$$(\forall x_1) \ldots (\forall x_n)[P(x_1, \ldots, x_n) \leftrightarrow \phi],$$

where ϕ is any wff of L, *predicative or nonpredicative* in nominalism's purely grammatical sense, in which no predicate variable has a free occurrence and in which $x_1, \ldots, x_n$ are all the distinct individual variables having free occurrences. (Note that if the same condition was allowed in nominalism, *viz.*, that ϕ might not be a first order wff, then by ($EG!_2$), i.e., the contrapositive of ($UI!_2$), the full unrestricted comprehension principle of standard impredicative second order logic would again be derivable, contrary to the notion of predicativity upon which nominalism is founded. Again, the same argument does not apply in constructive conceptualism because neither ($UI!_2$) nor ($EG!_2$) is derivable in the nonstandard predicative second order logic in question.)

Thus, where Σ is an applied theory of a language L, $P^n \notin$ L, and δ is a possible definition of P^n in terms of L in the above sense, then by ($S_c!$) it follows that (1) δ is *noncreative*, i.e.., for each wff ψ of L,

$$\text{if } \Sigma \cup \{\delta\} \vdash_c \psi, \text{ then } \Sigma \vdash_c \psi;$$

and (2) that P^n is provably *eliminable* by means of δ from every wff of L $\cup$ $\{P^n\}$, i.e.,

$$\Sigma \cup \{\delta\} \vdash_c \psi \leftrightarrow \psi[\phi/P(x_1, \ldots, x_n)],$$

where ϕ is the wff of L which is the definiens of δ. (The eliminability condition follows by applying ($S_c!$) to (LL_2), the second order analogue of (LL) described in Section 2, where the antecedent of (LL_2) is δ. The noncreativity condition follows by an inductive argument on the derivation of ψ from $\Sigma \cup \{\delta\}$.)

One important consequence of the difference in logistic role which predicate constants and predicate variables have in constructive conceptualism from that in nominalism is the irreducibility in constructive conceptualism of the concept of identity. In this regard it should be noted that while both nominalism and constructive conceptualism agree that any applied theory will contain at most finitely many predicate constants – though of course for both there are potentially infinitely many predicate constants that might be introduced in the development of that theory or in any of its definitional extensions – they will disagree as to whether indiscernibility with respect to those finitely many predicate constants amounts itself to a predicative wff, as well as to whether such indiscernibility suffices for full substitutivity; and in the latter regard they will disagree as to whether the identity sign is in principle eliminable in favor of such indiscernibility. In other words, as far as nominalism is concerned the identity sign can be eliminated in any applied context in favor of the indiscernibility wff regarding the finitely many predicate constants involved in the theory describing that context (cf. Quine [26], p. 63f).

This is not the case in constructive conceptualism, however; and in fact the indiscernibility wff regarding all of the predicate constants of a given context will not be predicative in constructive conceptualism's logico-grammatical sense unless all of the predicate constants in question are themselves predicative in this sense. Moreover, even if all of the predicate constants were assumed (e.g., by "meaning postulates") to stand for predicative concepts, such an indiscernibility wff will not suffice for the substitutivity of singular terms in wffs which are not predicative in constructive conceptualism's logico-grammatical sense. The identity sign, in other words, is not eliminable or otherwise reducible in constructive conceptualism. Moreover, since identity allows for the full substitutivity of singular terms even in impredicative contexts – and, in particular, since identity implies indiscernibility with respect to all predicative concepts – then an identity wff $(x = y)$ will not in general stand for a predicative concept, which is why the identity sign is not allowed to occur in instances of (CCP!). Thus, whereas

$$x = y \leftrightarrow (\forall F)[F(x) \leftrightarrow F(y)]$$

is provable in nominalism, the right-to-left direction of this same wff is in fact not provable in constructive conceptualism's nonstandard predicative second order logic.

Finally, let us now turn to an external set-theoretic semantics for our extensional version of constructive conceptualism. In this regard, we shall say that a structure of the form $\langle\langle D, f\rangle, X_n\rangle_{n\in\omega-\{0\}}$ is an L-*model structure* (*for constructive conceptualism*) if and only if (1) $\langle D, f\rangle$ is an L-model and (2) for each positive integer n, X_n is a countable set every member of which is a set of n-tuples drawn from D. Naturally, D is the domain of discourse of the applied context in question and, as in nominalism, $f(P^n)$, where $P^n \in$ L, is the set of n-tuples of members of D that P^n is true of in that context. Each of the sets X_n, moreover, is understood to consist of the extensions of the countably many n-ary predicative concepts that are constructible regarding that domain of discourse. We assume there to be only countably many such extensions because as a biologically based theory of universals constructive conceptualism is committed to there being at most only a potential infinity of concepts (and in our external semantics we ignore the difference between a potential and an actual infinity). (Nothing would be gained, at least as far as conceptualism is concerned, if uncountably many extensions were allowed in this semantics, since a Löwenheim-Skolem theorem would then be provable, permitting the restriction to countably many extensions.) It should be noted, incidentally, that we do not assume that $f(P^n) \in X_n$, for each $P^n \in$ L; i.e., it is not assumed that the predicate constants of the applied context in question stand for constructive or predicative concepts.

Because predicate quantifiers are now to be interpreted referentially and not substitutionally, we return to the notion (defined in Section 3) of a completely referential assignment in a given domain of discourse. Note that if $\mathfrak{A} = \langle\langle D, f\rangle, X_n\rangle_{n\in\omega-\{0\}}$ is an L-model structure for constructive conceptualism and A is a completely referential assignment in D, then it does not follow that $A(f^n) \in X_n$; i.e., relative to such an assignment, it is not assumed that *free* predicate variables stand for predicative concepts of the context in question (though of course *bound* predicate vriables must do so). Our semantics for constructive conceptualism, in other words, is free of existential presuppositions regarding both predicate constants and free predicate variables, which, as already indicated, is as it should be. The recursive definition of *satisfaction in* $\mathfrak{A}$ *by* A of an arbitrary wff of L proceeds now exactly as in the "standard" definition of satisfaction already given in Section 3 (for logical realism), except of course that the realist clause (7) for predicate quantifiers must now be redefined as follows:

(7) *A satisfies* $(\forall F^n)\phi$ *in* $\mathfrak{A}$ iff for all $K \in X_n$, $A(K/F^n)$ satisfies ϕ in $\mathfrak{A}$.

The definition of *truth* in an L-model structure is assumed defined in the usual way, i.e., as satisfaction in the structure by every completely referential assignment in the structure's domain of discourse. Such a structure is said to be *normal according to constructive conceptualism* if every instance of (CCP!) of the language in question is true in that structure, i.e., if constructive conceptualism's law of compositionality for concept-formation holds in the context in question. A wff ϕ is then said to be a *logical consequence according to constructive conceptualism* of a set of wffs Σ, in symbols $\Sigma \models_c \phi$, iff for any language L such that $\Sigma \cup \{\phi\}$ is a set of wffs of L, and for every completely referential assignment A in the domain of any L-model structure $\mathfrak{A}$ which is normal according to constructive conceptualism, if A satisfies every member of Σ in $\mathfrak{A}$, then A also satisfies ϕ in $\mathfrak{A}$. We say that ϕ is *valid according to constructive conceptualism*, in symbols $\models_c \phi$, if and only if it is a logical consequence of the empty set. The following completeness theorem now follows by means of the usual Henkin style proof.

COMPLETENESS THEOREM. $\Sigma \models_c \phi$ *if and only if* $\Sigma \vdash_c \phi$; *and therefore* $\models_c \phi$ *if and only if* $\vdash_c \phi$.

7. RAMIFICATION OF CONSTRUCTIVE CONCEPTUALISM

One important feature of all forms of conceptualism, it should be noted, is the assumption that concept-formation proceeds through developmental stages of increasing structural complexity where the later stages contain reconstructions of the earlier. These reconstructions are brought about by what Jean Piaget has called a process of reflective abstraction (cf. [22]). Such a process consists of a projection of previously constructed concepts onto a new plane of thought where they are reorganized in accordance with the closure conditions of new laws of concept-formation which are characteristic of the stage in question. It is essentially in just this sense of a reconstruction by a process of reflective abstraction, moreover, that holistic conceptualism is said to presuppose constructive conceptualism as an antecedent stage of concept-formation which it transcends by taking the predicative concepts of an applied context as the contents of new structures which it builds upon their basis.

Now the pattern of reflective abstraction by which holistic conceptualism transcends constructive conceptualism is neither direct nor immediate, but proceeds through a potentially infinite succession of intermediate stages where the concepts constructed or formed at one stage become the contents of the concepts constructed or formed at the next. The stages are intermediate insofar as there is always a next stage at which new concepts can be formed having the concepts of the prior stage as their contents; that is, insofar as at no such stage can concepts be formed whose content involves a referential concept referring to the totality of concepts formed at that stage. This reiterable pattern of reflective abstraction is finally concluded, according to holistic conceptualism, only by a new form of closure which is holistic with respect to all of the intermediate stages; i.e., by an idealized transition to a limit at which impredicative concept-formation is finally realized as a state of cognitive equilibrium (or stage of conceptual development) which can have no succedding stage under the pattern of reflective abstraction in question. Of course it is this holistic closure which applies to itself as well as to all of the intermediate stages which (unlike the closure that applies at any one such stage with respect to the immediately preceding stage) cannot be accounted for in terms of principles acceptable within constructive conceptualism. That is, it is precisely in this new form of closure or idealized transition to a limit that holistic conceptualism transcends constructive conceptualism.

Now without this holistic closure or idealized transition to a limit, it should be emphasized, the pattern of reflective abstraction which generates one intermediate stage from another can be accounted for by principles which are acceptable within constructive conceptualism. For that reason we shall call the framework characterizing the successive stages in question *ramified constructive conceptualism*. The title is especially appropriate, incidentally, because the system characterizing this framework stands to ramified second order logic in essentially the same way that constructive conceptualism's nonstandard predicative second order logic stands to nominalism's standard predicative second order logic. (Cf. Church [4], Section 58, for a formulation of standard ramified second order logic.)

In standard ramified second order logic, it should be noted, there are different types of (*n*-place) predicate variables corresponding to the different "levels" of predicate expressions that might be true or false of the same individuals of an applied theoretical context. This is because in standard predicative second order logic (which is the initial "level" of ramified second order logic) both bound and free (*n*-place) predicate variables have the same wffs as their legitimate substituends or "values". Thus, as understood in nominalism, where all predicate quantifiers are interpreted substitutionally, these different "levels" correspond to a potentially infinite cumulative succession of syntactical metalanguages beginning with an object language L, followed by its syntactical metalanguage ML, followed by the syntactical metalanguage MML (or M^2L) of ML, followed by the syntactical metalanguage MMML (or M^3L) of MML, and so on. Thus, where L_1 is the set of possible predicate constants definable (in nominalism) in terms of L together with those whose definiens contain quantifiers of ML ranging over all the possible predicate constants definable (in nominalism) in terms of L; and, proceeding inductively, where $L_{(j+1)}$, for each positive integer *j*, is the set of possible predicate constants definable in terms of L_j together with those whose definiens contain quantifiers of $M^{(j+1)}L$ ranging over all the possible predicate constants definable in terms of L_j; then, just as the substitutional interpretation of (*n*-place) predicate quantifiers in standard predicative second order logic corresponds to quantification in ML over all of the (*n*-place) predicate expressions of L, i.e., over all of the possible (*n*-place) predicate constants definable (in nominalism) in terms of L, so too the substitutional interpretation of (*n*-place) predicate quantifiers of "level" $(j + 1)$ in ramified second order logic corresponds to quantification in the $(j + 1)$th syntactical metalanguage $M^{(j+1)}L$ over all of the possible (*n*-place) predicate constants definable (in nominalism) in terms of L_j. Similarly, just as the (*n*-place) predicate variables of standard predicative second order logic have the first order wffs of L (with *n* free individual variables as subject position indicators) as their nominalistic values, so too the (n-place) predicate variables of "level" ($j + 1$) have the first order wffs of L_j (with n free individual variables as subject position indicators) as their nominalistic values. In such a substitutional semantics, in other words, free (n-place) predicate variables can have as their values only those nominalistic universals belonging to a certain "level" of the cumulative syntactical metalanguage hierarchy, and therefore each such predicate variable must then be restricted to that "level"; i.e., predicate variables must then be typed in the object language according to their "level" as well as their degree (or adicity).

In conceptualism, however, predicate quantifiers are interpreted referentially and not substitutionally, and, unlike bound predicate variables, free predicate variables can have arbitrary wffs as their legitimate substituends (barring "clashes" of variables). Moreover, introducing different types of (n-place) predicate variables is both unnecessary and unnatural in conceptualism since the (n-ary) concepts referred to at the different stages are all alike in their predicable nature and differ only in their conceptual priorities. Of course, in order to represent the different successive stages obtained by the type of reflective abstraction in question, we shall need to distinguish quantificational references over the concepts constructible at one stage of concept-formation from those constructible at another; i.e., we shall need to distinguish the different (second order) referential concepts regarding the concepts of the different cumulative stages of concept-formation determined by the type of reflective abstraction in question (and the initially given predicative concepts of an applied theoretical context). We shall do so here by introducing into our logical grammar a potential infinity of quantifier signs $\forall^1, \ldots, \forall^j, \ldots$ ($j \in \omega - \{0\}$) which are to be affixed only to predicate variables. (We of course shall understand $(\exists^j F^n)$ as $\sim(\forall^j F^n) \sim$.) Thus the predicate quantifier $(\forall^j F^n)$ is understood here to refer to the n-ary concepts of the jth stage of the pattern of reflective abstraction in question. Of course that regarding the first stage, namely $(\forall^1 F^n)$, is none other than our original predicate quantifier expression $(\forall F^n)$ of constructive conceptualism, i.e., the predicate quantifier originally taken as referring to the n-ary predicative concepts of the applied context in question. Accordingly, we now understand the set of (*ramified*) *wffs* of a language L to be the smallest set K consisting of all of the atomic wffs of L and such that $\sim\phi$, $(\phi \to \psi)$, $(\forall x)\phi$, $(\forall^j F^n)\phi$ are all in K whenever ϕ, ψ are in K and j, n are positive integers.

Beginning then with the original comprehension principle (CCP!) which represents the law of compositionality for predicative concepts, the pattern of reflective abstraction and subsequent closure which is in question here is represented by the following *ramified conceptualist comprehension principle* as applied at the jth stage (and which is none other than (CCP!) when $j = 1$):

$$\text{(RCCP!)}\quad (\forall^j G_1)\ldots(\forall^j G_k)(\exists^j F^n)(\forall x_1)\ldots$$
$$\ldots(\forall x_n)\,[F(x_1,\ldots,x_n)\leftrightarrow\phi],$$

where (1) ϕ is a *pure* (ramified) wff (i.e., one in which no non-logical constant occurs), (2) neither F^n nor the identity sign occur in ϕ, (3) no predicate variable is bound in ϕ by a quantifier of a stage $\geqslant j$ (i.e., for all $i \geqslant j$, $\forall^i$ does not occur in ϕ), and (4) $G_1, \ldots, G_k$ are all of the distinct predicate variables occurring free in ϕ and $x_1, \ldots, x_n$ are among the distinct individual variables occurring free in ϕ.

The axioms and inference rules of ramified constructive conceptualism are the same axioms and rules already described for constructive conceptualism, except of course for replacing (CCP!) by (RCCP!) for all positive integers j, and also for restating axioms (A1) and (A2) in terms of $\forall^j$ when applied to predicate variables. (Universal generalization with respect to predicate variables is of course understood to hold for all of the quantifiers $\forall^j$.) We do need one additional axiom corresponding to the cumulative nature of the stages of concept-formation in question; namely,

$$(\forall^j F^n)\phi \rightarrow (\forall^i F^n)\phi,\ \text{where } i \leqslant j.$$

Given this axiom, the cumulative nature of the different stages of concept-formation is now seen to be characterized in the following consequence of (RCCP!),

$$\vdash_{rc} (\forall^i G^n)(\exists^j F^n)(\forall x_1)\ldots(\forall x_n)[F(x_1,\ldots,x_n)\leftrightarrow$$
$$\leftrightarrow G(x_1,\ldots,x_n)],$$

where $i \leqslant j$. (We use '$\vdash_{rc}$' to stand for the derivability relation of ramified constructive conceptualism.)

The point of the above axiom schema (RCCP!) is that what is added at each stage $(j + 1)$ beyond the initial stage of predicative concepts are concepts whose construction involves the *totality* of concepts constructible at the jth stage (i.e., concepts that have referential concepts referring to the concepts constructible at the prior stages as part of their contents); and formally this is represented by allowing the occurrence of predicate quantifiers of the jth stage in any comprehending wff $\phi(x_1, \ldots, x_n)$ of an instance of (RCCP!) for the $(j + 1)$th stage. Thus, beginning with the predicative concepts of an applied theoretical context Σ in which not all of the wffs $\phi(x_1, \ldots, x_n)$ are predicative in constructive conceptualism's logico-grammatical sense, the constructive conceptualist is able to reconsider the nonpredicative wffs of Σ as representing constructible concepts of a new stage of concept-formation, though of course not as concepts which are initially predicative in Σ, since they are formed by a process of reflective abstraction which is based upon the predicative concepts of Σ, i.e., by a process which involves quantificational reference to the predicative concepts of Σ. The new concepts which are represented by the originally nonpredicative wffs of Σ are not themselves impredicative concepts, however, since they do not have themselves as part of their referential content, i.e., they are not impredicative in Σ with respect to the new stage of concept-formation; and, indeed, impredicativity in Σ in the logico-grammatical sense will now need to be modified with respect to the new totality of concepts, i.e., with respect to predicate quantifiers referring to the predicative concepts of Σ together with all those whose formation presupposes a quantificational reference to the predicative concepts of Σ. Wffs that are nonpredicative in Σ in the new sense can now of course be taken as representing constructible concepts of a yet further stage of concept-formation where concepts are formed by the same type of reflective abstraction but based now upon the new totality of concepts; and of course, strictly speaking, these concepts in turn will also not be impredicative, since they will not have themselves as part of their referential content. That is, impredicativity in Σ will again have to be modified with respect to the new totality, and again a new stage of concept-formation will result by the same pattern of reflective abstraction; and so on *ad infinitum*, with a genuine impredicativity in Σ coming about only by an holistic closure of all of the stages of this type of reflective abstraction.

In regard to a set-theoretic semantics for ramified constructive conceptualism we need only slightly modify the semantics already given for constructive conceptualism by replacing the notion of an L-model structure by that of an *extended L-model structure*; i.e., a structure of the form $\langle\langle D, f\rangle, X_{(j,n)}\rangle_{j,n\in\omega-\{0\}}$, where (1) $\langle D, f\rangle$ is an L-model, (2) for all j, $n \in \omega - \{0\}$, $X_{(j,n)}$ is a countable set of n-ary relations-in-extension over D, and (3) $X_{(i,n)} \subseteq X_{(j,n)}$, for all positive integers $i \leqslant j$. Satisfaction in such an extended structure $\mathfrak{A}$ is then assumed defined as in Section 6, except that clause (7) is replaced by the following:

(7) *A satisfies* $(\forall^j F^n)\phi$ *in* $\mathfrak{A}$ iff for all $K \in X_{(j,n)}$, $A(K/F^n)$ satisfies ϕ in $\mathfrak{A}$.

Truth in an extended model structure is defined as usual, and *validity* and *logical consequence* (according to ramified constructive conceptualism) must of course be restricted to those extended model structures which are *normal* in the sense that every instance of (RCCP!) formulable therein is true. We must restrict the semantics, in other words, to those extended model structures in which all of the laws of compositionality for concept-formation generated by the pattern of reflective abstraction in question can be said to be true. A completeness theorem is then proved in the usual way.

COMPLETENESS THEOREM: *If* $\Sigma \cup \{\phi\}$ *is a set of ramified wffs, then* $\Sigma \models_{rc} \phi$ *if and only if* $\Sigma \vdash_{rc} \phi$; *and therefore* $\models_{rc} \phi$ *if and only if* $\vdash_{rc} \phi$.

8. HOLISTIC CONCEPTUALISM

Impredicative concept-formation, according to holistic conceptualism, presupposes both the system of predicative concepts, i.e., the closure conditions characterizing the laws of compositionality for predicative concepts, and the holistic closure or idealized transition to a limit of the pattern of reflective abstraction represented by (RCCP!). However, because this abstraction is implicit in the distinction between predicative and impredicative concepts, we shall not explicitly include the detailed formalization of the intermediate stages involved in this abstraction in holistic conceptualism's formal theory of predication. Thus, for convenience we shall not resort to the many different predicate quantifiers involved in (RCCP!). Instead we shall distinguish between only two quantifier signs as applied to predicate variables, the first, $\forall!$, to refer to the totality of predicative concepts, and the second, $\forall$, to refer to all concepts, predicative or impredicative, of an applied context of holistic conceptualism. The set of wffs of a language L, accordingly, is now understood to be the smallest set K containing all of the atomic wffs of L and such that $\sim\phi$, $(\phi \rightarrow \psi)$, $(\forall x)\phi$, $(\forall! F^n)\phi$ and $(\forall F^n)\phi$ are all in K whenever ϕ, ψ are in K and n is a positive integer. (We understand $(\exists! F^n)$ and $(\exists F^n)$ to be $\sim(\forall! F^n)\sim$ and $\sim(\forall F^n)\sim$, respectively.)

The comprehension principle representing the laws of compositionality for concept-formation in constructive conceptualism is of course retained in holistic conceptualism, except that now it is expressed in terms of $\forall!$ and $\exists!$ as follows:

(CCP!) $(\forall! G_1) \ldots (\forall! G_k)(\exists! F^n)(\forall x_1) \ldots (\forall x_n)[F(x_1, \ldots, x_n) \leftrightarrow \phi]$, where (1) ϕ is a pure second order wff (i.e., one in which no non-logical constants occur), (2) neither F^n nor the identity sign occur in ϕ, (3) ϕ is predicative in nominalism's purely grammatical sense (i.e., no predicate variable has a bound occurrence in ϕ), and (4) $G_1, \ldots, G_k$ are all of the distinct predicate variables occurring (free) in ϕ and $x_1, \ldots, x_n$ are among the distinct individual variables occurring free in ϕ.

The comprehension principle representing the laws of compositionality for concepts in general, i.e., for impredicative as well as predicative concepts, is now formulated as follows:

$$\text{(CP)} \qquad (\exists F^n)(\forall x_1) \ldots (\forall x_n)[F(x_1, \ldots, x_n) \leftrightarrow \phi],$$

where ϕ is any wff in which F^n has no free occurrence and in which $x_1, \ldots, x_n$ are distinct individual variables occurring free (though not necessarily all such). The remaining axioms and inference rules are the same as those already described for constructive conceptualism, except for having universal generalization and axioms (A1) and (A2) extended so as to apply to $\forall!$ as well as $\forall$. Thus since (LL_2) and ($\exists/UI_2$) are provable in holistic conceptualism in the same way they are provable in nominalism and constructive conceptualism, then by (CP),

$$(\mathrm{UI}_2) \qquad \vdash_{\overline{h}} (\forall F^n)\phi \rightarrow \phi[\psi/F(x_1, \ldots, x_n)].$$

(We use '$\vdash_{\overline{h}}$' to represent the derivability relation of holistic conceptualism.) Accordingly, where ψ is $F(x_1, \ldots, x_n)$ itself, then by (UG) applied to (UI_2), together with tautologous transformations and (A1) and (A2), it follows that what is true of every n-ary concept in holistic conceptualism is thereby true of every predicative concept:

$$\vdash_{\overline{h}} (\forall F^n)\phi \rightarrow (\forall! F^n)\phi.$$

By (CP), it is clear, every wff $\phi(x_1, \ldots, x_n)$ stands for an n-ary concept in the sense of being a legitimate substituend of the bound n-place predicate variables; that is, every wff $\phi(x_1, \ldots, x_n)$ is "predicative" in holistic conceptualism's logico-grammatical sense. This means in effect that as applied to a given domain of discourse any open wff is "predicative' in the semantic sense of having a concept as the semantic grounds for its correct or incorrect application in that domain. Thus unlike the situation in constructive conceptualism there is no gap here between a wff's being grammatically predicative in a given context in the sense of being the definiens of a possible predicate constant in that context and its being semantically predicative in the sense of actually standing for a concept in that context, i.e., of its being predicative in that context in the logico-grammatical sense of provably being a legitimate substituend of the bound predicate variables. In constructive conceptualism, of course, it is precisely this gap between grammatical and semantical predicativity which leads to all of the intermediate stages of concept-formation in ramified constructive conceptualism; and even though more and more wffs that are only grammatically predicative at one stage become semantically predicative at later stages, nevertheless the gap remains and is finally closed only by the idealized transition to a limit, or holistic closure, which is represented by (CP). The conceptual drive for this closure, and in particular the conceptual tension in the gap which remains at each intermediate stage despite the closure of the preceding stage, is but one of the many mechanisms of this type, according to holistic conceptualism, which characterize the holistic nature of thought and representation.

Impredicative concept-formation, it should be noted, is a generalized capacity based not only on the intermediate stages which lead to its development but on the capacity humans have for language as well. That is, the capacity to give linguistic expression to predicative concepts and to the concepts of the intermediate stages is an essential component in the development of the holistic stage of concept-formation. It is the interconnected development of language and thought, in other words, which is the basis of impredicative concept-formation. In this sense, impredicative concept-formation is a mediated process, and language and the linguistic ability to use predicate expressions is the means used to master and direct such a process.

Finally, in regard to a set-theoretic semantics for holistic conceptualism we need again only slightly modify the definitions already given for constructive conceptualism. Thus, instead of the L-model structures for constructive conceptualism, we take the L-*model structures for holistic conceptualism* to be structures of the form $\langle\langle D, f\rangle, X_n, Y_n\rangle_{n\in\omega-\{0\}}$, where (1) $\langle D, f\rangle$ is an L-model, (2) for all positive integers n, $X_n \subseteq Y_n$, (3) Y_n is a countable set of n-ary relations-in-extension over D, and (4) $f(P^n) \in Y_n$, for all $P^n \in$ L. Each of the sets X_n is understood here to consist of the extensions of the countably many n-ary predicative concepts that are constructible in the domain of discourse, and each of the sets Y_n are similarly understood to consist of the extensions of the countably many n-ary concepts, predicative or otherwise, which can in principle be constructed in that domain. That $f(P^n) \in Y_n$, for all $P^n \in$ L, is a reflection of the condition of holistic closure, i.e., that every open wff, and therefore every predicate constant of the language in question, is to stand for a concept in the context in question.

Because of holistic closure, incidentally, completely referential assignments in a domain of discourse must be restricted to those whose values for predicate variables are always in the sets Y_n. That is, where $\mathfrak{A}$ is an L-model structure described as above, then A is an *assignment in* $\mathfrak{A}$ if and only if A is a completely referential assignment in D such that for all positive integers n and all n-place predicate variables F^n, $A(F^n) \in Y_n$. *Satisfaction in* $\mathfrak{A}$ *by* A is then defined as usual, except that clause (7) is now replaced by the following two clauses:

(7) *A satisfies* $(\forall! F^n)\phi$ *in* $\mathfrak{A}$ iff for all $K \in X_n$, $A(K/F^n)$ satisfies ϕ in $\mathfrak{A}$; and

(8) *A satisfies* $(\forall F^n)\phi$ *in* $\mathfrak{A}$ iff for all $K \in Y_n$, $A(K/F^n)$ satisfies ϕ in $\mathfrak{A}$.

Truth in a holistic model structure is of course defined as usual as satisfaction in the structure by every assignment in the structure. The structure is then said to be *normal according to holistic conceptualism* if every instance (of the language in question) of either (CCP!) or (CP) is true in the structure. Validity and logical consequence (according to holistic conceptualism) are then defined in the obvious way in terms of model structures that are normal according to holistic conceptualism. The following completeness theorem is proved in the usual way.

COMPLETENESS THEOREM: $\Sigma \vdash_{h} \phi$ *if and only if* $\Sigma \models_{h} \phi$; *and therefore* $\vdash_{h} \phi$ *if and only if* $\models_{h} \phi$.

9. LOGICAL REALISM VS HOLISTIC CONCEPTUALISM

Realism, like conceptualism, goes beyond nominalism in positing the existence of universals other than predicate expressions. Unlike conceptualism, however, realism posits universals which are said to exist not only independently of language but of the natural capacity humans have for thought and representation as well; that is, real universals are said to exist independently of all concept-formation and even the natural possibility for such. In keeping with tradition, we shall refer to real universals as properties and relations.

Now the sense in which properties and relations are said to exist is not the same in all forms of realism. We assume, however, that despite their differences otherwise all forms of realism fall under two general and exclusive types which we shall call *natural* and *logical realism*, respectively, Thus, for example, in all variants of natural realism properties and relations are assumed to exist only within nature as components of the causal structure of the world. These universals may exist independently of what is not causally or naturally necessary, and in that sense they may be said to exist independently of the world itself – at least to the extent that not everything which occurs in the world is necessary in this sense. But as components of the causal structure of the world, natural properties and relations cannot also exist independently of all causally possible worlds where the same laws of nature hold as hold in the actual world.

It is precisely this sort of dependence on the causal structure of the world, on the other hand, which all forms of logical realism reject. That is, all variants of logical realism maintain that real properties and relations exist independently not only of what is not determined by the causal structure of the world but also of that causal structure itself – and, indeed, even of whether there is a world at all! Logically real universals, in other words, exist independently of all forms of concrete existence and the causal grounds for such. They do not exist independently of one another, to be sure, but that dependence obtains only in a timeless, transcendent realm of being which, according to all variants of logical realism, is in fact the semantico-ontological ground for the logical truths we express in thought and language; and of course it is for just that reason that we refer to such a theory of universals as a variant of *logical* realism. Thus, unlike the situation in natural realism, it is assumed in all variants of logical realism that there is a real universal corresponding to every fully applied wff $\phi(x_1, \ldots, x_n)$, including those which are logically true or logically false of any individuals whatsoever. That is, in logical realism every instance of the full impredicative comprehension principle (CP) is logically true.

Now though the properties and relations of logical realism are sometimes called *concepts* (especially by Frege and early Russell), they are not concepts in the sense of cognitive capacities or cognitive structures otherwise based on the natural capacity humans have for concept-formation. Indeed, the only point of similarity between logically real universals and concepts in the sense intended here is that both are claimed (in their respective frameworks) to be the semantical grounds for the correct or incorrect application of predicate expressions. Moreover, since it is only in holistic conceptualism that every fully applied wff $\phi(x_1, \ldots, x_n)$ is assumed to stand for a concept, the similarity in question applies only to the concepts of holistic conceptualism.

Logical realism and holistic conceptualism are similar, in other words, insofar as both validate the full impredicative comprehension principle (CP). They are radically dissimilar, on the other hand, both in the nature or mode of being of their respective universals and in the grounds each provides for the validation of (CP) in terms of those universals. Thus, e.g., whereas the predicative-impredicative distinction is essential to the sense in which impredicative concepts are said to exist as holistic projections onto the structure of thought in holistic conceptualism, in logical realism there is no semantical or ontological basis whatsoever for the predicative-impredicative distinction. That is, there is no sense in logical realism in which the predicative-impredicative distinction applies to logically real universals in themselves. Of course such a distinction applies to wffs, but then it is only impredicativity in nominalism's purely grammatical sense which is in question. That is, unlike the situation in conceptualism, no wff $\phi(x_1, \ldots, x_n)$ can be impredicative in logical realism in anything other than the purely grammatical sense of containing bound predicate variables; and this is so, moreover, because logically real universals exist independently of how they are specified.

One important consequence of the difference between concepts on the one hand and logically real universals on the other is that whereas there can only be a potential infinity of concepts, in logical realism the number of universals can be of any transfinite cardinality. Indeed, there will in particular be at least as many properties as classes of individuals, since, according to logical realism, every class of individuals is the extension of some logically real universal. This in fact is *the fundamental thesis of logical realism* which is the basis of its validation of (CP) within its set-theoretic semantics and which differentiates it in that regard from both nominalism and conceptualism. Thus on the basis of the "standard" logical realist satisfaction clause for predicate quantifiers already described in Section 3, it is easily seen that if L is a language, $\mathfrak{A} = \langle D, f \rangle$ is an L-model, A is a completely referential assignment in D, and $\phi(x_1, \ldots, x_n)$ is any standard second order wff of L (in which F^n does not occur free), then A satisfies

$$(\exists F^n)(\forall x_1) \ldots (\forall x_n)[F(x_1, \ldots, x_n) \leftrightarrow \phi]$$

in $\mathfrak{A}$. There is a set $K \subseteq D^n$, in other words, namely, where $K = \{d \in D^n$: $A(d_1/x_1, \ldots, d_n/x_n)$ satisfies ϕ in $\mathfrak{A}\}$, such that $A(K/F^n)$ satisfies

$(\forall x_1) \ldots (\forall x_n)[F(x_1, \ldots, x_n) \leftrightarrow \phi]$ in $\mathfrak{A}$. Consequently, no instance of (CP) can be falsified in a model, and therefore every instance of (CP) must be logically true (according to logical realism).

This validation of (CP) and the related association of a logically real universal with each fully applied wff $\phi(x_1, \ldots, x_n)$ is not like the validation of (CP) and similar association of a concept with $\phi(x_1, \ldots, x_n)$ in holistic conceptualism, it should be noted. In particular, (CP) is not to be thought of in logical realism as characterizing the closure conditions or laws of compositionality regarding the *generation* of real universals. That is, unlike the situation in conceptualism, logically real properties and relations are not generated at all but rather exist timelessly in their transcendent realm outside of the causal nexus of the world. The validation of (CP) in this regard is but a consequence of the logical relationships obtaining in this transcendent realm between logically real universals; and the fact that there are more such universals than could in principle ever be represented in language (assuming an infinite domain of discourse) is but a sign of the deeper fact already noted in Section 4 that any formal theory of predication which can be associated with logical realism must be essentially incomplete.

10. POSSIBILISM AND ACTUALISM IN MODAL LOGICAL REALISM

Since logical realism is essentially incomplete as an extensional theory of predication, it can hardly be expected to be otherwise when considered as an intensional theory. Nevertheless, with the addition of the logical modalities there are certain distinctions and issues which arise in logical realism, or what we shall call *modal logical realism*, which do not arise in its extensional form. One such distinction is that between possibilism and actualism as variants of modal logical realism.

With $\Box$ as a new logical primitive (and $\Diamond$ defined as $\sim\Box\sim$), the set of wffs of a language L is now the smallest set K containing the atomic wffs of L and such that $\sim\phi$, $\Box\phi$, $(\phi \rightarrow \psi)$, $(\forall x)\phi$, $(\forall F^n)\phi$ are in K whenever ϕ, ψ are in K, x is an individual variable, and, for all positive integers n, F^n is an n-place predicate variable. Since the modalities in question in modal logical realism are the logical modalities, we return to the primary semantics for logical necessity described in Section 4 where, corresponding to a language L and a domain of discourse D, all of the possible worlds in question are represented by the L-models having D as their domain of discourse. Thus, given L and D, let $W(L, D) = \{\mathfrak{A}\colon \mathfrak{A}$ is an L-model having D as its domain of discourse$\}$.

In regard to a set-theoretic representation of logically real properties and relations, we note that their representation by their extensions in an extensional semantics readily generalizes in an intensional semantics to their representation by functions from possible worlds into the extensions these properties and relations have in those possible worlds. (This approach towards the representation of properties and relations was first taken by Rudolf Carnap in his 1955 'Notes on Semantics' which was published posthumously in *Philosophia*, 1972.) Thus, where L is a language and D is a domain of discourse, we can characterize the properties and relations representable in $W(L, D)$ as the functions in $(\mathscr{P}D^n)^{W(L,D)}$, for all positive integers n.

Where L is a language, $\mathfrak{A} = \langle D, f\rangle$ is an L-model, we now say that A is a *referential assignment in* $\mathfrak{A}$ if A is a function with the set of variables as domain and such that (1) $A(x) \in D$, for all individual variables x, and (2) for each positive integer n, $A(F^n) \in (\mathscr{P}D^n)^{W(L,D)}$, for all n-place predicate variables F^n. Where A is such an assignment, we define the *satisfaction by A in* $\mathfrak{A}$ of a wff of L exactly as in the "standard" semantics described in Section 3, except that now a clause for $\Box$ is added (which is essentially the same as that given in Section 4) and clauses (6) and (7) are redefined in a way that respects the intensional representation of properties and relations:

(6) *A satisfies* $F^n(x_1, \ldots, x_n)$ *in* $\mathfrak{A}$ iff $\langle A(x_1), \ldots, A(x_n)\rangle \in A(F^n)(\mathfrak{A})$;

(7) *A satisfies* $(\forall F^n)\phi$ *in* $\mathfrak{A}$ iff for all $K \in (\mathscr{P}D^n)^{W(L,D)}$, $A(K/F^n)$ satisfies ϕ in $\mathfrak{A}$; and

(8) *A satisfies* $\Box\phi$ *in* $\mathfrak{A}$ iff for all $\mathfrak{B} \in W(L, D)$, A satisfies ϕ in $\mathfrak{B}$.

The definitions of *truth* (in a model), *logical truth* and *logical consequence* (*according to modal logical realism*) are assumed defined in the usual way. We note that just as the extensional comprehension principle (CP) was validated in the extensional semantics of logical realism, so too its intensional counterpart,

$$(\Box\text{CP}) \quad (\exists F^n)\Box(\forall x_1) \ldots (\forall x_n)[F(x_1, \ldots, x_n) \leftrightarrow \phi],$$

where F^n does not occur free in ϕ, is validated in the present intensional semantics. For just as the n-ary relation-in-extension $\{d \in D^n: A(d_1/x_1, \ldots, d_n/x_n)$ satisfies ϕ in $\mathfrak{A}\}$ exists as a value of the n-place predicate variables in the extensional semantics of logical realism, so too does its intensional counterpart – namely, the function which assigns such a relation-in-extension to each possible world (model) in $W(\mathrm{L}, D)$ – exist in the intensional semantics of modal logical realism.

The principal assumption which leads to the essential incompleteness of logical realism, we have already noted, is the assumption that every class (of n-tuples) of individuals drawn from a given domain of discourse is the extension of some logically real universal. This assumption was validated in the extensional semantics by simply identifying or representing each property or relation with its extension in a given domain of discourse and then taking all of the classes of n-tuples of individuals drawn from that domain as the range of values of the n-place predicate variables. In the present semantics, on the other hand, where properties and relations are represented not by their extensions but by functions which assign such extensions to each possible world, the assumption is borne out by the fact that every class of n-tuples of individuals is the constant value of some such function. A property or relation which has the same extension in every possible world we say is *rigid*, a notion which can be represented as follows:

$$Rigid(F^n) =_{\mathrm{df}} (\forall x_1) \ldots (\forall x_n)[\Box F(x_1, \ldots, x_n) \vee \vee \Box \sim F(x_1, \ldots, x_n)].$$

(This notion was first formulated by Montague in [18], p. 132. It also occurs in Gallin [14], p. 77, where Gallin's principle of extensional comprehension is the counterpart of the principle of rigidity stated below.) Thus the assumption that every class of (n-tuples of) individuals is the extension of some property or relation is realized in the present semantics by the fact that every class of (n-tuples of) individuals is the extension of some rigid property or relation. It is this fact, moreover, which validates the following *principle of rigidity*:

$$\text{(PR)} \quad (\forall G^n)(\exists F^n)(Rigid(F^n) \mathbin{\&} (\forall x_1) \ldots (\forall x_n)[F(x_1, \ldots, x_n) \leftrightarrow \leftrightarrow G(x_1, \ldots, x_n)]).$$

Now it is clear of course that

$$(\forall F^n)\phi \rightarrow (\forall F^n)(Rigid(F^n) \rightarrow \phi)$$

is logically true in the present semantics; and therefore since the consequent of this wff is semantically equivalent in the present semantics to the antecedent under the extensional semantics for logical realism, then the same incompleteness theorem which applies in the latter semantics also applies in the present intensional semantics as well. Such a result, as already indicated, is as it should be in logical realism, with or without the logical modalities included.

Although the above semantics is a direct and natural generalization of the extensional semantics for logical realism, it does raise a question which does not arise in the latter semantics. Specifically, the question or problem has to do with how we are to understand the role of the domain of discourse as a fixed range of values of the bound individual variables. E.g., are we to assume that this domain consists only of the individuals that exist in the actual world and that other possible worlds are merely all of the logically possible redistributions of the properties and relations that apply to these individuals in the actual world? If so, then there would seem to be no way by which we might represent the purported logical possibility of there being either more or fewer individuals than there are in the actual world; and in that regard we would be unable to represent the purported logical possibility of there being individuals which do not exist in the actual world as well as of there being possible worlds in which individuals existing in the actual world do not exist.

One way of resolving this problem is the possibilist variant of modal logical realism. On this interpretation the fixed domain of discourse consists not just of the individuals existing in the actual world but also of the individuals existing in other possible worlds as well. First order quantifiers refer, in other words, not just to the individuals existing in a given possible world but to the individuals existing in other possible worlds as well. On this interpretation, following Bertrand Russell in his pre-1905 possibilist phase, we should distinguish *being* from *existence*, where *to be* is *to be the value of a bound individual variable* (i.e., *to be* is *to be something*) but *to exist* is to have a special property of existence which "is the perogative of some only amongst beings" (cf. [28], p. 449). The distinction between *existence* and *being*, according to this possibilist Russell, "is essential if we are ever to deny the existence of anything. For what does not exist must *be* something, or it would be meaningless to deny its existence; and hence we need the concept of being, as that which belongs even to the non-existent" (*ibid.*, p. 450).

We shall follow Russell here and assume that the special property of existence in question is such that it can be possessed only by concrete individuals; that is, only concrete individuals are such that they can exist in some possible worlds and fail to exist in others. Abstract individuals, if in fact there are any, only have being and do not exist in any possible world. Thus, according to this early possibilist Russell for whom properties were themselves individuals, the property of existence has being (as an individual) but does not itself exist (*ibid.*) That is, as abstract individuals whose being is independent of whether there is a world (of space, time and causality), logically real properties and relations are such that they necessarily do not exist (as concrete individuals). We shall not ourselves assume that logically real properties and relations are individuals; that is, we wish to leave open the choice between a Russellian and a Fregean form of modal logical realism (where in a Fregean form properties and relations are unsaturated entities and therefore cannot be individuals). For convenience, however, we shall adopt Russell's own later notation (which he applied only to definite descriptions) and use $E!$ as a 1-place predicate constant which, at least in modal logical realism, is assumed to stand for the property of (concrete) existence.

Note that in the possibilist framework quantification over existing individuals is understood as a restricted form of quantification over all individuals. Thus, new special quantifiers $\forall^e$ and $\exists^e$ for quantification over existing individuals can be contextually defined as follows:

$$(\forall^e x)\phi \ =_{\mathrm{df}} \ (\forall x)(E!(x) \rightarrow \phi),$$

$$(\exists^e x)\phi \ =_{\mathrm{df}} \ (\exists x)(E!(x) \ \& \ \phi).$$

We can also affix these new quantifiers to predicate variables, moreover, and take them when so affixed as referring to properties and relations which entail existence:

$$(\forall^e F^n)\phi \ =_{\mathrm{df}} \ (\forall F^n)(\Box(\forall x_1) \cdots (\forall x_n)[F(x_1, \ldots, x_n) \rightarrow$$
$$\rightarrow E!(x_1) \ \& \cdots \& \ E!(x_n)] \rightarrow \phi),$$

$$(\exists^e F^n)\phi \ =_{\mathrm{df}} (\exists F^n)(\Box(\forall x_1) \ldots (\forall x_n)[F(x_1, \ldots, x_n) \rightarrow$$
$$\rightarrow E!(x_1) \ \& \cdots E!(x_n)] \ \& \ \phi).$$

Now it is noteworthy that opposed to possibilism there is another variant of modal logical realism called *actualism*, the principal theses of which are (1) that quantificational reference to individuals can only be to individuals existing in the world in question, and (2) that no property or relation can be possessed in a given possible world except by the individuals existing in that world; and therefore, at least from the possibilist perspective, that the only properties and relations which are to be acknowledged in actualism are those which entail existence in the above sense. Actualism, in other words, is a variant of modal logical realism according to which the only legitimate forms of quantification are based on the quantifiers $\forall^e$ and $\exists^e$ (or the quantifiers $\forall$ and $\exists$ with the restricted semantics given $\forall^e$ and $\exists^e$). In actualism, of course, these quantifiers must be taken as primitives rather than as contextually defined above; and in that case the satisfaction clauses (5) and (7) of the preceding semantics must be formulated as follows:

(5^e) *A satisfies* $(\forall^e x)\phi$ *in* $\langle D, f\rangle$ iff for all $d \in f(E!)$, $A(d/x)$ satisfies ϕ in $\langle D, f\rangle$; and

(7^e) *A satisfies* $(\forall^e F^n)\phi$ *in* $\langle D, f\rangle$ iff for all $K \in (\mathscr{P}D^n)^{W(\mathrm{L}, D)}$, if $K(\langle D, g\rangle) \subseteq g(E!)^n$, for all $\langle D, g\rangle \in W(\mathrm{L}, D)$, then $A(K/F^n)$ satisfies ϕ in $\langle D, f\rangle$.

In actualism, of course, it is logically true that every individual exists; but as viewed from the possibilist perspective this only amounts to the trivial claim that every existing individual exists: $(\forall^e x)E!(x)$. For the possibilist the claim that every individual exists, *viz.*, $(\forall x)E!(x)$, is not logically true; and if abstract individuals are acknowledged at all, then it is even logically false. What is logically true in both possibilism and actualism, on the other hand, is the claim that *to exist* is *to possess an existence entailing property*; that is,

$$E!(x) \leftrightarrow (\exists^e F)(x)$$

is logically true *simpliciter* in modal logical realism, whether of the possibilist or actualist form. Perhaps it is this equivalence, incidentally, which is what is meant by the claim that something exists if and only if there is a fact about it. Of course, the related claim, at least according to actualism, that only existing individuals have properties or stand in relations to other individuals, i.e., formally,

$$(\forall^e F^n)[F(x_1, \ldots, x_n) \rightarrow E!(x_1) \,\&\, \cdots \,\&\, E!(x_n)],$$

is also logically true. But then in possibilism we must distinguish an existence entailing fact from a fact *simpliciter*. Thus, according to possibilism, whereas Pegasus's non-existence is a fact about Pegasus, it is not an existence entailing fact.

Note that although it is logically true in both possibilism and actualism that individuals are identical if and only if they have all the same properties, the interpretation each gives this claim is not the same. Thus in possibilism the claim amounts to the logical truth of

$$(\forall x)(\forall y)(x = y \leftrightarrow (\forall F)[F(x) \leftrightarrow F(y)]);$$

but in actualism it amounts to the logical truth of

$$(\forall^e x)(\forall^e y)(x = y \leftrightarrow (\forall^e F)[F(x) \leftrightarrow F(y)]).$$

One way of seeing the difference here is by dropping the initial quantifiers. Thus in possibilism,

$$x = y \leftrightarrow (\forall F)[F(x) \leftrightarrow F(y)]$$

is also logically true; but in actualism,

$$x = y \leftrightarrow (\forall^e F)[F(x) \leftrightarrow F(y)]$$

is not also logically true and in fact will be false (in its right-to-left direction) for certain values of x and y which do not exist. Actualism, in other words, is free of existential presuppositions regarding singular terms (and regarding predicates as well, incidentally). What is logically true in actualism, on the other hand, is the related open wff,

$$\sim E!(x) \;\&\; \sim E!(y) \rightarrow (\forall^e F)[F(x) \leftrightarrow F(y)],$$

which, according to actualism, accounts for the difference in question. In possibilism, however, the closure of this wff states the valid thesis that non-existing individuals are one and all indiscernible in their existence entailing properties (of which they have none) – regardless whether they are also indiscernible *simpliciter* (and therefore identical). According to actualism (as a logic free of existential presuppositions) what the above wff shows is that non-existing individuals are one and all indiscernible and that therefore since they have no properties it is pointless to acknowledge such individuals as further constituents of the world. Of course, for possibilism this argument is fallacious insofar as it construes indiscernibility as indiscernibility with respect only to existence entailing properties and assumes that a world is constituted only by the existence entailing facts that obtain in that world and not also by facts about non-existing individuals as well.

Note that whereas the principle (UI_1) of (first order) universal instantiation is logically true in possibilism, only the qualified form

$$(\exists/UI_1^e) \quad (\exists^e z)(y = z) \rightarrow [(\forall^e x)\phi \rightarrow \phi(y/x)],$$

is logically true in actualism. But then since individuals can have no possible mode of being in actualism other than possible (concrete) existence, $\Diamond(\exists^e z)(y = z)$ is also logically true in actualism. (It will not be logically true in a possibilism which admits abstract individuals.) In other words, since actuality is none other than concrete existence, then in actualism the domain of discourse of any world system must be none other than the sum of the sets of concretely existing objects of the different possible worlds in that system. In that case, by ($\exists$/UI$_1^e$) and obvious principles of modal logic, the following modal universal instantiation law,

$$(\Box\text{UI}_1^e) \quad \Box(\forall^e x)\Box\phi \rightarrow \phi(y/x),$$

is easily seen to be logically true in actualism. (The same law will be valid in possibilism only when $(\forall x)\Diamond E!(x)$ is assumed.) Finally, it should also be noted that whereas the second order principle (UI$_2$) is logically true in possibilism, only the qualified form,

$$(\exists/\text{UI}_2^e) \quad (\exists^e F^n)\Box(\forall^e x_1)\Box \cdots \Box(\forall^e x_n)\Box[F(x_1, \ldots, x_n) \leftrightarrow \psi] \rightarrow$$
$$\rightarrow [(\forall^e F^n)\phi \rightarrow \phi[\psi/F(x_1, \ldots, x_n)]],$$

is logically true in actualism.

Another way possibilism differs from actualism, incidentally, is in the generality of the comprehension principle validated in each. In particular, whereas the full unrestricted modal comprehension principle ($\Box$CP) is validated by the universals in possibilism, this is not the case in actualism even when the initial predicate quantifier $(\exists F^n)$ is replaced by $(\exists^e F^n)$. This is because the complement of an existence entailing property or relation is not itself an existence entailing property or relation, and therefore, as far as actualism is concerned, there is no such property or relation as the complement of an existence entailing property or relation.

Now instead of the unrestricted modal comprehension principle ($\Box$CP), what is validated in actualism is a kind of *Aussonderungsaxiom* for existence entailing properties and relations. This principle is schematically described as follows:

$$(\Box\text{CP}^e) \quad (\forall^e G^k)\Box(\exists^e F^n)\Box(\forall^e x_1)\Box \cdots \Box(\forall^e x_n)\Box[F(x_1, \ldots, x_n) \leftrightarrow$$
$$\leftrightarrow G(x_1, \ldots, x_k) \;\&\; \phi],$$

where $n \leqslant k$, F^n, G^k are distinct predicate variables which do not occur free in ϕ, and ϕ is an actualist wff (i.e., a wff in which the possibilist quantifier $\forall$ does not occur). As an *Aussonderungsaxiom* for existence entailing properties and relations what this principle maintains (when $n = k$) is that although there will not in general be an existence entailing property or relation corresponding to an arbitrary wff $\phi(x_1, \ldots, x_k)$, nevertheless there must be such a property or relation corresponding to the restriction of $\phi(x_1, \ldots, x_k)$ to any antecedently given k-ary existence entailing property or relation. (For $n < k$, the principle also allows that the n-ary subrelation of a k-ary existence entailing relation is itself in existence entailing relation.)

A related principle of actualism, incidentally, has the form of a special case of an unrestricted modal comprehension principle. This principle states that, for each positive integer n, the n-ary property or relation of standing in some n-ary existence entailing property or relation is itself an existence entailing property or relation. That is,

$$(\mathbb{E}!) \qquad (\exists^e F^n)\Box(\forall^e x_1)\Box \cdots \Box(\forall^e x_n)\Box[F(x_1, \ldots, x_n) \leftrightarrow \\ \leftrightarrow (\exists^e G^n)G(x_1, \ldots, x_n)]$$

is logically true in actualism. Of course, where $n = 1$, what ($\mathbb{E}$!) asserts is that the property of existence, when analysed as the property of possessing some existence entailing property, is itself an existence entailing property. Note that by ($\Box$CP^e), ($\mathbb{E}$!) and ($\exists$/UI$^e_2$), the following comprehension principle,

$$(\Box\text{CP}^e_2) \quad (\exists^e F^n)\Box(\forall^e x_1)\Box \cdots \Box(\forall^e x_n)\Box[F(x_1, \ldots, x_n) \leftrightarrow \phi \ \& \\ \& \ (\exists^e G^n)G(x_1, \ldots, x_n)],$$

is also logically true in actualism. And by ($\Box$CP$^e_2$) so is the following simpler actualist comprehension principle:

$$(\Box\text{CP}^e_3) \ (\exists^e F^n)\Box(\forall^e x_1) \ldots (\forall^e x_n)[F(x_1, \ldots, x_n) \leftrightarrow \phi].$$

By ($\Box$CP$^e_3$), incidentally, it can easily be seen why

$$(\forall^e x)(\forall^e y)(x = y \leftrightarrow (\forall^e F)[F(x) \leftrightarrow F(y)])$$

is logically true in actualism. For if, by *reductio*, a and b are *distinct* existing individuals which are indiscernible in their existence entailing properties, then by ($\Box$CP$^e_3$) there is an existence entailing property which necessarily is possessed by all and only those existing individuals that are distinct from a. That is,

$$(\exists^e F)\Box(\forall^e x)[F(x) \leftrightarrow x \neq a].$$

But, by hypothesis, b exists and is distinct from a, and so therefore b has the property in question. But then so does a, since a is indiscernible from b with respect to existence entailing properties; that is, then $a \neq a$, which of course is impossible.

11. LOGICAL REALISM AND ESSENTIALISM

As variants of modal logical realism, the notion of necessity involved in possibilism and actualism is that of logical necessity. It is this modality, in other words, together with the totality of logically real properties and relations (and, in some forms of logical realism, propositions as well), which, according to modal logical realism, is the real ontological ground of logical truth and logical consequence. It was for that reason that we returned to the semantics of $\square$ given in Section 4; i.e., to what we there called the primary semantics of logical necessity. Of course, in doing so we continue to validate the modal thesis of anti-essentialism. That is, whether our variant of modal logical realism is possibilism or actualism, in either case the modal thesis of anti-essentialism is logically true (which is as it should be if we are to retain the "standard" definition of logical truth in the extensional semantics of logical realism).

In possibilism this thesis takes the same form as that already described in Section 4, except that the wffs in question are now second order modal wffs in which no predicate variable has a free occurrence. That is, if ϕ is a wff of possibilist modal logical realism in which no predicate variable has a free occurrence and $x_1, \ldots, x_n$ are all of the pairwise distinct individual variables occurring free in ϕ, then in possibilism *the modal thesis of anti-essentialism* maintains that

$$(\text{A–E}_2)\quad (\exists x_1)\ldots(\exists x_n)[ID_j(x_1, \ldots, x_n)\ \&\ \square\phi] \rightarrow (\forall x_1)\ldots(\forall x_n)$$
$$[ID_j(x_1, \ldots, x_n) \rightarrow \square\phi]$$

is logically true. The thesis takes the same form in actualism, moreover, except that the quantifiers $\forall$ and $\exists$ must have the content given $\forall^e$ and $\exists^e$ instead. That is, in actualism the modal thesis of anti-essentialism maintains that

$$(\text{A–E}_2^e)\quad (\exists^e x_1)\ldots(\exists^e x_n)[ID_j(x_1, \ldots, x_n)\ \&\ \square\phi] \rightarrow$$
$$\rightarrow (\forall^e x_1)\ldots(\forall^e x_n)\ [ID_j(x_1, \ldots, x_n) \rightarrow \square\phi]$$

is logically true whenever $x_1, \ldots, x_n$ are as above and ϕ is a wff in which no predicate variable has a free occurrence and in which the possibilist quantifiers ∀ and ∃ do not occur. (The proof of the logical truth of (A–E_2) and (A–E_2^c) is by means of an obvious extension of the lemma described in Section 4 as the basis of the proof for (A–E); cf. Cocchiarella [10], Section 3.10.)

Now although the above results seem to be in order insofar as the modality in question is logical necessity, nevertheless there are variants of modal logical realism which reject the modal thesis of anti-essentialism and in which there are some properties and relations (other than those based upon identity and difference) which are necessarily possessed by some individuals but not necessarily possessed by all individuals (relative to a given identity-difference condition). Thus, for example, if properties and relations are individuals (as was claimed in Russell's pre-1905 form of possibilism), or if there are any abstract individuals otherwise, and the property of existence represented by *E*! is in fact the property of *concrete existence*, then whereas it is true in possibilism that some individuals (namely, the abstract individuals) necessarily do not exist (as concrete individuals), nevertheless it is not true in possibilism that all individuals necessarily do not exist (as concrete individuals). In such a variant of possibilism, in other words, the modal thesis of anti-essentialism is not logically true, and in fact there are properties (other than identity and difference) which some individuals have essentially but which other individuals do not have at all.

The thesis is validated in the semantics of Section 10, it should be noted, only because we assumed that relative to a language L and a domain of discourse *D* the logically possible worlds are *all* of the L-models in *W*(L, *D*); and therefore for any individual in *D* there will always be a possible world (model) in which that individual is in the extension assigned to *E*!. Of course, to remedy this particular example we need only exclude from *W*(L, *D*) those models in which abstract individuals are assigned to the extension of *E*!. But then, these need not be the only models which should be excluded if there are other properties and relations besides non-existence which are necessarily possessed by some individuals but not necessarily possessed by all; and, presumably, in an essentialist variant of modal logical realism there may be indefinitely many such properties and relations.

We shall not attempt to describe here any general principles for an essentialist variant of modal logical realism (and in fact, beyond specific examples, we know of no such general laws like those for natural kinds given in Section 14 below for an essentialist form of natural realism). Instead, we shall assume that if essentialist logical realism is a coherent framework at all, then some metaphysical scheme regarding essential properties and relations is in principle available and that at least the general form of its set-theoretic representation can be given as follows. (Our presentation allows for a simultaneous characterization of possibilism and actualism in which the modal thesis of anti-essentialism is not validated.)

Accordingly, if D is a non-empty domain of discourse and W is a non-empty set (the members of which are presumed to represent a relevant totality of possible worlds), then an indexed pair $\langle D, E_i \rangle_{i \in W}$ is said to be a *world system* if and only if $\bigcup_{i \in W} E_i \subseteq D$. For $i \in W$, the set E_i is assumed to be the set of individuals having concrete existence in the world i. We do not assume that the totality of concretely existing individuals exhausts D, needless to say, since we do not want to validate the thesis that every individual *can exist* (as a concrete individual): $(\forall x)\Diamond E!(x)$. We continue to represent logically real properties and relations as functions from possible worlds to the extensions which these properties and relations have in those possible worlds. For possibilism, these functions will draw their extensions from D regardless of whether all of the individuals in the extension exist in the world in question. For actualism, on the other hand, the extension which such a function assigns to a given possible world $i \in W$ will be drawn exclusively from the set E_i.

Where L is a language, we now say that $\langle D, E_i, f \rangle_{i \in W}$ is a *modal model structure* for L if and only if $\langle D, E_i \rangle_{i \in W}$ is a world system and f is a function with L as domain such that for all $P^n \in$ L, $f(P^n) \in (\mathscr{P}D^n)^W$. Note that if for some $i \in W$, $f(P^n) \nsubseteq E_i^n$, then, as viewed from within actualism, P^n fails to stand for an (existence entailing) n-ary relation. In actualism, in other words, both prediates and singular terms are free of existential presuppositions.

By a referential *assignment* in a modal model structure $\langle D, E_i, f \rangle_{i \in W}$ for a language L we now understand a function A with the set of variables as domain and such that $A(x) \in D$, for each individual variable x, and $A(F^n) \in (\mathscr{P}D^n)^W$, for each n-place predicate variable F^n. Where $i \in W$, $\mathfrak{A} = \langle D, E_i, f \rangle_{i \in W}$, and A is an assignment in $\mathfrak{A}$, then the *satisfaction in* $\mathfrak{A}$ *at i by A* is recursively defined as follows:

(1) *A satisfies* $(x = y)$ *in* $\mathfrak{A}$ *at i* iff $A(x) = A(y)$;

(2) *A satisfies* $P^n(x_1, \ldots, x_n)$ *in* $\mathfrak{A}$ *at i* iff $\langle A(x_1), \ldots, A(x_n) \rangle \in f(P^n)(i)$;

(3) *A satisfies* $F^n(x_1, \ldots, x_n)$ *in* $\mathfrak{A}$ *at* i iff $\langle A(x_1), \ldots, A(x_n)\rangle \in A(F^n)(i)$;

(4) *A satisfies* $\sim\phi$ *in* $\mathfrak{A}$ *at* i iff A does not satisfy ϕ in $\mathfrak{A}$ at i;

(5) *A satisfies* $(\phi \rightarrow \psi)$ *in* $\mathfrak{A}$ *at* i iff either A does not satisfy ϕ in $\mathfrak{A}$ at i or A satisfies ψ in $\mathfrak{A}$ at i;

(6) *A satisfies* $(\forall x)\phi$ *in* $\mathfrak{A}$ *at* i iff for all $d \in D$, $A(d/x)$ satisfies ϕ in $\mathfrak{A}$ at i;

(7) *A satisfies* $(\forall F^n)\phi$ *in* $\mathfrak{A}$ *at* i iff for all $K \in (\mathscr{P}D^n)^W$, $A(K/F^n)$ satisfies ϕ in $\mathfrak{A}$ at i; and

(8) *A satisfies* $\Box\phi$ *in* $\mathfrak{A}$ *at* i iff for all $j \in W$, A satisfies ϕ in $\mathfrak{A}$ at j.

In the case of actualism, where $\forall^e$ is taken as a logical primitive (or where $\forall$ is given the semantic content of $\forall^e$), clauses (6) and (7) are replaced by:

(6^e) *A satisfies* $(\forall^e x)\phi$ *in* $\mathfrak{A}$ *at* i iff for all $d \in E_i$, $A(d/x)$ satisfies ϕ in $\mathfrak{A}$ at i; and

(7^e) *A satisfies* $(\forall^e F^n)\phi$ *in* $\mathfrak{A}$ *at* i iff for all $K \in (\mathscr{P}D^n)^W$, if for all $j \in W$, $K(j) \subseteq E_j^n$, then $A(K/F^n)$ satisfies ϕ in $\mathfrak{A}$ at i.

We also say that a wff ϕ of L is *true in* $\mathfrak{A}$ at i if ϕ is satisfied in $\mathfrak{A}$ at i by every assignment in $\mathfrak{A}$. Similarly, we say that ϕ is *valid in* $\mathfrak{A}$ if and only if ϕ is true in $\mathfrak{A}$ at every $i \in W$; and finally we say that ϕ is *universally valid* if and only if ϕ is valid in every modal model structure for any language of which ϕ is a wff.

Note that where $\mathfrak{A} = \langle D, E_i, f\rangle_{i \in W}$ is a modal model structure for a language L, we can correlate each $i \in W$ with a unique L-model having D as its domain of discourse; namely, the L-model $\langle D, \langle f(P)(i)\rangle_{P \in L}\rangle$. The present semantics, in other words, is really a restricted form of the semantics described for modal logical realism in Section 10, where the restriction amounts to allowing arbitrary "cut-downs" on the original totality of logically possible worlds in the satisfaction clause for $\Box$. It is these "cut-downs" or restrictions of the original totality, needless to say, which allows for the denial of the modal thesis of anti-essentialism, since now some necessities may obtain in a world system that might not obtain with respect to the full totality of logically possible worlds of which that world system is a part. Of course, it is precisely these sorts of "cut-downs" which allows for a formulation of essentialism according to logical realism.

Despite these "cut-downs", however, logical realism remains essentially incomplete, which is as it should be since the incompleteness is a consequence of the logical realist interpretation of predicate quantifiers and not of whether the modal thesis of anti-essentialism is valid or not. Thus, for example, the principle of rigidity, (PR), is universally valid in the present semantics just as it was logically true in the semantics of modal logical realism without these "cut-downs". Of course, since

$$(\forall F^n)\phi \rightarrow (\forall F^n)(Rigid(F^n) \rightarrow \phi)$$

is also universally valid in possibilism, and, furthermore, since the consequent of this wff is semantically equivalent in the present semantics with the antecedent in the extensional semantics of logical realism, then, once again, the same incompleteness theorem which applies in the latter semantics also applies in the present semantics for possibilism as a variant of modal logical realism.

The principle of rigidity is not universally valid in actualism (when $\forall$ is interpreted as $\forall^e$), on the other hand, since none of the individuals existing in one possible world need exist in all possible worlds, and therefore no existence entailing property or relation need have the same (rigid) extension in every possible world. In actualism, however, the assumption that every class (of n-tuples) of individuals is the extension of some logically real property or relation is semantically represented by the fact that for all world systems $\langle D, E_i \rangle_{i \in W}$, all $i \in W$, all positive integers n, and all $J \subseteq E_i^n$, there is a function $K \in (\mathscr{P}D^n)^W$ such that for all $j \in W$, $K(j) \subseteq E_j^n$, and $K(i) = J$. And this semantical fact, it turns out, is sufficient to yield the following lemma from which follows the essential incompleteness of actualism as a variant of modal logical realism.

LEMMA. *If ψ is a modal free pure second order wff of actualism which is satisfiable, but only in a world system at a possible world in which there are infinitely many existing individuals, ϕ is a modal and identity free pure wff in which no predicate variable has a bound occurrence (i.e., ϕ is predicative in nominalism's purely grammatical sense), and $F_1, \ldots, F_m, x_1, \ldots, x_n$ are all of the variables occurring free in ϕ, then $[\psi \rightarrow \sim(\forall^e F_1) \ldots (\forall^e F_m)(\forall^e x_1) \ldots (\forall^e x_n)\phi]$ is universally valid iff ϕ is not logically true in accordance with the "standard" extensional semantics of logical realism (as described in Section 3).*

What this lemma shows, in other words, is that we can express in terms of the universally valid wffs of actualism the nonlogical truths of pure modal and identity free wffs which have only first order wffs as their fully applied instances. Accordingly, since the latter are not recursively enumerable, then neither are the universally valid pure wffs of actualism (or of possibilism, since actualism is a subsystem of possibilism).

THEOREM. *The set of wffs of either actualism or possibilism which are universally valid according to modal logical realism is not recursively enumerable; and therefore no (recursive) set of axioms can be complete for either actualism or possibilism, essentialist or otherwise, as variants of modal logical realism.*

12. POSSIBILISM AND ACTUALISM WITHIN CONCEPTUALISM

Although possibilism and actualism are essentially incomplete as variants of modal logical realism, they are not incomplete when considered within the framework of conceptualism. The problem here, however, is that actualism seems not to be fully coherent as an independent form of conceptualism. This is because in both constructive and holistic conceptualism the complement of any given concept is itself a concept; but in actualism the complement of an existence entailing concept (i.e., a concept under which only existing individuals can fall) is not in general also an existence entailing concept. In other words, by restricting itself to concepts which can have only existing objects falling under them, actualism excludes the full comprehension principles of both constructive and holistic conceptualism; and since these principles represent fundamental laws of compositionality for concept-formation it is dubious that actualism can stand as a fully coherent independent form of conceptualism. (Such a problem does nor arise in logical realism, incidentally, because the universals of logical realism have their being independently of all conceptual activity and are not subject to laws of compositionality. Whatever comprehension principles are validated in logical realism, in other words, are validated on the basis of the logically necessary relationships that obtain between logically real properties and relations and not because they can be composed or formed in thought.)

Notwithstanding its dubious status as an independent form of conceptualism, however, actualism is nevertheless an important part of any form of conceptualism in which modal concepts are acknowledged as a further category; and in fact, as we shall see, it may even be given a separate axiomatization. Since actualism can best be described in holistic terms, the formulation we give below will be of actualism as part of holistic conceptualism. For convenience, concepts which entail existence in the sense that only existing objects can fall under them (in any given possible world) will be referred to hereafter as *e-concepts*.

Now in regard to modality, we have already pointed out that no variant of conceptualism can account for the primary semantics of logical necessity. That is, since the latter implicitly involves reference to nondenumerably many universals when applied to an infinite domain, and since conceptualism can account only for a potential infinity of concepts, then conceptualism cannot provide a coherent account of the primary semantics of logical necessity. In this much at least conceptualism and nominalism agree: namely, that there can be no such modality as logical necessity as described in its primary semantics. Of course, since nominalism is committed to the thesis of anti-essentialism, and the latter can be validated only in the primary semantics of logical necessity, then what follows is nominalism's commitment to its second general thesis, *viz*., the thesis of extensionality. Conceptualism is not committed to the thesis of extensionality, on the other hand, and this is precisely because conceptualism is not committed to the thesis of anti-essentialism.

Conceptualism, in other words, does not exclude essentialism, and an essentialism based upon natural realism may in fact be fundamental to whatever ontology is presupposed in any given form of conceptualism. That is, as a socio-biologically based theory of the human capacity for systematic concept-formation, conceptualism may in fact be committed to the existence of natural (and in particular biological) properties and relations as the causal grounds of the laws of compositionality for concept-formation. Of course, the kind of modality involved in natural realism is a natural or causal necessity and not logical necessity, essentialist or otherwise. This is because logical necessity, even under the secondary semantics of Section 11, presupposes a framework of properties and relations which is independent of the causal structure of the world.

This does not mean that conceptualism is incompatible with logical realism, on the other hand; and in fact there is at least one variant of conceptualism, which we may call *conceptual platonism*, which posits logically real properties and relations as the (saturated) individuals denoted by nominalized predicates (cf. Cocchiarella [11], Section 17). (As unsaturated cognitive structures, concepts cannot themselves be the denotata of nominalized predicates, and in conceptual platonism the assumption is made that most (even if, on pain of Russell's paradox otherwise, not all) nominalized predicates denote logically real properties and relations. Of course, in an Abelardian form of conceptualism, all nominalized predicates will be assumed to be denotationless singular terms – cf. Cocchiarella [11], Section 14.) But aside from being only one variant of conceptualism, conceptual platonism presupposes a stage of concept-formation which is prior to the systematic use of nominalized predicates (especially since the latter usually occurs only in post-adolescence), and such an antecedent stage is independent of the logical realism implicit in the notion of logical necessity even under its secondary semantics. Since we are not dealing with the logic of nominalized predicates at all in this essay, we shall assume that the notion of necessity involved in our present discussion of conceptualism is not logical necessity.

Now besides a natural or causal necessity, most variants of conceptualism are committed to a variety of pragmatic modalities, including especially the temporal modalities corresponding to the use of tenses in natural language. Thus, e.g., the truth-conditions of an assertion about what was, is, or will be the case will in general depend on the time at which the assertion is made, and this indexicality or dependence on context of assertion is what characterizes these modalities as pragmatic. By combining temporal and causal considerations, moreover, we can easily formulate the concept of a pragmatic modality which covers not only the causally possible past in the sense of what could have been the case (given the laws of nature) but also the causally possible future in the sense of what might possibly come to be (given the same laws of nature). And a modality based upon such temporal and causal considerations is more than enough for the purposes of both conceptual possibilism and conceptual actualism.

For convenience, we shall continue to use $\Box$ for necessity, but understood now as a modality which covers both the causally possible past and the causally possible future, and which for brevity we shall simply call *causal necessity*. Because we shall not introduce tense and other modal operators here in terms of which causal necessity in the sense intended might be constructed, however, we shall not concern ourselves with the internal details of the temporal-causal structures involved in the semantics for this notion of necessity. (These structures involve more branching alternatives than the structures for the causal tense operators described in Cocchiarella [9], Section 15, it might be noted, since the latter are based only on the *actual(ized) causal connections* between the world-lines constituting the universe.) In other words, for purposes of brevity we shall continue to use the simple notion of a world system $\langle D, E_i \rangle_{i \in W}$ as described in Section 11, except that now the "worlds" in W are taken as representing the momentary configurations of the world-lines of a temporal-causal structure. The set of wffs of a language L involved in this semantics, it should perhaps be noted, is the smallest set K containing the atomic wffs of L and such that $\sim\phi$, $\Box\phi$, $(\phi \rightarrow \psi)$, $(\forall x)\phi$, $(\forall !F^n)\phi$, $(\forall^e F^n)\phi$, $(\forall F^n)\phi$ are all in K whenever ϕ, ψ are in K.

In conceptual possibilism, incidentally, the actualist quantifier $\forall^e$ must be taken as a new primitive logical constant (at least as applied to predicate variables). This is because in conceptualism, and in holistic conceptualism in particular, the concept of (concrete) existence is *impredicatively formed* as the concept of falling under some existence entailing concept:

$$E!(x) \quad =_{\mathrm{df}} \; (\exists^e F(x).$$

That is, as an *e*-concept *par excellence*, the concept of (concrete) existence is impredicatively formed in terms of a reference to the totality of *e*-concepts. It is the impredicativity of the concept of (concrete) existence, moreover, which explains its peculiarities and why it seems so different from the more mundane predicative concepts under which only existing individuals can fall. In logical realism, on the other hand, whether $E!$ is defined as above or taken as a primitive predicate constant is irrelevant, since the purported logically real property in question in no way depends on how it is specified or represented in thought. (Of course in constructive conceptualism, it should be noted, there are no impredicatively formed concepts. But still, unlike the situation in nominalism, a definiens with bound predicate variables is meaningful in constructive conceptualism, and the fact that $E!$ defined as above fails to stand for a predicative concept might well explain why some philosophical logicians have preferred to speak of $E!$ as a syncategorematic sign.)

Now where L is a language, we shall say that $\mathfrak{A} = \langle D, E_i, f, X_n, Y_n, Z_n \rangle_{i \in W, n \in \omega - \{0\}}$, is a *possibilist/actualist model structure* for L if (1) $\langle D, E_i \rangle_{i \in W}$ is a world system (as defined in Section 11), (2) for all positive integers n, $X_n \subseteq Y_n$, (3) Y_n is a countable subset of $(\mathscr{P}D^n)^W$, (4) $Z_n = \{K \in Y_n\text{: for all } i \in W, K(i) \subseteq E_i^n\}$, and (5) f is a function with L as domain such that for each n-place predicate constant $P^n \in$ L, $f(P^n) \in Y_n$. As in the extensional semantics for conceptualism, the predicative concepts of the structure are understood as beng represented by the members of X_n, while the members of Y_n represent all of the structure's concepts, predicative or otherwise. The members of Z_n, of course, represent the existence entailing concepts. An assignment A in such a structure in an assignment such that $A(x) \in D$, for all individual variables x, and $A(F^n) \in Y_n$, for all n-place predicate variables F^n.

Where $\mathfrak{A}$ is as described above, A is an assignment in $\mathfrak{A}$, and $i \in W$, we recursively define the *satisfaction by A in* $\mathfrak{A}$ *at i* exactly as in Section 11, except for clauses (7) and (7^e), and a new clause (7!) for $\forall!$, which are given as follows:

(7) *A satisfies* $(\forall F^n)\phi$ *in* $\mathfrak{A}$ *at i* iff for all $K \in Y_n$, $A(K/F^n)$ satisfies ϕ in $\mathfrak{A}$ at i;

(7!) *A satisfies* $(\forall! F^n)\phi$ *in* $\mathfrak{A}$ *at i* iff for all $K \in X_n$, $A(K/F^n)$ satisfies ϕ in $\mathfrak{A}$ at i; and

(7^e) *A satisfies* $(\forall^e F^n)\phi$ *in* $\mathfrak{A}$ *at i* iff for all $K \in Z_n$, $A(K/F^n)$ satisfies ϕ in $\mathfrak{A}$ at i.

We assume that truth and validity in $\mathfrak{A}$ are defined as in Section 11, and we say that $\mathfrak{A}$ is *normal* with respect ($\Box$CP) and ($\Box$CCP!) if every wff of L which is an instance of ($\Box$CP) or of ($\Box$CCP!) is valid in $\mathfrak{A}$. (By ($\Box$CCP!) we mean the modal counterpart of (CCP!), i.e., where the constraints are the same as in (CCP!) except that $(\exists! F^n)$ is now replaced by $(\exists! F^n)\Box$.) The restriction to semantic structures which are normal in this sense is of course a necessary constraint in conceptualism, since it is a restriction to structures which validate the laws of compositionality for concept-formation in holistic and constructive conceptualism. The requisite constraints for e-concepts is that given by the *Aussonderungs axiom* ($\Box$CP^e) and the special axiom ($\mathbb{E}$!) for the (n-ary) concept of existence described in Section 10. That is, $\mathfrak{A}$ is said to be *normal* with respect to ($\Box$CP^e) and ($\mathbb{E}$!) if every wff of L which is an instance of ($\Box$CP^e) or ($\mathbb{E}$!) is valid in $\mathfrak{A}$.

Now if $\Sigma \cup \{\phi\}$ is a set of wffs, then ϕ is said to be a *logical consequence of* Σ *according to conceptual possibilism*, in symbols, $\Sigma \models_P \phi$, if and only if for every language L, every possibilist/actualist model structure $\mathfrak{A}$ for L which is normal with respect to ($\Box$CP), ($\Box$CCP!), ($\Box$CP^e) and ($\mathbb{E}$!), every possible world (or momentary configuration) $i \in W$, and every assignment A in $\mathfrak{A}$, if $\Sigma \cup \{\phi\}$ is a set of wffs of L and for all $\psi \in \Sigma$, a satisfies ψ in $\mathfrak{A}$ at i, then A satisfies ϕ in $\mathfrak{A}$ at i. A wff ϕ is then said to be *universally valid according to conceptual possibilism*, in symbols $\models_P \phi$, if and only if the empty set $\models_P \phi$.

In conceptual actualism, even if only as a fragment of a larger conceptual framework, the only possible mode of being individuals can have is that of possible (concrete) existence. That is, according to actualism, where concrete existence is none other than actuality, there can be no abstract individuals, since, in such a framework, only what can *exist* can *be*. An *actualist model structure* for a language L, accordingly, is a possibilist/actualist model structure for L in which $D = \bigcup_{i \in W} E_i$. (It would seem that each E_i must also be non-empty in actualism, but we shall leave that issue open here.) Accordingly, if $\Sigma \cup \{\phi\}$ is a set of actualist wffs (i.e., wffs in which the only quantifer that occurs is $\forall^e$), then we say that ϕ is a *logical consequence of* Σ *according to conceptual actualism*, in symbols $\Sigma \models_e \phi$, if and only if for every language L, every actualist model structure $\mathfrak{A}$ for L which is normal with respect ($\Box CP^e$) and ($\mathbb{E}$!), every possible world $i \in W$, and every assignment A in $\mathfrak{A}$, if $\Sigma \cup \{\phi\}$ is a set of wffs of L and, for all $\psi \in \Sigma$, A satisfies ψ in $\mathfrak{A}$ at i, then A satisfies ϕ in $\mathfrak{A}$ at i. Universal validity according to conceptual actualism is then defined as logical consequence from the empty set.

By an *axiom of conceptual actualism* we now mean any actualist wff which is either (1) tautologous; (2) an instance of an axiom of the modal propositional logic S5; (3) an instance of (LL), ($\Box CP^e$), or ($\mathbb{E}$!); (4) or an instance of one of the following forms (where u is a predicate or individual variable):

($A1^e$) $(\forall^e u)[\phi \to \psi] \to [(\forall^e u)\phi \to (\forall^e u)\psi]$,

($A2^e$) $\phi \to (\forall^e u)\phi$, if u is not free in ϕ,

($A3^e$) $(x = x)$,

($A4^e$) $(\forall^e x)(\exists^e y)(x = y)$,

($A5^e$) $\Diamond(\exists^e y)(x = y)$,

($A6^e$) $(\exists^e F)F(x) \leftrightarrow (\exists^e y)(x = y)$.

As *inference rules* of conceptual actualism we continue to take *modus poneus* (MP) and *universal generalization* (UG) (of an individual or predicate variable); but in addition we also need the rule of *modal generalization*,

(MG) if $\vdash_e \phi$, then $\vdash_e \Box\phi$,

and a special rule for an embedded universal generalization of an individual variable within the scope of iterated occurrences of $\Box$:

($\Box$UG) if $\vdash_e \phi_1 \rightarrow \Box(\phi_2 \rightarrow \cdots \rightarrow \Box(\phi_n \rightarrow \Box\psi)\ldots)$, and x is not free in $\phi_1, \ldots, \phi_n$, then $\vdash_e \phi_1 \rightarrow \Box(\phi_2 \rightarrow \cdots \rightarrow \Box(\phi_n \rightarrow \Box(\forall^e x)\psi)\ldots)$.

The actualist wffs derivable from a given set of premises by means of these axioms and rules can now be proved to coincide with the wffs which are logical consequences of these premises according to conceptual actualism. (A proof of completeness can be found in Cocchiarella [10], Section 3.11.)

COMPLETENESS THEOREM. *If $\Sigma \cup \{\phi\}$ is a set of actualist wffs, then:*

(1) $\Sigma \models_e \phi$ *if and only if* $\Sigma \vdash_e \phi$*; and*
(2) $\models_e \phi$ *if and only if* $\vdash_e \phi$.

Note that except for (A5^e) every axiom of actualism, even when extended to nonactualist wffs, is universally valid in conceptual possibilism (and even (A5^e) would be valid in a form of possibilism which agreed with actualism that there are no abstract individuals). With additional possibilist axioms, however, (A3^e) and (A4^e) become redundant. Accordingly, by *an axiom of conceptual possibilism* we shall mean any wff (actualist or otherwise) which is either (1) an instance of an actualist axiom schema other than (A3^e)–(A5^e); (2) an instance of ($\Box$CP) or ($\Box$CCP!); (3) an instance of one of the axiom schemas (A1)–(A3) (as described in Section 2); or (4) an instance of one of the following schemas (where u is a predicate or individual variable):

$$(\forall u)\phi \rightarrow (\forall^e u)\phi$$

$$(\forall F^n)\phi \rightarrow (\forall! F^n)\phi.$$

As *inference rules* of conceptual possibilism we need only take *modus ponens* (MP), universal generalization (UG) for $\forall$ (as applied to individual or predicate variables) and modal generalization (MG). The rule (UG) for $\forall^e$ and $\forall!$ is now derivable by the two schemas indicated above. The special rule ($\Box$UG) of conceptual actualism is also derivable, and so it need not be taken as a primitive rule. Using $\models_p$ for the derivability relation of conceptual possibilism we now have the following completeness theorem (whose proof is essentially that given in Cocchiarella [5]).

COMPLETENESS THEOREM. *If* $\Sigma \cup \{\phi\}$ *is a set of wffs of conceptual possibilism, then:*

(1) $\Sigma \models_{\overline{p}} \phi$ *if and only if* $\Sigma \vdash_{\overline{p}} \phi$*; and*
(2) $\models_{\overline{p}} \phi$ *if and only if* $\vdash_{\overline{p}} \phi$.

13. NATURAL REALISM AND CONCEPTUALISM

Unlike the universals of logical realism, a property or relation of natural realism is not assumed to be the semantical grounds for the correct or incorrect application of a predicate expression. That is, although it is a theory of universals, natural realism is not also a theory of predication in the sense of providing a direct semantical account of the notion of predication in language and thought in the form of the universals it posits. In this regard natural realism does not validate (as a logical truth) any comprehension principle regarding the existence of real universals. Indeed, according to natural realism any posit of the form,

$$(\exists^n F^j)\Box(\forall x_1)\ldots(\forall x_j)[F(x_1,\ldots,x_j) \leftrightarrow \phi],$$

where F^j does not occur free in ϕ, $\Box$ stands for a natural or causal necessity, and $(\exists^n F^j)$ is taken as referring to a natural property or relation, always amounts at best to a scientific hypothesis; and understood as such, it is never immune from falsification.

In order for such an hypothesis even to be formulable, however, natural realism must assume that it can provide an account of its background logic in its own terms; that is, that it can provide a naturalistic account of the nature of predication in language and thought. Of course, since nominalism rejects as false or meaningless all hypotheses of the above form when predicate quantifiers are interpreted referentially, then it is not to nominalism that natural realism must turn for such an account. Logical realism must also be excluded, needless to say, since it would validate as logically true all instances of the full impredicative comprehension principle; and it would do so, moreover, on the basis of a metaphysical assumption which natural realism denies, namely, that properties and relations exist independently of nature and its causal nexus.

Conceptualism, on the other hand, is not only compatible with natural realism but, as already indicated, may even be described as having natural properties and relations as its causal ground. That is, as a socio-biologically based theory of the human capacity for systematic concept-formation, conceptualism may be described in precisely the naturalistic terms demanded of a theory of predication for natural realism. The difference between constructive and holistic conceptualism will then depend on certain assumptions about the biological basis of concept-formation. E.g., if the capacity for systematic concept-formation is embodied in the human genotype in a unitary manner, then it is likely that many of the regulatory processes that are constitutive of concept-formation at the later stages of its development are "preformed" in the regulatory processes of the earlier stages; and if that is so then what have been called *a priori* forms of thought would ultimately be "due to hereditary differentiations of the central nervous system which have become characteristic of the species, producing hereditary dispositions to think in certain forms" (Lorenz [17], p. 183). In other words, where the capacity for concept-formation is embodied in the human genotpye in a unitary manner, and in particular in the functioning of the central nervous system, it would appear that the compositional laws for systematic concept-formation are strongly constrained, and, in particular, that they will exclude the possibility for impredicative concept-formation. That is, there can be no compositional laws and subsequent closure conditions that can accommodate impredicative concept-formation if these conditions are to be reducible, even if only in principle, to dispositions of the central nervous system that are determined by hereditary differentiations.

Jean Piaget, on the other hand, would disagree with this assessment of the extent of the antecedent programming of the biological mechanisms that are prior to psychogenesis. These mechanisms, according to Piaget, are to "be regarded as arising from self-regulations of a more general and more fundamental nature than that of hereditary transmissions" ([22], p. 52). That is, the generic capacity humans have for systematic concept-formation is not embodied in the human genotype in a unitary manner but is rather the result of endogeneous self-regulations "whose effects are not innate" (*ibid.*, p. 57). As the most general mechanism common to organic and cognitive reactions, self-regulations are the basis of processes constitutive of organic structures in general rather than of structures that are "ready-made" and within which others may be found in a preformed state. Indeed, "organic regulations," according to Piaget, "already provide us with the picture of unlimited reconstructions from level to level, without the higher forms being contained in advance in the lower ones, the link between them being merely an analogy of functioning" (*ibid.*, p. 61). The reconstructions that occur on the conceptual level, moreover, are no different in this regard; that is, later stages of concept-formation are not in general reducible or otherwise predictable on the bases of the earlier, and in particular they will not be "preformed" in the phylogenetic programming of the human brain. The constraints on the powers of concept-formation that apply on the assumption of such preformations, accordingly, will not apply when the conceptual structures that are constructed at the later stages are no longer so reducible. Indeed, once the human child has reached the stage of adolescence of what Piaget calls formal operational thought, the powers for concept-formation are such as to allow for idealized transitions to a limit through the (idealized) closure of a pattern of reflective abstraction (such as that already described in Section 7–8) which can accommodate impredicative concept-formation. Of course it must be emphasized that according to both constructive and holistic conceptualism the maturation of the nervous system in adolescence "can do no more than determine the totality of possibilities and impossibilities at a given stage" ([15], p. 337). That is, "a particular social environment remains indispensable for the realization of these possibilities" (*ibid.*); and in fact "their realization can be accelerated or retarded as a function of cultural and educational conditions. This is why the growth of formal thinking as well as the age at which adolescence itself occurs . . . remain dependent on social as much as and more than on neurological factors" (*ibid.*). This is also why, according to holistic conceptualism, the post-adolescents of primitive societies appear to have few, if any, impredicative concepts as part of their repertoire. In other words, "the history of formal structures is linked to the evolution of culture and collective representations as well as their ontogenetic history" (*ibid.*).

We shall assume, accordingly, that natural realism can provide an explanatory causal account of the development of concept-formation and that, depending on what biological assumptions are made in such an account, natural realism may take as its background logic or formal theory of predication either a constructive or a holistic form of conceptualism. The positing of natural properties and relations, in other words, can be viewed as a referential act from within some form of conceptualism.

Now it should perhaps be noted in this context that concepts, as cognitive capacities or structures otherwise based upon such capacities, are not themselves real properties and relations. That is, notwithstanding their intersubjectivity, concepts do not exist independently of the natural capacity humans have for thought and representation. In this regard, accordingly, any formulation of natural realism with either a constructive or holistic conceptualism as its logical background must distinguish predicate quantifiers that refer to natural universals from those that refer to concepts – just as in holistic conceptualism predicate quantifiers that refer to predicative concepts must be distinguished from those that refer to concepts in general. Of course, each of the different types of predicate quantifiers involved in such a formulation will themselves stand only for (second order) referential concepts; but then just as different sortal and common noun quantifiers stand for (first order) referential concepts that refer to different kinds of individuals, so too different kinds of predicate quantifiers will stand for (second order) referential concepts that refer to different kinds of (unsaturated) universals; namely, concepts as intelligible universals on the one hand, and natural properties and relations as causally real universals on the other. For this reason we shall use $\forall^n$ as a new quantifier sign which when affixed to predicate variables in wffs in understood to refer to natural properties and relations. (Of course we take $(\exists^n F)$ as an abbreviation of $\sim(\forall^n F)\sim$.)

Some abstraction is permissible here, however. That is, where our concern is with natural realism as a theory of universals we may ignore those wffs which contain predicate quantifiers that refer to concepts, thereby leaving open the question whether the background is constructive or holistic conceptualism. In either case, however, we shall continue to use $\Box$ for the temporal-causal notion of necessity which covers both the causally possible past in the sense of what could have been the case (given the laws of nature) and the causally possible future in the sense of what might possibly be the case (given the same laws of nature).

Now the central thesis of natural realism, we have said, is the claim that natural properties and relations cannot exist independently of the causal structure of the world. This means not only that such properties and relations cannot exist independently of all concrete existence but also that somehow their being as universals is contained within nature's causal matrix. It does not mean, on the other hand, that a natural property or relation must be instantiated (at some time or other in the history of the actual world) but only that such an instantiation is causally possible. Formally, in other words, *the fundamental thesis of natural realism* may be stated as follows:

$$\text{(NR)} \quad (\forall^n F^j)\Diamond(\exists^e x_1)\ldots(\exists^e x_j)F(x_1, \ldots, x_j).$$

The implication here, it should be noted, is that a natural property or relation can hold only between concretely existing individuals. That is,

$$(\forall^n F^j)[F(x_1, \ldots, x_j) \rightarrow E!(x_1) \,\&\, \cdots \,\&\, E!(x_j)]$$

is understood to be universally valid according to natural realism; and therefore the quantifier prefix $(\exists^e x_1) \ldots (\exists^e x_j)$ could just as well be replaced by $(\exists x_1) \ldots (\exists x_j)$. We shall avoid the possibilist quantifier in the context of natural realism, however, if only so as to emphasize the fact that we are concerned here only with concrete existence which in the present restricted context we may assume is defined as follows:

$$E!(x) =_{df} (\exists^e y)(x = y).$$

A methodological corollary of the thesis (NR), incidentally, might well be the claim that with sufficient technology and unlimited energy resources instances of any natural property or relation can in principle be produced, i.e., caused to come to be, even if in fact no such instances are ever produced or some to be. E.g., when nature finished with the earth as we know it today, it contained the atoms of only ninety-two chemical elements, with uranium being the heaviest; and whether in fact there are atoms of transuranic elements existing in the universe outside of the earth is for the most part unknown. Nevertheless, even if atoms of some of these elements do not, nor ever will, exist, the elements themselves are so well known today that atoms of some, but not all, of these elements have actually been produced in accelerators in recent years. Of course, those which have had atoms produced in this way will trivially satisfy the thesis (NR) of natural realism; but still the fact that some of the other transuranic elements may never satisfy (NR) in no way shows that they do not exist as causally determinate structures. A natural property or relation, in other words, has its being as a universal not in its actually having instances (at some time or other in the history of the actual world) but in its being a structure which is causally realizable.

Now as causally determinate structures which need not have any instances in the world, natural properties and relations are contained within the causal structure of the world not as individuals but as unsaturated components of that structure. The argument for this claim is simply that if a natural property or relation were an individual, then it would be *an abstract individual* and as such it would transcend even the causal structure of the world insofar as its existence as an abstract individual does not depend on its having any instances in the world. Such a universal, in other words, would be a platonic form or logically real universal and not a causally determinate structure after all. Whether the unsaturatedness of such a structure consists only in its being the nexus of a causally possible state of affairs (the totality of which determine what could have or might yet come to be the case, and in terms of which the truth-conditions for $\Box$

are to be given) or, on the other hand, whether it has some other ontological ground for its explanation (such as the sense in which the laws of nature can be said to involve such structures), we shall leave to different forms of natural realism to decide. We might note, however, that although the nature of this unsaturatedness is not the same as that of concepts as cognitive capacities or cognitive structures otherwise based upon such capacities, nevertheless the sense in which the latter are said not to exist independently of the generic capacity humans have for concept-formation is not entirely unlike the sense in which natural properties and relations are said not to exist independently of nature and its causal matrix. In particular, just as the laws of compositionality for concept-formation can be said to characterize the logical structure of the generic capacity humans have for concept-formation, so too the laws of nature as laws regarding the causal connections between natural properties and relations can be said to characterize the causal structure of the world.

Finally, it should be noted that although natural properties and relations are not in general "identical" when they have the same extension, nevertheless as causally determinate structures natural properties and relations are "identical" when it is causally necessary that they have the same instantiations. This of course is another feature which shows how different natural properties and relations are from logically real properties and relations. For convenience, we shall abbreviate this sort of "identity" as follows:

$$F^j \equiv G^j =_{\mathrm{df}} \Box(\forall^e x_1)\Box \cdots \Box(\forall^e x_j)\Box[F(x_1, \ldots, x_j) \leftrightarrow \\ \leftrightarrow G(x_1, \ldots, x_j)].$$

Of course, although natural properties and relationds do not exist independently of the causal structure of the world, they nevertheless retain their "identity" as causally determinate structures in all causally possible contexts. That is, according to natural realism,

$$(\forall^n F^j)\Box(\exists^n G^j)(F \equiv G),$$

is universally valid, for all positive integers j.

14. ARISTOLELIAN ESSENTIALISM AND THE LOGIC OF NATURAL KINDS

Despite the occasional confusion of Aristolelian essentialism with the essentialism of logical realism (cf. Quine [25], p. 173f), the form of essentialism advocated by Aristotle was an essentialism of natural kinds within a framework of natural realism (cf. Bolton [2]). Moreover, unlike the situation in logical realism, the underlying logic of this form of essentialism is relatively unproblematic and can be formulated in terms of a few basic laws of natural kinds. We shall include these laws here in our formulation of natural realism; but it should be noted that since no particular natural kinds are actually posited as existing in these laws, the formulation we give below will also apply to a natural realism which rejects the essentialism of natural kinds. All one need do in that case is add an extra axiom to the effect that there are no natural kinds. That is, since natural kinds are special types of properties, we may assert that there are no natural kinds as follows,

$$(\forall^n F^1) \sim (\exists^k G^1)(F \equiv G),$$

where $(\exists^k G^1)$ is a predicate quantifier understood as referring to natural kinds. Such a predicate quantifier, in other words, stands for a (second order) referential concept which purports to refer to natural kinds.

As already indicated, because we are considering natural realism here only as a theory of universals, we may ignore wffs which contain predicate quantifiers that refer to concepts as intelligible universals and retain just those that refer to natural properties and relations as causally real universals, including now those that refer to natural kinds as well. The set of wffs of a language L, accordingly, is the smallest set K containing the atomic wffs of L and such that $\sim\phi$, $\Box\phi$, $(\phi \to \psi)$, $(\forall^e x)\phi$, $(\forall^n F)\phi$, $(\forall^k G^1)\phi$ are in K, whenever ϕ, ψ are in K, x is an individual variable, G^1 is a 1-place predicate variable, and F is an arbitrary predicate variable. (The existence predicate $E!$ is assumed defined as in Section 13.)

As with natural properties and relations in general, needless to say, which concepts, if any, have natural kinds corresponding to them is always an empirical question. That is, an assertion of the form,

$$(\exists^k F)\Box(\forall^e x)[F(x) \leftrightarrow \phi],$$

where F^1 does not occur free in ϕ, is always construed at best as a scientific hypothesis and never as a logical or conceptual truth. Though not all natural properties are natural kinds, nevertheless, we do assume that natural kinds are special types of properties. Minimally, in other words, the following is assumed to be universally valid:

(K1) $\quad (\forall^n F^1)\phi \to (\forall^k F^1)\phi.$

It follows, accordingly, by the fundamental thesis (NR) of natural realism that every natural kind is causally realizable:

$$(\forall^k F)\Diamond(\exists^e x)F(x).$$

Of course, it also follows that natural kinds are natural properties in all causally possible contexts; but then natural kinds fulfill an even stronger condition than that. That is, as the causal structures grounding the law-like behavior of things belonging to a natural kind, natural kinds retain their "identity" as natural kinds in all causally possible contexts. That is,

(K2) $(\forall^k F)\Box(\exists^k G)(F \equiv G)$,

is also universally valid according to natural realism.

The most fundamental law regarding natural kinds, however, and which is more to the point of their role as essences, is the thesis that an individual can belong to a natural kind only if being of that kind is essential to it; i.e., only if it *must* be an individual of that kind whenever it exists:

(K3) $(\forall^k F)(F(x) \rightarrow \Box[E!(x) \rightarrow F(x)])$.

An important principle which is derivable from (K3) and the fact that natural kinds are existence entailing properties (i.e., that only existing objects can belong to a natural kind) is the thesis that any individual which can belong to two natural kinds necessarily belongs to one when and only when it belongs to the other:

$$(\forall^k F)(\forall^k G)(\Diamond F(x) \;\&\; \Diamond G(x) \rightarrow \Box[F(x) \leftrightarrow G(x)]).$$

Now we shall speak of the principles (K1)–(K3) (together with a distribution law (K0) for $\forall^k$ over conditionals) as constituting the minimal principles of a theory of natural kinds. It is these principles, in other words, which are assumed in all forms of essentialist natural realism. The remaining principles cited below were indeed all adopted in one form or another by Aristotle and are implicit in our common sense view of natural kinds as the various genera and species of plants and animals. But then whether or to what extent they also apply to natural kinds of substances (in the chemical sense), i.e., substances such as gold, water, oxygen, etc., is not entirely clear; and for that reason we shall leave them as principles which may or not be adopted in various forms of essentialist natural realism. For convenience of expression, however, we shall adopt the following abbreviatory notation:

$$F \leqslant G \;=_{\text{df}}\; \Box(\forall^e x)[F(x) \rightarrow G(x)]$$

$$F < G \;=_{\text{df}}\; (F \leqslant G \;\&\; G \nleqslant F).$$

The first and most plausible principle beyond those of the minimal system, accordingly, is what we shall call *the partition principle for natural kinds*. On this principle, if two natural kinds are not necessarily disjoint, then one must be subsumed under the other:

(K4) $(\forall^k F)(\forall^k G)(\Diamond(\exists^e x)[F(x) \,\&\, G(x)] \rightarrow F \leqslant G \vee G \leqslant F)$.

The idea here is that the family of natural kinds to which an individual belongs determines a chain of subordination of these kinds one to another. The partition determined by this chain of subordination is stated in the following consequence of (K4) that natural kinds which are subsumed under the same *immediate* genus are either "identical" or necessarily disjoint:

$$(\forall^k F)(\forall^k G)(\forall^k H)(F < H \,\&\, G < H \,\&\, (\forall^k H')[F < H' \rightarrow$$
$$\rightarrow H \leqslant H'] \,\&\, (\forall^k H')[G < H' \rightarrow H \leqslant H'] \rightarrow$$
$$\rightarrow \sim \Diamond(\exists^e x)[F(x) \,\&\, G(x)] \vee [F \equiv G]).$$

An important ontological thesis in essentialist natural realism, incidentally, is *the summum genus principle* that if an individual belongs to any natural kind at all, then it belongs to a natural kind which is a summum genus in the sense that it has subordinate to it every natural kind to which that individual belongs:

(K5) $(\forall^k F)(F(x) \rightarrow (\exists^k G)[G(x) \,\&\, (\forall^k H)(H(x) \rightarrow H \leqslant G)])$.

The idea here is that unless there are summum genera in nature as posited in (K5), there can be no real individuation of natural kinds of things. (In conceptual natural realism, incidentally, individuation is an ontological process, and identification is a conceptual process. In logical realism, these two process tend to coalesce together.) In other words, the rationale for (K5) is that it is supposed to be causally impossible for things belonging to a natural kind to be individuated through an infinitely ascending chain of natural kinds as more and more generic causal structures, and that therefore there must be a most generic natural kind or causal structure determining the real individuation of any natural kind of thing. Given (K4), incidentally, (K5) can be equivalently stated as follows:

$$(\forall^k F)(\exists^k G)[F \leqslant G \,\&\, \sim(\exists^k H)(G < H)].$$

But even without (K4) this principle is still at least a consequence of (K5).

The dual of the summum genus principle, at least as far as natural kinds are concerned, is *the infima species principle* that if an individual belongs to any natural kind then it belongs to a natural kind which is subordinate to every natural kind to which that individual belongs and which also has no natural kinds subordinate to it:

(K6) $(\forall^k F)(F(x) \rightarrow (\exists^k G)[G(x) \,\&\, (\forall^k H)(H(x) \rightarrow G \leqslant H) \,\&$

$\& \sim(\exists^k F)(F < G)])$.

A consequence of (K6) is another form of the infima species principle:

$$(\forall^k F)(\exists^k G)[G \leqslant F \,\&\, \sim(\exists^k H)(H < G)].$$

It should perhaps be noted that if it is possible for an individual to belong to but one natural kind, then it is possible for the same natural kind to be both an *infima species* and a *summum genus*. Such in fact would seem to be the case with the different kinds of elementary particles posited in microphysics.

Finally, in connection with the partition principle, another candidate for the logic of natural kinds is the principle that every genus is the sum of its species:

(K7) $(\forall^k F)[(\exists^k G)(G < F) \rightarrow$

$\rightarrow \Box(\forall^e x)(F(x) \leftrightarrow (\exists^k G)[G(x) \,\&\, G < F])]$.

Of course this principle can also be formulated so as to apply only to genera that have *immediate* species and which are then posited as being the sum of these immediate species; i.e., where $(G < F)$ in (K7) is replaced in each of its occurrences by $(G < F \,\&\, (\forall^k H)[G < H \rightarrow F \leqslant H])$.

In regard to a set-theoretic semantics for natural realism, let us say that $\langle D, E_i, f, X_d, Y_n \rangle_{i \in W, d \in D, n \in \omega - \{0\}}$ is a *natural kind model structure* for a language L if (1) $\langle D, E_i, f \rangle_{i \in W}$ is a modal model structure for L (as defined in Section 11) such that $D = \bigcup_{i \in W} E_i$; (2) for all positive integers n, $Y_n \subseteq \{K \in (\mathscr{P}D^n)^W$: for all $i \in W$, $K(i) \in E_i^n\}$, i.e., every natural property or relation in the structure is an existence entailing property or relation; (3) for all $d \in D$, $X_d \subseteq Y_1$, i.e., every natural kind to which an individual d belongs is a natural property; (4) for all $K \in X_d$, for some $i \in W$, $d \in K(i)$, i.e., a property is a natural kind to which d belongs only if d *can* have that property; (5) for all $d' \in D$, for all $K' \in X_{d'}$, if $d \in K'(i)$, for some $i \in W$, then $K' \in X_d$, i.e., every natural kind to which d can belong is a natural kind of d; and (6) for all $K \in X_d$, for all $i \in W$, if $d \in E_i$, then $d \in K(i)$, i.e., d belongs to each of its natural kinds whenever it exists. Note that although Y_n is the set of n-ary natural relations, no stipulation is made that Y_n is non-empty. That is, whether there are any n-ary natural relations in the world or not is a factual (or rather causal) and not a logical matter; and therefore it is left as a matter for specific scientific theories as applied forms of natural realism to decide. Similarly, whether there are any natural kinds in the world or not, i.e., whether for any $d \in D$, X_d is non-empty, is also left open; and the fact that Y_1 might be non-empty, i.e., that there are natural properties after all, would not of itself decide the matter for natural kinds. Note, however, that since not every concretely existing object need belong to a natural kind, then X_d might be empty for some $d \in D$, whereas for others X_d could well be non-empty.

Now by an *assignment* A in a natural kind model structure $\mathfrak{A}$ defined as above we mean a referential assignment A in the modal model structure $\langle D, E_i, f\rangle_{i \in W}$ exactly as defined in Section 11. Where A is such an assignment and $i \in W$, we define the *satisfaction by* A *in* $\mathfrak{A}$ *at* i of an arbitary wff of the language in question exactly as in Section 11, except for the clauses for the predicate quantifiers which are now defined as follows:

(7^n) *A satisfies* $(\forall^n F^j)\phi$ *in* $\mathfrak{A}$ *at i* iff for all $K \in Y_j$, $A(K/F^j)$ satisfies ϕ in $\mathfrak{A}$ at i; and

(7^k) *A satisfies* $(\forall^k F^1)\phi$ *in* $\mathfrak{A}$ *at i* iff for all $d \in D$, for all $K \in X_d$, $A(K/F^1)$ satisfies ϕ in $\mathfrak{A}$ at i.

Truth and validity in a natural kind model structure are assumed defined in the usual way (as, e.g., in Section 11); and so too *logical consequence* and *universal validity according to natural realism*. Note that no closure condition for either natural properties and relations or for natural kinds is universally valid according to natural realism. Also, just as, for each positive integer j,

$$(\exists^n F^j)(G^j \equiv F^j)$$

is not universally valid according to natural realism, neither is

$$(\exists^k F^1)(G^1 \equiv F^1).$$

The logic of natural realism, in other words, is free of existential presuppositions regarding whether the concepts expressed by predicate expressions have either a natural property or relation or a natural kind corresponding to them. Of course, because the logic is based on the actualist quantifier $\forall^e$, it is also free of existential presuppositions regarding singular terms.

Now by an *axiom of natural realism* we shall mean any wff (as presently defined) which is either (1) tautologous; (2) an instance of an axiom of the modal propositional logic S5; (3) an instance of (LL) or of one of the schemas ($A1^e$)–($A5^e$) described in Section 12; (4) an instance of (K0)–(K3) (where (K0) is the distribution law for $\forall^k$ over conditionals); or an instance of one of the following schemas:

$$(\forall^n F^j)[\phi \rightarrow \psi] \rightarrow [(\forall^n F^j)\phi \rightarrow (\forall^n F^j)\psi],$$

$$\phi \rightarrow (\forall^n F^j)\phi, \text{ when } F^j \text{ is not free in } \phi,$$

$$(\forall^n F^j)\Diamond(\exists^e x_1) \ldots (\exists^e x_j)F(x_1, \ldots, x_j),$$

$$(\forall^n F^j)[F(x_1, \ldots, x_j) \rightarrow E!(x_1) \;\&\; \cdots \;\&\; E!(x_j)],$$

$$(\forall^n F^j)\Box(\exists^n G^j)(F^j \equiv G^j).$$

As *inference rules* of natural realism we take *modus ponens*, universal generalization (of an individual or predicate variable), and the rule ($\Box$UG) described in Section 12 for an embedded universal generalization of an individual variable within the scope of iterated occurrences of $\Box$. A proof of the following completeness theorem is easily constructed in terms of the proof of completeness for modal natural realism given in Cocchiarella [10], Section 3.7. (We use $\models_n$ for the relation of logical consequence according to natural realism, and $\vdash_n$ for the derivability relation based upon the above axioms and rules.)

COMPLETENESS THEOREM: *If* $\Sigma \cup \{\phi\}$ *is a set of wffs* (*as presently defined*), *then:*

(1) $\Sigma \models_n \phi$ *if and only if* $\Sigma \vdash_n \phi$; *and*
(2) $\models_n \phi$ *if and only if* $\vdash_n \phi$.

Completeness theorems can also be established for the systems obtained by adding the principles (K4), (K5), (K6) and (K7), respectively, to the above axiom set, and by then appropriately constraining the natural kind model structures with respect to which universal validity is to be defined. E.g., for (K4), a natural kind model structure must also satisfy the condition that for all $d \in D$, for all K, $K' \in X_d$, either for all $i \in W$, $K(i) \subseteq K'(i)$ or for all $i \in W$, $K'(i) \subseteq K(i)$; and for (K5) it must satisfy the condition that if X_d is non-empty, for any $d \in D$, then for some $K \in X_d$, for all $K' \in X_d$, for all $i \in W$, $K'(i) \subseteq K(i)$. We omit stating the corresponding conditions for (K6) and (K7) (but for details see Cocchiarella [8]).

In closing, we should perhaps emphasize once again how different natural realism, essentialist or otherwise, is from logical realism, essentialist or otherwise. There is some similarity, of course, between natural realism and actualism as a form of logical realism; but that similarity cannot be the basis of a defense of actualism over possibilism, or at least not without confusing logical realism with natural realism. The more relevant similarity, moreover, is really between natural realism and actualism as a form of conceptualism, since in fact it is the construction of concepts that have natural properties, and natural kinds in particular, corresponding to them which leads to the systematic formation of *e*-concepts, and thereby to the impredicative concept of concrete existence. Within conceptual natural realism, in other words, actualism is a natural outcome of our experience of a world causally conditioned by natural properties and relations, including especially our experience of things belonging to a natural kind.

Indiana University

REFERENCES

[1] Beth, E. W.: 1960, 'Extension and intension', *Synthese* **12**, 375–379.

[2] Bolton, R.: 1976, 'Essentialism and semantic theory in Aristotle', *Phil. Review* **85**, 514–544.

[3] Carnap, R.: 1946, 'Modalities and quantification', *J. of Symbolic Logic* **11**, 33–64.

[4] Church, A.: 1950, *Introduction to Mathematical Logic*, Princeton Univ. Press, N. J.

[5] Cocchiarella, N. B.: 1969, 'A completeness theorem in second order modal logic', *Theoria* **2**, 81–103.

[6] Cocchiarella, N. B.: 1975, 'On the primary and secondary semantics of logical necessity', *J. of Philosophical Logic* **4**, 13–27.

[7] Cocchiarella, N. B.: 1980, 'Nominalism and conceptualism as predicative second order theories of predication', *Notre Dame J. of Formal Logic* **21**, 481–500.

[8] Cocchiarella, N. B.: 1976, 'On the logic of natural kinds', *Philosophy of Science* **43**, 202–222.

[9] Cocchiarella, N. B.: 1983, 'Philosophical perspectives on quantification in tense and modal logic', in *Handbook of Philosophical Logic*, Vol. 2, eds. D. Gabbay and F. Guenthner, D. Reidel Publ. Co., Dordrecht.

[10] Cocchiarella, N. B.: 1986, *Logical Investigations of Predication Theory and the Problem of Universals*, Vol. 2 of *Indices*, Bibliopolis Press, Naples.

[11] Cocchiarella, N. B: 1986, 'Frege, Russell and logicism: A logical reconstruction', *Frege Synthesized*, L. Haaparanta and J. Hintikka, eds., D. Reidel Publ. Co., Dordrecht, 197–252.

[12] Doets, H. C.: 1969, 'Novak's result by Henkin's method', *Fund. Math.* **64**, 329–333.

[13] Flanagan, T. B.: 1976, 'A new finitary proof of a theorem of Mostowski', in G. Müller (ed.), *Sets and Classes: On the Work of Paul Bernays*, North-Holland.

[14] Gallin, D.: 1975, *Intensional and Higher-Order Modal Logic*, North-Holland.

[15] Inhelder, B. and Piaget, J.: 1958, *The Growth of Logical Thinking from Childhood to Adolescence*, Basic Books Inc.

[16] Kanger, S.: 1957, *Provability in Logic*, Univ. of Stockholm.
[17] Lorenz, K.: 1975, 'Kant's doctrine of the *a priori* in the light of contemporary biology', in *General Systems*, Vol. III (1962), eds. von Bertalanffy, L. and A. Rapaport; reprinted in *Konrad Lorenz: The Man and His Ideas*, Evans, R. I., Harcourt Brace Jovanovich, N. Y., pp. 181–217.
[18] Montague, R. M.: 1974, *Formal Philosophy*, Yale Univ. Press, New Haven.
[19] Novak, I. L.: 1951, 'A construction for models of consistent systems', *Fund. Math* **37**, 87–110.
[20] Parsons, C.: 1971, 'A plea for substitutional quantification', *Journal of Philosophy* **68**, 231–237.
[21] Parsons, T.: 1969, 'Essentialism and quantified modal logic', *Phil. Review* **78**, 35–52.
[22] Piaget, J.: 1972, *The Principles of Genetic Epistemology*, Routledge and Kegan Paul, N.Y.
[23] Piaget, J..: 1977, *The Development of Thought: Equilibration of Cognitive Structures*, Viking Press, N.Y.
[24] Quine, W. V.: 1960, *Word and Object*, M.I.T. Press, Cambridge.
[25] Quine, W. V.: 1966, *Ways of Paradox*, Random House, N.Y.
[26] Quine, W. V.: 1970, *Philosophy of Logic*, Prentice Hall Inc., Englewood Cliffs.
[27] Quine, W. V.: 1968, 'Replies', in *Essays in Honor of W. V. Quine, Synthese* **19**.
[28] Russell, B.: 1903, *The Principles of Mathematics*, Norton, N.Y.
[29] Shoenfield, J. R.: 1954, 'A relative consistency proof', *J. of Symbolic Logic* **19**, 21–28.

CHAPTER IV.4

MASS EXPRESSIONS

by Francis Jeffry Pelletier and Lenhart K. Schubert

INTRODUCTION

Problems associated with mass expressions[1] can be divided into the following general areas: (1) distinguishing a class of mass expressions, (2) describing the syntax of this class, (3) describing the formal semantics of this class, (4) explicating the ontology such a class of expressions presupposes, and (5) accounting for various epistemological issues involving our perception of the ontology. Permeating all these areas are questions of the role of pragmatics (construed broadly so as to encompass questions of how people typically use mass expressions to describe a state of affairs) in formal semantic descriptions. For example, if people normally use expression x in such-and-such circumstances (or normally take a use of x to imply y), should that have some reflex in the syntax, the semantics, the presupposed ontology, etc.? The "problems of mass expressions" therefore provides a rich area in which to formulate and test various logico-linguistic theories. In this article we indicate (what we take to be) the more important problems and puzzles, survey some suggested solutions, indicate what we think are outstanding difficulties with those approaches, and make a few not-well-defended suggestions of our own to these problems and puzzles. A bibliography of most work on mass expressions prior to 1978 can be found in Pelletier (1979b). Further additions to it are mentioned in the bibliography at the end of this article.

D. Gabbay and F. Guenthner (eds.), Handbook of Philosophical Logic, Volume IV, 327–407.

WHAT IS A MASS EXPRESSION?

Most works do not address this basic issue but rather concentrate attention on certain examples which they take to distinguish mass expressions from count expressions. *Water*, *dirty water*, and *water used in washing yesterday's dishes* are central examples of mass expressions, it is claimed; while *person*, *tall person*, and *person who washed yesterday's dishes* are central examples of count expressions. On the syntactic side, the usual view claims, mass expressions occur with the quantifiers *much* and *little*, with the unstressed article *some*, are susceptible to measurement phrases like *liters of* and *amount of*, and do not exhibit a singular/plural distinction. On the other hand, count expressions occur with the quantifiers *each*, *every*, *many*, *several*, *few*, and the stressed *some*, they use the indefinite article *a*(*n*), they are susceptible to counting phrases like *five*, *a score of*, and do exhibit a singular/plural dichotomy manifested in the count noun phrase itself and in agreement with the verb phrase.[2]

From the semantic point of view, count expressions and mass expressions are supposed to be differentiated by the way in which they refer to something. A count expression is supposed to refer to a discrete, well-delineated group of entities, whereas a mass expression refers without making it explicit how its referent is to be individuated or divided into objects. This feature of mass reference gives rise to the test of "cumulative reference" – any sum of parts which are *M* is also *M* (Quine [1960], p. 91), and to "distributed (or divided or divisive or homogeneous) reference" – any part of something which is *M* is also *M* (Cheng [1973], pp. 286–287).

Correlated with the semantic and syntactic conditions is a set of pragmatic features. (Perhaps the syntactic and semantic features follow from the pragmatic ones, or perhaps vice versa). These pragmatic features have to do with the way we differentially *use* various expressions of the language. Most relevant to the present discussion are the concepts of "individuating" and "identifying".

Traditional philosophic wisdom has it that some subset of the count expressions "individuate" the world in the sense that, given a space appropriate to such a expression *C*, there is an answer to the question "How many *C*s are in that space?"[3] Such count expressions are traditionally called "sortal expressions". Thus, given a building room, it makes sense to ask how many people are in that room – making *person* a sortal expression; given a solar system, it makes sense to ask how many planets are in that solar system – making *planet* a sortal expression. Mass expressions are supposed to fail this test, as are certain count expressions like *thing*, *object*, *entity*, etc. (This last because there is no definite answer to the question "How many things are in this room?", unless *thing* has antecedently been given some contextual understanding.) "Individuate" is supposed to mark off one instance of a count expression from another – and it is claimed that *thing*, *water*, etc., do not do this. "Individuate" should be distinguished from "identify". Presumably *water* will identify or characterize a certain portion of the environment and distinguish it from other portions – just as *person* does. However, we presume, *thing* does not identify any more than it individuates. And equally, we suppose, words like *stuff* and *quantity* do not identify anything. (Or rather, *thing*, *stuff*, etc., require some "pragmatic" background to do it.)[4]

There is no question that the standard examples of mass expressions are nouns or noun phrases; but various authors have made claims about classifying expressions other than nouns and noun phrases as being count or mass. The three main types of expressions that have been suggested for this extended use are: adjectives (and adjective phrases), verbs (and verb phrases), and adverbs (and adverb phrases). We shall consider these each in turn. Moravcsik [1973], Quine ([1960], p. 104), and Bunt [1980] suggest that some adjectives such as *spherical* be classified as count because they are only true of objects which can be counted. We note that very few adjectives have this property. Perhaps the numeral adjectives, like *one*, *two*, *many*, *several*, . . . , are the best candidates for + count adjectives. (Of course it is hard to make the argument for the plural ones, since one could simply say that, e.g., **two furniture* violates a number agreement – but even ?*one furniture* seems odd). Moravcsik also suggests that adjectives which are divisive should be considered as + mass. For example *light*, but not *heavy*, *short* but not *tall*. Of course, if one used the cumulativity test (as Quine does) for determining "massness", one would get *heavy* but not *light*, *tall* but not *short*. Which test shall we take? Bunt offers us this: Only adjectives which are homogeneous (both divisive and cumulative) in their reference can combine with mass nouns ([1981], p. 229; see also his [1980]). This becomes revised to exclude non-restrictive and generic and collective modification of mass nouns. In the course of this revision, we learn (for a variety of reasons, mostly having to do with polar opposite adjectives where one is divisive but not cumulative and the other is cumulative but not divisive – as in the above examples) that adjectives of size or amount, or of shape or other aspects of outward appearance, are *not* mass adjectives. It seems to us that this at least partially contradicts our intuitive starting point: for very few adjectives will be +mass.[5]

It is perhaps true that such expressions ought to be somehow differentially marked – so that *spherical* can only modify a noun which is judged count, or so that only nouns which are judged mass can be modified by *rancid*. But three things should be noted. First, this need not be a syntactic classification. If one takes the view that count/mass is a semantic well-formedness constraint, then *no* expressions will receive a syntactic marking for mass or count. It will rather be a feature of the semantic representation of *spherical*, say, that it can only combine with semantic representations that are true of individual objects, but it will not be a syntactic classification. If it is a semantic point being made, as it seems to be since the initial test was one of looking for "cumulativity" and "divisiveness", then the combination of *spherical* with, say, *water* will be well-formed (syntactically) but simply will not be given a coherent semantic representation. Second, even if it were called syntactic, one should note that the notions of mass and count are here being used in an extended way: it is not really that *spherical* is +count, but rather that *spherical* has a co-occurrence restriction to the effect that it can only combine with expressions judged to be "really" count – that is, with certain nouns. Third, the co-occurrence restrictions that have been offered seem suspect. Are

> Bring me the spherical ice and leave the irregularly-shaped ice there.
> Look at the fist-sized, spherical hail.
> Mary has box-shaped furniture.

really ungrammatical?[6]

Some people have been briefly tempted (e.g., Moravcsik [1973]) to classify verbs as mass or count, depending on what sort of subject they can take. For example, *flow* might be +mass (taking +mass subjects) while *think* might be +count (taking +count subjects). But they are quickly disabused upon considering that not only does water flow, but so do rivers; and even thinking is not beyond question – not only do people think, but possibly so does computer software (and *computer software* is normally assigned +mass status). More careful theorists have focussed on the semantic criterion of "divisiveness" in giving an intuitive account of why certain verb phrases should or should not be considered mass/count. Leech ([1969], p. 134ff), Mourelatos [1978], L. Carlson [1977], Åqvist and Guenthner [1978], Gabbay and Moravcsik [1979], Taylor [1977] and

Hoepelman [1976], all following Vendler [1967], suggest that verbs denoting processes be marked +mass while those denoting achievements be marked +count. The idea is that an event is the primitive verb-phrase denotation, and events can be part of larger events or contain subevents themselves. Some of these events, the ones which are processes like *to eat* and *to run*, have parts that are events denoted by the same verb. Others, the ones which are achievements like *to prove* and *to prepare*, are naturally bounded in the sense that they describe actions that involve change toward a final goal. So they do not have parts that are events denoted by the same verb. This is, of course, to apply the "divisiveness" criterion to verb phrases. Implicit in these discussions is also the possibility of carrying the "cumulativity" criterion to verb phrases.

An informative discussion of verbs (and adverbs) as mass expressions is in ter Meulen ([1980], Chapt. 4). She points out that either type of verb phrase can take either mass or count subjects. More interesting, she finds that the (direct) objects of the verb can be of either type, but which one it is determines whether the entire verb phrase is count or mass. Thus *eating cake* is +mass, *eating a cake* is +count; *preparing dinner* is +mass, *preparing a dinner* is +count. And thus the object dominates the simple verb (*eating* was +mass, *preparing* was +count).[7] Following Hoepelman [1976], [1978], ter Meulen considers certain adverb phrases to be either mass/count. Generally this means that the spatial or temporal extension or duration of the adverb is unbounded (or bounded). For instance, the adverb phrase *for hours* is temporally unbounded; the phrase *along the road* is spatially unbounded; the phrase *in an hour* is temporally bounded; the phrase *to the city* is spatially bounded. Generally, ter Meulen suggests, one should look for a final state that will be reached through the action described by the verb phrase in question; if so, then the adverb is +count. ter Meulen notes that any verb (count/mass) can take either type of adverb phrase (count/mass); but it appears that the feature of the adverb phrase dominates the one of the verb. The next question is whether the feature of the entire verb phrase generated by using some count/mass direct object will combine correctly with either type of adverb. The results are a bit complicated and the judgments involved seem to us to be somewhat unreliable, but the answers seem to be this: The entire verb phrase is dominated by the adverb phrase; regardless of whether the verb and direct object are mass or count. Two possible exceptions to this are: (a) When the verb and the direct object are −count and the adverb phrase is +count, we get some sort of anomaly: *John ate cake in an hour* – one needs some special understanding to interpret it, and (b) when the verb and adverb are +count but the object is −count, one gets what ter Meulen calls an iterative reading which is +count. Her example here is not well conceived: *Henry solved puzzles in an hour* has perhaps a +count verb and adverb, but it is unclear that *puzzles* is −count. What she says ([1980], p. 129) is: "it is interpreted as saying that Henry could provide solutions to puzzles in an hour, which is obviously [an achievement] rather than [a process]." A better example of this phenomenon might be *Henry repaired furniture in an hour*.

We should note that, whatever else is true of this claim about the count/mass distinction amongst verbs and adverbs, it is founded on a semantic understanding of the distinction. We are told to look for the denotation of the terms and see whether they are divisive (or cumulative). So all the evidence gathered here might find its proper home in an account of the semantic plausibility of certain kinds of combinations, rather than as a syntactic point. With regard to the claim that verbs denoting processes are mass while those denoting achievements are count we might remark that according to the sense which is being used, it is rather the noun phrase *Sandy's eating cake* which is +mass and *Kim's winning a race* which is +count, and not the mere verb phrase, *eat cake* or *win a race*.

As the preceding discussion suggests, there are two senses in which the count/mass distinction might carry over to categories other than nominals. (1) If the count/mass distinction for nominals is a syntactic distinction, then it might carry over to other phrase types as an agreement feature, just as the singular/plural distinction carries over to verb phrases, even though the distinction might be said to derive from a distinction inherent in noun phrases (or their denotations). Carry over of gender to adjectives (from nouns) in many languages is another example of this phenomenon. Or, (2) it might be that there are distinctions among verb phrases (and perhaps other non-nominal categories) which are in some sense *analogous* to the count/mass distinction for nominals, and raise similar problems as to whether the distinctions are syntactic, semantic, or pragmatic. For instance, we have seen the claim that certain verb phrases semantically are true of any part of what they are true of. This would be to claim that the mass/count distinction amongst nouns is a semantic distinction, and that verb phrases have an analogous character. We have given reasons to believe that there are no syntactic agreement features which would justify holding (1). As far as position (2) goes, we have seen that the only reasons for holding it are semantic, having to do with the denotation of the verb phrases in question. However true the analogy might be, we think the intuitions behind it are more clear in the nominal case and it ought to be studied there before being generalized to other categories. We therefore think it better to restrict the mass/count classification to nouns and noun phrases, regardless of whether in the latter case it be a syntactic or a semantic classification. We think that the intuitions being appealed to by those who wish to extend the classification beyond nouns and noun phrases simply do not apply except in a stretched sense. It is really always by considering the *objects* which are characterized by the adjectives that gives us any reason to call the adjectives mass or count – so it is not syntactic, and in fact even the semantic characterization only applies to the *things*. But *things* are what are designated by nouns and noun phrases. (Cp. Pelletier [1975], pp. 9–10 on calling *red* a sortal term). Again with the verbs, verb phrases, and adverb phrases, it is a matter of looking to the *events* denoted by the nominalized verb phrases, or to the kinds of events that can be modified by certain adverb phrases, which makes one want to call verbs, etc., mass or count. But then, just as with the adjectives, we should not call them count/mass, but rather call the noun phrases generated from them by nominalization either count or mass. For these reasons we shall restrict our further discussion to nouns and noun phrases.

Two of the issues that must be addressed at the outset are, first, the question of whether one is going to posit a mass/count distinction for *expressions*, for *senses of expressions*, or for *occurrences of expressions*. And second, one must address the question of how to classify expressions (or their senses or their occurrences) which do not immediately satisfy the syntactic/semantic/pragmatic conditions cited above. We mention these issues together because how one answers the one might influence how one answers the other.

Consider sentences like

(1) Kim put an apple into the salad

(2) Kim put apple into the salad

Most theorists would claim that *apple* in (1) is count, whereas in (2) it is mass. If it is the expressions themselves one is going to call or mass, one is faced with a slight, but not insuperable, difficulty, since the expression cannot be both. However, one could claim that there are two separate words here, spelled and pronounced the same, one of which is mass and the other of which is count.[8] (This would be akin to our normal procedure of counting *bank* as at least three separate words.[9]) Another way out for the "expression approach" to classification would be to claim that *apple* is always count and that sentence (2) has a deleted *bits of an* left out. Alternatively one might claim that *apple* is always mass and that (1) uses *an* to construct this mass into a well-defined object. So, with respect to the "dual life" words like *apple*, one might say (a) that *all* such expressions were "really" count, or (b) that *all* such expressions were "really" mass, or (c) that some were "really" count and others were "really" mass. One might go even further and say (d) that *every* expression, not just the dual life ones, is "really" mass, or (e) that every expressions is "really" count. (Presumably (d) and (e) would be to adopt (a) or (b) but to claim that all expressions led a "dual life"'). So according to (a), there are some apparently "dual life" expressions, but they should all be represented as count, and a rule of "massification" sometimes occurs. According to (b), some expressions lead a "dual life", but they are all to be represented as mass, and a process of "countification" sometimes occurs. According to (c), some "dual life" expressions are represented as count while others are represented as mass, and there are two processes at work. According to (d), every expression is to be represented as mass, and the process of countification has rather a lot of work to do. And according to (e), every expression is to be represented as count, and the process of massification has rather a lot of work to do. We shall consider these approaches shortly.

The difference between these "expression approaches" and a "sense approach" is difficult to explicate in the absence of a general theory of the syntax (and semantics) of the language which would show the relationship between the lexical representation of individual words (or senses) and the syntactico-semantic representation of a sentence in which it occurs. As an overview, however, let us just indicate that we have in mind a lexicon which for each word contains some base form of that word, a statement of its membership in a syntactic class (potentially construed as a bundle of features), perhaps some idiosyncratic information about it (e.g., irregular plural), and finally a semantic representation of that word. We furthermore view the syntactic theory as providing some method of analyzing the structure of a series of words in such a way as to show how the semantic representations of the words are combined to form a semantic representation of the whole series.[10]

On this account of an overall theory, the "expression approach" to the mass/count distinction would associate with each entry of the lexicon membership in a syntactic class. Presumably being a mass expression or a count expression are two such classes. The syntactic rules, in analyzing a longer sequence of words, might assign such a feature to these longer sequences, and presumably the syntactic rules will be sensitive to its presence in a subphrase when they analyze a longer sequence. For words like *apple* (as above), there might be two separate entries – one marked +count and the other +mass. Call this the "dual expression approach". Or there might be just one (marked either as +mass or +count in accordance with the discussion of (a)–(e) above) and some syntactic rule might assign the phrase *an apple* to the category +count. Call these approaches "unitarian expression approaches". The "senses approach", on the other hand, will have only one lexical entry but will have alternate "senses", some being senses appropriate to mass uses and others being appropriate to count uses.

It seems to us that there is more to the difference between a "dual expression approach" and a "senses approach" than merely the cosmetic issue of having one entry in the lexicon with many different semantic representations or having many entries in the lexicon each with one semantic representation. The crucial difference has to do with whether +count and +mass are perceived as *syntactic* features (or classifications), or whether whatever differences there may be between them are to be accounted for entirely within the semantic representations. *Any* "expression approach" explicitly assigns the features or categories +count and +mass to some elements of the lexicon. We take this to be syntactic information about the expression in question. Therefore, unless it is a completely superfluous classification, some syntactic rules must make explicit reference to these features. Now, if the "senses approach" were to assign to the different senses of a lexical entry the *syntactic* features +count and +mass, then it *would* just be an alternative notation for the "dual expression approach", and there would be nothing (other, perhaps, than computational ease) to choose between them. But we take it that central to a "sense approach" is the claim that +count and +mass are *not* syntactic features or categories, but rather are a description of the semantic representation of the expression. That is, whether a *sense* is mass or count is something to be discovered after examining the semantic representation of the expression in question.[11] In this approach, *no* syntactic rules refer to +count or +mass (since these are not syntactic objects). Rather, sentences like (1) and (2) each have (at least) two semantic representations – one using a mass representation of *apple* and one using a count representation of *apple*. Presumably, the semantic representation of (1) using the mass sense and the semantic representation of (2) using the count sense are both (semantically) incoherent – they violate some (semantic) type constraint, or violate some other semantic well-formedness rule – but they are (syntactically) well-formed.

Let's now briefly distinguish an "occurrence approach" from the "expression approach" and the "senses approach". An "occurrence approach" might take +count and +mass either to be syntactic features (or classifications) or to be a semantic characterization. (Call these "syntactic occurrence approach" and "semantic occurrence approach" respectively). Underlying the "occurrence approaches" is the conviction that the lexical items are *not* characterized in any way as being count or mass – they have neither a syntactic feature nor independent semantic representations which can be classified as being count or mass. Indeed, the "occurrence approach" says, it is only in the context of a longer sequence of words that such a classification can be determined. In the lexicon a "neutral" semantic representation for, say, *apple*, is given. As the expression is seen to occur in longer phrases, its representation is then transformed so as to fall into one or the other of count or mass.[12] If the transition to a larger phrase, e.g., from *apple* to *an apple*, is taken to assign to *an apple* the syntactic category or feature +count, then this is presumably because some syntactic rule refers to "+count": This is the "syntactic occurrence approach". On the other hand, the theory might hold that the transition does not assign any such syntactic category or feature to our longer phrases, but rather adopts the sort of semantic view espoused by the "senses approach" – one tells that it is a count occurrence by looking at the semantic representation of this longer phrase. This is the "semantic occurrence approach".

As representatives of these various approaches, we might point to Quirk *et al.* ([1978], p. 128) and Quine ([1960], p. 91f) as proponents of the "dual expression approach" to such words as *apple*. Such theories recommend that these dual life words be given separate lexical entries, one with the syntactic feature +mass and the other with the syntactic feature +count.

There seem to be no representatives of the view that some ordinary expressions are +count and others are +mass, but that the dual life ones are always +mass. However, Sharvy [1978] is a proponent of the "unitarian expression approach" which takes *all* expressions to be underlyingly +mass in the lexicon, using a "deletion of nominal measure word" rule to account for apparent count expressions. In brief his argument runs thus: Sentences like

(3) Give me three beers

do *not* use *beers* as a count expression. Rather there is some (unrecoverable) nominal measure – such as *glasses of*, *kinds of*, *liters of*, etc. – which has been deleted from (3). At least, he claims, a language just like English in all other respects except for this would at least be "Quine-Indiscernible" from our English. But Sharvy applies this to more than just the examples which, like *apple*, appear to exhibit both a count and a mass use. His remarks can be generalized, he claims, to show that all apparent count nouns can be analyzed as a deleted nominal measure and the resulting language would be "Quine-Indiscernible" from English. So a sentence like

(4) There are three horses in the yard

does *not* have a +count term *horse* in it nor a +count expression like *three horses* or *three horses in the yard* in it. Rather there is a deleted nominal measure phrase, such as *head of*. As an example of a natural language which does not even *appear* to contain any count expressions (unlike English), Sharvy cites Mandarin Chinese.

On the other side of the "expression approach" are those who take our apparently dual life words such as *apple* to be underlyingly +count. (Although they have in addition some expressions being simply +mass). Here we might point to Bunt ([1981], Chap. 18) where such words are "included in the lexicon as singular count nouns", and there is a "syntactic rule transforming them into ground nouns" ([1981], p. 248). This is done "to do justice to the observation that many nouns, like *apple*, *onion*, *cake*, and *rope* occur both as singular count nouns and as mass nouns ("ground nouns")" (*ibid.*). Some of these unitarian expression theories even go to the opposite extreme from Sharvy and say that *all* nouns are underlying +count. Perhaps in this category is Allan (1980) with his "degrees of countability", but even clearer are textbook accounts of translation from English to predicate logic. Such texts will translate (5) as (5′)

(5) Beavers are mammals

(5′) $(Ax)[Beaver(x) \rightarrow Mammal(x)]$

and by analogy will translate (6) as (6′)

(6) Snow is white

(6′) $(Ax)[Snow(x) \rightarrow White\ (x)]$

Now, the move from (5) to (5′) requires understanding *beavers* as being true of a set of individuals (i.e., as a count expression), so that the quantifier makes sense ("For each thing, if it is a beaver then it is a mammal"). Similarly the move from (6) to (6′) requires understanding *snow* as being true of a set of individuals (like count nominals), so that the quantifier makes sense ("For each thing, if it is (a? sm?) snow, then it is white"). Textbook accounts do not discuss the conceptual and linguistic difficulties here, but it is significant to note that people who have seriously advocated a predicate logic account have shied away from the representation of (6) as (6′).[13] As an example of a natural language which does not even *appear* to have mass expressions (unlike English). Whorf ([1939], pp. 140–143) cites Hopi. (Although the work of Malotki [1983] casts considerable doubt on Whorf's claim.)

As representatives of the "senses approach" we might point to dictionaries that have entries like

> **lamb** (n.) la. A young sheep. 3a. The flesh of a lamb used as food.

(But dictionaries are just as likely to have

> **lamb** (n.) 1. A young sheep or its flesh used as food.)

Pelletier ([1975], p. 2) says that "we need to distinguish not between mass and count *nouns* but between mass and count *senses* of nouns." He goes on to say that the fact that speakers can isolate the cause of the ambiguity in a sentence like (3) – three bottles of beer, three kinds of beer, three servings of beer – as being due to *beers*, shows that the ambiguity is lexical, and not a matter of the operation of a syntactic deletion rule. For if it were due to the operation of such a rule, then we should be willing to give different syntactic analyses of the sentence, which we do not. With some justification one might accuse Pelletier [1975] of confusion on two distinctions: that between a "senses approach" and an "occurrences approach" on the one side, and that between the role of +count and +mass as syntactic features or as a description of semantic representations on the other side. McCawley [1975] is apparently also a representative of this "approach", although he sometimes can be seen to suggest a "unitarian sense approach" which uses "lexical redundancy rules" to generate the other sense. And such an approach would be difficult to distinguish from a "unitarian expression approach".

In Ware [1975], Bunt ([1981], Chapt. 2) and sometimes in Pelletier ([1974], fn. 1), it is claimed that the notions count and mass are to be applied to *occurrences* of expressions in a sentence. As mentioned above, this entails that the individual lexical entries do not have any such feature in their lexical representation, but rather that such classifications arise as the expressions are seen to occur in longer sequences of words. (For this reason it is probably better to say that it is the entire phrase *an apple*, rather than the individual word *apple*, which occurs here as count.) Bunt, in this localized discussion,[14] wishes to assign such phrases the *syntactic* marker +count or +mass. His procedure here seems suspicious. He delineates a class of (syntactic) contexts like our standard syntactic tests in which an occurrence is definitely said to be one of either mass or count. For occurrences which do not get classified by these criteria (e.g., sentence (2) above), he applies this test: "If '*E*' can be paraphrased by an expression 'E_1' in which [the expression in question] occurs in such a syntactic configuration that it can be classified as a mass (count) noun [occurrence], then [the expression] occurs as a mass (count) noun [occurrence] in 'E' on the reading 'E_1'" ([1981], p. 15). And in his ([1979], p. 250) he says of such occurrences: "construct an expression . . . which does contain such a criterion [as listed above] and in which the noun is used in the same sense as in [the original]. The classification obtained from [this new construction] is also the classification of the noun as it occurs in [the original]." This seems questionable along three dimensions. First, (as Bunt notes) it may not be possible to find any such construction. Second (and more importantly) is there any reason to believe that a mass use can never be paraphrased as a count use? If this can sometimes happen, then perhaps there are some so far unclassified (but "really" mass) uses which would be paraphrased (and hence classified) as count. Third (and we think most importantly), if it is an *occurrence* which is to be classified, the move to evaluating some *other* occurrence as mass or count would seem to have little if any relevance to the enterprise. (Along these same lines, compare the criticism in Pelletier ([1975], pp. 9–10) of calling *red* a sortal expression.) Finally, we might note that this method is unmechanizable. How would a parsing procedure ever *discover* the classification of an occurrence as mass or count? Certainly it is hard to imagine a mechanical procedure which generates the required paraphrase and checks if it means the same as the original!!

As the quotes from Bunt in the previous paragraph make clear, he would want to assign a *syntactic* feature of +count or +mass to the individual expression when it occurs in the appropriate type of context. It is not exactly clear why he would want do do this since there are no syntactic rules which can make use of such a feature.[15] Since the syntactic clues which can be used to classify the expression are exactly the ones for which the information (+count or +mass) might be sintactically useful, it is not obvious what is the reason for wanting to classify the noun at all. There may be a reason to classify the entire phrase, if there were some rules which were sensitive to a phrase's being +count or +mass, but not the expression itself.[16] One might wonder whether there could be any syntactic reason for classifying an entire phrase +count or +mass; that is, could there be any syntactic rules which make use of an entire phrase's being so classified, but not on whether an individual expression is so classified? Possibly rules like plural agreement with a verb phrase, but Bunt's grammar does not make use of features such as +mass/+count when applied to entire phrases. Instead the rules are entirely semantic in their effect, forcing a certain (semantic) interpretation on certain combinations.

Ware [1975] proposes an "occurrence approach" according to which one can sometimes (semantically) determine that a phrase is either mass or count. His tests have to do with "what sort of evidence we recognize as being relevant to the truth of falsity of the sentence." thus our sentence (1) would be counted false if less than one apple were added to the salad or possibly if three-quarters each of two apples were added.[17] Such a test should make the occurrence of *an apple* in (1) be semantically count. Ware is very wary of this test however. He thinks that (almost?) every occurrence can be taken to be mass or to be count, because there are (normal!) circumstances which classify it each way. And indeed, he claims, for a vast number of circumstances the speakers simply do not have any intentions one way or the other. The main point of his paper is to show that no "pragmatic approach" – one based on speaker's intentions, beliefs, and desires – will suffice to characterize the distinction. Nonetheless, he believes that there is really a distinction there.

Let us return now to an evaluation of which of the various "approaches" outlined above seems best to account for the mass/count phenomena. At the outset we should say that we think that a completely-worked-out grammar and semantics (perhaps even a computer program) which used the mass/count dichotomy in any one of the above-mentioned ways would convince us that that was the preferred way to view the dichotomy, so long as it was "sufficiently elegant" and provided "insight" into the observed syntactic, semntic, and (perhaps) pragmatic phenomena. However, we are not in that position; rather, we have some grammars and semantics which are incompletely worked out (e.g., Montague [1973], Bennett [1977], ter Meulen [1980], Chierchia [1982a]) and some proposals which are little more than hints toward a semantic theory (e.g., Quine [1960], Cartwright [1965], [1970], Parsons [1970], Burge [1972], Moravcsik [1973], Pelletier [1974], Laycock [1975], Link [1981]). The only implemented grammar we know of is Bunt [1981], and we do not feel that it satisfies the "elegance" and "insight" criteria adequately. Thus we hope to be excused for an *a priori* statement of what appears to be the most intuitively satisfying account of the dichotomy.

Short of writing a complete grammar and semantics for all the mass/count phenomena, what evidence can be brought to bear on which of the above "approaches" we should take? It seems to us that the following evidence is relevant. Later, in discussing various of the specific proposals, we shall marshall further evidence in the form of various classes of sentences; here we merely list and discuss overall features of the language.

(a) The various criteria proffered for distinguishing count and mass expressions or occurrences (syntactic criteria, semantic criteria, pragmatic criteria) differ in how they classify.
(b) Within the same language, one and the same object can be referred to by means of an (apparent) count expressions or an (apparent) mass expression.
(c) (Almost?) every expression which can be used in one of the mass or count ways can be used in the other way.
(d) For a great number of occurrences, one cannot tell whether it is being used in a count manner or in a mass manner.
(e) Languages differ in what entities are referred to by their (alleged) mass expressions.
(f) The role of +count and +mass as syntactic features or classifications is either non-existence or at least very slight.

With respect to (a), that our different criteria classify differently, we might note that syntactic criteria (occurrence with numerals, etc.) make *thing* and *object* be count whereas the semantic criteria (cumulativity and divisiveness) make them be mass. And this is only the tip of the iceberg, for different versions of the semantic criteria will make one and the same expression or occurrence be mass or count. Under some versions of divisiveness, for example, *unicorn* in mass (since every "part" of a unicorn is a unicorn – there are no parts) whereas under other versions it is count. Similarly, some tables are made of smaller tables; a tree branch can consist of smaller tree branches; some animals live symbiotically; and lumps of coal can be counted but also divided into smaller lumps of coal. Although he was talking about the notion of a sortal expression, Feldman [1973] is an excellent source of counterexamples to the claim that the different criteria amount to the same classification. This seems to indicate that we should choose *one* of the criteria as our touchstone and not try to use a variety in concert. Thus Bunt's ([1981], Chapt. 2) method of setting up syntactic criteria to separate some mass from some count occurrences and then using a semantic paraphrase relation to find others, would be ruled out.[18] Such a method would be bound to cross-classify expressions.

A part of the problem is that the paraphrase test cannot work if point (b) is correct in holding that the same entity can be referred to in a count or in a mass manner. For, if (b) were true then a paraphrase could go in either direction. But point (b) is true. Consider whether there is really any paraphrase difference between *much more data* and *many more data*, between *much more justification* and *many more justifications*, between *much more groceries* and *many more groceries*, between *too much blues* and *too many blues*, between *lots of kindness* and *many kindnesses*, or between *a lot of difference* and *many differences*. Many times we have different words which refer to the same thing where one is mass and the other count. It seems to us that there is simply no semantic or pragmatic fact which could be used to account for the belief that, in the following list, the former are mass and the latter count. There just is no difference [and no differences] between *change* and *coins*, between *clothing* and *clothes*, between *shit* and *turds*, between *footware* and *shoes*, between *furniture* and *furnishings*, between *fuzz* and *cops*, or between *gin and orange juice* and *orange blossoms* which could be used.[19] And as further evidence that no account which appeals to "what the expression is about" can be correct, we might point to the fact that even when two expressions refer to the same material in what (naively) seems the same way, one might be count and the other mass. *Knowledge* is mass, *belief* count; *success* is mass, *failure* count; *fruit* is mass, *vegetable* count; *garlic* is mass, *onion* count; *baklava* is mass, *brownie* count; *spaghetti* is mass, *noodle* count; *flu* is mass, *cold* count; *rice* is mass, *bean* count.

But the cross-pervasiveness of mass and of count goes even further. In Pelletier [1975] a "universal grinder" was proposed: into one end is inserted an object of which some count expression is true, and from the other end spews forth the finely-ground matter of which it is composed. So a hat is entered into the grinder and after a few minutes there is hat all over the floor. This is so in spite of the fact that we could have also said that there is felt all over the floor, using an "ordinary" mass expression. Examples like this show that many count expressions can be seen to already have within them a mass sense or a mass use.[20] Universal objectifiers also come to mind, converting sm finely silted mud into a fine mud (cf. Pelletier [1975], p. 6, following Richard Grandy; and Bunt [1981], p. 11). Any stuff for which there are standard portions used for whatever purposes will immediately become countified: *three beers*, *an ice cream*, *an entertainment*, etc. To top this off, there seems to always be a count sense (or use) for every (alleged) mass expression *M* which means *kind of M*.

We now turn our attention to those cases wherein it seems unlikely that either mass or count is even intended by a speaker. Consider the italicized nouns in their uses here[21]

> I want more *eggs*
> Some people like *data* better than *theory*
> While in *prison*, Lee felt at *home* only when he was in (the) *hospital*
> The *water meter* reader hit the *snow* man which held the *tape* recorder
> Lee ran into the *brick* wall while she was sniffing *nose* drops
> Sandy worried about the *justification* for the *difficulty*
> I like *smelt* (*Beethoven*/*candy*/etc.)

We think that one would be hard pressed to find criteria according to which one would want to call any of these italicized expressions either count or mass in these uses. Perhaps the best one can do is look to the communicative intensions of the speaker; perhaps the speaker wants whole, entire eggs, and perhaps that makes *eggs* count. Perhaps the speaker believes that there is some one, particular justification – making *justification* count. Perhaps generally, speakers distinguish individualizing (counting) and amassing (measuring). When they individualize, the expression being used is count (regardless of how it may appear to an outsider) and when they amass, the expression used is mass (regardless of how it might otherwise appear). But even this concession to pragmatics seems insufficient, for it is just not true that we use so-called count expressions when individualizing and so-called mass terms when we are not. As Ware ([1975], pp. 26–27) puts the point

> . . . we can use a count noun for beans without individualizing and a mass noun for toast while individualizing. When we use 'consideration' and 'difference' as count nouns or

as mass nouns there is no reason to believe that we are either counting or measuring. In some contexts it is not appropriate to ask whether we are talking about it as stuff or as things. . . .

When I ask that the justification was for something, I can be totally devoid of intentions that involve individualizing or amassing, and there is no need to answer either in terms of a justification or in terms of justification. . . . If one Beethoven sonata was played and I say that I liked the Beethoven I don't have to be either using a count noun or a mass noun, or individualizing or amassing. That is something that may not be determined by my communication intentions, thus leaving a communicative gap. There can be a communicative gap with respect to mass and count occurrences when I say that I like smelt. I don't have to determine whether it is the stuff or the things.

Along these same lines, both the National Broadcasting Corporation (USA) and the Canadian Broadcasting Corporation news, in reporting [July 11, 1984] on the devastation caused by a tornado in Ohio, used the phrases *count the damage* and *count up the damage* more than once. It seems clear that there are no intentions on the part of the newscasters to be amassing damage and then counting it (them?), nor individuating the damage(s?) and then amassing the sum.

Obviously, if in the same language the same physical reality can be referred to by means of an (alleged) mass and an (alleged) count noun, then one should expect that different languages will use the distinction differently. One expects that some languages will describe some stuff by means of a mass noun whereas another language will do it by means of a count noun.[22] Sharvy has already been seen to state that Chinese has only mass nouns (whereas, ordinarily, we see English as having both); and Whorf claims that Hopi only has count nouns. Even within the same language family, we see that English *dandruff* is normally taken as +mass while French *les pellicules* is +count; English *furniture* is +mass while French *les meubles* is +count; English *dish*(*es*) is +count while French *la vaisselle* is +mass. We doubt that the French do anything different with their dishes or furniture than do the English, nor do we think that they conceive of them differently. Other examples from many different languages and language groups come to mind, but we take the point as having been clearly made by just these few examples.

We wish now to consider whether a grammar for English should have +mass and +count as syntactic categories or syntactic features. This means that we shall look to whether there is any syntactic construction which makes use of such information in combining simpler expressions together to form larger ones. Corresponding to our "expression approaches" is the view that lexical items are in some way marked

+mass or +count, and this information is used to rule in or rule out (as syntactically well formed) some longer expressions.[23] Corresponding to our "(syntactic) occurrence approach" is the view that longer stretches of text, for instance complete noun phrases, can be seen as +count or +mass (but this information is not due to the embedded noun's being count or mass). Here, certain properties of the embedded parts (e.g., a determiner *each*) might force the entire noun phrase to be marked +count, but have no effect on the head noun itself. But, of course, to be a *syntactic* occurrence approach, this information +count must be used in describing syntactic well-formedness of some still larger part – e.g., a sentence.

For these two types of approaches, different evidence is relevant. The expression approach will try to use such evidence as: *man* is +count and therefore *a man*, *each man*, *few men* [+plural], etc., are well formed; furthermore **sm man*, **much man*, **an amount of man* are not well formed. The occurrence approach will instead wish to focus on evidence relating an entire noun phrase to the rest of the sentence. They will say that, since *a man* and *the man in the room* are +count, they can be pluralized and this allows for plural agreement with the verb phrase. Since *much water* and *little wine*, etc., are +mass they cannot be pluralized.

Part of the reason for constructing universal grinders and objectifiers, for considering "kind of" meanings for mass terms, and generally for inspecting a wide range of sentences of the sort that we have considered, is to convince one that there is no point to an expression approach. *Every* noun – even *hole* and *pore* – sometimes occurs in noun phrases which we would intuitively call +mass. And every noun sometimes occurs in noun phrases we would intuitively call +count.[24] It simply seems that there is no construction to be ruled out by treating mass/count as features of lexical entries. Doesn't this provide a strong reason for discounting the (syntactic) expression approach, of any variety? Well, perhaps one could argue as follows in its favour. The normal usage of *book*, *bookshelf*, *hold*, *building*, *woman*, *man*, *lawyer*, *sky*, *dilemma*, and the like (cf. the last note), have been "stretched". This kind of stretched or extended use ought to be distinguished from ordinary use. If we do not make a distinction generally, we will in other cases be forced to deny, say, a noun/verb distinction due to such examples as *The weasel treed the squirrel for the kill*, *Leslie tricycled down the driveway*, or *Kim doorknobbed Sandy* (Kim is a practical joker who rigs doorknobs to deliver electrical shocks).[25] Rather than concluding that *tricycle* and *doorknob* double as verbs, or even that there *is* no noun/verb distinction but rather only a "predicate" category, we perhaps ought to say that strictly, or conventionally, *tricycle* and *doorknob* are nouns, not verbs, and that there are rules for "stretching the lexicon" – lexical extension rules – which enable us to use nouns as verbs unconventionally. In such a view of the lexicon, the lexical extension rules would tell us something about how to use (and how not to use) the semantic translation (and meaning postulates, and contingent properties) associated with the lexical entry. For example, knowing that *Sandy doorknobbed Kim* and *Leslie tricycled* involve a verb obtained by extension from a noun, tells us that we cannot just translate these as **Doorknob**′(s, k) or as **Tricycle**′(l), but rather that the "derived verb" translates as something like "x does something to y involving a doorknob", etc. Note that this requires that we specify the conventional classification of the term in the lexicon. So, if one wishes to hold on to the syntactic expression approach, the onus is put on the theory to come up with the lexical extension rules which convert mass nouns (in the lexicon) to "derived" count nouns, and rules which convert lexical count nouns to "derived" mass nouns. We think that there is some promise in such a view and will investigate it further at the end of this paper. For now we shall continue with the other approaches to mass/count nouns.

Regardless of what one thinks about the dispute over the syntactic expression approach states in the last paragraph, the arguments against it do not tell against the (syntactic) occurrence approach. In the examples of the last footnote, such an approach says, we wish to call *more book, bookshelf, hole, building* – taken as entire noun phrases, not just the lexical entries – +mass; and we wish to call *a fine wine* and *a joy to drink* +count. This judgment is not due to the lexical nouns' being so marked, but rather is a function of such subparts as *more*, *a*, etc.

There is something right and something wrong with this suggestion. What is right is that the interpretation of the entire constructions *more book* and *a book*, or *white wine* and *a fine wine* will be different in "massiness". There really is a difference in interpretation – which difference is perhaps called a mass/count difference. What we deny is that it ever plays a role in well-formedness. In the occurrence approach, what restriction does the alleged +mass or +count feature of the noun phrase supply in determining whether any longer stretch of language is well-formed? Two possibilities have been suggested. One is pluralization and the related claim of verb agreement. The other is as direct object of a verb phrase. But these are spurious. A grammar already has a [NUMBER] feature, necessary for distinguishing singular from plural within so-called +count noun phrases. If the so-called +mass noun phrases are marked as singular, the pluralization and verb agreement automatically works without recourse to any extra +mass feature. Above we discussed Hoepelman and ter Meulen, about the interaction of certain verbs and their direct objects. It will be recalled that their claim was that any verb can take a direct object of either +mass or +count, but that the interpretations will be different (yielding a process or an achievement as an interpretation). Thus the feature is *not* necessary as a syntactic marker – it would rule out nothing. Rather it is a *semantic* feature instructing us on how to interpret the result of an interpretation of certain constructions.

We conclude, then, that there is no rule of syntax which can be seen as necessitating features or classifications like +count or +mass. The best that can be said is that one way of looking at such matters as language change by lexical extension might be easier to express with a mass/count expression approach. And the view which is suggested by this would have us assign "usual" mass/count to the lexical items and attempt to state a variety of lexical extension rules. Any of the syntactic approaches – the various expression approaches and the syntactic occurrence approach – have this difficulty, though: since no syntactic construction is ruled out by their presence or absence, it seems incorrect to call them syntactic. (However, the "extension" approach suggested above can claim to rule out mass uses of count nouns, such as *There was hat all over the floor*, and count uses of mass terms such as *A fluffy snow fell*, re-admitting them only at a "semi-grammatical" level though the lexical extension rules).

In this section we have canvassed a number of ways that the mass/count distinction might be drawn, and (some of) the consequences each way has in trying to construct an overall grammar-semantics for a natural language. There are a number of ways we might tabulate the various moves that can be made, some more illuminating than others. Here is one way of tabulating our discussion.

The mass/count distinction might be drawn as a distinction of

1. syntax
   a. All nouns[26] are count or all nouns are mass [unitarian expression approaches]
      (1) All nouns are +count (first order logic, Hopi)
      (2) All nouns are +mass (Sharvy, Chinese)
   b. Some nouns are +count and others +mass
      (1) The "dual life" nouns are always +count (Bunt, Chap. 18) [unitarian expression approach]

[2] The "dual life" ones are always +mass [unitarian expression approach]
[3] Some "dual life" ones are +count, others are +mass, but each has exactly one representation [unitarian expression approach]
[4] The "dual life" ones are to be represented twice in the lexicon (Quirk, Quine) [dual expression approach]

c. No (lexical) nouns are +count or +mass, but only longer noun phrases (Bunt, Chap. 2) [(syntactic) occurrence approach][27]

2. Semantics
   a. +mass and +count distinguish between different senses of a lexical entry (Pelletier [1975], McCawley) [senses approach]
   b. +mass and +count are not semantic properties of lexical items, but of longer phrases [(semantic) occurrence approach] (Ware, Pelletier [1974])
3. Pragmatics – the mass/count distinction is neither a matter of syntactic well-formedness nor a description of the semantic representations (or what they are true of), but rather to be accounted for by the intentions, etc., of the speaker. (What Ware argues against).

Conclusion to this section: If there were any syntactic rules which used +mass or +count, then this would call either for some expression approach or for a syntactic occurrence approach: and what is (and isn't) mass or count would be dictated by these rules. In this case it seems unlikely that any "across-the-board unitarian" approach will work, since there are just too many different ways expressions will be converted in syntactically different circumstances; there seems to be no general procedure underlying the conversions. Rather, what's called for is some "unitarian" approach which appeals to "normal" usage, and has lexical extension rules of various sorts. It still might seem unlikely that such an approach will work but it might be given a chance, as we shall do later in this paper. If no unitarian approach works, perhaps we should favour a "dual expression" approach? No, this would be a pointless proliferation of lexical entries; and if we can avoid it, so much the better. There seems to be little motivation for any of the rules one can think of that might use +mass or +count, except possibly for the formation of noun phrases (or maybe other nouns) from nouns. but even these putative rules do not seem to do any work; nor do the rules which would motivate a syntactic occurrence approach. So we would be left with either a "senses approach", a "semantic occurrences approach", or a "pragmatic approach." There seems to be no point to having different senses of lexical entries – it leads only to proliferation of items in the lexicon. And in any case what we are interested in is the *overall* interpretation of the sentences, not in the interpretation of individual words of the sentence (except insofar as they aid in the overall interpretation). But there just is no semantic rule which requires information about the mass or count status of a constituent to compute that information about the entire noun phrase of which it is a part. We take Ware's critique of "pragmatic approaches" to be decisive – most times there is nothing that the speaker has in mind which would allow us to classify the use of a noun phrase as either +count or +mass. Hence we are left with an occurrence approach which takes entire noun phrases as being the bearers of the properties mass and count, and treats them as semantic properties of that noun phrase without comment on whether they have any semantic correlate in the individual lexical items involved. At the end of this paper we shall consider two theories for mass nouns: a syntactic expression approach which appeals to lexical extension rules and a semantic occurrence approach. These two theories seem to us to be the only viable alternatives to the puzzles about mass terms.

A HISTORY OF PROPOSALS FOR MASS TERMS[28]

Different theorists, focussing on different types of examples, have come to different conclusions as to what the "semantics" of mass expressions is and what the "logical form" of sentences containing mass expressions is. In the earliest of these, Quine (1960: 97–98), when considering sentences like

This puddle is water

says

> We can view the mass term in these contexts as general terms, reading 'is water' . . . as 'is a bit of water'. . . . In general a mass term in predicative position may be viewed as a general term which is true of each portion of the stuff in question, excluding only the parts too small to count.

The reason for the restriction on size is the view that for every mass term (e.g., *water*) there are parts of it (in the ordinary sense of 'part') of which the mass term is not true (e.g., the individual molecules of water). However, when focussing on examples like

Water is liquid

Quine says (p. 98)

> . . . the mass term is much on a par with the singular term of 'Mama is big' . . . A mass term used in subject position differs none from such singular terms as 'mama' . . . , unless the scattered stuff that it names be denied the status of a single sprawling object.

This proposal has been criticized in a number of writings. Burge [1972] remarks that the proposal is incomplete in that it does not assign a role to *snow* in

Sandy threw snow on Kim

Burge thinks it natural to assign *snow* here the status of a singular term, but then points out that it would give the wrong truth conditions, under the assumption that Sandy did not throw the totality of snow on Kim. Bunt ([1981], p. 27) recommends instead that this be treated as a position calling for the predicative interpretation of *snow*.

Pelletier [1974] pointed out that various logical properties of the regimented versions do not correspond to the natural language from which they were derived. For example, letting W be the predicate *is water*, w be the singular term *water*, and p name this puddle, the argument

This puddle is water, water is wet, therefore this puddle is wet

would be regimented as

$W(p)$, WET(w), therefore WET(p)

which is invalid. Furthermore, sentences like 'Water is water' and 'Dirty water is water' cannot be shown to be analytic without further specification of a relation between the predicate and the constant.

The reason for the difficulties is that "one must treat mass terms as being either predicates or individual constants but not both, on pain of failing to account for the logical relations binding different sentential occurrences together" (Burge [1972], p. 267). Of course these difficulties could be gotten around if one added a "meaning postulate" of the following sort for each mass term (where '$<$' is the mereological "part of" relation, M is the mass predicate, and m is the corresponding mass singular term).

For all x, $M(x)$ iff $x < m$

but the "minimal parts" view of mass terms prevents this from being acceptable, for it does not hold from right-to-left. And the other direction is insufficient, by itself, for avoidance of the logical difficulties.

Moravcsik [1973] gives two mereological approaches which attempt to treat all syntactic positions as referring to mereological wholes. The idea behind one of these approaches is (p. 283)

> For any mass term '*F*', to say '*x* is *F*' is to say that '*x*' is a part of that part of *F* that has the structural properties *SP*.

So, *F-sp* is that part of the mereological whole *F* which happens to have the appropriate structural properties (perhaps having to do with size) relevant to being *F*. It too is a mereological entity. Water-*sp* would be the restricted mereological whole made up by all water-parts of at least the size of a molecule. To say that something is water would be to say that it is a part of Water-*sp*. The "water puzzle" is then analyzed as (*water*, in subject position, retains its role as denoting the entire water totality):

$p <$ Water-*sp*

Water $<$ Wet-*sp*

Water-*sp* $<$ Water

therefore, $p <$ Wet-*sp*

Given the transitivity of '$<$', the argument is valid. However the proposal does not account for the analyticity of 'Water is water', which would come out

Water $<$ Water-*sp*

which is false under the proposal. Perhaps the thing to do here is eliminate the proposal that subject terms get the unrestricted Water, but rather assign them Water-*sp*.

But this proposal is not well conceived in any case. Suppose we are given that $a <$ water-*sp*, for example that a is a certain mereological entity like a single water molecule. Since it is such an entity, it has parts – atoms, perhaps. Maybe b is one such atom, so $b < a$. But now given the transitivity of '$<$', it follows that $b <$ Water-*sp*, contrary to the intended interpretation. As Bunt ([1981, p. 30), from whom this criticism is taken, puts it

> Generally speaking, the notion of a mereological object having as parts *only* those entities that have certain properties *SP* is not a well-defined notion.

Moravcsik's second version of mereology consists in leaving the mass expression denotations untouched – they remain the entire mereological entity – but in putting the *SP*-restrictions into the part-whole relation. For every mass term *M*, with structural properties *SP*(*m*), a part-whole relation $< sp(m)$ is introduced. So sentences like "*a* is *M*" get represented as "$a < sp(m)M$". Such a proposal perhaps accounts for the analyticity of 'Water is water'

$$W < sp(w)W$$

and perhaps also for the analyticity of 'Dirty water is water' (although this last is far from certain)

$$D \cdot W < sp(w)W$$

where $D \cdot W$ is the "overlap" of the mereological wholes Dirty and Water. (It is unclear because we do not know whether an overlap will retain the *SP*'s of the wholes.) However, it fails as an attempt to solve the "puddle puzzle"

$$p < sp(w)W$$

$$W < sp(\text{wet})\ \text{WET}$$

$$\text{therefore, } p < sp(\text{wet})\ \text{WET}$$

is not a valid argument because there is no uniform '<' relation to attribute transitivity to.

Bunt [1979], [1981] endorses a fully mereological account called "Ensemble Theory" for the interpretation of mass expressions. There are a number of nice details of the overall theory having to do with its integration of mereological entities and set theoretic constructs, but these are perhaps not germane to the present discussion.[29] Instead we concentrate on the proposals specific to the interpretation of mass expressions.

We have seen that adopting the Quinean "minimal parts view" (which says that for each mass expression *M* there is a lower size limit less than which *M* does not apply, even though there may be *parts* of *M*-things which are smaller) has led the Movavcsik position to impossibility. How then, one might ask, does Bunt's Ensemble Theory avoid these problems? Quite simply, he denies the minimal parts view. Of course, he does not deny that it is a *scientific* truth that, e.g., water has such a minimal size, But, he says, this is not reflected in our linguistic usage.

> ... mass nouns provide the possibility of talking about things as *if they do not consist of discrete parts*. ... Since we are dealing with the construction of a linguistic semantic theory, which should only account for *linguistic* facts, ... we take the viewpoint that a linguistic semantic theory should take into account that the use of a mass noun is a way of talking about things as if they were homogeneous masses, i.e., as having some part-whole structure but without singling out any particular parts and without any commitments concerning the existence of minimal parts. ... We believe that ... mass nouns are semantically different from count nouns. The difference is not in the structure of the entities that mass nouns and count nouns refer to, but in the way in which they refer to them. (pp. 47–48)

Similar views can be found in various authors, e.g. ter Meulen ([1980], pp. 67–68), Cartwright [1970], and Cook [1975]. But we should ask, is this really true? Is the homogeneous reference principle plausible? Bunt mentions apparent counterexamples

> ... like 'furniture', 'footware', and 'computing equipment', [about which] it is unlikely that any speaker would really believe that the mass noun refers to something non-discrete.

So it is not true that speakers have no beliefs in the matter. One *might* say that the discovery of water molecules is a "late development" in language, and so should not be counted as within the domain of semantics. But it seems clear that even in Urlanguage 'furniture' was a mass term of which all users knew there were minimal parts. So how is this to be accommodated here? The answer is that Bunt constructs a "multi level semantics" according to which the distinction between the two types of mass expressions is reflected *after* the "purely" semantic interpretation. In the "purely" semantic interpretation of sentences, there is no difference – other than which constants to use – between sentences containing *water* and those containing *furniture*; they are both interpreted as having no minimal entities. Indeed, the whole issue of what ensemble theory constants have minimal parts and which do not is viewed as not being an issue for "pure" semantics, but rather for some later stage of interpretation. Apparently this is because such a distinction is viewed as generally being a matter which calls for "information about the world", and hence is not "purely" semantic. We have a certain sympathy for this view of "multi stage" interpretation of sentences – although we would not like to call it all "semantics".[30] But we think that this general view of semantics should not interpret mass terms as mereological wholes but rather as predicates (at the level of nouns) and as abstract kinds (at the level of bare noun phrases). We shall later outline such a position.

Other theorists to make use of a mereological theory are Burge [1972], Cartwright (at least once, in [1975b]), and Cocchiarella [1976]. Burge takes the denotation of a mass expression, e.g., *gold*, to be the fusion (in the sense of the calculus of individuals) of the set of elements which are gold. A fusion of a set of elements is itself an individual of the same ontological type as the elements themselves were. This is obviously the same as the mereological wholes used in other throries, and gives us another way of defining them in terms of the entities which satisfy some predicate like 'is gold.' In his [1975], Burge argues against a form of a predicative approach to mass terms, which he calls the relational account. Most of his arguments have to do with the fact that on the relational account of mass terms, the basic individuals are stages of objects. In this account, sentences like

This ring is now gold

are represented as

Gold(r, now)

Below we shall discuss Chierchia [1982a], which in this respect anyway is a relational account of mass terms; Grandy's [1975] theory is also along these lines. Burge's own theory he calls the *C*-account because the relationship between the basic objects and mass terms is one of *Constitution*. The above sentence would be represented as

(Ex)(Gold(x) & $C(x, r$, now)

In this paper the favoured account is given as predicational (mass terms are predicates true of certain bits of stuffs), but we presume that it would be merged with the [1972] account of the denotation of mass substance names as fusions. One immediate difficulty with this *C*-account is in the representation of sentences which are not asserting a relation of constitution, such as

Rice is food.

In Cartwright [1975b] it is suggested that mass terms like *water* by themselves denote the natural kind or substance water. All other occurrences of mass terms in sentences are to be reduced to this kind by means of a translation using an amount function on the quantities of the substance. So predicative mass term occurrences use an amount function which takes sets of quantities into a set of linearly ordered "applied amounts." As in Moravcsik, the basic unit of the measurement of the amount is dependent on the structural properties of the substance in question. This account of the denotation of mass expressions is criticized by ter Meulen ([1980], pp. 52–53) on the grounds that it confuses "which item is constituted by what", and "what item has an amount." But many of the details of Cartwright's account are adopted by ter Meulen (in [1980], Chap. 3).

Cocchiarella ([1976], p. 212) analyzes mass terms in their occurrences that designate natural kinds of stuff as mereological sums, which, as in Burge [1972], are "derived" from predicative mass expressions. These nominalized predicates (except when occurring in identity statements) refer to some object or thing correlated with the concept or properties designated by these same predicates in a predicative position. So Cocchiarella distinguishes individuals that are "concept-correlates" from individuals that are ordinary objects. In his [1978] he gives a formal account of how the logical syntax of such a second-order system goes. Cocchiarella allows nominalized mass terms to denote individuals that can be values of free variables, but they cannot be values of bound variables, and hence cannot be quantities of the same substance they designate. So although the substance *water* is an individual just as quantities of water are, only the quantities can have *is water* predicated of them; the concept-correlate – that which is designated by the nominalized predicate *water* – cannot. So we have here another version of Quine's "dual approach", only now we can explicitly define the singular term occurrences by means of the predicative occurrences. ter Meulen ([1980], p. 55) mentions that the mereological interpretation of the nominalized predicates is not essential to the proposal, and for her own part has it denote something quite different. Both of these authors, Cocchiarella and ter Meulen, believe that far from being analytic, such sentences as

Water is water

are not even well formed, since a concept-correlate (the denotation of a nominalized predicate) cannot be in the extension of itself taken as a predicate.[31] We disagree with this judgment, and our own theory will count it as analytic. (See below.)

Parsons [1970] offers a view according to which every occurrence of a mass noun is taken as a name. But it is different from the mereological approaches in that it does not make use of any of the apparatus from the calculus of individuals. Rather these are taken as names within an (almost) ordinary first order logic. Thus, for example, *g* names "the substance Gold, which is to be taken in the chemist's sense, to stand for the *material.*"

In addition to the ontological realm of substances, there are are two other realms: physical objects and bits of matter. Objects are related to substances by being *constituted* by the substance; so

My ring is gold

becomes

rCg

where C is that relation between objects and substances which is true just in case the matter of the object is a *quantity* of the substance. The relation Q, being a quantity of, holds between bits of matter and the substance that they are a quantity of. By analogy to the well known lambda-abstraction and set-abstraction operators, Parsons introduces a substance abstraction operator μ, which is designed to obtain the substance by abstraction over the quantities of the substance. The formula $\mu x[xQg]$ denotes the substance whose quantities are all quantities of g (Gold); alternatively one could have denoted this substance merely by g. So the above example could have been expressed.

$rC\mu x[xQg]$

In fact, Parsons allows for substance-abstraction over any complex predicate, so that complex mass terms like *dirty water* are represented as $\mu x[xQw \,\&\, Dx]$. So Parsons' language contains a primitive name for the substance, a primitive relation of being a quantity of a substance, a primitive substance abstraction operator which forms individuals from sets of quantities of stuff, and a primitive relation of being constituted holding between physical objects and substances. Some theorists have objected to this ontological wealth (e.g., Burge [1972], Moravcsik [1973], Pelletier [1974], ter Meulen [1980], Bunt [1981]), and Parsons has provided a defense in Parson [1975]. Other objections that have been made to Parsons' theory include the objection that not all relations of physical objects to substances can be analyzed as "being constituted of", e.g.,

Hamburgers are food

This armchair is the furniture in this room

(from ter Meulen [1980], p. 58). Hamburgers are constitued of all kinds of ingredients, but they are not constituted by food. Various relations hold between objects and sets of quantities (or substances) other than just the

C-relation. And various relations can obtain between an object and bits of matter. In a Parsons-style approach we would need a new primitive relation for each one of these relations. ter Meulen ([1980], pp. 58–60) brings up other considerations against Parsons, such as the fact that the μ operator makes substances out of any sets of quantities of stuff, so that Muddy Water would be a substance, as would Leaded Tin That Has Been Annealed, and so on. But, she says, if substances are abstract individuals they cannot be muddy. (Cf. Moravcsik [1973] and Pelletier [1974]). Further, ter Meulen claims that some mass term occurrences are "rigid designators", but Parsons has no method of distinguishing amongst those that are (such as $\mu x[xQg]$) and those that are not (such as $\mu x[xQw \,\&\, Mx]$).

It must be admitted, however, that Parsons' proposal gives a satisfying solution to the "logical form" difficulties we mentioned earlier. The "puddle puzzle" becomes

$$(Ax)[xQw \rightarrow \text{WET}(x)]$$

$$pQw$$

$$\text{therefore, WET}(p)$$

which is valid. And the puzzles about the analyticity of *Water is water* and *Dirty water is water* are eliminated, since they are translated respectively as the analytic

$$(Ax)[xQw \rightarrow xQw]$$

$$(Ax)[(xQw \,\&\, Dx) \rightarrow xQw]$$

(We shall later mention some other sentences and arguments for which Parsons' does not work so well).

One problem with Parsons' approach which deserves to be mentioned here has to do with the interpretation of Q. Parsons claims ([1970], p. 366n) "what 'parts' of x are quantities of x depends on x, and not some abstract notion of 'part'." He says this because he is aware of Quine's objections to analyzing the copula as *is a part of* – that is, Parsons wishes to uphold the view that there are "minimal parts" of any substance. The Q relation, therefore, behaves either like Moravcsik's $<sp$ relation or else the substances themselves have to be interpreted like Moravcsik's *SP*-wholes. In either case, as we have seen, the proposal does not seem logically coherent. (Cf. the remarks in Pelletier [1974], p. 100 and Bunt [1981], p. 36).[32]

In this article Parsons introduced an argument which has been cited as showing that no mereological approach can be correct (e.g., by Pelletier [1974], [1979a]). Suppose all the furniture in the world is made of wood, and all the wood has been made into furniture. Then, in a mereological theory we would have

$$\text{Wood} = \text{Furniture}$$

and, since identity of wholes is defined as containing all the same parts, we would have

$$(Ax)(x < \text{Wood iff } x < \text{Furniture}]$$

But this is false, even in the imagined circumstances. The leg of this chair is wood but is not furniture. The reason for this is that they have different minimal parts.[33] This argument is closely related to the points made against Moravcsik's conception of mereological wholes as being restricted in some way and against the possibility of having differing part-whole relations relevant to different mass expressions. Of course if one denies that there are minimal parts of any mass noun, this argument will not seem to carry much force (e.g., Bunt [1981], pp. 48–50); but, we feel, it *should* be bothersome. Any proposal which flies in the face of the "primary semantic data" (here: claiming that the leg of the chair *is* furniture) like this should be adopted only after deep and careful analyses of the alternatives show them to be even more wanting. Indeed, there is another class of problems with the notion that mass noun phrases like *furniture* or *water* denote mereological wholes. If, for instance, one takes a subject occurrence of *water* as denoting all the water in the world, there will be a "paraphrase" problem, since presumably *all the water in the world* also denotes this entity. But *All the water in the world weighs billions of tons* is intelligible and true, yet **Water weighs billions of tons* seems semantically odd. This example, along with Parsons' furniture-wood case, and sentences like *Element 130 is identical with element 131* and *Atoms of element 130 are identical with atoms of element 131* (said of physically possible but unrealized elements – so *element 130* and *element 131* denote the empty totality, and have the same parts namely the empty part), strongly suggest that the denotations of bare *NP*'s are intensional.[34]

There have been a number of attempts to invoke sets of one variety or another as the interpretation of mass nouns.[35] In Strawson [1959], in Quine's 1964 review of Geach [1962], and in Clarke [1970], there are suggestions that uses of mass terms should be understood as elliptical for some more complex phrase in which there is an explicit "individuating standard" (or count phrase) by means of which we can give sense to there being a certain *number* of things of which the mass term is true. Thus, *is water* might, in certain circumstances, be elliptical for *is a body of water* or, in other circumstances, for *is a kind of water*; *is gold* might be short for *is a vein of gold* or *is a nugget of gold*. Such a position avoids the Parsons anti-mereological argument by pointing out that there are "individuating standards" applicable to wood which are not applicable to furniture; e.g., the set of pieces of wood is distinct from the set of pieces of furniture, and so *wood* and *furniture* do not denote the same sets. Such a view has been elucidated and criticized in Cartwright [1965].[36] A few of the difficulties are that first, some sentences will not receive a representation; second that the representation of some sentences will not have the right truth value; and third that the appeal to varying "standards" prevents us having a *uniform* translation procedure. As for the first, consider

What Kim spilled is the same coffee as Sandy wiped up

Since this is an identity claim, the same "individuating standard" must be relevant to both sides. But what could it be? It cannot be 'puddle of coffee' for that cannot be spilled. It cannot be 'cup of coffee' for that is not the kind of thing to wipe up. As for the second, consider sentences like

The sugar here is the same sugar as that which was on the boat

(said after the sugar which was on the boat was melted before it got here). The extension of 'this sugar here' contains no lumps, grains, etc., so there is no longer a set of these things (which there was on the boat). Thus the purported set equality fails, yet the English sentence may very well be true. For the third objection, we merely note that in a given particular case we may have no way to tell which of the various "individuating standards" that are true is applicable. For instance, sameness of shipment of sugar is often sameness of grains and of lumps of sugar. How do we know which one to use? We finally note that statements of constitution will not receive a representation. For instance, in

This shipment is constituted by the grains of sugar

it does not seem that any "individuating standard" is applicable. Rather it looks as if we need a use of *is sugar* that does not depend on the individuating standards of 'lump', 'shipment', or 'grain'.

It seems therefore that we should not call into play context dependent "individuating standards". If one wishes to call sets into play at all, what is required is the same set for all occurrences of the mass term. One plan, used by Laycock [1972] and Bacon [1973], is to let the mass term denote the set which contains all and only the *smallest* (minimal) elements of which the mass term is true.[37] One should note that this proposal also avoids the Parsons argument: the set of minimal wood pieces is distinct from the set of minimal furniture pieces. Nonetheless, the proposal has not seemed very attractive. It first seems to involve various "empirical" hypotheses into the meaning of mass terms (e.g., that part of the meaning of *water* is that is comes in molecules of H_2O). Secondly, it seems that it is often impossible to actually determine a set of minimal elements, even for "normal" mass terms, such as *garbage*. And thirdly, if we are to include abstract mass nouns like *speed*, *information*, and *time*, we shall have to invent a whole host of new "minimal elements".[38]

Another proposal which assigns a set uniformly to all occurrences of a mass noun can be found in Cartwright [1965], [1970] and Grandy [1973]. In this account, the mass noun *M* denotes the set of *all* quantities of M. It is also similar to the proposals of Montague [1973], Pelletier [1974], Bennett [1977], and ter Meulen [1980] wherein a mass term denotes the function from indices (possible worlds) to such sets. Such an approach obviously does not suffer from the defects of the "varying standards" approach nor, apparently, from the defects of the "minimal elements" approach. But there do seem to be some problems with it (ignoring those problems having to do with its lack of an intensional component). In Pelletier ([1974], pp. 96–98) various details of Cartwright's view are criticized. First, we should note that Cartwright distinguishes an *amount* from a *quantity*. We could have the same amount of water but not have the same quantities of water; non-identical quantities may be the same amount, and to bring this out Cartwright adopts the terminology of saying that a quantity *contains* a certain amount of it (rather than that it *is* that amount). It is also important to mention that Cartwright's notion of the amount contained in a quantity is *not* dependent upon a choice of measure, nor upon the conditions of measurement. Cartwright then attempts to show that there is a "strong analogy" between a *set* of ordinary objects and a *quantity* of some stuff. As she puts the analogy ([1970], p. 29)

> D1: *x* is a set of *A* if and only if, for some *y*, *x* and *y* are comparable with respect to the number of *A* each contains, and *x* contains nothing other than *A* . . .

D2: x is a quantity of B if and only if, for some y, x and y are comparable with respect to the amount of B each contains, and x contains nothing other than B.

The point of this analogy is to show that, just as we understand quantification in the "set" case by understanding that 'x is a person' is true if and only if x denotes a member of the set denoted by 'person', so too we can understand quantification in the "mass" case if we can understand what are the permissible values of x in 'x is *sm* water'. But the analogy goes wrong in its details. In the "set" case, one assigns something *contained in* the set to x. In her ([1965], p. 485) Cartwright claims that the values of x in 'x is *sm* water' are quantities of water. But by the above analogy this would be like claiming that it is *sets* of men which satisfy 'x is a man'. In her ([1970], p. 39) Cartwright claims that it is what is contained in a quantity of water which satisfies 'x is *sm* water'. But by the analogy, what is contained in a quantity of water is an amount of water; and *that* is not what we want to quantify over. (Cartwright had already said this in her [1965]). Further every quantity contains exactly one amount, yet sets contain many elements. So one is not quantifying over the quantities, for that corresponds to the set; and one is not quantifying over the amounts, for that corresponds to the number (measure) of the set. What is it that corresponds to elements of a set?

Laycock [1972], [1975] offers more "philosophical" arguments against the utility and coherence of the notions of quantity and amount as employed by Cartwright. In turn, the notions are defended against Laycock's attack by Cook [1975]. Bunt ([1980], pp. 41–43) offers this attack against Cartwright, which he apparently also thinks applies to all the authors mentioned in this part.[39] Consider, says Bunt, a noun phrase like

The gold on the table

There are two ways of understanding the present proposal concerning amounts and quantities, corresponding to

The thing which is gold and is on the table

$(ix)[x \in G \,\&\, \mathrm{ON}(x, t)]$

or as

The things which are gold and are on the table

$\{x\colon x \in G \,\&\, \mathrm{ON}(x, t)\}$

The first alternative is incorrect since if there is *sm* gold on the table, it is generally wrong that there is exactly one quantity of gold on the table. Furthermore, since one can quantify this noun phrase, it seems incorrect to treat it as if it were a definite description

All the gold on the table

But if the noun phrase is taken in the second way, then there are difficulties with measurement because we shall be "counting the same quantity many times". Consider sentences like

The gold on the table weighs 2 grams

Under this proposal we are to sum the weights of the quantities of gold on the table and have it result in 2 grams. Bunt gives this formula as a possible representation

$$\text{SUM}(\{y\colon (Ez)(z \in \{x\colon x \in G \;\&\; \text{ON}(x, t)\} \;\&$$
$$\&\; y = \text{Weight}(z)\} = 2\text{g}.$$

Certainly this representation is incorrect since the set $\{x\colon x \in G \;\&\; \text{ON}(x, t)\}$ contains many overlapping elements whose weight will be counted many times. Obviously we wish to first sum the elements and then take their (its) weight. These considerations once again appear to lead to mereology as opposed to a set-theoretic approach; for, on the one hand we must be able to refer to the many quantities that go into one physical object, but on the other hand we must be able to refer to them as a group in a way that does not involve the same quantity being in two distinct quantities (or else we shall not be able to discuss such properties as the weight of a quantity).

The final proposal we shall consider is that mass nouns should be analyzed as predicates. Recall that we have distinguished this proposal from the one where mass terms are taken to denote sets; in this latter proposal a mass expression is interpreted extensionally, whereas in the former proposal a mass expression is interpreted as denoting a function from points of reference (perhaps to sets). The difference between the two is not merely a matter of what happens in intensional contexts,[40] but also involves the possibility of stating an intuitively plausible relationship between occurrences of mass expressions that appear to characterize things ("predicative mass expressions", as in *This ring is gold*) and those occurrences which seem to name something ("nominal mass terms", as in *Water is wet*). As representatives of this group we might consider

Cocchiarella [1976], Montague [1973], Pelletier [1974], Bennett [1977], ter Meulen [1980], Carlson [1977], and Chierchia [1982a].

We have already had occasion to mention Cocchiarella and his view that the nominal mass expressions name a mereological entity. One can explain his overall theory as follows:[41] A predicative mass expression denotes a function on possible worlds which picks out a set in each possible world. Nominal mass expressions are systematically related to these predicative mass expressions in that they denote a function on possible worlds which in each possible world picks out the mereological fusion of the set picked out in that world by the predicative mass expression. (Such a view is obviously closely related to that of Burge [1972], [1975] – with the exception of the intensional aspect of it). Since we have already argued against the mereological-sum (-fusion, -whole) interpretation of nominal mass expressions, we shall not discuss Cocchiarella further.

Montague [1973] presented an outline of a theory of mass nouns in the general framework of "Montgue Grammar" which, however, (as he puts it, p. 173) "is not within the elaborately inflated framework of [Montague 1973a]". In this account "a mass term will denote that function on possible worlds which takes as its value for a given world the set of all samples (or, to give synonyms, portions or quantities (in the sense ascribed to Parsons) or "parts with the correct structural properties") of the substance in question in that world." So goes Montague's account for occurrences "standing alone in normal substantive positions", that is, in such positions as *portion of* α and α *is a liquid.* To account for other contexts, Montague remarks that standing alone in predicative position it should be synonymous with *portion of* α; that with a quantifier, a demonstrative, or an adjective phrase the mass noun should there denote the extension of the property usually denoted; and that as an adjective (e.g., *iron bed*) it denotes the extension of the property usually denoted by the mass term. Montague (pp. 174–175) also proposes to treat phrases as *the gold in my ring* as being synonymous with *the gold constituting my ring*. Here *constituting* is taken to denote the set of maximal portions of gold in the ring, and hence the *the* is proper: it denotes a unit set. In this account, *Water is liquid* would be elliptical for *all water is liquid* and hence quite different from *Water is a liquid.* In Montague's account, as opposed to Parsons', quantities or portions of water (say), are taken to be ordinary individuals but the mass noun *water* denotes (at a given possible world) the set of these individuals. In Parsons' account the mass noun also denotes an individual, a *substance*, and there is a primitive (intensional) relation of *being a quantity of*. Montague took a virtue of his analysis to be that he had given an analysis of substances in terms of their (possible) quantities, and thereby had given an analysis of *being a quantity of*.

Bennett [1977] develops an account also within the "Montague Grammar" framework, but his account diverges from Montague's in that, for him, mass terms are not regarded as denoting properties of individuals. In fact he regards mass terms as Parsons does – as denoting individuals, by making *substances* be "ordinary individuals". He then follows Parsons in introducing an intensional relation *being a quantity of*, that tells us the quantities of any substance at a point of reference. The difference here between Bennett and Parsons is that, unlike Parsons, who uses his substance abstraction operator to yield substance names from any open formula. Bennett allows only simple mass nouns to denote substances.

A little terminology here will help, not only to understand Bennett's proposal, but also below in the discussion of ter Meulen. Bennett uses the terminology "count noun (phrase)", CN, as a syntactic classification of such phrases as *man*, *tall man*, *man who is a spy*; and by analogy uses "mass noun (phrase)", MN, as a syntactic classification of such phrases as *water*, *blue water*, *water that is pure*. These are to be distinguished from (the syntactic category) "terms". Terms also come in two types: "count terms", CT, are such phrases as *John*, *a man*, *every man*; "mass terms", MT, are such phrases as *water*, *little water*, *all water*. CNs and MNs semantically denote sets of individuals. Montague (whom Bennett follows here) gives all CTs the same sort of denotation, so that *John* and *every man* could be regarded as being in the same semantic category. Bennett would to the same with all the MTs, although in his [1977] he wants to "forget Montague's complications." MNs are the sort of expression that can be modified by adjectives and relative clauses, while MTs are the sort of expression that, when combined with a verb phrase can form a sentence. In Bennett's account, the basic MTs (such as *water* and *gold*) denote substances. From these basic MTs, one can form a sentence when they are combined with a verb phrase of the right sort. Thus, for example, we can form

Water is a liquid.

But there is also a syntactic rule which converts the MT *water* into a count noun phrase (CN) by combining it with *quantity of*. After having formed the count expression *quantity of water*, one can perform ordinary quantification over it to form a term. For example, *All water flows* get translated as

$$(Ax)[(\mathbf{quantity\text{-}of}'\ (\lambda \mathbf{P}\ \mathbf{P}(\mathbf{water}')))(x) \rightarrow \mathbf{flow}'(x)]$$

This view is obviously very close to Parsons', the only difference having to do with the absence of the substance abstraction operator (and the consequent inability to have complex substance names.) *Blue water is a liquid* would be ill-formed, since *blue*, when it combines with *water*, forms a MN rather than at MT and so cannot be the subject of this sentence; *is a liquid* is taken here as a property of substances, *is liquid* is a property of quantities, so *Blue water is liquid* is well-formed in this grammar. What, then, about such sentences as *Water is liquid*? In brief, Bennett's answer is that these are elliptical for *All water is liquid*, where as usual *all water* is analyzed as above in terms of quantities.

In spite of this slight difference between Bennett and Parsons, it seems that there are enough similarities that various of the objections to Parsons' account will also apply to Bennett's, in particular those objections having to do with the understanding of basic mass terms as denoting abstract substances. ter meulen ([1980], pp. 61–63) mentions three further objections to Bennett's account. We ourselves are less than taken by these arguments, but mention them here to lead us into ter Meulen's account.[42] First, the relation of being a quantity of does not always correspond to an element in a syntactic derivation of the sentence. thus, *This ring is gold* does not contain *quantity of*, although in Bennett's translation the relation does occur. This, says ter Meulen, goes against a fundamental tenet of Montague Grammar. She herself claims that the sentence should be translated in such a way that it is logically equivalent (given a meaning postulate which says that every entity in the denotation of a predicative mass term necessarily has the property of being a quantity of some substance) to *This ring is a quantity of gold*, but not identical. According to ter Meulen, a consequence of taking the substance to be a first order individual is that there is no way to distinguish between substances and their quantities – they all can have the same properties. Especially, she says, the substance *m* and the quantities of *m* can all have the property of being *m* – a feature which ter Meulen finds "contradictory and hence undesirable." Finally, ter Meulen criticizes the "nominalistic outlook" of Bennett's theory (presumably the fact that a substance is a first order object) as being "absolutely alien to the ontological wealth of Montague Grammar."

ter Meulen's theory ([1980], see also her [1981]), is actually quite similar to Bennett's, the major differences being the ones just noted. A crucial feature of her account is the existence of two distinct syntactic categories of basic terms: nominal mass nouns and predicative mass nouns. These basic expressions are such as *gold*, *water*, *gas*, *furniture*, and *food*. All these basic lexical entries are given twice: once as nominal (NMNs) and once as predicative (PMNs). The idea is that any of these simple mass nouns might be used to indicate a substance (by using it as a nominal mass noun and converting it to a nominal mass term – recall Bennett's use of 'term'). In this use it can be the subject of a sentence where one predicates a rather special property of it – the type of property which holds of substances, such as *is a liquid*, *is an element*, *has atomic number n*. On the other hand these simple mass nouns might be used to say something about the quantities of the substance, as in *Water is wet*. In this case the PMN *water* is converted to a predicative mass term, which is translated as $\lambda P[\mathbf{water}'(^\wedge P)]$ which is supposed to reduce, by a meaning postulate to $\lambda P(Ax)[\mathbf{water}'(x) \rightarrow P(x)]$.[43] So, in ter Meulen's account, nominal mass terms and predicative mass terms form distinct syntactic categories and cannot be viewed as deriving one from the other. This is, she says (e.g., [1980], pp. 6, 193; [1981], p. 105), the most fundamental innovation in her system. A number of important observations have been made by her in support of this distinction. Four of these observations are: (a) terms constructed from nominal mass nouns exhibit backwards pronominalization behaviour similar to that of proper names, (b) nominal mass terms bind only pronouns that are interpreted as denoting the substance, whereas predicative mass terms bind only pronouns that are interpreted as denoting a set of quantities, (c) nominal mass terms semantically behave like rigid designators, unlike predicative mass terms, and (d) nominal mass terms take wide scope over other terms and intensional operators whereas predicative mass terms take narrow scope over quantifiers and intensional operators. As for the first: in the sentence

The man she loves betrays Sarah

it is possible to interpret *she* as anaphoric to *Sarah*. However, when a general term, e.g., *some women*, is substituted for *Sarah*, the resulting sentence is impossible to read with an anaphoric *they*:

The men they love betray some women

This behaviour is mirrored by mass terms:

Its chemical formula defines water

allows for an anaphoric relation between *its* and *water*, but

The person who finds it sells some gold

does not allow for *it* to be anaphoric with *some gold.* As for (b): consider sentences like *Water is* H_2O and *Water is muddy*. Intuitively in the former reference is made to a kind, whereas in the latter reference is made to quantities of water. To prove that such an interpretation is correct, ter Meulen considers sentences lie

Water is H_2O and it is muddy

Water is muddy and it is H_2O

Both of these sentences are deviant, she says, because the anaphoric bindings in the sentences are deviant. In the first, substances (being abstract entities) cannot be muddy; and in the second, quantities cannot have properties of substances, such as being defined by their atomic structure. As for (c), ter Meulen claims that once the reference of a nominal mass term is fixed in an interpretation, it remains the same at all reference points. "What water is in the sense of what the substance water is, is once and for all established in an interpretation of the language. But what is water, in the sense of what quantities of water there are, is a contingent matter. . . . Substances determine what their quantities are, but the quantities do not make up the substance." ([1981], p. 108). And as for (d), most sentences with double quantifiers are considered ambiguous. E.g., *Every man loves some woman* is usually thought to be ambiguous between every man having some women or other whom he loves, and there being some specific woman (Aphrodite, perhaps) whom every man loves. This ambiguity is described by saying that in the first interpretation, *every man* has wide scope over *some woman,* whereas in the second interpretation *some woman* has wide scope over *every man.* If, in these sentences, *some woman* were replaced by a proper name, e.g., *Aphrodite*, then only the second sort of interpretation is possible – *Aphrodite* is given wide scope over *every man.* ter Meulen thinks that predicative mass terms "preferably have narrow scope" and that nominal mass terms "usually take wide scope", over other terms. E.g.,

Every child drinks some milk

Mary found little gold but Jane found it in abundance

illustrate the view. It is very difficult to give *some milk* a wide scope over *every child*, even in the passive version (*Some milk is drunk by every child*). In the second sentence, it is very difficult to give *little gold* a wide scope over *Mary* and *Jane* (for then they would have found the same gold). Yet the nominal mass terms, as with proper names, the wide scope is usual.

Every child drinks milk

Mary found gold and Jane found it too

These sentences show *milk* and *gold* having wide scope over *every child* and *Mary*, *Jane* respectively. Similar remarks could be made for the relative scope of terms and intensional operators. In

John believes that gold has atomic number 79

we have wide scope for the nominal mass term *gold* over *believes*, as evidenced by its equivalence to *Gold is such that John believes that it has atomic number 79*. But with predicative mass terms, such as *some gold* in

John believes that some gold is in the box

we do not have the paraphrase *Some gold is such that John believes that it is in the box*.

How convincing are these data? We think that they do not demonstrate the thesis ter Meulen wishes them to. There are two ways one might attack her data: first, show that it simply is incorrect or that it is correct but for some reason other than what she claims; or second, one might concede the data but give some other account of them than that nominal mass terms are names of some entity (a substance) which cannot be defined in terms of anything else (such as the quantities). We will not attempt the former way, although it does seem to us that the data are not so rigid as ter Meulen makes out.[44] Instead we shall take issue with the interpretation ter Meulen places on this data. Her picture of the situation is this: a predicate like *water* denotes a function on possible worlds which has as its value in a possible world the set of quantities of water in that world. So far we have two types of objects: a semantic one (the function) and a metaphysically real one (the set of quantities). In addition to this, ter Meulen thinks, there is another metaphysically real object, the substance water. This object bears a certain relation to the other real objects, namely the relation of being exemplified by (the converse of being a quantity of). Since the substance is a metaphysically real object, we must have some way to name it – hence the nominal mass terms. The nominal mass term has as its value the object water, which is unchanged when considered from any possible world. Why not, one might ask, just identify this substance with the intension of the predicative mass term? After all that object, the function on possible worlds to sets, does not change from possible world to possible world. Isn't the introduction of a substance just a needless proliferation of entities? Doesn't this function do all the necessary work of picking out what is and what isn't water in each possible world? Aren't we just duplicating this relationship when we add an "exemplification" relation? ter Meulen's answer would presumably be that the function is a semantic object, and not a metaphysically real one. Any attempt to pretend that they were the same would be a cross-matching of types from different categories. Semantic objects are conveniences in describing the world; metaphysical objects really exist. Underneath it all, it seems to come to ter Meulen's realist feeling that substances exist and are not in any way a construct from more basic entities (such as the quantities). We think (and perhaps ter Meulen does too) that all her data could be accounted for by identifying her substances with the intensions of predicative mass terms. But she would nonetheless object that we have ignored a certain piece of reality: the substance. It would be, to her, rather like denying that people were primitive existents and instead "constructing" or "defining" them in terms of person-stages. Maybe it can be done formally and account for all the syntactic and semantic data, but it would just be incorrect as an account of reality.[45] The one argument we find in ter Meulen against this identification is the following. *Any* predicative mass term, even a complex one, has an intension. But only *simple* mass terms can function as a nominal mass term, so in any case we have to distinguish between intensions of predicative mass terms and nominal mass terms. Her evidence here comes from counterfactuals. For example, the sentence *This muddy water might not be muddy* is true; but *This water might not be water* is false.[46] According to her then, complex mass terms cannot be rigid designators, but simple mass terms can be; and so, she concludes, it is not possible to define substances as the intensions of predicative mass terms. In fact though, this does not follow: the simple syntactic maneouver of having a category of simple mass nouns to which various syntactic rules apply will do the trick. One rule is to directly convert it into a nominal mass term (and the corresponding semantic rule then would have it denote the intension). Another rule allows it to form predicative mass nouns either by modification (for *muddy water*) or simply by an identity mapping (for use in sentences like *water is wet*). In this kind of case the predicative mass term will designate the family of sets containing all quantities of [say] water.[47] Once a simple mass noun has been converted to a predicative mass noun, it can no longer be converted to a nominal mass term.

In short, while we find ter Meulen's data interesting and parts of her solution intriguing, we do not find the fundamental starting point – a syntactic distinction between nominal and predicative mass nouns – to be well motivated. We think we can account for the data (what hard data there is, anyway) differently, and will try to do so later.

Carlson [1977] is concerned to present a unified account of mass expressions and "bare plurals". It had been noted before that bare plurals (e.g., *dogs* in *Dogs are barking outside* and *Dogs are mammals*) resemble mass nouns in a variety of ways, both syntactic and semantic. Carlson's central idea is that both the formation of bare plurals and of nominal mass terms are best regarded as the transformation of an expression that semantically designates a property into something like a proper name. So all unquantified noun-phrases are to be analyzed as names of first order individuals. Carlson gives a wide variety of data, some of which overlaps that mentioned above as coming from ter Meulen, for this conclusion. He argues that we can make sense of all these data by assuming that bare plurals act as names of kinds of things, and that the different readings (sometimes entailing "all" as in *dogs are mammals*, sometimes entailing "some" as in *dogs are barking outside*, sometimes having neither sort of entailment as in *dogs are numerous*) that arise in different contexts depend upon the properties of classes of predicates used in the rest of the sentence and upon the aspect of the verb phrase. One class of verb phrases is said to denote properties of *stages* of (ordinary) individuals, e.g., *be drunk*, *be on the roof*, *be eating a cake*. Other verb phrases, such as *be intelligent*, *be a person*, *know Greek*, denote properties of (whole, ordinary) objects. The "stage level" verb phrases select "existential" readings of bare plurals, while object level verb phrases select the "universal" reading. Carlson also introduces two relations, R and R′, which connect individuals to their stages and kinds to their instances. These are introduced by meaning postulates. The result is that sentences like *This ring is gold*, *Dogs are intelligent*, and *Dogs are barking* will be translated as

$$(Ex))[R(x, g) \ \& \ \mathbf{ring}'(x)]$$

$$(Ax)[\mathbf{Dog}'(x) \rightarrow \mathbf{intelligent}'(x)]$$

$$(Ex)[R(x, d) \ \& \ \mathbf{barking}'(x)]$$

(The second sentence is the result of applying a meaning postulate to the initial translation according to which the property of being intelligent was attributed to the kind, Dog.) Chierchia [1982a] and ter Meulen ([1980], pp. 63–64, 99–100) criticize this account. ter Meulen's criticism is mainly that "properties of kinds and of their members cannot be distinguished, so kinds can realize themselves," It is not clear that this in an undesireable consequence; Carlson himself thinks it is true and opposes ter Meulen's evidence about "type restricted binding" with such sentences as

Gold is an expensive metal, but John buys it regularly

ter Meulen's response to this is merely the question-begging "apparent counterexamples to the type restricted binding of mass terms can usually be shown to hinge on a failure to recognize that substance-properties can only be inherited by quantities of the substance" ([1980], p. 100). Apparently what she means is that the present sentence is to be interpreted as meaning "All quantities of gold are quantities of a metal and are expensive, and John buys them regularly".[48]

Chierchia [1982a] gives a more thorough development of Carlson's ideas, especially as to how they account for mass nouns. The main points of divergence are: a kind is identified as the intension of a common noun (contra Carlson and ter Meulen), and the notion of a "stage" is replaced by "quantities of matter" (the latter interpreted broadly so as to include "stages" of ordinary objects – for details see Chierchia [1982a]). Common nouns, and in general characterizing properties, are treated as sets of individual concepts (in the Montague sense). Episodic properties denoted by verb phrases like *run* and *hit* are analyzed as sets of portions of stuff. So it is suggested that the difference between states (expressed by certain stative verb phrases) on the one hand, and processes and events (expressed by non-stative verb phrases) on the other, is merely a difference concerning the kinds of entities involved. A kind is the intension of a property, and for an ordinary thing to instantiate a kind is simply for it to have the corresponding property. Moreover, ordinary objects are individual concepts that have stages as their values; so for a stage to realize an object will simply be for it to be the object's value at a given possible world. And the definition of what it is for a stage to realize a kind is that the stage is the (mereological) fusion of stages of one or more objects belonging to that kind. Chierchia [1982a], [1982b] gives an explicit model theoretic account of all this employing Cocchiarella's [1979] theory which has no semantic types. The meaning postulates that are relevant to the interpretation have the following consequences:

1. if Σ is a count noun or a stative verb phrase, and Σ is true of x, then x is an object.

2. if Σ is a non-stative verb phrase and Σ is true of x, then x is a quantity of matter ("stage").

3. if Σ is a stative transitive verb (e.g., *love*) and Σ is true of the pair $\langle x, y \rangle$, then x is not a quantity of matter.

4. Σ is a non-stative verb phrase (e.g., *hit*) true of the ordered pair $\langle x, y \rangle$ and x is not a quantity of matter if and only if there is some realization of x which stands in the Σ relation to y.

Much of the work is to be done by the aspectual system. We will be given (at least) three classes of verb phrases: stative verb phrases that select for kinds, ordinary stative verb phrases, and non-stative verb phrases. The translations of an example of each are (after the application of meaning postulates):

1a. Gold is an element.
b. **element′**(^**gold′**)

2a. Gold is yellow.
b. $(Ax)[R(x,$ ^**gold′**$) \rightarrow$ **yellow′**$(x)]$

3a. Water is flooding the city.
b. $(Ey)[R(y,$ ^**water′**$)$ & **flood-the-city′**$(y)]$

The idea here is that in the syntactic analysis of the sentence we will notice whether a verb phrase is stative or non-stative. In the latter case the translation will proceed as in 3. In the former case, the translations will proceed as in 1. After this initial translation is complete, the application of certain meaning postulates is considered. Those verb phrases which do directly attribute qualities to the "kind" – as for example, *is an element*, *are numerous*, *is getting expensive*, and the like – do not have a meaning postulate which allows the property to be "inherited downward" to all the quantities ("stages") of the kind. Thus, 1 remains as initially translated;

but 2 allows the application of such a meaning postulate, and the property of being yellow is inherited by all the realizations of gold.

Chierchia concludes that there is a "true generalization" here which his theory is able to capture. The name-like properties of bare plurals and of mass nouns follow simply from their being nominalizations of common nouns. Exactly the same explanations given for the behaviour of bare plurals goes through for mass terms without additional stipulations.

The range of evidence given especially in Chierchia [1982b] is quite impressive. Nonetheless, we think that there are a variety of problems not handled in this approach. Below we shall list some of these data and indicate our own solutions to them.

AN UNDERLYING MISTAKE

It seems to us that an underlying flaw in all the proposals thus far canvassed – both those accounts of what a mass expression is and those accounts of what the "logical form" of sentences containing mass expressions is – stems from a failure to adequately distinguish among the semantics associated with different levels of syntactic description.[49] For example, consider a sentence like

> The dirty water used in washing yesterday's dishes is still in the sink.

When one asks: what is the semantic value of *water*, the question is ambiguous at least between the following. What is the semantic value of *water*, considered merely as an entry in the lexicon? vs. What is the semantic value of *water*, as it occurs in this sentence? And we should like to add further that even in this sentence *water*, *water used in washing yesterday's dishes, dirty water used in washing yesterday's dishes*, and *the dirty water used in washing yesterday's dishes* will all be given distinct semantic values. Furthermore, these values might be so different that (for example) *water* could be a predicate while *the dirty water used in washing yesterday's dishes* could be a name. There are a variety of possible syntactic analyses of this sentence, dependent on various grammatical theories. For definiteness we mention an X-bar theory along the lines of Gazdar and associates (cf. Gazdar *et al.* [1985]). In this theory there are, as it were, different "levels" of being a noun phrase or a verb phrase. We indicate these levels by superscript. (So N^2 is the third level of noun phrase.) An analysis of the sentence in this theory might be [50]

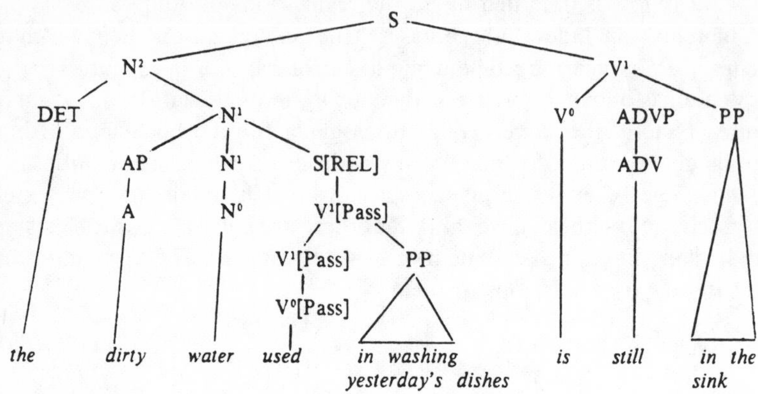

Here we see that there are many different nodes of the syntactic analysis tree, and that each one might be associated with radically different semantic values. Following the methodology of the Gazdar (ultimately Montague) approach, the only restrictions here are (1) for each node other than the terminals, its semantic value is some function of the semantic values of the immediately dominated nodes, and (2) the semantic value of the topmost node is a truth value. Perhaps our point, that the semantic values corresponding to our N^2, N^1, N^0 nodes in this tree are distinct, seems obvious for this example; but now consider such examples as

Gold is expensive

Water is widespread

Again the question "What is the semantic value of *water*?" is ambiguous at least between the questions "What is the semantic value of *water*, considered as an element in the lexicon?" and "What is the semantic value of *water* in this sentence?" And again, there is no reason for them to be the same. There might be various answers to the second question even in this example. Perhaps a sentence is formed by the combination of an N^2 and a V^1; but since the lexical item *water* is simply an N^0, it cannot be combined with the V^1 to form a sentence. Perhaps one expansion of N^2 is as N^1 and one expansion of N^1 is N^0. Then the structure of the sentence

would be

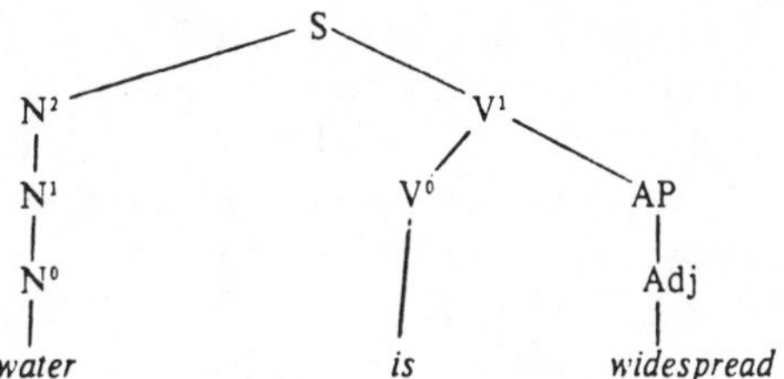

In such a case, the question "What is the semantic value of *water* in this sentence?" is itself ambiguous amongst asking it of *water* considered as N^0 (its value in the lexicon), of *water* considered as an N^1, and of *water* considered as an N^2.

We think that much of the disagreement and confusion in the whole area is due to inattention to this fact. The proper investigation of mass expressions will proceed by first finding a suitable type of value for the lexical entries so that, when combined to form "higher" node-values, these too will have the correct sort of values. A condition of adequacy then is that all values corresponding to a particular syntactic category will have the same semantic type. If, for example, N^2's are sometimes names, then all N^2's are names. If some combination function sometimes combines (say) two predicates to form another predicate, then that function cannot be used to combine (say) a predicate and a name to form a predicate. And so on.

We suggested that there may be several "levels" of being a noun phrase or a verb phrase. The theory that we cited employs two levels of each, in addition to the basic N^0 and V^0 level. (The V^2 level is the entire sentence). Semantically speaking, one needn't distinguish all three "levels", and for the simple grammar we shall present later we will distinguish only two "levels". We shall call these two levels CN and NP, identifying NP with our earlier N^2, and CN ambiguously with N^0 and N^1. The (semantic) justification for this simplification is that we can safely ignore mass nouns with subcategorized complements (*cream of wheat*, *salt of epson*), so that for the rest N^0 reduces to N^1 without semantic change. So we shall make a distinction between a lexical noun (and other CN's), e.g., *water*, and its noun phrase occurrence in a sentence, especially as a "bare noun phrase" – as in *water is wet*. From now on, we shall mark this difference as *water : CN* and *water : NP*. Similarly, we distinguish *water : CN* from the predicate *is water : VP* – an entire verb phrase. This makes the basic problem that of providing an analysis of the semantics of mass (and plural) CN's. The semantics of bare NP's, quantified NP's (e.g., *this wine*, *some dogs*) and predicates (e.g., *is water*, *are dogs*) must build on this basis. For example, if *water : CN* is claimed to express a predicate over quantities of matter while *water : NP* expresses the name of a substance, we would like to know how the former is transformed (syntactically and semantically) into the latter. Similarly, we would like to know how the meaning of the sentential predicate *is water : VP*, in such contexts as *the stuff in this bottle is water*, and *mineral water is water*, is derived from the meaning of *water : CN*.

We mention here three desiderata: (1) A syntactic-semantic homomorphism. Since we are looking at semantics within a Montague-style framework, we cannot be satisfied with just any "philosophically satisfying explication" of mass term and bare plural semantics. (In particular we want to avoid *ad hoc* "reconstruals" of sentences containing bare noun phrases.) Instead, we require in addition that there be semantic rules paired with syntactic rules of formation, delivering the logical translations of the syntactic constituents. In the current context, this entails at least that phrases of the same syntactic category should be of the same semantic category. This makes it desirable, for example, that *water : NP* should denote the same thing in the sentences

Water covers most of the globe.

Water is a liquid.

Water is dripping from the faucet.

(2) Uniform treatment of bare mass noun phrases and bare plural noun phrases. There is plenty of linguistic evidence (cf. Carlson [1977] and Chierchia [1982a]) for taking these two types of noun phrases as being of the same semantic category, despite the fact that they carry the SING and PLUR feature respectively.[51] Any semantic analysis should explain the systematic connection between the two types of bare noun phrases. (3) Substitutivity of alleged paraphrases. Often, semantic analyses of mass and plural CN's (and the corresponding NP's and VP's) are accompanied by informal (i.e., English) paraphrases or elucidations that indicate how these constituents are to be understood intuitively. For example, *water : CN* might be said to denote "(is a) quantity of water"; similarly *water : NP*

might be said to denote "All the water in the world", or perhaps "The substance, water". We should take such paraphrases seriously, and hence require that (a) substitution of alleged paraphrases for the original phrases should map an originally intelligible sentence into another intelligible sentence synonymous with the original – apart, perhaps, from pragmatic defects such as unwanted implicatures. (b) It should be possible to analyze the paraphrase formally, showing that it leads to a translation logically equivalent to the translation of the original CN, NP or VP. In other words, it should be possible to formalize (a).

There seem to be at least five intuitively distinct roles that mass terms can assume, and the problem then is to try to determine which of them is semantically primitive, i.e., which corresponds to the meaning category of lexical mass nouns such as *water : CN*.

1. As names: This seems to be motivated by examples like *Water covers most of the globe*, *Water is oxidized hydrogen*, *Methane is abundant on Titan*, together with the tacit assumption that the denotations of CN's and the corresponding bare NP's are the same.

2. As predicates over quantities/portions of matter: This is apparently motivated by such sentences as *The contents of this bottle is* (*are*) *water*, *John drank* (*sm*) *water*. It is also motivated by quantified phrases such as *sm water* and *the gold on the table*, which can plausibly be read as *some quantity of water* and *the* (*largest*) *quantity of gold on the table*.

3. As predicates true of objects: This is motivated by such examples as *This ring is gold*, *This sandwich is food*, and *This boy is skin and bones*.

4. As predicates over kinds/substances: This is motivated by such examples as *Claret is wine*, *Rice is food.*

5. As predicate modifiers: This is motivated by such examples as *She wore a gold anklet*. It would be akin to treating mass nouns as adjectives.

6. As predicates which combine the kinds of predicates mentioned in (2), (3), and (4). – So there are four such possible combinations: (2)–(3), (2)–(4), (3)–(4), (2)–(3)–(4).

We now give some general problems for (some of) these approaches. With respect to (1), that mass (and plural (CN's) are names, we might consider such sentences as *Pure lemonade is lemonade* and *Fake fur is not fur*. It seems pretty clear in general that adjectives map predicates into predicates. For example, *red* and *fake* map *flower* (whose extension is the set of flowers) into *red flower* and *fake flower* respectively (whose extensions are the set of red flowers and the set of fake flowers respectively). But if *lemonade : CN* and *fur : CN* are names, then we require that adjectives also map names into names, yielding such new names as *pure lemonade : CN* and *fake fur : N* (the extensions of names being individuals rather than sets of individuals). Attributing two sorts of functions to adjectives seems inelegant at best, and it is hard to see how to axiomatize the latter function. Both Bunt and Parsons try to get around the problem by analyzing [ADJ CN] : NP, where CN is a mass noun,[52] as *the substance whose quantities are quantites of CN and are ADJ* (where for Bunt "the substance . . ." corresponds to mereological supremum while for Parsons it remains his unanalyzed substance-abstraction operator). Thus *pure lemonade* comes out as *the substance whose quantities are of lemonade and are pure*. But this is wrong, for whether a quantity of lemonade is "pure" depends on whether it is viewed as lemonade (lemon juice plus water plus sugar) or as something else, such as water (with lemon and sugar impurities), or lemon juice (diluted with sugar water). In other words, *pure* is *essentially* a predicate operator, which cannot be construed conjunctively.[53] Finally, it is well known that mass CN's, like count CN's, are premodifiable by nouns (*rabbit fur*), attributive adjectives (*white fur*), quantifying adjectives (*much fur*), ordinals and determiners (*the first snow*), and can occur as predicates in sentences like *White wine is wine* where the *is* is not the "*is* of identity". These NP roles indicate that mass CN's are to be treated on a par with count CN's. We conclude that position (1), that the mass CN's should be assigned the status of names, is highly implausible, and appears to be the product of precisely the kind of confusion between the levels of noun phrase structure that we have drawn attention to.

Are mass (and plural) CN's then to be predicates over quantities of matter, as in ter Meulen [1980], Chierchia [1982a], Montague [1973]?[54] A problem with this approach is finding a way to translate *quantity of fur : NP* so that its intension will differ from the intension of *fur : CN*, and hence, that *quantity of fake fur*, *quantity of small furniture*, etc., will have intensions different from *fake quantity of fur*, *small quantity of furniture*, etc.[55] Another problem is to show how this choice of primitive role for mass expressions can account for the role of mass NP's as predicates over (ordinary) objects and predicates true of kinds/substances. It should be noted that, even if one chooses this role of mass NP's to be primitive and assigns it to the CN's, one can still translate the NP as something other than a predicate. For example, Chierchia's [1982a] translation of *dew worms* would be the "name" $^\wedge$**dew-worm**′, and Link's [1981] translation would be $(Ex)(\oplus$**dew-worm**′(x) & . . . x . . .). (For Link $\oplus$**dew-worm**′ is true of $\oplus$-sums of individual dew worms (at least two).)

There seem to be no theorists who allow role 3, that mass terms are predicates true of ordinary objects, to be their primitive role assigned to mass CN's. However some theorists (e.g., Pelletier [1974], Chierchia [1982a], [1982b] whose formal semantics does not distinguish quantities of matter (in the informal sense) from "things", and Link [1981] (see especially his example 45) combine roles 2 and 3, allowing that mass nouns are at root predicates over quantities of matter, lumps, nuggets, slices, helpings, etc., of stuff, as well as over ordinary things. The main problem with such an approach is its apparent inability to handle mass NP's which exemplify role 4. There is also a paraphrase problem here analogous to that which we cited for position 2: intuitively, a sample of M, where M is a mass NP, is true of both quantities of M and things exemplifying M (e.g., a given quantity of furniture is a sample of furniture, as is any piece of furniture). Yet a small sample of furniture is not the same as a sample of small furniture.

Role 4 is that mass nouns are true of substances, or kinds, or varieties.[56] Examples of this role are such sentences as *Claret is wine* where *wine* appears to classify or describe or be true of kinds/species/varieties. Such uses are often paraphrasable as *wine of a certain kind/variety* or *variety/kind of wine* or *a wine*. In other words, this use of mass NP's in forming VP's comes close to, or coincides with, the use of the putative mass noun as a count noun. It is tempting to try to paraphrase these uses of mass NP's as a quantification of some sort. E.g., ter Meulen [1980], and Pelletier [1974] would paraphrase this (after the application of meaning postulates) as *All* (*the quantities of*) *claret are* (*quantities of*) *wine*; and possibly some theories would paraphrase it as *Some* (*quantities of*) *claret are* (*quantities of*) *wine*. But sentences like *Claret is wine and is currently out of stock* (in response to a liquor store request. "I need some claret, whatever that is") thwarts this move. Obviously this conjunctive sentence has the same term *claret* as subject of both conjuncts, and yet no quantificational reading can work for both conjuncts. Other examples of this role are *Muddy water is water but is not potable*, *Fake fur is not fur yet sells well*.

If this is chosen as the meaning of mass CN's, it would entail that all mass terms derive from count nouns (expressing predicates over substances/kinds). Corresponding to roles 1, 2, and 3 above we would have to introduce into our semantics certain operators that work on these basic meanings to form names of the substance, to form predicates over quantities of matter, and to form predicates over ordinary objects. For example, we might have (for one of these mass CN's M)

$$\mu M \quad qM \quad cM$$

where μ is a name-forming operator, giving the top (most general kind) of a "is a kind of" semilattice of M-kinds, q maps substance-predicates to quantity-of-matter predicates, and c maps substance-predicates to thing-predicates (things, unlike substances, being concrete). If we wish to conflate the q and c operators along the lines suggested earlier, we might introduce

$$pM$$

which forms a predicate that is true of quantities of matter and ordinary objects, given a substance predicate. An objection to this approach is that there is *already* a category of nouns generally regarded as denoting substance-predicates, and these are intuitively quite unlike mass nouns. Examples of this category are *element*, *substance*, *metal*, *liquid*, *stuff*, *gas*, *material*, etc. Why are these not transformable to mass terms by applying μ, q, c, and p? The answer, possibly, is that these terms *are* so transformable – consider *traces of yellow gas*, *lots of material*, *molten metal in large vats*, and the like. At least some of these kinds of "substance predicates" seem to be convertible to mass NP's. It is worth noting that position 4 is not as susceptible to the "paraphrase problem" as positions 2 and 3 were; e.g., *This hamburger is greasy food* seems roughly paraphrasable as *This hamburger is a greasy kind of food*, and similarly for *This shipment is heavy furniture* to *This shipment is a heavy kind of furniture*.

As far as the 5th role of mass terms goes, the predicate modifier role, the only person we know to have suggested that it is basic was Aristotle. We shall consider it no further as a candidate for primitiveness.

Re position (6) – combinations of (2)–(4). We have already seen that a paraphrase problem arises not only for position (2) (that mass CN's denote quantities of matter) but also for (2) and (3) in combination. There is a similar difficulty for *all* combinations of (2), (3), and (4). For example, if we want to claim that mass CN's are predicates over quantities and varieties of matter, it seems that we should be able to paraphrase *is CN* as *is a CN-quantity or CN-variety*. Yet *is a small furniture quantity or furniture variety* is not synomymous with *is small furniture*, since the former is necessarily true of any small furniture quantity, while the latter is not. Since one of the theories we shall develop later appeals to a combination, we shall let it have its say about this paraphrase argument. Briefly, it denies the inference

1. CN′ denotes quantities and varieties of CN therefore,
2. CN is paraphrasable as *CN-quantity or CN-variety*

Its reason for denying this inference might be that the explicit English *or* in 2 is always read *distributively* in context (e.g., *x* is *a small P or Q* is always read as *x is a small P or x is a small Q*), whereas in *small CN*, where CN is mass, *small* acts on the entire intension of CN′.

TWO THEORIES FOR MASS NOUNS

In this section we offer two theories to account for the data we have mentioned earlier in this paper. One theory, the *p*-theory, is a semantic occurrence approach. In this theory MASS and COUNT are not syntactic categories or features, but are rather a description of the sort of semantic translations generated by the Gazdar-style syntactic rules. The other theory, the *s*-theory, is a syntactic expression approach. In this theory MASS and COUNT are explicitly syntactic features which attach to lexical nouns. The *s*-theory invokes the lexical extension rules (mentioned above) to describe certain phenomena. These two theories seem to us to be the most plausible candidates for a unified theory of mass nouns. The fragment of English that these two fragments are concerned with is roughly those sentences with a copular verb and either a mass, a count, or an adjectival expression as predicate and whose subjects are either bare noun phrases or quantified noun phrases.

The two theories share much – in particular, they are in agreement as to the range of data to be accounted for. They agree, for example, that sentences like the result of the universal grinder are to be given a representation, that sentences using a putative mass term with the indefinite article (e.g., *a wine*) are to be given a representation, and in general they agree on what semantic representation is appropriate to the entire range of example sentences to be considered. The ontological underpinnings of both theories are that "reality" contains two sorts of items: (1) "ordinary objects" such as rings, sofas, puddles (and including here what many theorists have called "quantities/portions of matter").[57] (2) "kinds of stuff", that is, "varieties", "substances", etc. (we have in mind here such items as wine, claret, red wine, and the like) and "kinds of portions", i.e., kinds of standard or conventionally recognized servings. (*This is a beer* might be true because *this* denotes a quantity of beer or a berry object – type (1) objects; but it might be true because *this* denotes a kind of stuff, say pilsner, or because it denotes a standard kind of serving, e.g., a bottle of beer). We wish to make no special metaphysical claims about the relationships that might hold between "ordinary objects" and "kinds" instead we content ourselves with describing how such an ontology leads to a simple and natural description of various of the facts concerning mass (and possibly plural) expressions.

Linguistically, that is semantically, we take there to be two distinct types of predicates: (a) those which apply only to "kinds", e.g., *is a substance*, *is a wine*, *is a kind of wine*,[58] *is scarce*, *is abundant*, and (b) those which can apply to both "kinds" and "objects". In this last group we have in mind mass predicates such as *is wine*, *is furniture*, *is food*, *is computer software*, and the like on the one hand, and count predicates such as *is a dog*, *is a puddle*, *is a quantity of gold*, and the like, on the other.[59] We hope that the discussion of the last section has made it plausible that mass VP's can apply to both "kinds" and "objects" as required by (b). We also take it to have been established in that section that these "kinds" are abstract rather than "scattered individuals".[60]

Both of these theories take it that *is wine* is true of the (abstract) kind claret in addition to an individual quantity such as the contents of this glass. Moreover, they take *is wine* to be true of an object such as a drop or puddle of wine, occupying the same region as some quantity of wine. (*This ring is gold* or *This hamburger is food* are clearer examples of the application of mass predicates to objects.) Generally speaking, the theories view the kinds of *M* as forming an upper semilattice of kinds with *M* at the top. This is a "formal" semilattice in that the union of any two elements of it is a member of the semilattice, and we view *is wine* as being true of any of these formal kinds. So a sentence like *Cheap wine is wine* will be true, since *cheap wine* names an element of the semilattice. Predicates like *is a wine* are true of "conventional" kinds (*Claret is a wine* is true) but not of every "formal" kind since, e.g., *Cheap wine is a wine* is not true.[61] (*Sauterne mixed with claret is a wine* is also not true, showing that *is a wine* is not true of unions of elements of the semilattice). These predicates are not only true of the conventional kinds but also of individual servings such as the bottle of wine on the table or the 250 ml in this glass and of conventional kinds of servings of wine (such as pints in a British pub, or 7 oz in Alberta bars). When a bare mass noun phrase (or indeed other bare noun phrases, although we shall not dwell on them here) is used as a subject (or object, but again we shall not consider that here), it is taken to name the kind. So in *Cheap wine is wine*, the subject *cheap wine* names a kind; and since the sentence is true it must name a "formal kind" so that *is wine* can be predicated of it. But since *Cheap wine is a wine* is not true, the formal kind cannot be a conventionally recognized kind (nor, for that matter, a conventional kind of serving nor an individual quantity).

Both theories hold that mass CN's should be translated into the semantics as predicates. Strictly this is not required: for, all we have given direct evidence for is that mass VP's be translated as predicates with a mixed object/kind extension. It could be the case that mass CN's are quite different, yet in the formation of a mass VP the entire VP gets assigned a mixed, predicate denotation. Still, it would be simple, and in keeping with much philosophical and linguistic analysis, to assume coincidence of CN and "is CN" denotations (at least when tense is ignored, as here).

With just this much of the theories sketched, we can see that they overcome various of the difficulties that plagued other theories. For example, it is most unclear that any other theory can adequately translate sentences like

Tap water is water

This puddle is water

without doing one of the following implausible things: Make *water* in the two sentences designate different predicates (one over kinds, the other over objects); make *tap water* be an implicit universal quantification over quantities of tap water,[62] or make *tap water* be an implicit existential quantification over quantities of tap water.[63] Consider also sentences like

All wine is wine

wherein the subject *all wine* seems to quantify over both kinds of wine and quantities of wine, covering both *White wine is wine* and *The litre of wine in this bottle is wine*, for example. It seems to us that no other theories allow for this comprehensiveness. An even clearer example of such comprehensive denotation is

Everything edible is food

from which both of

Rice is food

This sandwich is food

follow, given that rice is edible and this sandwich is edible. (Note also the comprehensive denotation of *edible*). No other theories we know of can account for the validity of these two arguments, although it is clear that (both of) ours can, at least if we assign correct denotations to the subject NP's.

According to these two theories, and following most work on mass terms, bare NP's in subject position are taken to be names. We use μM to form the name of the kind/substance/etc. M.[64] Thus we have the (a) sentences translated as (b):

a. Water is wet
b. **Wet′**(μ **Water′**)

a. Tap water is water
b. **Water′**(μ(**Tap Water**)′)

Universally quantified mass NP's, as these theories see them, are comprehensive in their extension: sentences like *All water* . . . assert that . . . is true of all varieties and of all quantities.

a. All water is wet
b. (Ax)[**Water′**(x) → **Wet′**(x)]

Existentially quantified NP's, as these theories see them, using the stressed *some*, are not comprehensive: rather they are true only of "kinds". Our two theories distinguish these from occurrences of the unstressed *sm*, which they see as being true only of "quantities". Yet, given the assumed comprehensive extension of many of our predicates, they can be translated identically. Thus both

Some rice is tasty

Sm rice is tasty

would be translated as

(Ex)[**Rice′**(x) **&** **Tasty′**(x)]

Given the comprehensive extension of *Rice* (and of *Tasty*), the first sentence is true because, say, *Rice* is true of the kind "brown rice" (which is tasty), and the second sentence is true because *Rice* is true of, say, this bowl of brown rice (which is tasty). According to this "bare bones" version of the two theories, the difference between our two sentences is not a matter of semantics, since they have the same representation. However, should we choose to treat this difference as a semantic phenomenon, these theories could be augmented with operators β and p which respectively take a mass CN M and form a predicate true only of M-kinds/varieties or a predicate true only of M-objects/quantities. Our two sentences would then be translated respectively as

$$(Ex)[\beta\ \mathbf{Rice}')(x)\ \&\ \mathbf{Good}'(x)]$$

$$(Ex)[(p\ \mathbf{Rice}')(x)\ \&\ \mathbf{Good}'(x)]$$

(Of course these β and p operators would be introduced by the syntactic-semantic rules by translations of *some* and *sm*, not by ambiguously translating *rice* sometimes as (β **Rice**′) and sometimes as (p **Rice**′).) A sentence which is unquestionably ambiguous is

This is a beer

which the p-theory translates as

$$(\beta\ \mathbf{Beer}')(t) \vee (\gamma\ \mathbf{Beer}')(t)$$

and the s-theory translates as either

$$(\beta\ \mathbf{beer}')(t)$$

$$(\gamma\ \mathbf{Beer}')(t)$$

making it refer ambiguously to a kind of beer or a quantity/portion/etc. of beer. ((γ **Beer**′) is true of conventional portions and kinds of portions of beer.)

Our grammars will cover only certain stative sentences. However, in view of the importance of the stative/non-stative distinction in the interpretation of sentences containing mass nominals, we should indicate briefly how we would treat mass nominals in an extended grammar covering non-stative verb phrases. The translations of the following sentences would be as indicated:

a. Snow is falling
b. $\mathbf{Falling}'(\mu\ \mathbf{Snow}')$

a. John threw snow at mary
b. $\mathbf{Threw\text{-}at}'[j, (\mu\ \mathbf{Snow}'), m]$

a. Sm snow is falling
b. $(Ex)[(p\ \mathbf{Snow}')(x)\ \&\ \mathbf{Falling}'(x)]$

a. John threw sm snow at Mary
b. $(Ex)[(p\ \mathbf{Snow}')(x)\ \&\ \mathbf{Threw\text{-}at}'(j, x, m)]$

Note that the bare mass terms in the first two sentences have been interpreted as kind-denoting, despite the non-stative context, as in the Carlson-Chierchia approach. The implication of the first sentence, that sm snow is falling, would be obtained by a meaning postulate to the effect that if x is falling and x is a kind, then some "object" y of kind x is falling. Similarly for the implication of the second sentence that John threw sm snow at Mary. In this way the first two sentences lead to consequences which coincide with the translations of the third and fourth sentences, without actually being synonymous with those sentences at the level of logical form.[65]

Our examples have invoked the predicate-restricting operators β, γ, and p, either in the translations of certain English sentences, or in the meaning postulates applicable to the translations. In any case, whether we choose to do it in the direct translations or in the meaning postulates, both of our theories will want to be able, in the semantics, to form predicates which are true of kinds, or of servings, or of individuals, given a predicate which has comprehensive extension. So, for example, from the predicate **water′** which is assumed to be true of quantities, servings, and kinds, we shall want to be able to form (β **water′**) which is true of conventional kinds of water, to form (γ **water′**) which is true of conventional portions (and kinds of portions) of water, and to form (p **water′**) which is true of quantities of water and objects which coincide with such quantities. Conversely, if we have a predicate which is true of individuals and kinds, we shall want to form a predicate true of *all* the entities that mass predicates are true of – quantities of stuff, kinds of stuff, conventional kinds of servings, and objects coincident with quantities of stuff. For example, if **man′** is a predicate true of objects and kinds, then (Σ **man′**) is the comprehensive predicate formed therefrom.

Having seen where they agree, let us look at where they differ. The p-theory takes all predicates, *animal* as well as *water*, to have comprehensive extension – to be true of kinds (of stuff or things), true of conventional kinds of servings, true of quantities (of stuff), and true of objects coinciding with quantities of stuff. It makes no use of syntactic features +mass and +count, and instead interprets certain occurrences of entire noun phrases as being (semantically) mass or count – depending upon whether the predicate as used therein is true of stuff or of things and kinds. (This is determined by the syntactic constructions.) On the other hand, while the s-theory agrees with the p-theory on the basic extension of mass predicates like *water*, assigning them a comprehensive extension, it distinguishes these from the count predicates which it takes to be true only of individual objects and kinds of individual objects. It marks this difference in the lexicon with the syntactic features +mass and +count respectively. In certain sentences a count noun is used as a mass noun; to generate these, the *s*-theory employs a "lexical extension" rule which (a) changes the mass/count feature, (b) adds a +EXT (for "extended usage") marker, and (c) suitably alters the semantic representation of the original noun to make it fit the extended usage. Overall, a sentence in which some term is marked +EXT is judged "semi-grammatical".

We shall close by giving an indication of how the two theories handle a variety of constructions involving mass expressions. We do not give a detailed statement of either the syntax or the semantics but rather just state the relevant rules of syntactic combination and the associated semantic representation.[66] The two theories use a Gazdar-style grammatical theory and an associated intensional logic semantics. Here we content ourselves with an informal presentation of the associated notions; in particular it is part of our goal to stay as "close" to a traditional first-order theory as we can (augmented by certain operators).[67] Finally, we here ignore certain grammatical niceties, such as the feature system which handles number and person agreement. Rules of these theories are stated by two parts: a context-free parse and the associated semantic representation, separated by a comma. We use a prime to indicate the semantic representation of some syntactic item. We allow ourselves complex syntactic nodes, for example [N +ADJP] is a noun which has the feature of being premodified by an adjective phrase, and [ADJP +INT] is an adjective phrase which is intersective.

The rules for the relevant portion of our two theories are (with tacit restriction to the *singular* throughout):

1. S → NP VP, **VP′(NP′)**

2. VP → [V +be] PRED, **PRED′**

3p. PRED → N, **N′**
3s. PRED → [N +MASS], **N′**

4p. PRED → [DET +*a*] N, $(\lambda x)[(\beta\ \mathbf{N}')(x) \vee (\gamma\ \mathbf{N}')(x)]$
4s. PRED → [DET +*a*] [N +COUNT], **N′**

5. PRED → NP, $(\lambda x)(x = \mathbf{NP}')$

6. PRED → ADJP, **ADJP′**

7p. NP → N, (μ **N′**)
7s. NP → (N +MASS], (μ **N′**)

8p. NP → DET N, ⟨**DET′ N′**⟩[68]
8s. NP → [DET SING] [N SING], ⟨**DET′ N′**⟩ (where SING may particularize to +MASS or +COUNT)[69]

9. [N + ADJP] → [ADJP +INT] N, (λx)(**ADJP″**(x) & **N′**(x))

10. [N +ADJP] → [ADJP ¬INT] N, **ADJP′(N′)**

The *s*-theory distinguishes in the lexicon MASS from COUNT nouns. And it has what might be called "lexical extension" rules to give us the "stretched" meaning of nouns that we have earlier talked about. For example, it has lexical entires

[N +COUNT] → *sofa*, *man*, *substance*, *wine*, . . .

[N +MASS] → *wine*, *water*, . . .

and extension rules

[N +EXT +COUNT] → [N −EXT +MASS], (β **N′**)

[N +EXT +COUNT] → [N −EXT +MASS], (γ **N′**)

[N +EXT +MASS] → [N −EXT +COUNT], (Σ **N′**)

So, for the minimal grammar considered here, a sentence is a noun phrase and a verb phrase. A verb phrase is a copula followed by a PRED, which in turn is either an adjective, or a bare noun (as in *Claret is wine* or *This puddle is man* – the latter said after an application of the universal grinder), or an *a* followed by a noun (as in *John is a man* or *Claret is a wine*), or is an entire noun phrase (as in *John is the man most likely to succeed* or *Claret is my favourite red wine*). A noun phrase is either a bare noun (as in *Claret is a dry red wine* or *Dogs are barking outside*) or else is a quantified term (as in *All men are mortal* or *Sm red wine is tasty* – we include as determiners *this*, *all*, *some*, *sm*, *much*, *little*, *each*, *every*, and the numeral quantifiers). Nouns may themselves be either an adjective-phrase noun combination, or just a noun.[70] We consider here two cases of adjective modification: intersective and non-intersective. For the former we have in mind such adjectives as *red*, while for the latter we think of such adjectives as *fake*.[71] The rules which give alternatives, such as (3p) vs. (3s) are those rules which are different for the two theories of mass terms. The *p*-rules are for the semantic occurrence approach while the *s*-rules are for the syntactic expression approach.

Now, both of these theories can give the correct semantic representation to a wide range of sentences involving mass terms, given certain meaning postulates. (The two theories do it slightly differently, as might be expected since they have somewhat different semantic understandings of the lexical nouns. For example, the *s*-theory takes *man* to be true of individual men and of kinds of men, while the *p*-theory takes it to be true also of the stuff of which men are made. In the *p*-theory, when a sentence uses *a* – as in *a man* – then the semantic operators convert this "basic" meaning into one that is true of individual men and of kinds of men. The *s*-theory rather has a lexical redundancy rule which will convert the lexical count noun *man* into one which is a mass noun and is true of the stuff of which men are made and quantities thereof. They will also take a different tack on what quantified terms designate, although that has been hidden in rule 8 above by assigning the same logical form to both theories. Nonetheless, the meaning postulates of the two theories will differ for these.) Let us state, then, the translations for three of our DET's into our "first order-ish semantics" for the two theories:

11. **sm′** $= (\lambda P)\langle E\ (\mu\ P)\rangle$

12p. **some′** $= (\lambda P)\langle E\ (\lambda x)[(\gamma\ P)x\ \vee\ (\beta\ P)x]\rangle$

12s. **some′** $= (\lambda P)\langle E\ \mathbf{P}\rangle$ (where *some* is [DET NUMB])

13. **all′** $= (\lambda P)\langle A\ P\rangle$[72]

So, in the *p*-theory, the predicates of the lexicon are comprehensively true of anything that any use might dictate. But in certain contexts, such as being in a quantified noun phrase of a certain sort, the semantic representation of the entire noun phrase is sometimes (e.g., when the quantifier is *some* or *a*) such that we can say that it is true of individuals (kinds) and sometimes (e.g., when the quantifier is *sm*) such that we can say that it is true of stuff. So, by looking at the semantic representation, we can say whether it is a count noun phrase or a mass noun phrase – but this is all in the semantics and holds of entire noun phrases (not individual nouns), as befits a semantic occurrence approach. The *s*-theory will already have COUNT and MASS as features on the nouns and use this information to induce the correct, allowable combinations and meanings of quantified terms.

These rules, and the informal remarks preceding them, indicate how the theories would translate the usual run of sentences encountered in discussions of the semantics of mass expressions. For example, using the rules just given,

Sm Einsteinium is in the lab

is translated as

$(Ex)[(p\ \mathbf{Einsteinium}')(x)\ \&\ \mathbf{In\text{-}Lab}'(x)]$

making the sentence assert that some quantity (ingot, puddle, etc.) of Einsteinium is in the lab. In contrast, the rules assign to

All Einsteinium is in the lab

the translation

$(Ax)[\mathbf{Einsteinium}'(x) \rightarrow \mathbf{In\text{-}Lab}'(x)]$

where the values of x can be both quantities (ingots, puddles) of Einsteinium and also the kinds of Einsteinium. And then the sentence asserts that they all are in the lab.[73]

One might further wish to inquire about the relationship between quantities and kinds. For example, if a quantity of Einsteinium is in the lab, is there also a kind of Einsteinium in the lab, and vice versa? If all quantities are in the lab, are all kinds, and vice versa? Our two theories answer these questions negatively, in general. While it may seem plausible in the example just given, the following slightly different examples show that these four propositions are false.

1. If a kind of *M* has property *P*, then some quantity of *M* has property *P*.
2. If a quantity of *M* has property *P*, then some kind of *M* has property *P*.
3. If all kinds of *M* have property *P*, then all quantities of *M* have property *P*.
4. If all quantities of *M* have property *P*, then all kinds of *M* have property *P*.

Consider, respectively, that while some kinds of wine are scarce, no quantity is; that this quantity of water is a puddle but no kind of water is a puddle; that all kinds of wine are kinds of wine, but no quantity is a kind;

and finally, it might be the case that all the quantities of Einsteinium are in the lab, but not all the kinds (maybe there are two kinds, but currently only one of them is instantiated by any quantities – all of which are in the lab).

Of course, *sometimes* the inferences are valid, and for those cases the two theories use meaning postulates. Recall that these theories have two types of predicates: those true only of kinds, and those true of both kinds and objects/quantities. Our theories hold that if the predicate is *not* exclusively a kind predicate, and if it holds of a kind M, then it (1) holds of some (perhaps all) subkinds of M (recall that the theories take kinds of M to form an upper semilattice), and (2) holds of at least some (perhaps all) quantities of any such subkind. Sometimes there are relationships between the different types of predicates. For example *is a liquid* is true only of kinds while *is liquid* is true of both.[74] The two theories take the view that in such cases the kind property entails the comprehensive property. Thus (as indicated earlier) the (a) sentences entail the (b) sentences.

(a) Water is a liquid
(b) Water is liquid

(a) Claret is a wine
(b) Claret is wine

(a) Gold is a metal
(b) Gold is metal[lic]

When the predicate in question is one of the comprehensive ones, the theories take the view that if the kind has it, then at least some quantity of the subject has it. Thus the following inference from (a) to (b) (and hence to (c)) is taken to be valid.

(a) Water is dripping from the faucet
(b) Sm water is dripping from the faucet
(c) Some quantity of water is dripping from the faucet.

Furthermore, following Chierchia [1982a], [1982b], the theories take the position that stative VP's of the comprehensive sort, such as *is liquid*, *is wine*, etc., induce a "universal" reading on the sentence. Non-stative VP's of the comprehensive sort, such as *is dripping from the faucet* or *is lying on my desk*, are assumed to induce an "existential" reading on the sentence. Thus while *water is liquid* is equivalent to *all water is liquid* and entails *all quantities of water are liquid*, the non-stative sentence *gold is lying on my desk* entails *some quantity of gold is lying on my desk*. These, and similar inferences, can be handled in the theories either by means of meaning postulates or directly in the translations (or by a combination of both, as in Chierchia who handles the universal import in stative sentences via meaning postulates but handles the existential import of non-stative VP's directly through a semantic rule for non-stative VP's).[75]

It will be recalled that both theories take a PRED formed with *a* plus *M* to denote the subset of conventional kinds of *M* belonging to the semilattice of kinds of *M*. (For present purposes, we can ignore the alternative denotation consisting of conventional portions and kinds of portions of *M*.) So *is an M* is a predicate true of all the subkinds of *M* (whereas *M* as a complete noun phrase, as when it is the bare subject of a sentence, names the uppermost element of the semilattice). This gives our theories' account of why

Claret is a wine
Sauterne is a wine
Burgundy is a wine

and the like are all true (given that *Claret*, *Sauterne*, and *Burgundy* are, in fact, conventional kinds of wine). It is also our theories' reason for saying that *All wines are wines* is true. Although we have not discussed our theories' handling of plurals, we will assert that *are wines* is the plural VP corresponding to *is a wine*, and hence the sentence will be true if *all wines* designates a class of conventionally recognized subkinds of wine – which it does according to these theories, namely the class of all conventionally recognized subkinds of wine. This, together with the previously mentioned meaning postulate, is the reason *All wines are wine* is true – *are wine* is the (plural) comprehensive predicate corresponding to the kind predicate *are wines*. Since *All wines are wines* is analytically true, the meaning postulate will have it entail *All wines are wine*. *All wine* (as opposed to *all wines*), it will be recalled, designates not only the subkinds of wine, but all quantities of wine, so a sentence like *All wine is a wine* will be false (or semantically anomolous) because of the mismatch between the subject designator and the predicate. Similarly, *Wine is wine* and *All wines are wine* will be true (and analytic), but *Wine is a wine* will be false.[76]

With these meaning postulates and semantic theory, the two theories can account for a variety of sentences and inferences, by generating them and giving them the correct logical form, which are beyond the scope of other theories.[77] Correct logical form is assigned to these sentences.

> Wine is wine (two readings, both analytic)
> Wine is a wine (false)
> Chilled wine is a wine (false)
> All wine is wine (analytic)
> *All wine is a wine (semantically anomalous)
> Water is dripping from the faucet (entails: sm water is dripping from the faucet)
> Water is a liquid (entails: water is liquid)

Consider these inferences:

1. Claret is a wine, wine is a liquid, so claret is a liquid
2. Claret is a wine, wine is a liquid, so claret is liquid
3. Claret is a wine, wine is liquid, so claret is liquid

These three are valid inferences, according to the theories, because: the first premise of each asserts that Claret is a kind (of wine), and in (1) and (2) we are also told that wine is a kind (of liquid), hence in (1) we get from the semilattice of kinds[78] that claret is a kind (of liquid) and by our meaning postulate we get (2) that claret is liquid. In (3) we are told that wine is liquid, so everything which is wine is liquid and hence claret is liquid.

Now consider

4. Claret is a wine, wine is liquid, so claret is a liquid
5. Claret is wine, wine is a liquid, so claret is a liquid
6. Claret is wine, wine is a liquid, so claret is a liquid
7. Claret is wine, wine is liquid, so claret is a liquid
8. Claret is wine, wine is liquid, so claret is liquid

(4) is invalid because although claret is a conventionally recognized wine-kind, it may not be a conventionally recognized liquid-kind. In (5)–(8) we are not given in the first premise that claret is a kind – for all we are told, it might be a name or description of a quantity or object – and so the arguments mentioned above no longer work to show any of these to be valid arguments. Of course, with an added premise to the effect that claret *is* a conventional kind (5), (6) and (8) would be valid; but we are currently interested in which of the inferences are valid without this added information. Obviously, if it is not given that claret is a conventional kind, then the conclusions of (5) and (7) might be false, so they cannot be valid

arguments. The second premise of (6) entails, by the meaning postulates, the second premise of (8); thus, showing that (8) is valid will imply that (6) is also. And indeed (8) is valid, since *is liquid* and *is wine* are stative VP's and hence induce a universal reading upon the sentences, as in *All claret is wine* and *All wine is liquid*.

We know of no other theories which can do all these things. Yet the two theories are radically different: one has a mass/count distinction in the syntax and the other doesn't, and they have different extensions assigned to the lexical items. So the question naturally arises – which is better? What can be said against the two theories? There is not space in a survey to go into the issues involved here in detail, so we shall content ourselves with just hurling the main charge that each one directs against the other. Briefly, the *p*-theory charges the *s*-theory with pretending to use syntactic features +mass and +count but allowing them to do no syntactic work. For every sentence which has a mass term in a given location, there is another sentence which has a count term in that position. No constructions are ruled out (although some are marked as +EXT, i.e., as involving lexical extension); the only use of the +mass/+count features is in directing the semantic translation process. And that suggests that the features should all along have been semantic. The *s*-theory charges the *p*-theory with being unable to give coherent meaning or factual postulates because of its commitment to a comprehensive extension to the lexical terms. For example, suppose one wanted to give as a factual postulate that *A lamb has fur*. The *s*-theory can do this without difficulty: **lamb′** is true of individual lambs and the postulate says of each of them that they have fur. But the *p*-theory cannot easily do this: **lamb′** is true of stuff, among other things, so the predicate must be converted to one which is true only of individuals. But there is no provision in the *p*-theory for doing this – the closest that it could come is with a predicate that is true of both conventional kinds and "conventional portions" (i.e., ordinary lambs or servings of lamb).

So, which theory is to be preferred? That is a topic for further research. We hope to have shown where we think the issues lie and where future efforts should (and should not) focus.

CONCLUSION

It will be recognized that the two theories draw from many other theories. Many of the remarks about kinds – although not such ones as "only simple lexical expressions can denote kinds" and "no sentence of the form

M is *M* can be well-formed" – can be found in ter Meulen [1980]. Some of the remarks about semilattices can be found in Link [1981] – although he does not consider a semilattice of kinds as we do. The use of stative vs. non-stative VP's to induce an universal vs. existential reading on sentences is due to Chierchia [1982a], [1982b]. Various of the informal understandings of the notion of a quantity can be found in Cartwright [1970]. The conflation of "ordinary" objects and quantities into a single category can be found in Pelletier [1974], Chierchia (*op. cit.*), and Link [1981]. And the underlying conception of a grammatical theory is due, of course, ultimately to Montague [1973a] as filtered by the works of Gazdar *et al.* [1985]. Perhaps what is most novel here is the attention paid to the denotation of mass CN's as opposed to mass NP's, and the view that they are predidates true not only of objects and quantities, but also of kinds.

One thing should be clear from this survey of the work done on the topic of the semantics (and syntax) of mass expressions: The time for studies of mass expressions with only causal reference to the syntax and semantics of language is past. Only systematic attempts to account for large classes of mass expressions within formal syntactic-semantic-pragmatic frameworks can hope to resolve the remaining issues.

ACKNOWLEDGEMENTS

The authors gratefully acknowledge the help of members of the "Logical Grammar Study Group" at the University of Alberta, especially Matthew Dryer (Linguistics), David Justice (Linguistics), and Bernard Linsky (Philosophy). Their stimulating discussions each week during 1983 and 1984, and various memos they wrote on particular topics involving mass expressions, allowed us to more carefully formulate and test our ideas. Many of the things they brought up have been incorporated into this work without explicit acknowledgment. Further, we thank Harry Bunt, James McCawley, and Alice ter Meulen for their numerous astute comments on an earlier draft of this article, and hope that the revisions we have made address, to some extent, the issues they have raised.

The authors also acknowledge the aid of NSERC Grant A5525 (FJP), NSERC Grant A8818 (LKS), and Grant R0245/12 from the Deutsche Forschungs Gemeinschaft under the auspices of Christian Rohrer in the summer of 1983 (FJP).

University of Alberta

NOTES

[1] There is a terminological problem involved in even discussing the topic in a neutral manner, for most of the natural ways of saying things have been pre-empted by theory. For example, we would like a neutral manner of discussing the question of what is properly called mass. 'Mass word' here is incorrect, for some theories apply 'mass' to longer stretches of discourse. 'Mass term' is likely to engender confusion, since some theories (e.g., those inspired by Montague) reserve 'term' for the sort of phrase which can be the (complete) subject of a sentence. 'Mass noun' or 'mass noun phrase' is incorrect since some theories allow other things to be mass. 'Mass phrase' seems wrong since 'phrase' seems to require a sequence of words and further seems to presuppose that words have been already judged as forming some syntactic group. It is with some hesitation that we have adopted 'mass expression', 'count expression', and 'expression' as theory-independent descriptive phrases. It is intended that an expression can be any word, group of words, or other (perhaps smaller) meaningful unit of any syntactic category. Then we can ask such questions as "What are mass expressions?" or "Is it the senses of expressions which are mass?", and so on.

[2] The usual view, as expressed in the last paragraph, acknowledges that there are many quantifiers which both mass and count expressions accept: *all*, *most*, *a lot of*, and others. Perhaps it should be mentioned that some theorists, e.g., McCawley ([1981], p. 435) have contemplated the view that *many* and *few* (the apparent count quantifiers) and *much* and *little* (the corresponding apparent mass quantifiers) are, respectively, really the same quantifiers. Some support for this view comes from the fact that some languages use the same word for *much* as for *many*. (German *viel*, French *beaucoup*, Japanese *takusan*). (This is a peculiar argument. Usually the conslusion follows in the other direction: since language *X* distinguishes these meanings by different words, the single word is language *Y* must represent two different meanings. Let us be clear in remarking that McCawley does not endorse this position.)

It is also usually admitted that some apparently pluralized expressions are used in a mass-like manner, e.g., *beans*, *brains*, *suds*, and *physics*. These are often viewed as syntactic irregularities: *brains* just is separately lexicalized as a (unpluralizable) mass expression. (As in *Joe doesn't have much brains*). In what follows we shall regularly use *sm* for the unstressed *some*, following usage popularized by Cartwright (1965). Contrast the intonations of *Some linguists are smart* and *I want sm water*.

These tests, which we below call "the standard syntactic tests", are of course geared to test nouns (or noun phrases) and do not directly apply to other syntactic categories. We shall later investigate whether the distinction is to be extended to these other categories. For now we just note that the *central* examples come from mouns and noun phrases.

[3] Often the test is "Does it make sense to ask how many . . . ?" or "Can you count them?", etc.

[4] The issues of the last number of paragraphs are more fully discused in Pelletier [1975] and Bunt ([1981], Chap. 2).

[5] This is somewhat unfair to Bunt's theoretical exposition in ([1981], Chap. II). Here the idea was *semantic*: when a +count adjective modifies a noun, a certain interpretation is forced upon the resulting noun phrase, and similarly for when a +mass adjective modifies a noun. The discussion in the text here is really an argument against taking count/mass to be a *syntactic* well-formedness constraint on adjective-noun combinations.

[6] Appendix A of Bunt [1981] reports an experiment claiming that ordinary speakers of English (and Dutch) do react oddly to some adjective/noun combinations. One might ask,

however, whether these reactions have to do with the assumed features +mass and +count, or perhaps just to other oddities of the situation described by the experimental questions – such as the (assumed) mass noun being liquid at normal temperatures while the (assumed) count adjective presupposing a shape. (thus *round water* would be judged peculiar because water has no independent shape, not because *water* is a mass expression and *round* a count adjective).

[7] J. D. McCawley (personal communication) points out that the mass:count distinction for verb phrases also depends on the verb: *It takes two hours to bake/*eat bread*; he suggests that the difference lies in whether the verb denotes a process that consumes the object a bit-at-a-time or affects the object all at once.

[8] One might even back up a bit and point out that it is not entirely plain that *apple* in (2) need be classified as mass, since none of the standard syntactic criteria listed above directly apply to it.

[9] The issue of when a word is to be entered into the lexicon with multiple senses and when it is to be considered separate homographs is too deep to go into here. Such matters as etymology are involved, as is the more "practical" matter of "large" meaning differences giving homographs and "smaller" meaning differences giving sense distinctions. We take it to be a sufficient condition of homographs that the different words belong to different syntactic categories. Thus, for us *fast* (ADJ) is a different word from *fast* (ADV). So if +count and +mass are syntactic features or classifications, then *apple* (+count) and *apple* (+mass) would be different words. (We realize that there are borderline problems for this sufficient condition: are *threaten* (_+NP) and *threaten* (_+[VP INF]) separate words? Further, the test says nothing about how many homographs *bank* represents.)

[10] This viewpoint – one associated with Montague Grammar and related approaches – is not the only one possible, and not even the only one which will be discussed in this article. Perhaps in philosophical logic it is even more common to forego an explicit syntactic theory and to give intuitive "translations" into first order predicate logic. While we are ourselves committed to the approach outlined in the text, it is no doubt unfair in a survey to insist that all other accounts follow this methodology. In any case, much of what we intend to discuss is independent of the particular format chosen.

[11] And perhaps even later, after examining meaning postulates associated with the representation or its elements; or even, only after examining "factual axioms" (contingent properties) of the representation or its elements.

[12] Rather, it *sometimes* is seen to fall into one or the other. The "occurrence approach" can admit that with some sentences one just cannot tell whether the occurrence is mass or count. The other approaches would have to count such sentences as ambiguous: syntactically ambiguous according to the "expression approach", semantically ambiguous according to the "senses approach". The "occurrence approach" can claim that such a sentence is *un*ambiguous, but has a broad range of interpretations.

[13] Quine ([1960], p. 91ff), Davidson ([1967], fn. 9).

[14] But his final implementation in Chap. 18 is, as observed earlier, an "expression approach" rather than an "occurrence approach."

[15] In the actual implementation of his Chap. 18 there is a point to having the features +count and +mass; namely, certain syntactic rules recognize various combinations of expressions on this basis. Thus if *apple* is +count, then there will be a rule so that the syntactic parser can recognize *six apples* as well-formed. But as we indicated earlier, *apple* is marked +count already in the lexicon – this is an "expression approach" rather than the "occurrence approach" being advocated in his Chap. 2.

[16] Except for semantic reasons having to do with how to translate the sentence in question; but then +count and +mass would not be *syntactic* features. Such features ought to be marked as being part of the semantics of the expression; a syntactic feature is one which determines the syntactic properties of an expression – and that means that they should be used in some syntactic rules.

[17] Ware is quick to point out however that this test is highly context-dependent. In certain contexts, putting in half each from two apples might count as putting in an apple, but putting in half each from three apples would not.

[18] Not that he really used it anyway. As we have seen, he takes a "expression approach" in his actual grammar of Chap. 18. And even in the theoretical discussion of Chap. 2, he is willing to allow the semantic criteria to override his syntactic criteria. For example, he claims that being preceded by "an expression like 'a dozen'. 'a gross'" ([1981], p. 14) establishes a count occurrence. So in *a dozen eggs* or *a gross of wieners*, *eggs* and *wieners* are +count. Yet in *a ream of paper*, *paper* is supposed to be +mass. Similarly, when faced with *I have a great admiration for Bette*, he retracts the natural attributation of +count status to *admiration* on the grounds that it is paraphrasable as *I have much admiration for Bette*, in which he thinks – for whatever reason – *admiration* occurs in a +mass manner.

Bunt (personal communication) has told us that in his Bunt [1985] he withdraws the paraphrase test, and instead uses the expression approach of his ([1981], Chap. 18) exclusively. He also says he would prefer a syntactic occurrence approach to this syntactic expression approach but does not seen how to implement it.

[19] Differences there are, for instance change seems to be money in coin form (compare *ancient Roman coins* with ?*ancient Roman change*); but none seem to be relevant to the count/mass distinction. These examples, and the ones following, are culled from Ware [1975] and McCawley [1975].

[20] Similar remarks have been made by Gleason ([1965], pp. 136–137) and Sampson ([1975], pp. 546–547). Various people have also pointed out that certain expressions, such as *hole*, *number*, *pore*, *dilemma*, *sky*, *noun*, etc., do not seem susceptible to the grinder. See the discussion in Pelletier ([1975], pp. 6–7) and Ware ([1975], p. 19).

[21] These examples are from Ware [1975] who gives many other reasons to believe that there may just be nothing, whether in the utterance or in the speaker's mind, to determine whether an occurrence is mass or count.

[22] Of course, we have also offered evidence that everything can be referred to by either a mass or a count noun even in the same language. But this sometimes calls for non-ordinary measures, such as a universal grinder. Here we just point to ordinary circumstances.

[23] We do *not* consider those cases where this +mass or +count information is used to choose one or another "interpretations" of the longer expression. That use of +mass/+count is *semantic*, and then these features would not be syntactic. We are here interested in their usefulness as *syntactic* features, determining whether the longer expression is *syntactically* well-formed.

[24] Try it once again: *He has more book than bookshelf*, *This site has more hole than building*, *She's all woman*, *What a hunk of man*, *Donald Macdonald, six feet five inches' worth of Toronto lawyer, . . . ,* [*MacLean's*, 24 Jan. 1983], *A fine wine is a joy to drink*, *Prairie artists like to put lots of sky in their pictures*, *He claims to be caught on the horns of a dilemma but I see no horns nor much dilemma in his situation*. Given the all-pervasiveness of such constructions, why would one wish to posit that *book*, *bookshelf*, *hole*, *buildi 'g*, *woman*, *man*, *lawyer*, *sky*, *dilemma* are +count; or posit that *wine* and *joy* are +mass?

[25] Cf. Clark and Clark [1979].
[26] We restrict our discussion here to nouns, but this general classification of approaches is wider than that. One could put in here the names of all categories for which one thinks a mass/count distinction appropriate.
[27] We note that under certain theories of grammar, e.g., generalized phrase structure grammar of Gazdar *et al.* [1985], this is not possible. Any head feature (other than "bar level") of a node is inherited downward by the daughter node which is the "head". In the present example, a Noun Phrase marked +mass would have the feature +mass transmitted to the daughter Noun node (by their "head feature convention"). Perhaps this is just a counterexample to the head feature convention.
[28] The discussion of this section generally follows the more detailed critiques of Pelletier [1974] and Bunt ([1981], Chap. 3).
[29] However, on issues having to do with the completeness of theories which merge set theoretic constructs (that is, sets [or predicates] which contain [or are true of] discrete individuals, or mereological entities with "atoms") with pure mereology, see Hendry [1982].
[30] See Schubert and Pelletier [1982].
[31] Of course in such a sentence it is open to them to claim that we have an identity. This would then seem to leave them open to the charge that they cannot account for the (alleged) analyticity of *Dirty water is water*. But the charge is not necessarily well taken, for first *dirty water* may not be a nominalized predicate of the right sort, and second they may claim that in such a sentence we are *not* asserting a nominalized predicate to be in its own extension (*dirty water* and *water* are, after all, different expressions). It is only if these do not hold in their systems that they would have to fall back on denying that it is well formed. (We presume that ter Meulen would take the first way: *dirty*, when applied to *water*, does not yield a nominalized predicate.)
[32] One might also wonder whether the notion of substance abstraction is coherent: might one not get into Russell-like difficulties of the form 'the substance all of whose quantities are not quantities of themselves'? Or perhaps 'the substance all of whose quantities are substances'?
[33] A similar argument can be found in Sharvy [1975] where it is put in terms of predicates with different extensions having identical mereological fusions.
[34] The objections we have listed to the interpretation of mass noun phrases as mereological wholes do not preclude other possible roles for mereology in a theory of mass expressions; e.g., mereology may still be the right tool to account for cumulativity and divisiveness (to the extent that these hold for the referents of predicative mass expressions), or the right basis for an account of measure phrases (see Bunt [1981] and Cartwright [1975]).
[35] This is to be distinguished from the use of predicates as the interpretation of mass expressions. The latter are to be understood as *properties*, which are the kind of thing that admit of an intensional interpretation (at least in some contexts). However, in extensional contexts, the interpretation of predicates may well be a set. Thus some authors (e.g., Pelletier [1974], pp. 106–107, Bunt [1981], p. 36) often treat them together. But it should be noted that various of the logical properties of an intensional interpretation cannot be mirrored in a set-theoretic interpretation – namely, those properties which have to do with intensional contexts, such as "natural kinds", "rigid designators", and perhaps even (as we shall investigate) the feature of being alternately viewed as a predicate proper and as a (complex) singular term. For this reason we treat them separately here.
[36] See also Pelletier ([1974], pp. 92–94) and Bunt ([1981], pp. 39–40).

[37] Laycock has changed his mind on this. In his [1975] he outlines a theory wherein the use of a mass noun in applying to [undifferentiated] matter is prior to *any* use of it as being true of objects, whether by a singular or plural count noun (cf. his [1975], p. 118 n. 8). The attribution of the theory to Bacon is somewhat more shakey.

[38] These criticisms and others can be found in Pelletier ([1974], pp. 94–95) and Bunt ([1981], p. 40).

[39] It is most unclear that it really applies against any of the intensional accounts, although it seems to work against Grandy's [1973] account. It is for this reason (amongst others) that Grandy [1975] rejects the [1973] account as inadequate.

[40] Nor in related areas such as the representation of natural kind expressions, counterfactuals, and natural kinds as rigid designators.

[41] Actually his theory is stated in a rather idiosyncratic formalism. We here "translate" it into the intensional logic idiom.

[42] Later we will exhibit various classes of sentences that Bennett's (and ter Meulen's) accounts do not handle.

[43] In fact we think ter Meulen's rules do not generate either of these sentences. As for the first, which she uses as an example ([1980], pp. 183–184), her rule S10 will not operate on the top level unless the "second order terms" are a subset of the terms (which is not stated). Presumably she wishes to use S11 here, but this requires *is an element* to be a second order intransitive verb phrase. There are no basic such phrases and none of her rules generates one. As for the second similar remarks hold. Unless predicative mass terms are terms S10 does not apply. Finally, as she points out, the closest sentence to the second sentence that her rules analyze is *water is a* [are?] *wet entity*[ies].

[44] For instance some sentences using a predicative mass term do allow for backward pronominalization. *Whoever finds it can keep the gold* (cf. ter Meulen [1981], n. 2). As for the "type restricted binding", consider *Snow is always piling up in his driveway and that's why John hates it*. One reading of this sentence (indeed the preferred one) has it that John hates snow, the stuff, and not just that he hates the quantities which pile up in his driveway. Further counterexamples due to J. D. McCawley (personal communication) are: *Water is H_2O. It is often muddy*, and *Water, which is H_2O, is often muddy*. And finally, is it so clear that it is due to the predicative mass term *some milk* that, in a sentence like *every child drinks some milk*, *some milk* is given narrow scope? Cannot it be due to idiosyncracies of the verb *drink*? After all, it is very difficult for more than one child to drink the same quantity of milk. Compare *Every third world country wants some new military hardware which France is exporting – namely the Exoret II*, in which the opposite phenomenon is observed even though *new military hardware* . . . is a predicative mass term. J. D. McCawley suggests *Every philosopher believes some nonsense* as having a clear scope ambiguity.

[45] The doctrine we here attribute to ter Meulen has been stated in various places, but put like this it most closely resembles the remarks made by Kaplan ([1975], p. 518, n. 31).

[46] Or at least so ter Meulen thinks. For us, we find this evidence which relies on intuitions about possible worlds less than convincing. Just as there is a possible world in which this muddy water is not muddy, isn't there a possible world in which this water (i.e., the entity designated by the noun phrase *this water*) is not water but rather, say aqua regia? (Because in that world the fiendish waiter whom I asked for some water poured acid instead).

[47] We note here that ter Meulen's meaning postulate (6i) ([1980], p. 180) makes all bare predicative mass terms in subject position be given a universal interpretation. This seems incorrect as sentences like *Muddy water is found on mars* illustrates.

[48] This is question begging since no one would ever try to paraphrase the sentence unless convinced of ter Meulen's "type binding restriction" in the first place.

[49] Exception here should be made for those explicit theories of Bennett, ter Meulen, and Chierchia. For them we acknowledge that they have recognized the general point, but we disagree with their specific accounts.

[50] We do not wish these analyses to be taken as if they were fixed in stone, immune from revision, and clear on all details. The point rather is that any serious linguistic theory will make certain kinds of distinctions amongst "levels" of analysis of (say) being a noun phrase. Here these distinctions are put: an N^2 is the kind of syntactic entity which can be the subject of a sentence (for example), an N^1 is that N^2 considered without its determiner (e.g., *the tall man* considered without the *the*), and possibly without certain of its modifiers (e.g., *fast flight to London that we took* considered without *fast* and without *that we took*), and an N^0 is the noun as it occurs in the lexicon (here, it is that part of the N^1 *flight to London* without the prepositional phrase *to London* – i.e., the N^0 is merely *flight*). We take it that the addition of premodifying adjectives and optional postmodifying prepositional phrases, relative clauses, etc., leave the semantic category of the nominal unaltered, e.g., *flight* and *fast flight that we took* map into the same semantic category. We therefore do not distinguish them here (unlike some theories). Different linguistic theories might make somewhat different claims about these categories, but they all have to make some such distinctions, and that is the point being made here.

[51] We think that the differences in "discreteness" (atomicity) should probably be handled via meaning postulates, rather than differences in translation.

[52] The [ADJ CN]: NP means an NP formed by premodifying a CN by an adjective. We intend this discussion also to include the postmodification of a CN by a relative clause.

[53] Theoretically it is open to Bunt and Parsons to read [ADJ CN]: NP as *The substance whose quantities are ADJ quantities of CN*. Thus *pure lemonade* is the substance whose quantities are pure quantities of lemonade. But this is not very plausible. It seems to us that "quantity of ADJ CN" in general differs in meaning from "ADJ quantity of CN". For example, a "fake quantity of fur" may be a bag of cocaine topped off with (real) fur, rather than being a quantity of fake fur. Similarly, quantities of coarse sand, cube sugar, and small furniture are not the same as coarse quantities of sand, cube (cubical?) quantities of sugar, and small quantities of furniture, respectively.

[54] Possibly this role of bare NP's is also exemplified in sentences with bare NP subjects and a stative VP, e.g., (*Sm*) *Water is dripping from the faucet*, (*Some*) *Dogs are barking*.

[55] Or show how to treat *fake*, *small*, etc., as operating on the two-place relator *quantity of*, without positing new logical forms and meanings for them, distinct from those in contexts such as *fake flower*, *small chair*.

[56] We shall ignore here certain implicatures about such uses of mass NP's to the effect that it be a *natural* or a *conventionally recognized* or a proper *sub*-kind.

[57] We take no stand on the issue of whether an object is identical with the bit of matter that constitutes it. We do insist, however, that there are both types of individual; thus our lack of a position here amounts to taking no stand on whether two objects can occupy exactly the same spatial location. (E.g., the ring and the bit of gold of which it is made).

[58] This use of *kind* is to be carefully distinguished from its use as a "hedge": *This is kind of a wine*. Contrast *A zebra is a kind of horse*, *A zebra is kind of a horse*, and **A zebra is a kind of a horse*.

[59] We think there are no predicates which apply only to "objects". the usual examples one can think of, such as *is a sofa* or *is a dog*, can in reality apply also to kinds, as in *An ottoman is a sofa* and *The Beagle is a dog*. (Which we take to *imply* statements about individual ottomans and dogs, but not to be equivalent to them. They seem to us to rather be equivalent to *The ottoman is a kind of sofa* and *The beagle is a kind of dog*, making *is a sofa* and *is a dog* be used as a kind designator.) Even *is a puddle* and *is a quantity of gold* can be used in this way: consider *This nugget is a quantity of gold*, which uses the predicate to be be true of an "object", and *A nugget is a quantity of gold*, which uses it to be true of the kind. (It may be inaccurate, but it is so used). In the sequel we shall make no assumption that any English predicate is true only of individuals, but will instead introduce into the formal semantics an operator which forces such a reading in certain contexts. We take the view that most English predicates have the mixed, comprehensive extension; and that some few others are true only of kinds.

[60] Established by the *Water weighs billions of tons* examples (in addition to the evidence that the denotation of a mass NP must be intensional).

[61] The appropriate sense of "conventional" appears to be "bearing a proper name". Thus *Sauterne*, being a proper name, designates a conventional kind while *red wine* does not.

[62] Implausible because it incorrectly predicts that *Tap water is water and is in short supply* or *Fake fur is not fur, yet sells well* will get a universally quantified subject. But their meanings are not preserved when *all tap water* or *all fake fur* are substituted for the respective subjects.

[63] More plausible perhaps for sentences like *Water is dripping from the faucet*, but consider *Water, though scarce in this part of the world, is dripping from the faucet*, wherein *water* cannot be replaced by *some water*.

[64] This is reminiscent of Parsons' substance abstraction operator, although it is not defined in terms of quantities as his is. We also do not take a position on the question of whether these are to be distinguished from the intension of M, i.e., from $^\wedge\mathbf{M}$.

[65] It remains unclear to us whether non-stative verb phrases should be viewed as predicates which apply to stages rather than to things, as Carlson and Chierchia claim. While the claim seems plausible for progressive verb phrases, it seems much less plausible for non-stative verb phrases in the simple past tense, such as *threw snow at Mary* or *melted and then evaporated*; such phrases appear to apply to temporal segments of things which are longer than mere "instantaneous" stages.

[66] For example, we follow our practice of earlier and only distinguish CN's from NP's, rather than other intermediate levels of NP's.

[67] For details see Schubert and Pelletier [1982].

[68] This can be thought of as a quantifier **DET**′ which is *restricted* by the predicate **N**′. This restricted quantifier is assigned a scope in a post-parsing phase. For further details on how these theories handle quantified terms, see Schubert and Pelletier [1982]. Here we forgo discussion of "quantifier scoping" and the like, but rather just assert that the stated translations actually come out as given. The quantified terms are those in angle brackets and are treated specially.

[69] The *s*-theory takes SING to be a node of an agreement feature hierarchy, with root NUMB, with disjoint successors SING and PLUR, and with SING in turn subdivided into MASS and COUNT. Thus COUNT applies to singular count nouns and determiners only. A node agrees with any of its descendants, including itself.

[70] This is one of the places where a fuller grammar would distinguish different "levels" of noun phrases.

71 We do not wish to enter the dispute as to whether there really are any truly intersective adjectives. Cf. Kamp [1975] and Siegel [1979] for statements of the views. On the one hand, even the colour terms do not seem to be clear examples; for example, white wine is hardly white! On the other hand, we *can* take many (perhaps most) adjectives as truly intersective, as long as we recognize that their meanings are context-dependent. Thus, white wine is indeed white (note that we do say things like *This wine is white*), but in a sense appropriate to wines. After all, even nouns can show such context dependence. *This hay is food* has a truth value dependent on whether we are talking about food for people or food for cattle, for example. Moreover, the interaction of an attributive adjective with the noun it modifies does not exhaust its context-dependence: the truth of *He is a tall man* still depends on whether we are talking about horse racing or basketball.

72 Where in the *s*-theory, *all* is [DET +MASS] (when restricted to the singular).

73 For the more difficult quantifiers, such as *most*, the formal translation would be much like those in 11–13, viz.,

$$\textbf{most}' = (\lambda P)\langle \text{MOST } P\rangle.$$

However, a direct reduction to standard first-order format, as for *A* and *E*, is not possible for MOST. (Cf. Barwise and Cooper [1981]). Rather, a semantics would have to be supplied for the unreduced, predicate-restricted, scoped quantifier; alternatively, some sort of set theory could be introduced to allow translation of statements involving such quantifiers into first-order statements about set cardinality. For discussion along these lines, see McCawley ([1981], pp. 425–433).

74 The two theories get this result in different ways: *liquid* is an adjective which has comprehensive extension, and is introduced by rule 10, in both theories. *a liquid* is a PRED and introduced by rule 4 from the noun *liquid*. In the *p*-theory, the noun *liquid* has comprehensive extension, but *a liquid* gets its meaning restricted to kinds/servings by rule 4p. In the *s*-theory, *liquid* is marked as MASS, and a lexical extension rule converts this to a COUNT noun with the meaning "kind of liquid"; and finally rule 4s says that the PRED has the meaning of this derived COUNT noun.

75 Perhaps it should be noted that many VP's have both a stative and non-stative reading, for example in *Gold glitters in the mines* or in *orang-utans inhabit the jungles of Borneo*.

76 Since both theories take *is a wine* to be a predicate true only of conventional subkinds of wine, *wine* – as a "formal" kind but not a conventionally recognized subkind of wine – will not fall in the extension of *is a wine*. For the same reason, *Chilled wine is a wine*, *Cheap wine is a wine*, *Claret mixed with Sauterne is a wine*, etc., are false.

77 Of course we believe to have already shown that our theories can account for the analyticity of certain sentences, the anomoly of certain sentences, the validity of certain inferences, and so on, which no other theory can account for. Here we give further examples.

74 In particular, the following is a valid inference in the semilattice: $A(B)$, $B(C)$ therefore $A(C)$. Replace *A* by *liquid*, *B* by *wine*, and *C* by *claret* for our inference.

BIBLIOGRAPHY

Allan, K.: 1977, 'Classifiers', *Language* **53**, 285–311.

Allan, K.: 1980, 'Nouns and countability', *Language* **56**, 451–467.

Allan, K.: 1981, 'Review of U. Weinreich *On Semantics*', *Language* **57**, 941–948. (esp. 947).

Åqvist, L. and Guenthner, F.: 1978, 'Fundamentals of a theory of verb aspect and events within the setting of an improved tense-logic', in F. Guenthner and C. Rohrer (eds.), *Studies in Formal Semantics* (North Holland: Amsterdam), 167–199.

Bacon, John: 1973, 'Do generic descriptions denote?' *Mind* **82**, 331–347.

Bahm, A.: 1976, 'Degrees and scales' *ITA Humanidades* **12**, 67–73.

Barwise, J. and R. Cooper: 1981, 'Generalized quantifiers and natural language' *Linguistics and Philosophy* **4**, 159–219.

Bennett, Michael: 1977, 'Mass nouns and mass terms in Montague grammar' in S. Davis and M. Mithun *Linguistics, Philosophy, and Montague Grammar* (Austin: Univ. of Texas Press), 263–285.

Bunt, H. C.: 1976, 'The formal semantics of mass terms', *Papers from the IIIrd Scandinavian Conference of Linguistics*, Helsinki.

Bunt, H. C.: 1978, 'A formal semantic analysis of mass terms and amount terms', in J. Groenendijk and M. Stokhof (eds.), *Amsterdam Papers in Formal Grammar*, Vol. II.

Bunt, H. C.: 1979, 'Ensembles and the formal semantic properties of mass terms' in F. J. Pelletier (1979), 249–277.

Bunt, H. C.: 1980, 'On the why, the how, and the whether of a count-mass distinction among adjectives' in J. Groenendijk and M. Stokhof (eds.), *Formal Methods in the Study of Language*, (Mathematical Centre: Amsterdam), 51–77.

Bunt, H. C.: 1981, *The Formal Semantics of Mass Terms*, Dissertation, University of Amsterdam.

Bunt, H. C.: 1985, *Mass Terms and Model Theoretic Semantics*, (Cambridge Univ. Press: Cambridge).

Burge, Tyler: 1972, 'Truth and mass terms', *Journal of Philosophy* **69**, 263–282.

Burge, Tyler: 1975, 'Mass terms, count nouns, and change', *Synthese* **31**, 459–478. Reprinted in Pelletier (1979), 199–218. Page references to reprint.

Carlson, G.: 1977, 'A unified analysis of the English bare plural', *Linguistics and Philosophy* **1**, 413–456.

Carlson, G.: 1982, 'Generic terms and generic sentences', *Jour. Philosophical Logic* **11**, 163–182.

Cartwright, H. M. R.: 1965, 'Heraclitus and the bath water' *Philosophical Review* **74**, 466–485.

Cartwright, H. M. R.: 1970, 'Quantities' *Philosophical Review* **79**, 25–42.

Cartwright, H. M. R.: 1975, 'Amounts and measures of amounts' *Noûs* **9**, 143–164. Reprinted in Pelletier (1979), 179–198. Page references to reprint.

Cheng, C-Y.: 1973, 'Response to Moravcsik', in Hintikka *et al.* 286–288.

Chierchia, G.: 1982a, 'Bare plurals, mass nouns and nominalization' in D. Flickinger, M. Macken, and N. Wiegand (eds.), *Proceedings of the First West Coast Conference on Formal Linguistics*, 243–255.

Chierchia, G.: 1982b, 'Nominalization and Montague grammar. A semantics without types for natural language', *Linguistics and Philosophy* **5**, 303–354.

Chierchia, G.: 1983, 'On plural and mass nominals and the structure of the world' in Borowsky, T. and D. Finer (eds.), *Univ. of Massachusetts Occasional Papers, VIII*, Amherst: GLSA.

Clark, H. and E. Clark: 1979, 'When nouns surface as verbs' *Language* **55**, 767–811.

Clarke, D. S.: 1970, 'Mass terms as subjects', *Philosophical Studies* **21**, 25–29.

Cocchiarella, N.: 1976, 'On the logic of natural kinds', *Philosophy of Science* **43**, 202–222.

Cocchiarella, N.: 1977, 'Sortals, natural kinds, and re-identification', *Logique et Analyse*, 439–474.

Cocchiarella, N.: 1978, 'On the logic of nominalized predicates and its philosophical interpretations' *Erkenninis* **13**, 339–369.

Cook, K.: 1975, 'On the usefulness of quantities' *Synthese* **31**, 443–457. Reprinted in Pelletier (1979), 121–135. Page references to reprint.

Davidson, D.: 1976, 'Truth and meaning', *Synthese* **17**, 304–323.

Feldman, F.: 1973, 'Sortal predicates' *Noûs* **7**, 268–282.

Gabbay, D. and Moravcsik, J. M. E.: 1979, 'Verbs, events and the flow of time', in C. Rohrer (ed.), *Time, Tense and Quantifiers* (Niemeyer: Stuttgart), 59–84.

Gazdar, G., E. Klein, G. Pullum, and I. Sag: 1985, *Generalized Phrase Structure Grammar* (Harvard UP: Cambridge, Mass.).

Geach, P.: 1962, *Reference and Generality* (Cornell UP: Ithaca).

Gleason, H. A.: 1965, *Linguistics and English Grammar* (New York: Holt, Rinehart & Winston), esp. 134–137.

Grandy, R.: 1973, 'Reply to Moravcsik' in Hintikka *et al.* (1973), 295–300.

Grandy, R.: 1975, 'Stuff and things', *Synthese* **31**, 479–485. Reprinted in Pelletier (1979), 219–225. Page references to the reprint.

Hansen, C. D.: 1976, 'Mass nouns and 'A white horse is not a horse'', *Philosophy East and West* **26**, 189–209.

Hendry, H.: 1982, 'Complete extensions of the calculus of individuals' *Noûs* **16**, 453–460.

Hestevold, H. S.: 1978, *A Metaphysical Study of Aggregates and Continuous Wholes*, Ph.D. dissertation, Brown Univ.

Hintikka, J., J. M. E. Moravcsik, and P. Suppes (eds.): 1973, *Approaches to Natural Language* (D. Reidel: Dordrecht).

Hoeksema, J.: 1983, 'Plurality and conjunction', in A. ter Meulen (ed.), *Studies in Model-theoretic Semantics* (Foris: Dordrecht), pp. 63–83.

Hoepelman, J.: 1976, 'Mass nouns and aspects, or: Why we can't eat gingercake in an hour', in J. Groenendijk and M. Stokhoff (eds.), *Amsterdam Papers in Formal Grammar*, Vol. I (Math. Centrum: Amsterdam).

Hoepelman, J.: 1978, 'The treatment of activity verbs in a Montague-type grammar, a first approximation' in F. Guenthner and C. Rohrer (eds.), *Studies in Formal Semantics* (North Holland: Amsterdam), 121–165.

Hoepelman, J. and C. Rohrer: 1980, 'On the mass-count distinction and the French imparfait and passé simple', in C. Rohrer (ed.), *Time, Tense and Quantifiers* (Niemeyer: Tübingen), 85–112.

Kamp, J.: 1975, 'Two theories about adjectives', in E. Keenen (ed.), *Formal Semantics of natural Language* (Cambridge UP: London), 123–155.

Kaplan, D.: 1973, 'Bob and Carol and Ted and Alice', in Hintikka *et al.* (1973), 490–518.

Klooster, W.: 1972, *The Structure Underlying Measure Phrase Sentences* (D. Reidel: Dordrecht).

Laycock, H.: 1972, 'Some questions of ontology', *Philosophical Review* **81**, 3–42.

Laycock, H.: 1975, 'Theories of matter', *Synthese* **31**, 411–442. Reprinted in Pelletier (1979), 89–120. Page references to the reprint.

Leech, G. N.: 1969, *Towards a Semantic Description of English* (Longman: London).

Link, G.: 1981, 'The logical analysis plurals and mass terms: A lattice-theoretical approach' in Bäuerle, R., C. Schwartze, and A. von Stechow (eds.), *Meaning, Use and Interpretation of Language* (de Gruyter: Berlin), 302–323.

Malotki, Ekkehart: 1983, *Hopi Time: A Linguistic Analysis of the Temporal Concepts in the Hopi Language*, (Mouton Pub.: Berlin).

Massey, G. J.: 1976, 'Tom, Dick, and Harry, and all the king's men', *American Philosophical Quarterly* **13**, 89–107.

McCawley, J.: 1975, 'Lexicography and the count-mass distinction', *Proceedings of the First Annual Meeting of Berkeley Linguistics Society*, 314–321.

McCawley, J.: 1981, *Everything the Linguists Always Wanted to Know About Logic* (Univ. Chicago Press: Chicago).

Mellama, G.: 1981, 'On quantifiers and mass terms', *American Philosophical Quarterly* **18**, 165–170.

Montague, R., 1973, 'Reply to Moravcsik' in Hintikka *et al.* (1973), 289–294, Reprinted in Pelletier (1979), 173–178, as 'The proper treatment of mass terms in English'.

Montague, R.: 1973a, 'The proper treatment of quantifiers in English', in Hintikka *et al.* (1973), 221–242.

Moravcsik, J. M. E.: 1973, 'Mass terms in English' in Hintikka *et al.* (1973), 263–285.

Mourelatos, A. P. D.: 1978, 'Events, processes, and states' *Linguistics and Philosophy* **2**, 415–434.

Mufwene, S. S.: 1980, 'Number, countability and markedness in Lingala LI-/MA- noun class', *Linguistics* **18**, 1019–1052.

Mufwene, S. S.: 1981, 'Non-individuation and the count mass distinction', *Chicago Linguistic Society Papers* **20**, 221–238.

Mufwene, S. S.: 1983, 'The proper name/common noun distinction', paper presented at Winter Meeting of Linguistic Society of America, Minneapolis.

Mufwene, S. S.: 1984, 'The count mass distinction and the English lexicon', *Chicago Linguistics Society: Parasession on Lexical Semantics.*

Nola, R.: 1980, ''Paradigms lost, or, the world regained' – An excursion into realism and idealism in science', *Synthèse* **45**, 317–350.

Noonan, H.: 1978, 'Count nouns and mass nouns', *Analysis* **38**.

Parsons, T.: 1970, 'An analysis of mass and amount terms' *Foundations of Language* **6**, 363–388. Reprinted in Pelletier (1979), 137–166, Page references to the reprint.

Parsons, T.: 1975, 'Afterthoughts on mass terms' *Synthese* **31**, 517–521. Reprinted in Pelletier (1979), 167–171. Page references to the reprint.

Pelletier, F. J.: 1974, 'On some proposals for the semantics of mass terms' *Journal of Philosophical Logic* **3**, 87–108.

Pelletier, F. J.: 1975, 'Non-singular reference: Some preliminaries', *Philosophia* **5**. Reprinted in Pelletier (1979), 1–14. Page references to the reprint.

Pelletier, F. J. (ed.): 1979, *Mass Terms: Some Philosophical Problems* (D. Reidel: Dordrecht).

Pelletier, F. J.: 1979a, 'A bibliography of recent work on mass terms' in Pelletier (1979), 295–298.

Quine, W. V.: 1960, *Word and Object* (MIT Press: Cambridge).

Quine, W. V.: 1964, 'Review of [Geach 1962]', *Philosophical Review* **73**, 100–105.

Quirk, R., S. Greenbaum, G. Leech, and J. Svartvik: 1972, *A Grammar of Contemporary English* (Longman: London).

Reeve, C. D.: 1980, *Mass, Quantity and Amount*, Ph.D. dissertation, Cornell University.

Roeper, P.: 1983, 'Semantics for mass terms with quantifiers', *Noûs* **17**, 251–265.

Sampson, G.: 1975, 'Review of Hintikka *et al.* (1973)', *Foundations of Language* **12**, especially pp. 546–547 (review of Moravcsik 1973).

Scha, R.: 1980, 'Distributive, collective and cumulative quantification', in J. Groenendijk, T. Janssen, and M. Stokhof (eds.), *Formal Methods in the Study of Language*, Vol. II, 483–512.

Schubert, L. K. and F. J. Pelletier: 1982, 'From English to logic: Context-free computation of 'conventional' logical translation', *American Journal of Computational Linguistics* **8**, 26–44.

Sharvy, R.: 1975, 'The indeterminacy of mass predication', in Pelletier (1979), 47–54.

Sharvy, R.: 1978, 'Maybe English has no count nouns: Notes on Chinese semantics', *Studies in Language* **2**, 345–365.

Siegel, M.: 1977, 'Measure adjectives in Montague grammar', in Davis, S. and M. Mithun (eds.), *Linguistics, Philosophy, and Montague Grammar* (Univ. Texas Press: Austin), 223–262.

Simons, P.: 1981, 'Plural reference and set theory' in B. Smith (ed.), *Parts and Moments* (Philosophia: Munich), 199–260.

Smith, R.: 1978, 'Mass terms, generic expressions, and Plato's theory of forms', *Jour. History of Philosophy* **16**, 141–153.

Strawson, P.: 1977, 'Tense and continuity', *Linguistics and Philosophy* **1**, 199–220.

ter Meulen, A.: 1980, *Substances, Quantities and Individuals*, Ph.D. Dissertation, Stanford University. Available through Indiana University Linguistics Club.

ter Meulen, A.: 1981, 'An intensional logic for mass terms', *Philosophical Studies* **40**, 105–125.

Vendler, Z.: 1967, *Linguistics in Philosophy* (Cornell UP: Ithaca), Chapter 4.

Ware, R. X.: 1975, 'Some bits and pieces' *Synthese* **31**, 379–393. Reprinted in Pelletier (1979), 15–29. Page references to the reprint.

Whorf, B.: 1939, 'The relation of habitual thought and behavior to language' in J. Carrol (ed.), *Language, Thought and Reality: Selected Writings of Benjamin Lee Whorf* (MIT Press: Cambridge), 1956, 134–159.

Works on Mass Expressions which have been drawn to our attention since this article was prepared (1984).

Bricker, P.: 1988, 'Review of [Bunt, 1985]', *Journal of Symbolic Logic* **53**, 653–656.

Heintz, J.: 1985, 'Do lambs have fur?' Paper read at Canadian Philosophical Association meetings, Montreal.

Laycock, H.: 1985, 'Mass terms again'. Paper read at Canadian Philosophical Association meetings, Montreal.

Pólos, L.: 1985, 'Mass nouns: A semantic representation' (in Hungarian), *Tertium non Datur* **2**.

Pólos, L.: 1987a, 'Mass nouns, plurals, events' (in Hungarian), *Tertium non Datur* **4**.

Pólos, L.: 1987b, 'Mass terms: A found key to Parmeuides' (in Hungarian) *Tertium non Datur*, **4**.

CHAPTER IV.5

REFERENCE AND INFORMATION CONTENT: NAMES AND DESCRIPTIONS*

by Nathan Salmon

Language is primarily a medium of communication, particularly the communication of information. (Of course, language is more than just a device for communication.) Declarative sentences are the primary vehicles for the communication of information. Declarative sentences have various semantic attributes. They have truth value, either truth or falsehood. They have semantic *intensions*, i.e., correlated functions from possible worlds to truth values, and so on. But the fundamental semantic role of a declarative sentence is to encode a piece of information.[1] This fundamental semantic role of encoding information lies behind many of the other semantic and pragmatic characteristics of declarative sentences. For example, since declarative sentences encode information, they may be used not only to communicate information to others, but also to record information for possible future use, and perhaps even to register information with no anticipation of any future use. A declarative sentence may be said to *contain* the piece of information it encodes, and that piece of information may be described as the *information content* or *cognitive content* of the sentence. The piece of information encoded by a sentence is what philosophers generally mean when they speak of the "proposition" expressed by the sentence.

D. Gabbay and F. Guenthner (eds.), Handbook of Philosophical Logic, Volume IV, 409–461.

Some of the central and most puzzling questions in the philosophy of language arise from these simple observations. To begin with, what is the nature of information, and what is the nature of the semantic connection between a declarative sentence and its information content?

A preliminary philosophical investigation reveals that pieces of information are ontologically complex; they have components. This is apparent from consideration of distinct pieces of information having components in common. The information that Socrates is wise has some component in common with the information that Socrates is snub-nosed, since both of these are directly about Socrates, and some other component again in common with the information that Plato is wise, since both of these directly ascribe wisdom. These two components of the information that Socrates is wise are separately semantically correlated with the two components that make up the sentence 'Socrates is wise' – the name 'Socrates' and the predicate 'is wise'. Together, these two information components exhaust the information encoded by the sentence.

Let us call the information component semantically correlated with an expression the *information value* of the expression. The information value of the name 'Socrates' is that which the name contributes to the information encoded by such sentences as 'Socrates is wise' and 'Socrates is snub-nosed', and similarly the information value of the predicate 'is wise' is that which the predicate contributes to the information encoded by such sentences as 'Socrates is wise' and 'Plato is wise'. The systematic method by which it is secured which information is semantically encoded by which sentence is, very roughly, that a sentence semantically encodes that piece of information whose components are the information values of the sentence parts, with these information values combined in a manner parallel to that in which the sentence parts are themselves combined to form the sentence.[2] In order to analyze the information encoded by a sentence into its components, one simply decomposes the sentence into its information-valued parts, and the information values thereof are the components of the encoded information. In this way, declarative sentences not only encode, but also codify, information.

For simplicity of exposition we may suppose that the meaningful expressions that make up a language – for example English, or some rich fragment thereof – are of the following logical categories:

Singular terms, that is, expressions that, when used in a particular context, are to be taken as *referring to* (denoting, designating), i.e., as standing for, a single individual or thing. This category includes proper names, demonstratives, certain other single-word indexicals and pronouns such as 'you', 'he', etc., and individual variables. Usually regarded as singular terms are singular definite descriptions, i.e., singular noun phrases formed from the definite article 'the' or from a possessive adjective (such as 'the oldest living American', 'John's first wife', etc.). Also included may be abstract nouns (such as 'happiness') and 'that'-clauses (such as 'that all men are created equal'). The *extension* of a singular term (with respect to such semantic parameters as an assignment of values to variables, a context of utterance, a possible world, or a time) is just its referent (with respect to those parameters).

(*First-order*) *n-place predicates* ($n > 0$), which *attach to* an *n*-ary sequence of singular terms to form a sentence. A *monadic* predicate is one-place, e.g., 'is wise'; a *dyadic* predicate is two-place, e.g., 'is taller than'. A *polyadic* predicate is an *n*-place predicate where $n > 1$, e.g., '______ gives ______ to ______'. An *n*-place predicate *applies to*, or fails to apply to, an *n*-ary sequence of objects. Thus, the predicate 'is taller than' applies to an ordered pair $\langle x, y\rangle$ if and only if *x*'s height is greater than *y*'s. The *extension* of a (typical) *n*-place predicate (with respect to semantic parameters) is the class of *n*-ary sequences to which the predicate applies (with respect to those parameters).

n-place sentential connectives ($n > 0$), which combine with an *n*-ary sequence of sentences to form a sentence, e.g., 'if and only if'. The *extension* of a truth-functional *n*-ary connective (with respect to semantic parameters) is an *n*-ary function from truth values to truth values.

(*First-order*) *n-place quantifiers* ($n > 0$), which attach to an *n*-ary sequence of (first-order) predicates (or to an *n*-ary sequence of pairs consisting of a variable together with an open or closed sentence) to form a sentence, e.g., 'a man', 'something', 'for everyone', 'there are three books which', etc. A quantifier may be *restricted*, e.g., 'some cat', or *unrestricted*, e.g., 'something'. A quantifier of the form ⌜every ϕ⌝, ⌜each ϕ⌝, ⌜any ϕ⌝, etc. is *universal*, whereas ⌜a ϕ⌝, ⌜some ϕ⌝, ⌜at least one ϕ⌝, ⌜there is a ϕ which⌝, etc. are *existential*. Quantifiers may be regarded as *second-order predicates*, i.e., predicates attachable to (sequences of) first-order predicates (or variable/sentence pairs) rather than to (sequences of) singular terms.

Other *operators*, which attach to (sequences of) singular terms, predicates, or sentences, to form singular terms, predicates, or sentences, e.g., 'necessarily', tense, and possibly adverbs. This category includes the *definite–description operator*, i.e., the definite article 'the' in the sense of 'the one and only', which attaches to a monadic predicate (or to a variable together with an open or closed sentence) to form a singular term, and the 'that'-operator, which attaches to a sentence to form a singular term referring to the information content of the sentence.

Sentences. We consider here only *declarative sentences*, open or closed, where an *open* sentence is an expression of the grammatical form of a sentence but for the presence of a free variable, e.g., '*x* is wise'. Sentences are the encoders of information, and when used in a particular context, are to be taken as being either true or false (with respect to semantic parameters, under an assignment of values to their free variables). The *extension* of a sentence (with respect to semantic parameters) is its truth value (with respect to those parameters).

The central philosophical question concerning information content can now be stated as follows: What are the information values of the various expressions that make up a language? What, for example, is the information value of a singular term, such as 'Socrates'? What is the information value of a predicate, such as 'is wise'? What is the information value of a sentential connective? Similarly for the remaining categories. The present article is concerned primarily with the question of the information value of singular terms, particularly proper names and singular definite descriptions.

THE NAIVE THEORY

One natural and elegantly simple theory of information value identifies the information value of a singular term, as used in a particular context, with its *referent* in that context, that is, with the object or individual that the term refers to with respect to that context. Gilbert Ryle called this *the 'Fido'-Fido theory*. Elements of this theory can be traced to ancient times. Likewise, the information value of a predicate, as used in a particular context, might be identified with the semantically associated attribute with respect to that context (or something similar) – that is, with the corresponding property in the case of a monadic predicate, or the corresponding *n*-ary relation in the case of an *n*-place polyadic predicate. Thus for example, the information value of the predicate 'is tall' might be identified with the property of being tall and the information value of the predicate 'is taller than' might be identified with the binary relation of being taller than. (A sophistocated version of this theory identifies the information value of a predicate, as used on a particular occasion, with a corresponding temporally indexed (and spatially indexed, if necessary) attribute, for example the property of being tall at *t*, where *t* is the time of utterance. This yields a more plausible notion of information.)[3] On this theory, the information encoded by a simple atomic subject-predicate sentence such as 'Socrates is wise', as used in a particular context, is what David Kaplan has called a *singular proposition* – a complex abstract entity consisting partly of things like properties, relations, and concepts, and partly of individuals, such as the man Socrates. By contrast, a (*purely*) *general proposition* is made up entirely of the former sorts of entities (in a certain way). On this theory, a sentence is a means for referring to its information content, by specifying the components that make it up. A sentential connective may be construed on the model of a predicate. The information value of a connective would thus be an attribute (property if monadic, relation if polyadic) – not an attribute of individuals such as Socrates, but an attribute of pieces of information, or propositions. For example, the information value of the connective 'if and only if' might be identified with the binary equivalence relation between propositions having the same truth value. Similarly, the information value of a quantifier might be identified with a property of properties of individuals. For example, the information value of the unrestricted universal quantifier 'everything' may be the second-order property of being a universal first-order property, i.e., the property of being a property possessed by every individual. The information value of a sentence, as used in a particular context, is simply its information content, the proposition made up of the information values of the information-valued sentence components.

This may be called *the naive theory* of information value. Its central theses may be listed as follows:

(I) (Declarative) sentences encode pieces of information, called *propositions*. The proposition encoded by a sentence, with respect to a given context, is its *information content* with respect to that context.

(II) The information content, with respect to a given context, of a sentence is a complex, ordered entity (something like a sequence) whose constituents are semantically correlated systematically with expressions making up the sentence, typically the simple (noncompound) component expressions. Exceptions arise with quotation marks and similar devices.

(III) The information value (contribution to information content), with respect to a given context *c*, of any singular term is its referent with respect to *c* (and the time of *c* and the possible world of *c*).

(IV) Any expression may be thought of as *referring*, with respect to a given context, time, and possible world, to its information value with respect to that context (and time).

(V) The information value, with respect to a given context, of a first-order *n*-place predicate is an *n*-place attribute (a property if $n = 1$, an *n*-ary relation is $n > 1$) – ordinarily an attribute ascribed to the referents of the attached singular terms. Exceptions arises in connection with quotation marks and similar devices.

(VI) The information value, with respect to a given context, of an *n*-place sentential connective is an attribute, ordinarily of the sorts of things that serve as referents for the operand sentences.

(VII) The information value, with respect to a given context, of an *n*-place quantifier or second-order predicate is an *n*-ary attribute, ordinarily of the sorts of things that serve as the referents for the operand first-order predicates.

(VIII) The information value, with respect to a given context, of an operator other than a predicate, a connective, or a quantifier is an appropriate attribute (for sentence-forming operators), or operation (for other types of operators), ordinarily an attribute of or operation on the sorts of things that serve as referents for its appropriate operands.

(IX) The information value, with respect to a given context, of a sentence is its information content, the encoded proposition.

For all its naiveté, the naive theory is a powerful and extremely compelling theory. It is a natural result – perhaps *the* natural result – of a preliminary investigation into the nature and structure of information. Both of the two greatest figures in the philosophy of language, Gottlob Frege and Bertrand Russell, came to the subject with an initial presupposition of some rudimentary form of the naive theory. The theory yields a plausible rendering of the observation that the information that Socrates is wise is information *about* or *concerning* Socrates: the proposition is about Socrates in the straightforward sense that Socrates is an individual constituent of it. The naive theory gives substance to the oft-repeated slogan that to give (or to know, etc.) the semantic content or "meaning" of a sentence or statement is to give (know, etc.) its truth conditions. Its notion of information content is exemplary of the kind of notion of proposition that is needed in connection with questions of so-called *de re* modality: If I utter the sentence 'Socrates is wise', I assert something such that it is true if and only if the individual, Socrates, has the property, wisdom. Moreover, what I assert is such that it is true with respect to an arbitrary possible world *w* if and only if that same condition, the very individual Socrates having wisdom, obtains *in w*. It is not enough, for instance, that someone in *w* who resembles or represents the actual Socrates in a certain way be wise in *w*, or that someone in *w* who fits a certain conceptual representation of the actual Socrates be wise in *w*. It must be the very individual Socrates. The naive theory also yields a straightforward notion of *de re* belief, and other *de re* propositional attitudes: to believe *p* is to believe the proposition *p*. So to believe *of* or *about* Socrates that *he* is wise is to believe the proposition *of* or *about* Socrates that *he* is wise, that is, the piece of information consisting of Socrates and wisdom. Perhaps the most important thing to be said for the naive theory is its cogency and intuitive appeal as a theory of assertion. When I utter 'Socrates is wise', my speech act divides into two parts: I pick someone out, Socrates, and I ascribe something to him, wisdom (at *t*). These two component speech acts – reference and ascription – correspond to two components of *what I assert* when I assert that Socrates is wise. My asserting that Socrates is wise consists in my referring to Socrates and my ascribing wisdom (at *t*) to him; so too, that Socrates is wise (what I assert) consists of Socrates (what I refer to) and wisdom (at *t*) (what I ascribe to him).

THE MODIFIED NAIVE THEORY

The naive theory rests upon two central ideas. The first is the identification of the information value of a singular term with its referent, i.e., the 'Fido'-Fido theory. By analogy, the referent of a predicate, connective, quantifier, etc. is identified with its information value. The second major idea is that the information value of a sentence, as uttered on a particular occasion, is made up of the information values of its information-valued parts. Unfortunately, these two ideas come into conflict in special cases, specifically the case of definite descriptions. According to the naive theory, the information value of a definite description such as 'the individual who wrote *the Republic*' is simply its referent, Plato. Consequently, the information encoded by such a sentence as 'The individual who wrote *the Republic* is wise' is to be the singular proposition about Plato that he is wise. But the definite description is a phrase which, like a sentence, has parts with identifiable information values – for example the dyadic predicate 'wrote' and the singular term (book title) '*the Republic*', as well as the monadic predicate 'wrote *the Republic*'. These information-valued components of the definite description are *ipso facto* information-valued components of the containing sentence. If the information value (= information content) of a sentence is made up of the information values of its information-valued parts, the information values of these description components must also go in to make up part of the information that the author of *the Republic* is wise. And if the information value of a sentence is something made up of the information values of its information-valued parts, it stands to reason that the information value of a definite description, which is like a sentence at least in having information-valued parts, should also be something made up of the information values of those parts. Thus, instead of identifying the information value of 'the individual who wrote *the Republic*' with its referent, one should look instead for some complex entity made up partly of something like the relational property of having written *the Republic* at some time earlier than t (which in turn is made up of the relation of *having written earlier than* t, and the work *the Republic*), where t is the time of utterance, and partly of whatever serves as the information value of the definite–description operator 'the'. On this modified version of the naive theory, the information that the author of the Republic is wise is not the singular proposition about Plato that he is wise, but a different piece of information, one that does not have Plato as a component and has in his place something involving the property of authorship of *the Republic*.

One extremely important wrinkle in the modified naive theory is that a definite description ⌜the ϕ⌝, unlike other sorts of singular terms, is seen as having a two-tiered semantics. On the one hand, there is the description's referent. This is the individual to which the description's constitutive monadic predicate ϕ applies, if there is only one such individual, and is nothing otherwise. On the other hand, there is the description's information value. This is a complex made up, in part, of the information value of the predicate ϕ. By contrast, a proper name or other simple singular term is seen as having a one-tiered semantics: its information value (with respect to a particular context) is just its referent (with respect to that context). From the point of view of the modified naive theory, the original naive theory errs by treating definite descriptions on the model of a proper name. Definite descriptions are not single words but phrases, and therefore have a richer semantic structure.

Given the modification of the naive theory considered thus far, other sorts of simple expressions, such as single-word connectives and quantifiers, may be treated on the model of a proper name, but any expression other than a simple singular term is, at least in principle, capable of having a two-tiered semantics. For example, though the information value of a sentence is its information content, sentences might be regarded as referring to their truth values, or to facts, or to states of affairs, etc. The modified naive theory, as defined so far, is tacit on the question of the referents of expressions other than singular terms (sentences, predicates, connectives, quantifiers, and other operators), and may be properly extended in any one of several directions. An extremely compelling argument, due to Alonzo Church and independently to Kurt Gödel, however, seems to show that if a notion of reference is to be extended to expressions other than singular terms, then an expression of any variety is best regarded as referring to its semantic extension.[4] Thus, an n-ary predicate is best regarded as referring to the set of n-tuples to which it applies, a truth-functional sentential connective to its associated function from truth values to truth values, a quantifier or other second-order predicate to the class of classes of individuals to which it applies, a sentence to its truth value, etc. From the following three assumptions a proof may be given that if an expression of any variety is a referring expression, then any two co-extensional such expressions are also co-referential: (*i*) a definite description ⌜the ϕ⌝ refers to the unique individual to which the constitutive predicate (or open formula) ϕ applies, if there is exactly one such individual; (*ii*) trivially logically equivalent referring expressions are co-referential; and (*iii*) barring such devices as quotation marks, the referent of a compound referring expression is preserved when a component singular term is replaced by one that is co-referential with it. Since each of these assumptions represents a fundamental feature of singular term reference, any attempt to extend the notion of reference to other sorts of expressions ought to respect these assumptions.

Accordingly, the central theses of the modified naive theory may be given as follows:

(I) (Declarative) sentences encode pieces of information, called *propositions*. The proposition encoded by a sentence, with respect to a given context, is its *information content* with respect to that context.

(II) The information content, with respect to a given context, of a sentence is a complex, ordered entity (something like a sequence) whose constituents are semantically correlated systematically with expressions making up the sentence, typically the simple (noncompound) component expressions. Exceptions arise in connection with quotation marks and similar devices.

(III′) The information value (contribution to information content), with respect to a given context c, of any simple singular term is its referent with respect to c (and the time of c and the world of c).

(IV′) Any expression may be thought of as *referring*, with respect to a given context, time, and possible world, to its extension with respect to that context, time, and possible world.

(V′) The information value, with respect to a given context c, of a simple first-order n-place predicate is an n-place attribute (a property if $n = 1$, an n-ary relation if $n > 1$) – ordinarily an attribute ascribed to the referents of the attached singular terms. Exceptions arise in connection with quotation marks and similar devices.

(VI′) The information value, with respect to a given context, of a simple n-place sentential connective is an attribute, ordinarily of the sorts of things that serve as referents for the operand sentences.

(VII′) The information value, with respect to a given context, of a simple n-place quantifier or second-order predicate is an n-ary attribute, ordinarily of the sorts of things that serve as referents for the operand first-order predicates.

(VIII′) The information value, with respect to a given context, of a simple operator other than a predicate, a connective, or a quantifier is an appropriate attribute (for sentence-forming operators), or operation (for other types of operators), ordinarily an attribute of or an operation on the sorts of things that serve as referents for its appropriate operands.

(IX′) The information value, with respect to a given context, of a typical compound expression, if any, is a complex, ordered entity (something like a sequence) whose constituents are semantically correlated systematically with expressions making up the compound expression, typically the simple (noncompound) component expressions. Exceptions arise in connection with quotation marks, and similar devices (and perhaps with compound abstracted predicates, which introduce further complications). The information value, with respect to a given context, of a sentence is its information content, the encoded proposition.

The modified naive theory, with its theses (IV′) and (V′), involves a significant departure from the original naive theory with regard to simple predicates, such as 'is a man'. This is in addition to its departures concerning definite descriptions with its theses (III′) and (IX′). The naive theory identifies the information value of 'is a man' with its referent, and both with the property of being a man (at *t*). Instead, the modified naive theory casts the class of all men as the analogue in the case of 'is a man' to reference in the case of singular terms, and reserves the property of being a man for the separate role of information value. Thus the modified naive theory attributes a two-tiered semantics even to single-word predicates as well as to definite descriptions. The theory retains the principle that the information value of a complex expression is typically made up of the information values of its information-valued parts, but by attributing a two-tiered semantics to single-word predicates, this involves abandoning the original naive-theoretical principle that the information value of a complex expression is made up of the referents of its single-word components. The information value of the definite description 'the tallest man' involves the property of being a man taller than any other, but this property, according to the modified naive theory, is not constructed from what 'is a man' refers to, i.e., the class of Tom, Dick, Harry, etc. Thus, while the modified naive theory resembles the original naive theory as regards simple (noncompound) singular terms, it departs significantly from the original naive theory as regards simple predicates, and as a consequence it abandons some of the spirit of the naive theory as regards multi-worded phrases built from predicates (including definite descriptions).

THE PUZZLES

There are at least four well-known puzzles that arise on both the original and the modified naive theories. Most of the philosophical literature in the theory of meaning is concerned in one way or another with one or more of the four puzzles. Each of the puzzles has been put forward as a refutation of the theory on which it arises.

First there is Frege's puzzle about the informativeness of identity sentences. The sentence 'Hesperus is Phosphorus' (or 'The Evening Star is the Morning Star') is informative; its information content apparently extends knowledge. The sentence 'Hesperus is Hesperus' (or 'The Evening Star is the Evening Star') is uninformative; its information content is a "given." According to both the original and the modified naive theories, the information content of 'Hesperus is Hesperus' consists of the planet Venus, taken twice, and the relation of identity (or the relation of identity-at-*t*, where *t* is the time of utterance). Yet the information content of 'Hesperus is Phosphorus', according to these theories, is made of precisely the same components, and apparently in precisely the same way.[5] Assuming a plausible principle of compositionality for pieces of information – according to which different pieces of information having the same structure and mode of composition must differ in some components – the original and modified naive theories ascribe precisely the same information content to both sentences. This seems to fly in the face of the fact that the two sentences differ dramatically in their informativeness.

The second puzzle is, from one point of view, simply a generalization of the first, though it can arise on any of a wide variety of semantic theories. This is the apparent failure of substitutivity of co-referential names or other (single-word) singular terms in certain contexts, especially in propositional–attitude contexts. If Jones learns the names 'Hesperus' and 'Phosphorus' but is unaware that they refer to the same heavenly body, he may sincerely and reflectively assent to the sentence 'Hesperus appears in the evening' and sincerely and reflectively dissent from the sentence 'Phosphorus appears in the evening', while understanding both sentences perfectly, i.e., while fully grasping their information content. It seems, then, that Jones believes the information encoded by the first sentence – i.e., Jones believes that Hesperus appears in the evening – but does not believe the information encoded by the second – i.e., Jones does not believe that Phosphorus appears in the evening. This presents a serious problem for any semantic theory, since it appears to violate a classical logical rule of inference, commonly called *the Substitutivity of Equality* or *Leibniz's Law*. This inference rule permits the substitution of an occurrence of any singular term β for an occurrence of any singular term α in any sentence, given $\ulcorner\alpha = \beta\urcorner$. The sentences 'Jones believes that Hesperus appears in the evening' and

'Hesperus is Phosphorus' are both true, yet we seem prohibited from substituting 'Phosphorus' for 'Hesperus' in the first sentence. Of course, classical Substitutivity of Equality is subject to certain well-known restrictions. Most notably, the inference rule does not extend to contexts involving quotation marks and similar devices: the sentences 'The expression 'Paris' is a five-letter word' and 'Paris is the capital of France' are both true, whereas the sentence 'The expression 'the capital of France' is a five-letter word' is false. However, failure of substitutivity in propositional-attitude contexts poses an especially pressing difficulty for the original and modified naive theories. These theories are unable to accommodate the apparent fact about Jones that he believes the information encoded by the sentence 'Hesperus appears in the evening' but does not believe the information encoded by the sentence 'Phosphorus appears in the evening', since the theories ascribe precisely the same information content to both sentences. Hence, the original naive theory requires the validity of Substitutivity of Equality in propositional-attitude contexts, and the modified naive theory requires the validity of a restricted, but apparently equally objectionable, version of the same.

Failure of substitutivity involving definite descriptions in temporal and modal contexts presents a similar difficulty. For example, although the sentences 'In 1978, Ronald Reagan was a Republican' and 'Ronald Reagan is the U.S. President' are both true, the sentence 'In 1978, the U.S. President was a Republican' is false, since in 1978 Jimmy Carter was President, and a Democrat. An analogous example involving modality, due to W. V. O. Quine, effectively refutes the original naive theory: the sentences 'It is necessary that 9 is odd' and 'The number of planets is 9' are both true, but the sentence 'It is necessary that the number of planets is odd' is false, since there might have been 10 planets rather than 9. Though the singular terms '9' and 'the number of planets' refer to the same number, the sentence '9 is odd' encodes a mathematically necessary truth, whereas the sentence 'The number of planets is odd' encodes information which, though true, is not necessarily so. Modality creates no like difficulty for the modified naive theory, which requires the validity of substitutivity of co-referential names and other simple singular terms in modal contexts (e.g., 'It is a necessary truth that . . .'), but does not require the validity of substitutivity of *all* co-referential singular terms, including definite descriptions, in such contexts.

Thirdly, there is the puzzle of true negative existentials, such as 'The present king of France does not exist' (Russell) and 'Hamlet does not exist'. These sentences are true if and only if the singular term in subject position does not refer to anything. Yet on any of a variety of semantic theories, a sentence involving a singular term can be true only if the term has a referent. This appears to be a consequence of the naive theory in particular, since on that theory, a sentence involving a singular term is true if and only if the individual referred to by the term has the property semantically correlated with the monadic predicate formed by the remainder of the sentence. In the case of a negative existential, the correlated property is that of nonexistence (at t). This property is such that it is impossible that there should exist something (at t) that has it. A similar problem arises on the modified naive theory in connection with such negative existentials as 'Hamlet does not exist' involving a nonreferring proper name or other simple singular term.

In fact, on the (modified) naive theory the information content of any sentence involving a (simple) singular term will lack a necessary component if any contained (simple) singular term lacks a referent. This presents a fourth and more general puzzle for the theory concerning any meaningful sentence involving nonreferring singular terms, such as Russell's problematic sentence 'The present king of France is bald' and sentences from fiction, e.g., 'Hamlet killed Polonius'. The theory seems to require that such sentences have no content, yet they clearly seem to have content. How is it, then, that a sentence involving a nonreferring (simple) singular term can have content? It seems clear, moreover, that such a sentence as 'The present king of France is bald' cannot be counted true. Should it then be counted false? If so, one should be able to say that the present king of France is *not* bald, yet this is no better than saying that he is bald. There seems to be a violation of the classical Law of Excluded Middle: either p or not-p. We may not assert 'Either the present king of France is bald, or the present king of France is not bald'. (Russell: "Hegelians, who love a synthesis, will probably conclude that he wears a wig.")

FREGE'S EARLY THEORY

In his early work, *Begriffsschrift* (1879), Gottlob Frege proposed to solve his puzzle about the informativeness of identity sentences by reading the identity predicate, '=' or the 'is' of identity ('is' in the sense of 'is one and the very same as'), as covertly metalinguistic: It was held that the sentence 'Hesperus is Phosphorus' encodes information not about the referent of the names 'Hesperus' and 'Phosphorus' (i.e., the planet Venus), but about the very names 'Hesperus' and 'Phosphorus' themselves, to the effect that they are co-referential. It is, of course, trivial that 'Hesperus' is co-referential with itself, but then this is transparently different information from the information that 'Hesperus' is co-referential with 'Phosphorus', since the latter concerns the name 'Phosphorus'. It is no wonder, therefore, that the sentences 'Hesperus is Hesperus' and 'Hesperus is Phosphorus' differ in informativeness.

There are a number of serious difficulties with this account, taken as an account of identity sentences in ordinary language, and Frege himself later came to reject it. Frege's ground for rejecting the account was that it misrepresents the information encoded by the sentence 'Hesperus is Phosphorus' as something which, if true, is made so entirely by virtue of arbitrary linguistic convention or decision, whereas the information that Hesperus is Phosphorus is actually made true by virtue of a certain celestial state of affairs, quite independently of human convention. There are also technical difficulties with the account. It renders the identity predicate an oblique device, similar to quotation marks. This makes quantification in impossible, or at least highly problematic, as for example in 'For every x and every y, if x and y are both natural satellites of the Earth, then $x = y$'. Moreover, the account fails to solve the general problem of which Frege's puzzle is only a special case: whereas we are told how the sentence 'Hesperus is Hesperus' can differ in informativeness from the sentence 'Hesperus is Phosphorus', unless the theory is only part of a much more sweeping proposal concerning the information values of all expressions (as is the theory of John Stuart Mill) and not just that of the identity predicate, we are given no explanation for the analogous difference in informativeness between such pairs of sentences as 'Hesperus is a planet if Hesperus is' and 'Hesperus is a planet if Phosphorus is'. In any event, the account does not even address the remaining problems of the apparent failure of substitutivity and of nonreferring singular terms.

MILL'S THEORY

The theory proffered by John Stuart Mill in *A System of Logic*, Book I, Chapter II, 'Of Names,' and Chapter V, 'Of the Import of Propositions' (1843) has several points of contact with the modified naive theory outlined above. Mill drew a distinction between what he called 'concrete general names' and 'singular names' or 'individual names'. The former are generally monadic predicates, or more accurately, common nouns (such as 'man'), adjectives (such as 'wise'), and perhaps intransitive verbs (such as 'flies'). The latter are what are here called singular terms, and include definite descriptions. Mill also distinguished two semantic attributes of a term, which he called 'denotation' and 'connotation', though not all terms which have one, according to Mill, have the other as well. In the case of singular terms, Mill's use of 'denotation' corresponds with the term 'referent', as used here. In addition, a "general name" is said to "denote" the various individuals to which the term applies, or alternatively the single class thereof, i.e., the extension of the term. The connotation of a term, on the other hand, consists of attributes or properties, and forms part of the information value of the term. In effect, Mill uses the term 'connotation' for a special kind of information value, namely one involving attributes. All concrete general names were held by Mill to have connotation. For example, 'man' was held to connote the properties of "corporeity, animal life, rationality, and a certain external form, which for distinction we call the human." Thus general names have both denotation and connotation. Among singular terms, according to Mill, proper names, such as 'Plato', (typically) have denotation but never connotation, whereas definite descriptions, such as 'the author of *the Republic*', have connotation, and (typically) denotation as well. A definite description connotes "some attribute, or some union of attributes, which being possessed by no object but one, determines the name exclusively to that individual." This contrasts sharply with Mill's account of proper names. As if forestalling Russell's theory that ordinary proper names are semantically equivalent to definite descriptions, as well as Frege's later theory, Mill argued on the contrary that ordinary proper names "are simply marks used to enable . . . individuals to be made subjects of discourse. . . . Proper names are attached to the objects themselves, and are not dependent upon . . . any attribute of the object." A proper name, he wrote, is " a word which answers the purpose of showing what thing it is we are talking about, but not of telling anything about it." He adds, "When we predicate of anything its proper name; when we say, pointing to a man, this is Brown or Smith, or pointing to a city, that it is York, we do not, merely by so doing, convey to the hearer any information about them, except that those are their names."

As regards singular terms, Mill's theory resembles the modified naive theory in according definite descriptions and simple predicates a disjointness between referent and information value that is denied to proper names. There is an essential difference, however. Mill did not identify the information value of a proper name simply with its referent or denotation. Rather, he held a complex theory of information value, according to which the information encoded by such a sentence as 'Socrates is wise' has at least two components: the proposition about Socrates that he has the property of wisdom, and the metalinguistic proposition about the expressions 'Socrates' and 'wise' that the individual denoted by the former has the property connoted by the latter, and is therefore among the things denoted by the latter. This means that any term itself forms part of its own information value. In the special case of an identity sentence, such as 'Hesperus is Phosphorus', Mill held that the first component was null, so that the information encoded reduces to the metalinguistic truth that the name 'Hesperus' denotes the same thing as the name 'Phosphorus'. Thus Mill solved Frege's puzzle in a manner similar to Frege's early theory. Mill went beyond the early Frege, though, generalizing the metalinguistic solution to Frege's puzzle in such a way as to make a similar solution available for the other three puzzles. (It is doubtful that this was part of Mill's motivation in propounding his dual-component theory of information value, though it doubtless would have been seen as lending independent support to the theory.) By the same token, the objections to Frege's early view apply *a fortiori* to the metalinguistic component of Mill's theory of information value.

Even ignoring its metalinguistic component, Mill's theory involves the same significant departures from the original naive theory as those of the modified naive theory. These departures constitute an important anticipation of Frege's later theory of meaning. Mill's theory of singular terms, on the other hand – especially his semantic distinction between proper names and definite descriptions – constitutes an important anticipation of a theory diametrically opposed to Frege's later theory, the theory of direct reference.

RUSSELL'S THEORY

Inspired by his new theory of logical analysis, Bertrand Russell's theory of information value (post 1904) was explicitly promoted by Russell on the grounds of its alleged ability to handle the four puzzles. Russell's theory more or less retains (and augments) the central theses of the original naive theory. The primary departure from the central theses of the naive theory is the replacement of attributes with *propositional functions*: Russell took the information value of a predicate, a connective, or a quantifier to be a function that assigns to any object(s) in its domain, a singular proposition concerning that (those) object(s). For example, the propositional function *being wise*, which is alleged to be the information value of the predicate 'is wise', assigns to any person x the singular proposition that x is wise.

Russell handled the apparent inconsistency of the naive theory in the case of definite descriptions and other such phrases (which he called 'denoting phrases') by supplementing the naive theory with his so-called Theory of Descriptions. Russell's Theory of Descriptions has both a special and a general form.

The General Theory of Descriptions concerns the logico-semantic status of so-called indefinite descriptions, that is, restricted universal or existential quantifier phrases such as 'every woman', 'some logician', 'an instructor', etc. We consider first restricted universal sentences of the form $\ulcorner\Pi(\ldots, \text{every } \phi, \ldots)\urcorner$ (or $\ulcorner\Pi(\ldots, \text{any } \phi, \ldots)\urcorner$, $\ulcorner\Pi(\ldots, \text{each } \phi, \ldots)\urcorner$, $\ulcorner\Pi(\ldots, \text{all } \phi, \ldots)\urcorner$, etc.) built from an n-place predicate Π together with the indefinite description $\ulcorner\text{every } \phi\urcorner$ instead of a singular term in one or more of the n singular-term positions. On the General Theory, such a sentence is analyzed as

$$(\beta)[\phi_\beta \supset \Pi(\ldots, \beta, \ldots)] \text{ (For every } \beta\text{, if } \beta \text{ is } \phi\text{, then } \Pi(\ldots, \beta, \ldots)).$$

(We may let β be the first variable, in some ordering of the variables, that does not occur in $\ulcorner\Pi(\ldots, \text{every } \phi, \ldots)\urcorner$.) Likewise, a restricted existential sentence of the form $\ulcorner\Pi(\ldots, \text{some } \phi, \ldots)\urcorner$ (or $\ulcorner\Pi(\ldots, \text{a } \phi, \ldots)\urcorner$) built from an n-place predicate together with the indefinite description $\ulcorner\text{some } \phi\urcorner$ instead of a singular term in one or more of the n singular-term positions is analyzed as

$$(\exists\beta)[\phi_\beta \,\&\, \Pi(\ldots, \beta, \ldots)] \text{ (For some } \beta\text{, } \beta \text{ is } \phi \text{ and } \Pi(\ldots, \beta, \ldots)).$$

One important aspect of Russell's General Theory of Descriptions concerns the matter of *scope*. Quantificational constructions generally, such as 'every man', 'some logician', etc., often yield syntactic scope ambiguities in surface structure when properly coupled with predicates and embedded within any sentential operator or connective, as for example 'Every man is not an island' or 'Some treasurer must be found'. The two readings of the latter sentence correspond to two ways of extending the Russellian analysis of $\ulcorner\Pi(\ldots, \text{some } \phi, \ldots)\urcorner$ to $\ulcorner O\Pi(\ldots, \text{some } \phi, \ldots)\urcorner$, where O creates a sentential context (as is done for example by a sentential operator or connective). One way is simply to embed the analysis of ⌜Π(. . . , some ϕ, . . .)⌝ within the context O, as in

$$O(\exists\beta)[\phi_\beta \mathbin{\&} \Pi(\ldots, \beta, \ldots)],$$

for example, 'it must be that: some treasurer is found'. On this reading the indefinite description 'some treasurer' is given *narrow scope*, and is said to have its *secondary occurrence*. The other way of extending the Russellian analysis is to imitate the treatment for the atomic case, treating the extended singular term context ⌜$O\Pi$(. . . , ______, . . .)⌝ as if it were a simple, unstructured unit (monadic predicate):

$$(\exists\beta)[\phi_\beta \mathbin{\&} O\Pi(\ldots, \beta, \ldots)],$$

for example, 'Some treasurer is such that he or she must be found'. On this reading the indefinite description 'some treasurer' is given *wide scope*, and is said to have its *primary occurrence*. Further embeddings within sentential contexts yield further scope ambiguities, one additional reading for each additional context embedding. For example, the sentence 'Some treasurer must not be poor' will have three readings: 'It must not be that: some treasurer is poor' (narrow scope); 'It must be that: some treasurer is such that he or she is not poor' (intermediate scope); 'Some treasurer is such that he or she must not be poor' (wide scope).

Russell's General Theory of Descriptions is relatively uncontroversial – though its consequences for the theory of information value should render the theory more controversial than it is. Construing second-order predicates (or quantifiers) on the model of a first-order predicate, the original naive theory may take the information value of the indefinite description 'some man' to be the second-order property of being a first-order property possessed by at least one man.[6] Using Russell's preferred notion of a propositional function, the description 'some man' might be regarded as having as its information value the second-order propositional function f that assigns to any one-place first-order propositional function F the proposition that at least one man instantiates F (where an individual x is said to *instantiate* a one-place propositional function F if the proposition obtained by applying F to x is true). The proposition that some man is wise would then be regarded as made up of the second-order propositional function f and the first-order propositional function *being wise*.[7] Russell did not identify the information value of an indefinite description with its corresponding second-order propositional function, however. Instead, he claimed that indefinite descriptions, and restricted quantificational constructions generally ("denoting phrases"), "have no meaning in isolation." This claim may seem puzzling, since for any such quantificational construction, Russell's theory of higher-order propositional functions presents clearly identifiable entities (such as the function f) that could serve as the contribution of the construction to the information encoded by the sentences in which it figures.

Russell was initially impressed with the fact that a restricted quantificational construction, such as 'a man', may serve as the grammatical subject of a sentence, e.g., 'A man is wise'. He may have feared that if 'a man' was assigned a "meaning in isolation" – for example the second-order propositional function f described above – then the sentence 'A man is wise' would have to be construed on the model of 'Socrates is wise', ascribing the property of wisdom to f, the "meaning" ("in isolation") of the grammatical subject (see Note 2). In denying that restricted quantificational constructions have meaning, it is possible that Russell meant only to emphasize that though descriptions are noun phrases that may serve as the grammatical subject of a sentence and that are replaceable by a genuine proper name, their logical form, as opposed to their grammatical form, is not that of a singular term to which the first-order predicate attaches, but rather that of a second-order predicate that attaches to the first-order predicate. Since Russell separates logic from grammar, however, this observation is insufficient grounds to deprive restricted quantificational constructions of information value.

There is a far better reason for Russell to have denied that restricted quantifier phases have information value. On the General Theory of Descriptions, the phrase 'some man' in the sentence 'Some man is wise' corresponds to the result of deleting the first-order predicate 'is wise' from the sentence 'For something x, x is a man and x is wise'. The phrase 'some man' thus corresponds to the essentially incomplete string 'for something x, x is a man and x ______', or more formally, '$(\exists x)$ [x is a man & x ______]'. Indeed, Russell called restricted quantifier phrases 'incomplete symbols'. It is natural and plausible on this theory to regard the simple quantifier 'for something' as a second-order predicate, while regarding the remainder of the sentence 'x, x is a man and x is wise' as a closed compound first-order predicate, synonymous with 'is both a man and wise', or roughly synonymous with 'is a wise man' (see Note 2). The sentence 'Some man is wise' would thus be construed as having the logical form of 'Something is both a man and wise', whose information content is made up of the second-order propositional function that is the information value of the unrestricted quantifier 'something' and the first-order propositional function that is the information value of the compound first-order predicate 'is both a man and wise'. There is no place in this proposition for the second-order propositional function f described above. Hence, there is nothing that the phrase 'some man' contributes on its own to the proposition that some man is wise, i.e., the proposition that something is both a man and wise – although the determiner 'some' may be regarded as contributing its information value and the information value of 'man' figures indirectly in the construction of the information value of 'is both a man and wise'. Exactly analogous results obtain in connection with restricted universal quantifier phrases such as 'every man'. On the General Theory of Descriptions, the determiner component of an indefinite description – the word 'every' or 'some', etc. – may be regarded as making a separate contribution to the information content of the sentence in which the description occurs. The rest of the description makes no contribution of its own but joins with the surrounding sentential context to yield something that does. In this way, indefinite descriptions are justifiably regarded on the General Theory of Descriptions as not forming semantically self-contained, information-valued units. Instead, they form part of a larger construction whose information-valued components overlap with the description.

Russell's Special Theory of Descriptions concerns singular definite descriptions, treating them as indefinite descriptions of a particular kind in accordance with the General Theory of Descriptions, rather than as semantically self-contained singular terms. A sentence having the surface structure of a subject-predicate sentence $\ulcorner\Pi(\ldots, \imath\alpha\phi_\alpha, \ldots)\urcorner$ consisting of an n-place predicate and containing a complete definite description $\ulcorner\imath\alpha\phi_\alpha\urcorner$ among its n occurrences of singular terms, is analyzed into a conjunction of three sentences:

(*i*) $(\exists\alpha)\phi_\alpha$ (*The existence condition*: There is at least one ϕ);

(*ii*) $(\alpha)(\beta)[\phi_\alpha \,\&\, \phi_\beta \supset \alpha = \beta]$ (*The uniqueness condition*: There is at most one ϕ); and

(*iii*) $(\beta)[\phi_\beta \supset \Pi(\ldots, \beta, \ldots)]$ (*The subsumption condition*: $\Pi(\ldots$, every $\phi, \ldots)$).

(We may let β be the first variable, in some ordering of the variables, that does not occur in $\ulcorner\Pi(\ldots, \alpha\phi_\alpha, \ldots)\urcorner$. ϕ_β is the same formula as ϕ_α except for having (free) occurrences of β wherever ϕ_α has free occur rences of α. $\ulcorner\Pi(\ldots, \beta, \ldots)\urcorner$ is the same formula as $\ulcorner\Pi(\ldots, \imath\alpha\phi_\alpha, \ldots)\urcorner$

except for having occurrences of β wherever $\ulcorner\Pi(\ldots, \imath\alpha\phi_\alpha, \ldots)\urcorner$ has occurrences of $\ulcorner\imath\alpha\phi_\alpha\urcorner$. If $\ulcorner\Pi(\ldots, \imath\alpha\phi_\alpha, \ldots)\urcorner$ is constructed from a polyadic predicate and contains a further definite description $\ulcorner\imath\gamma\psi_\gamma\urcorner$ – so that $\ulcorner\Pi(\ldots, \imath\alpha\phi_\alpha, \ldots)\urcorner$ is of the form $\ulcorner\Pi'(\ldots, \imath\alpha\phi_\alpha, \ldots, \imath\gamma\psi_\gamma, \ldots)\urcorner$ – sentence (*iii*) is to be analyzed further by iteration of the method just given, treating the string $\ulcorner(\beta)[\phi_\beta \supset \Pi'(\ldots, \beta, \ldots, \underline{\qquad}, \ldots)]\urcorner$ as if it were a simple and unstructured monadic predicate. The new subsumption condition is $\ulcorner(\delta)(\psi_\delta \supset (\beta)[\phi_\beta \supset \Pi'(\ldots, \beta, \ldots, \delta, \ldots)])\urcorner$. The process is to be repeated until all definite descriptions have undergone an application of the procedure.) A definite description $\ulcorner\imath\alpha\phi_\alpha\urcorner$ is said to be *proper* if conditions (*i*) and (*ii*) both obtain (i.e., if there is exactly one individual that answers to it), and is said to be *improper* otherwise. Thus, according to the Special Theory of Descriptions, the propriety of a definite description $\ulcorner\imath\alpha\phi_\alpha\urcorner$ is a necessary condition for the truth of $\ulcorner\Pi(\ldots, \imath\alpha\phi_\alpha, \ldots)\urcorner$. The only other requirement is that whatever satisfies ϕ_β also satisfies $\ulcorner\Pi(\ldots, \beta, \ldots)\urcorner$.

None of the sentences making up the analysans is logically subject-predicate; each is a quantificational generalization containing no definite descriptions. The Russellian analysis is thus said to be a method of "eliminating" definite descriptions, replacing an apparently subject-predicate sentence by a conjunction of quantificational generalizations.

Equivalently, $\ulcorner\Pi(\ldots, \imath\alpha\phi_\alpha, \ldots)\urcorner$ may be analyzed by means of a single complex generalization:

$$(\exists\beta)[(\alpha)(\phi_\alpha \equiv \alpha = \beta) \;\&\; \Pi(\ldots, \beta, \ldots)],$$

or less formally, $\ulcorner$For something β, β and only β is ϕ and $\Pi(\ldots, \beta, \ldots)\urcorner$. This version of the analysis illustrates the central tenet of the Special Theory of Descriptions: A complete definite description such as 'the author of *the Republic*' is regarded as semantically equivalent to the corresponding uniqueness-restricted existential quantifier, 'some unique author of *the Republic*' or 'a unique author of *the Republic*', which falls under the purview of the General Theory of Descriptions. This central tenet of the Special Theory of Descriptions has several important consequences.

First, the Special Theory of Descriptions predicts scope ambiguities in cases where definite descriptions are embedded within sentential contexts, one additional reading for each additional embedding. For example, the three predicted readings of the sentence 'The U.S. Treasurer must not be poor' correspond exactly to those of the sentence 'Some unique U.S.

Treasurer must not be poor' on the General Theory of Descriptions: 'It must not be that: some unique U.S. Treasurer is poor' (narrow scope); 'It must be that: some unique U.S. Treasurer is such that he or she is not poor' (intermediate scope); 'Some unique U.S. Treasurer is such that he or she must not be poor' (wide scope).

Most important for present purposes are the consequences of conjoining the Special Theory of Descriptions with the naive theory of information value. The central tenet of the Special Theory of Descriptions distinguishes it from semantic theories that accord definite descriptions the logico-semantic status of singular terms rather than that of restricted quantifiers. The Special Theory of Descriptions conflicts with the first assumption used in the Church-Gödel argument for the replacement of thesis (IV) of the naive theory by thesis (IV′) of the modified naive theory, thus enabling Russell to maintain both of the original naive-theoretical theses (III) and (IV). This distinguishing characteristic of Russell's theory is obscured by the fact that he misleadingly and artificially extended Mill's term 'denotation' to the semantic relation that obtains between a uniqueness-restricted existential quantifier ⌜some unique ϕ⌝ (which the theory regards as not forming a semantically self-contained unit) and the individual, if any, to which ϕ uniquely applies (i.e., the individual that uniquely satisfies ϕ). For example, although the definite description 'the author of *the Republic*' does not refer to Plato on Russell's theory, and indeed is not any kind of referring expression on Russell's theory, it is nevertheless said by Russell to "denote" Plato. Thus Russell verbally mimics Mill's version of the modified naive theory, while retaining the term 'meaning' for the semantic notion here called 'information value'.

Although the Theory of Descriptions was put forward by Russell as a supplement to the original naive theory, its consequences concerning the information contents of sentences involving definite descriptions are very similar to those of what is here called 'the modified naive theory' – the chief difference being that the Special Theory of Descriptions regards definite descriptions as incomplete symbols, whereas the modified naive theory (which is strictly silent concerning the logico-semantic status of definite descriptions) is intended to be coupled with a treatment of definite descriptions as semantically self-contained singular terms. On both theories, a definite description such as 'the author of *the Republic*' is regarded as not contributing the author of *the Republic*, i.e., Plato, as its information value. As with the modified naive theory, on Russell's theory such a sentence as 'The author of *the Republic* is wise' does not encode the singular proposition about Plato that he is wise. Plato does not "occur as a constituent" of the encoded proposition; the sentence is only indirectly about him. Rather, the sentence is regarded as being directly about the propositional function (or property), *being both a unique author of the Republic and wise*, to the effect that it is instantiated. Thus, the sentence does not directly concern any individual, but only a certain propositional function. This is critical to Russell's solutions to the four puzzles that arise in connection with the naive theory.

The four puzzles arise primarily from the identification of the information value of a singular term with the term's referent. Since the modified naive theory and the Special Theory of Descriptions do not identify the information value of a definite description with the individual that uniquely answers to it, they are able to solve the puzzles in the special case where the singular terms involved are all definite descriptions, but the problems remain in the case where the terms involved are proper names, demonstratives, or other single-word indexicals. Russell handled these remaining difficulties by combining his Special Theory of Descriptions with the thesis that terms ordinarily regarded as proper names or indexicals are ordinarily used not as "genuine names" (singular terms) for individuals, but as "disguised," "concealed," "truncated," or "abbreviated" definite descriptions. This thesis has the effect of "reducing to the previous case" the special problems that arise in connection with proper names and indexicals.

For example, let 'Evening Star' or 'ES' be an abbreviation for the descriptive conditions associated with the name 'Hesperus', and let 'Morning Star' or 'MS' abbreviate the descriptive conditions associated with the name 'Phosphorus'. Then the sentence 'Hesperus is Hesperus' is a truncated formulation of 'The ES is the ES', which, according to the Special Theory of Descriptions, analyzes into (something logically equivalent to) 'There is a unique ES'. The sentence 'Hesperus is Phosphorus', on the other hand, abbreviates 'The ES is the MS', which analyzes into (something logically equivalent to) 'Something is both a unique ES and a unique MS'. Thus Russell solves Frege's puzzle by reading the sentence 'Hesperus is Hesperus' as encoding the simple proposition about the propositional function that is the information value of 'ES' that it is uniquely instantiated, while reading the sentence 'Hesperus is Phosphorus' as encoding the much richer proposition about the two propositional functions that are the information values of 'ES' and 'MS' to the effect that they are each uniquely instantiated, and in addition co-instantiated.

Russell's treatment of the puzzle of the failure of substitutivity in propositional-attitude contexts is complicated by the fact that on Russell's theory, a propositional-attitude attribution such as 'Jones believes that Hesperus appears in the evening' is ambiguous. It has a wide-scope reading, 'Some unique ES is such that Jones believes: that it appears in the evening'. It also has a narrow-scope reading, 'Jones believes: that some unique ES appears in the evening'. On the wide-scope reading, the sentence attributes to Jones a belief of the singular proposition about a unique ES (i.e., about the planet Venus) that it appears in the evening. Given the further premise 'The ES is the MS' – i.e., given that something is both a unique ES and a unique MS – it does indeed follow that a unique MS is such that Jones believes: that it appears in the evening. This is the wide-scope reading of the sentence 'Jones believes that Phosphorus appears in the evening'. Thus on the wide-scope reading of propositional-attitude attributions, Substitutivity of Equality is valid. It is only on the narrow-scope reading that the inference fails. Russell's theory solves the puzzle of failure of substitutivity in narrow-scope propositional-attitude contexts by reading the embedded 'that'-clause, 'that Hesperus appears in the evening', as referring to a proposition about the propositional function that is the information value of 'ES', while reading the other 'that'-clause, 'that Phosphorus appears in the evening', as referring to the corresponding proposition about the propositional function that is the information value of 'MS'. Since these are distinct pieces of information concerning different propositional functions, it is to be expected that one may be believed without the other. It is consonant with Russell's theory not to extend Substitutivity of Equality to substitutions involving a definite description. Substitutivity of Equality licenses the substitution of co-referential genuine singular terms (including individual variables). According to Russell's theory, definite descriptions, concealed or not, have the logical status of restricted quantifiers rather than that of singular terms, and hence the traditional rule of Substitutivity of Equality does not apply to them. However, when it is given that two propositional-function expressions (predicates) ϕ and ψ apply to (are satisfied by) exactly the same individuals, restricted quantifiers of the same sort constructed from ϕ and ψ, such as $\ulcorner$some $\phi\urcorner$ and $\ulcorner$some $\psi\urcorner$, will usually be interchangeable on other logical grounds. In the special case where it is given that there is something that uniquely satisfies ϕ and also uniquely satisfies ψ, the uniqueness-restricted existential quantifiers $\ulcorner$some unique $\phi\urcorner$ and $\ulcorner$some unique $\psi\urcorner$, and hence the definite descriptions $\ulcorner$the $\phi\urcorner$ and $\ulcorner$the $\psi\urcorner$, will be interchangeable in most contexts on logical grounds (which include the Substitutivity of Equality as applied to individual variables). It is for this reason that substitutivity is upheld in the wide-scope reading of propositional-attitude attributions. But since ⌜some unique ϕ⌝ and ⌜some unique ψ⌝ may still contribute differently to the information contents of sentences in which they occur, they need not be interchangeable when occurring within the scope of operators, such as those of propositional attitude, that are sensitive to information value. Exactly analogous solutions to the problems of failure of substitutivity in modal and temporal contexts are available.[8]

The remaining two puzzles are solved in a similar manner, by reading sentences involving improper definite descriptions as encoding propositions about the corresponding propositional functions. In particular, a negative existential, such as 'The present king of France does not exist', is ambiguous on the Special Theory of Descriptions. It has a true narrow-scope reading, 'There is no unique present king of France', and a contradictory wide-scope reading, 'There exists a unique present king of France who does not exist'. An unmodified sentence involving an improper definite description (concealed or not), such as 'The present king of France is bald', will in general be false, since part of what it asserts is that there is a unique present king of France. Its apparent negation, 'The present king of France is not bald', is ambiguous on the Special Theory of Descriptions. Its narrow-scope reading yields the genuine negation of the original sentence, i.e., 'There is no unique present king of France who is bald', and on this reading the sentence is true. Its wide-scope reading, 'Some unique present king of France is such that he is not bald', is perhaps the more natural reading, but on this reading the sentence is false, as is the original sentence. The Law of Excluded Middle is preserved in the case of 'Either the present king of France is bald, or the present king of France is not bald', provided both occurrences of the description 'the present king of France' are given narrow scope. Other readings of this ambiguous sentence are not proper instances of the law.

Russell's thesis that proper names and indexicals are ordinarily used as disguised definite descriptions, together with his Special Theory of Descriptions, has the effect of purging (closed) simple singular terms from the language, and replacing them with restricted quantifiers. It might seem, therefore, that Russell ultimately solves the philosophical problems that beset the naive theory by denying the existence of all singular terms other than individual variables. However, Russell acknowledged the possibility of (closed) "genuine names" or "proper names in the strict, logical sense," which are semantically simple and unstructured, and which therefore function semantically in accordance with the naive theory. The class of possible information values for genuine names was severely limited by Russell's Principle of Acquaintance: every proposition that one can grasp must be composed entirely of constituents with respect to which one has a special sort of intimate and direct epistemic access, "direct acquaintance." Predicates, connectives, and quantifiers contribute the right sort of entities to propositions, since their information values are propositional functions, and these were held by Russell to be directly epistemically accessible. Because the information value of a genuine name is to be its referent, the only genuine names of individuals that one could grasp, according to Russell, were generally the demonstrative 'this', used deictically by a speaker to refer to mental items presently (or at least very recently) contained in his or her consciousness, and perhaps the first-person pronoun 'I' used with introspective deictic reference to oneself. Uses of genuine names for oneself or one's own mental items, however, were held to be rare, since the singular proposition encoded by a sentence involving such a name would be apprehended by the speaker of the sentence only very briefly, and never by anyone else. Even if a speaker were to use a genuine name, his or her audience would be forced to understand the name as a disguised definite description for the intended referent. Since communication using genuine names must be circumvented in this way, even when speaking about oneself or one's own present experiences one might typically employ definite descriptions, disguised or not, in lieu of genuine names. Genuine names of individuals are expedient only when conversing with oneself about oneself.

Though Russell acknowledged the possibility of genuine names, for which information value coincides with reference, his restriction on admissible referents seems sufficient to prevent the four puzzles from arising. True identity sentences involving genuine names for an item of direct acquaintance are all equally uninformative, and all co-referential genuine names are validly intersubstitutable in propositional-attitude contexts. Russell did not countenance genuine names lacking a referent; thus the remaining two puzzles are also blocked. (Russell claimed that singular existential or negative existential statements involving genuine names are without content, in part, because the information value of the unrestricted existential quantifier, 'there exists something that', is seen to be a higher-order propositional function that applies only to propositional functions of individuals, and not to the individuals themselves. This observation overlooks the fact that the unrestricted existential quantifier together with the identity predicate defines (something equivalent to) a first-order existence predicate, 'there exists something identical with', whose information value would be the corresponding universally true propositional function of individuals. It would have been better to say that singular existentials and negative existentials involving genuine names have content, but are always trivially true and trivially false, respectively.)

FREGE'S THEORY OF SENSE AND REFERENCE

Frege's later theory of meaning, though superficially similar to Russell's in certain respects, is fundamentally different. In his *Begriffsschrift* (1879), Frege was generally sympathetic to something like the naive theory, proposing an *ad hoc* reinterpretation of the identity predicate to avoid the problem of differing informativeness among true identity sentences concerning the same individual. By the time he wrote his classic *Über Sinn und Bedeutung* (1892), Frege no longer advocated any version of the naive theory. Posing the problem of the informativeness of identity sentences as a refutation of the naive theory, Frege now proposed abandoning that theory in favor of a different and richly elegant philosophy of semantics. Whereas the modified naive theory and the theory of Mill attribute a two-tiered semantics to definite descriptions and predicates ("concrete general names"), Frege further extended two-tiered semantics across the board, to include all singular terms, predicates, connectives, quantifiers, operators, and even sentences. Frege distinguished between the *Bedeutung* of an expression and its *Sinn*. The former corresponds in the case of singular terms to what is here called the 'referent'. The latter, standardly translated by the English 'sense', is the expression's information value. Frege's conception of sense is similar to Mill's notion of connotation, except that all meaningful exressions, including proper names, are held to have a sense. Frege explained his notion of the sense of an expression as something "wherein the mode of presentation is contained" and that "is grasped by anybody who is sufficiently familiar with the language or the totality of designations to which it [the expression] belongs; but with this the *Bedeutung*, in case it is available, is only one-sidedly illuminated." The sense of an expression is something like a purely conceptual representation, by means of which a referent for the expression is secured.[9] It is a conception of something, and the referent of the expression is whoever of whatever uniquely fits the concept. Since the sense of a singular term secures the term's referent, *strictly synonymous* expressions (i.e., expressions having the very same sense) must have the same referent – although different expressions having the same referent may differ in sense. An expression is said to *express* its sense, and its sense, in turn, (typically) determines an object. The reference relation is simply the relative product of the relation of *expressing* between an expression and its sense, and the relation of *determining* between a sense and the object that uniquely fits it. An expression "*expresses* its *Sinn*, *designates* its *Bedeutung*. By means of a sign we express its *Sinn* and designate its *Bedeutung*." In the special case of a sentence, Frege called its sense a 'thought' (*Gedanke*).

The clearest examples of expressions exhibiting something like Frege's distinction between sense and referent are certain definite descriptions. (Like Mill and unlike Russell, Frege counted definite descriptions as genuine singular terms.) One of Frege's illustrations involves descriptions in the language of geometry: if *a*, *b*, and *c* are the three medians of a triangle, then the expressions 'the point of intersection of *a* and *b*' and 'the point of intersection of *b* and *c*' refer to the same point – the centroid of the triangle – but they do so by presenting that point to the mind's grasp in different ways, by means of different aspects of the point. The descriptions thus share a common referent, but differ in sense. These senses are the information values of the two expressions, and it is in virtue of this difference in sense that the sentence 'The point of intersection of *a* and *b* is the point of intersection of *b* and *c*' encodes different, and more valuable, information than that encoded by the sentence 'The point of intersection of *a* and *b* is the point of intersection of *a* and *b*'.

Anticipating (though perhaps not advocating) Russell's thesis that ordinary proper names are concealed definite descriptions, Frege wrote:

> As concerns an actual proper name such as 'Aristotle' opinions as to the sense may differ. One could take as such, for example: the pupil of Plato and teacher of Alexander the Great. Anybody who does this will attach to the sentence 'Aristotle was born in Stagira' a different sense than will one who takes as the sense of this name: the teacher of Alexander the Great who was born in Stagira (the second footnote to "*Über Sinn und Bedeutung*").

Here and elsewhere, Frege illustrates what the sense of a proper name is by means of carefully chosen definite descriptions. The observation that proper names have this sort of conceptual content as well as a referent, together with Frege's doctrine that this conceptual content and not the referent serves the role of information value, immediately solves the puzzle about the informativeness of identity sentences involving two names for the same individual. The distinction between sense and referent also immediately solves the problem of how sentences involving nonreferring singular terms can have content. Since a singular term contributes its sense, rather than its referent, to the information content of any sentence containing the term, lack of a referent will not matter as regards a containing sentence's capacity to encode information.

Crucial to Frege's theory are a pair of principles concerning the referent and sense of complex expressions. These are the *Principle of Compositionality* (*Interchange, Extensionality*) *of Reference* and the analogous *Principle of Compositionality* (*Interchange*) *of Sense*. They hold that the referent or sense of a complex expression is a function only of the referents or senses, respectively, of the constituent expressions. In the latter case Frege often spoke (explicitly metaphorically) of the sense of a constituent expression as a *part* of the sense of the complex expression. Thus, if a constituent expression is replaced by one having the same referent but differing in sense, the referent of the whole is preserved, but not the sense. If a constituent expression is replaced by something strictly synonymous, both the sense and the referent of the whole are preserved. In particular, Frege held as a special case of the Compositionality of Reference that a compound expression having a nonreferring part (as for example 'the wife of the present king of France') must itself be nonreferring. Relying on the Compositionality of Reference, Frege argued that what he called the 'cognitive value' (*Erkenntniswert*) of a sentence, i.e., its encoded information, is not the referent of the sentence, but its sense or "thought" content, and that the referent of a sentence is simply its truth value, either truth or falsehood ("the True" or "the False"). (The Church–Gödel argument mentioned earlier was inspired by Frege's arguments, and was offered independently by both Church and Gödel on Frege's behalf.)

The Compositionality of Reference thus solves the problem of the truth value of sentences involving nonreferring singular terms. Since a sentence refers to its truth value, either truth or falsehood, and a sentence involving a nonreferring singular term itself refers to nothing, such a sentence as 'The present king of France is bald' is neither true nor false. It lacks truth value. The same holds for its negation, 'The present king of France is not bald'. Frege held that (the "thoughts" expressed by) these sentences do not assert, but merely *presuppose*, that there is a unique present king of France and that the expression 'the present king of France' has a referent. Thus Frege's theory does not preserve an unrestricted Law of Excluded Middle, nor any other law of logic. The laws of logic must be restricted to sentences whose presuppositions are fulfilled.

Whereas the Compositionality of Reference solves one of the puzzles, it also issues in the problems of failure of substitutivity and of true negative existentials. Since on Frege's theory sentences refer to their truth values, and the referent of a sentence is a function of the referents of its constituents, the theory requires the universal validity of Substitutivity of Equality – an inference pattern apparently violated in quotational, propositional-attitude, temporal, and modal contexts. Furthermore, since on Frege's theory sentences involving nonreferring singular terms are neither true nor false, the theory appears unable to accommodate the truth of the negative existential 'The present king of France does not exist'.

Frege explicitly considered the problems of failure of substitutivity in quotational and propositional-attitude contexts, treating both in a like manner. Quotation marks and the sentential operator 'that' associated with propositional-attitude operators (e.g., 'Jones believes that'), according to Frege, create an *oblique context*, in which expressions take on a different referent from their customary referent. Whereas the expression 'Hesperus' customarily refers to the planet Venus, when occurring within quotation marks, as in the sentence 'The expression 'Hesperus' is a sequence of eight letters', it instead refers to itself. Such is the case when someone's remarks are quoted in "direct discourse," that is, when reporting the very words used by the speaker, as in 'Jones said 'Hesperus appears in the evening''. Analogously, when occurring in a 'that'-clause in a propositional-attitude attribution, as in 'Jones believes that Hesperus appears in the evening', the name 'Hesperus' refers neither to its customary referent nor to itself, but to its customary sense. Similarly, the entire embedded sentence 'Hesperus appears in the evening', when occurring within the 'that'-operator, refers to its customary sense rather than to its customary referent. Such is the case when someone's remarks are quoted in "indirect discourse," that is, when reporting the content of his or her remarks rather than the very words used, as in 'Jones said that Hesperus appears in the evening'. It is not the truth value of the sentence 'Hesperus appears in the evening' that Jones is said to believe or to have asserted, but its information or "thought" content, the information that Hesperus appears in the evening. The principle of Compositionality of Reference is to be understood as requiring the validity of substituting for the name 'Hesperus' in such a position any expression having the same referent as 'Hesperus' *in that position*. This validates the substitution of any expression having the same customary sense as 'Hesperus' (e.g., perhaps 'the Evening Star'); it does not validate the substitution of an expression merely having the same customary referent. Similarly, the Compositionality of Reference does not validate any substitution within quotation marks. Thus Frege's theory gives central importance to a relativized semantic notion of an expression *e* referring to an object *o* (or expressing a sense *s*) as occurring in a particular position *p* within a sentence (or equivalently, to a notion of an expression occurrence (within a sentence) referring to an object or expressing a sense), rather than the more standard notion in contemporary semantics of an expression referring to (denoting) an object ("in isolation"). The latter is what Frege called 'customary reference', and might be defined in Fregean theory as reference in a particular favored position (for example as one of the terms of an identity sentence).

Since all reference on Frege's theory is mediated by sense, an occurrence of an expression standing within a single occurrence of a propositional-attitude operator (e.g., the occurrence of 'Hesperus' in 'Jones believes that Hesperus appears in the everning') must refer to its customary sense by expressing some further sense that is also associated with the expression and that determines the expression's customary sense in the usual way that sense determines reference. This Frege called the 'indirect sense' of the expression. He explained his notion of indirect sense by observing that the indirect sense of a sentence such as 'Socrates is wise' is just the customary sense of the phrase 'the thought that Socartes is wise'. Unfortunately, this device cannot be extended to other varieties of expressions, since the 'that'-operator attaches only to sentences. In order to give the indirect sense of an arbitrary expression, a device less restrictive than the 'that'-operator must first be introduced into the language. For example, we may decide to use the lower case italic letter '*s*' in the manner of a quotation mark, to quote not an expression but its customary sense.[10] Thus for example, the expression "^{s}Socrates^{s}" is a singular term that refers to the customary sense of the name 'Socrates'. (These sense-quotation marks, when applied to a complete sentence, function in exactly the same way as the 'that'-operator.) Then the indirect sense of 'Socrates' is just the customary sense of "^{s}Socrates^{s}", i.e., $^{s\,s}$Socrates$^{s\,s}$. Just as an occurrence of a single propositional-attitude operator induces a shift in the reference of expression occurrences standing within its scope from customary referent to customary sense, so it stands to reason that an occurrence of an expression standing within the embedding of one occurrence of a propositional-attitude operator within another (e.g., the occurrence of 'Hesperus' in

'Smith doubts that Jones believes that Hesperus appears in the evening') refers not to its customary sense but to its indirect sense. This would have to be accomplished by means of some yet third sense associated with the expression – a doubly indirect sense – which determines the expression's (singly) indirect sense. Since there is no limit in principle to the number of allowable embeddings of operators ('Brown realizes that Smith doubts that Jones believes that . . .'), Frege admitted the existence of an infinite hierarchy of senses associated with each meaningful expression. The doubly indirect sense of 'Socrates' is the customary sense of '^{s} ^{s}Socrates^{s} ^{s}', the triply indirect sense the customary sense of '^{s} ^{s} ^{s}Socrates^{s} ^{s} ^{s}', etc. Using such a device as sense-quotation marks, one may formulate a simple rule for computing the *n*-fold indirect sense of an expression, for arbitrary *n*.[11]

Frege did not explicitly consider the problem of failure of substitutivity in modal contexts, nor that of true negative existentials. It is in the spirit of Frege's theory, though, to regard modal operators as creating further oblique contexts in which expressions refer to their customary sense, as in the sentence 'It is necessary that 9 is odd', since it is not the truth value of the embeded sentence '9 is odd' that is said to be necessary, but rather its information or "thought" content, the information that 9 is odd. Similarly, Frege could regard the predicate 'exists' as creating an oblique context in which a singular term refers to its customary sense (in which case the sentence 'The present king of France does not exist' means the same as '^{s}the present king of France^{s} determines nothing'), or perhaps itself (' 'the present king of France' refers to nothing').

Frege explicitly considered certain temporal contexts – specifically the phenomenon of tense as well as such temporal indexicals as 'yesterday'. From his treatment of these it is possible to extract a solution to the problem of failure of substitutivity in temporal contexts. A tensed or temporally indexical sentence, according to Frege, is incomplete and must be supplemented by a time specification before it can properly express a thought and refer to a truth value. Whenever such a sentence is uttered, the very time of utterance is relied upon as the needed time specification, a specification of itself. "The time of utterance is part of the expression of the thought." Presumably, the tensed sentence itself has the logico-semantic status of a monadic predicate that attaches to a singular term referring to a time, and the time of utterance serves as a singular term referring to itself. Analogously, a definite description whose referent may vary with the time of utterance may be regarded as an operator that forms a complete singular term only when joined with a time specification, as may be provided by the time of the utterance itself. Thus, although the definite description 'the U.S. President', supplemented by the time of the writing of the present article, refers to the same individual as the name 'Ronald Reagan', the description 'the U.S. President' cannot be substituted for the name 'Ronald Reagan' in the sentence 'In 1978, Ronald Reagan was a Republican', since this sentence already includes a verbal time specification, 'in 1978', which supersedes the time of utterance in completing any expressions occurring within its scope in need of completion by a time specification. (A similar solution is possible for such complex constructions as 'When I lived in Princeton, Ronald Reagan was a Republican' and for quantificational temporal operators, such as 'always', in place of specific time indicators.) The solution here is significantly different from that for quotational and propositional-attitude contexts. The time specification 'in 1978' is seen not as creating an oblique context inducing a reference shift, but as providing a component needed to complete the singular term so that it may properly refer to an individual, relative to its position in the sentence. Once completed by the time specification, its referent, as occurring in that position, is just its customary referent.

The present article concerns singular terms. A full account of Frege's theory of meaning, not undertaken here, would require consideration of Frege's further doctrines concerning functions and their role in the semantics of predicates, connectives, quantifiers, and operators.

THE ORTHODOX THEORY

The theories of Russell and Frege have been extremely influential in contemporary philosophy. Although they are significantly dissimilar in that Russell's is essentially a supplement to the naive theory whereas Frege's involves a total abandonment of the naive theory (and any modification thereof), there is considerable common ground concerning singular terms, especially in regard to ordinary proper names. This area of agreement between the two theories has ascended to the status of orthodoxy. The orthodox theory can be explained as follows. Let us say that an expression α, as used in a particular possible context, is *descriptional* if there is a set of properties semantically associated with α in such a way as to generate a semantic relation, which may be called 'denotation' or 'reference', and which correlates with α (with respect to such semantic parameters as a possible world *w* and a time *t*) whoever or whatever uniquely has all (or at least sufficiently many) of these properties (in *w* at *t*), if there is a unique such individual, and nothing otherwise.[12] A descriptional term is one that denotes by way of properties. It is a term that expresses a way of conceiving something, and its "denotation" (with respect to a possible world and a time) is secured indirectly by means of this conceptual content. Definite descriptions, such as 'the author of *the Republic*', are descriptional. A nondescriptional singular term is one whose reference is not semantically mediated by associated conceptual content. The paradigm of a nondescriptional singular term is the individual variable. An individual variable is a singular term that refers (or "denotes") *under an assignment of values to individual variables.* The referent or "denotation" (with respect to a possible world and a time) of a variable under such an assignment is semantically determined directly by the assignment, and not by extracting a conceptual "mode of presentation" from the variable.

The descriptional theory of proper names is the theory that ordinary proper names, as used in a particular context, are descriptional. Frege and Russell held a strong version of this theory. On their view, if a name such as 'St. Anne' is analyzable as 'the mother of Mary', it must be in some sense analyzable even further, since the name 'Mary' is also supposed to be descriptional. But even 'the mother of the mother of Jesus' must be in this sense further analyzable, in view of the occurrence of the name 'Jesus', and so on.

Let α be a nondescriptional singular term referring to Socrates. Then the definite description $\ulcorner$the wife of $\alpha\urcorner$, though descriptional, is not thoroughly so. The property expressed is not one like that of being married to the philosopher who held that such-and-such. Rather, it is an intrinsically relational property directly involving Socrates, the property of being *his* wife. We may say that the description is only *relationally descriptional*, and that it is *descriptional relative to* Socrates. A *thoroughly descriptional* term, then, is one that is descriptional but not relationally descriptional.[13]

The orthodox theory is the theory that proper names, demonstratives, and such indexical terms as 'you', 'here', etc., as used in a particular possible context, are either thoroughly descriptional or descriptional relative only to items of "direct acquaintance," such as sensations, visual images, and the like.

Frege held the very strong version of this theory that proper names, demonstratives, and other indexical singular terms, as used in a particular context, are all thoroughly descriptional. Only if a term is thoroughly descriptional can there be something that counts as a genuine Fregean *Sinn* for the term. The reason for this is that the Fregean conception of *Sinn* is a compilation or conflation of at least three distinct linguistic attributes. First, the sense of an expression is a purely conceptual mode of presentation. Individuals that are not themselves senses, such as persons and their sensations, cannot form part of a genuine Fregean sense. Second, the sense of a singular term is the mechanism by which its referent is secured and semantically determined. Third, the sense of an expression is its information value. Nothing counts as the sense of a term, as Frege intended the notion, unless it is all three all at once. It is supposed that the purely conceptual content of any singular term is also its information value, which also secures its referent. This three-way identification constitutes a very strong theoretical claim. A descriptional singular term is precisely one whose mode of securing a referent is its descriptive content, which also serves as its information value. Only if the term is thoroughly descriptional, however, can this be identified with a *purely* conceptual (or a *purely qualitatively* descriptive) content. Strictly speaking, even a Russellian term descriptional relative only to items of direct acquaintance (if there is any such term) does not have a genuine Fregean sense. Some contemporary orthodox theorists (Leonard Linksy and John Searle, for example) have proposed various refinements of Frege's original doctrines concerning the senses of proper names, but typically these do not involve a departure from the thesis that names are thoroughly descriptional. Any departure from this thesis would constitute a rejection of fundamental Fregean theory.

THE THEORY OF DIRECT REFERENCE

In the 1960s and 1970s the orthodox theory was forcefully challenged by some philosophers, most notably by Keith Donnellan, David Kaplan, Saul Kripke, Ruth Barcan Marcus, and Hilary Putnam.[14] Instead they held the opposing theory that ordinary proper names and single-word indexical singular terms are nondescriptional. Since they deny that the reference of names and single-word indexicals is mediated by a descriptive concept, their view has come to be called 'the theory of direct reference'. This title may be misleading, however, since it suggests the obviously false thesis that reference is entirely unmediated. Also misleading, though literally correct, is the characterization of the direct-reference theory as the doctrine that names and indexicals have reference but not sense. In denying that proper names are descriptional, the direct-reference theory is not denying that a use of a particular proper name may exhibit any or all of the three aspects of a Fregean sense mentioned in the previous section. It is not denied that a use of a name typically evokes certain concepts in the minds of those who have learned the name, perhaps even purely conceptual or purely qualitative concepts. Nor is it denied that there is some means by which a referent for the name is secured. Nor is it necessarily denied that there is something other than the name's referent that serves as its information value. What the direct-reference theory denies is that the conceptual content is what secures the referent. Since a genuine Fregean sense would have to be, in addition to the information value, both the conceptual content and what secures the referent, it follows that names do not have Fregean sense. It is clearer and more direct, however, to state the central thesis by saying that names are nondescriptional, in the sense used here. Thus, for example, the direct-reference theory would hold that a proper name such as 'Shakespeare' is not shorthand for any description or cluster of descriptions, such as 'England's greatest bard', 'the author of *Hamlet*', etc. But the central thesis of the direct-reference theory is significantly stronger than a simple denial of Russell's doctrine that ordinary names are concealed definite descriptions. (It has been argued that, despite his illustrations, even Frege did not endorse this Russellian doctrine.) The direct-reference theory holds that ordinary names are not only not synonymous with definite descriptions, they are not even similar. According to the orthodox theory, ordinary names are either thoroughly descriptional or descriptional relative only to private sensations and other items of "direct acquaintance." Against this, the direct-reference theorists argue that names and single-word indexical singular terms, as ordinarily used, are not descriptional at all. An immediate consequence is that a great many definite descriptions fail to be thoroughly descriptional, or descriptional relative only to items of direct acquaintance, since so many contain proper names or indexicals referring to ordinary individuals.

THE ARGUMENTS

A number of arguments have been advanced in favor of the central thesis of the direct-reference theory. Although the arguments are many and varied, most of them may be seen as falling under one of three main kinds: modal arguments, epistemological arguments, and semantic arguments.

The modal arguments are due chiefly to Kripke. Consider the name 'Shakespeare' as used to refer to the famous English dramatist. Consider now the properties that someone might associate with the name as forming its conceptual content on a particular occasion. These properties might include Shakespeare's distinguishing characteristics, or the criteria by which we identify Shakespeare – such properties as that of being a famous English poet and playwright of the late 16th and early 17th Centuries; authorship of several classic plays including *Hamlet*, *Macbeth*, and *Romeo and Juliet*; partnership in the Globe Theatre; and so on. Suppose then that the name 'Shakespeare' simply means "the person, whoever he or she may be, having these properties," or for simplicity, "the English playwright who wrote *Hamlet*, *Macbeth* and *Romeo and Juliet*." Consider now the following sentences:

> Shakespeare, if he exists, wrote *Hamlet*, *Macbeth*, and *Romeo and Juliet*.
>
> If anyone is an English playwright who is sole author of *Hamlet*, *Macbeth*, and *Romeo and Juliet*, then he is Shakespeare.

If the orthodox theory of names is correct, then by substituting for the name its longhand synonym we find that these two sentences taken together simply mean: Someone is the English playwright who wrote *Hamlet*, *Macbeth*, and *Romeo and Juliet* if and only if he is the English playwright who wrote *Hamlet*, *Macbeth*, and *Romeo and Juliet*. That is, if the orthodox theory is correct, the sentences displayed above should express *logical truths* – indeed they should be *analytic* sentences in the traditional sense – and should therefore express necessary truths, propositions true with respect to all possible worlds. But surely, the argument continues, it is not at all necessary that someone is Shakespeare if and only if he is an English playwright who wrote *Hamlet*, *Macbeth*, and *Romeo and Juliet*. In the first place, it might have come to pass that Shakespeare elected to enter a profession in law instead of becoming a writer and dramatist. Hence, the first sentence displayed above does not express a necessary truth. Furthermore, assuming Shakespeare had gone into law instead of drama, it could have come to pass that some Englishman other than Shakespeare – say, Francis Bacon – should go on to write these plays. That is, it is not impossible that someone other than Shakespeare should write these plays. Hence even the second sentence displayed above expresses only a contingent truth. It follows that the name 'Shakespeare' is not descriptional in terms of the properties mentioned.

The intuition that the two sentences displayed above are false with respect to certain possible worlds supports and is supported by a complementary intuition concerning reference: that the name 'Shakespeare' continues to refer to the same person even with respect to counterfactual situations in which this individual lacks all of the distinguishing characteristics that we actually use to identify him. In particular, the name 'Shakespeare' continues to refer to the same individual even in discourse about a counterfactual situation in which not he but some other Englishman wrote *Hamlet*, *Macbeth*, and *Romeo and Juliet*, whereas the definite description 'the English playwright who wrote *Hamlet*, *Macbeth*, and *Romeo and Juliet*' will refer in such discourse to the other Englishman. Consequently, the two sentences displayed above must be false in such discourse. Thus the main intuition behind the modal arguments is intimately connected with a related linguistic intuition concerning the reference of proper names and indexical singular terms, in contrast to definite descriptions, with respect to other possible worlds. The orthodox theory comes into conflict with this intuition, whereas the direct-reference theorists offer an alternative that conforms with this intuition. One important consequence of the theory of direct reference is that such expressions as proper names and single-word indexical singular terms are *rigid designators* (Kripke). An expression is a rigid designator if it designates the same thing with respect to every possible world in which that thing exists, and does not designate something else with respect to any possible world.

The epistemological arguments against the orthodox theory – also due chiefly to Kripke – are similar to the modal arguments. Consider again the two sentences displayed above. Assuming that the orthodox theory is correct, these sentences are analytic in the traditional sense, and hence they should encode information that is knowable *a priori*, i.e., knowable solely by reflection on the concepts involved and without recourse to sensory or introspective experience. But it is not difficult to imagine circumstances in which it is discovered that, contrary to popular belief, Shakespeare did not write *Hamlet*, *Macbeth*, *Romeo and Juliet*, or any other work commonly attributed to him. Since this possibility is not automatically precluded by reflection on the concepts involved, it follows that the first sentence displayed above encodes information that is knowable only *a posteriori*, i.e., knowable only by recourse to sensory or introspective experience. One can even imagine circumstances in which it is discovered that we have been the victims of a massive hoax, and that, though Shakespeare is not responsible for any of these great works, some other Englishman (say, Bacon) wrote every one of the plays and sonnets commonly attributed to Shakespeare. This means that even the second sentence displayed above is not analytic or true by definition, as alleged, but encodes genuine *a posteriori* information.

The most persuasive and direct of the three kinds of arguments for the direct-reference theory are the semantic arguments offered by Donnellan, Kaplan, Kripke, and Putnam. One example is Donnellan's argument concerning Thales. Consider the set of properties that might be associated with the name 'Thales' according to the orthodox theory. Suppose, for example, that the sense or conceptual content of 'Thales' is determined by the description 'the Greek philosopher who held that all is water'. On the orthodox view, the name refers to whoever happens to satisfy this description. Suppose now that, owing to some error or fraud, the man referred to by writers such as Aristotle and Herodotus, from whom our use of the name 'Thales' derives, never genuinely believed that all is water. Suppose further that by a very strange coincidence there was indeed a Greek hermit-philosopher who did in fact hold this bizarre view, though he was unknown to them and bears no historical connection to us. To which of these two philosophers would our name 'Thales' refer? This is a clear semantic question with a clear answer: The name would refer to the first of the two. Our use of the name would bear no significant connection to the second character whatsoever. It is only by way of a comical accident that he enters into the story at all. He happens to satisfy the misdescription associated with the name 'Thales', and no one else does.

This example is not to be confused with the corresponding modal or epistemological arguments ("Thales might not have been the Greek philosopher who held that all is water"). In the modal and epistemological arguments, the main question is what the truth value of such a sentence as 'Thales is the Greek philosopher who held that all is water', which is alleged to be analytic, becomes when the sentence is evaluated with respect to certain imagined circumstances that are possible, in either a metaphysical or an epistemic sense. The strategy in the semantic arguments is more direct. The issue here is not whom the name *actually* refers to *with respect to* the imagined circumstances; the issue is whom the name *would* refer to if the circumstances described above *were to obtain*. The modal arguments are indirectly related to the question of what a particular term refers to *with respect to another possible world*; the semantic arguments are directly concerned with the nonmodal question of reference *simpliciter*. The key phrase in the definition of a descriptional singular term is not 'correlate with respect to a possible world', but 'whoever or whatever uniquely has the properties'. On any descriptional theory of names, precisely whom a name refers to or "denotes" depends entirely on whoever happens to have certain properties uniquely. The theory predicts that, if these circumstances were to obtain, the name would refer to or denote the hermit instead of Thales. But here the theory is simply mistaken. The existence of the hermit-philosopher would be irrelevant to the reference or denotation associated with our use of the name 'Thales'.

CONTEXTUAL FACTORS IN REFERENCE

The theory that proper names and single-word indexical singular terms are entirely nondescriptional should not be understood as involving the thesis that no descriptive concepts or properties are ever semantically associated with names or indexicals. Indeed, proponents of the direct-reference theory allow that some nondescriptional terms may be "defined," or introduced into a language or idiolect, by way of descriptional expressions. In this special kind of definition the descriptional expression serves only to assign a referent to the term being introduced, and does not simultaneously bequeath its descriptionality to the new term. To use Kripke's apt phrase, the descriptional expression is used only to "fix the reference" of the nondescriptional term. It does not provide the term with a Fregean sense.

This admission that there are sometimes descriptional elements at work in fixing the referent of a term of the sort in question is generally coupled with the observation that there are almost always nondescriptional contextual elements at work as well. The semantic arguments reveal something about the nature of what secures the referent of a name. The way its referent is determined is not a purely conceptual matter; external factors enter into it. The surrounding settings in which speakers find themselves are crucial to determining the referents of the names and other terms they use. This is true not only of the extra-linguistic setting in which the referent is to be found, but also of the linguistic setting in which the term is used or was learned by the speaker, i.e., the history of the use of the name leading up to the speaker's learning it. In a word, the securing of a referent for a proper name is a *contextual* phenomenon. Donnellan and Kripke have provided accounts of the securing of a referent for a proper name by means of such historical chains of communication. Putnam has given a similar account of certain terms designating something by means of a "division of linguistic labor" and a "structured cooperation between experts and nonexperts." In virtue of these accounts the theory of direct reference is often called the "causal" theory of reference. But causal factors are not the only external factors involved in securing a referent – or at any rate, it is an independent philosophical thesis that all of these external factors in reference are reducible to causal phenomena. It would be better to call the theory the *contextual* theory of reference. The contextual accounts provided by direct-reference theorists are usually sketchy and incomplete. Though there have been attempts to work out the details of how the referent of a proper name or demonstrative, as used in a particular context, is secured, there is much that remains to be done in this area.

OPEN ISSUES

The direct-reference theory is concerned primarily to distinguish two of the three aspects of a name conflated by the Fregean conception of sense: the conceptual content and what secures the referent. As noted above, if the conceptual representation associated with a use of a particular name is not what secures the name's referent, it follows that the name lacks genuine Fregean sense. We are still left, then, with the question concerning the third aspect of sense with which the present article began: What is the information value of a proper name? If the direct-reference theory is correct, it cannot be the sense of the name, for there is none. What then is the information value?

A tempting answer is that the information value of a name is simply its associated conceptual content, its "mode of presentation" of its referent. The identification of information value with associated conceptual or descriptive content does not require the further identification of this with the manner of securing a referent. Thus, this idea preserves a good deal (but not all) of Frege's point of view concerning information value without positing full-blown Fregean senses for names.

The theory that the information value of a name is simply its conceptual or descriptive content misrepresents the information content of sentences involving a name. Many of the considerations that count against the orthodox theory – the modal arguments for example – extend also to this simpler theory. Or to use a variant of the semantic argument, there could be two distinct individuals, *A* and *B*, such that the descriptive content individual *C* associates with *A*'s name is exactly the same as that associated by individual *D* with *B*'s name. (This could happen for any number of reasons. Perhaps *A* and *B* are very much like one another, or one or both of *C* and *D* is mistaken, etc.) Hence according to the proposed theory, *A*'s name has the same information value for *C* that *B*'s name has for *D*. This conflicts with the fact that when *C* uses *A*'s name to ascribe something to *A* – say, that he weighs exactly 165 pounds – and *D* uses *B*'s name to ascribe the very same thing to *B*, *C* asserts accurate information concerning *A* (and does not assert any information concerning *B*) whereas *D* asserts misinformation concerning *B* (and does not assert any information concerning *A*).

This suggests that conceptual content cannot be the whole of information value for proper names. It might be proposed, then, that the information value of a name is constituted partly by associated descriptive or conceptual content, and partly by something else, say the context that secures the referent, or perhaps the referent itself. But any proposal that identifies the information value of a name even only partly with descriptive or conceptual associations faces some of the same difficulties as the orthodox theory. Suppose, for example, that the descriptive content associated with the name 'Shakespeare' – one's concept of Shakespeare – includes some particular property as a central or critical element, say the authorship of *Romeo and Juliet*. If the proposed theory of information value is correct, the information encoded by the sentence 'If Shakespeare exists, then he wrote *Romeo and Juliet*' must be knowable *a priori*. But as was noted in the epistemological arguments against the orthodox theory, the information content of this sentence is *a posteriori*. There is always a possibility of error and inaccuracy in conceptual or descriptive associations. The descriptive content one associates with the name 'Plato' may be riddled with misattribution and misdescription, enough so as to befit someone else, say Aristotle, far better than Plato. Nevertheless, the sentence 'Plato wrote *the Republic*' encodes information that is entirely accurate and error-free. Hence, the descriptive associations that attach to the name cannot be even only a part of the name's information value.[15]

The primary philosophical question concerning proper names, demonstratives, and other single-word indexical singular terms – what is their information value? – is yet to be solved. The question was pressing for Frege and Russell because the answer provided by the naive theory – that the information value of a name is simply its referent – seemed to be refuted by the existence of the four puzzles of reference. The question is all the more pressing today, since the answer provided by the orthodox theory and its variants – that the information value of a name is given, at least partly, by its descriptive or conceptual associations – appears to fail as well. Among contemporary philosophers of reference, there are a number who attempt to defend some version of the orthodox theory against the criticisms of direct-reference theorists. In addition, there are those who offer some alternative theory of information value, and others who reject the question of information value on philosophical grounds. There are also some (including this author) who seek to revive some version or variant of the naive theory. Amid this controversy, a deeper understanding is needed of the four puzzles that gave rise to the present situation in the theory of reference, as well as a re-examination of the modified naive theory in light of this deeper understanding. Until this is achieved, it is premature to reject the modified naive theory solely on the basis of the puzzles.

University of California, Santa Barbara

NOTES

* Portions of my books *Reference and Essence* (1981) and *Frege's Puzzle* (1986) have been incorporated into the present article by permission of Princeton University Press/Basil Blackwell and the MIT Press, respectively.

[1] A word of clarification is needed concerning my use of the semantic predicates 'encode' and 'information'. The term 'information' is used here in a broad sense to include misinformation and even pieces of information that may be neither true nor false. Throughout this article we shall be concerned with discrete units of information that are specifiable by means of a 'that'-clause, e.g., the information that Socrates is wise. These discrete units are *pieces of information.* We shall generally use the mass noun 'information' as if it were shorthand for the count noun phrase 'piece of information', i.e., as a general term whose extension is the class of pieces of information. Thus, we shall write 'information that is such-and-such' to mean "pieces of information that are such-and-such," 'the same information' to mean "the same piece(s) of information," 'different information' to mean "different pieces of information," and so on. We shall use the verb 'encode' in such a way that an unambiguous declarative sentence encodes (with respect to a given possible context *c*) a *single* piece of information, which is referred to (with respect to *c*) by the result of prefixing 'the information that' to the sentence and which is to be called 'the information content' of the sentence (with respect to *c*). A declarative sentence may encode (with respect to a given context) two or more pieces of information, but if it does so, it is ambiguous. Pieces of information encoded by the proper logical consequences of an unambiguous sentence are not themselves encoded, in this sense, by the sentence. The (piece of) information that snow is white and grass is green is different information (a different piece of information) from the

(piece of) information that snow is white, though intuitively the latter is included as part of the former. The sentence 'Snow is white and grass is green' encodes only the former, not the latter. This constitutes a departure from at least one standard usage, according to which the information content of a sentence is perhaps something like a class of pieces of information, closed under logical consequence.

We shall not be concerned in this article with a notion of an *amount* of information, which arises in the mathematical theory of communication or information. The information *that snow is white and grass is green and Socrates is Socrates* may be no more or less information than the information *that both snow is white if and only if grass is green and either snow is white or grass is green.* Nevertheless general considerations involving Leibniz's Law strongly suggest that they are numerically distinct pieces of information. For instance, the first concerns Socrates whereas the second does not.

[2] The latter clause is needed in order to distinguish 'Bill loves Mary' from 'Mary loves Bill', where the sequential order of composition is crucial. This succinct statement of the rule connecting sentences and their information contents is only an approximation to the truth. A complicated difficulty arises in connection with the latter clause of the rule and with such quantificational locutions as 'someone' in 'Someone is wise'. Grammatically the sentence 'Someone is wise' is analogous to 'Socrates is wise', though logically and semantically they are disanalogous. In 'Socrates is wise', the predicate 'is wise' attaches to the singular term 'Socrates'. This situation is reversed in 'Someone is wise', wherein the restricted quantifier 'someone' attaches to the predicate 'is wise'. Thus, whereas grammatically 'someone' is combined with 'is wise' to form the first sentence in just the same way that 'Socrates' is combined with 'is wise' to form the second sentence, the information values of 'someone' and 'is wise' are combined very differently from the way the information values of 'Socrates' and 'is wise' are combined.

A perhaps more important qualification to the general rule arises in connection with quotation marks and similar devices – since the information value of the numeral '9' is no part of the information content of the sentence 'The numeral '9' is a singular term'. Yet another important qualification concerns overlaid quantifiers. It is necessary to distinguish between the information contents of such constructions as:

(A) For everyone x there is someone y such that x loves y.

and

(B) For everyone x there is someone y such that y loves x.

One method, due to Alonzo Church and pointed out to me by David Kaplan, employs Russellian propositional functions, i.e., functions from an individual to a singular proposition involving that individual. (Church himself applies the general method in such a way as to invoke only Fregean functions from pure concepts of individuals to Fregean purely general propositions, although the general method can also accommodate anti-Fregean theories by invoking propositional functions.) A variant of this method (closer to the spirit of the naive and modified naive theories defined below) invokes attributes (properties and relations) in place of these functions. The information values of the quantifiers 'for everyone' and 'there is someone such that' are certain higher order properties. (Specifically, they are properties of classes of individuals.) The former proposition may be regarded as made up of the property of being an individual x such that x loves someone or other y, together with the information value of 'for everyone'. The latter proposition may be regarded as made up of the property of being an individual x such that someone or other y loves x, together with the information value of 'for everyone'. (Alternatively, the property of loving someone may be replaced by that of loving someone at *t*, where *t* is the time of utterance, and the property of being loved by someone by that of being loved by someone at *t*.) The information content of sentence (A) may thus be assimilated to the following complex proposition: that the class of individuals who love someone or other includes everyone whatsoever. Likewise, the information content of sentence (B) is assimilated to the proposition that the class of individuals loved by someone or other includes everyone whatsoever. (For more details, see my *Frege's Puzzle*, Appendix C, pp. 143–151.)

This method need not assign any information value to an open sentence such as '*x* loves *y*', except relative to an assignment of values to its free variables. Unlike some other possible proposals, it thus generates no counterexamples to a certain principle of compositionality (or interchange) for information value, commonly attributed to Frege, according to which the information value of a compound expression such as (A) or (B) is a function solely of the information values of its information-valued components. (See Church, 'Intensional Isomorphism and Identity of Belief', *Philosophical Studies* **5**, No. 5 (October 1954), pp. 65–73, for a similar but more sharply articulated principle.) However, the method does generate counterexamples to a stronger compositionality principle, also attributable to Frege, according to which the information value of a compound expression is a complex made up entirely of the information values of its information-valued components. If the first element of the information content of sentence (A) is the information value of any component of the sentence – for example, the component '*x* there is someone *y* such that *x* loves *y*' – then the information value of that component is not made up of the information values of its information-valued components.

[3] For further details, see by *Frege's Puzzle*, Section 2.2, pp. 24–43; and 'Tense and Singular Propositions,' in J. Almog, J. Perry, and H. Wettstein (eds.), *Themes from Kaplan* (Oxford University Press, forthcoming).

[4] See Alonzo Church, 'Review of Carnap's *Introduction to Semantics*,' *The Philosophical Review* **52** (1943), pp. 298–304, at pp. 299–301; and Kurt Gödel, 'Russell's Mathematical Logic', in P. A. Schilipp (ed.), *The Philosophy of Bertrand Russell*, The Library of Living Philosophers, Evanston, Illinois (New York: The Tudor Publishing Company, 1944), at pp. 128–129. The general argument is applied to the special case of monadic predicates in my *Frege's Puzzle*, at pp. 22–23, and in greater detail to the special case of common nouns in my *Reference and Essence*, at pp. 48–52.

[5] See Note 2 above. It has been argued, however, that the information content of a sentence is a function not only of the information-values and the sequential order of the information-valued parts but also of the very logical structure of the sentence as a whole, and that therefore, since the two identity sentences differ in logical structure the mode of composition of the information values of their parts are different from one another. See Hilary Putnam, 'Synonymity, and the Analysis of Belief Sentences', *Analysis* **14**, No. 5 (April 1954), pp. 114–122, at pp. 118–119, and especially Note 8. For response, see Alonzo Church, 'Intensional Isomorphism and Identity of Belief', *loc. cit.*, at pp. 68–71; and my *Frege's Puzzle*, at pp. 164–165, Note 4.

[6] Cf. Richard Montague, 'The Proper Treatment of Quantification in Ordinary English', in his *Formal Philosophy*, R. Thomason, ed. (New Haven: Yale University Press, 1974), pp. 247–270.

[7] Alternatively, the determiner 'some' is plausibly regarded as a two-place quantifier, so that the phrase 'some man' would be seen as an incomplete string that is formed by attaching a

two-place quantifier to a single first-order predicate, and that stands in need of completion by a second first-order predicate. The General Theory of Descriptions also conflicts with this extremely plausible theory concerning the logico-semantic status of 'some man'.

[8] See Arthur F. Smullyan, 'Modality and Description', in L. Linsky (ed.), *Reference and Modality* (Oxford University Press, 1971), pp. 35–43.

[9] In characterizing the sense of an expression as a purely conceptual entity, I intend the word 'concept' with a more or less ordinary meaning and not with that of Frege's special use of '*Begriff*'. Senses are neither empirically observed (as are external, concrete objects) nor "had" in the way that sensations or other private experiences are had, but are abstract entities that are 'grasped" or "apprehended" by the mind. In addition, I intend the term 'pure' to exclude concepts that include nonconceptual elements as constituents. A genuine sense may involve reference to an object, but it must do so by including a conceptual representation of the object in place of the object itself. This will be clarified below.

[10] Cf. David Kaplan's "meaning marks," introduced in 'Quantifying In', in D. Davidson and J. Hintikka (eds.), *Words and Objections: Essays on the Work of W. V. Quine* (Dordrecht: D. Reidel, 1969), pp. 206–242, at pp. 213–215.

[11] This tends to militate against the unlearnability objection of Donald Davidson in 'Theories of Meaning and Learnable Languages', in Y. Bar-Hillel (ed.), *Logic, Methodology, and Philosophy of Science: Proceedings* (Amsterdam: North-Holland, 1965), pp. 383–394, wherein he complains (at p. 393) that "the problem [with Frege's doctrine of reference-shifting from n-fold indirect sense to $(n + 1)$-fold indirect sense] is not how the individual expressions that make up a sentence governed by 'believes', given the meanings they have in such a context, combine to denote a proposition; the problem is rather to state the rule that gives each the meaning it does have."

[12] Frege and Russell wrote before the advent of modern intensional semantics, and consequently neither spoke of reference or truth *with respect to a possible world* or *with respect to a time*, but only of reference ("meaning") or truth (in a language) *simpliciter*. The parenthetical phrase 'with respect to a possible world and a time' indicates the natural and usual extensions of their account to modal and temporal semantics. However, both Frege and Russell treated the phenomenon of tense and other temporal operators differently from the usual treatment today, and neither clearly distinguished tense from the distinct phenomenon of indexicality.

[13] See my *Reference and Essence*, pp. 14–21, 43–44, and 54–55, for a more detailed discussion of these notions.

[14] Keith Donnellan, 'Proper Names and Identifying Descriptions', in D. Davidson and G. Harman (eds.), *Semantics of Natural Language* (Dordrecht: D. Reidel, 1972), pp. 356–379, and 'Speaking of Nothing', *The Philosophical Review* **83** (January 1974), pp. 3–32; David Kaplan, 'Demonstratives', in *Themes from Kaplan*; Saul Kripke, 'Identity and Necessity', in M. Munitz (eds.), *Identity and Individuation* (New York University Press, 1971), pp. 135–164, and 'Naming and Necessity' in Davidson and Harman (eds.), *op. cit.*, pp. 253–355, 763–769, and as a book (with an additional, substantive preface), *Naming and Necessity* (Harvard University Press and Basil Blackwell, 1980); Ruth Barcan Marcus, 'Modalities and Intensional Languages', *Synthese* **13**, No. 3 (September 1961), pp. 303–322; Hilary Putnam, 'Meaning and Reference', *Journal of Philosophy* **70** (November 8, 1973), pp. 699–711, and 'The Meaning of "Meaning"', in K. Gunderson (eds.), *Minnesota Studies in the Philosophy of Science VII: Language, Mind, and Knowledge* (Minneapolis: University of Minnesota Press, 1975), pp. 131–193. See also my *Reference and Essence*, pp. 9–75, 93–157.

[15] For further difficulties with this and other proposed alternatives to the original and modified naive theories and the orthodox theory, see my *Frege's Puzzle*, pp. 63–75.

SELECTED BIBLIOGRAPHY

Barwise, J. and J. Perry.: 1983, *Situations and Attitudes* (Cambridge, Mass.: MIT Press).

Blackburn, S.: 1975, 'The identity of propositions', in Blackburn (ed.), *Meaning, Reference and Necessity* (Cambridge University Press), pp. 182–205.

Bradley, R. and N. Swartz,: 1979, *Possible Worlds: An Introduction to Logic and Its Philosophy* (Indianapolis: Hackett Publishing Company).

Burge, T.: 1977, 'Belief *De Re*', *The Journal of Philosophy* **69**, 338–362.

Burge, T.: 1978, 'Belief and synonymy', *Journal of Philosophy* **75**, 119–138.

Burge, T.: 1979, 'Individualism and the mental', in P. French, T. Uehling, and H. Wettstein (eds.), *Midwest Studies in Philosophy IV: Studies in Metaphysics* (Minneapolis: University of Minnesota Press), pp. 73–121.

Cappio, J.: 1981, 'Russell's philosophical development', *Synthese* **46**, 185–205.

Carnap, R.: 1947, *Meaning and Necessity: A Study in Semantics and Modal Logic* (University of Chicago Press).

Cartwright, R.: 1966, 'Propositions', in R. Butler (eds.), *Analytical Philosophy*, First Series (Oxford: Basil Blackwell), pp. 81–103.

Church, A.: 1943, 'Review of Carnap's *Introduction to Semantics*', *The Philosophical Review* **52**, 298–304.

Church, A.: 1950, 'On Carnap's analysis of statements of assertion and belief', *Analysis* **10**, No. 5, 97–99; also in L. Linsky (ed.), 1971, pp. 168–170.

Church, A.: 1951, 'A formulation of the logic of sense and denotation', in Henle, Kallen, and Langer (eds.), *Structure, Method and Meaning: Essays in Honor of Henry M. Sheffer* (New York: Liberal Arts Press), 3–24.

Church, A.: 1954, 'Intensional isomorphism and identity of belief', *Philosophical Studies* **5**, No. 5, 65–73; also in N. Salmon and S. Soames, eds., *Propositions and Attitudes* (Oxford University Press, 1988), pp. 159–168.

Church, A.: 1956, *Introduction to Mathematical Logic 1* (Princeton University Press).

Church, A.: 1973, 'Outline of a revised formulation of the logic of sense and denotation', Part I, *Noûs* **7** (March), 24–33; Part II, *Noûs* **8** (May), 135–156.

Church, A.: 1982, 'A remark concerning Quine's paradox about modality', Spanish translation in *Analisis Filosofico* **2**, Nos 1–2 (May–November 1982): 25–34; in English in N. Salmon and S. Soames, eds., *Propositions and Attitudes* (Oxford University Press, 1988), pp. 58–65.

Cresswell, M. J.: 1985, *Structured Meanings: The Semantics of Propositional Attitudes* (Cambridge, Mass.: MIT Press).

Davidson, D.: 1965, 'Theories of meaning and learnable languages', in Y. Bar-Hillel, ed. *Logic, Methodology, and Philosophy of Science*: *Proceedings* (Amsterdam: North-Holland), pp. 383–394.

Devitt, M.: 1981, *Designation* (New York: Columbia University Press).

Donnellan, K.: 1966, 'Reference and definite descriptions', *The Philosophical Review* **75** (July), 281–304.

Donnellan, K.: 1972, 'Proper names and identifying descriptions,' in D. Davidson and G. Harman (eds.), *Semantics of Natural Language* (Dordrecht: D. Reidel), pp. 356–379.
Donnellan, K.: 1974, 'Speaking of nothing', *The Philosophical Review* **83**, 3–31; also in Schwartz, 1977, pp. 216–244.
Donnellan, K.: 1983, 'Kripke and Putnam on natural kind terms', in C. Ginet and S. Shoemaker (eds.), *Knowledge and Mind* (Oxford University Press), pp. 84–104.
Dummett, M.: 1973, *Frege: Philosophy of Language* (New York: Harper and Row, 1973).
Dummett, M.: 1978, *Truth and Other Enigmas* (Cambridge, Mass.: Harvard University Press).
Fitch, G. W.: 1985, 'On the logic of belief', *Noûs* **19**, No. 2 (June), 205–228.
Fitch, G. W.: 1986, 'Belief ascription', *Philosophical Studies* **49**, 271–280.
Frege, G.: 1972, '*Begriffsschrift*', in English in Frege's *Conceptual Notation and Related Articles* (Oxford University Press); also in Frege, 1952, pp. 1–20.
Frege, G.: 1984, '*Funktion und Begriff*,' in English in Frege, 1984, pp. 137–156; also in Frege, 1952, pp. 21–41.
Frege, G.: 1893, '*Über Sinn und Bedeutung*,' *Zeitschrift für Philosophie und Philosophische Kritik* **100** (1893): 25–50; in English in Frege, 1984, pp. 157–177; also in Frege, 1952, pp. 56–78.
Frege, G.: 1984, '*Über Begriff und Gegenstand*,' in English in Frege, 1984, pp. 182–194; also in Frege, 1952, pp. 42–55.
Frege, G.: 1952, *Translations from the Philosophical Writings*, translated by P. Geach and M. Black (Oxford: Basil Blackwell).
Frege, G.: 1984, '*Der Gedanke*,' in English in Frege, 1984, pp. 351–372; also in Frege, 1977, pp. 1–30; also in N. Salmon and S. Soames, eds., *Propositions and Attitudes* (Oxford University Press, 1988), pp. 33–55.
Frege, G.: 1977, *Logical Investigations* (New Haven: Yale University Press, 1977).
Frege, G.: 1979, 'Logic,' in Frege, 1979, pp. 126–151.
Frege, G.: 1979, *Posthumous Writings*, H. Hermes, F. Kambartel, and F. Kaulbach, eds., translated by P. Lond and R. White (Chicago: University of Chicago Press).
Frege, G.: 1980, *Philosophical and Mathematical Correspondence*, G. Gabriel, H. Hermes, F. Kambartel, C. Thiel, and A. Veraart, eds., abridged by B. McGuinness, translated by H. Kaal (Chicago: University of Chicago Press, 1980).
Frege, G.: 1984, *Collected Papers on Mathematics, Logic, and Philosophy*, B. McGuinness, ed., translated by M. Black, V. H. Dudman, P. Geach, H. Kaal, E.-H. W. Kluge, B. McGuiness, and R. H. Stoothoff (Oxford: Basil Blackwell, 1984).
French, P., Uehling, T., and Wettstein, H. (eds.): 1979, *Contemporary Perspectives in the Philosophy of Language* (Minneapolis: University of Minnesota Press).
Gödel, K.: 1946, 'Russell's mathematical logic', in P. A. Schlipp (ed.), *The Philosophy of Bertrand Russell*, The Library of Living Philosophers, Evanston, Illinois (New York: The Tudor Publishing Company): pp. 125–153.
Kaplan, D.: 1969, 'Quantifying in', in D. Davidson and G. Harman (eds.) *Words and Objections: Essays on the Work of W. V. Quine* (Dordrecht: D. Reidel), pp. 206–242: also in Linsky, 1971, pp. 112–144.
Kaplan, D.: 1978, 'Dthat,' in P. Cole, ed., *Syntax and Semantics 9: Pragmatics* (New York: Academic Press): pp. 221–243; also in French *et al.*, 1979, pp. 383–400.
Kaplan, D.: 1973, 'Bob and Carol and Ted and Alice,' in J. Hintikka, J. Moravcsik, and P. Suppes (eds.), *Approaches to Natural Language* (Dordrecht: D. Reidel), pp. 490–518.

Kaplan, D.: 1979, 'On the logic of demonstratives,' in French *et al.*, pp. 401–412; also in N. Salmon and S. Soames, ed., *Propositions and Attitudes* (Oxford University Press, 1988), pp. 66–82.

Kaplan, D.: 1975, 'How to Russell a Frege–Church,' *Journal of Philosophy* **72** (November 6), 716–729; also in M. Loux (ed.), *The Possible and the Actual* (Ithaca: Cornell University Press, 1979), pp. 210–224.

Kaplan, D. 1977, 'Demonstratives (Draft #2),' unpublished manuscript, U.C.L.A. Department of Philosophy, 1977; published in J. Almog, J. Perry, and H. Wettstein, eds., *Themes from Kaplan* (Oxford University Press, forthcoming).

Kaplan, D. 'Opacity,' in L. E. Hahn and P. A. Schilpp (eds.), *The Philosophy of W. V. O. Quine* (La Salle, Ill.: Open Court), pp. 229–289.

Kripke, S.: 1971, 'Identity and necessity', in M. Munitz (ed.), *Identity and Individuation* (New York: New York University Press), pp. 135–164; also in Schwartz, 1977, pp. 66–101.

Kripke, S.: 1972, *Naming and Necessity* (Harvard University Press and Basil Blackwell, 1972, 1980); also in D. Davidson and G. Harman (eds.), *Semantics of Natural Language* (Dordrecht: D. Reidel, 1972), pp. 253–355, pp. 763–769.

Kripke, S.: 1979, 'Speaker's reference and semantic reference,' in French *et al.*, 1979, pp. 6–27.

Kripke, S.: 1979, 'A puzzle about belief,' in A. Margalit (ed.), *Meaning and Use* (Dordrecht: D. Reidel, 1979): pp. 239–275; also in N. Salmon and S. Soames, eds., *Propositions and Attitudes* (Oxford University Press, 1988), pp. 102–148.

Linsky, L. (ed.): 1971, *Reference and Modality* (Oxford University Press).

Linsky, L. (ed.): 1977, *Names and Descriptions* (Chicago: University of Chicago Press).

Linsky, L. (ed.): 1983, *Oblique Contexts* (Chicago: University of Chicago Press).

Marcus, R. B.: 1961, 'Modalities and intensional languages,' *Synthese* **13**, No. 3 (September), 303–322.

Mates, B.: 1950, 'Synonymity,' *University of California Publications in Philosophy*, 25 also in L. Linsky, ed., *Semantics and the Philosophy of Language* (University of Illinois Press, 1952), pp. 109–136.

Mill, J. S.: 1843, *A System of Logic* (New York: Harper Brothers).

McGinn, C.: 1982, 'The structure of content,' in A. Woodfield (ed.), pp. 207–258.

McKay, T.: 1981, 'On proper names in belief ascriptions,' *Philosophical Studies* **39** 287–303.

Montague, R.: 1974, 'The proper treatment of quantification in ordinary English,' in Montague, 1974, pp. 247–270.

Montague, R.: 1974, *Formal Philosophy*, R. Thomason, ed. (New Haven: Yale University Press).

Moore, G. E.: 1962, 'Facts and propositions,' *Aristotelian Society, Supplementary Vol. 7*, (1927); also in Moore's *Philosophical Papers* (New York: Collier Books), pp. 60–88.

Perry, J.: 1977, 'Frege on demonstratives,' *The Philosophical Review* **86,** pp. 474–497.

Perry, J.: 1979, 'The problem of the essential indexical,' *Noûs* **13,** pp. 3–21, also in N. Salmon and S. Soames, eds., *Propositions and Attitudes* (Oxford University Press, 1988), pp. 83–101.

Perry, J.: 1980, 'Belief and acceptance,' in P. French, T. Uehling, and H. Wettstein, eds., *Midwest Studies in Philosophy V: Studies in Epistemology* (Minneapolis: University of Minnesota Press), pp. 533–542.

Perry, J.: 1980, 'A problem about continued belief,' *Pacific Philosophical Quarterly* **61,** pp. 317–332.

Plantinga, A.: 1978, 'The Boethian compromise,' *American Philosophical Quarterly* **15** (April), pp. 129–138.

Plato: 1961, *Sophist*, in E. Hamilton and H. Cairns (eds.), *The Collected Dialogues of Plato* (Princeton University Press), pp. 957–1017.

Price, H. H.: 1969, *Belief* (London: George Allen and Unwin Ltd.).

Prior, A. N.: 1971, Objects of Thought (Oxford: Oxford University Press).

Putnam, H.: 1954, 'Synonymity, and the analysis of belief sentences,' *Analysis* **14**, No. 5 (April), 114–122; also in N. Salmon and S. Soames, eds., *Propositions and Attitudes* (Oxford University Press, 1988), pp. 149–158.

Putnam, H.: 1973, 'Meaning and reference,' *The Journal of Philosophy* **70** (November 8), 699–711; also in Schwartz, 1977, pp. 119–132.

Putnam, H.: 1975, 'The meaning of 'meaning',' in K. Gunderson (ed.), *Minnesota Studies in the Philosophy of Science VII: Language, Mind, and Knowledge* (Minneapolis: University of Minnesota Press), also in Putnam's *Philosophical Papers II: Mind, Language, and Reality* (Cambridge University Press, 1975): pp. 215–271.

Putnam, H.: 1979, 'Comments [on Kripke, 1979],' in A. Margalit (ed.), *Meaning and Use* (Dordrecht: D. Reidel), pp. 285–288.

Quine, W. V. O.: 1953, 'Reference and modality,' in Quine, pp. 139–159; also in L. Linsky, ed., 1971, pp. 17–34.

Quine, W. V. O.: 1953, *From a Logical Point of View* (New York: Harper and Row).

Quine, W. V. O.: 1956, "Quantifiers and propositional attitudes," *Journal of Philosophy* **53**, No. 5 (March 1): 177–187; also in Quine's *The Ways of Paradox* (New York: Random Hous, 1966): pp. 183-194; also in L. Linsky, ed., 1971, pp. 101–111.

Quine, W. V. O.: 1960, *Word and Object* (Cambridge, Mass.: MIT Press).

Quine, W. V. O.: 1979, 'Intensions revisited,' in P. French, *et al.*, 1979, pp. 268–274.

Reddam, J. P.: 1981, 'Van Fraassen on propositional attitudes,' *Philosophical Studies* **42**, 1 (July), pp. 101–110.

Richard, M.: 1981, 'Temporalism and eternalism,' *Philosophical Studies* **39**, 1–13.

Richard, M.: 1982, 'Tense, propositions, and meanings,' *Philosophical Studies* **41**, 337–351.

Richard, M.: 1983, 'Direct reference and ascriptions of belief,' *Journal of Philosophical Logic* **12**, 425–452; also in N. Salmon and S. Soames, eds., *Propositions and Attitudes* (Oxford University Press, 1988) pp. 169–196.

Russell, B.: 1903, *Principles of Mathematics* (New York: W. W. Norton and Co.).

Russell, B.: 1905, 'On denoting,' *Mind* **14** (October): 479–493; also in Russell, 1956, pp. 41–56; also in Russell, 1973, pp. 103–119.

Russell, B.: 1911, 'Knowledge by acquaintance and knowledge by description,' Chapter X of Russell's *Mysticism and Logic and Other Essays* (London: Longmans, Green and Company), pp. 209–232; also in N. Salmon and S. Soames, eds., *Propositions and Attitudes* (Oxford University Press, 1988) pp. 16–32.

Russell, B.: 1912, *The Problems of Philosophy* (Oxford: Oxford University Press).

Russell, B.: 1919, "Descriptions," Chapter 16 of Russell's *Introduction to Mathematical Philosophy* (London: George Allen and Unwin): pp. 167–180.

Russell, B.: 1921, *The Analysis of Mind* (George Allen and Unwin).

Russell, B.: 1956, 'The philosophy of logical atomism,' in Russell, 1956, pp. 177–281.

Russell, B.: 1956, *Logic and Knowledge*, R. C. Marsh, ed. (London: George Allen and Unwin).

Russell, B.: 1957, 'Mr Strawson on referring,' *Mind* **66** (July), 385–389; also in Russell, 1973, pp. 120–126.

Russell, B.: 1973, *Essays in Analysis* D. Lackey, ed. (New York: George Allen and Unwin).

Sainsbury, R. M.: 1983, 'On a Fregean argument for the distinctness of sense and reference,' *Analysis* **43** (January), 12–14.

Salmon, N.: 1979, 'Review of Linsky, 1977,' *Journal of Philosophy* **76** (August), 436–452.

Salmon, N.: 1981, *Reference and Essence* (Princeton University Press and Basil Blackwell).

Salmon, N.: 1983, 'Assertion and incomplete definite descriptions,' *Philosophical Studies* **42**, 37–45.

Salmon, N.: 1984, 'Fregean theory and the four worlds paradox,' *Philosophical Books* **25**, No. 1 (January), 7–11.

Salmon, N.: 1986, *Frege's Puzzle* (Cambridge, Mass.: MIT Press/Bradford Books).

Salmon, N.: 1986, 'Reflexivity,' *Notre Dame Journal of Formal Logic* **27**; also in N. Salmon and S. Soames, eds., *Propositions and Attitudes* (Oxford University Press, 1988), pp. 240–274.

Salmon, N.: 1988, 'Tense and singular propositions,' in J. Almog, J. Perry, and H. Wettstein, eds., *Themes from Kaplan* (Oxford University Press, forthcoming).

Schiffer, S.: 1978, 'The basis of reference,' *Erkenntnis* **13** (July), 171–206.

Schiffer, S.: 1983, 'Indexicals and the theory of reference,' *Synthese* **49**, 43–100.

Schwartz, S.: 1977, *Naming, Necessity, and Natural Kinds* (Cornell University Press).

Searle, J.: 1983, *Intentionality* (Cambridge University Press).

Sellars, W.: 1955, 'Putnam on synonymity and belief', *Analysis* **15**, 117–120.

Smullyan, A. F.: 1948, 'Modality and descriptions', *The Journal of Symbolic Logic* **13**, 31–37; also in L. Linsky (ed.), 1971, pp. 35–43.

Soames, S.: 1988, 'Direct reference, propositional attitudes, and semantic content' in N. Salmon and S. Soames, eds., *Propositions and Attitudes* (Oxford University Press, 1988), pp. 197–239.

Sosa, E.: 1975, 'Propositional attitudes *De Dicto* and *De Re*,' *Journal of Philosophy* **71** (December), 883–896.

Stalnaker, R.: 1976, 'Propositions,' in A. MacKay and D. Merrill, eds., *Issues in the Philosophy of Language* (New Haven: Yale University Press), pp. 79–92.

Stalnaker, R.: 1978, 'Assertion,' in P. Cole, ed., *Syntax and Semantics 9: Pragmatics* (New York: Academic Press), pp. 315–332.

Stalnaker, R.: 1981, 'Indexical belief,' *Synthese* **49**, 1 (October), 129–151.

Stalnaker, R.: 1984, *Inquiry* (Cambridge, Mass.: MIT Press).

Strawson, P. F.: 1950, 'On referring', *Mind* **59**, No. 235, 320–344; also in Strawson (1971), pp. 1–27.

Strawson, P. F.: 1971, *Logico-Linguistic Papers* (London: Methuen and Company).

Tienson, J.: 1984, 'Hesperus and Phosphorus,' *Australasian Journal of Philosophy* **62**, 1 (March), 16–25.

Tye, M.: 1978, 'The puzzle of Hesperus and Phosphorus,' *Australasian Journal of Philosophy* **56**, No. 3 (December), 219–224.

Wettstein, H.: 1979, 'Indexical reference and propositional content,' *Philosophical Studies* **36**, 91–100.

Wettstein, H.: 1981, 'Demonstrative reference and definite descriptions,' *Philosophical Studies* **40**, 241–257.

Whitehead, A. N. and Russell, B.: 1927, *Principia Mathematica*, Second Edition (Cambridge University Press).

Wiggins, D.: 1968, 'Identity-statements,' in R. J. Butler, ed., *Analytical Philosophy*, Second Series (Oxford: Basil Blackwell), pp. 40–71.

Wiggins, D.: 1976, 'Frege's problem of the Morning Star and the Evening Star,' in M. Schirn, ed., *Studies on Frege II: Logic and the Philosophy of Language* (Stuttgart: Bad Canstatt), pp. 221–55.

Wittgenstein, L.: 1922, *Tractatus Logico-Philosophicus* (London: Routledge and Kegan Paul).

Wittgenstein, L.: 1953, *Philosophical Investigations*, translated by G. E. M. Anscombe (New York: The Macmillan Company).

Woodfield, A. (ed.): 1982, *Thought and Object* (Oxford University Press).

Yagisawa, T.: 1984, 'The pseudo-Mates argument,' *The Philosophical Review* **93**, No. 3, 407–418.

CHAPTER IV.6

INDEXICALS

by Graeme Forbes

1. THE TWO PROBLEMS: SEMANTICS AND LOGIC

The central examples of indexical expressions are the English pronouns 'I', 'you', 'it', and temporal and spatial expressions like 'now', 'tomorrow', 'here' and 'there'; these indexicals are syntactically simple, and can be used to form complex indexical expressions such as 'my father', 'over there', and 'the day after tomorrow'. An indexical expression is, roughly, one whose associated object is not given once for all, as it is for a proper name. Rather, the associated object changes in some systematic way from context of use to context of use, and one task of any semantics of indexicals is to explain or articulate the system which is actually at work here. In the literature, it is also usual to include with indexicals other kinds of expression which have a salient semantic feature that changes as the context of use changes; for instance, demonstratives, such as 'that man', and tense operators, such as 'it will be the case that'; so we will count these as indexicals too.[1]

What are the problems in philosophical logic to which indexicals give rise? Looking at the examples just given, it appears that the function of such expressions is to pick out objects of a certain sort. This remark, at least for the case of 'I', 'here' and 'now', would be regarded as superficial by those who doubt that there are such objects as selves, places and times, but for the sake of making some initial progress, we will indulge outselves in a little metaphysical naïvety. So indexicals are amongst the singular terms of natural language, along with proper names and pure definite

D. Gabbay and F. Guenthner (eds.), Handbook of Philosophical Logic, Volume IV, 463–490.

descriptions (these latter are expressions of the form 'the *F*' where '*F*' does not itself contain a name or an indexical – 'the inventor of the wheel' as opposed to 'the girl next door'). Hence they need to be accommodated by any general theory of singular reference, such as the theories described by Salmon in his essay on names and descriptions in this volume. For instance, Russell at one point held that ordinary names abbreviate definite descriptions (which themselves describe, rather than refer) and that the phenomenon of genuine reference is rare; where then do indexicals fit in his scheme? Within this area, we should also investigate the question of how indexicals are related one to another. For example, we said that 'I', 'here' and 'now' are syntactically simple, but it does not follow that they are conceptually simple. Perhaps 'here' means 'where I am now located', or maybe it is 'I' which is complex, meaning something like 'the self now here'. A third possibility is that the three central indexicals reduce to demonstratives picking out experiences: 'I' would have the content of 'the subject of *this* experience'. A more radical, eliminative, proposal, is that the content of indexicals can be spelt out without ultimately relying on *any* indexical resources. We shall return to these matters of semantic analysis below.

Another kind of question indexicals raise concerns the proper way of extending the set-theoretic notion of a formal interpretation for a language when indexicals are added to that language, so that we can characterize and investigate the notion of logical validity for indexical languages. What new apparatus is required to deal with 'I', 'here', 'now' and 'that *F*' over and above what we already have for languages with names, descriptions, tenses and modal adverbs such as 'possibly' and 'necessarily'? In this connection, we have some pretheoretic intuitions to which any adequate proposals should do justice. For instance, it appears to be a contingent truth that I am here now (Scotland, July 1984), for I could have remained in California for this month, rather than travelling here. So the interpretative apparatus must admit a possible world in which I am not here now, but am rather in California. On the other hand, there is a sense in which no reasonable interpretation can result in the sentence 'I am here now' being assigned the truth-value false; so although we want to allow worlds where I am not here now, we do not want models in which 'I am here now' is falsified. An elegant formal apparatus has been devised by David Kaplan which satisfies these desiderata (Kaplan [1977]), and we shall explain his "logic of demonstratives", probably the single most influential contribution to our topic, in some detail. Of course, the problem of analysis and the problem of logic are interconnected in ways which will emerge as we pursue our discussion.

2. INDEXICALITY IN A FREGEAN CONTEXT

Descartes was perhaps the first philosopher to interest himself in, or at least to exploit, the phenomenon of indexicality, and we might expect a semantic analysis of indexicals to throw some light on the peculiarities of his famous premise "I think". However, it is not unreasonable, here as elsewhere in philosophy of language, to regard the modern phase of the subject as beginning with Frege. As is well known (see Salmon's essay for details) Frege held that each expression of an interpreted language has both a reference and a sense, the sense, in general, being a mode of presentation of, or a way of thinking of, the reference. He subscribed to two compositionality principles, that the reference of a complex is wholly determined by the references of its constituents, *mutatis mutandis* for senses, and two dependence principles, that a complex lacks a reference if it has a constituent which lacks a reference, *mutatis mutandis* for senses. Examples of types of expression which have reference and sense are names and sentences. The reference of a name is the object it stands for, while its sense is some associated way of thinking of that object; the reference of a sentence is its truth-value, while the sense of a sentence is the proposition or thought it expresses, which is said to be a way of thinking of the truth-value.

The specific details of the nature of the sense of different types of expression are still a matter of controversy amongs present-day Fregeans, but it would be widely agreed that the following criterion of identity of sense is roughly correct for the categories of expression to which it is applicable: expressions S and S^* of such a category have distinct senses if an object-language sentence using these expressions to effect the claim that they have the same reference is potentially informative. Suppose S is 'Marilyn Monroe' and S^* is 'Norma Jean Baker'; then these names have different senses since

(1) Marilyn Monroe = Norma Jean Baker

is informative. Suppose S is 'Given any family F of pairwise disjoint non-empty sets, there is a set Z which contains exactly one member of each set Y in F' while S^* is 'Given any set X there is a total ordering R on X such that each subset Y of X has an R-least member'. S, the Axiom of

Choice, and S^*, the Well Ordering Principle, are equivalent as a matter of mathematical necessity, but the object-language sentence relevant to the test for sameness of sense, the biconditional 'S iff S^*', is clearly informative, so S and S^* are (express) different propositions. Indeed, more than one author has pointed out that by this criterion, S and S^* may differ in sense even if they are tokens of the same linguistic type:

(2) That telephone = that telephone

is informative in circumstances where one is faced with an array of telephones, each of which both rings and flashes. The sense of the first 'that telephone' is the auditory mode of presentation, of the second, the visual, and the judgement is again clearly informative.[2] The verdicts of informativeness we are making here are quite intuitive. Whenever a sentence of the kind in our examples is informative, it will be epistemically possible that that sentence is false, but this remark is not very illuminating, since for a Fregean, an explanation of epistemic possibility would employ the notion of sense.

This example indicates how Frege's two-tiered semantic theory appears to accommodate demonstratives rather well. What of the central indexicals 'I', 'here' and 'now'? Frege made the following observation about 'I' [1977, pp. 11–13]:

> Dr. Gustav Lauben says, 'I was wounded' . . . Now everyone is presented to himself in a special and primitive way, in which he is presented to no-one else. So, when Dr. Lauben has the thought that he was wounded, he will probably be basing it on this primitive way in which he is presented to himself. And only Dr. Lauben himself can graps thoughts specified in this way. But now he may want to communicate with others. He cannot communicate a thought he alone can grasp. Therefore, if he now says 'I was wounded', he must use 'I' in a sense which can be grapsed by others, perhaps in the sense of 'he who is speaking to you at this moment' . . .

That each individual is presented to himself in a way in which he cannot be presented to anyone else is not implausible: consider the proposal already mentioned, that 'I' means 'the subject of *this* experience', and the fact that only the subject is in a position to use demonstratives directly to pick out his own experiences. But this raises a consequent difficulty for communication, as Frege notes, although he does not deal with it very successfully. It is awkward to say that 'I' in Dr. Lauben's mouth changes its sense according to whether he is addressing someone else or not; for instance, this view has the implication that it is impossible to eavesdrop comprehendingly on someone talking out loud about himself to himself.

Further, the specific proposal about the sense of 'I' in its other-directed use does not work in Fregean terms, since 'I am he who is speaking to you at this moment' could be informative for Lauben in unusual circumstances.

One solution of the difficulty here would be to introduce an entity intermediate between the sense and the reference of Lauben's utterance, for example the sequence of the referents of the components of the utterance. In the case of Lauben's 'I was wounded', this sequence could be the ordered pair of Lauben himself and the property of having been wounded, or perhaps the ordered triple of Lauben, that property, and the time of Lauben's utterance; indeed, on some views, some such entity as this, rather than the Fregean sense, *is* the content of what Lauben says (Kaplan [1977], Perry [1979]). The proposal is then that all that is required for successful communication between Lauben and his audience is that the audience should graps this sequence as a result of hearing Lauben speak. However, this proposal would not help solve the problem if the audience has to arrive at the sequence *through* its grasp of the sense of Lauben's utterance, for the difficulty is to see how the audience *can* grasp that sense. So the proposal implicitly rejects Fregean sense as a vital ingredient in the theory of communication.

Perhaps we will in the end be forced to make such a move, but we should not abandon Fregean semantics lightly, since the Fregean notion of sense has much explanatory power that alternative approaches seem unable to equal (see Salmon's account of the four puzzles). Fortunately, there are other aspects of Frege's semantics, introduced to deal with other problems, which may yield a Fregean solution to the problem of first-person communication. Although Frege held that expressions have sense unless they are meaningless, and reference except in special cases like 'Vulcan' in Leverrier's mouth and nowadays 'the present King of France', he did not hold that at any given time, for a fixed language, every use of an expression of that language at that time has the same sense and reference as every other. Rather, there are certain operators in natural language within the scope of which the senses and references of expressions alter in a systematic way. To see how one is led to this view, consider

(3) Smith believes snow is white.

It is natural to regard (3) as asserting that a particular two-place relation, that of believing, holds between Smith and a certain item. *Propositions* are the kinds of thing which are believed, and so on this natural view (3) has the general form '*Rab*', where '*R*' is the relation of believing, '*a*' refers to the believer and '*b*' to the proposition believed. Hence just as the first word in (3) is a name of the first item in the instance of the belief relation (3) reports, so the last three words seem to function as a complex name of the second item, the believed proposition. However, when 'snow is white' is considered in isolation, Frege would say that these words refer to the truth value *true*; the proposition that snow is white is rather the *sense* of 'snow is white'. In that case, what 'snow is white' refers to as those words occur in (3) is the entity which is their sense when they occur in isolation (their 'customary sense'); so the verb 'believes' changes the reference of the sentence which follows it from that sentence's customary reference, a truth value, to its customary sense. Moreover, since expressions always have sense if they have reference, 'snow is white' in (3) must have a sense, which by the general account of sense will be a mode of presentation of the referent, which is now the customary sense, i.e. the proposition that snow is white (whatever the detailed account of this new sense, one would expect it to differ from the customary sense, since this latter is a mode of presentation of a truth-value, not a proposition, but if we take the next step up the hierarchy, to senses of senses of senses, the same reason for expecting a still different entity does not apply: see T. Parsons [1981] for an investigation of what Frege is committed to). So the verb 'believes', as well as having the effect of cancelling the normal reference of the sentence which follows, substituting the customary sense as the new reference, also induces (at least in the simplest case) a new sense for that sentence. Any propositional attitude verb will have the same effect; these verbs are so-called because they stand for attitudes to propositions, attitudes such as hoping-true, fearing-true, trying-to-make-true (sometimes their presence requires analysis to uncover, e.g. as in Kaplan's example 'Schliemann sought the site of Troy', which should probably be rendered 'Schliemann tried to make it true that he himself finds the site of Troy'). The reader may refer to Section 3 of Salmon's essay for an account of how Frege's general analysis of propositional attitude verbs nicely resolves the threat of counterexample to Leibniz's Law in their logical behaviour, e.g. in the failure of 'Smith expects Marilyn Monroe to win an Oscar' and (1) to entail 'Smith expects Norma Jean Baker to win an Oscar'; for this makes it look as if Marilyn has a property Norma Jean lacks, that of being expected by Smith to win an Oscar.

Returning to our problem of communication in the first person, we now have some materials with which to attack it. First, in discussing the introduction of the intermediate entity as a solution to the problem, we were assuming that for a speaker to communicate with his audience, the audience had to grasp something given just by the speaker's utterance. But it seems to be a better account of communication that what is required of the audience is that it come to know, of the speaker, *that he is saying such-and-such*; this is what 'grasping what the speaker says' amounts to. So Lauben will succeed in communicating that he was wounded only if his listener comes to know (4), as a result of hearing Lauben's words:

(4) he (Lauben) is saying that he was wounded;

(knowledge of (4) is said just to be necessary not because there are extra propositions which also have to be known, but because the way in which (4) comes to be known must be further constrained). The point of our digression into Frege's account of propositional attitude verbs is that the 'says' of (4) is itself an attitude verb, so Frege's theory applies here. The proposition which Lauben's listener comes to know has the following constituents; the sense of 'he' as the listener uses it to refer to Lauben, that is, the listener's mode of presentation of Lauben; then the sense of 'is saying that'; and then the sense of 'he was wounded' in the 'saying that' context. The question is what the senses and reference of the occurrence of 'he' in the propositional attitude context should be, given its relationship to the 'I' which Lauben uttered, for we would like to assign this 'I', as sense, that way of thinking Lauben which only Lauben can employ. A proposal within the spirit of Frege's treatment of propositional attitude verbs is that the second occurrence of 'he' in (4) should have as its reference that special way of thinking of Lauben, and as its sense the listener's way of thinking of that way of thinking. This proposal solves the problem of first-persom communication by revealing how Lauben's communicating with his audience does not require the audience to have the thought Lauben has; but unlike the 'intermediate entity' proposal, Fregean sense still plays a central role.

It might be objected to this that the listener is in no position to refer to the sense of Lauben's 'I', since he, the listener, cannot employ it himself. But it is hard to see why this should make it not a possible object of reference for him; after all, it is at least *prima facie* plausible that the listener can refer to Lauben's experiences, for instance the present hurting of his old war wound, though he cannot *have* those experiences. Moreover, other accounts of the function of the second 'he' in (4) are less satisfactory. It cannot simply refer to Lauben and have as sense the listener's way of thinking of Lauben, i.e. function just as the first 'he' functions, unless there is no cogency at all in the Fregean account of attitude verbs we sketched in connection with (3). Nor can the second occurrence of 'he' in (4) have as reference the sense of the first occurrence, if the details of that account of attitude verbs are right; for then the clause after the verb should refer to the proposition Lauben himself believes, and that proposition has as constituents the senses *Lauben* employs, not those employed by his listener in referring to him; in particular, it does not have as constituent the sense of the first 'he' in (4), so this sense should not be the reference of the second 'he'.

The problem which Frege raises for 'I' also recurs with other indexicals, for instance, perceptual demonstratives. From the example of (2), it seems clear that although there is no logical difficulty in two subjects assigning the same sense to 'that telephone is ringing', in fact it will usually be the case that they do not. Perhaps they both identify the telephone in question visually, but from different viewpoints; in general, a Fregean must admit that such differences in viewpoint lead to differences in sense for 'that telephone', for it will sometimes be informative that that telephone is that telephone, in a situation where the difference is just one of viewpoint. So even here it would be too much to demand of successful communication that the hearer come to employ in thought the very same proposition that the speaker expresses by 'that telephone is ringing'. By the token of our treatment of the 'I' case, all that we would require is that he comes to know (5):

(5) the speaker is saying that that telephone is ringing

where the sense of 'that telephone' has as reference the speaker's sense for 'that telephone'.

Nevertheless, there is some intuitive force in the objection that when the hearer thinks that the speaker is saying that that telephone is ringing, the constituent of the hearer's thought corresponding to 'that telephone' refers to the telephone in question, rather than to a constituent of the speaker's thought about it (Loar [1972]). This intuition appears to challenge the whole Fregean analysis of propositional attitude verbs, according to which the sentence following the verb should refer to the proposition to which the subject of the verb has the attitude, the components of that sentence referring to the components of the proposition. The only way to bring the intuition into line with this theory would be to adopt the view that the proposition is after all the 'intermediate entity', the sequence of references, for then 'that telephone' could refer both to the first member of the sequence and to the telephone, these being the same; but this, again, is to abandon the Fregean notion of sense as individuated by the criterion of informativeness.

The Fregean must do some justice to the intuition, which he might try to accomodate by distinguishing two readings of (5). One reading is the *de dicto* reading, the one Frege himself analysed: (5) reports a relation between a speaker and a proposition. But there is also a *de re* reading of (5), on which it could be pedantically reformulated as

(6) the speaker is saying, of that telephone, that it is ringing.

(6) reports a relation between the speaker, the telephone, and the constituent of the speaker's thought corresponding to 'is ringing'.[3] The Fregean can then say that we use the sentence (5) ambiguously, to express both readings at once, and the intuition that 'that telephone' has its customary reference is evidence that the *de re* reading is in some way more salient, perhaps because in the normal situation, the hearer is himself in a position to refer to the thing to which the speaker refers. If some such response to the intuition as this is adequate, it seems the problem of communication is not one on which Fregean semantics need founder.

3. SENSE AND CONSTITUTIVE ROLE

In Frege's general semantic theory, sense and reference are not independent; rather, the sense of an expression *determines* its referent, and recent critics of Frege have attacked his theory on the grounds that when the senses of expressions for referring to objects are sufficiently spelt out so as to guarantee that a unique referent, if any, is determined, the results lead to incorrect predictions about the status of the propositions certain sentences express (see Salmon's essay for an account of these criticisms). But the objections standardly assume that the spelt out version of the Fregean sense of e.g. a name '*NN*', will be encapsulated in some definite description 'the *F*', which will be *analytically equivalent* to the name. A name and a description may be said to be analytically equivalent if and only if no matter what context *C* the identity '*NN* = the *F*' is embedded in, the resulting sentence '*C*(*NN* = the *F*)' is a truth (it may be reasonable to exclude iterated attitude contexts from the class of admissible instances of *C*). Such objections, advanced most forcefully by Kripke [1972] (in whose writings *C* is usually 'necessarily'), are therefore relatively ineffective against more sophisticated modern theories of non-descriptive sense for indexicals, such as the theories of Burge [1977], Evans [1983], McGinn [1983], Peacocke [1983] and Searle [1983].

Certainly, as Burks [1948], and more recently Castañeda [1968] and Perry [1979] have argued, there is little likelihood that any description free of indexicals could encapsulate the sense of an indexical. The standard objections also seem to work against the descriptive proposal we have occasionally used for illustrative purposes, that when a subject uses 'I', its sense in his mouth is given by 'the subject of *this* experience', where the indexical 'this experience' refers to some experiential state he is then in. For it is a contingent fact that he is the subject of that experience, even if experiences are individuated (transworld as well as intraworld) partly in terms of the subject to whom they belong; that is, the description alleged to give the sense of a fixed use of 'I' is not analytically equivalent to 'I' in that use, because 'Necessarily I am the subject of this experience' is false in the relevant situation, for the reason that it is not necessary that I have that experience. On the other hand, since there is no room for a notion of error such that the subject mistakenly believes that it is he who is the subject of the (presently-occurring) experience, the Fregean theory will not wrongly determine the reference of 'I' to be some other person who 'coincidentally' satisfies the description; so what Salmon, in his contribution to this volume, calls the 'strongest' objection to the descriptive semantics for names, does not apply to indexicals. Similar considerations hold of the natural analogue for the sense of 'now', viz. 'the time at which this experience is occurring', and for the sense of 'here', perhaps 'the place where this experience is (now) occurring', or better, 'the place where I am now located': there is no resisting the point that it is contingent that the place where I am now located is here, since it could have been California instead (however, this remark raises issues which will not be tackled until our discussion of 'I am here now' in the context of Kaplan's formal semantics for indexical languages). Much the same also appears to be true of perceptual demonstratives; 'that *F*' might be associated with the description 'the *F* causing, in the way needed for perception, this experience as of an *F*' (Peacocke [1983], p. 113). There is no analytic equivalence, since it is contingent that that *F* causes any experiences, but (arguably) the description could hardly lead us wrong about the referent on any particular occasion of use of 'that *F*', *modulo* a problem about the case where the demonstrated object is not an *F*.

These examples show that Russell was incorrect to hold that indexicals (which he called 'egocentric particulars') could all be *defined* using the demonstrative 'this', if definition requires the production of an analytically equivalent expression, in the sense explained two paragraphs back (Russell [1940], p. 102). Nevertheless, the impossibility of our descriptions leading us to the *wrong* referents suggests that there is an intimate relationship between them and their associated indexicals which is of theoretical significance. Perhaps this relationship is what Kripke ([1972], p. 276) calls 'fixing the reference'. For instance, we can introduce a new name '*MM*' into the language by stipulating that it is to refer to whoever is the *F*, if there is exactly one *F*, and otherwise to be referentless. It would then be a requirement of understanding the name, at least in the absence of any subsequent encounter with whoever or whatever is the *F*, that one knows that its reference is so fixed. However, from the Fregean perspective, the suggestion that e.g. 'the subject of this experience' gives the sense of a use of 'I' in virtue of fixing its reference, is unacceptable, for senses are supposed to be the expressing, and so the proposal implausibly requires that anyone competent with the first person pronoun has the concept of experience and of a subject of experience. This objection could also be urged by the non-Fregean, so long as it is held that understanding uses of 'I' requires knowledge that that is how the reference is fixed on each occasion of use; it lapses only if the description plays no cognitive role whatever for the subject, but then it is hard to see how the overall proposal could be part of an explanation of the semantic functioning of indexicals.

However, Peacocke has advanced an account of the relationship between the indexicals and the descriptions which fits well with the Fregean apparatus. We have already remarked that the intuitive notion of informativeness which we employ in making judgements of distinctness of sense cannot be explained in terms of epistemic possibility, on pain of an unilluminating circle of explanation. So what is it which accounts for the informativeness of, say, a biconditional, even when its two sides are necessarily equivalent? One answer is that the informativeness arises from a difference in the possible evidential grounds for each side. This answer, however, is vulnerable to the objection that one cannot assign fixed confirmation and disconfirmation conditions to propositions which distinguish them appropriately, for propositions are confirmed or disconfirmed only in the context of theories, and in such a context, anything can be evidence for anything, if enough adjustments are made throughout the theory. To escape this objection, Peacocke introduces the notion of *canonical* evidence (Peacocke [1983], p. 116): although anything can be evidence for anything, *canonical* evidence for a proposition is such that a subject who does not take it as providing at least *prima facie* confirmation of the proposition thereby manifests a lack of comprehension of the proposition. Thus a biconditional will be informative provided its two sides have distinct canonical evidence conditions; and the coreferentiality of subsentential components, such as names or predicates, will be informative provided intersubstitution does not preserve canonical evidence conditions.

Peacocke's idea is then to explain Frege's notion of the sense of an expression in terms of the pattern of canonical evidential sensitivity associated with it; for instance, it will be constitutive of someone's employing an expression as the first person personal pronoun that judgements he makes with it are sensitive to evidence about his own states in a certain way that makes it clear that for him, such evidence is canonical for those judgements. Evidently, such a proposal about the nature of Fregean sense does not imply that there must be some analytically equivalent articulation of the sense of 'I', so the proposal is not vulnerable to the Kripke-style objections discussed above. However, it would unquestionably be highly complex to spell out in detail exactly what pattern of evidential sensitivity is constitutive of the sense of any given indexical, and rather than leave the details of such senses quite mysterious, Peacocke ([1983], p. 110) introduces the idea of the *constitutive role* of such senses, which can be spelt out using concepts more sophisticated than those required for employment of the senses in question themselves. The constitutive role of a sense can be embodied in some description such that someone with the more sophisticated concepts who employed those descriptions in place of the associated indexicals would manifest the same pattern of canonical evidential sensitivity as someone who employed only the indexicals. And unsurprisingly, the descriptions which give the constitutive roles turn out to be the ones we have been considering: 'the subject of *this* experience' for 'I', 'the *F* causing, in the way required for perception, this experience as of an *F*' for 'that *F*', etc.

The emphasis to this point has been on Fregean approaches, which, in view of the various avowedly non-Fregean approaches in the literature, may seem unbalanced. But on closer inspection of these alternatives, the emphasis seems not unjustifiable, since some such alternatives have the appearance of terminological variants of the Fregean account. For example, on the approach of Perry ([1979], pp. 16–20), there is a distinction between what is believed, a proposition, and a belief state, a way of believing that proposition. Perry writes (*op. cit*., p. 19): 'Anyone can be in the belief state classified by the sentence "I am making a mess". But only

I can have that belief by being in that state.' So the ordered pair of the proposition (Perry, the property of making a mess) and the belief state corresponding to that sentence is an entity as finely discriminated as the Fregean sense of "I am making a mess" in Perry's mouth. And the belief state, being something in common to all who think "I am making a mess", seems to correspond to the pair of the sense of 'making a mess' and the *type* of way of thinking of oneself of which each individual's sense for 'I' in his own mouth is a token, the type of way of thinking articulated by the constitutive role 'the subject of this experience', where the reference of 'this' at a fixed time varies from subject to subject.[4] None of this is to deny the intrinsic interest of Perry's theory of indexicals and indexical belief, only to suggest that it is not so far removed from a Fregean approach as its author imagines. It is for this reason that our discussion of the semantic functioning of indexicals has been pursued in Frege's framework.

4. KAPLAN'S LOGIC

The two-tier structure of Fregean semantics is capable, up to a point, of being given an abstract set-theoretic model: each expression of a formal language can be assigned a pair of entities, meant to represent its reference and its sense, so that the Fregean compositionality principles, according to which the senses and references of complex expressions are determined respectively by the senses and references of their constituents, are cashed our rigourously in terms of the notion of functional application. To see how this works, let us consider for a moment just a simple first-order language with no indexicals and no non-extensional operators such as 'necessarily' or 'in the future'. Suppose this language includes a non-logical one-place predicate '*F*', and a proper name 'α', so that we can form the atomic sentence 'Fα'. On Frege's view, the sentence has a referent, its truth-value, which is determined by the referents of its parts. For a formal representation of this idea, then, we want to assign the predicate '*F*' and the name 'α' some formal entities which determine a truth-value. It is natural to assign an object to a proper name as its referent, so we must assign to the predicate something which, given an object, determines a truth-value. The most obvious candidate for such a thing is a function from objects to truth-values; so '*F*' will be assigned such a function. We now have a clear account of what it means to say that the referents of '*F*' and 'α' determine the referent of '$F\alpha$'; for the truth-value of the atomic sentence is what is yielded when the function from objects to truth-values which is assigned to 'F' is applied to the object which is assigned to 'α'. Less formally, suppose we take the natural language sentence 'Socrates is wise'; then in terms of the above, the truth-value of this sentence is what is given when the function from objects to truth-values which is assigned to 'wise' – the function which maps x to truth if and only if x is wise – is applied to the object assigned to 'Socrates', viz., Socrates.

We have spoken of functions from objects to truth-values as the referents of atomic predicates, but let us now use the Fregean terms more cautiously. Granting that expressions do have both sense and reference, it is a separate question how good some formal apparatus is for modelling Fregean semantics, given Fregean views about reference and sense. So, following an ancient tradition (Carnap [1947]), we will call the pair of entities such an apparatus assigns to an expression the *extension* and the *intension* of that expression. It will then be a further question how well these extensions and intensions reflect Fregean views about, respectively, reference and sense.

It is worth developing our account begun two paragraphs back of the references, or extensions as we are now calling them, of the expressions of an ordinary first-order language. There are many ways of doing this, and what follows is only intended to give the flavour (see C. Parsons [1982] for a recent version of the intensional type hierarchy). One-place predicates were assigned functions from objects to truth-values as their extensions; so two-place predicates will be assigned functions from pairs of objects to truth-values, three-place predicates from triples of objects, and so on. Let us write 'D' for the set of objects, '$\{A \rightarrow B\}$' for the set of functions with domain A and range B, and 'Ext[S]' for the extension assigned to the expression S; for example, if 'G' is a two-place predicate of our language, it is assigned a function Ext[G] which maps pairs of objects to truth-values. That is,

(7) $\quad \text{Ext}[G] \in \{D \times D \rightarrow \{T, F\}\}.$

We can simplify our treatment somewhat by supposing that, whatever set D is, the language always has a name for each member of D (and each name in the language denotes some member of D). Using such names, we can form atomic sentences, such as 'Gbc', and we have

(8) $\quad \text{Ext}[Gbc] = \text{Ext}[G](\langle \text{EXT}[b], \text{Ext}[c]\rangle),$

i.e. the extension of the sentence is determined by applying the extension of the predicate to those of the names. However, a first-order language also has sentential connectives and quantifiers, which must also be assigned extensions. In a sentential language, the extension of e.g. '&' is a function on truth-values (sentence-extensions), the function given by the familiar truth-table, but in a first-order language, the propositional connectives often connect expressions which are not complete sentences, e.g. 'Fx' and 'Gx' in '$(\exists x)(Fx \& Gx)$'. However, our assumption that every object has a name enables us to reduce this kind of occurrence to a sentential one, because it allows a functional treatment of quantifiers itself based on truth-values. Since every object has a name, the following evaluation clauses for the quantifiers suffice:

(9) $(\forall v)\phi(v)$ is true iff for every name t, $\phi[t/v]$ is true.

(10) $(\exists v)\phi(v)$ is true iff for some name t, $\phi[t/v]$ is true.

It follows from this that the extensions of the universal and existential quantifiers are functions on the power set of $\{T, F\}$. For instance, Ext[$\forall$] is a function on the set of truth-values of all instantiations of the matrix it governs; if this set is (T), then Ext[$\forall$] yields T, otherwise it yields F (the case where its argument is the empty set never arises).

So much for the extensions of expressions. What of their intensions? The standard way of defining the intension of an expression is to think of it as a function which determines the extension in a variety of different possible circumstances, or possible worlds. As things actually are, the extension of 'green' maps all and only green objects to T, but if different objects had been green, then there would have been a corresponding difference in the extension of 'green'. Hence, given any particular way things could have been (any possible world) in which the green things are not the same as those which are actually green, the intension of 'green' should take that possible world as argument and yield the extension for 'green' which maps all and only the green things of *that* world to T. In the simplest case, then, the intension of an expression is a function from the members of some set W of possible worlds to extensions. It is usual to assign some expressions constant intensions, i.e. intensions which yield the same extension at every world. The intensions of quantifiers and sentential connectives are *independently* constant: they are the same regardless of what sets W and D (possible objects, i.e. all the objects which exist at some world or other in W) are chosen. The intensions of names may be said to be *dependently* constant: a name is assigned a particular referent from D, but that assignment constitutes its extension at every world. This is what is meant by saying that names are "(strongly) rigid designators".

Finally, once we have introduced intensions and possible worlds, we can augment our language by adding intensional operators, that is, operators whose extensions are functions of the intensions of the expressions they govern. For instance, the operator 'necessarily' or '$\Box$' has the following (simplified) evaluation clause:

(11) $\Box A$ is true iff A is true at every world

which means that the extension of '$\Box$' is a function which maps the intension of A (we write 'Int[A]') to T iff Int[A] maps every w in W to T. Int[$\Box$] itself assigns '$\Box$' this extension at every world, though the particular nature of Int[$\Box$] depends on what intensions there are, since these make up its domain.

Kaplan has proposed the following apparatus for adding indexicals to the formal language. What is characteristic of an indexical is that there is a notion of context such that the extension of the indexical varies with the context. The context determines the intension of the expression, which in turn determines an extension given a possible world, or in Kaplan's full framework, given a *circumstance*, which is an ordered pair of a world and a time. So we begin with a set C of contexts and a 'grid' of circumstances:

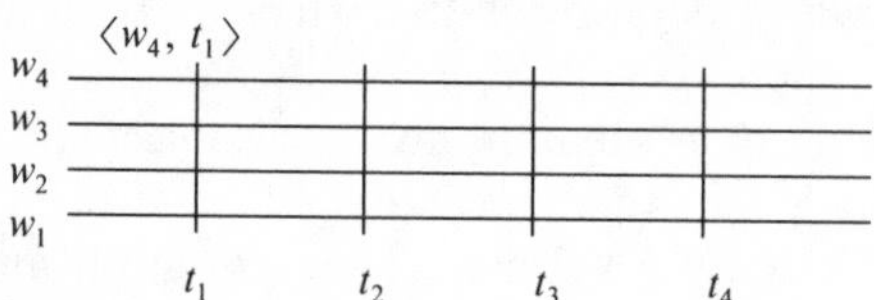

In this grid, horizontal lines are the histories of possible worlds and vertical lines are instants of time in these histories: a circumstance is an intersection of a horizontal and a vertical line, and intensions are functions on these intersections, i.e. on world-time pairs, which assign each expression an extension at such a pair; e.g., Int[green] will assign to $\langle w_4, t_1 \rangle$ the function which maps to T exactly those members of D which are green in world w_4 at time t_1.

The nature of the contexts depends on the range of indexicals to be added to the language. Suppose we are concerned just with 'I', 'here' and 'now'. Then a context should include a speaker (or more generally, an agent), a place and a time. It is also useful to include possible worlds in contexts, whether or not one thinks that there are indexicals which refer to worlds, since we can then define the notion of a *proper* context: a sequence $c = \langle c_a, c_p, c_t, c_w \rangle$ is a proper context iff c_a (the agent of c) is located in c_p (the place of c) at c_t (the time of c) in c_w (the world of c). The set of contexts C which goes along with the grid of circumstances is a set of proper contexts, whose components are themselves members of chosen sets W (worlds), T (times), P (places) and D (possible objects, including agents).

Intensions are functions from circumstances to extensions. But before we can determine the intension of an indexical expression, we need to know the context; for indexicals are governed by rules which yield intensions, given a context. Kaplan calls such rules the *characters* of the indexicals, and they are related to commonsense principles about indexicals, such as that in any context, the reference of 'here' is the place of the context. However, in terms of the present framework, this rule gives an extension, rather than an intension, for 'here', so we will think of characters as functions from contexts to intensions. Since indexicals are name-like, the intensions will be constant, as name-intensions are. So for instance we have:

(12) $\text{Char}[\text{I}](c) \;=\;$ the function f such that for every circumstance $\langle w, t\rangle, f(\langle w, t\rangle) \;=\; c_a$.

An expression S is said to have a *stable* character iff the intension of S is the same in every context, i.e. iff $\text{Char}[S]$ is a constant function, and a stable intension (Kaplan's term is 'content') iff the extension of S is the same at every $\langle w, t\rangle$, i.e. iff $\text{Int}[S]$ is a constant function. Expressions free of any indexical component all have stable character, while indexicals, names, and sentential connectives all have stable intension.

To increase familiarity with this apparatus, let us work again through the case of intensional operators (remember that an intension is now a function on circumstances, i.e. world-time pairs). The intension in a context c of a suitable operand for an intensional operator is a function from world-time pairs to truth-values, i.e. a member of

(13) $\{(W \times T) \rightarrow \{1, 0\}\} \quad (=\alpha).$

The extension at a world-time pair of any intensional operator is therefore a function from functions of the sort displayed in (13) to truth-values, i.e. it is a member of

(14) $\{\alpha \rightarrow \{1, 0\}\}. \quad (=\beta).$

The intension at a context c of an intensional operator is therefore a function which assigns to each world-time pair an extension of the sort displayed in (14), that is, it is a member of

(15) $\{(W \times T) \rightarrow \beta\}$ $(=\gamma)$.

Finally, then, the character of an intensional operator is a function which assigns to each context c an intension of the sort displayed in (15), that is, it is a member of

(16) $\{C \rightarrow \gamma\}$ $(=\delta)$.

Let us take the operators '**F**' (in the future') and '$\Diamond$' ('possibly') as examples. At any world-time pair $\langle w, t\rangle$, the extension of '**F**' at $\langle w, t\rangle$ will take a member of α and map it to truth if that member of α maps to truth some pair $\langle w, t'\rangle$, w as before and t' later than t: the future tense directs us to the future of the given world. For i in α, we have:

(17) $$\begin{aligned} \text{Ext}_{\langle w,t\rangle}[\mathbf{F}](i) &= 1 \text{ if } \exists t' > t\text{: } i(w, t') = 1 \\ &= 0 \text{ otherwise.} \end{aligned}$$

It follows that the intension of '**F**' is unstable, since the extension it assigns to a world-time pair will vary with the time, even with the world held fixed. For instance, let t^* be the last time in w at which P is true, and let i be the intension of P. Then if $t < t^*$, we have:

(18) $\text{Ext}_{\langle w,t\rangle}[\mathbf{F}](i) = 1$

while

(19) $\text{Ext}_{\langle w,t^*\rangle}[\mathbf{F}](i) = 0$.

In fact, since we are taking extensions relative to both a world and a time, we can make the same point without exploiting the ordering on the times, since the truth-value to which i is mapped will clearly depend on the world. But the character of '**F**' is stable, for the intension of the future tense is not sensitive to the context in which it is used.

With the modal operator '$\Diamond$' there is a slight complication. The extension of the future tense at $\langle w, t\rangle$ looks to the future of t in w alone; concomitantly, the extension of '$\Diamond$' at $\langle w, t\rangle$ might be held to look at all the worlds at t alone; alternatively, it might be held to look at all times within every world. The first, narrower, interpretation, gives us:

(20) $$\begin{aligned} \text{Ext}_{\langle w,t\rangle}[\Diamond](i) &= 1 \text{ if } \exists w'\text{: } i(w', t) = 1 \\ &= \text{ otherwise.} \end{aligned}$$

The second, broader, interpretation, is:

$$(21) \qquad \mathrm{Ext}_{\langle w,t\rangle}[\Diamond](i) = 1 \text{ if } \exists w', t': i(w', t') = 1$$
$$= 0 \text{ otherwise.}$$

Since Kaplan remarks ([1979], p. 408, rmk. 5) that both tense and modal operators would have unstable content (= intension) on his approach, it is (20) he has in mind.[5] For it is clear from the statement of (21) that the relativization of the extension of '$\Diamond$' to $\langle w, t\rangle$ is redundant. But suppose that it is not possible that P be true at any time other than t^*. For $t \neq t^*$, w a world where P holds and i the intension of P, (20) then gives us:

$$(22) \qquad \mathrm{Ext}_{\langle w,t\rangle}[\Diamond](i) = 0$$

but

$$(23) \qquad \mathrm{Ext}_{\langle w,t^*\rangle}[\Diamond](i) = 1.$$

So by (20), the extension of '$\Diamond$' at $\langle w, t\rangle$ depends on t.[6] Finally, the character of '$\Diamond$' is stable, like that of '**F**'; regardless of the context, its intension stays the same.

The set-theoretic apparatus underlying Kaplan's approach has its limitations as a device for articulating Frege's conception of how language works, though for Kaplan, from his non-Fregean point of view, this is hardly a penetrating objection to it. The basic problem is that intensions are insufficiently finely individuated to represent Fregean senses. Two mass terms, such as 'water' and 'H_2O', evidently have different senses, by the criterion of informativeness, but whatever the precise details of the semantics of mass terms, on any view on which water and H_2O are the same substance at every circumstance (whether or not they are the same across circumstances), the terms will have the same intension; more generally, there is no reason to believe that every cognitive distinction must be encapsulable by some modal distinction (recall also the example in Section 2 of the telephones). But even from Kaplan's own viewpoint, the apparatus has its drawbacks. This is because Kaplan works simultaneously with two conceptions of the thought, or proposition, expressed by a sentence, which he calls its *content*. One conception is the functional one, on which the set-theoretic apparatus yields an adequate, though non-Fregean, model of contents, which are not discriminated as finely as senses. The other conception is the structural conception, which he combines with a view of the functioning of indexicals as "directly referential': the content expressed by a sentence is a structured entity, and an indexical serves to introduce its referent directly into the structure of the content. So the content Jones expresses by 'I am hot' would be an ordered pair consisting of Jones and the property of being hot; that is, the first constituent is the extension of 'I' in the given context, rather than its intension. Besides the Fregean objections to this position, it raises a doubt as to whether the set-theoretic framework is the right one for Kaplan to employ in presenting a systematic version of his theory of content, if the structural conception of contents is the primary one.

A major point of Kaplan's work is to characterize the notion of logical validity for indexical languages. There is a general technique for constructing definitions of validity: one formulates some notion of model for the language and defines truth in a model, and the valid formulae are then those true in every model. With intensional languages, it is usual to define truth in a model as a special case of some other truth property relativized to the special entities needed to deal with the intensional operators; for instance, in a first-order modal language, truth in a model is truth at the actual world of the model, and the general notion of truth of which this is a special case is that of truth at an arbitrary world in a model. In Kaplan's semantics, we have circumstances, i.e. world-time pairs, instead of worlds, and the extra ingredient of a set of contexts, which makes his procedure slightly more complex. A model, as might be expected, consists in a set C of contexts, a set W of worlds, a set T of times, a set P of places, and a set U of individuals, together with a function which assigns appropriate extensions at each world-time pair to the names and atomic predicates of the language. The most general truth-property Kaplan defines is that of truth (for some model) in a context c relative to a circumstance $\langle w, t\rangle$, this being exactly the truth-property for which the apparatus of character, intension and extension is designed: given a sentence θ and a context c, the character of θ determines its intension in c, which can then be applied to $\langle w, t\rangle$ to yield the truth-value of θ with respect to (wrt) $\langle w, t\rangle$. This property is inductively defined in a standard way by the assignment and evaluation clauses for the language's primitive predicates, terms and operators. Once this is done, the restricted truth-property of truth in a context c can then be defined as truth in c relative to the circustance $\langle c_w, c_t\rangle$ comprising the world and time of the context. We can then define validity as truth in every context in every model (on this definition, such a formula as '$P \leftrightarrow \text{Now}(P)$' is valid, so the kind of validity in question is the 'weak' validity of two-dimensional tense logic).

Precisely which formulae come out valid on this definition depends on the details of the evaluation clauses. Since the most characteristic feature of Kaplan's logic is its solution of the problem we mentioned at the outset about such sentences as 'I exist' or 'I am here now', let us focus on these examples for illustrative purposes. Kaplan treats 'now' as an operator, 'I' and 'here' as referring terms; so he must define the referents of 'I' and 'here' at a context c wrt a circumstance $\langle w, t\rangle$, and say how the extension of 'Now(ϕ)' in c wrt $\langle w, t\rangle$ depends on that of ϕ in c wrt $\langle w, t\rangle$. Unsurprisingly, he stipulates:

(24) the referent of 'I' ('here') in c wrt $\langle w, t\rangle$ is $c_a(c_p)$, the agent (place) of c.

(25) 'Now(ϕ)' is true in c wrt $\langle w, t\rangle$ iff ϕ is true in wrt $\langle w, c_t\rangle$.

(24) and (25) exhibit how the nature of the context enters non-trivially into the evaluation clauses for indexicals. By contrast, the evaluation clause for the non-indexical operator '$\Box$' ('necessarily') does not refer to a context:

(26) '$\Box(\phi)$' is true in c wrt $\langle w, t\rangle$ iff for every w' in W, ϕ is true in c wrt $\langle w', t\rangle$.

To see how these clauses work, consider 'I am here now', regimented in Kaplan's system as

(27) N(Located(I, here)).

It is stipulated that, in any model, the interpretation of 'Located' includes each pair $\langle c_a, c_p\rangle$ for every context c in C. Let c^* be an arbitrary context; then by the definition of 'true in c' as 'true in c wrt $\langle c_w, c_t\rangle$', (27) is true in c^* iff it is true in c^* wrt $\langle c^*_w, c^*_t\rangle$, iff (by (25)) 'Located(I,here)' is true in c^* wrt $\langle c^*_w, c^*_t\rangle$, iff (by (24)) $\langle c^*_a, c^*_p\rangle$ is in the interpretation of 'Located', which, as remarked, is always the case. So in any model, (27) is true in all contexts, and is therefore logically valid.

However, Kaplan's logic has the unusual feature that the validities are not closed under necessitation, for although (27) is valid,

(28) $\Box[N$(Located(I, here))]

is not. For (28) to be true in c^* requires that it be true in c^* wrt $\langle c^*_w, c^*_t\rangle$, which requires that for any w' in W, (27) is true in c^* wrt $\langle w', c^*_t\rangle$, which in turn requires the very strong condition that in *every world* w' at the time c^*_t, the agent of the particular context, c^*_a, be located in w' as he is in the particular world c_w^* at c_t^*; but there is no reason at all why he should be so located in *every* world.

How can something, viz. (27), which is true in virtue of logic alone, fail to be necessary? That such a category is admissible has some plausibility in the light of arguments by Kripke ([1972], pp. 273–5) for the category of the *contingent a priori.* According to Kripke, it is one thing whether a proposition can be known independently of experience, another whether or not it holds at every possible world: this is just another case of failure of the epistemic and modal status of a proposition to mirror one another. So Kaplan can say that truth in virtue of logic is to be likened to, or is a species of, *a priori* truth, a view encouraged by the plausibility of analyses of the concept of a logical constant which employ the notion of *a priori* knowledge (Peacocke [1976]). So, provided Kripke is right about the possibility of contingent *a priori* truths, there is a region of conceptual space to which Kaplan's logic elegantly assigns some occupants.

5. INDEXICALS AND THE CONTINGENT *A PRIORI*

The existence of a category of propositions which are both contingent and *a priori* is controversial, for although it cannot be denied that such sentences as 'I am here now' and 'I exist' are special, there is at least one other reasonable account of their special feature. On this account, advanced, for instance, by Donnellan [1979] (though not specifically in connection with indexicals), we must sharply distinguish epistemic relations to the propositions sentences express from epistemic relations to the sentences themselves. On Donnellan's view, there is no single thing which is both contingent and *a priori*; in the case of 'I am here now', on a Donnellan-style approach, the proposition expressed is undeniably contingent, but the speaker does not know it *a priori.* What he knows *a priori*, just on the basis of his mastery of the language, is that in every context, the sentence 'I am here now' expresses a truth; the speaker may also know which true proposition is expressed, but for this to be the case, he must satisfy conditions inconsistent with his knowing *a priori*, concerning that proposition, that it is true. Such a condition might be that he has to know *where he is*, in some sense which entails dependence on experience. Indeed, Kaplan himself seems to endorse such a view, when he writes ([1977], p. 72):

> The bearers of logical truth and contingency are different entities. It is the *character* (or, the sentence, if you prefer) that is logically true, producing a true content in every context. But it is the *content* (the proposition, if you will) that is contingent or necessary.

But it is not clear that these remarks are consistent with a univocal interpretation of the assertion '*A* is valid' in Kaplan's logic; for either 'is valid' is ambiguous, according to whether or not *A* contains indexicals, or the source of the validity of *A* is mislocated when *A* is indexical-free. For in that case, validity is not an intrinsic property of *A*'s character, which will be stable whether or not *A* is valid, but is rather a property of *A*'s intension, which maps every circumstance to truth, this fact translating into A's validity precisely because *A*'s being indexical-free renders all features of all contexts irrelevant. So there is some interest in the question whether or not the Donnellan-style account of the situation is correct.

A less controversial case of knowing that a sentence expresses a truth but not knowing which truth is expressed might involve an untutored subject opening a physics textbook and reading that the relativistic 3-force on a particle is the product of its rest mass and proper acceleration. He believes, of the sentence he is reading, that it is true, and believes this for reasons which, in the circumstances, make his belief knowledge (though the nature of the reasons would of course make it *a posteriori* knowledge). But since he has no understanding at all of such concepts as 'proper acceleration' he can fairly be said not to know the truth the sentence expresses, in this case the relativistic analogue of Newton's second law. The question before us is whether a subject who says 'I am here now' and all of whose knowledge associated with this statement is *a priori* is in a similar position.

The most likely kind of case involves a subject who is lost. If such a person says 'I am here now' then perhaps he does not know what truth he is expressing since he does not know where he is; so what he says, his content, is contingent but not known *a priori*. To know what proposition he expresses would involve finding out where he is, by exploring, or using a map, say; but then he comes to know the proposition he expresses by an *a posteriori* route, which undermines the thought that his knowledge concerning it, that it is true, is a clear case of *a priori* knowledge.

This reasoning locates the course of the difficulty with 'here', but implicit in it is too strong a condition for proper employment of this indexical. Someone who is lost can still think 'it's dangerous here', and know it *a posteriori*. Or at the very least, if it is a consequence of accepting the Donnellan-style analysis that he only knows *a posteriori* that 'it's dangerous here' expresses a truth, then that analysis loses the initial plausibility it has for the case of 'I am here now'. Note, for instance, that just by giving this subject knowledge of where he is, we would not, in the absence of special circumstances, expect to alter behaviour motivated by the belief associated with 'it's dangerous here', which suggests that the belief his behaviour is manifesting is the same, whether or not he has that knowledge; the ways in which his behaviour would change would trace to the new belief 'I am at such-and-such a place'.

However, we can defend Donnellan's account by utilizing Evans' notion of egocentric space (Evans [1982], pp. 153–154):

> The subject conceives himself to be in the centre of a space (at its point of origin) with its coordinate axes given by 'up' and 'down', 'left' and 'right', and 'in front' and 'behind' . . . 'here' will denote a more or less extensive area which centres on the subject.

Given egocentric space, we can introduce an egocentric use of 'here', not to denote a position in public space, but rather with the constitutive role of 'the origin of my egocentric space'. The point in defence of Donnellan is not that 'here' in the mouth of the man who is lost is analytically equivalent to 'the origin of my egocentric space', but rather that the subject carries one and the same egocentric space around with him – that is what makes it egocentric. This point about the *nature* of egocentric space means that if 'here' has the egocentric use, then what the subject says in uttering 'I am here' is a proposition he can know even if he is lost, and furthermore, it is a proposition which is *necessary*, not contingent (*modulo* worlds where he does not exist),[7] for his egocentric position is one and the same through time and across worlds: concerning the origin of his egocentric space, he must always be at *it*. So the objection to 'I am here now' as contingent *a priori* is that if the content is knowable *a priori*, then 'here' (and analogously 'now') is being used egocentrically, hence the content is necessary.

Nevertheless, Evans himself would not have accepted this argument. He wrote ([1982], p. 157): ' . . . when I speak of information "specifying a position in egocentric space" I am not talking of information about a special kind of space, but of special information about space . . . a public, three-dimensional space'. For Evans, the man who is lost will use 'here' egocentrically, but that does not prevent him from referring to a point in objective space. Evans' view was that such reference is secured by a general ability the subject has, to superimpose his egocentric space on his 'cognitive map' of objective space *if* he is presented with evidence about his objective spatial relations to familiar landmarks (his 'frame of reference'). The lost man's problem is only that he is short of evidence of this kind, not that his use of 'here' takes on a special, non-public reference; so what he says is genuinely contingent.

If defence of Donnellan's position in the case of indexicals demands an expansion of ontology, by treating egocentric space as a new entity, then the defence seems weak. Consider the analogous case of 'I exist'. Corresponding to the lost man who says 'I am here now' is the amnesiac who says 'I exist'; the former does not know where he is, the latter, who he is. More specifically, to be lost is not to know one's approximate spatial relations to relatively stable public landmarks which constitute one's frame of reference. To be an amnesiac (from a time *t*) is not to know which impersonally specified life history (up to *t*) is one's own. But it would be unintuitive to hold that the amnesiac does not know what proposition he expresses if he thinks 'I'm in danger', and equally unintuitive to say that if he does know this, then his use of 'I' refers to some special egocentric self.

What, then, is the source of the initial plausibility of the view that the lost subject can at most know that he is sure to speak truly if he says 'I am here now'? Is the main source just that one can infer merely from the characters of the words that any utterance of them will be true, regardless of the particular context and circumstances? This would certainly make for *a priori* knowledge, whatever it is knowledge of, but it is not sufficient motivation for Donnellan's view. For the rules governing indexicals relevant here ('"I" refers to the speaker') may be likened to definitional introduction of a new word, e.g. 'bachelor', into the language, and we are not inclined to say that we do not know *a priori* that bachelors are male, but only that 'bachelors are male' expresses a truth. The extra premise needed is that this is not the correct parallel: the definition is meaning-giving, and what gives the meaning, in the same sense, of an utterance of 'I am here now', is the Kaplanian content in the context of utterance. Therefore, if we can know that any utterance of *S* is true without ever inspecting the associated content and comparing with the circumstances, we must be exploiting 'extraneous' principles, just as the layman who comes to know that '3-force is rest mass times proper acceleration' is true exploits extraneous principles about the reliability of textbooks. So in the former case as much as the latter, only metalinguistic knowledge results, the *a priori*/*a posteriori* difference between the cases being explained by the principles in the former case being rules of language.

But these considerations do not show that the subject cannot know *a priori both* the metalinguistic fact *and*, concerning the proposition he expresses, that it is true. To show (within Kaplan's framework) that *all* he can know *a priori* is the metalinguistic fact, we need an argument that a subject who is lost cannot grasp a Kaplanian content involving his location, and it is difficult to see what such an argument could be. Perhaps grasp of such a content should bring with it more information about the place it involves than a lost man has. However, it is interesting to note that on a Fregean approach, we are not tempted to explain the subject's lack of information, and the corresponding uninformativeness of 'I am here now' for him, by saying that it is only a metalinguistic fact that he knows. On the Fregean analysis, to say that the subject knows, on some occasion, the proposition he would express by 'I am here now', is to say that he stands in the knowledge relation to the complex sense such an utterance would have. And his being lost does not put him out of touch with the sense of 'here'. For this sense is a mode of presentation of his location, and the most likely constitutive role for it, 'where these experiences are occurring', does not advert to the spatial relations between the subject's location and the landmarks of his frame of reference. The uninformativeness of the proposition the subject would express by 'I am here now' is explained not by his knowledge being knowledge of an elementary metalinguistic fact, but by the component senses of his thought interlocking in such a way that no question of evidence, canonical or otherwise, for or against the proposition, arises. It seems plausible that a proposition is informative for a subject if he lacks canonical evidence for it and does not believe it on non-canonical grounds; but, spelling out the constitutive roles, grasp of 'I am here now' requires sensitivity to 'evidence' that the subject of these experiences is located at the place, and exists at the time, of having them. So at least for someone to whom it is self-evident that a subject cannot be at a spatio-temporal distance from his own experiences, his thought 'I am here now' would be no great discovery.[8]

Tulane University, New Orleans

NOTES

[1] See Section 4 for what is intended here in the case of the tense operator.

[2] The example is Peacocke's. See also Kaplan [1977].

[3] The saying relation, and other propositional attitude relations, have no fixed degree here; they are multigrade.

[4] See Davies [1982] for an account of the interrelationships amongst various approaches.

[5] A modal operator can have an unstable intension in the manner of the example involving the tense operator if there is an accessibility relation on worlds analogous to the ordering relation on times.

[6] Whichever of (20) or (21) we choose, Kaplan's logic has the resources to define an operator satisfying the other; with (20) as basic, 'possibly sometimes' satisfies (21), while with (20) as basic, 'possibly now' satisfies (21).

[7] It would be strategically unwise of Donnellan's opponent to insist that the contingency of my being at p^*, the origin of my egocentric space, which arises in virtue of my possible nonexistence, suffices to prove his point, for this contingency could be disputed. The Falsehood Principle dictates that I am not located at p^* at worlds where I and p^* fail to exist, but in this case application of the Principle could be disputed, just as one can argue about the truth-value to assign to 'Socrates $\in$ {Socrates}' at worlds where Socrates does not exist. Certainly, it may be that there is both a weak and a strong notion of membership, the holding of the strong definable in terms of the possibility of the holding of the weak, and analogously with spatial relations between subjects and points in their egocentric spaces, the strong predicates yielding contingent *a priori* truths. But it seems that the dispute between Donnellan and his opponent should be pursued independently of these metaphysical issues about modality.

[8] There is no counterexample to this self-evident principle in the predicament of the narrator of 'Where Am I?' (Dennett [1978]). His problem is not that his experiences are occurring at one place while his self is located at another, but rather, that there are two places such that it is unclear which is the location of both his self and the occurrences of the experiences. The truth of the principle is most neatly explained by the view that the self is constituted by the stream of experiences which "belong" to it, its self-evidence by the view that that view is the one which guides our commonsense thinking about personal identity.

SELECTIVE BIBLIOGRAPHY

Barwise, J. and Cooper, R.: 1982, 'Generalized quantifiers and natural language', *Linguistics and Philosophy* **4**, 159–219.

Burge, T.: 1979, 'Sinning against Frege', *The Philosophical Review* **88**, 318–432.

Burks, A. W.: 1948, 'Icon, index and symbol', *Philosophy and Phenomenological Research* **9**, 673–89.

Carnap, R.: 1947, *Meaning and Necessity*, Chicago University Press.

Castañeda, H.-N.: 1968, 'On the logic of attributions of self-knowledge to others', *Journal of Philosophy* **65**, 439–56.

Chisholm, R.: 1970, *Person and Object*, Open Court.

Church, A.: 1973, 1974, 'Outline of a revised formulation of the logic of sense and denotation', *Noûs* **7**, 24–33, **8**, 135–56.

Davies, M.: 1982, 'Individuation and the semantics of demonstratives', *Journal of Philosophical Logic* **11**, 287–310.

Dennett,. D.: 1978, *Brainstorms*, Bradford Books.

Donnellan,. K.: 1979, 'The contingent *a priori* and rigid designators', in P.. French *et al.* (eds.), *Contemporary Perspectives in the Philosophy of Language*, Minnesota University Press, pp. 45–60.

Dummett, M.: 1973, *Frege: Philosophy of Language*, Duckworth.

Evans, G.: 1982, *The Varieties of Reference*, Oxford University Press.

Frege, G.: 1977, *Logical Investigations*, Blackwell.

Goodman, N.: 1977, *The Structure of Appearance*, 3rd. edition, D. Reidel.

Husserl, E.: 1970, *Logical Investigations Volume One*, Routledge and Kegan Paul.
Kaplan, D.: 1977, *Demonstratives*, Draft #2, mimeo, UCLA.
Kaplan, D.: 1979, 'The logic of demonstratives', in P. French *et al.* (eds.), *Contemporary Perspectives in the Philosophy of Language*, Minnesota University Press, pp. 401–412.
Kamp, H.: 1971, 'Formal properties of "now"', *Theoria* **37**, 227–73.
Kripke, S.: 1972, 'Naming and necessity', in D. Davidson and G. Harman (eds.), *Semantics of Natural Language*, D. Reidel, pp. 253–355.
Lewis, D.: 1980, 'Index, context and content', in S. Kanger and S. Ohman (eds.), *Philosophy and Grammar*, D. Reidel, pp. 79–100.
Loar, G.: 1972, 'Reference and propositional attitudes', *The Philosophical Review* **81**, 43–62.
McGinn, C.: 1981, 'The mechanism of reference', *Synthese* **49**, 157–86.
McGinn, C.: 1983, *The Subjective View*, Oxford University Press.
Montague, R.: 1974, *Formal Philosophy*, Yale University Press.
Parsons, C.: 1982, 'Intensional logic in extensional language', *Journal of Symbolic Logic* **47**, 289–328.
Parsons, T.: 1981, 'Frege's hierarchies of indirect senses and the paradox of analysis', in P. French, *et al.* (eds.), *Midwest Studies in Philosophy Vol. VI: The Foundations of Analytic Philosophy*, Minnesota University Press, pp. 37–58.
Peacocke, C.: 1976 'What is a logical constant?', *Journal of Philosophy* **73**, 221–40.
Peacocke, C.: 1983, *Sense and Content*, Oxford University Press.
Peirce, C. S.: 1932, *Collected Papers Volume II: Elements of Logic*, Harvard University Press.
Perry, J.: 1977, 'Frege on demonstratives', *The Philosophical Review* **86**, 474–97.
Perry, J.: 1979, 'The problem of the essential indexical', *Noûs* **13**, 3–31.
Reichenbach, H.: 1948, *Elements of Symbolic Logic*, MacMillan.
Russell, B.: 1940, *An Inquiry into Meaning and Truth*, George Allen and Unwin.
Searle, J.: 1983, *Intentionality*, Cambridge University Press.
Stalnaker, R.: 1978, 'Assertion', in P. Cole, (ed.), *Syntax and Semantics Vol. 9: Pragmatics*, Academic Press, pp. 315–32.
van Fraassen, B.: 1977, 'The only necessity is verbal necessity', *Journal of Philosophy* **74**, 71–85.

CHAPTER IV.7

PROPOSITIONAL ATTITUDES

by Rainer Bäuerle and M. J. Cresswell

1. THE PROBLEM

The phrase "propositional attitude" was used by Russell[1] to cover such "mental" things as beliefs, hopes, wishes, fears and the like. One can even use the phrase more widely to refer to indirect discourse in general. The grammatical mark of an expression for a propositional attitude in English is that it can take a *that*-complement. Russell's example was

(1) Othello believes that Desdemona loves Cassio

and the problem is the individuation of the object of Othello's belief. To believe is to believe something, but since Desdemona does not love Cassio, there is no such thing as "Desdemona's loving Cassio" to be the object of Othello's belief. It seems plausible, however, to analyse the meaning of (1) by requiring it to be true if and only if Othello stands in whatever relation is the meaning of 'believe' to the entity which is the meaning of

(2) Desdemona loves Cassio

and that this be an entity which still exists, even though Desdemona does not in fact love Cassio.

This way of putting it is of course controversial. Indeed one of the problems about propositional attitudes is that there is not even agreement about the framework within which the issue should be raised. In this article we must make one or two assumptions about the general nature of semantic theory, assumptions which may not be accepted by all theorists.

D. Gabbay and F. Guenthner (eds.), Handbook of Philosophical Logic, Volume IV, 491–512.

We shall in fact assume that semantics is both model-theoretic and truth-conditional. We shall propose, at least initially, that the task of a semantic theory is to assign to every well-formed expression of a language an appropriate semantic entity, its meaning. This task is guided by two important principles:

(a) *The Frege Principle*[2]. The meaning of a complex expression is a function of the meanings of its parts.
(b) *The "Most Certain Principle"*. For two sentences α and β, if α is true and β false, α and β must have different meanings.

The MCP only claims that truth-conditions are a necessary ingredient of accounts of meaning. Whether or not they are also sufficient might be of no importance to the logician, who may only be interested in questions of truth, were it not for the need to account for the truth-conditions of sentences about propositional attitudes. Consider

(3) John believes that $2 + 2 = 4$

(4) John believes that $\sqrt[2]{144} = 12$.

It is easy to imagine a situation in which (3) is true and (4) false. The paradox of propositional attitudes arises here because of the inconsistency which results from an apparently plausible strengthening of the MCP to

(b′) for any two sentences α and β, if α and β have the same truth-values in all circumstances, then α and β have the same meaning.

Actually no one would make a claim quite as strong as (b′). In the first place, the claim would have to be qualified to take account of context dependence. And in the seond place, there are all kinds of properties which many would regard as semantic which seem to have nothing to do with truth-conditions. For instance questions about illocutionary force, presuppositions and the like seem to be aspects of meaning not covered by (b′). When properly stated, (b′) would claim rather that there is an autonomous core of semantics which is concerned solely with the truth-conditions, in a context, of the sentence in question. In particular, questions of illocutionary force, presuppositions and the like do not enter into the determination of the truth-conditions of a sentence by the meanings of its parts. The correct statement of (b′) would therefore be somewhat complicated. Luckily *those* complexities do not affect the problems this article is concerned with, and so we shall use the unqualified version.

Principles (a) and (b′) are incompatible with the fact that (3) and (4) differ in truth value, for the following reason: according to the "Most Certain Principle", (3) and (4) differ in meaning, because they differ in truth values; according to the Frege Principle, this difference in meaning is due to a difference in the meaning of their parts; and the only parts which are different in (3) and (4) are the embedded sentences (5) and (6).

(5) $2 + 2 = 4$

(6) $\sqrt[2]{144} = 12.$

Therefore (5) and (6) must be different in meaning, although they are true in exactly the same situations, viz. all situations. But this contradicts (b′).

We can now classify approaches to the problem of propositional attitudes according to which they reject of the premisses which lead to the inconsistency.

Approach I: On this approach, we deny the Frege Principle. The meaning of a complex expression may thus be a function of the meaning of the parts and something else, or perhaps just "something else"; i.e. propositional attitude operators are sensitive to more than the meaning of a sentence. On this approach we face the problem of individuating the "something else"-argument for propositional-attitude operators.

Approach II: This approach rejects the strengthened version (b′) of the "Most Certain Principle" and says that although meanings may determine truth-conditions, they are not to be identified with them. Thus the approach accepts the Frege Principle as formulated above and buys the problem of the nature of the meaning of sentences.

Of course an even stronger version of approach I would reject even the weaker form of the "Most Certain Principle" and claim that semantics has nothing to do with truth-conditions. This seems to be the position adopted by J. J. Katz [1972].

Approach III: This approach says that for the purposes of a viable semantics we must treat (3) and (4) as if they had the same truth value. It argues as follows: (5) and (6) are true in exactly the same situations (more precisely: in all possible situations), and as logical equivalents they have the same meaning. Therefore embedding either of them in the context of "John believes . . ." must result in the same truth value for the complex sentences.

But given the case that (3) may be true where (4) is false, we should be aware that, on this approach, we either see ourselves as giving an account of what one ought to believe rather than of what one does believe, or else we have to offer an explanation how apparent counterexamples like (3) and (4) can nevertheless be accommodated in such a framework.

It is clear that a key question in evaluating these three approaches is our attitude to (b′): how central are truth-conditions to meaning? We have already mentioned that there may well be non-truth-conditional aspects of meaning, e.g. illocutionary force. But these features seem not to be at issue in the difference between (5) and (6). The argument for (b′) is that a very large amount of semantics, as shown by the work of Richard Montague [1974] and others (see, e.g. Partee [1976]) can be done by treating meaning purely truth-conditionally. Since this is so, it is useful to have some terminology which makes it clear when we are talking about truth-conditions. This will have the additional advantage that we do not have to beg any questions about whether meanings are truth-conditions. Let us call the truth-conditions of a sentence its *intension*.[3] Expressions in other syntactic categories can also have intensions which are related in specifiable ways to the intensions of sentences. Some semanticists of course want to avoid even intensions and deal only with extensions (i.e. with merely the truth value of a sentence and with merely the set of things which actually satisfy a predicate).

Using this terminology, approach II says that intensions do not individuate meanings finely enough, which is quite plausible in view of the fact that such an approach seems to require that we say that sentences true in all possible circumstances, such as (5) and (6), have in fact the same meaning. Approach I, on the other hand, has it that intensions are fair enough renderings of the meaning of a sentence, but claims at the same time that certain contexts are sensitive to aspects of language different from intensions pure and simple. Again this is quite plausible, considering

(7) He told me to keep my mouth shut

(8) He told me, in those very words, to keep my mouth shut.

In a situation in which the same message has been conveyed in words different from those reported in (7) and (8), (7) will still be true, but not (8).

The common core, however, of approaches I and II is that some contexts are obviously sensitive to more than either the extension or the intension of a sentence. Therefore the problem can be referred to as the problem of hyperintensional[4] contexts. We shall sometimes use the above terminology in going through various theories about the proper analysis of the semantics of propositional attitudes.

2. THEORIES RELATED TO APPROACH I

Approach I says that propositional attitudes are sensitive to more than the meaning of a sentence. Now there clearly are some contexts which are sensitive to more than meaning, e.g. quotational contexts. Where Cicero and Tully are the same person, we clearly have an equivalence of truth value in the following sentences:

(9) Cicero is the author of 'De Senectude'

(10) Tully is the author of 'De Senectude'.

Quine called the occurrence of "Cicero" and "Tully" in such sentences *referentially transparent*. He contrasted this with quotational contexts which he called *referentially opaque*.[5] Thus we do not have an equivalence of truth value in the following case:

(11) 'Cicero' has six letters

(12) 'Tully' has six letters.

The difference between (9) and (11) is obvious; "Cicero" in (9) denotes an individual: Marcus Tullius Cicero; "'Cicero'" in (11) denotes a name.

Now propositional attitude sentences are similar to these quotational contexts in that they do not allow for the substitution of logical equivalents, because (13) may be true and (14) false:

(13) Arabella believes that (9)

(14) Arabella believes that (10).

But it is not obvious that they can be analysed *purely* quotationally in the way (11) and (12) can. Such an approach would have it that (9) and (10) denote intensions all right, but as embedded sentences in (13) and (14) they are names for sentences (the name of a sentence being constituted by the sentence itself), i.e.

(13′) Arabella believes that "Cicero is the author of 'De Senectude'"

(14′) Arabella believes that "Tully is the author of 'De Senectude'".

But there are problems. In some cases the embedded sentence does not in any obvious way seem to be a suitable entity. Certainly Jack may have a consistent belief about Rockefeller which can be correctly reported by saying

Jack believes that Rockefeller is richer than he actually is

but if Jack believes the sentence

Rockefeller is richer than he actually is

he certainly has an inconsistent belief. And also in

Jack believes that this is a valid formula

the sentence named is

This is a valid formula.

And here it becomes obvious that the belief is not about the word "this", but rather it is its referent on a particular occasion of its use that makes me believe the sentence or not.

However, the main objection to a quotational approach is that neither the particular expression nor the particular language used to express the belief are decisive, in that the belief-relation only too obviously involves meaning. Jack's belief can be expressed by the sentence "Rockefeller is a millionaire" because that sentence means what in fact it does mean. And as that sentence could mean quite different things under different meaning assignments (in different languages or different historical stages of a language, if you prefer), "believe" cannot just be a relation between a person and a sentence. Just imagine a language in which "being a millionaire" means what "being a New Zealander" means in ordinary English, and see how it affects John's belief.[6] This is not to deny that there may be a quotational element in many propositional-attitude sentences, particularly in indirect discourse. As Barbara Partee has pointed out, the complement of a *manner*-verb of communication (such as "hiss", "scream", "giggle" etc.) always seems to be a near-quote:

Ted hollered that them brown cows was back in the cornpatch again. (Partee [1973], p. 326)

Still less is it to deny that certain features of the structure of the embedded sentence may be relevant to the truth-value of a propositional attitude sentence.

A theory which looks quotational, but isn't really, is Donald Davidson's analysis of indirect discourse in "On Saying That". Davidson [1969] analyzes an utterance of the sentence

Galileo said that the earth moves

as consisting of the following two separate utterances:

(a) The earth moves

(b) There is an utterance x by Galileo, such that x and my last utterance make us samesayers.

This avoids the objection to a purely quotational approach, because whereas sentences may have varying meaning-assignments, the meaning of an utterance of a particular sentence in a particular language at a particular time and context is unique. The proposed solution depends entirely on accepting "samesaying" (which is a relation among utterances which corresponds to sameness of meaning among sentences) as a primitive relation.

It is arguable, however, that if this is to be a semantic relation, it should be explainable from truth-conditions. And if, in addition, we have to take care of possible utterances as well as actual ones, we get all the ingredients of a hyperintensional semantics. Furthermore, it is far from clear how this can be applied to attitudes other than those expressed by verbs of communication. Samesayings are relations between utterances, but what are the relata in the utterance of a belief sentence? If we are to make any solution of a Davidsonian type work, it would seem that we would need to postulate something like an internal counterpart of an utterance.

Actually, though Davidson might not welcome such entities, Jerry Fodor [1978] has put forward just such a solution to the problem of the semantics of belief. Fodor suggests that we get into trouble with a quotational approach only if we take the objects of propositional attitudes to be sentences of a public language. And he urges us to construe the belief-relation as a relation between the believer and an "internal sentence" in a "language of thought" (see also Fodor [1975] and Moore and Hendrix [1982]). We have no quarrels with Fodor's contention that such an internal representation is crucial to psycholinguistic problems, but the question remains, whether such a representation is relevant or necessary for truth-conditional semantics. For it could turn out that the semantics can be developed autonomously; without being committed to discover the exact nature of these internal representations. In fact, Fodor explicitly claims that the view that propositions (and not their internal representations) are to be the objects of attitudes does not conflict with his proposal, for there is nothing wrong with saying that internal representations express propositions.

Now it is doubtless true that we cannot stand in certain kinds of relations to a proposition except by standing in a relation to some token of an (internal) formula which expresses that proposition. But does that solve any of the semantic problems? Let us suppose that I ascribe to John the belief that Rockefeller is a millionaire. Let us suppose in addition that there is an internal representation R in John's mind which represents this belief. Unfortunately it is I who am ascribing the belief and so it is the meaning of *my* sentence which is at issue. Let us suppose that my representation of Rockefeller's being a millionaire is S. R and S then are sentences in the language of thought. There is no reason to suppose that they are identical. Indeed in Fodor's account they cannot be, since R plays a causal role in John's behaviour, whereas S does not, except perhaps incidentally. What is crucial is that R and S must have the same meaning, because otherwise my sentence would not be correctly reporting John's belief. Let us call this meaning μ. Then the sentence

John believes that Rockefeller is a millionaire

will be true if and only if John stands in the belief-relation to an internal representation of μ. What we notice now is that R and S have become completely irrelevant. The thing whose nature we are interested in is μ. Our problem is when are two *meanings* the same or different. And no amount of discrimination among the things that represent meanings will help in the individuation of the meanings they represent.

Meanings cannot be private entities. Two people may have the same belief, and the belief they have in common must be something that we can make public reference to. One advantage which the regular quotational approach has over Fodor's approach is that the object of the attitude is a sentence in a public language. As we saw, however, the sentence itself will not do, because the same sentence can have quite different meanings. What we are looking for, if we take this line, is an entity which has an intimate connection with the sentence, yet is not able to be given a different meaning. This suggests that we might think of meanings as things which involve intensions, but in some way also reflect the structure of the sentences which express them. A solution along these lines is perhaps closer to approach II than approach I in that it seems to be postulating a notion of meanings which is finer than intensions. For that reason it will be given a section on its own.

3. STRUCTURED MEANINGS

In working out the intension of a sentence in a semantics based on the principle of compositionality, we need to have, on the way, intensions for all the symbols of the sentence. If we now replace every separate symbol by its intension, we get a structured meaning, which is, on the one hand, a "public entity" in the sense that its structure depends on the sentence used in the belief-ascription, and which allows, on the other hand, different internal or external representations. Where $\sim$ and $\neg$ are both (classical) negation operators "$\sim p$" and "$\neg p$" will have the same structured meaning; and as structured meanings are completely determined by intensions, together with the structure of the sentence, we have only made use of things that we have already used anyway.

This account of structured meanings has been worked out by David Lewis [1970], who did not specifically think of propositional attitudes, when he argued for a distinction over and above intensions of sentences. His starting point was that we probably would not want to accept that all tautologies, being true in all possible worlds, have the same meaning. Intensionally equivalent sentences may still have different structured meanings.

The difference is that in calculating the intensions of sentences we apply the intensions of the functor categories to the values of their arguments, whereas structured meanings are just the ordered sequence of the values of functors and arguments. On a very simple account, in

Cicero sings

Cicero will be a name whose semantic value I(**Cicero**) is Cicero, and **sings** will have as its value I(**sings**) the function ω such that for any object a, $\omega(a)$ is the set of worlds in which a sings. The intension of the whole sentence is arrived at by the rule that I(**Cicero sings**) is the set of worlds which is the value of the function I(**sings**) for the argument I(**Cicero**). The structured meaning, however, is something like the ordered pair

(I(**sings**), I(**Cicero**))

and reflects, in a way, the steps that enter our calculation of the intension of the sentence.

Lewis' structured meanings were prompted by the notion of *intensional isomorphism* developed by Rudolf Carnap [1947]. Roughly speaking, two sentences are intensionally isomorphic, if and only if they are built in the same way out of designators such that any two corresponding designators are logically equivalent. And Bertrand Russell [1912, pp. 124–130] argues that when Othello believes that Desdemona loves Cassio, he must not have before his mind a single object (such as "Desdemona's love for Cassio" or "that Desdemona loves Cassio"), but rather stand in a relation to "Desdemona", "loving", and "Cassio". To see how an approach like Russell's would solve the problem of propositional attitudes consider the following example:

Both "2 + 2 = 4" and "7 is prime" express the necessary proposition, so if I were to believe the first but not the second, my belief would violate the laws of identity. However, starting from Quine's [1956] relational or de re-analysis of belief (which was designed to avoid quantification into intensional contexts), we might say that I believe of the pair (2, 2) that its terms sum to four, and that I do not believe of 7 that it is prime. Under such an analysis, different properties ("being prime", "summing to 4") are ascribed to different entities (7, (2, 2)), and there is no violation of the law of identity. Clearly such a solution requires us to take into account the internal structure of the belief-ascription (see Cresswell and von Stechow [1982]). The idea that not every belief is propositional and that belief can be analyzed as the ascription of a property to a res has been put forward in Lewis [1979]. Lewis argues that if Heimson believes that he is Hume, he does not believe the impossible proposition "Heimson is Hume", but self-ascribes the property of being Hume.

However, a treatment of propositional attitudes along the lines of Lewis [1970] needs a few adjustments. First of all, the intension of the attitude operator itself would still be something that takes individuals and propositions as arguments, i.e.

Dale believes that Cicero sings

would be

I(**believes**)(I(**Dale**),I(**Cicero sings**))

The intension of **Cicero sings** is, of course, the same as that of any logically equivalent sentence. But as we do not want interchangeability of equivalents to hold, we have to assume that $\omega_{\textbf{believe}}$ is a function that takes individuals and structured meanings as arguments:

I(**believes**)(I(**Dale**), ⟨I(**sings**), I(**Cicero**)⟩)

But this adjustment brings the major problem of the theory with it: the hyperintensional definition of the attitude operator introduces the set-theoretic impossibility that the intension of **believes** takes its own intension as an argument when the attitude operator is iterated. The structured meaning of

Earnest believes that Dale believes that Cicero sings

would now become

I(**believes**)(I(**Earnest**), (I(**believes**), I(**Dale**), ⟨I(**sings**), I(**Cicero**)⟩))

One way to get around this impossibility is to allow flexibility in how much of the structure is taken into consideration. This could be achieved by complicating the semantics of **THAT**. **THAT** may simply form a name out of a sentence, but in the case of a de re-belief, it seems to operate on names and predicates. The general idea is that a systematically ambiguous **THAT** (or a family of **THAT**s of different syntactic categories) makes a name out of expressions which, put together, can form a sentence, i.e. out of an *n*-place property and its arguments. And thus, although the belief-context is sensitive to more than the intension of the sentence following the clause-initial **THAT** in the surface structure, the **THAT** operates only on intensions (the intensions of the parts of the sentence (see Cresswell and von Stechow [1982])). If there really are cases in which "believe" appears to operate on itself, then one would probably have to introduce a hierarchy of **believe**-operators: let a type-0-operator $\textbf{believe}_0$ operate on intensions, a type-1-operator $\textbf{believe}_1$ operate on Lewis-meanings that do not contain a $\textbf{believe}_n$ (where $n \geqslant 1$), and in general, let any $\textbf{believe}_n$ operator operate on Lewis-meanings containing at most an operator $\textbf{believe}_m$ $(m < n)$. The problem is now with the infinity of operators. We have to look for a way to reverse the effects of hyperintensionality and to make the semantics of all these operators dependent on a basic $\textbf{believe}_0$. Other ways to cope with this set-theoretic difficulty are discussed in Cresswell [1975] and Bigelow [1978] but they bring other complications with them. This particular problem is in fact symptomatic of a deep worry about structured meanings. For Richmond Thomason [1977], [1980b] has produced a very general argument to show that there can be no operators satisfying certain apparently plausible requirements which have all meanings as their range. He uses a technique based on the work of Gödel [1931], Tarski [1935], and Montague [1963]. If we assume that there is an effective one-to-one correspondence between sentences and meanings, then Thomason shows that there can be no operator on (all) propositions which means "it is true that" (here he adapts the work on Tarski), and he also shows that there can be no operator which means "a knows that" (here he adapts the work of Montague). How to avoid the undesirable consequences of Thomason's argument is one of the most important tasks for a defender of structured meanings. A sustained exposition of the structural meanings approach is found in Cresswell [1985].

But even if we satisfy ourselves with the development so far, the theory is not yet capable, as Mates [1950] notes, of handling propositional attitudes. Lewis-meanings follow the structure of the sentence, and thus they can discriminate between logically equivalent sentences of a different syntactic make-up. As they stand, they cannot cope with equivalent sentences that only differ in lexical items. Clearly the intensions of "New Zealand" and "Aotearoa" are the same. And so the sentences

Kupe landed in New Zealand

and

Kupe landed in Aotearoa

do not only have the same intension, but also the same structured meaning, for I(**NZ**) = I(**Aotearoa**) = α:

⟨I(**landed-in**), I(**Kupe**),α⟩

But obviously there are people who believe that Kupe landed in New Zealand, but do not believe that Kupe landed in Aotearoa (and the other way round).

Here we encounter the Frege-problem about Hesperus and Phosphorus. We are not only concerned with the res New Zealand (or Aotearoa), but also with how the res is presented. If we ascribe the property that Kupe has landed there to a place, we have a de re-attitude only if the reporting sentence does not claim to specify how the subject represents the res; but in our example, "New Zealand" and "Aotearoa" occur essentially, for it is not a consistent belief to say that the res New Zealand/Aotearoa has the property that Kupe landed there and at the same time that this res has the property that Kupe did not land there. And such an attitude *de expression* is not only distinct from the de re-attitudes, but also from the quotational attitudes already discussed, for my belief that Hesperus and Phosphorus are identical may not involve the names "Hesperus" and "Phosphorus" at all – after all I may know them only under their German names "Morgenstern" and "Abendstern".

And also, as Lewis-meanings reflect the structure of the sentence they derive from, there arises a problem if sentences can express more than one de re-belief. And indeed, a belief that

Arabella has met Clarissa

may state

of Arabella and Clarissa that they have met

of Arabella that she has met Clarissa

of Clarissa that Arabella has met her

according to the question the sentence is meant to answer: Who met Clarissa? Whom did Arabella meet? and so on. Cresswell and von Stechow [1982] therefore suggest that the structured meaning does not depend on syntax alone, but also on the division of the sentence into topic and focus (theme/rheme). One thing, however, can certainly be held in favour of all the variants of the structured meaning approach: they do not force us to abandon the established analysis of extensional and intensional contexts. Preserving intensions, they look for structures defined – at least partly – in terms of these. Other versions of approach II look for completely new entities.

4. OTHER THEORIES RELATED TO APPROACH II

Following approach II, one could adopt the point of view that truth in the same possible worlds is a criterion for logical equivalence, but not for propositional identity. And that propositional identity could be accounted for by adding more worlds to the logically possible ones.

There is an intuitively plausible argument: to believe in square circles and to believe in square triangles is certainly to have different beliefs. And as both are impossible objects, there seem to be different impossible worlds. And could we not postulate that p and $\sim\sim p$ have the same truth value in all possible worlds, which is why they are logically equivalent, and yet have different values in different impossible worlds, which is why they are not identical?.

But anything goes in impossible worlds: both p and $\sim\sim p$ may be true in some and false in others, and they may have opposite truth values in still others. Remember now that approach II was designed to be able to stick to the Frege Principle. But how are we to explain how the meaning of $\sim\sim p$ is determined by the meaning of $\sim$ and p? In the logically possible worlds, $\sim$ simply maps a set of worlds onto its set-theoretical complement. But obviously the meaning of a negated sentence is not so predictable from the unnegated sentence in impossible worlds. There does not even seem to be a unique way of predicting the meaning of negated sentences. If we allow $\sim$ to do something different in impossible worlds from what it does in possible worlds, then there are, in the absence of any restrictions on impossible worlds, many ways of doing "something different" – which of these represent negation? And even apart from this difficulty, there is a more fundamental objection: whether or not $\sim$ maps sets of worlds to their complements is a matter of the meaning of $\sim$, and not of the particular world we are looking at. If so, we have arrived at the point where p and $\sim\sim p$ are different in some worlds, because $\sim$ does not stand for negation in the usual sense. But why should we need an impossible world for *that* to happen? Surely it is logically possible that $\sim$ does not mean what it does in fact mean. There are possible worlds where $\sim$ is taken to be something different from set-theoretical complementation. The argument here is not about ways the world can or cannot be, but about whether words may not mean something different from what they actually do mean. So that on at least one way of looking at this solution the extra worlds are not logically impossible at all.

What solutions of this kind are *really* saying is that in the worlds of our beliefs, words do not mean what they do in the actual world (it is a bit like Frege's view that in certain contexts words do not denote whay they ordinarily do). We have imagined that no restrictions at all have been placed on their interpretations in these worlds – this is what caused the embarrassment. One might try to solve this problem by suggesting that the non-standard meaning of an expression is not arbitrary, but governed by a non-standard logic. This seems to be the view taken by Hintikka [1975] when he commends the use of structures called "urn models" in the logic of belief. Urn models were developed by Veiko Rantala [1975] and amount in essence to a non-standard quantification theory in which the domains are allowed to vary depending on the position of the quantifier in the formula. Urn models provide a non-classical quantification theory. Possibly we would need in addition a non-standard propositional logic, perhaps e.g. the kind of logic that Belnap ([1977a], [1977b]) thinks a computer should use when reasoning from inconsistent information. But it is hard to see how any approach of this kind can guarantee that we have enough impossible worlds, for could there not be two sentences that are true in the same set of possible and impossible worlds, the one being believed by me, but not the other? If so, we have not found a qualitative solution to our problem.

Some philosophers try to avoid the dilemma by arguing that propositions should not be identified with sets of worlds at all, either possible worlds or impossible worlds. Instead, one might try to choose propositions or properties as primitive, and to define "worlds" as sets of propositions, (possible worlds then being maximal consistent sets of propositions "impossible worlds" simply inconsistent sets). Propositions as primitives can be found in Thomason [1980a], properties (or relations) as primitives in Rescher [1975] and also in Castañeda [1974], Parsons [1974] and in Barwise and Perry [1983].

Then of course there are approaches which descend from Carnap's [1947] notion of a *state description* and take worlds to be certain sets of formulae. The possible worlds then become the consistent sets.[7]

But the problem now is that this is not just a matter of choosing a new set of ontological primitives – we also need a new notion of consistency, which is not easily found unless explained by possible worlds right at the beginning. We may of course have a set of – primitive – possible worlds as well and postulate the truth of the proposition in some of these, but not others. But the relation between a proposition and its truth in a world becomes perfectly arbitrary in such an approach, and thus it contradicts the principle we have committed ourselves to right at the begining (Most Certain Principle): that semantics has to be based on the notion of truth. It seems, therefore, that whatever the seeming merits of these approaches for a semantics of propositional attitudes may be, pursuing these approaches further would involve a loss in explanatory power in other respects and pose as many problems as are supposedly solved.

5. THEORIES RELATED TO APPROACH III

Nothing that has been said about approaches I and II proves that they cannot be made to work in the sense that the entities used there provide us with the necessary distinctions in a formal way. The question has been thoughout what they tell us intuitively about the objects of belief. The problem seemed to arise from the fact that there are different expressions which determine the same intension. The problem is that we do not always realize this, and that it can be a significant discovery that two sentences do in fact have the same intension. Adding more and more "worlds" to the logically possible ones enables us to say of more and more expressions that are equivalent in logically possible worlds that they do not express the same hyperintensional meaning and are therefore, on closer inspection, not expressions for the same objects of belief. But then there may still be different expressions for hyperintensional equivalents, and so, no matter how many more worlds we "discover", the original problem may still be there. On the other hand, the more we make our objects of belief similar to the actual sentence used in the belief-statement, the more difficult it gets to explain just how these objects of belief relate to the way the world is or is believed to be. The extreme case here would be to base the explication of belief totally on the linguistic expression of it. But as we have seen, some of our attitudes (attitudes in the broadest possible sense), and the epistemic ones among them, seem to require an analysis in terms of our understanding of the linguistic expression.

What is so attractive about taking propositions in the possible-worlds sense to be the objects of belief is that this also gives us an intuitively persuasive account of what it is to believe: when we believe that α, we base our rational actions on the assumption that these would in fact achieve our aims in a situtiom in which α is the case. And if α and β are true in exactly the same situations, they determine the same base for figuring out our actions, and there is no need, or even possibility, to distinguish between them: they denote, on the possible worlds view, the same set of worlds. Thus, whoever believes a proposition, believes it under whatever representation. And furthermore, nobody believes an impossible proposition, for if to believe a proposition is to believe that our world is a member of a certain set of worlds, then it does not make sense to assume that it is a member of a set without members.

But if this is not to be a semantics of belief for a community of perfect reasoners only (it has to be a community, for as Barbara Partee [1979] pointed out, even perfect reasoners' semantics must involve hyperintensionality if they are to talk about humans, i.e. imperfect reasoners), we have to offer an explanation for what happens when people say that they believe that but not that β, where α and β do in fact express the same proposition. Well, perhaps they don't know what proposition their sentence expresses. For instance, a person might say on Thursday

(15) Tomorrow is a holiday

thinking that he is speaking on Friday. In that context, (15) expresses the proposition that Friday is a holiday, but the utterer thinks he is expressing the proposition that Saturday is a holiday. Certain remarks made by Robert Stalnaker [1976] suggest that he thinks that this fact might enable us to keep the objects of propositional attitudes simply sets of worlds and explain away the apparent counter examples discussed at the beginning of this article.

This approach has some plausibility when applied to certain allegedly necessary sentences involving proper names. The most famous example is Frege's

(16) The morning star is the evening star

This is actually not a very good example, since it seems plausible to take a phrase like "the morning star" as a definite description meaning something like "star which appears in the morning". So taken, there is no temptation to think of (16) as a necessary truth. Russell's theory of descriptions (as applied to modality by Arthur Smullyan [1948]) shows us how it can be true in some worlds but not in others. The case is different, however, if they are treated as names. For if we assume that the function of a name is purely to refer to something, then (16), if true, becomes simply an instance of the law of identity and so true in all possible worlds. Is it then possible for (16) to be nevertheless informative? One might think that the informativeness of (16) resides in the fact that "the morning star" is always a description. But other examples with genuine names are easy to come by, e.g. we might follow Kripke ([1972], p. 256) and use "Phosphorus" and "Hesperus" or we might use

(17) Tully is Cicero

It seems clear in this case that there is no world in which the thing which is in fact named by "Tully" fails to be identical with the thing in fact named by "Cicero". And yet it seems equally clear that someone may fail to know or believe (17).

(16) and (17) are instances of necessary propositions whose truth can only be known a posteriori. If so, knowledge of or belief in (16) and (17) depends on empirical evidence which may not be available even to a perfect reasoner. But this means that there must be some possible worlds in which what is said with (16) or (17) is false. Whatever we learn when we learn that Tully is Cicero, it must be a contingent proposition which rules

out a number of possible worlds as compatible with the actual world. The worlds ruled out cannot be worlds in which Tully is not Cicero, for there are none of these, they must be worlds in which the proposition expressed by somebody who says that Tully is Cicero is different from the one expressed by that sentence in the actual world: it must be an empirical proposition misleadingly expressed by a sentence whose meaning is in fact a necessary proposition.

Now all these examples involve knowledge *a posteriori*, by empirical investigation. Of course we can imagine a world in which somebody else wrote under the *nom de plume* "Mark Twain", in which something different from water flows from the tap and quenches thirst, and in which some other star shines in the evening in the very place where Venus appears in the actual world. Problems of this kind are discussed in Kripke [1972] and Putnam [1975], and attempts to put them into the framework of a more or less formal semantical theory have been made by Stalnaker [1978], Kaplan [1979], van Fraassen [1979] and Klein [1978].

But promising as these attempts are it is very doubtful whether logical or mathematical beliefs can be explained in this way. For in the case of a false belief of this kind there just is no world which is compatible with our beliefs. Stalnaker [1985] has admittedly tried to dispute this by suggesting that there are such worlds, *viz.* worlds in which the mathematical sentence has a different meaning and therefore a different truth value. But it seems very implausible to suppose that when someone mistakenly believes that $14 + 23 = 47$ the belief world of that person is a world in which this expression has a different meaning and expresses a truth. For one may well believe *that* without believing that $14 + 23 = 47$. If one believes that the sentence "pigs fly" is true because one believes that "pigs" is the word for birds it cannot be concluded that one believes that pigs fly.

6. POSTSCRIPT

At the beginning of the article we mentioned how little agreement there was even about the scope or nature of the problem of the proper analysis of sentences of propositional attitudes. We have concentrated on the problem of contexts which do not admit the replacement of logically equivalent expressions and have had very little to say about the rather more general questions of the whole matter of intensional semantics itself. Thus, for instance, we have hardly mentioned the problem of quantifying in to such contexts despite the work of people like Quine, Hintikka,

Kaplan and others.[8] In the end of course a philosophically satisfactory solution to the problem of propositional attitudes will have to take us into the question of the nature of knowledge itself. And if we have solved the major problems of epistemology, then no doubt we shall also have solved the problem of quantifying in to epistemic contexts and many more problems as well. The editors of this handbook have however assured us that we have not been expected to solve the major problems of epistemology.

Universität Konstanz
Victoria University of Wellington

NOTES

[1] The phrase is used in Russell [1940] in several places. It appears to be Russell's own and was attributed to him by Quine in *Word and Object* [1960]. Russell certainly discussed the problems as early as [1912] (pp. 124–130) and it is in here that the Othello example is used, though he did not there use the expression "propositional attitude". The phrase almost emerges on pp. 241–243 of *The Analysis of Mind* [1921].

[2] This principle seems a guiding principle in Frege's work though the name is not used here for any principle which Frege states explicitly. The "most certain principle" is stated in Cresswell [1982].

[3] This term has a long history in logic. C. I. Lewis [1918] uses it in describing modern logic from Leibniz on. The distinction between extension and intension, in our sense, probably owes most to its adoption by Carnap in [1947].

[4] The world "hyperintensional" comes from Cresswell [1975].

[5] Quine [1953], p. 142. Quine introduces the phrase "referentially opaque" and says in a footnote: "This term is roughly the opposite of Russell's 'transparent' as he uses it in his Appendix C to *Principia*, 2nd ed., Vol I" (Whitehead and Russell [1910], p. 665).

[6] This argument is set out more fully in Cresswell [1979] and Bigelow [1978]. It has very close links with what is often known as Church's Translation argument. The translation argument begins with Church [1950]. A well documented discussion is found in Burge [1978].

[7] This approach has been worked out most fully by Hintikka from [1962] on. In ([1962], p. 41) Hintikka says of what he calls 'model sets' (in a propositional logic) that "they constitute, in the absence of logical constants other than propositional connectives, a very good formal counterpart to the informal idea of a (partial) *description of a possible state of affairs*." (Hintikka's italics.) Model set on this account are consistent but need not be maximal. Now, whether a set of sentences is consistent or not depends upon the logic used, but with respect to a given logic we can have sets that are consistent but not maximal, consistent and maximal or are maximal but not consistent. (If the logic is anything like a classical one and if the set is deductively closed, then an inconsistent set must be maximal.) Hintikka does not, in this passage, actually *identify* possible worlds with maximal consistent sets of sentence but he does seem to be suggesting that we use such sets as a formal explication of a possible world, and his discussions in later works of whether we know all the logical consequences of what we know (e.g. [1970] and Chapter 7 of [1973]) seem best understood in terms of the logical steps used in deducing one formula from another. In [1973] he regards

Rantala's work as important in showing how to give a semantical definition of certain sets whih are not classically consistent but which may correspond to someone's belief "world".
[8] Quine's worries about quantifying in date from at least as early as [1953]. Authors such as Hintikka (from [1962] on) and Kaplan [1969] have attempted to show what kind of sense can be made of these questions and the whole logic has a large and increasing literature of its own. For an excellent discussion consult Reeves [1973].

BIBLIOGRAPHY

This is not a complete list of works dealing with propositional attitudes and contains only items referred to in the article. We have tried to make reference to a reasonably wide variety of approaches to the subject and we hope that in the works listed here together with the works listed in *their* bibliographies the reader who wants to take the subject further will find sufficient material.

Barwise, J. and Perry, J.: 1983, *Situations and Attitudes*, Cambridge Mass. MIT Press.

Belnap, N. D.: 1977a, 'How a computer should think', *Contemporary Aspects of Philosophy*, in G. Ryle (ed.), Oxford, Oriel Press, pp. 30–56.

Belnap, N. D.: 1977b, 'A useful four-valued logic', *Modern Uses of Multiple-Valued Logic*, (in J. M. Dunn and G. Epstein (eds.), Dordrecht, D. Reidel, pp. 8–37.

Bigelow, J. C.: 1978, 'Believing in semantics', *Linguistics and Philosophy* **2**, 101–144.

Burge, T.: 1978, 'Belief and synonymy', *The Journal of Philosophy* **75**, 119–138.

Carnap, R.: 1947, *Meaning and Necessity*, Chicago, University of Chicago Press (2nd edition with appendices, 1956).

Casteñeda, H.-N.: 1974, 'Thinking and the structure of the world', *Philosophia* **4**, 3–40.

Church, A.: 1950, 'On Carnap's analysis of statements of assertion and belief', *Analysis* **10**, 97–99.

Cresswell, M. J.: 1975, 'Hyperintensional logic', *Studia Logica* **34**, pp. 25–38.

Cresswell, M. J.: 1980, 'Quotational theories of propositional attitudes', *Journal of Philosophical Logic* **9**, pp. 17–40. Reprinted as chapter 6 of *Semantical Essays: Possible Worlds and their Rivals*, Dordrecht, Kluwer 1988, pp. 78–103.

Cresswell, M. J.: 1982, The autonomy of semantics. *Processes, Beliefs, and Questions* in S. Peters and E. Saarinen (eds.), Dordrecht, D. Reidel, pp. 69–86.

Cresswell, M. J.: 1985, *Structured Meanings: The Semantics of Propositional Attitudes*, Cambridge Mass., MIT Press.

Cresswell M. J. and A. von Stechow: 1982, 'De re belief generalized', *Linguistics and Philosophy* **5**, pp. 503–535.

Davidson, D.: 1969, 'On saying that', *Words and Objections* in D. Davidson and K. J. J. Hintikka (eds.), Dordrecht, D. Reidel, pp. 158–174.

Fodor, J. A. 1975: *The Language of Thought*. New York, Crowell.

Fodor, J. A.: 1978, 'Propositional attitudes', *The Monist* **61**, pp. 501–523.

van Fraassen, B. C.: 1979, 'Propositional attitudes in weak pragmatics', *Studia Logica* **38**, pp. 365–374.

Godel, K.: 1931, 'Über formal unentscheidbare Sätze der *Principia Mathematica* und verwandter Systeme I', *Monatshefte für Mathematik und Physik* **38**, 173–198. (Translated by B. Meltzer with an introduction by R. B. Braithwaite, Edinburgh, Oliver and Boyd).

Hintikka, K. J. J.: 1962, *Knowledge and Belief*, Ithaca, Cornell University Press.

Hintikka, K. J. J.: 1970, 'Knowledge, belief and logical consequence', *Ajatus*, Vol 32, pp. 32–47 (reprinted in Hintikka [1975, pp. 179–191].)

Hintikka, K. J. J.: 1975, *The Intentions of Intentionality and Other New Models for Modalities*, Dordrecht, D. Reidel.

Hintikka, K. J. J.: 1973, *Logic, Language Games and Information*, Oxford, Oxford University Press.

Hintikka, K. J. J.: 1975, 'Impossible possible worlds vindicated'. *Journal of Philosophical Logic* **4**, 475–484.

Kaplan, D.: 1969, Quantifying in. *Words and Objections* (ed. D. Davidson and K. J. J. Hintikka) D. Reidel Dordrecht, pp. 206–42.

Kaplan, D.: 1979, On the logic of demonstratives. The *Journal of Philosophical Logic* **8**, 81–98.

Katz, J. J.: 1972, *Semantic Theory*, New York, Harper and Row.

Klein, E. H.: 1978, On sentences which report beliefs, desires and other mental attitudes, Ph.D. Thesis, University of Cambridge.

Kripke, S. A.: 1972, 'Naming and necessity', *Semantics of Natural Language*, in D. Davidson and G. H. Harman (eds.), Dordrecht, D. Reidel, pp. 253–355.

Lewis, C. I.: 1918, *A Survey of Symbolic Logic*, Berkeley, University of California Press.

Lewis, D. K.: 1970, 'General semantics', *Synthese* **22** (1970) pp. 18–67. (Reprinted in *Semantics of Natural Language*, (ed. D. Davidson and G. H. Harman), Dordrecht, D. Reidel, pp. 169–218, and in Partee [1976] pp. 1–50).

Lewis, D. K.: 1979, 'Attitudes de dicto and de se', *Philosophical Review* **8**, 513–43.

Mates, B.: 1950, 'Synonymity', *University of California Publications in Philosophy* Vol 25, pp. 201–226 (reprinted in *Semantics and the Philosophy of Language*, ed. L. Linsky, University of Illinois Press 1952, pp. 111–136).

Montague, R. M.: 1963, Syntactical treatments of modality. *Acta Philosophic-Fennica, Modal and Many-valued Logics* pp. 153–166 (Reprinted in Montague (1974) pp. 286–302).

Montague, R. M.: 1974, *Formal Philosophy*, in R. H. Thomason (ed.), New Haven, Yale University Press.

Moore, R. C. and G. G. Hendrix: 1982, 'Computational models of beliefs and the semantics of belief sentences', in Peters, P. S. and E. Soarinen (eds.), *Processes, Beliefs, and Questions*, Dordrecht, Reidel, pp. 107–27.

Parsons, T.: 1974, A Prolegomenon to Melinongian semantics. *Journal of Philosophy* **71**, 561–80.

Partee, B. H.: 1973, 'The semantics of belief sentences', *Approaches to Natural Language* (ed. K. J. J. Hintikka, J. M. E. Moravcsik, and P. Suppes) Dordrecht, Reidel, pp. 309–336.

Partee, B. H.: 1979, 'Semantics: Mathematics or Psychology?', *Semantics from Different Points of View* (ed. R. Bäuerle, U. Egli, A. von Stechow) Berlin, Springer, pp. 1–14.

Putnam, H.: 1975, 'The meaning of "meaning"', *Language, Mind and Knowledge* (ed. K. Gunderson) Minnesota Studies in the Philosophy of Science, Vol 7, University of Minnesota Press, Minneapolis, 1975, reprinted in *Mind, Language and Reality: Philosophical Papers Vol 2*, Cambridge, Cambridge University Press, pp. 215–271.

Quine, W. V. O.: 1953, 'Reference and modality', *From a Logical Point of View*, Cambridge Mass., Harvard University Press (second edition 1961), pp. 139–159.

Quine, W. V. O.: 1956, 'Quantifiers and propositional attitudes', *The Journal of Philosophy* **53**, 177–187.

Quine, W. V. O.: 1960, *Word and Object*. Cambridge Mass., MIT Press, 1960.

Rantala, V.: 1975, 'Urn models: A new kind of non-standard model for first-order logic', *Journal of Philosophical Logic*, **4**, 455–474.

Reeves, A. L.: 1973, 'In defence of a simple solution', *Australasian Journal of Philosophy* **51**, 17–38.

Rescher, N.: 1975, *A Theory of Possibility*, Oxford Blackwell.

Russell, B. A. W.: 1912, *The Problems of Philosophy*, London, Oxford University Press.

Russell, B. A. W.: 1921, *The Analysis of Mind*, London, George Allen and Unwin.

Russell, B. A. W., 1940, *An Enquiry into Meaning and Truth*, London, George Allen and Unwin, 1940.

Smullyan, A. F.: 1948, 'Modality and description', *The Journal of Symbolic Logic* **13**, 31–37.

Stalnaker, R. C.: 1976, 'Propositions', *Issues in the Philosophy of Language*, in A. F. Mackay and D. Merill (eds.), New Haven, Yale University Press, pp. 79–91.

Stalnaker, R. C.: 1978, 'Assertion', *Syntac and Semantics Vol 9*, in P. Cole (ed.), New York, Academic Press, pp. 315–332.

Stalnaker, R.: 1985, *Inquiry*, Cambridge Mass., Bradford Books, MIT Press.

Tarski, A.: 1935, 'Der Wahrheitsbegriff in den formalisierten Sprachen', *Studia Philosophica* **1**, 261–405. Translated in *Logic, Semantics and Metamathematics*, Oxford, Oxford University Press, 1956, pp. 151–278).

Thomason, R. H.: 1977, 'Indirect discourse is not quotational', *The Monist* **60**, 340–354.

Thomason, R. H.: 1980a, 'A model theory for propositional attitudes', *Linguistics and Philosophy* **4**, 47–70.

Thomason, R. H.: 1980b, 'A note on syntactical treatments of modality', *Synthese* **44**, 391–395.

Whitehead, A. N. and B. A. W. Russell: 1910, *Principia Mathematica*, Cambridge, Cambridge, Cambridge University Press, 3 Vols, 1st edition 1910–1913, second edition 1923–1927.

CHAPTER IV.8

TENSE AND TIME[1]

by Steven T. Kuhn

I. INTRODUCTION

The semantics of tense has received a great deal of attention in the contemporary linguistics, philosophy and logic literatures. This is probably due partly to a renewed appreciation for the fact that issues involving tense touch on certain issues of philosophical importance (viz., determinism, causality, and the nature of events, of time and of change). It may also be due partly to neglect. Tense was noticeably omitted from the theories

D. Gabbay and F. Guenthner (eds.), Handbook of Philosophical Logic, Volume IV, 513–552.

of meaning advanced in previous generations. In the writings of both Russell and Frege there is the suggestion that tense would be absent altogether from an ideal or scientifically adequate language. (Tense was not the only blemish of natural language to be removed, of course, but tense is a more serious omission than, say, pronouns because it is present in every sentence of the language.) Finally, in recent years there has been a greater recognition of the important role that all of the so-called "indexical" expressions must play in an explanation of mental states and human behavior. Tense is no exception. Knowing that our friend *died* is cause for mourning, knowing that he *dies* is just another confirmation of a familiar syllogism.

This article will survey some attempts to make explicit the truth conditions of English tenses. The restriction to English does not stem from any parochial bias on the part of the author, but rather reflects the fact that English is the language that has been studied the most systematically.

II. EARLY WORK

1. *Jespersen.* The earliest comprehensive treatment of tense and aspect with direct influence on contemporary writings is that of Otto Jespersen. Jespersen's *A Modern English Grammar on Historical Principles* was published in seven volumes from 1909 to 1949. Grammar, for Jespersen, is comprised of *morphology*, and *syntax*. In morphology one begins with linguistic forms and studies their meaning whereas in syntax one begins with meanings and investigates how they can be expressed by linguistic forms. Thus Jespersen's grammar includes much of what we would call semantics and (since he seems to accept some kind of identification between meaning and use) a good deal of pragmatics as well. The aims and methods of Jespersen's semantic investigations, however, are not quite the same as ours.[2]

First, Jespersen is more interested than we are in cataloging and systematizing the various uses of a particular English construction and less interested in trying to characterize its meaning in a precise way. This leads him to discuss seriously uses we would consider too obscure or idiomatic to bother with. For example, Jespersen notes in the *Grammar* that the expressions of the form *I have got A* and *I had got A* are different than other present perfect and past perfect sentences. *I have got a body*, for example, is true even though there was no past time at which an already existent me received a body. Jespersen suggests *I have in my possession* and *I had in my possession* as readings for *I have got* and *I had got*. And this discussion is considered important enough to be included in his *Essentials of English Grammar*, a one volume summary of the *Grammar*.

The interest in cataloging and systematizing also leads Jespersen to distinguish uses we might lump together. For example, Jespersen includes a special chapter on tense in indirect discourse, in which he points out that the tenses appropriate for indirect discourse are "backshifted" versions of those appropriate in direct discourse. *He said that he was glad to see me* is a back-shifted version of *He said*, "*I am glad to see you*" and *He said that he had seen her on Tuesday* is a back-shifted version of *He said*, "*I saw her on Tuesday*." But the analysis of Reichenbach makes it possible to see the past perfect following *He said that* as the same linguistic phenomenon as an ordinary past perfect following the sentence *Yesterday was uneventful*.

Jespersen clearly does *not* see his task as being merely to collect and classify obscure English sentences in the way a classical botanist might collect and classify rare flora. He criticizes Henry Sweet, for example, for a survey of English verb forms that includes such paradigms as *I have been being seen* and *I shall be being seen* on the grounds that "they are so extremely rare that it is better to leave them out of account altogether." Nevertheless there is an *emphasis* on cataloging, and this emphasis is probably what leads Jespersen to adhere to a methodological principle that we would ignore; viz., that example sentences should be drawn from published literature wherever possible rather than manufactured by the grammarian. Contemporary linguists and philosophers of language see themselves as investigating fundamental intuitions shared by all members of a linguistic community. For this reason it is quite legitimate for them to produce a sentence and assert without evidence that it is well-formed or ill-formed, ambiguous or univocal, meaningful or unmeaningful. This practice has obvious dangers. Jespersen's methodological scruples, however, provide no real safety. On the one hand, if one limits one's examples to a small group of "masters" of the language one will leave out a great deal of commonly accepted usage. On the other hand, one can't accept *anything* as a legitimate part of the language just because it has appeared in print. Jespersen himself criticizes a contemporary by saying of his examples that "these three passages are the only ones adduced from the entire English literature during nearly one thousand years."

A final respect in which Jespersen differs from the other authors discussed here is his concern with the recent history of the language.

Although the *Grammar* aims to be a compendium of contemporary idiom, the history of a construction is recited whenever Jespersen feels that such discussion might be illuminating about present usage. A good proportion of the discussion of the progressive form, for example, is devoted to Jespersen's thesis that *I am reading* is a relatively recent corruption of *I am a-reading* or *I am on reading*, a construction that survives today in expressions like *I am asleep* and *I am ashore*. This observation, Jespersen feels, has enabled him to understand the meaning of the progressive better than his contemporaries.[3] In discussing Jespersen's treatment of tense and aspect, no attempt will be made to separate what is original with Jespersen from what is borrowed from other authors. Jespersen's work is obviously part of a long tradition. Discussion of tense and aspect occurs already in the writings of the Latin grammarian Varro in the first century B.C. Discussions of English tense and aspect can be found, for example, in Ward, Brown and Sweet.[4] Furthermore there is a long list of grammarians contemporaneous with Jespersen who independently produced analyses of tenses. See, for example, Curme, Kruisinga and Poutsma. Jespersen, however, is particularly thorough and insightful and, unlike his predecessors and contemporaries, he continues to be widely read (or at least cited) by linguists and philosophers. Jespersen's treatment of tense and aspect in English can be summarized as follows.

Time. It is important to distinguish *time* from *tense*. Tense is the linguistic device which is used (among other things) for expressing time relations. For example, *I start tomorrow* is a present tense statement about a future time. To avoid time-tense confusion it is better to reserve the term *past* for time and to use *preterit* and *pluperfect* for the linguistic forms that are more commonly called past tense and past perfect. Time must be thought of as something that can be represented by a straight line, divided by the present moment into two parts: the past and the future. Within each of the two divisions we may refer to some point as lying either before or after the main point of which we are speaking. For each of the seven resulting divisions of time there are *retrospective* and *prospective* versions. These two notions are not really a part of time itself, but have rather to do with the perspective from which an event on the time line is viewed. The prospective present time, for example, "is a variety of present which looks forward into the future." In summary, time can be pictured by the diagram below. The three divisions marked with *A*'s are past; those marked with *C*'s are future. The short pointed lines at each division indicate retrospective and prospective times.

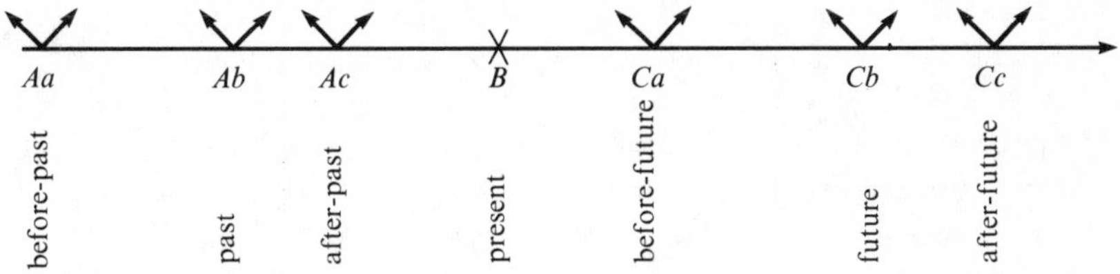

Aa	*Ab*	*Ac*	*B*	*Ca*	*Cb*	*Cc*
before-past	past	after-past	present	before-future	future	after-future

Tense morphology. The English verb has only two tenses proper, the present tense and the preterit. There are also two "tense phrases," the perfect (e.g., *I have written*) and the pluperfect or anteperfect (e.g., *I had written*). (Some verbs, including *can*, *may*, *must*, *ought*, *shall* and *will* cannot form perfects and pluperfects.) Corresponding to each of the four tenses and tense phrases there is an *expanded* form. For example, *had been writing* is the expanded pluperfect of *write*. It is customary to admit also future and future perfect tenses, as in *I will write* and *I shall have written*. But these constructions lack the "fixity" of the others. On the one hand, they are often used to express nontemporal ideas (e.g., volition, obstinacy) and on the other hand future time can be indicated in many other ways.

The present tense is primarily used about the present time, by which we mean an interval containing the present moment whose length varies according to circumstances. Thus the time we are talking about in *He is hungry* is shorter that that in *None but the brave deserve the fair*. Tense tells us nothing about the duration of that time. This same use of the present is found in expressions of intermittent occurrences (*I get up every morning at seven*) and (*Whenever he calls, he sits close to the fire*). Different uses of the present occur in statements of what might be found at all times by all readers (*Milton defends the liberty of the press in his Areopagitica*) and in expressions of feeling about what is just happening or has just happened (*That's capital!*). The present can also be used to refer to past times. For example, the *dramatic* present can alternate with the preterit: *He perceived the surprise, and immediately pulls a bottle out of his pocket, and gave me a dram of cordial*. And the present can play the same role as the perfect in subordinate clauses beginning with *after*: *What happens to the sheep after they take its kidney out?* Present tense can be used to refer to future time when the action described is considered part of a plan already fixed: *I start for Italy on Monday*. The present tense can also refer to future events when it follows *I hope*, *as soon as*, *before* or *until*.

The perfect is actually a kind of present tense that "seems to connect the present time with the past." It is both a "retrospective present, which looks upon the present as a result of what happened in the past" and an "inclusive present, which speaks of a state that is continued from the past into the present time" (or at least one that has "results or consequences bearing on the present time").

The preterit differs from the perfect in that it "refers to some time in the past without telling anything about its connection with the present moment." Thus *Did you finish?* refers to a past time while *Have you finished?* is a question about present status. It follows that the preterit is appropriate with words like *yesterday* and *last year* while the perfect is better with *today*, *till now* and *already*. *This morning* requires a perfect tense when uttered in the morning and a preterit in the afternoon. Often the correct form is determined by context. For example, in discussing a schoolmate's Milton course *Did you read Samson Agonistes?* is appropriate, whereas in a more general discussion *Have you read Samson Agonistes?* would be better. In comparing past conditions with present the preterit may be used (*England is not what it was*), but otherwise vague times are not expressed with the preterit but rather by means of the phrase *used to* (*I used to live at Chelsea*). The perfect often seems to imply repetition where the preterit would not. (Compare *When I have been in London*, with *When I was in London*.) Finally, the preterit can be used after *after*, *as soon as*, *before* and *until* to indicate a before-past time. The perfect can be used in similar circumstances to indicate a before-future time.

The pluperfect serves primarily to denote before-past time or retrospective past, two things which cannot easily be kept apart. (An example of the latter use is *He had read the whole book before noon*.) After *after*, *when* or *as soon as*, the pluperfect is interchangeable with the preterit.

The expanded tenses indicate that the action or state denoted provides a "temporal frame" encompassing something else described in the sentence or understood from context. For example, if we say *He was writing when I entered*, we mean that his writing (which may or may not be completed now) had begun, but was not completed at the moment I entered. In the expanded present the shorter time framed by the expanded time is generally understood to be *now*. In the expanded perfect it is generally considered to be *very recently*. The expanded tenses also serve some other purposes. In narration simple tenses serve to carry a story forward while expanded tenses have a retarding effect. In other cases expanded tense forms may be used in place of the corresponding simple forms to indicate that a fact is already known rather than new, that an action is incomplete rather than complete or that an act is habitual rather than momentary. Finally, the expanded form is used in two clauses of a sentence to mark the simultaneity of the actions described. (In that case neither really frames the other.)

In addition to the uses already discussed, all the tenses can have somewhat different functions in passive sentences and in indirect speech. They also have uses apparently unrelated to temporal reference. For example, forms which are primarily used to indicate past time are often used to denote unreality, impossibility, improbability or non-fulfillment.

Tense syntax. In the preceding discussion we started with the English tense forms and inquired about their meanings. Alternatively we can start with various temporal notions and ask how they can be expressed in English. If we do, several additional facts emerge:

(1) The future time can be denoted by present tense (*He* ***leaves*** *on Monday*), expanded present tense (*I* ***am dining*** *with him on Monday*), *is sure to, will, shall, come to* or *get to*.

(2) The after-past can be expressed by *would, should, was to, was destined to*, expanded preterit (*They* ***were going out*** *that evening and when he came back from the club she* ***was dressing***) or *came to* (*In a few years he came to control all the activity of the great firm*).

(3) The before-future can be expressed by *shall have, will have* or present (*I shall let you know as soon as I* ***hear*** *from them* or ***Wait*** *until the rain stops*).

(4) The after-future is expressed by the same means as the future (*If you come at seven, dinner* ***will soon*** *be ready*).

(5) Retrospective pasts and futures are not distinguished in English from before, pasts and before, futures. (But retrospective presents, as we have seen, are distinct from pasts. The former are expressed by the past, the latter by the perfect.)

(6) Prospectives of the various times can be indicated by inserting expressions like *on the point of, about to* or *going to*. For example, *She is about to cry* refers to a prospective present.

2. *Reichenbach.* In his general outlook Reichenbach marks a sharp and deliberate break with the tradition of grammarians like Jespersen. Jespersen

saw himself as studying the English languge by any means that might prove useful (including historical and comparative investigations). Reichenbach saw himself as applying the methods of contemporary logic in a new arena. Thus, while Jespersen's writings about English comprise a half dozen scholarly treatises, Reichenbach's are contained in a chapter of an introductory logic text. (His treatment of tense occupies twelve pages.) Where Jespersen catalogs dozens of uses for an English construction, Reichenbach is content to try to characterize carefully a single use and then to point out that this paradigm does not cover all the cases. While Jespersen uses, and occasionally praises, the efforts of antecedent and contemporary grammarians, Reichenbach declares that "the state of traditional grammar is hopelessly muddled by its two-millennial ties to a logic that cannot account even for the simplest linguistic forms."

Despite this difference in general outlook, however, the treatment of tenses in Reichenbach is quite similar to that in Jespersen. Reichenbach's chief contribution was probably to recognize the importance of the distinction between what he calls the *point of the event* and the *point of reference* (and the relative unimportance and obscurity of Jespersen's notions of prospective and retrospective time). In the sentence *Peter had gone*, according to Reichenbach, the point of the event is the time when Peter went. The point of reference is a time between this point and the point of speech, whose exact location must be determined by context. Thus Reichenbach's account of the past perfect is very similar to Jespersen's explanation that the past perfect indicates a "before past" time. Reichenbach goes beyond Jespersen, however, in two ways.

First, Reichenbach is a little more explicit about his notion of reference times than is Jespersen about "the time of which we are speaking." He identifies the reference time in a series of examples and mentions several rules that might be useful in determining the reference time in other examples. Time determiners like *yesterday*, *now* or *November 7, 1944*, for example, are said to refer to the reference point. Similarly, words like *when*, *after* and *before* relate the reference time of a relative clause to that of the main clause. And if a sentence does not say anything about the relations among the reference times of its clauses, then every clause has the same point of reference.

Second, Reichenbach argues that the notion of reference time plays an important role in *all* the tenses. The present perfect, for example, is distinguished by the fact that the event point is before the point of reference and the point of reference coincides with the point of speech. (So

I have seen Sharon has the same meaning as *Now I have seen Sharon*.) In general, each tense is determined by the relative order of the point of event (E), the point of speech (S), and the point of reference (R). If R precedes S we have a kind of past tense, if S precedes R we have a kind of future tense and if R coincides with S we have a kind of present. This explains Jespersen's feeling that the simple perfect is "a variety of present." Similarly the labels anterior, posterior and simple indicate that E precedes, succeeds or coincides with R. The account is summarized in the following table.

Structure	New Name	Traditional Name
E ___ R ___ S	Anterior past	Past perfect
E, R ___ S	Simple past	Simple past
R ___ E ___ S		
R ___ S, E	Posterior past	———
R ___ S ___ E		
E ___ S, R	Anterior present	Present perfect
S, R, E	Simple present	Present
S, R ___ E	Posterior present	Simple future
S ___ E ___ R		
S, E ___ R	Anterior future	Future perfect
E ___ S ___ R		
S ___ R, E	Simple future	Simple future
S ___ R ___ E	Posterior future	———

Each of the tenses on this table also has an expanded form which indicates, according to Reichenbach, "that the event covers a certain stretch of time."

Notice that the list of possible tenses is beginning to resemble more closely the list of tenses realized in English. According to Jespersen there are seven divisions of time, each with simple, retrospective and prospective versions. This makes twenty-one possible tenses. According to Reichenbach there are twelve possible tenses, corresponding to the twelve orderings of E, S, and R. Looking more closely at Reichenbach, however, we see that the *tense* of a sentence is determined only by the relative order of S and R, and the *aspect* by the relative order of R and E. Since there are three possible orderings of S and R, and independently three possible orderings of R and E, there are really only nine possible complex tenses (six of which are actually realized in English).

Finally, Reichenbach acknowledges that "actual language" does not always keep to the schema set forth. The expanded forms, for example, sometimes indicate repetition rather than duration: *Women are wearing larger hats this year*. And the present perfect is used to indicate that the event has a certain duration which reaches up to the point of speech: *I have lived here for ten years.*

3. *Montague.* Despite Reichenbach's rhetoric, it is probably Montague, rather than Reichenbach, who should be credited with showing that modern logic can be fruitfully applied to the study of natural language. Montague actually had very little to say about tense, but his writings on language have been very influential among those who do have something to say. Two general principles underlie Montague's approach.

(1) Compositionality. The meaning of an expression is determined by the meaning of its parts.

(2) Truth conditions. The meaning of a declarative sentence is something that determines the conditions under which that sentence is true. Neither of these principles, of course, is original with Montague, but it is Montague who shows how these principles can be used to motivate an explicit account of the semantics of particular English expressions.

Initially, logic served only as a kind of paradigm for how this can be done. One starts with precisely delineated sets of *basic expressions* of various categories. *Syntactic rules* show how complex expressions can be generated from the basic ones. A class of permissible *models* is specified, each of which assigns interpretations to the basic expressions. *Rules of interpretation* show how the interpretation of complex expressions can be calculated from the interpretations of the expressions from which they are built.

The language of classical predicate logic, for example, contains predicates, individual variables, quantifiers, sentential connectives and perhaps function symbols. Generalizations of this logic are obtained by adding additional expressions of these categories (as is done in modal and tense logic) or by adding additional categories (as is done in higher order logics). It was Montague's contention that if one generalized enough, one could eventually get English itself. Moreover, clues to the *direction* this generalization should take are provided by modal and tense logic. Here sentences are interpreted by functions from possible worlds or times, or in general from "indices" to truth values. English, for Montague, is merely an exceedingly baroque intensional logic. To make this hypothesis plausible,

Montague constructed, in [1970], [1970a] and [1973], three "fragments" of English of increasing complexity. In his final fragment, commonly referred to as PTQ, Montague finds it convenient to show how the expressions can be translated into an already-interpreted intensional logic rather than to specify an interpretation directly. Thus logic assumes a more explicit role in Montague's program. The goal is now to find a translation procedure by which every expression of English can be translated into a (comparatively simple) intensional logic.

We will not attempt here to present a general summary of PTQ. (Readable introductions to Montague's ideas can be found in Montague [1974] and Dowty *et al.*) We will, however, try to describe its treatment of tense. To do so requires a little notation.

Montague's intensional logic contains tense operators **W** and **P** meaning roughly "it will be the case that" and "it was the case that." It also contains a special operator $\hat{}$, that makes it possible to refer to the intension of an expression. If A is an expression denoting the object **a**, then $\hat{A}$ denotes the function that assigns **a** to every possible word w.

Among the expressions of English are *terms* and *intransitive verb phrases*. If A is an intransitive verb phrase then the translation A' of A is an expression of intensional logic of category $\langle e, t\rangle$. (That is, A' denotes a function from entities to truth values.) The translation B' of a term B is an expression of category $\langle s, \langle e, t\rangle\rangle$. (That is, B' denotes a function from possible worlds to functions from entities to truth values.) Tense and negation in PTQ are treated together. There are six ways in which a term may be combined with an intransitive verb phrase to form a sentence. These generate sentences in the present, future, present perfect, negated present, negated future and negated present perfect forms. The rules of translation corresponding to these six constructions are quite simple. If B is a term with translation B' and A is an intransitive verb with translation A' then the translations of the six kinds of sentences that can be formed by combining A and B are just $A'(\hat{}B')$, $\mathbf{W}A'(\hat{}B')$, $\mathbf{P}A'(\hat{}B')$, $-A'$, $-\mathbf{W}A'(\hat{}B')$ and $-\mathbf{P}A'(\hat{}B')$.

This treatment is obviously crude and incomplete. It was probably intended merely as an illustration of how tense *might* be handled within Montague's framework. Nevertheless, it contains the interesting observation that the past tense operator found in the usual logics corresponds more closely to the present perfect tense than it does to the past. In saying *John has kissed Mary* we seem to be saying that there was *some* time in the past when *John kisses Mary* was true. In saying *John kissed Mary*, we seem to be saying that *John kisses Mary* was true at *the* time we happen to be talking about. This distinction between *definite* and *indefinite* past times was pointed out by Jespersen, but Jespersen does not seem to have thought it relevant to the distinction between present perfect and past. Jespersen's idea was that the indefinite past is expressed by *used to.* This seems accurate for his *I used to live in Chelsea.* But in general it does not. *John used to kiss Mary* does not mean *At some time in the past John kissed Mary. John has kissed Mary*, on the other hand, does seem capable of expressing this. There are, as we shall see, some good reasons to reject the analysis of present perfect as indefinite past. But the view does have certain attractions, particularly when combined with Reichenbach's view that the difference between the present perfect and the past has to do with whether the event time or the reference time is past.

One of the differences between Reichenbach's reference time and event time seems to be that the former is often fixed by an adverbial clause or by contextual information, whereas the latter is less often so fixed. So it is approximately correct to say that the reference time is determinate whereas the event time is indeterminate. This may help explain the frequent remarks that only two times are needed to specify the truth conditions of all the tenses. In one sense these remarks are wrong. S, R and E all play essential roles in Reichenbach's explanation of the tenses. But only S and R ever need to be extracted from the context. All that we need to know about E is its position relative to R and this information is contained in the sentence itself. Thus a tense logic following Reichenbach's analysis could be two-dimensional, rather than three-dimensional. If s and r are the points of speech and reference, for example, we would have $(s, r) \models$ PASTPERFECT A if and only if $r < s$ and, *for some* $t < r$, $t \models A$.

Whatever the correct analysis turns out to be, it seems clear that the past tenses can not be added to PTQ without adding something like Reichenbach's point of reference to the models. Moreover adherence to the idea that there should be a separate way of combining tenses and intransitive verb phrases for every negated and unnegated tense would be cumbersome and would miss important generalizations. Montague's most important legacies to the study of tense were probably his identification of meaning and truth conditions, and his high standards of rigor and precision. It is striking that Jespersen, Reichenbach and Montague say successively less about tense with correspondingly greater precision. A great deal of the contemporary work on the subject can be seen as an attempt to recapture the insights of Jespersen without sacrificing Montague's precision.

III. CONTEMPORARY VIEWS

Explicit truth conditions for the tensed sentences of English are given in Bennett-Partee,[6] Vlach [1980][7], Kuhn [1979], and [1979], Parsons [1980],[8] Tichý and Bäuerle-von Stechow.[9] A number of other authors have suggested truth conditions for particular tenses or discussed the tense system in such a way that the contribution of tense to truth conditions can easily be extracted. There is, of course, a great deal of overlap and agreement among these accounts. In this section we will concentrate on the issues on which they seem to disagree.

A. *Simple Present and Present Progressive*

In many contemporary accounts the semantic analysis of the present underlies that of all the other tenses.[10] But despite this allegedly fundamental role, the only use of the present that seems to have been treated formally is the "reportive" use, in which the sentence describes an event that is occurring or a state that obtains at the moment of utterance.[11] The preoccupation with reportive sentences is unfortunate for two reasons. First, the reportive uses are often the less natural ones. (Consider the sentence *John walks to work*, for example.) Second, if the present tense is taken as fundamental, the omission of a reading in the present tense can be transferred to the other tenses. (*John walked to work* can mean that John habitually walked to work.) The neglect is understandable, however, in view of the variety of uses the present can have and the difficulty of analyzing them. One encounters immediately, for example, the issue discussed below.

4. *Statives and non-statives.* There is a discussion in the philosophical literature beginning with Aristotle about the kinds of verb phrases there are and the kinds of things verb phrases can describe. Details of the classification and terminology vary widely. One reads about events, processes, accomplishments, achievements, states, activities and performances. The labels are sometimes applied to verb phrases, sometimes to sentences and sometimes to "eventualities." There seems to be general agreement, however, that some kind of classification of this kind will be needed in a full account of the semantics of tense. In connection with the present tense there is a distinction between verb phrases for which the reportive sense is "easy" (e.g., *John knows Mary*, *The cat is on the mat*,

Sally is writing a book) and those for which the reportive sense is "difficult" (e.g., *John swims in the channel, Mary writes a book*). This division almost coincides with a division between verb phrases that have a progressive form and those that do not. (Exceptions – noted by Bennett and Partee – include *John lives in Rome* and *John resides in Rome*, both of which have easy reportive uses but common progressive forms.) It also corresponds closely to a division of sentences according to their behavior in the presence of *when* clauses. The sentence *The cat was on the mat when John came home* implies that the cat was on the mat for some time preceding and extending up to John's coming home. In contrast the sentence *Mary swam in the Channel when John came home* implies that Mary began swimming only *after* John's arrival. In general, the sentences for which appended *when* clauses can be paraphrased using *just after* correspond to those with difficult reportive uses and common progressive forms. The sentences for which appended *when* clauses can be paraphrased using *still at the time* correspond to those with easy reportive uses and no common progressive forms. (Possible exceptions are *I knew her when I saw her* and *I hated him when he told the joke*.)

The correspondence among these three tests suggests that they reflect some fundamental ways in which language users divide the world. The usual suggestion is that sentences in the first class (easy reportive readings, no progressives and *when* = *still at the time*) describe *states*. States are distinguished by the fact that they seem to have no temporal parts. The way Emmon Bach puts it is that it is possible to imagine various states obtaining even in a world with only one time, whereas it is not possible to imagine events or processes in such a world. Sentences that describe states are *statives*; those that do not are *non-statives*.

There is some disagreement about whether sentences in the progressive are statives. The fact that Harry is building a house, for example, can go on at discontinuous intervals and the fact that Mary is swimming in the Channel is composed of a sequence of motions, none of which is itself swimming, lead Moravcsik and Gabbay to the conclusion that present progressives do not denote states. But according to the linguistic tests discussed above progressives clearly do belong with the state sentences. For this reason, Vlach, Bach and Bennett all take the other side. It is as if, as English speakers, we pretend that there is some state that obtains steadily exactly while Mary makes the swimming motions and another state that obtains exactly while Harry does the house building activities. Perhaps we are adopting some kind of determinism according to which whatever happens does so because the world is in the state of its happening.

5. *Intervals and instants.* Those who wrote about the truth conditions of English tenses in the 1960's assumed that sentences were to be evaluated at instants of time. Montague [1968] and Dana Scott each suggested a treatment of the present progressive according to which *Mary is swimming in the Channel* is true at an instant t if *Mary swims in the Channel* is true at every instant in an open interval that includes t. This account has the unfortunate consequence of making the present progressive form of a sentence imply its (indefinite) past. For a large class of sentences this consequence is desirable. If John is swimming in the Channel he did, at some very recent time, swim in the Channel. On the other hand there are many sentences for which this property does not hold. *John is drawing a circle* does not imply that John drew a circle. *Mary is climbing the Zugspitze* does not imply that Mary climbed the Zugspitze.

In Bennett–Partee, Vlach [1980] and Kuhn [1979] this difficulty is avoided by allowing some present tense sentences to be evaluated at extended intervals of time as well as instants. *John is drawing a circle* means that the present instant is in the interior of an interval at which *John draws a circle* is true. The present instant can clearly be in such an interval even though *John drew a circle* is false. Sentences like *John swims in the Channel*, on the other hand, are said to have what Bennett and Partee label the *subinterval* property: their truth at an interval entails their truth at all subintervals of that interval. This stipulation guarantees that *Mary is swimming in the Channel* does imply *Mary swam in the Channel.*

The idea that some sentences should be evaluated at extended intervals has been criticized recently by Terence Parsons on the grounds that there is no way to test the truth of sentences at long intervals. Parsons' idea is that we can test whether a sentence is true at an instant, or a very short interval, by imagining that we utter the sentence at that time. This method clearly cannot be used to test whether a sentence is true at a long interval. One must be a little careful, however, how one understands the notion of truth "at" a time. In Parsons' fragment it is assumed that a single time supplies all the information needed to evaluate the truth of a sentence. In a treatment along the lines of Reichenbach we would expect sentences to be true or false with respect to both an utterance time and a reference time. It is plausible to interpret the defenders of the interval approach as maintaining that some sentences should be evaluated at extended reference times. Parsons' test, on the other hand, seems to determine whether a sentence is true at an utterance time. How, after all, do we check whether *Hilary climbed Everest* is true when uttered with reference to 12:30 p.m.,

July 23rd, 1954? Certainly not by imagining uttering *Hilary climbed Everest* at that time.

Parsons suggests further that in some cases it is *impossible in principle* to test whether a sentence is true at a specific long interval. This is a reasonable point. It does seem impossible in principle to determine whether *Bob plays ball* is true at the open interval between three and four o'clock yesterday afternoon rather than, say, the closed interval beginning with three and ending with four. But the argument from testability seems to weigh just as heavily against the instant approach in this case as it does against the interval approach. If we could test whether *Bob plays ball* was true at three o'clock and four o'clock then we could test whether the interval at which Bob played was open or closed. In fact, one might argue that this order of evaluation is reversed. To find out whether *Bob is playing ball* was true at three o'clock yesterday we check whether *Bob played ball* would be true if it referred to some interval containing that time.

As Parsons acknowledges, however, the question of whether a semantic theory renders truth "testable" is of less importance than the question of whether it gives a better overall explanation of our linguistic intuitions. If the interval approach led to an account of the progressive with fewer counterintuitive consequences than the instant approach that would itself be good evidence in its favor. As will be seen below, the verdict in this case is not yet in.

6. *Unfinished progressives and inertia worlds.* According to the Bennett-Partee account of progressives, *John was building a house* does not imply that John built a house. It does, however, imply that John will eventually build a house. Yet it seems perfectly reasonable to say *John was building a house when he was killed.* One attempt to modify the account to handle this difficulty is outlined in Dowty [1979]. Dowty's proposal is that we make the progressive a *modal* notion.[12] The progressive form of a sentence A is true at time t in world w just in case A is true at an interval containing t in all worlds w' such that w' and w are exactly alike up to t and the course of events in w after t "develops in the way most compatible with past events." The w'-worlds mentioned are referred to as "inertia worlds." *John was building a house when he died* means that *John builds a house* is eventually true in all the worlds that are inertia worlds relative to ours at the time of John's death.

If an account like this is to be useful, of course, we must have some understanding of the notion of inertia world independent of its role in making progressive sentences true. The idea of a development maximally compatible with past events may not be adequate here. John's death and consequent inability to finish his house may have been natural, even inevitable, at the time he was building it. In Kuhn [1979] the suggestion is that it is the expectations of the language users that are at issue. But this seems equally suspect. It is quite possible that because of a bad calculation we all mistakenly expect a falling meteor to reach earth. We would not want to say in this case that the meteor *is* falling to earth.

7. *Instantaneous events and gappy processes.* Other objections have been made to the Bennett–Partee analysis and the inertia worlds modification of it. These have to do with the application of the analyses to two special classes of sentences. The first class comprises sentences that cannot plausibly be said to be true at extended intervals, but that do have progressive forms. Vlach, following Gilbert Ryle, calls these achievement sentences. We will follow Gabbay-Moravcsik and Bach in calling them instantaneous event sentences. They include *Baltimore wins*, *Columbus reaches North America*, *Columbus leaves Portugal* and *Mary starts to sweat*. It seems clear that instantaneous event sentences fail all the tests for statives. But if they are really true only instantaneously then the interval analysis would predict that they would never form true progressives.

The second class contains just the sentences that are immune from the imperfective paradox (because their present progressive does imply their indefinite past). According to Bennett–Partee these are subinterval sentences. To avoid begging the question of whether they really do have the subinterval property we follow Vlach and others in calling them *process* sentences. The Bennett–Partee analysis and its inertia worlds variation have the consequence that process sentences can't have "gappy" progressives. If *I sit in the front row of the Jupiter theater* was true at the interval from two o'clock to four o'clock last Saturday afternoon, then *I am sitting in the front row of the Jupiter theater* was true at all instants between those times including, perhaps, some instants at which I was really buying popcorn. This according to Vlach, Bennett and Moravcsik-Gabbay, is a conclusion that must be avoided.[13]

8. *Optimism and pessimism.* Vlach's solution to the problems of instantaneous events and gappy processes is to give up the idea that a uniform treatment of the progressive is possible. For every non-stative sentence *A*, according to Vlach, we understand a notion Vlach calls the *process of A*

or, simply proc(A). The present progressive form of A simply says that our world is now in the state of proc(A)'s going on.

The nature of proc(A), however, depends on the kind of sentence A is. If A is a process sentence then proc(A) is "the process that goes on when A is true."

For the other non-stative sentences, proc(A) is a process that "leads to" the truth of A, i.e., a process whose "continuation . . . would eventually cause A to become true." In fact, Vlach argues, to really make this idea precise we must divide the non-process, non-stative sentences into at least four subclasses.

The first subclass contains what we might (following Bach) call extended event sentences. Paradigm examples are *John builds a house* and *Mary swims across the Channel.* If an extended event sentence is true at an interval i then proc(A) starts at the beginning of i and ends at the end of i. For the second subclass (*John realizes his mistake*, *Mary hits on an idea*) proc is not defined at all. For the third class (*Mary finishes building the house*, *Columbus reaches North America*) the progressive indicates that the corresponding process is in its final stages. For the fourth class (*Max dies*, *The plane takes off*) proc must be a process that culminates in a certain state.

Vlach's account is intended only as a rough sketch. As Vlach himself acknowledges, there remain questions of clarification concerning the boundaries of the classes of sentences and the formulation of the truth conditions. Furthermore, Vlach's account, like Dowty's, introduces a new theoretical term. If the account is to be really enlightening we would like to be sure that we have an understanding of proc that is independent of, but consistent with, the truth conditions of the progressive.

Even if all the questions of clarification were resolved, Vlach's theory might not be regarded as particularly attractive because it abandons the idea of a uniform account of the progressive. Not even the sources of irregularity are regular. The peculiarity of the truth conditions for the progressive form of a sentence A are explained sometimes by the peculiarity of A's truth conditions, sometimes by the way proc operates on A and sometimes by what the progressive says about proc(A). In this sense, Vlach's account is *pessimistic*. Two attempts have recently been made to give a more uniform account of the progressive. One is by Michael Bennett (Bennett [1981]) and one by Terence Parsons (Parsons [1983]).

The accounts of Vlach, Bennett and Parsons (and presumably anyone else) must distinguish between statives and non-statives because of the differences in their ability to form progressives. Non-statives must be further divided between processes and events if the inference from present progressive to past is to be selectively blocked. But in the treatments of Bennett and Parsons all the differences among these three kinds of sentences are reflected in the untensed sentences themselves. Tenses and aspects apply uniformly.

Bennett's proposal is extremely simple.[14] The truth conditions for the present perfect form of *A* (and presumably all the other forms not involving progressives) require that *A* be true at a *closed* interval with the appropriate location. The truth conditions for the progressive of *A* require that *A* be true in an open interval with the appropriate location. Untensed process sentences have two special properties. First, if a process sentence is true at an interval, it is true at all closed subintervals of that interval. Second, if a process sentence is true at every instant in an interval (open or closed) then it is true at that interval. Neither of these conditions need hold for event sentences. Thus, if *John is building a house* is true, there must be an open interval at which *John built a house* is true. But if there is no *closed* interval of that kind, then *John has built a house* will be false. On the other hand, *Susan is swimming* does imply *Susan has* (*at some time*) *swum* because the existence of an open interval at which *Susan swims* is true guarantees the existence of the appropriate closed intervals.

If this proposal has the merit of simplicity, it has the drawback of seeming very *ad hoc* – "a logician's trick" as Bennett puts it. Bennett's explanatory remarks are helpful. Events have a beginning and an end. They therefore occupy closed intervals. Processes, on the other hand, need not. But a process is composed, at least in part, of a sequence of events. If Willy walks then there are many positions *A* and *B* such that the events described by *Willy walks from A to B* occupy portions of the period of his walking. And *Willy walks* is also true at these intervals. Events, however, need not be decomposable in this way.

The account offered by Parsons turns out to be similar to Bennett's account. Parson's exposition seems more natural, however, because the metaphysical underpinnings discussed above are exposed. Parsons starts with the assumption that there are three kinds of eventualities: *states*, *processes* and *events*. Eventualities usually have *agents* and sometimes *objects*. An agent may or may not be *in* a state at a time. Processes may or may not be *going on* at a time. Events may or may not be *in development* at a time. In general, if *e* is an eventuality, we say that *e holds* at time *t* if the agent of *e* is in *e* at *t* or *e* is going on at *t* or *e* is in development at *t*. In addition, events can have the property of *culminating* at a time. The set of times at which an event holds is assumed to be an open interval and the time, if any, at which it culminates is assumed to be the least upper bound of the times at which it holds.

The structure of language mirrors this metaphysical picture. There are three kinds of untensed sentences: statives, process sentences and event sentences. Tensed sentences describe properties of eventualities. Stative and process sentences say that an eventuality *holds* at a time. Event sentences say that an eventuality *culminates* at a time.

The treatment of progressives is remarkably simple. Putting a sentence into the progressive has no effect whatsoever, other than changing the sentence from a non-stative into a stative. This means that, for process sentences, the present and progressive are equivalent. *John swims* is true at *t* if and only if *John is swimming* is true at *t*. *John swam* is true at *t* if and only if *John was swimming* is true at *t*. For event sentences, the change in classification does affect truth conditions. *John swam across the Channel* is true at *t* if the event described *culminated* at *t*. *John was swimming across the Channel*, on the other hand, is true at *t* if the state of John's swimming across the Channel *held* at *t*. But this happens if and only if the event described by *John swims across Channel* was *in development* at *t*. So it can happen that *John was swimming across the Channel* is true even though John never got to the other side.

It seems likely that, with the proper understanding of primitives, Parsons, Vlach and Bennett could be seen as saying very similar things about the progressive. Parsons' exposition seems simpler than Vlach's, however, and more natural than Bennett's. These advantages may have been won partly by reversing the usual order of analysis from ordinary to progressive forms.

Vlach's account proceeds from *A* to proc(*A*) to the state of proc(*A*)'s holding. In Bennett's, the truth conditions for the progressive of *A* are explained in terms of those for *A*. If one compares the corresponding progressive and non-progressive forms on Parsons' account, however, one sees that in the progressive of an event sentence, something is *subtracted* from the corresponding non-progressive form. The relations between the progressive and non-progressive forms seem better accommodated by viewing events as processes plus culminations rather than by viewing processes as eventualities "leading to" events.

On the other hand the economy of Parsons' account is achieved partly by ignoring some of the problems that exercise Vlach. The complexity of

Vlach's theory increases considerably in the face of examples like *Columbus reaches America* and *Max dies.* To accommodate these examples Parsons has two options. He can say that these are ordinary event sentences that are in development for a time and then culminate, or he can say that they belong to a new category – "achievement" – of sentences that culminate but never hold. The first alternative is unrealistic. (We don't really want to make *John swam across the Channel at 5:01* and *Max dies at 5:01* equally acceptable.) The second requires us to say that the progressive of these sentences, if it can be formed at all, involves a "change in meaning." But the progressive *can* be formed and spelling out the details of the meaning changes involved will certainly spoil some of Parsons' elegance.

B. *Past Tense*

9. *Definite and indefinite time.* Every account except those of Dowty and Parsons accommodates in some way the notion that past tense sentences are more definite than the usual tense logic operators. Even Dowty and Parsons, while claiming to treat the more fundamental use of the past tense, acknowledge the strength of the arguments that the past can refer to a definite time. Both cite Partee's example:

> When uttered, for instance, half way down the turnpike such a sentence [as *I didn't turn off the stove*] clearly does not mean that there exists some time in the past at which I did not turn off the stove or that there exists no time in the past at which I turned off the stove.

There are, however, some sentences in which the past does seem completely indefinite. We can say, for example, *Columbus discovered America* or *Oswald killed Kennedy* without implying or presupposing anything about the date those events occurred beyond the fact that it was in the past. It would be desirable to have an account of the past that could accommodate both the definite and indefinite examples. One solution, employed in Bennett–Partee and Kuhn [1979], is that we interpret the past as a quantifier over a set of possible reference times.[15] *I left the oven on* is true now only if the oven was left on at one of the past times I might be referring to. The context serves to limit the set of possible reference times. In the absence of contextual cues to the contrary the set comprises *all* past times and the past is completely indefinite. The suggestion that the context determines a set of possible reference times seems more realistic than the suggestion that it determines a unique such time. In addition the reference set accounts facilitate treatment of coordinate conjunctions whose clauses refer to different times.

There is still something a little suspicious, however, about the notion that context determines a reference interval or a range of reference times for past tense sentences to refer to. One would normally take the "context of utterance" to include information like the time and place the utterance is produced, the identity of the speaker and his audience, and perhaps certain other facts that the speaker and his audience have become aware of before the time of utterance. But in this case it is clear that *Baltimore won the Pennant* and *Columbus discovered America* uttered in *identical* contexts would have *different* reference times.

A way out of the dilemma might be to include the sentence itself as a part of its own context of utterance. This would ensure that *Baltimore won the Pennant* and *Columbus discovered America* could never be uttered in the same context. Such a move has been defended for other reasons in Lewis. It is not clear, however, exactly how a past tense sentence would contribute to the determination of its own reference time. A full explanation of the mechanism might require taking into account what *new* information is being communicated by the sentence. For example, when a teacher tells her class *Columbus discovered America*, teacher and class probably already share the assumption that America was discovered. The teacher is communicating the fact that the discovery was made by Columbus. When the teacher says *Bobby discovered the solution to problem number seven*, teacher and students probably know that Bobby was trying to solve problem number seven. The new information is that he succeeded. In those cases it is plausible to suppose that possible reference times would be the time of America's discovery and the times after which Bobby was believed to have started working on the problem. (As support for the latter claim consider the following scenario. Teacher assigns the problems at the beginning of class period. At the end she announces *Bobby discovered the solution to problem seven*. Suzy objects *No, he didn't. He had already done it at home.*)

In general we can view a declarative sentence as having two functions. First, it identifies the relevant part of our mutual knowledge. Second, it supplies a new piece of information to be added to that part. It is the first function that helps delimit possible reference times. Previous discourse and non-linguistic information, of course, also play a role. When I say *Baltimore won the Pennant* it matters whether we have just been talking about the highlights of 1963 or silently watching this week's "Monday Night Baseball."

It is a consequence of this view that one can *refer* to a time without knowing it. A student, for example, who says *Columbus discovered America* may be unknowingly referring to a date in 1492. In itself this is not problematic. Someone can, for example, unknowingly refer to the number 666 by saying *the sum of the numbers on a roulette wheel.* But one must be a little bit careful. Suppose Ann tells Betty *James Watt called William Robinson a cripple*. It would be wrong to say that the range of reference times include only those times at which James Watt called William Robinson a cripple. (There may be no such time.) Perhaps the appropriate restriction is that the range of reference times must be denoted by descriptions that speaker and audience believe, prior to the utterance, to be non-vacuous.[16] So *the time America was discovered* and *the time since the class began work on the problem* are suitable descriptions, but *the time James Watt called William Robinson a cripple* is not.

What *is* the range of possible reference times for *James Watt called William Robinson a cripple*? In the absence of any further clues, it is probably the times during which James Watt is alive and after which William Robinson was born. But in more realistic contexts the speaker would probably be saying something a little more definite. Possible occurrences of the incident when Watt and Robinson were in school do not count.

10. *Frequency*. Bäuerle and von Stechow point out that interpreting the past tense as a quantifier ranging over possible reference times (or over parts of the reference time) makes it difficult to explain the semantics of frequency adverbs. Consider, for example, the sentence *Angelika sneezed exactly three times*, uttered with reference to the interval from two o'clock to three o'clock yesterday morning. We might take the sentence to mean that there are exactly three intervals between two and three with reference to which *Angelika sneezed* is true. But if *Angelika sneezed* means that she had the sneezing property at least once during the time referred to, then whenever there is one such interval there will be an infinite number of them. So *Angelika sneezed exactly three times* could never be true. Alternatively we might take the sentence to mean that there was at least one time at which Angelika had the sneezing-three-times property. But the intervals when Angelika sneezed three times will contain subintervals

during which she sneezed twice. So in this case *Angelika sneezed exactly three times* would imply *Angelika sneezed exactly twice.*

This problem leads Bäuerle and von Stechow to insist that the past tense itself indicates simply that the eventuality described occupies that part of the reference time that lies in the past. On this interpretation, it does make sense to say that *Angelika sneezed three times* means that there were three times with reference to which *Angelika sneezed* is true. Tichý, using a different framework, arrives at a similar analysis. Unfortunately, this position also has the consequence that the simple sentence *Angelika sneezed*, taken literally, would mean that Angelika's sneeze lasted for the full hour between two and three. Bäuerle–von Stechow and Tichý both suggest that past tense sentences without explicit frequency operators often contain an *implicit* "at least once" adverb. In a full treatment the conditions under which the past gets the added implicit adverb would have to be spelled out, so it is not clear how much we gain by this move. But the alternative would seem to be to insist that the "at least once" qualification is a normal part of the meaning of the tense which is dropped in the presence of frequency adverbs. This seems little better.

Vlach handles the frequency problem by allowing sentences to be true either "in" or "at" a time interval. *Angelika sneezed exactly three times* is true *at* the reference interval if it contains exactly three subintervals *at* which Angelika sneezes. On the other hand *Angelika sneezed* would normally be taken to assert that Angelika sneezed *in* the reference interval, i.e., that there is at least one time in the interval at which she sneezed. Again, a complete treatment would seem to require a way of deciding, for a given context and a given sentence, whether the sentence should be evaluated in or at the reference time. (The possibilities are actually even more numerous. See *Varieties of truth* below.)

We might argue that *all* the readings allowed by Vlach (or Bäuerle–von Stechow) are always present, but that language users tend to ignore the implausible ones – like those that talk about sneezes lasting two hours . But the idea that ordinary past tense sentences are riddled with ambiguities is not appealing.

11. *Location of reference time.* There appears to be some disagreement among the contemporary accounts about whether the reference time of a present tense sentence must wholly precede the utterance time or whether it may merely overlap the utterance time. This difference is probably largely terminological. Consider the sentence *Billy forgot his mittens today.*

If one takes the "reference time" to be the time at which Billy forgot his mittens, then clearly this must precede the utterance. If one takes the reference time to be the time denoted by *today*, however, then overlap will apparently suffice. One author for whom the difference may be more than terminological is Vlach, who writes: "I take it that *Max arrived today* is semantically anomalous unless there is some further contextual information as to when Max arrived."

C. *Present Perfect*

12. *Definite and indefinite time.* Nearly every contemporary writer has abandoned Montague's position that the present perfect is a completely indefinite past.[18] The current view seems to be that the time to which it refers (or the range of times to which it might refer) must be an interval of time that begins in the past and includes the moment of utterance. This is plausible. When we say *Pete has bought a pair of shoes* we normally do not mean just that a purchase was made at some time in the past. Rather we understand that the purchase was made recently. The view also is strongly supported by the observation that the present perfect can always take temporal modifiers that pick out intervals overlapping the present and never those that pick out intervals entirely preceding the present. We can say *Mary has bought a dress since Saturday*, but not *Mary has bought a dress last week*.

On the other hand the view seems to clash with Reichenbach's intuition that in a present perfect sentence the reference time coincides with the utterance time. In support of Reichenbach is the observation that one can add *now* to a sentence in the present perfect without changing its meaning. Furthermore, other temporal modifiers don't *always* serve to delimit the times at which the untensed sentence is true. *Today Asimov has written over three hundred books* does not mean that three hundred books were written since last midnight. It seems that, for this example at least, a Montagovian analysis does provide the correct interpretation: *It is true today that Asimov has the property of having written in the past three hundred books*.

There is something else a little odd about the contemporary analyses of the present perfect. The present perfect is frequently used when speaker and audience are well aware that the event or state alluded to does *not* obtain at the moment of utterance. It becomes very mysterious why I should ever say *I have bought a pair of shoes* when I am not in a shoe store, or *I have lived in Finland* when I am very obviously living in Germany. If the context determines *any* interval in those cases it would appear to supply a completely past interval.

13. *Existence presuppositions.* Jespersen's observation that the present perfect seems to presuppose the present existence of the subject in cases where the past tense does not has been repeated and "explained" many times. We are now faced with the embarrassment of a puzzle with too many solutions. The contemporary discussion begins with Chomsky, who argues that *Princeton has been visited by Einstein* is all right, but *Einstein has visited Princeton* is odd. James McCawley points out that the alleged oddity of the latter sentence actually depends on context and intonation. Where the existence presupposition does occur, McCawley attributes it to the fact that the present perfect is generally used when the present moment is included in an interval during which events of the kind being described *can* be true. Thus, *Have you seen the Monet exhibition?* is inappropriate if the addressee is known to be now unable to see it. (*Did you* is appropriate in this case.) *Frege has contributed a lot to my thinking* is appropriate to use even though Frege is dead because Frege *can* now contribute to my thinking. *My mother has changed my diapers many times* is appropriate for a talking two year old, but not for a normal thirty year old. *Einstein has visited Princeton* is odd because Einsteinean visits are no longer possible. *Princeton has been visited by Einstein* is acceptable because Princeton's being visited *is* still possible.

The explanation in Tichý is that, in the absence of an explicit indication of reference time, a present perfect generally refers to the life time of its subject. If this does not include the present, then the perfect is inappropriate.

In Kuhn [1983] the suggestion is that the explanation may be partly syntactic. Existence presuppositions can be cancelled when a term occurs in the scope of certain operators. Thus *Santa is fat* presupposes that Santa exists, but *According to Virginia, Santa exists* does not. There are good reasons to believe that past and future apply to sentences, whereas perfect applies only to intransitive verb phrases. But in that case it is natural that presuppositions that do hold in present perfect sentences fail in past and future sentences.

Guenthner requires that at least one of the objects referred to in a present perfect sentence (viz., the topic of the sentence) must exist at utterrance time.

The question of whether these explanations are compatible and whether they are equally explanatory, remains open.

D. *Other Topics*

14. *Future tense.* The architects of fragments of English with tense seem to have comparatively little to say about the future. Vlach omits it from his very comprehensive fragment, suggesting he may share Jespersen's view that the future is not a genuine tense. Otherwise the consensus seems to be that the future is a kind of mirror image of the past with the exception, noted by Bennett and Partee, that the times to which the future can refer include the present. (Compare *He will now begin to eat* with *He now began to eat.*)

There appears to be some disagreement over whether the future is definite or indefinite. Tichý adopts the position that it is ambiguous between the two readings. This claim is difficult to evaluate. The sentence *Baltimore will win* can mean *Baltimore will win next week* or *Baltimore will win eventually.* But this difference can be attributed to a difference in the set of possible reference times as easily as to an ambiguity in the word *will.* As in the case of the progressive, it is probably preferable on methodological grounds to adopt a uniform treatment if possible.

15. *Complex tenses.* On the simplest analyses all the tenses are built by appropriate combination of past, present, future, perfect and progressive. This idea is extremely attractive, but there is a difficulty in the case of past perfect. On the compositional analyses this is to be regarded as the past of the perfect form. Now consider the following sentences:

(1) At three o'clock, Ron had run four miles.

(2) Bill had finished the book at three o'clock.

The first seems to fit the proposed analysis quite well. The past tense sentence is to be evaluated with respect to three o'clock. At that time the perfect *Ron has run four miles* was true. The second sentence, however, seems to suggest that Bill finished the book at three o'clock. Because of restrictions on the use of temporal modifiers with the perfect, this cannot be paraphrased along the lines: *It was once the case that "Bill has finished the book at three o'clock" was true.*[19] One might try to salvage the simple analysis by insisting that the second sentence can be made to fit the same mold as the first: *At three o'clock "Bill has (just this moment) finished the book" was true*. (One could give pragmatic arguments to explain why the perfect so often refers to "nearby" events in these contexts.) But there are some cases for which this move is not plausible. Consider, for example, *Last week he had come on Monday*. At least some authors are driven to the view that the past perfect sometimes behaves more like the past of a past than the past of a perfect. Vlach manages to save at least the shell of the compositional account of the complex tenses by arranging that the perfect behaves differently according to the tense in whose scope it appears.

16. *Varieties of truth*. Perhaps the most radical suggestion in the contemporary literature is Vlach's idea that proper treatment of tenses requires recognition of three distinct and fundamental ways that a sentence can be true or false with respect to a time. Vlach's suggestion plays a role in his treatment of all the tenses and temporal modifiers, but the best evidence in its favor probably occurs in connection with the perfect. The idea can be conveyed by a few examples. *Max was a sailor* is true *at* three o'clock yesterday afternoon if Max was then a sailor. It is true *in* 1980 if he was a sailor at some time during that year. It is true *for* 1980 if he was a sailor at some time during that year. It is true *for* the month of December, 1980, if he was a sailor at intervals "scattered more or less all" over December, 1980. The variety of truth that is appropriate for evaluating a sentence can be indicated by prepositions in temporal modifiers. Often there are several appropriate varieties. For example, *Max worked in 1980* can mean that he worked *in* the year 1980 or *for* the year 1980. *Max has been running since Thursday* can mean that he has been running *in* the interval from Thursday to now or *for* that interval. In the absence of restraints imposed by such prepositions, present tense sentences can be evaluated *at* or *for* a time; present perfect sentences, *in*, *at*, or *for* a time.

This arrangement allows us to explain the observation in McCawley and elsewhere that the present perfect can have either an "existential" or "universal" force. It allows us to distinguish two kinds of present (e.g., *Max works at this minute*, which implies Max is awake, and *Max works this year*, which does not.)[20] It exposes ambiguities in sentences like *Nora was in New York yesterday*. And, as we have seen, it facilitates the treatment of frequency adverbs. But it does all this at the expense of a fundamental change in our conceptual framework.

It might be possible to get at least some of these benefits by attributing the ambiguities to expressions that pick out the reference times rather than to the variety of truth. For example, the ambiguity in *The cat has been on the mat since Saturday* can be explained by saying that *since Saturday* is ambiguous between *ever since Saturday* and *at least once since Saturday*. In the first case the prepositional phrase denotes a single interval and in the second it denotes a set of instants. Similarly, in *John was in New York yesterday*, *yesterday* could denote a single 24 hour period (which would entail that John spent the day in New York) or a set of instants (which would entail only that John spent some time in New York).

Where there are no temporal adverbs to help delimit the reference times we could (following Gabbay-Moravcsik) attribute the ambiguity to the basic verb phrase. In *Max works this year*, for example, *works* is to be given a "wide" reading that permits interruptions for eating, sleeping, and sufficiently short periods of illness and vacation. In *Max works at this minute* it is to be given a narrower reading that excludes such interruptions.[22]

Ultimately, however, it may turn out that the simplest adequate theory will use Vlach's varieties of truth. If this seems distasteful, we *could* evade it by making Vlach's varieties features of the index or sentence. Mathematically it doesn't matter whether we say that *Max works* is true for the time 1984 or that *Max works* is true with respect to (1984, *for*) or that *Max works for* is true with respect to 1984. Vlach's analyses could be formulated with any of these locations. But only Vlach's formulation seems to be capable of being understood without further explanation.

17. *Tense logics for natural language*. A general survey of tense logic is contained elsewhere in this Handbook [Chapter 11.2]. In this section we survey briefly some of the systems that claim affinity to the tense mechanisms in natural language. There are at least two motivations for the study of such systems. There is first the program of generalizing the language of classical logic so that arguments that hinge on tense can be represented in a form that resembles natural language. For example, to represent *Mount Vesuvius erupted* in classical logic by $\exists t(t < 1988 \,\&\, Emt)$ seems less natural than to represent it in a tense logic by $\mathscr{P}Em$. The "representation" in this case is supposed to provide a logical form for tensed sentences, so one would not expect the nonclassical operators to capture all the nuances of particular English tense constructions. On the other hand one *would* expect a certain economy in logical vocabulary and rules of inference. A second motivation is to provide a framework in which the semantics of particular tense constructions can be expressed precisely. Montague's

tensed intensional logic, discussed above, provides such a framework, but it is both too simple and too complex. It is too simple because the past and future of English don't have the truth conditions of the usual $\mathscr{P}$ and $\mathscr{F}$; too complex because the intensional part of the logic does not seem needed for the treatment of tense.[23]

The contemporary literature on tense logic begins with Arthur Prior who formulated a number of simple tense logic systems and, in a series of publications in the 1950's and 1960's, advocated, chronicled and contributed to their development, (These have been collected in Prior [1967] and Prior [1968].) Prior wrote about a variety of such systems, but most of them have nonclassical operators $\mathscr{P}$ for past and $\mathscr{F}$ for future. Furthermore, in nearly all of Prior's systems formulas can be thought of as being true or false *relative to a time* and the operators can be interpreted by the conditions:

$$t \models \mathscr{P}A \text{ iff } \exists s\,(s < t \;\&\; (s \models A))$$

$$t \models \mathscr{F}A \text{ iff } \exists s(t < s \;\&\; (s \models A))$$

(where $t \models A$ means A is true at time t and $s < t$ means time s is *before* time t). Linguistically minded successors to Prior have generalized this framework in two basic ways. First, they have suggested allowing formulas to be evaluated at temporal intervals as well as instants. Second, they have suggested allowing formulas to be evaluated at many-dimensional "indices" instead of single times. The first of these generalizations is discussed in Chapter II.2 of this Handbook, so we will only deal with the second.

Hans Kamp (Kamp [1971]) proposed that the logic of *now* could be represented in a system in which every formula was evaluated with respect to a *pair* of times, one of which "keeps track of" the time of speech and the second of which is used to evaluate expressions inside tense operators. More precisely the truth conditions for Kamp's operators can be written as follows:

$$\langle s, t\rangle \models \mathscr{P}A \text{ if and only if } \exists t' < t\ \langle s, t'\rangle \models A$$

$$\langle s, t\rangle \models \mathscr{F}A \text{ if and only if } \exists t' > t\ \langle s, t'\rangle \models A$$

$$\langle s, t\rangle \models \mathscr{N}A \text{ if and only if } \langle s, s\rangle \models A.$$

With Kamp's $\mathscr{N}$ operator we can represent the sentence *Officer Jones will cite everyone now speeding* (with the reading that they will all be cited at once) by $\mathscr{F}\forall x(\mathscr{N}Sx \rightarrow Cjx)$. Without it this reading could not be represented.

Frank Vlach (Vlach [1973]) realized that Kamp's *now* in a more general setting could be used to cause evaluation at moments other than the moment of utterance. More specifically, he added to Kamp's framework an *index* operator $\mathscr{I}$ with truth conditions:

$$\langle s, t\rangle \vDash \mathscr{I}A \text{ if and only if } \langle t, t\rangle \vDash A.$$

If an $\mathscr{N}$ occurs within the scope of an $\mathscr{I}$ it can be read as *then*. This allows, for example, the sentence *Officer Jones was once going to cite all those who were then speeding* to be represented as

$$\mathscr{P}\mathscr{I}\mathscr{F}\forall x(\mathscr{N}Sx \rightarrow Cjx).$$

In general, if A contains no occurrence of $\mathscr{I}$ the utterance time is "fixed" in the sense that the truth value of A at $\langle u, t\rangle$ depends on the truth values of its subformulas at pairs $\langle u, t'\rangle$. The occurrence of an $\mathscr{I}$ "shifts" the utterance time so that evaluating A at $\langle u, t\rangle$ may require evaluating the subformulas that are within the scope of the $\mathscr{I}$ at pairs $\langle u', t'\rangle$ for u' different than u.

Lennart Åqvist (Åqvist [1976]) pointed out that the two coordinates in the semantics of Kamp and Vlach could be identified with Reichenbach's points of utterance and reference. Åqvist and Franz Guenthner have tried to construct some general two- and three-dimensional logics within which tense operators with Reichenbach's interpretation could be defined. In one of the less elaborate versions, for example, there are operators with the following truth clauses:

$$\langle s, t\rangle \vDash \mathscr{O}A \text{ iff } \forall t' \, \langle s, t'\rangle \vDash A$$

$$\langle s, t\rangle \vDash \mathscr{X}A \text{ iff } \langle t, s\rangle \vDash A$$

$$\langle s, t\rangle \vDash \mathscr{P}A \text{ iff } \exists t' \, (t' < t \mathbin{\&} \langle s, t'\rangle \vDash A)$$

$$\langle s, t\rangle \vDash \mathscr{F}A \text{ iff } \exists t' \, (t' > t \mathbin{\&} \langle s, t'\rangle \vDash A)$$

Now let **Past** and **Future** be abbreviations for the operator sequences $\mathscr{P}\mathscr{X}\mathscr{O}$ and $\mathscr{F}\mathscr{X}\mathscr{O}$. Åqvist suggests that the Reichenbachian tenses simple past, simple future, anterior present, anterior past and anterior future can be represented by **Past**, **Future**, $\mathscr{P}$, **Past** $\mathscr{P}$, and **Future** $\mathscr{P}$, respectively. It is easy to see that

$$\langle s, t\rangle \vDash \textbf{Past}\, A \text{ iff } \exists t' \, (t' < t \mathbin{\&} \langle t', t'\rangle \vDash A), \text{ and}$$

$$\langle s, t\rangle \vDash \mathscr{P}A \text{ iff } \exists t' \, (t' < t \mathbin{\&} \langle s, t'\rangle \vDash A).$$

It follows that if A itself has no tense operators then the past and anterior present (our present perfect) are equivalent. Åqvist regards this fact as a virtue of his account, but in view of the earlier discussion about the present perfect, it is not clear that it is.[24]

In later versions, Åqvist and Guenthner use similar general many-dimensional logics to define a host of "aspectual" operators, like "remains false" or "is becoming more and more", that may be needed for a more detailed representation of sentences in natural language.

Dov Gabbay (Gabbay [1974], [1976]) has argued that certain other many-dimensional operators are needed to properly represent constructions in which tensed clauses are embedded within one another. Some of these suggestions are listed below.

1. The sentence *It has always been thought that Tomac the Great invented the wheel* can be represented $\mathscr{H}_2\mathscr{P}A$, where $\mathscr{P}$ is Aqvist's two-dimensional past operator and $\mathscr{H}_2$ has the truth conditions: $\langle u, t\rangle \models \mathscr{H}_2 A$ iff $\forall v > u$ $\langle v, t\rangle \models A$. (The index $\langle u, t\rangle$ in this case is supposed to represent a state of affairs in which the inhabitants of world u believe their world to be t.)

2. One reading of the sentence *John said he will come* can be represented $\mathscr{P}$ **Johnsaysthat** $\mathscr{F}^2 A$, where $\langle u, t\rangle \models \mathscr{F}^2 A$ iff $\exists s(u < s \,\&\, t < s \,\&\, \langle u, s\rangle \models A)$.

3. One reading of *John said he would come* (*John will say he would come*) can be represented $\mathscr{P}$ **Johnsaysthat** $\mathscr{F}_2 A$ ($\mathscr{F}$ **Johnsaysthat** $\mathscr{F}_2 A$), where $\langle u, t\rangle \models \mathscr{F}_2 A$ iff either $t < u$ and $\exists s(t < s < u \,\&\, \langle u, s\rangle \models A)$ or $u < t$ and $\exists s(u < s < t \,\&\, \langle u, s\rangle \models A)$.

These examples are not all completely convincing. It is not clear, for example, whether *John said he would come* really has a reading that implies that he said he would come before now, or whether we make that inference on the basis of some extra-linguistic information that if John were coming at all he would have done so by now. Nevertheless they are suggestive enough to have convinced Gabbay and others that a general study of such logics is worth pursuing. In the end Gabbay rejects the idea that the $\mathscr{F}_2$ and $\mathscr{F}^2$ operators in the above examples are adequate to represent the English "would" and "will" in all contexts. He suggests some considerably more complicated candidates, but eventually concludes that even these are inadequate because "we must keep record of the entire sequence of points [that figure in the evaluation of a formula] and not only that, but also keep track of the kind of operators used." Since many-dimensional logics can only keep track of finitely many points, it would seem that no such logic could do the job.

Esa Saarinen has suggested that Gabbay's problem can be solved in a one-dimensional tense logic with special "backwards looking" operators. Saarinen's system can be viewed as the result of introducing a special operator-transforming functor, **D**. For any operator $\square$, **D**($\square$) "looks back" to the time at which the preceding $\square$ was evaluated. For example, we can represent Saarinen's *Every man who ever supported the Vietnam War believes now that one day he will have to admit the he was an idiot then* by: $\forall x(x$ is a man $\rightarrow \mathscr{X}(x$ supported the Vietnam war $\rightarrow$ $\mathscr{N}(x$ **believesthat** $\mathscr{F}(x$ **hastoadmitthat** $(\mathbf{D}(\mathscr{N})(x$ is an idiot$)))))$. A little more generally, $w \models A \,\&\, (B \,\&\, \mathscr{P}(C \,\&\, \mathscr{F}(D \,\&\, \mathbf{D}(\mathscr{P})E)))$ iff $\exists x \exists y \exists z(w < x, y < x, y < z, w \models A, x \models B, y \models C, z \models D$ and $x \models E)$. In this example **D**($\mathscr{P}$) "looks back" to the time at which the preceding $\mathscr{P}$ was evaluated, namely, x. One might wonder how we could look back to x if there were *several* $\mathscr{P}$'s before the E. It turns out that Saarinen's **D** has the property that iterated **D**($\square$)'s look back to more distant $\square$'s, so that:

$$w \models A \,\&\, \mathscr{F}(B \,\&\, \mathscr{F}(C \,\&\, \mathscr{F}(D \,\&\, \mathbf{D}(\mathscr{F})\mathbf{D}(\mathscr{F})E))) \text{ iff}$$

$$\exists x \exists y \exists z(w < x < y < z, w \models A, x \models B$$

$$y \models C, z \models D \text{ and } x \models E).$$

But it is worth noting that **D** also provides another means to do this. Since $A \,\&\, \mathscr{F}B$ is equivalent to $\mathscr{F}(\mathbf{D}(\mathscr{F})A \,\&\, B)$ the formula above is equivalent to $\mathscr{F}(\mathbf{D}(\mathscr{F})A \,\&\, (B \,\&\, \mathscr{F}(C \,\&\, \mathscr{F}(D \,\&\, \mathbf{D}(\mathbf{D}(\mathscr{F}))E))))$ Here **D**($\mathscr{F}$) serves only as something unique to look back to. Since the **D**'s can be nested indefinitely there are infinitely many such signposts available. This suggests that we might keep track of times more directly by generalizing the Kamp-Vlach $\mathscr{N}$ and $\mathscr{I}$. Add to the standard tense logic connectives $\mathscr{N}_i$ and $\mathscr{I}_i$ for all natural numbers i. $\mathscr{I}_i A$ indicates that A is to be evaluated at the time referred to when $\mathscr{N}_i$ was encountered.

These new operators have something of the flavor of ordinary free variables. In fact, many of the attempts to fortify the Priorean tense logics so that they are more useful for the study of natural language seem to lead to logical systems that are just disguised versions of predicate logic. This observation has led some[25] to suggest that it might be better to forget about tense logics altogether. The fact that a tense logic is equivalent to predicate logic or some fragment thereof is not itself proof that the logic is uninteresting. We do not customarily use locutions like "there was a time t before the present such that . . . " to indicate indefinite past. The observation that some of the expressions we do use are equivalent to that locution would, if true, be interesting. The general task of laying bare the "logic" of English tenses seems a worthy, if somewhat vague, enterprise. Unfortunately, however, none of the systems discussed here comes close doing this. If our aim is to construct an "idealized" version of tensed discourse in which arguments can easily be represented and evaluated then most of the systems are too baroque. Vlach's $\mathscr{I}$, Gabbay's $\mathscr{H}_2$, Åqvist's $\mathscr{X}$, and Saarinen's **D** don't correspond to any expressions in natural language. On the other hand, if our aim is merely to construct a language in which the truth conditions of tensed sentences can be expressed, this *can* probably be done more perspicuously in ordinary predicate logic (or, more precisely, in a first order theory of temporal succession).

18. *Discourse representation structures.* There is also a third way that we might construe the task of finding the logic of tense. We might take our goal to be that of finding an algorithm by which the truth conditions for a sentence can be computed from the sentence itself (or more strongly that of finding the algorithm which native speakers actually use to make this computation). Montague's work can be seen as an attempt to do this, although as we have seen, Montague does not offer a particularly sophisticted account of tense. One might try to defend some of the more esoteric tense logics on the grounds that they provide *intermediaries* between natural language and truth conditions. Our goal is to get from English to truth conditions. The tense logician tells us how to get from his logic to truth conditions and gives us strong hints about how to get from English to his logic. Recently a more direct attack along these lines has been made by Hans Kamp. (See Kamp [1980] and [1983], and Kamp and Rohrer.) Kamp's ideas have not been worked out in the same kind of detail as Montague's. They are quite suggestive, however, and the treatment of tense was one of the *motivations* for their development, so it would be appropriate to give some account of them here.

Kamp focuses on the "discourse" as a unit of analysis rather than the sentence. This move would seem to be particularly helpful in dealing with tenses. Sentential analyses treat reference times either as completely indeterminate or as simply given by context. In fact the "context" that determines the time a sentence refers to may just be the sentences that were uttered previously. And it seems clear that the way the reference time is extracted from the linguistic context may depend on the tenses of the sentence and its predecessors.

The semantic theory that Kamp envisages would specify a set of rules for constructing from each (syntactically analyzed) discourse a *discourse representation structure* (henceforth DRS) and a second set of rules for determining from the DRS the conditions under which the discourse is true. The DRS corresponds to what a competent speaker has in mind when he understands the discourse. In the simplest case (say, discourse consisting of a single, present tense, "atomic" sentence) the DRS is simply a *partial model*, i.e., a set of objects referred to in the discourse and a set of properties and relations said to hold among them.[26] The discourse is *true* if the DRS is *correct*, i.e., if the representation can be embedded in a full model representing the way the world is. More complicated kinds of discourse lead to more complicated DRS's and more complicated truth definitions. For example, the DRS for a discourse consisting of a conditional sentence is a pair of partial models, the second of which "extends" the first. The discourse is true in a model M if the larger partial model can be embedded in M.

Discourse Representation Structures for tensed discourse will include *events*, *states* and *times* as objects in the universe of discourse and will specify relations of precedence and containment among them. For example,[27] the DRS for the discourse

> Pedro entered the kitchen. Mary was doing the dishes.
> Pedro went into the hall. He took off his coat.

will contain the event e_1 of Pedro entering the kitchen, the state s of Mary's doing the dishes, the events e_2 and e_3 of Pedro going into the hall and taking off his coat, and the time n at which the discourse is uttered. It will specify that s contains e_1, e_1 precedes e_2, e_2 precedes e_3 and that all three precede n.

The exact DRS construction rules that should be associated with each tense construction are not clear. Consider for example the discourse

> Mary did the dishes carefully. She filled the sink
> with hot water. She added a half cup of soap. Then
> she gently dipped each glass into the sudsy liquid.

In the previous example each past tense sentence marked an event that succeeded the one described in the previous sentence. In this example the last three sentenes describe events that *comprise* the event described in the previous sentence. Furthermore, it should not be expected that the introduction of the DRS framework will solve the problems of interpreting tense that have been described here. When it comes to specify rules for determining the truth conditions of a discourse with a given DRS these problems are likely to emerge again. Still, we may hope that raising the old problems in the same context as the new ones might facilitate solutions.

Department of Philosophy, Georgetown University

NOTES

[1] This paper was written in 1983 while I was in Tübingen on a fellowship from the Alexander von Humboldt foundation. I have benefitted from the help of Franz Guenthner and from the written comments of Frank Vlach and Rainer Bäuerle.

[2] By "ours" I mean those of the authors discussed in the remainder of this article. Some recent work, including that of F. Palmer and R. Huddleston, say, is more in the tradition of Jespersen than this.

[3] Interestingly, a similar claim is made in Vlach [1981]. For the most part, however, the history of English is ignored in contemporary semantics.

[4] Many of these old grammars have been reprinted in the series *English Linguistics: 1500–1800* (*A Collection of Facsimile Reprints*) edited by R. C. Alston and published by Scolar Press Limited, Menston, England in 1967.

[5] This is the case, for example, in Bennett-Partee who say they are following the lead of Montague. But not everyone agrees. Kuhn [1983], for example, argues that present, past and future should be (equally fundamental) modes of combining noun phrases with verb phrases to form sentences.

[6] See also Bennett [1977] and [1981].

[7] See also Vlach [1981].

[8] See also Parsons [1983].

[9] See also Bäuerle [1979a]. These works actually present truth conditions for German sentences, but the remarks about tense are clearly intended to apply to English as well.

[10] This is true, for example, of Bennett-Partee. But there is no consensus here. Kuhn [1983], for example, argues that past, present and future should be taken as (equally fundamental) modes of combination of noun phrases and verb phrases.

[11] Many authors restrict the use of the term "reportive" to event sentences.

[12] Dowty attributes this idea to David Lewis.

[13] This argument is not completely decisive. It would seem quite natural to tell a friend one meets at the popcorn counter *I am sitting in the front row*. On the other hand, if one is prepared to accept *I am not sitting in the front row* at popcorn buying time, then perhaps one should be prepared to accept *I sat in the front row before I bought the popcorn* **and again** *after*. This would suggest the process went on *twice* during the long interval rather than once.

[14] Bennett attributes the idea behind his proposal to Glen Helman.

[15] The proposal is made in these terms in Kuhn [1979]. In Bennett-Partee the idea is rather that the reference time is an interval over whose subintervals the past tense quantifies. Thus the main difference between these accounts has to do with whether the reference time (or range of reference times) can be discontinuous. One argument for allowing it to be is the apparent reference to such times in sentences like *John came on a Saturday*. Another such argument might be based on the contention in Kuhn [1979] that the possible reference times are merely the times that happen to be maximally *salient* for speaker and audience. Vlach [1980] goes Partee-Bennett one further by allowing the past to indicate what obtains *in*, *at*, or *for* the reference interval. This idea is discussed under "varieties of truth" below.

[16] One can imagine various refinements. Perhaps it is sufficient that speaker believes listener believes description is non-vacuous and listener believes speaker believes description is non-vacuous.

[17] This view is completely compatible with the frameworks of Bäuerle and Tichy. I suspect that, semantically, it does not matter whether words like *throughout* or *during* modify the proposition or the set of times at which the proposition is to be evaluated. I prefer the latter on the grounds that *John kissed Mary throughout* and *John kissed Mary during* are of dubious grammaticality. But Tichý is not likely to be moved by such arguments since his interest lies merely in providing semantic analyses of those expressions that *are* grammatical.

[18] One exception is Kuhn, who hereby recants (without thereby subscribing to the current orthodoxy).

[19] It is a little unfair to make the argument hinge on the availability of a suitable paraphrase. We might maintain that sentence 2 does indeed mean that at some understood time in the past Bill had the property of having finished the book at 3:00. Such a reading is consistent with the sentence's being interpreted as the past tense combination of a noun phrase with a perfect verb phrase. But this leaves it a mystery why we can't say *Bill has finished the book at 3:00* to mean that Bill *now* has the property of having finished the book at 3:00. Interestingly, *Bill has finished reading the book at 3:00 twice* and *This time, Bill has finished reading the book at 3:00* seem all right. Furthermore, analogues to the problematic example in other languages are fine. (*Guillaume a terminé la lecture du livre à trois heures Um drei Uhr hat Wilhelm das Buch fertig gelesen.*)

[20] The issue here is the size of the interruptions that are permitted while one "works." It is not clear whether this distinction coincides with the reportive/habitual distinction, or whether both sentences are varieties of the reportive present.

[21] In a full account, of course, we would have to evaluate sentences with respect to reference time, utterance time and other features of context.

[22] Actually this two-way division might be a bit oversimplified. Consider the sequence: *Max worked for a minute*, *Max worked for a year*, *Max worked for thirty years*. There appears to be no way to divide these sentences into just two groups according to the size of permitted interruptions. One might do better by formulating the distinction in terms of the *proportion* of non-work time permitted, or the *distribution* of such times, but the issue is not clear.

[23] Indeed, the availability of terms denoting, and quantifiers ranging over, intensions of arbitrary level does not seem to be needed to explain any linguistic phenomena.

[24] If we take Reichenbach to be saying that the anterior present is *indefinitely* anterior, then a natural "Reichenbachian" logic in which the simple past and anterior present are different

(even in the absence of other connectives) would have

$$\langle u, r\rangle \vDash \textbf{\textit{Past}}\ A \text{ iff } r < u \text{ and } r \vDash A\text{, and}$$

$$\langle u, r\rangle \vDash \textbf{Anteriorpresent}\ A \text{ iff } u = r \,\&\, \exists t(t < r \,\&\, \langle u, t\rangle \vDash A).$$

[25] See van Benthem and Massey. Van Bentham cites Needham as the source of his "awakening".

[26] This is what Kamp *says* a DRS would be in this case. The "format" in which the DRS's are presented in his examples, however, consists of a list of objects and a set of atomic sentences true of these objects. Since some of the atomic sentences are of the form **a** = **b** for different names **a** and **b**, it is clear that the DRS contains a little more information than would be provided by a partial model in even the simplest case.

[27] This example is intended merely as an illustration of how tense might be treated within Kamp's framework. It should not be taken too seriously. Kamp's own application had to do with French tenses. I have "translated" his examples into English only to preserve the uniformity of the survey.

REFERENCES

Åqvist, L.: 1976, 'Formal semantics for verb tenses as analyzed by Reichenbach', in van Dijk (ed.), *Pragmatics of Language and Literature*, North Holland, Amsterdam, pp. 229–236.

Åqvist, L. and Guenthner, F. (eds.): 1977, *Tense Logic*, Nauwelaerts, Louvain.

Åqvist, L. and Guenthner, F.: 1978, 'Fundamentals of a theory of verb aspect and events within the setting of an improved tense logic', in F. Guenthner and C. Rohrer (eds.), *Studies in Formal Semantics*, North Holland, pp. 167–199.

Bäuerle, R.: 1979, 'Tense logics and natural language' *Synthese* **40**, 226–230.

Bäuerle, R.: 1979a, '*Temporale Deixis, Temporale Frage*', Gunter Narr Verlag, Tuebingen.

Bäuerle, R. and von Stechow, A.: 1980, 'Finite and non-finite temporal constructions in German', in Rohrer [1980], pp. 375–421.

Bach, E.: 1983, 'A chapter of English metaphysics', manuscript, University of Massachussetts at Amherst.

Bach, E.: 1981, 'On time, tense and aspect: An essay in English metaphysics', in Cole (ed.), *Radical Pragmatics*, Academic Press, New York, pp. 63–81.

Bennett, M.: 1977, 'A guide to the logic of tense and aspect in English', *Logique et Analyse* **20**, 137–163.

Bennett, M.: 1981, 'Of tense and aspect: One analysis', in Tedeschi and Zaenen (eds.), pp. 13–30.

Bennett, M. and Partee, B: 1972, 'Toward the logic of tense and aspect in English', Systems Development Corporation Santa Monica, California, reprinted by Indiana University Linguistics Club, Bloomington.

Brown, G.: 1865, *The Grammar of English Grammars*, William Wood & Co., New York.

Bull, W.: 1968, *Time, Tense, and the Verb*, University of California Press, Berkeley.

Burgess, J.: 1984, 'Basic Tense Logic', in this Handbook, Vol. II, pp. 89–133.

Curme, G.: 1935, *A Grammar of the English Language* (vols III and II), D. C. Heath and Company, Boston.

Dowty *et al.*: 1981, *Introduction to Montague Semantics*, Kluwer, Boston.

Dowty, D.: 1977, 'Toward a semantic analysis of verb aspect and the English imperfective progressive', *Linguistics and Philosophy* **1**, 45–77.
Dowty, D.: 1979, *Word Meaning and Montague Grammar: The Semantics of Verbs and Times in Generative Grammar and Montague's PTQ*, D. Reidel, Dordrecht.
Gabbay, D.: 1974, 'Tense Logics and the tenses of English', in Moravcsik (ed.), *Logic and Philosophy for Linguists: A Book of Readings*, Mouton, The Hague, reprinted as Chapter 12 in Gabbay [1976a].
Gabbay, D.: 1976, 'Two dimensional propositional tense logics', in Kasher (eds.), *Language in Focus: Foundation, Methods and Systems – Essays in Memory of Yehoshua Bar-Hillel*, D. Reidel, Dordrecht, pp. 569–583. Reprinted as Chapter 10 in Gabbay [1976a].
Gabbay, D.: 1976a, *Investigations in Modal and Tense Logics with Applications to Problems in Philosophy and Linguistics*, D. Reidel, Dordrecht.
Gabbay, D. and Moravcsik, J.: 1980, 'Verbs, events and the flow of time', in Rohrer [1980].
Guenthner, F.: 1980, 'Remarks on the present perfect in English', in Rohrer [1980].
Hinrichs, E.: 1981, *Temporale Anaphora im Englischen*, Zulassungsarbeit, University of Tübingen.
Jespersen, O.: 1924, *The Philosophy of Grammar*, Allen & Unwin, London.
Jespersen, O.: 1933, *Essentials of English Grammar*, Allen & Unwin, London.
Jespersen, O.: 1949, *A Modern English Grammar Based on Historical Principles*, 7 vols, Allen & Unwin, London.
Kamp, H.: 1971, 'Formal properties of "now"', *Theoria* **37**, 237–273.
Kamp, H.: 1980, 'A theory of truth and semantic representation', in Groenendijk *et al.* (eds.), *Formal Methods in the Study of Language, Part I.*, Mathematisch Centrum, Amsterdam, pp. 277–321.
Kamp, H.: 1983, 'Discourse representation and temporal reference', manuscript.
Kamp, H. and Rohrer, C.: 1983, 'Tense in Texts', in R. Bäuerle *et al.* (eds.) *Meaning, Use and Interpretation of Language*, de Gruyter, Berlin, pp. 250–269.
Kruisinga, E.: 1932, *A Handbook of Present Day English*, 4 vols., P. Noordhoff, Groningen.
Kuhn, S.: 1979, 'The pragmatics of tense', *Synthese* **40**, 237–263.
Kuhn, S.: 1983, 'Where does tense belong?', manuscript, Georgetown University, Washington D.C.
Lewis, D.: 1979, 'Scorekeeping in a language game', *J. Philosophical Logic* **8**, 339–359.
Massey, G.: 1969, 'Tense logic! Why bother?', *Noûs* **3**, 17–32.
McCawley, J.: 1971, 'Tense and time reference in English', in Fillmore and Langendoen (eds.), *Studies in Linguistic Semantics*, Holt Rinehart and Winston, New York, pp. 96–113.
Montague, R.: 1970, 'English as a formal language', in Visentini, *et al.* (eds.), *Linguaggi nella Societa a nella Tecnica*, Milan, reprinted in Montague [1974].
Montague, R.: 1970a, 'Universal grammar', *Theoria* **36**, 373–398, reprinted in Montague [1974].
Montague, R.: 1973, 'The proper treatment of quantification in ordinary English', in Hintikka *et al.* (eds.), *Approaches to Natural Language*, D. Reidel, Dordrecht, reprinted in Montague [1974].
Montague, R.: 1974, *Formal Philosophy, Selected Papers of Richard Montague*, R. H. Thomason (ed.), Yale University Press, New Haven.
Needham, P.: 1975, *Temporal Perspective: A Logical Analysis of Temporal Reference in English*, *Philosophical Studies* **25**, University of Uppsala.
Nishimura, H.: 1980, 'Interval logics with applications to study of tense and aspect in

English', *Publications of the Research Institute for Mathematical Sciences, Kyoto University* **16**, 417–459.

Parsons, T.: 1980, 'Modifiers and quantifiers in natural language', *Canadian Journal of Philosophy* **6**, 29–60.

Parsons, T.: 1985, 'Underlying events in the logical analysis of English', in E. LePore and B. P. McLaughlin (eds.), *Actions and Events, Perspectives on the Philosophy of Donald Davidson*, Blackwell, Oxford, pp. 235–267.

Poutsma, H.: 1916, *A Grammar of Late Modern English*, 5 vols, P. Noordhoff, Groningen.

Prior, A.: 1967, *Past, Present and Future*, Oxford University, Oxford.

Prior, A.: 1968, *Papers on Time and Tense*, Oxford University Press, Oxford.

Reichenbach, H.: 1947, *Elements of Symbolic Logic*, MacMillan, New York.

Rohrer, C. (ed.): 1980, *Time Tense and Quantifiers: Proceedings of the Stuttgart Conference on the logic of Tense and Quantification*, Max Niemayer Verlag, Tuebingen.

Scott, D.: 1970, 'Advice on modal logic', in K. Lambert (eds.), *Philosophical Problems in Logic*, D. Reidel, Dordrecht, pp. 143–174.

Sweet, H.: 1898, *A New English Grammar, Logical and Historical*, 2 vols, Oxford University Press, Oxford.

Tedeschi, P. and A. Zaenen (eds.): 1981, *Tense and Aspect* (*Syntax and Semantics* **14**), Academic Press, New York.

Tichý, P.: 1980, 'The logic of temporal discourse', *Linguistics and Philosophy* **3**, 343–369.

van Benthem, J.: 1977, 'Tense logic and standard logic', in Åqvist and Guenthner [1977].

Vlach, F.: 1973, *Now and Then: A Formal Study in the Logic of Tense Anaphora*, Ph.D. dissertation, UCLA.

Vlach, F.: 1979, 'The semantics of tense and aspect', ms., University of New South Wales.

Vlach, F.: 1980, 'The semantics of tense and aspect in English', ms., University of New South Wales.

Vlach, F.: 1981, 'The semantics of the progressive', in Tedeschi and Zaenen (eds.), pp. 271–292.

Ward, W.: 1767, *An Account of the Principles of Grammar as Applied to the English Language*.

CHAPTER IV.9

PRESUPPOSITION*

by Scott Soames

1. BASIC QUESTIONS

To presuppose something is to take it for granted in a way that contrasts with asserting it. For example, if one assertively utters

(1a) It was Sam who broke the typewriter.

one presupposes that the typewriter was broken and asserts that Sam was the one who did it. Similarly, if one assertively utters

(2a) John is going to drop out of school again.

one presupposes that he has dropped out of school before and asserts that he will drop out in the future. In each case, the speaker commits himself both to that which he presupposes and to that which he asserts. However, there are important differences between the two.

D. Gabbay and F. Guenthner (eds.), Handbook of Philosophical Logic, Volume IV, 553–616.

One such difference is that commitments that are presupposed are highly heritable, whereas those that are only asserted are not. If assertive utterances of a sentence *S* are used to assert *A* and presuppose *P*, then assertive utterances of more complicated sentences containing *S* often presuppose *P* without carrying any commitment to *A*. This is illustrated by the examples in (1) and (2).

(1b) It wasn't Sam who broke the typewriter.

(1c) Maybe it was Sam who broke the typewriter.

(1d) It is unlikely that it was Sam who broke the typewriter.

(1e) If it was Sam who broke the typewriter, then he will have to fix it.

(P) Someone broke the typewriter.

(A) Sam broke the typewriter.

(2b) John isn't going to drop out of school again.

(2c) Maybe John is going to drop out of school again.

(2d) It is unclear whether John is going to drop out of school again.

(2e) Either John will pass the course or he is going to drop out of school again.

(P) John has dropped out of school before.

(A) John will drop out of school in the future.

Heritability is such a striking feature of presuppositions that they are often identified as those commitments that are inherited in the kinds of linguistic environments just mentioned. However, presuppositions are not inherited in all environments. For example, utterances of the sentences in (3) do not presuppose *P* even though they contain constituents that do.

(3a) If the typewriter was broken, then it was Sam who broke it.

(3b) The typewriter was broken and it was Sam who broke it.

(3c) Either the typewriter wasn't broken, or it was Sam who broke it.

(P) The typewriter was broken.

The contrast between these examples and those given above raises three basic questions which, in broadest terms, define the descriptive task for linguistic theories of presupposition.

DESCRIPTIVE QUESTIONS

(Q1) What presuppositions do various constructions give rise to?

(Q2) Which constructions allow utterances to inherit the presuppositions of their constituents and which do not?

(Q3) What do utterances of arbitrary sentences presuppose?

Although these questions are important, they are not the only ones that theories of presupposition are responsible for. In addition to the descriptive task of identifying the presuppositions of various utterances, a theory of presupposition should specify the kind of phenomenon presupposition is, and how it fits into general theories of the semantic encoding of information by sentences and the pragmatic exchange of information in communicative situations. In short, an adequate theory should answer not only descriptive questions about the scope of presupposition, but also foundational questions about its nature.

FOUNDATIONAL QUESTIONS

(Q4) What is presupposition – what does it mean to say that x presupposes y?

(Q5) Why are there linguistically expressed presuppositions at all – what functions do presuppositions have in the representation and communication of information?

(Q6) How are presuppositions of utterances affected by the semantic rules that determine the information encoded by a sentence relative to a context, and the pragmatic rules that specify the manner in which utterances increment sets of assumptions common among conversational participants?

Historically, three main approaches to presupposition have been developed corresponding to three different answers to (Q4). The first approach consists of theories of logical presupposition, deriving ultimately from the work of Gottlob Frege.[1] According to these theories, presupposition is, in its primary sense, a relation between propositions. A proposition P is said to logically presuppose a proposition Q iff the truth of Q is a necessary condition for P to be either true or false.

Logical Presupposition: A proposition P logically presupposes a proposition Q iff for all possible circumstances w, if P is true or false in w, then Q is true in w.

Sentences are said to bear logical presuppositions in a derivative sense: A sentence S logically presupposes a proposition Q (relative to a context C of utterance) iff S expresses a proposition P (in C) that logically presupposes Q.

The second approach derives from the work of Peter Strawson and consists of theories of what might be called "expressive presupposition".[2] According to these theories, presupposition is a relation between a sentence, or a use of a sentence, and a proposition. A sentence (or use of a sentence) S can be said to expressively presuppose a proposition P iff the truth of P is a necessary condition for S (or a use of S) to express a proposition. On this view, an assertive utterance of a sentence S in a context C fails to semantically express a proposition if one or more of the relevant presuppositions is false.

The third approach consists of theories of pragmatic presupposition in roughly the sense articulated by Robert Stalnaker.[3] According to these theories, presuppositions are requirements that sentences, or utterances of sentences, place on sets of common background assumptions built up among conversational participants. Typically, the requirement is that this set of assumptions contain a specific proposition, or some proposition from a limited range of alternatives. Presuppositions in this sense are essentially things taken for granted at a given point in a conversation. The sources of these pragmatic presuppositions vary from case to case, and theory to theory. For example, logical presuppositions, expressive presuppositions, conventional implicatures, conversational implicatures, and general pragmatic strategies of context incrementation have all been held to play important roles in determining the pragmatic presuppositions of utterances. On this view, theories of presupposition are neither exclusively semantic nor exclusively pragmatic, but rather require the integration of both kinds of information.

The leading ideas behind these three approaches can be made clearer by considering some paradigmatic examples in historical context.

2. THREE APPROACHES TO PRESUPPOSITION

2.1. *Fregean Examples of Logical Presupposition*

The most widely discussed (putative) examples of logical presupposition are so-called referential presuppositions, corresponding to uses of singular terms. The classical Fregean explanation of these examples relies on his bipartite semantics of sense and reference. For example, consider the positive version of (4).

(4) The queen of England is (isn't) popular.

(P) England has a (unique) queen.

According to Frege, the proposition expressed by this sentence can be broken down into two parts: one part consisting of the sense (or meaning) of the subject expression, and the other part consisting of the sense (or meaning) of the predicate. Each of these is a "mode of presentation" of a referent. Senses of predicate expressions present functions from objects to truth values (Truth and Falsity). Senses of singular terms present objects. The truth value of the proposition expressed by a simple subject-predicate sentence (as well as that of the sentence itself) is defined to be the value of the function referred to by the predicate at the argument referred to by the subject.

An important aspect of Frege's distinction between sense and reference is his recognition that some singular terms have senses that fail to present referents – for example, 'the Democrat elected President of the U.S. in 1980'.[4] It follows from Frege's semantics that sentences containing such terms (in environments in which they occur with their customary sense and reference) must lack truth values. One such sentence is the positive version of (5).

(5) The Democrat elected President of the U.S. in 1980 is (isn't) popular.

(P) A (unique) Democrat was elected President of the U.S. in 1980.

By definition, the truth value of this example is the value of the function referred to by the predicate 'is popular' at the argument referred to by the subject. But since the subject fails to refer, there is no such argument and hence no such truth value. Therefore, the example is truth valueless.

Negations are treated similarly. The propositions expressed by the negative versions of (4) and (5) are taken to consist of the propositions expressed by their positive counterparts together with with sense of a sentential negation operator. This sense presents a function *f* as referent from Truth and Falsity to Falsity and Truth. The truth value of each of these negative propositions is defined to be the value of the negation function f at the argument consisting of the truth value of the corresponding positive proposition. Where this positive proposition is truth valueless, there is no such argument and, hence, no truth value for the negation. Thus, corresponding positive and negative examples are either jointly truth valued or jointly truth valueless.

In both cases (4–5), the truth of P is a necessary condition for the positive and negative propositions to have a truth value. As a result, these propositions (and the sentences that express them) logically presuppose P. As such, they illustrate the more general point that negations share the logical presuppositions of their positive counterparts.

It is worth distinguishing those aspects of this analysis that are peculiar to Frege from those that are central to theories of logical presupposition in general. For Frege, claims about presuppositions are consequences of his compositional theory of sense and reference. His decision to take predicates to designate (total) functions from objects to truth values has the consequence that truth valuelessness arises from reference failure on the part of singular terms. His decision to take truth functional operators to denote functions from the truth values of their operands to the truth values of larger, compound sentences has the consequence that these sentences will be truth valued only if each of their truth functional constituents is. Neither of these decisions is constitutive of logical presupposition in general.

What is constitutive is an analysis of (4–5) in terms of the following pair of assumptions:

(i) The negative versions of (4–5), and the propositions they express, are (logically) negations of their positive counterparts.[5]

(ii) In each case, P is entailed (necessitated) by both the positive and negative propositions (sentences).[6]

It follows from these assumptions that the positive and negative examples in each case logically presuppose P; and, hence, that these examples are neither true nor false when P is untrue.

It so happens that the only logical presuppositions generated by Frege's explicit semantics are referential presuppositions (which express necessary conditions for the propositional constituents corresponding to singular terms to present referents). However, there is nothing essential in this either to theories of logical presupposition in general, or to Frege's basic semantic framework. Suppose one dropped the Fregean requirement that the functions designated by predicates be total. For example, the function designated by 'is forgetful' might be defined only on animate beings; and the function designated by the factive verb 'realize' might be defined only over pairs consisting of individuals and true propositions. On this analysis, the deviance of the examples in (6) could be traced to the falsity of the logical presupposition arising from the sortal restriction on the predicate; and the inferences from (7a) and (7b) to (7P) could be treated on a par with those in (4–5).

(6a) The speed of light is (isn't) forgetful.

(6b) The sum of 2 and 3 is (isn't) forgetful.

(7a) John realized that time was running out.

(7b) John didn't realize that time was running out.

(P) Time was running out.

A similar analysis could, in principle, be applied to the examples in (8).[7]

(8a) All graduate students in the class wrote term papers.

(8b) Not all graduate students in the class wrote term papers.

(P) There were graduate students in the class.

Since presupposition was not one of Frege's main concerns, he did not canvass natural language to determine the different kinds there might be. In particular, he did not consider nonreferential examples like those just illustrated. However, the ease with which his system can be extended to provide a unified account of referential and non-referential cases has made it an important model for later theories of logical presupposition.

Unfortunately, there is one respect in which Frege's theory is clearly mistaken. We have seen that Frege takes *n*-place truth functional operators to designate *n*-place truth functions; and that he defines the truth value of a truth functional compound to be the value of the relevant *n*-place truth function at the *n*-tuple of truth values of its truth functional constituents. As a result, the argument used to show that a negation is truth valueless iff its positive counterpart is truth valueless can be generalized to yield the conclusion that a truth functional compound is truth valueless iff one of its constituents is.[8] But this conclusion is incorrect, as is shown by the fact that (9a) is true, and by the fact that the examples in (9) do not presuppose (9P).[9]

(9a) Either there is no king of France or the king of France is in hiding.

(9b) If there is a king of France, then the king of France is one of the few remaining European monarchs.

(9c) There is a king of France and the king of France is wise.

(P) There is a (unique) king of France.

A more reasonable treatment of truth functional connectives that avoids the difficulties posed by (9) is given in (10).

(10a) *A* or *B*

A \ *B*	*T*	*F*	*
T	*T*	*T*	*T*
F	*T*	*F*	*
*	*T*	*	*

(10b) If *A*, then *B*

A \ *B*	*T*	*F*	*
T	*T*	*F*	*
F	*T*	*T*	*T*
*	*	*	*

(10c) *A* and *B*

A \ *B*	*T*	*F*	*
T	*T*	*F*	*
F	*F*	*F*	*F*
*	*	*	*

(10d) Not *A*

A	Not *A*
T	*F*
F	*T*
*	*

The consequences of this treatment of the connectives will be explored further in Section 3.3.2.

2.2. *Strawsonian examples of expressive presupposition*

The examples in (11) illustrate a different kind of presupposition.

(11a) He is wealthy.

(11b) This is a fine red one.

(11c) That little bug is harmless.

In each case, a singular term is used to refer to an entity which the rest of the sentence says something about. This entity is intimately involved in what is said by the utterance in a way that contrasts with examples like (12).

(12) The president of General Motors (whoever he may be) is wealthy.

For example, imagine a situation in which the pronoun in (11a) is used demonstratively to refer to a man *m* who, in fact, is the president of General Motors. Let *P* be the proposition expressed by such a use of (11a) and *Q* be the proposition expressed by a use of (12). Clearly, *P* and *Q* are different propositions. *P* is true in a possible circumstance of evaluation *w* iff *m* is wealthy in *w*. *Q* is true in *w* iff whoever happens to be the president of General Motors in *w* is wealthy in *w*. In effect, that which a use of 'he' contributes to the truth conditions of what is said by an utterance is its referent in the context: whereas that which a use of 'the president of General Motors' contributes is its descriptive sense. Thus, an utterance of (13a) is true just in case the (present) referent *m* of 'he' was poor 20 years ago; whereas (13b) is true just in case 20 years ago General Motors had a pauper for a president.

(13a) Twenty years ago he was poor.

(13b) Twenty years ago the president of General Motors (whoever he may have been) was poor.

The point seems to hold even in propositional attitude constructions. For example, the assertions made by utterances of the sentences in (14) seem to be the same.

(14a) Mary said that he was poor. (Uttered pointing at *m*.)

(14b) Mary said that I was poor. (Uttered by *m*).

(14c) Mary said that you were poor. (Uttered to *m*).

This suggests that the contribution of a demonstrative or indexical to the proposition expressed by an utterance of a sentence containing it is simply its referent in the context. Following David Kaplan, we may refer to this view as the thesis that demonstratives are directly referential.

What happens when a use of a directly referential term fails to semantically determine a referent? In such a case, it is natural to suppose that the sentence fails to semantically express a proposition relative to the context.

With this in mind, one can define a notion of expressive presupposition as follows:

Expressive Presupposition: A sentence *S* expressively presupposes a proposition *P* relative to a context *C* iff the truth of *P* is necessary for *S* to semantically express a proposition in *C*.

(11a–c) can then be characterized as expressively presupposing (15a–c).

(15a) There is a contextually salient male under discussion.

(15b) 'This' refers to something relative to the context.

(15c) There is a contextually salient little bug under discussion.

Although both logical and expressive presupposition have been motivated using examples in which the truth of a proposition is necessary for a (use of a) singular term to secure a referent, the two kinds of presupposition are conceptually quite different. If *S* logically presupposes *P* relative to a context *C*, then the proposition expressed by *S* in *C* must entail (necessitate) *P*. This is not so when *S* expressively presupposes *P* relative to *C* (even when *S* succeeds in expressing a proposition).

One can think of this difference as corresponding to two different stages in the semantic evaluation of a sentence. The first stage consists in associating the sentence with the proposition it expresses in the context. The semantic mechanisms responsible for this are what give rise to expressive presuppositions. The second stage consists in evaluating the truth or falsity of that proposition with respect to different (possible) circumstances of evaluation. The semantic mechanisms for determining the extensions (referents) presented by propositions and their various constituents are what give rise to logical presuppositions.

It is just this two stage conception of semantics that is needed to distinguish expressive and logical presuppositions involving directly referential and non-directly-referential singular terms.[10] However, there is a historical irony in this. The first presentation and discussion of expressive presuppositions is given in Peter Strawson's influential paper 'On Referring', long before the systematic development of two stage semantic theories by David Kaplan, and others.[11] As a result, Strawson's important insights were obscured and he was unable to successfully distinguish his new notion of expressive presupposition from the Fregean notion of logical presupposition. Since Strawson's work has been highly influential in bringing presupposition to the attention of semantic theorists, it may be worthwhile to say a word about this.

The central theses of "On Referring" are as follows:

Thesis 1: Meaning is a property of expressions; referring, being true or false, and saying something are properties of uses of expressions in contexts.

Thesis 2: A sentence is meaningful iff it could be used to say something true or false.

Thesis 3: To give (or know) the meaning of a sentence is to give (or know) a rule for determining the contexts in which it is used to say something true and the contexts in which it is used to say something false.

Thesis 4: The semantic function of a singular term (demonstrative, pronoun, name, definite description) in its primary referring use[12] is to refer to an entity which the rest of the sentence is used to say something about. The meaning of such an expression is a rule for determining its referents in different contexts.

Thesis 5: If a singular term *b* in a sentence ⌜*Fb*⌝ is used referringly in a context *C*, then this use of ⌜*Fb*⌝ in *C* says something true (false) in *C* iff in *C*, the referent of *b* has (doesn't have) the property *F* is used to express. If the use of *b* fails to refer to anything, then the use of ⌜*Fb*⌝ in *C* doesn't say anything true or false.

Thesis 6 (Definition): If the truth of *P* is a necessary condition for a use of *S* in *C* to say something true or false, then *S* presupposes *P* relative to *C*.

Thesis 7: Uses of ⌜*G*[the *F*]⌝, ⌜All *F*'s are *G*'s⌝, ⌜Some *F*'s are *G*'s⌝, ⌜No *F*'s are *G*'s⌝, and ⌜Some *F*'s are not *G*'s⌝, presuppose that which is expressed by ⌜There is at least one *F*⌝.

The key thesis is Thesis 3. However, there is a problem with it. As it stands, it does not rule out, and may even be taken to suggest, that the meaning of a sentence can be represented as a function from contexts of utterance to truth values. This is at variance with the two stage conception of semantics in which the meaning of a sentence is represented by a function from contexts to propositions, where the latter determine functions from (possible) circumstances of evaluation to truth values. Since the circumstance of the context is one of these possible circumstances, the meaning of a sentence determines the one stage mapping from contexts to truth values suggested in Strawson's Thesis 3. However, the latter does not determine the former. As a result, there are important semantic distinctions that Thesis 3 does not explicitly accommodate.

This is illustrated by (16a) and (16b).

(16a) I exist.

(16b) I am here now.

Any context in which one of these would express a truth (falsehood) is a context in which the other would as well. Nevertheless, they do not have the same meaning. If I were to assertively utter both in the present context, my utterance of (16a) would express the proposition that Scott Soames exists; whereas my utterance of (16b) would express the proposition that Scott Soames is in Santa Cruz on April 6, 1984. Since these are different propositions, the semantic contents of (16a) and (16b) must be distinguished. To make this explicit, Thesis 3 should be replaced with Thesis 3′.[13]

Thesis 3′: The meaning of a sentence is a rule for determining the propositions it expresses in different contexts. Each such proposition determines a rule for assigning truth values to (possible) circumstances of evaluation.

This requires corresponding changes in Theses 4 and 5. To make these changes, one must decide what singular terms contribute to the propositions expressed by uses of sentences containing them. Strawson's insistence that the semantic function of a singular term is to refer to an object, and his tendency to treat referring uses of demonstratives as prime examples of this function, suggest a reformulation in which all referring uses of singular terms are directly referential.[14]

Thesis 4′: The propositional constituent corresponding to a (referring) use of a singular term *b* in a context *C* is the referent of *b* in *C*. The meaning of a singular term is a rule for determining the propositional constituents corresponding to uses of the term in different contexts.

Thesis 5′: If a singular term *b* in a sentence ⌜*Fb*⌝ is used to refer to an object *o* in a context *C*, and if *F* is used in *C* to express the property *P*, then

⌜*Fb*⌝ expresses a proposition in *C* which is true (false) in a possible circumstance *w* iff *o* has (doesn't have) *P* in *w*. If *b* fails to refer to anything in *C*, then there is no propositional constituent corresponding to *b* in *C*, and ⌜*Fb*⌝ fails to semantically express a proposition in *C*.

The theory of presupposition that emerges from this reconstruction of Strawson's theses is a theory of expressive presupposition, as defined above. As we have seen, the combination of Theses 3′–5′ provides a plausible account of examples like those in (11) in which a pronoun, demonstrative, or demonstrative phrase is used referringly. However, it clearly produces incorrect results when extended to the range of cases mentioned in Thesis 7.

This extension also conflicts with Strawson's expressed intentions. In Chapter 6 of *Introduction to Logical Theory*, Strawson defines presupposition as follows:[15]

(17) A statement (proposition) *S* presupposes a statement (proposition) *S*′ iff the truth of *S*′ is a necessary condition for *S* to be true or false.[16]

It was apparently this definition that he had in mind when discussing presuppositions of examples of the kind mentioned in Thesis 7 (both in 'On Referring' and in *Introduction to Logical Theory*). Since (17) defines logical presupposition, Strawson's adoption of it belies any clear commitment to expressive presupposition, or any systematic analysis of the constructions mentioned in Thesis 7 along directly referential lines.

This points up a second possible reconstruction of Strawson's position. On this construal, his account of presupposition is basically the same as Frege's, without the compositional semantics, but with an explicit stipulation that propositions involving restricted quantifiers are bearers of presuppositions. This theory is potentially broad in scope and has been historically influential. However, its leading ideas are not original with Strawson.

As a historical point, it would be a mistake to attribute to Strawson either an account of presupposition that is systematically Fregean (logical) or an account that is systematically expressive. His major discussions include elements of both, the conflict being masked by his failure to articulate the crucial account of meaning given in Thesis 3′. Once this deficiency is corrected, Strawson's main original contribution to the study of presupposition lies in the reconstructed account of expressive presupposition suggested by his work. This account is attractive for examples involving various kinds of indexical elements. Some have suggested that it may be possible to extend it from examples involving demonstrative phrases like 'this little bug' to corresponding examples involving definite descriptions like 'the little bug'.[17] What it cannot be is a comprehensive theory of presupposition in general.

2.3. *Pragmatic presupposition*

Robert Stalnaker, and others, have argued that in order to arrive at a comprehensive theory, it is necessary to adopt a pragmatic account of what presupposition is. The account is based on the observation that sentences are used in communication to contribute to an already existing conversational record, which contains a set of common background assumptions built up among conversational participants. Because of this it is natural for speakers to develop conventional means of indicating what assumptions they are making about the common background to which their utterances contribute. In particular, it is understandable that certain words and constructions should come to be used for this purpose.

For example, the following (a)-constructions seem to be designed for use in conversations in which the information expressed by the (b)-constructions is already assumed.

(18a) It was *NP* that *VP*ed.

(18b) Something (someone) *VP*ed.

(19a) Even *NP VP*ed.

(19b) Others under consideration, besides *NP*, *VP*ed. Of those under consideration, *NP* was among the least likely to *VP*.

A sentence that indicates that such assumptions are being made can be thought of as putting requirements on the conversational record at the time of utterance – requirements that must be satisfied if the speaker's communicative intentions are to be fulfilled. Presuppositions, on this view, are just such requirements.

Suppose now that a speaker utters a sentence *S* which requires that the conversational record satisfy a certain condition – say that it contain a specific proposition *P* as part of the common background. Suppose further that *P* is not already part of the background, but that the conversational participants are ready to accept *P* as uncontroversial, at least for present conversational purposes. What sort of response would be reasonable on the part of hearers in such a case?

The legalistic response would be to object to the speaker's remark on the grounds that *P*, which was required by the remark, had not already been established prior to the utterance. The speaker could then ask whether his hearers were willing to accept *P*, and be told that they were. After adding *P* to the context, the speaker could repeat his original remark and continue along as before.

But there is really no point in this. Since the hearers are ready to accept *P* anyway, they might as well add it to the background and let the speaker go on without objection. In other words, the most efficient and cooperative response on the part of the hearers is to accommodate the speaker by updating the conversational record so that it meets the requirements of the speaker's utterance.

The reasoning leading to this strategy of accommodation is something that conversational participants can be expected to be familiar with, or to work out for themselves. Knowing this, a speaker can exploit the strategy by uttering sentences whose presuppositional requirements he knows are not already satisfied by the existing conversational record. So long as he takes the content of these requirements to be both recognizable and unlikely to provoke objections, he can rely on his hearers to accommodate him by incrementing the common background in the appropriate way. A speaker can use presuppositions in this way to introduce new information, as well as, in some cases, to subtly insinuate a point of view regarding what can be taken to be uncontroversial, and hence beyond further discussion.[18]

On this picture, the presuppositions of a sentence are conditions it imposes on conversational records. Often these conditions take the form of requirements that the common background assumptions present in the record contain a specific proposition. However, other types of presuppositional requirements are also possible. For example, a sentence may require that the common background contain at least one proposition from a specified range of propositions, it may require that it contain no propositions of a specified sort; or it may require that the topic of conversation at the time of the utterance be one thing rather than another. Once the presuppositional requirements of sentences have been determined, incrementation of the record occurs in accordance with the strategy of accommodation.

An important feature of this approach is its eclecticism regarding the factors that give rise to different presuppositional requirements. Proponents of pragmatic presupposition have suggested that these requirements might be derived from a variety of sources – including logical presupposition, expressive presupposition, conventional implicature, and non-conventional pragmatic facts.

For example, in discussing logical presupposition as a source of pragmatic presuppositional requirements, Robert Stalnaker says the following:

> The relation between the semantic [logical] notion of presupposition and the pragmatic notion of presupposition requirement is not, of course, just accidental. Among the reasons that a pragmatic presupposition might be required by the use of a sentence, by far the most obvious and compelling reason would be that the semantical rules for the sentence failed to determine a truth value for the sentence in possible worlds in which the required presupposition is false. Since the whole point of expressing a proposition is to divide the relevant set of alternative possible situations – the presupposition set – into two parts, to distinguish those in which the proposition is true from those in which the proposition is false, it would obviously be inappropriate to use a sentence which failed to do this. Thus, that a proposition is presupposed by a sentence in the technical semantic sense provides a reason for requiring that it be presupposed in the pragmatic sense whenever the sentence is used. This explains where the semantic notion gets its name, and why linguists and philosophers have been tempted to identify presupposition in general with this semantic relation.[19]

According to this picture, the exchange of information in a conversation increases the number of propositions in the common background against which the conversation takes place. Taken together, these propositions determine a set of possible worlds which, at any given moment, represent the alternatives compatible with everything that has been said or assumed in the conversation up to that point. The function of an assertive utterance of a sentence *S* is to further constrain these alternatives by eliminating the worlds in which the proposition expressed by *S* is false, while retaining those in which it is true. Note that if *S* logically presupposes a proposition *Q* which is not entailed by the propositions in the common background, then the set of conversationally alternative worlds will contain some members *w* in which the proposition expressed by *S* cannot be correctly characterized as either true or false. Stalnaker's contention is that in such a case the assertive utterance will fail in its primary purpose of determining a new set of alternative possibilities, since it will fail (barring accommodation) to determine what should be done with worlds in the old alternative set in which *Q* is false. Thus, Stalnaker maintains, logical presuppositions provide one kind of principled explanation of the pragmatic presuppositional requirements of sentences.

A similar story might be told regarding expressive presuppositions. Suppose that the truth of *P* is a necessary condition for a use of *S* to semantically express a proposition in a context *C*. Suppose further that speakers standardly intend to assert, and to be recognized as asserting, the proposition semantically expressed by the sentence assertively uttered, relative to the context of utterance. Then, a speaker will assertively utter *S* in *C* only when the truth of its expressive presupposition *P* can be taken for granted – only when *P* is either already part of the conversational record, or uncontroversial enough to be added by accommodation. Thus, expressive presupposition may be another source of pragmatic presuppositional requirements.[20]

Finally, proponents of pragmatic presupposition maintain that some presuppositional requirements arise from more straightforwardly pragmatic sources. For example, Lauri Karttunen and Stanley Peters [1979] argue that the pragmatic presuppositions of (18a) and (19a) are Gricean conventional implicatures that are carried by these sentences independently of the propositions they express. Particularly interesting from a pragmatic point of view are constructions which, though they don't give rise to presuppositions themselves, allow compound sentences to inherit the presuppositions of their constituents. Two examples of such constructions are conjunctions and indicative conditionals.[21] The presuppositional requirements of these constructions have been characterized as follows: A set *R* of common background assumptions satisfies the presuppositional requirements of ⌜*A* and *B*⌝ and ⌜If *A*, then *B*⌝ iff *R* satisfies the requirements of *A*, and the result *R′* of adding the proposition expressed by *A* to *R* satisfies the presuppositional requirements of *B*. Roughly speaking, ⌜*A* and *B*⌝ and ⌜If *A*, then *B*⌝ inherit all the presuppositional requirements of *A*, plus those requirements of *B* that are not automatically satisfied by the addition of *A* to the conversational record.

Different theorists have proposed different explanations of these requirements. Among the simplest is one suggested by Robert Stalnaker. Stalnaker's proposed explanation is based on two principles. First, whenever something is asserted or supposed, it is immediately added to the conversational record (in the case of supposition, the addition is often temporary). Second, someone who assertively utters a conjunction asserts the proposition expressed by its first conjunct prior to uttering the second conjunct; someone who assertively utters a conditional posits the proposition expressed by its antecedent as a supposition prior to uttering the consequent. Since the initial conversational record is augmented with *A* prior to the utterance of *B*, this guarantees that the only substantive presuppositional requirements arising from *B* will be those that remain unsatisfied after the addition of the proposition expressed by *A*. In this way, Stalnaker attempts to provide a non-semantic explanation of the presuppositional requirements of these compound sentences.

Whether or not this explanation proves to be correct, there is something both right and important about the pragmatic approach. Presupposition is, first and foremost, a matter of what is assumed or taken for granted. As such, linguistically expressed presuppositions should be described in terms of the beliefs and assumptions of language users. Although this makes presupposition a pragmatic notion, it does not rule out semantic explanations of pragmatic facts. What it does do is avoid conflating data about the commitments carried by various utterances with highly theoretical accounts of those commitments in terms of one or another kind of semantic presupposition.

This theoretical neutrality makes the pragmatic approach extremely useful in investigating the scope of presupposition in natural language. In recent years, linguists and philosophers have used this approach in pursuing the following three part strategy: First, the scope of pragmatic presupposition is described by specifying both the natural language constructions that give rise to presuppositions and those that allow larger compounds to inherit the presuppositions of their constituents. Second, an inquiry is made into the sources of pragmatic presuppositions in particular cases. Third, an attempt is made to determine whether semantic presuppositions are needed to explain pragmatic ones; or whether purely pragmatic mechanisms suffice to account for the data without appeal to special semantic assumptions.

3. RECENT DESCRIPTIVE WORK

3.1. *The scope of pragmatic presupposition*

As we have seen, traditional discussions of presupposition by philosophers have tended to focus on a small range of presupposition creating and inheriting constructions – the prime examples of the former being singular terms and restricted quantifiers, the prime example of the latter being negation. Recent descriptive work has greatly expanded both classes of cases.

Among the constructions that give rise to pragmatic presuppositions one finds a large variety of different types in addition to those discussed by Frege and Strawson; for example:

(20a)	Bill regrets that he lied to Mary.	(Factive)
(P)	Bill lied to Mary.	
(21a)	Ivan has stopped beating his wife.	(Aspectual)
(P)	Ivan has beaten his wife.	
(22a)	Harry managed to find the book.	(Implicative Verb)
(P)	Finding the book required some effort.	
(23a)	Andy met with the PLO again today.	(Iterative)
(P)	Andy has met with the PLO before.	
(24a)	It was in August that we left Connecticut.	(Cleft)
(P)	We left Connecticut sometime.	
(25a)	What John broke was his typewriter.	(Pseudo cleft)
(P)	John broke something.	
(26a)	Pat is leaving, too.	(Focus on 'Pat')
(P)	Someone other than Pat is leaving.	
(27a)	Even Sam passed the test.	
(P)	Others, besides Sam, passed the test; and of those under consideration Sam was among the least likely to do so.	

One also finds a variety of constructions that typically inherit the pragmatic presuppositions of their constituent clauses. These include, in addition to negation, epistemic modals, indicative conditionals, disjunctions, conjunctions, and sentences containing certain complementizable verbs.

(28a) It wasn't Jane who solved the problem.

(28b) Maybe it was Jane who solved the problem.

(28c) Either it was Jane who solved the problem, or they awarded the fellowship to the wrong person.

(28d) If the problem was as important as they indicated, then it was probably Jane who solved it.

(28e) That it was Jane who solved the problem isn't very likely.

One way of thinking about these sentences is to see the pragmatic presuppositions of the constituent clauses as projected onto, and hence inherited by, the larger sentences. In this respect, the sentences in (28) contrast with their counterparts in (29).

(29a) Jane didn't solve the problem.

(29b) Maybe Jane solved the problem.

(29c) Either Jane solved the problem, or they awarded the fellowship to the wrong person.

(29d) If the problem was as important as they indicated, then Jane probably solved it.

(29e) That Jane solved the problem isn't very likely.

The difference between these two sets of sentences is that those in (28) have the cleft sentence (30) as a sentential constituent, whereas those in (29) have its noncleft counterpart, (31).

(30) It was Jane who solved the problem.

(31) Jane solved the problem.

Although both (30) and (31) entail (32), only (30) presupposes it.

(32) Someone solved the problem.

Utterances of (28a–e) inherit this presupposition. Utterances of (29a–e) have nothing to inherit.

This sort of projection is all but ubiquitous. However, it is not universal, as was illustrated by (3a–c) in Section 1. The chief descriptive problem occupying presupposition theorists in the past several years – the projection problem – has been to determine which utterances inherit the presuppositions of their constituents, which do not, and why.

It is useful in approaching this problem to have a characterization of the pragmatic notion of an utterance presupposition. This is given below, using the notion of the conversational background to represent the background information, common among speakers and hearers, against which utterances are evaluated.

The Conversational Background: The conversational background at a time *t* is the set of propositions *P* such that at *t* the conversational participants believe or assume *P*; and recognize this about each other.

Utterance Presupposition: An utterance *U* presupposes *P* iff one can reasonably infer from *U* that the speaker *S* accepts *P* and regards it as uncontroversial, either because

(a) *S* thinks that *P* is already part of the conversational background at the time of *U*; or because
(b) *S* thinks that the conversational participants are prepared to add *P*, without objection, to the background.

The projection problem is the problem of determining the presuppositions of utterances of compound sentences in terms of presuppositions associated with their clausal constituents.

3.2. *The projection problem*

Standardly, presuppositions associated with a sentential constituent become presuppositions of an utterance of a compound sentence of which the constituent is a part. There are, however, three factors that can prevent this from happening – cancellation of the presupposition by propositions in the conversational background, cancellation by Gricean conversational implicatures, and suspension of the presupposition by "local context incrementation" utilizing other clauses in the compound sentence.

The first of these factors is illustrated by a discourse in which (33b) is uttered after (33a) (where the utterances may be by the same or different speakers).

(33a) There is no king of France.

(33b) Therefore the king of France isn't in hiding.

(33b) is the negation of a sentence that presupposes that France has a king. Although negations typically share the pragmatic presuppositions of their positive counterparts, this utterance of (33b) does not. The reason it doesn't is that the putative presupposition conflicts with a proposition already placed in the conversational background. When this happens, utterances of negative sentences may be felicitous, but they do not inherit the presuppositions of their positive counterparts.[22]

The second way in which presupposition inheritance can be blocked is illustrated by the contrast between (34) and (35).

(34) If I regret later that I haven't told the truth, I will confess it to everyone.

(A) I will regret later that I haven't told the truth.

(P) I haven't told the truth.

(35) If I realize later that I haven't told the truth, I will confess it to everyone.

(A) I will realize later that I haven't told the truth.

(P) I haven't told the truth.

In each case, the antecedent *A* presupposes *P*. Since utterances of indicative conditionals normally inherit the pragmatic presuppositions of their antecedents, one would expect utterances of these sentences to presuppose *P* as well. However, this is true only of (34). In the case of (35) the presupposition is blocked by the conversational implicature that the speaker doesn't know the antecedent of his statement to be true. (If he did, he could have made a stronger statement.) Since he doesn't know that he will later realize that he hasn't told the truth, one may conclude (in normal circumstances) that he doesn't now know that he hasn't told the truth. In this way, the normal presumption that the speaker is taking the presupposition of the antecedent for granted is defeated by a conversational implicature that indicates that he cannot be doing so.[23]

The third way in which presupposition inheritance can be blocked is illustrated by (36).

(36) If all the Smith brothers have children, then John Smith's children will probably inherit the family fortune.

(B) John Smith's children will probably inherit the family fortune.

(P) John Smith has children.

Here, *P* is not presupposed by an utterance of (36) even though it is presupposed by its consequent *B*. There is also no conversational implicature to the effect that the speaker is not assuming *P*. Rather the utterance is noncommital regarding his attitude toward *P* – it neither indicates that he takes it to be true, nor indicates that he does not take it to be true.

This can be explained using assumptions (i) and (ii).

(i) The pragmatic presuppositional requirement of *B* arising from its grammatical subject is that a certain contextually defined set of propositions entails *P*.

(ii) When *B* occurs as the consequent of an indicative conditional the contextually relevant set is the one that results from adding the proposition expressed by the antecedent to the conversational background prior to the utterance.

These assumptions guarantee that the relevant presuppositional requirement of *B* will be satisfied by an utterance of (36) no matter whether *P* is in the conversational background prior to the utterance or not. (We assume that it is part of the background that John Smith is one of the Smith brothers.) Hence, the utterance provides no indication whether the speaker regards *P* to be true, or whether he is unsure of its truth value.

The three factors capable of preventing presupposition inheritance have provided the basis for three different theories of presupposition projection. The first of these, presented in Karttunen [1974], is based entirely on the kind of presupposition suspension illustrated by (36). The theory takes the form of an inductive definition of a two place relation of admittance between sets of propositions (called "contexts") and sentences. It is assumed that a context *C* will admit (20a)–(27a) only if *C* entails (20P)–(27P), respectively. Clauses for compound sentences include the following:

(37) A context *C* admits a negation of a sentence *A* iff
C admits *A*.

(38) A context *C* admits ⌜Maybe *A*⌝, ⌜It is likely that *A*⌝,
⌜It is possible that *A*⌝, iff *C* admits *A*.

(39) A context *C* admits ⌜*A* and *B*⌝, ⌜If *A*, then *B*⌝ iff
C admits *A*, and
$C \cup [A]$ admits *B*.[24]

(40) A context *C* admits ⌜Either *A* or *B*⌝ iff
$C \cup [-A]$ admits *B*, and
$C \cup [-B]$ admits *A*, and
for all propositions *P*, if *P* is entailed by every context that admits both *A*, and *B*, then *C* entails *P*.[25]

Pragmatic presuppositional requirements are consequences of the inductive definition plus a general requirement that the conversational background prior to an utterance admit the sentence uttered. In cases in which the background does not admit the sentence, the strategy of accommodation is invoked to allow for utterance presuppositions to introduce new information.

This theory accounts for presupposition suspension both in simple cases like (36) and in more complicated examples like (41).

(41) If Martha buys a blue dress and Susan buys a blue dress too, then Martha will regret buying a dress that is the same color as one bought by Susan.

However, it cannot be accepted as it stands. The basic problem, as shown in Gazdar [1979], Soames [1979], and Soames [1982], involves a conflict between conversational implicatures and presuppositional requirements generated by the admittance conditions. The conflict arises in cases in which the latter make a prediction that something is being taken for granted, which the former deny. One example of such a case is (35). The Karttunen theory predicts that (35) requires the conversational background to entail *P*, and hence that utterances of (35) commit the speaker to *P*. The theory of conversational implicature, on the other hand, tells us that (normal) utterances of (35) conversationally implicate that the speaker is not assuming *P*. In every such case, the conversational implicature is genuine and the presupposition is nonexistent.[26]

The ability of conversational implicatures to prevent presupposition inheritance was noted independently and used to develop an alternative approach to presupposition projection in Gazdar [1979] and Soames [1979]. The basic ideas of the approach are the following:

(i) Presupposition creating constructions like those in (20–27) give rise to "potential presuppositions" (illustrated by 20P–27P).

(ii) Compound sentences inherit all the potential presuppositions of their constituents.

(iii) If *Q* follows from potential presuppositions $P_1, \ldots, P_n$ of *S*, then an utterance *U* of *S* presupposes *Q* unless
   (a) $P_1, \ldots, P_n$ are jointly incompatible with the conversational background; or
   (b) *U* conversationally implicates that the speaker is not taking $P_1, \ldots, P_n$ for granted.

In effect, utterances presuppose all the presuppositions of their constituents except those that are incompatible with the conversational background or cancelled by conversational implicatures.

This approach handles much of the data that motivated the Karttunen theory while accounting for counterexamples like (35).[27] However, there is a range of examples, specified in Soames [1979], [1982], that this approach cannot accommodate. Two such examples are (36) and (42).

(42) Maybe Bill proved the theorem and Mary proved it too.

(P) Someone other than Mary proved the theorem.

In each case, an utterance of the sentence as a whole fails to inherit the constituent presupposition P, even though there is no cancelling conversational implicature. The reason for this seems to be that the presupposition is suspended by the kind of "local context incrementation" proposed by Karttunen.

The upshot of this is that a proper theory of pragmatic presupposition projection must include both mechanisms for suspending presuppositions of the sort suggested by Karttunen and mechanisms for cancelling them of the kind suggested by Gazdar and Soames. This conclusion is drawn in Soames [1982] where two different methods of incorporating these mechanisms into a single theory are explored. According to one method, Karttunen-like devices are used to generate pragmatic presuppositional requirements of sentences that are cancellable by contextual and conversational means. According to the other method, cancellation first eliminates certain potential presuppositions, with the remaining uncancelled potential presuppositions providing the input for a computation of utterance presuppositions along Karttunen-like lines. Although the second method was ultimately selected in that paper, each method has its own advantages. The end result, though not entirely free of descriptive problems, extends the reach of descriptive theories of pragmatic presupposition to a significant range of data.[28]

3.3. *Foundational implications of descriptive work*

3.3.1. *Cancellation and accommodation.*

Although contextual and conversational cancellation seem to be indispensable parts of an adequate descriptive account of pragmatic presupposition, they also appear to threaten the original conception of pragmatic presupposition outlined in

Section 2.3. According to that conception, a theory of pragmatic presupposition can be thought of as a bipartite affair. The first part specifies the presuppositional requirements that various sentences place on conversational backgrounds. These requirements are illustrated by (37–40), and by requirements that (20P–27P) be entailed by the conversational backgrounds for (20a–27a). The second part of the theory specifies the role of presuppositional requirements in determining how conversational backgrounds are incremented on the basis of utterances. The crucial element here is the strategy of accommodating apparent violations. According to this strategy, an assertive utterance of a sentence *S* that pragmatically presupposes *P* will result in the addition of *P* to the conversational background unless the hearers object.

The combination of this foundational conception of pragmatic presupposition, together with the descriptive requirements just mentioned, encapsulates the Karttunen theory discussed in the last section. Since that theory is descriptively inadequate, some change in either foundational structure or descriptive requirements is needed. The introduction of contextual and conversational cancellation into this picture worked out in Soames [1982] can be thought of as a change in the former. The basic idea is to introduce a new kind of accommodation to supplement the Lewis–Stalnaker variety that we have been considering up to now.

The kind of accommodation discussed by Lewis and Stalnaker might be called "de facto accommodation". When a presuppositional requirement is not met because the conversational background does not entail the presupposed proposition, the law does not change (the requirement remains in force); rather, the conversational facts are adjusted to bring the speaker's performance into line with the law (the proposition is added). However, this is not the only kind of accommodation possible. Suppose a speaker utters a sentence that requires the conversational background to satisfy a certain condition – say, to entail the proposition *P*. Suppose also that something about the speaker's utterance makes it clear that the requirement is to be waived in this case. The hearers, recognizing this, will not add *P* to the background, but will go ahead with the process of incrementing it on the basis of the proposition asserted.

This sort of accommodation can be called "*de jure* accommodation". In this case, apparent violations of presuppositional requirements are accommodated not by adjusting the existing conversational facts to fit the requirements, but by adjusting the requirements to fit the facts. In effect, presuppositional requirements become defeasible. Unless there is an indication to the contrary, they remain in force. However, there are means available to cancel them in particular cases.

The addition of this kind of accommodation to the foundational conception of pragmatic presupposition increases not only its descriptive accuracy, but also its intuitive plausibility. It must be remembered that incrementation of the conversational background is a complex process involving the nature of the preceding background, the syntactic form and semantic content of the sentence uttered, the presuppositional requirements of the sentence, plus the conversational implicatures of the utterance. Without *de jure* accommodation, the interaction of general principles involving various aspects of this process could easily lead to communicative conflicts.

For example, suppose all sentences of a certain form require preceding backgrounds to contain some proposition *P*. Suppose further that in certain special cases the content of the sentence, plus Gricean conversational principles, and the conversational background, generate a conversational implicature to the effect that *P* is not being assumed. A pragmatic strategy allowing *de facto*, but not *de jure*, accommodation would give rise to a communicative impass, since the addition of *P* to the background would be both required by the rules governing presupposition and prohibited by a conflicting conversational implicature.

The existence of *de jure* accommodation allows speakers to avoid this kind of difficulty, while adhering to simple and general rules for determining presuppositional requirements. By allowing these requirements to be defeasible, one keeps the task of computing them manageable, with exceptions to one's general rules being clearly recognizable in virtue of other, independently needed, pragmatic principles. Thus, there is good reason why pragmatic presuppositions ought to be governed by *de jure*, as well as *de facto*, accommodation.

3.3.2. *The insufficiency of conventional implicature and logical presupposition as sources of pragmatic requirements.* We now have the outlines of a theory of pragmatic presupposition that combines significant descriptive content with plausible answers to a number of basic foundational questions, including the following elaborations of the original Q5 of Section 1.

(Q5a) Why should there be linguistically expressed pragmatic presuppositions at all?

(Answer) To provide speakers with conventional means of indicating what assumptions they are making about the conversational backgrounds to which their utterances contribute.

(Q5b) Why, given that there are such presuppositions, should conversational participants be ready to follow a policy of *de facto* accommodation when the presuppositional requirements of a sentence are not satisfied by the conversational background prior to the utterance?

(Answer) To allow speakers to use presuppositions to introduce new, but uncontroversial, information; and to avoid pointless objections involving propositions they are ready to accept.

(Q5c) Why should pragmatic presuppositions be defeasible?

(Answer) To avoid communicative conflicts with other pragmatic implicatures; and to allow the rules for computing the presuppositional requirements of sentences to be kept manageably simple.

If one were designing a language for use in communication one would presumably want it to incorporate the main elements of this account. Thus, it is not surprising that natural languages do.

This points up the explanatory attractiveness of the pragmatic theory. There is, however, a serious gap in this explanatory picture. Although we have explained why there ought to be pragmatic presuppositions in general, we have not explained why various sentences carry the particular presuppositional requirements they do.

In some cases, the requirements seem to be simply matters of linguistic meaning. For example, in the case of the word 'even' it is plausible to hold that its meaning consists in the presuppositions, illustrated by (27P), that it introduces. A similar point might be made about the cleft construction, illustrated by (24a) and (30). If this is right, then what the meaning of the cleft construction adds to the meaning of (30) (that is not contained in the meaning of its noncleft counterpart (31)) is a certain presuppositional requirement.

Lauri Karttunen and Stanley Peters have suggested that these presuppositional requirements are Gricean conventional implicatures that are independent of the propositions expressed by sentences bearing them.[29] This suggestion is generalized to cover the constructions in (20–27), as well as the presuppositions arising from definite descriptions, quantified phrases, negations, conditionals, conjunctions, and disjunctions.[30] For the constructions covered in Karttunen's 1974 theory, this amounts to taking the presuppositional requirements corresponding to (37–40) to be conventional implicatures arising from a recursive assignment of "non-truth-conditional" content to sentences that parallels the more familiar assignment of propositional content.[31]

Although the resulting system is elegant, it suffers from both explanatory and descriptive problems. On the explanatory side, pragmatic presuppositional requirements are reduced to arbitrary linguistic conventions associated with lexical items and constructions. This is plausible for examples like 'even' and clefts; but it is implausible for other cases, particularly the connectives. Surely, there is some connection between the truth conditional content of the connectives and the pragmatic presuppositions of sentences containing them.[32] We are not inclined to think that learning their meanings consists of two separate and unrelated tasks; nor do we expect to find natural languages containing connectives that share the truth conditional contents of their English counterparts while differing arbitrarily from them in their contributions to presupposition inheritance. This suggests some explanatory link between pragmatic presuppositions, truth conditional content, and general principles governing communication – a link that is missing from the Karttunen–Peters account.

There are also serious descriptive problems arising from the systematic identification of pragmatic presuppositions with conventional implicatures. As Grice, Karttunen, and Peters all have stressed, such implicatures are aspects of linguistic meaning, and are therefore uncancellable. Thus, the examples of presupposition cancellation cited in Gazdar [1979], Soames [1979], and [1982] refute the Karttunen–Peters account, along the original theory of Karttunen [1974].[33] Nor is the problem resolvable by simply changing the content of the implicatures.[34] Rather, it seems that some pragmatic presuppositions are not conventional implicatures after all.

A corresponding point can be made about logical presupposition. Suppose one wanted systematically to derive the pragmatic presuppositions of sentences from their logical presuppositions (including (20–27), negations, conjunctions, conditionals, disjunctions, and so on). To do this, one would need a nonbivalent semantics incorporating the truth tables in (10), together with the bridge principle (43) (justified by the Stalnaker argument given in Section 2.3).

(43) If *S* logically presupposes *P* relative to a context *C*, then an utterance of *S* in *C* pragmatically requires the conversational background to entail *P*.

According to this analysis, the logical presuppositions of negations, conjunctions, conditionals, and disjunctions are given in (P) – where ⌜Pre: *S*⌝ expresses a proposition whose truth is a necessary and sufficient condition for the proposition expressed by *S* to be true or false.

(Pa)	Pre: Not S	=	Pre: S
(Pb)	Pre: (A and B)	=	(Pre: A & ($A \rightarrow$ Pre: B))
(Pc)	Pre: (If A, then B)	=	(Pre: A & ($A \rightarrow$ Pre: B))
(Pd)	Pre: (A or B)	=	($A \vee$ Pre: B) & ($B \vee$ Pre: A)

Since these presuppositions parallel those predicted in Karttunen [1974] and Karttunen and Peters [1979], the examples of presupposition cancellation that falsify those theories falsify the present analysis as well.[35] It is shown in Soames [1979] that this problem cannot be solved by simply changing the nonbivalent truth tables of the connectives.[36]

The upshot of this is that pragmatic presuppositions cannot be systematically explained as arising from either logical presuppositions or conventional implicatures. We have not shown that no pragmatic presuppositions are logically presupposed, or conventionally implicated; only that, in each case, some are not. In one respect, this conclusion is not surprising. After all, pragmatic presuppositions were introduced, in part, to allow for radically different sources of presuppositional requirements. However, in another respect, the result is disquieting. As indicated earlier, there seem to be some expressions – including truth functional connectives – whose contributions to pragmatic presuppositions are not arbitrary matters of linguistic convention or pragmatic practice, but rather are linked in some way to their propositional content, plus general pragmatic principles governing cooperative communication. Until this link is spelled out precisely, we have no explanation of why various compound sentences bear the presuppositions they do.[37]

This explanatory difficulty is an important unsolved problem in the field that will be taken up again in Section 5. Before doing that, however, it is worthwhile to examine the more basic question of whether logical presuppositions ever provide the explanation of pragmatic ones. Is there any sentence *S* and proposition *Q* such that the reason that *S* pragmatically presupposes *Q* is that it logically presupposes *Q*?

4. PHILOSOPHICAL FOUNDATIONS OF NONBIVALENT ANALYSES

4.1. *Why truth value gaps don't explain presuppositions*

In order to derive pragmatic presuppositions from logical ones, a non-bivalent semantics plus the bridge principle (43) are needed. In order for such a derivation to have significant explanatory force, both the semantics and the bridge principle should be independently motivated. The point of the derivation is not simply to state the pragmatic requirements of various sentences, since (in most cases) we can do that directly. Rather, the point is to explain those requirements in terms of deeper semantic and pragmatic principles that are important for the explanation of other phenomena as well. Thus, if pragmatic presuppositions are to be ascribed to logical sources, we will need both a reason for abandoning bivalence and an account of (43) according to which it reflects something more than an arbitrary and unexplained correlation.

The main attempt to give the latter is the one by Robert Stalnaker discussed in Section 2.3 above. According to him, the link between the logical presuppositions of sentences and their pragmatic presuppositional requirements is provided by the function of assertions in incrementing conversational backgrounds. Stalnaker notes that the propositions in the background determine a set of conversationally alternative possible worlds compatible with everything established or assumed in the conversation at a given point. The function of an assertion is to further constrain these alternatives by eliminating some of the worlds and retaining the rest. The ones to be eliminated are those that are incompatible with the proposition asserted. These, Stalnaker maintains, are those in which the proposition expressed by the sentence uttered is false. The ones to be retained are those in which the proposition is true. Note, if *S* logically presupposes (relative to the context) a proposition *Q* which is not entailed by the propositions in the conversational background, then the set of conversationally alternative worlds will contain some members *w* in which the proposition *P*, expressed by *S*, cannot be correctly characterized as either true or false. Stalnaker's point is that in such a case the assertion will fail in its primary purpose of determining a new set of alternative possibilities, since it will fail, barring accommodation, to determine what should be done with worlds in the old alternative set in which *Q* is false. If this is right, it provides a straightforward explanation of why a sentence that logically presupposes Q (relative to a context) should give rise to a pragmatic requirement (subject to satisfaction via accommodation) that Q be entailed by the conversational background.

But is it right? Certainly, if Q is false in w, then there is no basis for retaining w, since retention is possible only if the world is one in which P is true. But if P is not true in a world w in which its presupposition fails, then, surely, w will be incompatible with P, and should be eliminated. For suppose that P is definitely not true in a world in which Q is false. Conversational participants who accept P should be in a position to remove the world from the set of alternative worlds among which the conversation has not decided. Since Stalnaker's argument assumes that w isn't eliminated on this basis, it requires a notion of radical presupposition in which a sentence or proposition with a false presupposition cannot be correctly characterized either as true or as not true. (This is implicit in his remark that a sentence with a false presupposition will fail to divide the set of conversationally alternative worlds into two exhaustive parts. Surely it would do so if the true and the not true were jointly exhaustive.)[38]

The idea, apparently, is that P cannot be correctly characterized either as being true in w or as being not true in w. Because of this, it fails to determine whether w should be retained or eliminated from the set of alternative possibilities among which it is the function of the assertion to discriminate. Thus, Stalnaker's argument for (43) depends on the view that the relevant presuppositions constitute not just necessary conditions for their bearers to be correctly characterized as true or false, but rather, necessary conditions for their bearers to be correctly characterized either as true or as not true.

Let us see if we can construct an example that exhibits these stronger conditions.[39] Imagine the predicate 'smidget' being introduced into a language by the following semantic stipulation.

(44) Smidget: Stipulative Definition

(i) Any adult human being under three feet in height is a smidget.

(ii) Any adult human being over four feet in height is not a smidget (or is such that it is not the case that he/she is a smidget.)

The stipulation consists of a sufficient condition for something to be a smidget and a sufficient condition for something not to be a smidget. On the basis of these conditions, there will be clear cases in which one is justified in characterizing someone as a smidget and clear cases in which one is justified in saying of someone that he is not a smidget. Moreover, these characterizations will convey information to other members of the linguistic community. An assertive utterance of 'Jack is a smidget' will convey to one's hearers the information that Jack is an adult under three feet tall and an assertive utterance of 'Jack is not a smidget' will convey the information that Jack is an adult over four feet tall. In short, 'smidget' will enter the language as a useful and meaningful predicate.

The interesting thing about the predicate is, of course, that the defining conditions for something to be a smidget, and for something to fail to be a smidget, are not jointly exhaustive. Adults between three and four feet tall cannot be correctly characterized either as being smidgets or as not being smidgets.

The same point can be made in certain cases in which the predicate is introduced ostensively. Rather than recite a stipulative definition, one might point to a number of adults under three feet tall and say "These people are smidgets", and a number of adults over four feet tall and say "These people are not smidgets" (or "It is not the case that these people are smidgets"). We may suppose that adults between three and four feet in height are extremely rare, and perhaps unheard-of in the linguistic community. Thus, the occasion may never arise during the period in which the term is being introduced to specify how such individuals are to be characterized. Language, being an institution designed to meet various practical contingencies, doesn't require linguistic conventions to be framed in terms of all logically, or metaphysically, possible circumstances.

Suppose now that the word 'smidget' has become entrenched in the language. We now consider the question of whether a man three feet six inches tall is or is not a smidget. This question might arise either because we finally encounter a man of that height, or because we wish to evaluate a counterfactual claim. Note, the question is not "Should the concept smidget be extended so as to include or exclude the individual in the relevant circumstance?"; rather it is, "Given the concept smidget as it already exists in the language, is the individual a smidget in the circumstance or is he not?". I suggest that our linguistic conventions provide no basis for answering this question. To characterize the man as not being a smidget would be just as unjustified as to characterize him as being a smidget. The concept is simply not designed for this case. As a result, neither (45a) nor (45b) can be accepted.

(45a) He is a smidget. (Said referring to the 3′ 6″ man)

(45b) He is not a smidget. (Said referring to the same man)

(45c) That he is (isn't) a smidget is true.

(45d) That he is (isn't) a smidget is not true.

If 'true' and 'apply' satisfy (i–iii), it will follow that (45c–d) cannot be accepted either. (In (i–iii) $\ulcorner A$ iff $B\urcorner$ holds whenever A and B have the same status – both true, both false, both neither true nor false, or both such that they cannot be correctly characterized as true or not true.)

(i) For any predicate P and term t, $\ulcorner Pt \urcorner$ is true (not true) iff P applies (doesn't apply) to the referent of t.

(ii) For any sentence S, $\ulcorner - S\urcorner$ is true (not true) iff S is not true (true).

(iii) The predicate 'red' applies (doesn't apply) to an object o iff o is (isn't) red.
The predicate 'smidget' applies (doesn't apply) to an object o iff o is (isn't) a smidget.

⋮

Under these assumptions, the radical partiality of 'smidget' results in the radical partiality of 'apply' and 'true'.[40]

This suggests that there are sentences and propositions of the kind required by Stalnaker's argument – sentences and propositions that cannot be correctly characterized as true or as not true unless certain conditions are met. What must be noted, however, is that the examples that have traditionally been analyzed as instances of logical presupposition are not of this kind.

For example, consider (46).

(46) The king of France is cultured (whoever he may be).

Although this example expresses a proposition, the proposition it expresses isn't true. Since there is no king of France, it can't be true. Unlike the smidget case, we don't feel that it would be just as wrong to say that (46) is not true as to say that it is true. Nor will one get an argument about this from traditional defenders of logical presupposition. The standard claim about such examples is that the falsity of their logical presuppositions leads to their being neither true nor false. But if (46) is neither true nor false, then it is not true, and hence distinguished from (45).

This means that Stalnaker's argument does not apply to the cases for which it was intended. In fact, the situation is worse. Examples like (45), to which Stalnaker's argument ostensibly applies, do not give rise to pragmatic presuppositional requirements on conversational backgrounds. If they did, then those requirements would be inherited in the normal way by larger presupposition inheriting constructions. However, examples like (47) do not bear the relevant pragmatic presuppositions.

(47a) Maybe Bill's uncle is a smidget.

(47b) It is unlikely that Bill's uncle is a smidget.

(47c) If Bill's uncle is a smidget, then he is probably entitled to special benefits from the government.

Although utterances of these sentences presuppose that Bill has an uncle, they do not presuppose that Bill's uncle is over four feet or under three feet tall. This suggests that Stalnaker's argument is not only inapplicable to standard cases of logical presupposition, but unsound as well.

Since the argument is unsound, no pragmatic presupposition can be explained by appeal to it. Thus, if any pragmatic presuppositions are to be derived from logical ones, a new argument is needed, which excludes smidget type cases.

Let us introduce some terminology. Following traditional discussions of logical presupposition we will reserve the term 'logical presupposition' for examples that satisfy the following definition.

Logical Presupposition: A proposition P logically presupposes a proposition Q iff

(a) for all possible circumstances w, if P is true or false in w, then Q is true in w; and

(b) for all possible circumstances w, if Q is not true in w, then P is neither true nor false in w (i.e. P is not true in w and P is not false in w).

Smidget type cases are instances of a different relation, in which the "presupposition" constrains acceptance or assertability rather than truth values directly.

Radical Presupposition: A proposition P radically presupposes a proposition Q iff (for all circumstances w) the proposition that P is true (in w) as well as the proposition that P is not true (in w) must be rejected if Q is not true (in w).

A more general relation encompassing both logical presupposition and radical presupposition is the following:

R-Presupposition: A proposition P R-presupposes a proposition Q iff (for all circumstances w) both the proposition that P is true (in w) and the proposition that P is false (in w) must be rejected if Q is not true (in w).

Derivative relations holding between sentences, contexts, and propositions can be defined in the usual way in each of these cases.

We have used the smidget examples to show that some radical presuppositions, and hence some R-presuppositions, are not pragmatically presupposed. Since R-presupposition is not enough to yield pragmatic presupposition, no pragmatic presupposition Q of S can be explained simply by claiming that S R-presupposes Q. If there is to be an argument explaining some pragmatic presuppositions in terms of logical ones, it must apply to logical presuppositions alone.

The difficulty in finding such an argument is illustrated by the following attempt. Suppose a speaker asserts a proposition P in a conversation in which one of its logical presuppositions, Q, is potentially controversial. Since Q is potentially controversial, the hearers might reject it, and thereby reject P. However, they cannot do this by assertively uttering the negation of the speaker's sentence, since on the preferred interpretation this sentence will express a proposition that commits them to Q. Thus if they are to reject P, they must say something more complicated – e.g. "Since it is not the case that Q, the proposition P cannot be accepted." Now, it might seem that a cooperative speaker should avoid putting his hearers in the position of having to go to such lengths. That is, it might seem that he should assert a proposition P only if he thinks that any objection to P on the part of his hearers can be expressed simply by negating his remark.

If this argument were correct, it would provide a rationale for the view that logical presuppositions give rise to pragmatic presuppositions. However, the argument cannot be correct. For the same line of reasoning applies with even greater force to smidget type cases in which P radically presupposes Q; and these don't give rise to pragmatic presuppositions.

Results like these present a challenge to the idea that pragmatic presuppositions can ever be explained in terms of logical presuppositions. There simply is no known reason why language use ought to be governed by the bridge principle (43) (rather than by analogous principles covering radical presuppositions). Indeed, it is hard to see how any explanation could be forthcoming. If there is no such explanation, then the fact that *S* pragmatically presupposes *Q* can never be explained simply by citing the (alleged) fact that *Q* is a logical presupposition of *S*.

This argument does not show that there is no such thing as logical presupposition; nor does it show that no pragmatic presuppositions are logically presupposed. Rather, it suggests that the connection between the two notions is at best indirect. If there are logical presuppositions, they must be motivated independently of pragmatic presuppositions. If pragmatic presuppositions are to be explained, their explanation can never rest simply on logical presupposition.

4.2. *Paradox, partiality, and presupposition*

The importance of the smidget example to the arguments just given might lead one to wonder whether there are non-artificial smidget type predicates in natural language. I believe there are. Many vague predicates, and predicates learned ostensively, share the kind of partial definedness that characterizes 'smidget'. So, I believe, do natural language truth predicates. Indeed, it was Kripke's theory of partially defined truth predicates that provided the model for 'smidget'.[41]

An important feature of this approach to truth lies in its avoidance of the strengthened liar paradox.

(48) Sentence (48) is not true.

A proper treatment of the paradox should explain the characteristics of the truth predicate that provide the basis for rejecting both the claim that (48) is true and the claim that (48) is not true. (Where to reject these claims is to refuse to accept them, without, of course, asserting their negations.)[42] This is what one gets if, following Kripke, one analyzes the truth predicate as partially defined in the manner of 'smidget'.[43]

What hasn't been sufficiently appreciated is that this kind of partiality is a fundamentally different phenomenon, semantically and pragmatically, from traditional examples of presupposition studied by philosophers and linguists.[44] If the argument in the previous section is correct, analyses of semantic (and other) paradoxes in terms of partially defined predicates must be sharply distinguished from theories of logical presupposition, expressive presupposition, and pragmatic presupposition generally.

4.3. *Three possible deviations from bivalence*

Think of a sentence obtaining a truth value as a result of the following process: First, the sentence is placed in a context and a proposition is determined by applying its meaning to the context. (The meaning is thought of as a function from contexts to propositions.) Next a possible circumstance is selected and the proposition is evaluated for truth value relative to it. This gives the truth value of the sentence relative to the context and circumstance.

There are three ways in which this process might lead one to reject a strong principle of bivalence for sentences. First, the meaning of the sentence might fail to be defined on the context in question, in which case the sentence will fail to express a proposition in the context. Second, the proposition expressed might contain a partially defined property that makes determinate evaluation impossible at the given circumstance. Third, the process of evaluating the proposition at the circumstance might yield a value of neither truth nor falsity.

The first of these ways corresponds to a failure of what I have called "expressive presupposition" and is definitely relevant to the study of pragmatic presupposition in natural language. The second way corresponds to smidget type cases, and is not. The third type of deviation is not as clear cut. If there are such cases, they can be used to characterize a notion of logical presupposition. However, it appears that this notion has no direct connection with presupposition in the primary sense of that which is taken for granted.

With this in mind, I now turn to another attempt to explain the basis for such presuppositions.

5. DISCOURSE SEMANTICS AND THE EXPLANATION OF PRAGMATIC PRESUPPOSITIONS

5.1. *A recent proposal*

Why do various sentences bear the pragmatic presuppositions that they do? In Section 3.3.2, it was suggested that pragmatic presuppositions arising from certain lexical items, like 'even', and certain syntactic constructions, like clefts, could be viewed as non-truth-conditional aspects of the conventional meanings of these elements. However, it was also argued that the presupposition inheriting characteristics of connectives, including 'and', 'or', and 'if, then', cannot be treated in this way. Since these characteristics are neither arbitrary, nor the result of special semantic stipulation, it ought to be possible to link them to the semantic contents of the connectives, plus general pragmatic principles governing communication. We have seen that this link is not provided by three valued truth tables that give rise to logical presuppositions. Thus, some other explanation is needed.

Recently, Irene Heim has attempted to provide such an explanation.[45] The key innovation is the development of a semantics for entire discourses, rather than individual sentences. Since pragmatic presupposition is itself a discourse phenomenon, it is natural to think that it might be linked to such a semantics.

Heim's semantic system is designed to assign (single) propositions to discourses, where propositions are taken to be sets of possible worlds and discourses are thought of as sequences of sentences. In this framework, the meaning of a sentence is given by a rule that determines its contribution to the propositions expressed by discourses containing it. These propositions are determined by starting with the set of all possible worlds and eliminating those incompatible with each of the sentences in turn. The end result is the "proposition" that the semantics assigns to the discourse as a whole.

To begin with, we can think of semantic-contexts as sets of possible worlds. The semantics consists of a recursive definition of a function, +, from sentences and semantic-contexts to semantic-contexts. For example, the clauses in the definition for negation, conjunction, and material conditionals are given in (49).

(49a) $C + \ulcorner A \text{ and } B \urcorner = (C + A) + B$

(49b) $C + \ulcorner \text{Not } A \urcorner = C \backslash (C + A)$

(49c) $C + \ulcorner \text{If } A, \text{ then } B \urcorner = C \backslash ((C + A) \backslash ((C + A) + B))$

($X \backslash Y$ is the intersection of X with the complement of Y.)

The discourse sensitivity of the semantics shows up in an analysis of indefinite noun phrases designed to handle discourses like (50).

(50) Mary met a man_i. He_i was F.

The problem illustrated by this example arises from two elementary observations. The first is that the initial sentence expresses an existential rather than a singular proposition. The second observation is that the pronoun in the second sentence is anaphoric with the indefinite *NP* in the initial sentence. This suggests that the semantic value of the pronoun in the context should be tied to that of the indefinite *NP*. If the two were in the same sentence, separated by 'and', this would be no problem for standard semantic accounts, since the pronoun could be seen as functioning as a variable bound by an existential quantifier. However, in (50) the pronoun and noun phrase are in different sentences. This creates a problem for standard systems in which variables can be bound within, but not across, sentences. It is this difficulty which motivates Heim to expand the scope of semantics from the single sentence to the entire discourse.

Her treatment of discourses containing indefinite *NP*'s consists of several elements: First, the semantic representation (51) is assigned to the discourse (50).

(51) x_i is a man, Mary met x_i, x_i was F.

Next, the notion of a semantic-context is enriched to make sense of the contribution of an open sentence to a discourse.

Enriched Notion of a Semantic-Context: A semantic-context is a set of pairs, $\langle g, w\rangle$, such that g is an infinite sequence of individuals and w is a possible world.

The Proposition Determined by a Semantic-Context: The proposition determined by a semantic-context C is the set of worlds w such that for some sequence of individuals g, $\langle g, w\rangle$ is a member of C.

Finally, the effect of incrementing a semantic-context with an open sentence is specified (informally) as follows.

Incrementation of a Semantic-Context by an Open Sentence: $C + Ax_i =$ the set of all $\langle g, w\rangle$ in C such that $g(i)$ is an A in w (where $g(i)$ is the ith element of the sequence g).

Applying this rule to the representation (51) of (50), we end up with a context that determines a proposition every member of which is a world in which Mary met a man who was F.

Intuitively, what is happening is this: When one uses an indefinite description one sets aside a certain variable for the rest of the discourse. One stipulates, in the case of (50), that from now on the variable x_i will be used only to express further constraints on men Mary met. This permanent setting aside of the variable means that constraints on these individuals can be built up piece by piece throughout the discourse. It is like letting the scope of an existential quantifier go across sentences, except the semantics is set up so that no explicit quantifier is needed.

In order to make this work, one needs to be careful about introducing variables. Whenever variables are bound by quantifiers or introduced by an indefinite description they must be distinct from all previously used free variables in the discourse. If this convention is observed, a universally quantified sentence ⌜Every A is B⌝ can be represented as ⌜Every x_i, Ax_i, Bx_i⌝, with the semantics given in (52).[46]

(52) C + ⌜Every x_i, Ax_i, Bx_i⌝ = The set of $\langle g, w\rangle$ in C such that for every individual a, if $\langle gi/a, w\rangle$ is a member of $C + Ax_i$, then $\langle gi/a, w\rangle$ is a member of $(C + Ax_i) + Bx_i$. (Where gi/a is the sequence that results from g by substituting a for its ith element.)

Using this definition, plus the above treatment of indefinites, Heim provides a semantics for problematic "donkey sentences" like (53).

(53) Every man who owns a donkey beats it.

Although the pronoun in this sentence is outside the scope of the relative clause containing its antecedent, Heim's theory succeeds in predicting that (53) is true iff every farmer who owns a donkey beats every donkey that he owns.

In addition to semantic predictions of this sort, Heim believes that her theory can be used to explain the semantic basis of pragmatic presuppositions. Her leading idea can be reconstructed in terms of the notion of a semantic discourse presupposition:

Semantic Discourse Presupposition: P is a semantic discourse presupposition of S iff for all semantic-contexts C, $C + S$ is defined only if the proposition determined by C is a subset of (i.e. entails) P.

In the case of clefts, Heim simply adds to the characterization of C + ⌜It was *NP* who *Fed*⌝ the condition that this operation is defined only if the proposition determined by C entails that someone has *F*ed. Here, the presuppositional requirement is a separate semantic stipulation, independent of other aspects of the meaning of the construction.

No such stipulations are needed to capture the inheritance characteristics of truth functional compounds. For example, consider conjunctions. It follows from (49a) that $C + \ulcorner A$ and $B\urcorner$ is defined only if $C + A$ is defined, and $(C + A) + B$ is defined. That will be the case only if the initial context C determines a proposition that entails the semantic discourse presuppositions of A, and, moreover, the result of incrementing C with A determines a proposition that entails the semantic discourse presuppositions of B. Note, this inheritance condition for semantic discourse presuppositions parallels the Karttunen [1974] condition for pragmatic presuppositions, later incorporated in modified form in Karttunen and Peters [1979] and Soames [1982]. A similar result holds for negations and indicative conditionals. Heim's point is that in her system the inheritance conditions carried by these constructions are direct consequences of their semantics. Thus, they are explained by the same devices that determine the truth conditional content of the connectives.

Of course, pragmatic presupposition is not the same as semantic discourse presupposition. The former has to do with requirements that utterances place on the conversations in which they occur; the latter has to do with requirements that one sentence in a sequence places on the semantic-context it increments in determining the proposition expressed by the entire sequence. Suppose that P is a semantic discourse presupposition of the *n*th member of a sequence of sentences. If the semantic context incremented by S_n does not determine a proposition that entails P, then the entire sequence fails to express any proposition. However, if $S_1, \ldots, S_n$ represent the sentences uttered in a conversation, and the conversational record incremented by S_n does not satisfy its pragmatic presuppositional requirements, then it is typically not the case that the conversation is empty of content. Thus, if Heim's system is to explain the pragmatic presuppositions of sentences, some way of relating the semantics and pragmatics of discourses must be found.

Although she is not fully explicit on this point, Heim seems to assume that assertive utterances increment contexts that contain all the semantic and pragmatic information contained in conversational backgrounds. Since the incrementation proceeds in accordance with the + function, requirements on the contexts in which it is defined become requirements on conversational backgrounds. Suppose now that a sentence is uttered in a situation in which the proposition determined by the preceding conversational background does not entail one of its semantic discourse presuppositions. In such a case, the + function will be undefined and the conversation will come to a halt, unless the hearers either object or accommodate the speaker by adding the needed presupposition to the background against which the speaker's remark is to be evaluated. In this way, Heim's semantic discourse presuppositions can be seen as giving rise to pragmatic presuppositions that trigger the Lewis–Stalnaker strategy of (*de facto*) accommodation.

Two further points are needed to fill out the basic picture. The first involves the interaction of presupposition and quantification illustrated by (54).

(54) Every student likes his advisor.

Heim takes this sentence to presuppose that every student has an advisor. She derives the presupposition by representing (54) as

(55) Every x_i, x_i is a student, x_i likes x_i's advisor.

and stipulating that

(56) $C + \ulcorner x_i$ likes x_i's advisor$\urcorner$

is to be defined only when every $\langle g, w \rangle$ in C is such that $g(i)$ has an advisor in w. The attractive feature of this is that the mechanisms for generating presuppositions arising from subsentential constructions like singular terms apply to open as well as closed expressions. Although this feature is not uncommon in semantic approaches to presupposition, it has proven difficult to incorporate into pragmatic approaches.

Finally, a word must be said about the different kinds of accommodation required by Heim. We have already noted the use she makes of the Lewis–Stalnaker strategy of (*de facto*) accommodation. In Sections 3.2 and 3.3, we demonstrated the need for another kind of accommodation in which pragmatic presuppositional requirements are cancelled by conversational implicatures and pre-existing contextual information (*de jure* accommodation). Since one can hardly use pragmatic facts to cancel definedness conditions on the + function, Heim needs to find another way to get the essential effects of *de jure* accommodation.

The simplest illustration of this point involves negation.

(57) There is no king of France. Therefore the king of France isn't in hiding.

First, the conversational background is incremented on the basis of the initial sentence in the discourse. The proposition determined by the resulting conversational context entails that France doesn't have a king. Next, this context is incremented using the second sentence of the discourse. Since the sentence is a negation, incrementation proceeds in accordance with (58).

(58) $C \backslash (C + A)$

(where in this case A = 'The king of France is in hiding')

However, this operation is defined only if the proposition determined by C entails that France has a king. But now we seem to be faced with a dilemma. If the hearers fail to accommodate the speaker, then the theory predicts that the conversation will be contentless. Yet if they do accommodate the speaker, the resulting context will determine the empty, or incoherent, proposition, and the theory will characterize the discourse as contradictory. Neither of these results is correct.

Heim avoids them by introducing a new kind of accommodation. The kind of accommodation countenanced up to now is one that takes the presupposition P of A, eliminates all non-P worlds from C (let us express this by 'C & P'), and then computes (59).

(59) $(C \,\&\, P) \backslash ((C \,\&\, P) + A)$

In the case of the second sentence of (57), this results in the empty set, which Heim doesn't want. However, there is another way of ensuring that + is defined in this case – namely by computing (60).

(60) $C \backslash ((C \,\&\, P) + A)$

Here accommodation occurs internally. Since the added proposition P is used only for computational purposes, it is not entailed by the proposition determined by the context that results from the utterance. Heim calls this "local accommodation" and contrasts it to the usual "global accommodation" in which the added proposition is entailed by the result of context incrementation.

In the case of (57) local accommodation in the evaluation of the second sentence of the discourse results in a context identical with the one resulting from the evaluation of the initial sentence. Thus, the discourse as a whole is assigned the same truth conditions as the initial sentence. In this way, local accommodation has the effect of cancelling the presuppositional requirement of the negation, and producing the right truth conditions.

This illustrates a general strategy of using local accommodation to account for the phenomena of contextual and conversational cancellation emphasized by Gazdar and Soames. In order to preserve empirical content in the theory, it is, of course, necessary to specify when global accommodation is to be employed and when local accommodation is to be used. Although Heim does not offer any precise proposals on this point, she does suggest that global accommodation is "strongly preferred" over local accommodation, except in cases – like the second sentence of (57) – in which global accommodation would result in an inconsistent context.[47] If this strategy is to be pursued, conversational implicatures that conflict with global accommodations must also be given precedence, thereby forcing local accommodation.[48] In this respect, Heim's proposal is like that of Soames [1982] in containing Karttunen-like inheritance conditions together with conversational and contextual devices capable of blocking, or cancelling, normal accommodation of presuppositional requirements. However, the overall theoretical frameworks of the two approaches are very different, as are some of the specific empirical results.

5.2. *Potential problems*

5.2.1. *Explanatory incompleteness.* A central goal of Heim's proposal is to explain the inheritance characteristics of truth functional operators in terms of an independently motivated semantics. Thus, the question "Why does operator *O* have inheritance property *P*?" is answered by citing the semantic characterization of the + function corresponding to *O*. In this way Heim provides a unified account of presuppositional requirements and semantic information.

Nevertheless, certain explanatory questions remain. Suppose one asks why a conjunction, ⌜*A* and *B*⌝, normally inherits a pragmatic presupposition corresponding to (61a) rather than (61b) or (61c)?

(61a) Pre: A & ($A \rightarrow$ Pre: B)

(61b) Pre: B & ($B \rightarrow$ Pre: A)

(61c) Pre: A & Pre: B

(Where ⌜Pre: *P*⌝ expresses the pragmatic presuppositions of *P*.)

It is not sufficient to answer that (61a) is a consequence of the semantics, (49a), for conjunction. For then one wants to know why (49a), rather than

(49a′) or (49a″), is the proper semantics for conjunction.

(49a) $(C + A) + B$

(49a′) $(C + B) + A$

(49a″) $C + A \cap C + B$

Aside from questions of presuppositions and accommodation, these alternatives are semantically equivalent in Heim's system. Thus, it would seem that the reason for selecting (a) over (a′) and (a″) is just that it does the better job in predicting presuppositions. But it cannot be that the reason that conjunctions inherit presuppositions in the way that they do, rather than in accord with (61b) or (61c), is because the semantics of conjunction are given by (49a); and, moreover, that the reason that (49a), rather than (49a′) or (49a″), gives the semantics of conjunction is that conjunctions inherit presuppositions in the way that they do.

This argument does not show that Heim's theory is incorrect; nor does it deny the attractiveness of deriving inheritance conditions from semantic specifications. What it shows is that Heim's theory leaves open some explanatory gaps of its own. For example, Heim criticizes the theory of Karttunen and Peters [1979] for having a structure that leaves open the possibility that "there could well be a lexical item – presumably not attested as yet – whose [truth conditional] content and presupposition properties are identical to those of 'if', while its heritage property [for presuppositions] is different".[49] However, the structure of Heim's theory allows this possibility to be realized (in a relatively small number of ways) through the selection of truth conditionally equivalent, but presuppositionally different, specifications of the + function. Unless some basis can be found for further limiting these alternative specifications, the aim of explaining the inheritance properties of truth functional connectives will not have been fully realized.

5.2.2. *Quantification and accommodation.* The second major goal of Heim's treatment of presupposition is to account for the interaction of presupposition and quantification by extending the mechanisms generating presuppositions to expressions containing free variables. Although Heim is reasonably successful in cases like (54), she notes that her theory faces problems with examples like (62a), which she represents as (62b).

(62a) Everyone who serves his king will be rewarded.

(62b) Every x_i, x_i serves x_i's king, x_i will be rewarded.

One of the steps in evaluating C + (62a) involves computing $C + \ulcorner x_i$ serves x_i's king$\urcorner$. In order for this to be defined every $\langle g, w \rangle$ in C must be such that $g(i)$ has a king in w. In Heim's system this means that the proposition determined by C must entail that everyone has a king.[50] So if (a) is uttered in a conversation in which it is not already assumed that everyone has a king, then (a) should fail to express a proposition unless the hearers accommodate the speaker. Global accommodation is expected here, since the proposition to be added may be consistent with everything else in the conversation. Thus, the theory incorrectly predicts that utterances of (a) pragmatically presuppose that everyone has a king.

The only way to block the prediction, within the overall framework of the theory, is to appeal to local rather than global accommodation. But this contradicts the general rule selecting global over local accommodation except in cases in which the former would result in inconsistency. Thus, the challenge posed by (62a) is to determine whether it is possible to modify this rule in some principled fashion.

A similar challenge is posed by (63).

(63a) A fat man was pushing his bicycle.

(63b) x_i was a fat man, x_i was pushing x_i's bicycle

Reasoning similar to that in (62) leads to the incorrect prediction that (63a) presupposes that every fat man had a bicycle. Heim notes that this result can be avoided if the context relevant for accommodation is not the context C preceding the utterance, but rather the internal context $C + \ulcorner x_i$ was a fat man$\urcorner$. If this context, C', is amended to a context C'' in which each $\langle g, w \rangle$ is such that $g(i)$ had a bicycle in w, then $C'' + \ulcorner x_i$ was pushing x_i's bicycle$\urcorner$ will be defined even though the proposition determined by the resulting context does not entail that every fat man had a bicycle. Heim contends that although this accommodation is, in a certain sense, "internal", it should also be regarded as global, since the accommodated information is not used simply for computational purposes, but rather "remains in the context for good."[51] Thus, she maintains that it does not threaten the principle that global accommodation takes precedence over local.

However, there are two problems here. First, even if the accommodation in question is global, it is a different sort of global accommodation from "external" accommodation on the initial context C. Unless principles are formulated specifying when each of the different varieties of "global accommodation" is to be available, the theory will fail to make empirical predictions in crucial cases. Second, by embedding (63a) in various constructions, one can create situations in which the accommodation Heim advocates must be classified as "local". For example, it is clear that (64) does not presuppose that every fat man had a bicycle.

(64) If a fat man pushed his bicycle across the flowers, you should have called a cop.

The only way for Heim to account for this is to require accommodation involving information that does not "remain in the context for good". Thus, some more precise statement of the principles governing accommodation is needed.

There are, of course, both descriptive and explanatory issues at stake. Principles of accommodation must be stated in order for the theory to make definite empirical predictions. If the explanatory goals of the theory are to be met, these principles should not be *ad hoc*, language particular stipulations, but general and independently motivated rules.

5.2.3. *Other constructions.* Many constructions important to theories of presupposition have so far not been treated in Heim's framework. Included among them are epistemic modals and propositional attitude verbs, which pose profound and familiar problems for semantic systems based on the notion of truth-supporting circumstances. However, important questions arise even with more elementary constructions.

Disjunctions are a case in point. Although Heim doesn't give an explicit semantics for disjunction, the most natural treatment in her framework is (65).

(65) $C + \ulcorner A \text{ or } B \urcorner \;=\; C + A \cup C + B$

Her account of presupposition will then predict that a disjunction inherits all the semantic discourse presuppositions of its disjuncts. Thus, disjunctions will not be assigned Karttunen-like inheritance conditions for presupposition, even though other truth functional connectives will.

This raises certain difficulties. For example, consider (66) and (67).

(66a) If France has an intelligent king, then the king of France is one of the few intelligent monarchs in Europe.

(66b) Either France doesn't have an intelligent king, or the king of France is one of the few intelligent monarchs in Europe.

(66c) Either the king of France is one of the few intelligent monarchs in Europe, or France doesn't have an intelligent king.

(P) There is a king of France.

(67) There has been some speculation that the projection problem has been solved, perhaps even by someone at the conference. Do you know anything about that?

(67a) If anyone at the conference solved the problem, then it was Susan who solved it.

(67b) Either no one at the conference solved the problem, or it was Susan who solved it.

(67c) Either it was Susan who solved the problem, or no one at the conference did.

(P) Someone solved the problem.

In each case, (a), (b), and (c) do not pragmatically presuppose *P*, even though they contain constituents that do. In the case of (a), Heim captures this by a semantic condition, (49c), which guarantees that the presuppositional requirement, *P*, of the consequent will be satisfied (without accommodation) no matter what context precedes the utterance. Since (65) does not have this character, the preference for global accommodation will incorrectly predict that (b) and (c) presuppose *P*.

The problem with the (b) sentences could be avoided by trading (65) for the more complicated condition (65′).

(65′) $C + \ulcorner A \text{ or } B \urcorner \;=\; C + A \cup (C + \ulcorner \text{Not } A \urcorner) + B$

However, the (c) sentences remain problematic. Nor can they be handled by the symmetric condition (65″), which, in the absence of unexplained local accommodation, gives rise to the same presuppositions as (65).

(65″) $C + \ulcorner A \text{ or } B \urcorner \;=\; (C + \ulcorner \text{Not } B \urcorner) + A$

$$\cup\ (C + \ulcorner \text{Not } A \urcorner) + B$$

Moreover, no other formulation of the clause for disjunction does any better.

It must be admitted that clear, unproblematic examples of the (c) type are somewhat unusual, and difficult to construct. However, unless an explanation of them is found within Heim's framework, they will pose a serious threat to the theory.

5.2.4. *What is said vs what is suggested.* It is customary to distinguish what is said by an assertive utterance of a sentence from what is merely suggested, implicated, or presupposed by the utterance. If *P* follows from what is said, then its truth is a necessary condition for the truth of the sentence, as used on that occasion. Otherwise, *P* may be part of the total information conveyed by the utterance, but it will not be part of the truth conditions of the sentence so used.

It is vital for a theory of pragmatic presupposition to respect this distinction. However, it is not clear how to do this in Heim's framework. For example, consider (68).

(68a) Even Bill likes Mary.

(P) Others besides Bill like Mary; and of the people under consideration, Bill is among the least likely to like Mary.

It has been argued, convincingly, in Horn (1969), Stalnaker (1973), and Karttunen and Peters (1979) that the pragmatic presupposition *P* of (68a) is not part of its truth conditions.

What is needed is a way of expressing this within Heim's framework. Let *C* be a context that does not contain the information *P* and therefore does not satisfy the definedness conditions on the + function corresponding to (68a). Let *C′* be the result of accommodating *C* so that the conditions are satisfied. One can't hold that what is said is determined by *C* + (68a), since there is no such set. Nor can one hold that what is said is determined by *C′* + (68a), since the proposition it determines entails *P*, which is not part of what is said. Perhaps what is said is determined by what *C* + (68a) would be if presuppositional requirements weren't definedness conditions on the + function. But if so, isn't this evidence that, in fact, at least some presuppositional requirements are not conditions of this kind?

6. VARIETIES OF PRESUPPOSITION: UNRESOLVED ISSUES

One of the most striking lessons of recent work is that there are many kinds and sources of presupposition; so many that there may be no single theory capable of incorporating them all. For example, it seems likely that expressive presuppositions and their pragmatic counterparts arise from a theoretical framework quite different from that which underlies presuppositions of the kind carried by 'even'. Thus, future progress in the field may call for the development of a number of circumscribed theories of well articulated types of presupposition, rather than one all encompassing model.

With this in mind, it may be useful to note certain kinds of variation that have not as yet been widely appreciated. One of these involves the cancellation of pragmatic presuppositional requirements by contextual and conversational means. The reality of this phenomenon is demonstrated in Gazdar [1979] and Soames [1979] and [1982], where many instances of cancellation involving compound sentences are cited – particularly negations, disjunctions and conditionals. However, these works fail to note that some pragmatic presuppositions cannot be cancelled in this way. For example, one cannot felicitously cancel the pragmatic presupposition *P* of (68a) in the manner in which the normal presupposition of (33b) is cancelled by the presence of (33a) in a discourse.

(33a) There is no king of France.

(33b) Therefore, the king of France isn't in hiding.

Karttunen and Peters were sensitive to this point, taking the non-cancellability of (68P) as showing that it is a conventional implicature of (68a). Unfortunately, they over-generalized in the opposite direction by extending the conventional implicature model to all sorts of pragmatic presuppositions, many of which turned out to be cancellable. What is needed at this point is a careful separation of cancellable and non-cancellable presuppositions, and principled explanations of each.

Another dimension along which pragmatic presuppositions vary involves the ease with which they allow the Lewis–Stalnaker variety of (*de facto*) accommodation. The standard case is illustrated by examples (69) and (70).

(69a) John managed to find the book.

(P) Finding the book required some effort.

(70a) Bill's son is (isn't) a genius.

(P) Bill has a son.

With these examples, accommodation works as expected. An utterance of a sentence in a conversation in which the common background does not already contain *P* results in the addition of *P* to the background, provided no objection is heard. Typically, there is no pretense that *P* has been part of the preceding background; nor is there any awkwardness or infelicity owing to the fact that it has not been.

By contrast, there are cases in which presuppositional requirements resist accommodation. For example, consider (71).

(71) The foreman was fired too. (Focus on 'The foreman')

This is often said to presuppose (72).[52]

(72) Someone other than the foreman was fired.

Suppose, however, that (71) were uttered in a conversation in which (72) was not already assumed. Even if the hearers were disposed to accept the suggestion that someone else had been fired, the remark would call for some further identification of the person or persons in question. Thus, (71) requires something other than (72).

I suggest that (71) requires the preceding conversational background to contain a set of propositions characterizing individuals both as being distinct from the foreman and as having been fired. If the sentence is uttered in a conversation containing the information in (73), but not that in (74), then the latter will typically be added by accommodation.

(73) John was fired.

Betty was fired.

The man from New Jersey was fired.

(74) John wasn't the foreman.

Betty wasn't the foreman.

The man from New Jersey wasn't the foreman.

However, if (71) is uttered in a conversation not containing information of the sort illustrated by (73), accommodation will generally not occur. The reason it won't is that the hearers may have no way of knowing how to accommodate the speaker, even if they desire to do so. The crucial point is that what (71) requires is not that the general proposition (72) be in the preceding background, but that one or more members of a set of more specific propositions be there.[53] If this requirement is not met, hearers will typically not know which propositions to add, and so will have to ask for clarification. Here, resistance to accommodation is explained by the nature of the requirement to be accommodated.[54]

A related example intermediate between (69) and (70), on the one hand, and (71), on the other, is (75).

(75) It was Mary who broke the typewriter.

This is often said to presuppose (76).

(76) Someone broke the typewriter.

Suppose, then, that it is uttered in a conversation that lacks this assumption. Accommodation may take place and the conversation might proceed without comment. However, there is something a bit odd about such a case – a kind of pretense that the (or a) topic of conversation prior to the remark was that of determining who broke the typewriter. For a speaker to utter (75) in a conversation in which this is not at issue is for him to reveal that his conception of the conversational plan differs from that of the other conversational participants.

This suggests that (75) pragmatically requires the (or a) topic of conversation prior to the utterance be that of determining who broke the typewriter. A conversation satisfying this requirement will be one in which (76) is entailed by the common background. However, it will also be one in which the conversational agenda is specified in a certain way.

These examples indicate that pragmatic presuppositions are more varied and complicated than is often assumed. Standard accounts have tended to view conversational records as sets of commonly assumed propositions, and pragmatic presuppositions as requirements that these sets contain certain specified propositions. The preceding examples illustrate two needed modifications of this picture. The first is the recognition that presuppositional requirements are not always requirements that some identifiable proposition be part of the preceding conversational background. Sometimes the requirement is that unspecified members of an identifiable set of propositions be included in the background. The second modification is the acknowledgement that there is more to conversational records, and requirements, than simply propositions. In particular, conversational records should be thought of as containing specifications of conversational topic and agenda.[55]

This more complex picture has the advantage that it allows a more natural treatment of various vague requirements arising from particular lexical items. For example ⌜*A* but *B*⌝ requires that the conversational background be such that the two conjuncts be seen as contrasting. This does not mean that the background must contain a proposition stating that there is some sort of contrast between the two conjuncts. What is required is just that the combination of conjuncts be unexpected in light of what is taken for granted.

One general lesson to be learned from all this is that the conception of pragmatic presuppositions as requirements on conversational records is a fruitful one that brings together a variety of different phenomena. A related moral is that the explanations of these different phenomena may require the conjunction of several different theories. In short, presupposition may not be a single phenomenon with a unitary explanation, but rather a domain of related issues involving the interaction of several semantic and pragmatic principles.

Department of Philosophy, Princeton University

NOTES

* This chapter was written in 1983–1984 while on leave from Princeton University on The Class of 1936 Bicenntenial Preceptorship, and while a guest of the Syntax Research Center at the University of California, Santa Cruz. (Minor revisions were made in 1986 prior to publication.) Portions of the chapter served as the basis for talks in the spring of 1984 at The Center for the Study of Language and Information at Stanford University. I would like to thank the Syntax Research Center at Santa Cruz for the use of their facilities; and also Joseph Almog, Saul Kripke, Julius Moravcsik, and Nathan Salmon for discussion of parts of the manuscript.

[1] See in particular, Frege [1891], 1892a], and [1892b].

[2] Strawson [1950], [1952].

[3] Stalnaker [1972], [1973], [1974].

[4] For Frege, this term results from applying the definite description operator to the complex predicate '______ is a Democrat and ______ was elected President of the U.S. in 1980'. The referent of the latter is a function from objects to truth values. The referent of the definite description operator is a second level function which takes a function *f* as argument and assigns the value *o* iff *o* is the unique object that *f* assigns the value Truth. In the case of 'the Democrat who was elected President of the U.S. in 1980' there is no such object. As a result, the definite description fails to refer.

Reference failure can also come about in simpler cases, e.g. 'the king of France'. Here the expression 'the king of ______' could be analyzed as referring to a function from countries to their kings, which is undefined at the argument France.

[5] The relevant negation operator is sentential, and the resulting negation is true (false) iff its corresponding positive counterpart is (false) true.

[6] If *R* and *S* are propositions, I say that *R* entails (necessitates) *S* iff there is no possible circumstance *w* such that *R* is true in *w* and *S* is not true in *w*.

[7] For Frege the unrestricted universal quantifier refers to a second level function that takes a function *f* (denoted by a complex predicate) as argument and assigns the value Truth iff *f* assigns Truth to every object; otherwise it assigns Falsity. Taking this as a model, one might

claim that the restricted quantifier 'all graduate students in the class' referred to a second level function that assigned Truth to *f* iff there were graduate students in the class and *f* assigned Truth to each of them; and assigned Falsity to *f* iff there were graduate students in the class and *f* assigned Falsity to at least one of them; and otherwise was undefined.

[8] When a truth functional constituent is truth valueless, there is no such thing as the *n*-tuple of truth values of the constituents; and hence, no such thing as the value of the relevant truth function at that *n*-tuple.

[9] This point was first noted by Bertrand Russell in 'On Denoting'. There Russell criticizes Frege as follows: "Or again consider such a proposition as the following: 'If *u* is a class which has only one member, then that one member is a member of *u*', or, as we may state it, 'If *u* is a unit class, the *u* is a *u*'. This proposition ought to be always true, since the conclusion is true whenever the hypothesis is true. But, 'the *u*' is a denoting phrase, and it is the denotation, not the meaning, that is said to be a *u*. Now if *u* is not a unit class, 'the *u*' seems to denote nothing; hence our proposition would seem to become nonsense as soon as *u* is not a unit class. Now it is plain that such propositions do not become nonsense merely because their hypotheses are false. The King in *The Tempest* might say, 'If Ferdinand is not drowned, Ferdinand is my only son'. Now, 'my only son' is a denoting phrase, which, on the face of it, has a denotation when, and only when, I have exactly one son. But the above statement would nevertheless have remained true if Ferdinand had been in fact drowned" (Russell [1905], p. 484).

If 'truth valueless' is substituted for 'nonsense' in the above quotation, then the passage correctly diagnoses the central difficulty with Frege's theory of presupposition inheritance. Further, implicit, criticism of Frege is contained in Russell's recognition that 'the king of France is not bald' is sometimes understood in such a way that it is true (if there is no king.)

[10] The two stage conception of semantics needed to distinguish logical from expressive presuppositions has important theoretical consequences for the familiar conception of semantic theory as consisting of a definition of truth relative to a context and a circumstance of evaluation. Although such a definition doesn't mention propositions, it does associate sentence/context pairs with functions from circumstances to truth values. If those functions are identified with propositions, then an analysis of an example like 'The largest prime number is odd' as logically presupposing a necessary falsehood will assign the sentence the degenerate function which is undefined on all circumstances. A problem arises when one notices that such a theory will assign the same "proposition" to 'This is a fine red one' in a context in which 'this' fails to refer. (Since 'this' has no referent relative to the context *C*, there is no circumstance *E* such that the referent of 'this' relative to *C* and *E* is a member of the extension or anti-extension of 'is a fine red one' in *E*.) Thus, a semantic theory of this familiar sort will miss the distinction between expressing a proposition that lacks a truth value in every circumstance, and failing to express a proposition at all. This distinction is captured in the semantic frameworks of Salmon [1986] and Soames [1987], where propositions are not identified with functions from circumstances to truth values, but rather are assigned to sentence/context pairs prior to evaluation for truth value.

[11] The seminal work on the two stage conception of semantics is Kaplan [1977]. Kamp [1968] is a significant precursor. The conception is developed further in Salmon [1986] and Soames [1987], where semantic theories assign sentence/context pairs structured Russellian propositions which determine, but are not determined by, functions from circumstances of evaluation to truth values.

[12] In 'On Referring', Strawson says that a (uniquely) referring use of a singular term is one in which the term is used to mention some particular individual. This is intended to rule out the use of 'the whale' in the generic claim (i).

(i) The whale is a mammal.

Strawson also indicates that predicative uses of singular terms are not (uniquely) referring uses. This stipulation is intended to exclude the use of 'the greatest French soldier' in (ii).

(ii) Napolean was the greatest French soldier.

The idea in both cases is to exclude uses in which what is grammatically a singular term is not functioning semantically as a singular term.

It is important not to confuse Strawson's notion of a (uniquely) referring use of a singular term with Keith Donnellan's notion of a referential use of a singular term, Donnellan [1966]. For Strawson, any use of a term which is genuinely singular and non-predicative would seem to qualify as a (uniquely) referring use. There is no recognition in Strawson [1950] or [1952] that some such uses might work in an essentially Fregean fashion, while others might work demonstratively.

[13] The pattern of argument here follows Kaplan [1977].

[14] See also the beginning of section iv of Strawson [1950], where he claims that to state a fact about an individual one must perform the (uniquely) referring task and the attributive task. To use an expression to perform the first of these tasks is, he says, to use it in a uniquely referring way. However, one can perform the same task without using any expression. He illustrates this by examples in which the object is presented directly, without linguistic mediation, as when one paints the words 'unsafe for lorries' on a bridge, or ties the label 'first prize' on a vegetable marrow. The suggestion here seems to be that in all these cases – linguistic and nonlinguistic – the statement made can be thought of as consisting of the object secured by the uniquely referring task, together with the property attributed to it.

[15] Strawson [1952], p. 175.

[16] Strawson distinguished statements from sentence types, sentence tokens, and acts of uttering a sentence in a context. The statement made by an utterance of a given sentence was supposed to be that which was said or asserted by the utterance. Although this made statements sound like propositions, the situation was complicated by Strawson's insistence that there really were no such entities as statements or propositions. This policy of countenancing talk about statements/propositions, without taking them seriously, was an important contributing factor to his failure to distinguish between expressive and logical presupposition.

Had he taken statements or propositions seriously, he might have been led to ask what sorts of things uses of various expressions contributed to them. Where the contributed elements were missing, it might have been natural to describe the situation in terms of expressive presupposition. Where the contributed elements were present, but it still seemed as if the statement or proposition failed to be true or false, logical presupposition would have been the relevant notion.

[17] For discussion see Donnellan [1978], Kripke [1979], Wettstein [1981], Salmon [1982], Barwise and Perry [1983], and Soames [1986].

[18] The strategy of accommodation, implicit in Stalnaker [1973], is explicitly formulated in Lewis [1979].

[19] Stalnaker [1973], p. 452.

[20] A related phenomenon involves cases in which the speaker has an object *o* in mind that he wants to say something about. In order to identify the object for his hearers, he may use a descriptive or demonstrative phrase that the conversational participants presume applies to *o*. For example, a speaker at a party might assertively utter (a) or (b) with the intention of asserting the singular proposition that he could have expressed by (impolitely) pointing at *o* and uttering (c).

(a) The man in the corner drinking champagne is famous.

(b) That man in the corner drinking champagne is famous.

(c) He is famous.

Such an utterance would presuppose that *o* is a man in the corner drinking champagne – i.e. it would indicate that the speaker regards that proposition to be either in the conversational record already, or evident enough to be added by accommodation. The important point about this proposition is not that it be true, but that the conversational participants accept it. If they do, then the speaker may succeed in saying something true about *o* even though the proposition presupposed by his utterance is false – because *o* is in fact drinking seltzer.

It should be noted that the true proposition asserted by the speaker may not be semantically expressed by his sentence relative to the context of utterance. Speakers may assert many things instead of, or in addition to, the propositions semantically expressed by their utterances. One of the factors determining what they do assert seems to be the pragmatic presuppositions of their utterances.

[21] Karttunen and Peters maintain that in addition to inheriting presuppositions from their constituents, indicative conditionals carry pragmatic presuppositions that their antecedents are not known to be false. In the interest of simplicity, I am ignoring this for present purposes.

[22] See Gazdar [1979] for further discussion.

[23] This point is made in Stalnaker [1974].

[24] In stating these conditions, I am using [*S*] to stand for the set whose only member is the proposition expressed by *S*.

[25] (40) is not the condition for disjunction given in Karttunen [1974], but rather a modification formulated in Soames [1979]. See pp. 629–630, 636–640, and footnote 24 of the latter for further discussion.

[26] Further examples of this type are given in (i) and (ii).

(ia) Either Bill regrets voting for Reagan or he regrets not voting for Reagan.

(b) If Bill regrets voting for Reagan or he regrets not voting for Reagan, then he is probably unhappy.

(c) It may be that either Bill regrets voting for Reagan or he regrets not voting for Reagan.

(PRE:A) Bill voted for Reagan.

(PRE:B) Bill didn't vote for Reagan.

(iia) If Mary's boss doesn't have children, then it wasn't his child who won the fellowship.

(A) Mary's boss doesn't have children.

(Pre:A) Mary has a boss.

(Pre:B) Mary has a boss and Mary's boss has a child and someone won the fellowship.

It follows from the definition of the Karttunen admittance relation that a context *C* admits (ia), (ib), or (ic) only if *C* entails the proposition expressed by (ia). Thus, the requirement that the conversational background prior to an utterance admit the sentence uttered incorrectly predicts that (ia), (ib) and (ic) pragmatically presuppose the proposition expressed by (ia). In fact, speakers who utter these sentences conversationally implicate that they are not presupposing this.

In the case of (ii), the admittance conditions require the conversational background to entail (iii), which in this case is equivalent to (iv).

(iii) (Pre: A & ($A \rightarrow$ Pre: B))

(iv) Mary's boss has a child.

Thus, the theory wrongly predicts that utterances of (iia) presuppose (iv), and fails to predict that they presuppose (v).

(v) Someone won the fellowship.

In fact what happens in this case is that the constituent presupposition (iv) is cancelled by a conversational implicature to the effect that the speaker doesn't know the truth value of the antecedent of his statement.

See Soames [1979], [1982], and Gazdar [1979] for further discussion of these and other examples.

27 Plus those mentioned in the previous footnote.

28 One descriptive problem with the account given in Soames [1982] involves conditionals in which the presuppositions of the consequent (together with the common background) entail the antecedent, but the antecedent (together with the common background) does not entail the presuppositions of the consequent.

An example of this kind, given in Heim [1983], is (i).

(i) If John has children, then Mary will not like his twins.

The second method of combining cancellation with Karttunen-like inheritance conditions fails to predict that utterances of (i) intuitively presuppose (ii).

(ii) John has children → John has twins.

(See, however, Soames [1982], pp. 502–504 for complicating factors.)

Further descriptive problems are noted and dealt with in Section 6 below.

29 Karttunen and Peters [1979].

30 Karttunen and Peters do not explicitly mention all the constructions that their theory is meant to apply to. For example, they do not mention 'stop', 'regret' (though they do explicitly include similar factives such as 'realize'), or 'again' (though they do explicitly include 'too', and 'also'). Nothing in the present discussion relies crucially on any items not explicitly treated by Karttunen and Peters.

31 Karttunen and Peters adopt a version of Montague semantics in which propositions are identified with functions from circumstances of evaluation to truth values.

[32] This point is made forcefully in Heim [1983]. However, it is also present, in various forms, in Stalnaker [1974], Gazdar [1979], and Soames [1982].
[33] For example, Karttunen and Peters claim that the conventional implicatures (pragmatic presuppositions) of conjunctions, conditionals, and disjunctions are given in (i). (Where *S* is a sentence, ⌜Imp: *S*⌝ is an expression that represents the conventional implicatures of *S*, and ⌜Ex: *S*⌝ is an expression that represents its truth conditions.)

(ia) Imp: (*A* and *B*) = Imp: *A* & (Ex: *A* → Imp: *B*)

(ib) Imp: (If *A*, then *B*) = Imp: *A* & (Ex: *A* → Imp: *B*)

(ic) Imp: (*A* or *B*) = (Ex: *A* ∨ Imp: *B*) & (Ex: *B* ∨ Imp: *A*) & (Imp: *A* ∨ Imp: *B*)

(See pp. 636–640 of Soames [1979] for discussion of (ic).) Applying these conditions to (35), as well as the sentences in Note 26, results in the same incorrect predictions that were shown to falsify the Karttunen (1974) account.
[34] See in particular Soames [1979], pp. 640–650; and Soames [1982], pp. 499–501.
[35] If anything, the problem is worse, since (43) and (LP) lead not only to false predictions about pragmatic presuppositions, but also to false claims about the truth conditions of examples like (33), (35), and the sentences in Note 26.

In the case of (33), the standard response by proponents of logical presupposition has been to claim that negative sentences in natural language are lexically ambiguous between a reading in which negation preserves neither-truth-nor-falsity and a reading in which neither-truth-nor-falsity is mapped onto truth. This response, under increasing attack in Wilson [1975], Kempson [1975], Atlas [1977], and Gazdar [1979], is insufficient, if one's goal is to derive pragmatic presuppositions from logical ones. For example, if the negation in the consequent of (iia) in Note 26 is claimed to be the (logical) presupposition preserving kind, then the conditional is wrongly characterized as entailing the denial of its antecedent. However, if the negation is claimed to be the (logical) presupposition blocking kind, then one gives up any hope of using logical presuppositions to derive the pragmatic presuppositions that do arise from the consequent. In effect, avoiding incorrect predictions about truth conditions requires explaining a significant range of pragmatic presuppositions independently of logical ones.
[36] Pp. 644–646.
[37] Stalnaker [1974] and Soames [1982] suggest that the link is pragmatic. However, they do not spell out mechanisms that apply to the full range of cases that need to be explained.
[38] Stalnaker does not explicitly call for a notion of radical presupposition according to which presupposition failure makes it impossible to correctly characterize the bearer of the presupposition either as true or as not true. However, without this notion his argument loses its force.
[39] The idea that there are predicates of the type illustrated by (44) was suggested to me by Nathan Salmon in a discussion of Kripke [1975].
[40] This point can be reinforced by considering a case in which the stipulative definition of 'smidget' is expanded to include (Siii), and the clauses for 'true' and 'apply' are formulated using (i′)–(iii′).

(Siii) For all *x*, if *x* is a smidget or is not a smidget, then *x* is an adult human being.

(i′) For any predicate P and term t, $\ulcorner Pt \urcorner$ is true (false) iff P positively applies (negatively applies) to the referent of t.

(ii′) For any sentence S, $\ulcorner - S \urcorner$ is true (false) iff S is false (true). Any sentence that is false is not true.

(iii′) The predicate 'red' positively applies (negatively applies) to an object iff it is (isn't) red. The predicate 'smidget' positively applies (negatively applies) to an object iff it is (isn't) a smidget.

⋮

Given this expansion, one can correctly say of a child that it is not true that he is a smidget. (Since it is not true that the child is an adult human being, we can conclude from (Siii) that it is not true that the child is a smidget or is not a smidget, from which the result follows.) However, the point about adults between three and four feet tall remains. Neither the claim that they are smidgets nor the claim that it is not true that they are smidgets can be accepted on the basis of the definition. Rather, the definition provides a reason for rejecting both claims without asserting their negations.

[41] Kripke [1975]. The idea of using smidget type examples to illuminate Kripke's conception of truth originated with Nathan Salmon.

[42] See Parsons [1984] for a discussion of relevant issues.

[43] See in particular p. 701, where Kripke indicates his intention of capturing the intuition that 'true' can be explained to someone along the lines of (i) and (ii).

(i) For any sentence S, one is entitled to assert that S is true in exactly those circumstances in which one is entitled to assert S.

(ii) For any sentence S, one is entitled to assert that S is not true in exactly those circumstances in which one is entitled to assert the negation of S. (In the interest of simplicity, we equate a sentence's being not true with its being false, and hence with its negation being true. The basic picture could be reconstructed so as to allow sentences that are not true and not false, but nevertheless grounded in Kripke's sense – however, there is no need to do so here.)

If 'true' is introduced in this way, then assertability conditions for claims to the effect that something is or is not true will standardly be grounded in assertability conditions of other claims, and ultimately in assertability conditions for sentences not involving truth at all. However, in "ungrounded" cases like (iii), the directions (i) and (ii) are silent.

(iii) Sentence (iii) is true.

Just as adults between 3 and 4 feet tall are not covered by the instructions governing 'smidget', so ungrounded examples like (iii) are not covered by (i) and (ii).

[44] Even Kripke likens ungrounded sentences to sentences which, for Strawsonian reasons, fail to express propositions. I believe this to be a mistake. One can, I think, construct ungrounded sentences that fail to express propositions. However, the phenomenon of ungroundedness is independent of this. Thus, I take Kripke's comments on pp. 699–700 to be a misstatement of the central philosophical insight underlying his analysis.

Parsons [1984] is, I think, guilty of a different mistake. While correctly noting the radical partiality of the truth predicate on Kripke-like analyses, he wrongly assimilates traditional examples of logical presupposition to this kind of partiality. This is just the converse of the widespread error in truth-gap solutions to the paradoxes of assimilating what are in fact instances of radical partiality to instances of logical presupposition.

These issues are discussed in greater detail in Soames [1985].

45 Heim [1982], [1983].

46 In order to assign the right truth conditions, the system must be set up so that the input context C to (52) satisfies (i) and (ii).

(i) If the proposition P determined by C contains a world w such that for some object o in w, o "is an A" in w but "is not a B" in w, then for some g and j, $g(j) = o$ and $\langle g, w \rangle$ is a member of C. (C contains all relevant "witnesses" from the set of worlds it determines.)

(ii) For any sequences g and g' that differ at most in their ith member, and for any world w, $\langle g, w \rangle$ is a member of C iff $\langle g', w \rangle$ is a member of C. (All "witnesses" can be found in the ith place of some sequence.)

C will satisfy (i) and (ii), if the semantic evaluation of sequences of sentences always starts with a context that satisfies them (the set of all $\langle g, w \rangle$), and the convention on introducing new variables is observed.

47 Heim [1983], p. 120.

48 Although Heim doesn't mention cancellation by conversational implicatures, her proposal must be formulated so as to include it – if it is to handle the data in Gazdar [1979], Soames [1979], and [1982].

49 Heim [1983], p. 115.

50 If it didn't entail this, then for some world w in the proposition determined by C and some individual o in w, o would lack a king in w. It would then follow by (i) and (ii) of Note 46 that for some $\langle g, w \rangle$ in C, $g(i) = o$. But this contradicts the hypothesis that every $\langle g, w \rangle$ in C is such that $g(i)$ has a king in w.

51 Heim [1983], p. 124.

52 Karttunen and Peters [1979], Soames [1979], [1982].

53 The theoretical significance of this point is illustrated by the fact that, in their present form, the theories of Gazdar [1979], Karttunen and Peters [1979], and Heim [1983] cannot incorporate this sort of requirement.

54 A number of facts bearing on this analysis have been brought to my attention by Saul Kripke. For example, consider the sentences in (i).

(ia) Herb's wife will come and Francis will come too.

(ib) If Herb's wife comes, then Francis will come too.

(ic) If Herb and his wife both come, then Francis will come too.

(id) Sam's wife will come. If Herb and his wife come, then Francis will come too.

Kripke observes that the presuppositions of these examples should include the following:

Pre: (ia) Herb's wife will come → Francis is not Herb's wife.

Pre: (ib) Herb's wife will come → Francis is not Herb's wife.

Pre: (ic) Herb and his wife both come → Francis is not Herb and Francis is not Herb's wife.

Pre: (id) Herb and his wife come → Francis is not Sam's wife, and Francis is not Herb, and Francis is not Herb's wife.

He also observes that these presuppositions would not be forthcoming if the pragmatic presupposition arising from 'too' in (ii) were the general proposition expressed by (iii).

(ii) NP VP's too. (Focus on NP)

(iii) Someone(thing) other than NP VP's.

These observations fit the analysis given above. However, Kripke goes further, noting that similar examples can be produced with other constructions.

(iv) If Reagan criticizes Hart in his radio talk, then he will criticize him again in his press conference.

(v) If Bill watches the opera at 2 o'clock, he will stop watching it when the Redskins' game begins.

As before, Kripke points out that the presuppositions of these sentences seem to include:

Pre: (iv) Reagan criticizes Hart in his radio talk → the radio talk will take place before the press conference.

Pre: (v) Bill watches the opera at 2 o'clock → the Redskins' game will begin after 2 o'clock.

These presuppositions would not be forthcoming if the presuppositions of (iv) and (v) arising from 'again' and 'stop' were (vi) and (vii).

(vi) Reagan criticizes Hart in his radio show → Reagan will have criticized Hart prior to the press conference.

(vii) Bill watches the opera at 2 o'clock → Bill will have been watching the opera prior to the beginning of the Redskins' game.

Kripke suggests that in all of these cases, the content of the presupposition of a sentence or clause containing 'too', 'again', or 'stop' may vary with, and be dependent upon, the preceding discourse or conversational context. The idea is that these presupposition creating elements may, in some way, be anaphoric with other elements in the discourse or context.

[55] See Lewis [1979] and Thomason (unpublished manuscript, Department of Philosophy, University of Pittsburgh) for further discussion.

REFERENCES

Atlas, J. D.: 1977, 'Negation, ambiguity, and presuppositions', *Linguistics and Philosophy* **1**, 321–336.

Barwise, J. and J. Perry: 1983, *Situations and Attitudes*, M.I.T. Press, Cambridge.

Donnellan, K. S.: 1966, 'Reference and definite descriptions', *The Philosophical Review* **75**, 281–304.

Donnellan, K. S.: 1978, 'Speaker reference, descriptions and anaphora', in P. Cole (ed.), *Syntax and Semantics 9: Pragmatics*, Academic Press, New York, pp. 47–69.

Frege, G.: 1891, 'Function and concept', in P. Geach and M. Black (eds.) [1952].
Frege, G.: 1892a, 'On concept and object', in P. Geach and M. Black (eds.) [1952].
Frege, G.: 1892b, 'On sense and reference', in P. Geach and M. Black (eds.) [1952].
Gazdar, G.: 1979, *Pragmatics: Implicature, Presupposition, and Logical Form*, Academic Press, New York.
Geach, P. and M. Black (eds.): 1952, *Translations from the Philosophical Writings of Gottlob Frege*, Basil Blackwell, Oxford.
Heim, I.: 1982, *The Semantics of Definite and Indefinite Noun Phrases*, Doctoral dissertation, University of Massachusetts, Amherst.
Heim, I.: 1983, 'On the projection problem for presuppositions', in M. Barlow, D. P. Flickinger, and M. T. Wescoat (eds.), *Proceedings of the West Coast Conference on Formal Linguistics*, Volume 2, Stanford Linguistics Association, Stanford, California, pp. 114–125.
Horn, L.: 1969, 'A presuppositional analysis of 'only' and 'even'', in R. Binnick, A. Davison, G. Green, and J. Morgan (eds.), *Papers from the Fifth Regional Meeting of the Chicago Linguistics Society*, Chicago Linguistics Society, pp. 98-107.
Kamp, H.: 1968, *Tense Logic and the Theory of Linear Order*, Doctoral dissertation, U.C.L.A.
Kaplan, D.: 1977, 'Demonstratives', unpublished manuscript, U.C.L.A., Dept. of Philosophy.
Karttunen, L.: 1974, 'Presupposition and linguistic context', *Theoretical Linguistics* **1**, 181–194.
Karttunen, L. and S. Peters: 1979, 'Conventional implicature', in Ch.-K. Oh and D. Dineen (eds.), *Syntax and Semantics 11: Presupposition*, Academic Press, New York, pp. 1–56.
Kempson, R.: 1975, *Presupposition and the Delimitation of Semantics*, Cambridge University Press, Cambridge.
Kripke, S.: 1975, 'Outline of a theory of truth', *The Journal of Philosophy* **72**, 690–716.
Kripke, S.: 1979, 'Speaker's reference and semantic reference', in P. French, T. Uehling, and H. Wettstein (eds.), *Contemporary Perspectives in the Philosophy of Language*, University of Minnesota Press, Minneapolis, pp. 6–27.
Lewis, D.: 1979, 'Scorekeeping in a language game', *Journal of Philosophical Logic* **8**, 339–359.
Parsons, T.: 1984, 'Assertion, denial, and the liar paradox', *Journal of Philosophical Logic* **13**, 137–152.
Russell, B.: 1905, 'On denoting', *Mind* **14**, 479–493.
Salmon, N. U.: 1982, 'Assertion and incomplete definite descriptions', *Philosophical Studies* **42**, 37–45.
Salmon, N. U.: 1986, *Frege's Puzzle*, M.I.T. Press, Cambridge.
Soames, S.: 1979, 'A projection problem for speaker presuppositions', *Linguistic Inquiry* **10**, 623–666.
Soames, S.: 1982, 'How presuppositions are inherited: A solution to the projection problem', *Linguistic Inquiry* **13**, 483–545.
Soames, S.: 1985, 'Truth, paradox, and partially defined predicates', unpublished manuscript Princeton University, Dept. of Philosophy.
Soames, S.: 1986, 'Incomplete definite descriptions', *Notre Dame Journal of Formal Logic* **27**, 349–375.
Soames, S.: 1987, 'Direct reference, propositional attitudes, and semantic content', *Philosophical Topics* **15**, 47–87.

Stalnaker, R.: 1972, 'Pragmatics', in D. Davidson and G. Harman (eds.), *Semantics of Natural Language*, Reidel, Dordrecht, pp. 380–397.

Stalnaker, R.: 1973, 'Presuppositions', *Journal of Philosophical Logic* **2**, 447–457.

Stalnaker, R.: 1974, 'Pragmatic presuppositions', in M. K. Munitz and P. K. Unger (eds.), *Semantics and Philosophy*, New York University Press, New York, pp. 197–214.

Strawson, P. F.: 1950, 'On Referring', *Mind* **59**, 320–344.

Strawson, P. F.: 1952, *Introduction to Logical Theory*, Methuen, London.

Thomason, R. H.: 1983, 'Accommodation, conversational planning, and implicature', unpublished manuscript, Dept. of Linguistics, University of Pittsburgh.

Wettstein, H.: 1981, 'Demonstrative reference and definite descriptions', *Philosophical Studies* **40**, 241–257.

Wilson, D.: 1975, *Presuppositions and Non-Truth-Conditional Semantics*, Academic Press, New York.

CHAPTER IV.10

SEMANTICS AND THE LIAR PARADOX

by Albert Visser

ON THE CHOICE OF MATERIAL

The semantical paradoxes are not a scientific subject like Inductive Definitions, Algebraic Geometry or Plasma Physics. At least not yet. On the other hand the paradoxes exert a strong fascination and many a philosopher or logician has spent some thought on them, mostly in relative isolation. The literature on the paradoxes is vast but scattered, repetitive and disconnected. This made it impossible to give a presentation in which all ideas in the literature receive their due.

The chapter consists of two parts. In the first I give a rather labyrinthine, admittedly far from complete, survey of problems and ideas for 'solution' connected with the liar. In Part II I concentrate on ideas mainly of

D. Gabbay and F. Guenthner (eds.), Handbook of Philosophical Logic, Volume IV, 617–706.

Kripke, Herzberger and Gupta, to be found in Kripke [1975], Gupta [1982], Herzberger [1982a, b]. Some reasons for this choice are: above papers are connected to important earlier work like Herzberger [1967], [1970a], [1970b]; the treatment in Kripke's paper uses inductive definitions. A good presentation of the two way connection of his solution with inductive definability is still lacking in the literature. Moreover similar constructions occur a.o. in Aczel [1980], Aczel & Feferman [1980], Cantini [1979a, b], Feferman [1967], [1975a, b, c], [1976], [1977], [1979], [1984], Fitch [1948], [1963], [1980], Gilmore [1974], [1980], Martin & Woodruff [1975], Nepeǐvoda [1973], Scott [1975]; these papers contain work that is technically explicit and fruitful; these papers share an interest in the meaning of selfreferential sentences in addition to the more mathematical interest to exploit the extra expressive possibilities obtained by adding a truth predicate to a given model.

I: THE LABYRINTH OF THE LIAR

1. THE SEMANTICAL PARADOXES AND THEIR INTEREST

Ramsey, following Peano (see Ramsey [1925]) distinguished the Semantical from the Set Theoretical Paradoxes, or in his own terminology the Linguistical from the Logical or Mathematical Paradoxes. The Semantical Paradoxes are those involving concepts like truth, satisfaction, denotation, definition, concept, proposition and the Set Theoretical Paradoxes those involving notions like set, element, number.

Such a distinction involves two theses: (i) The paradoxes are paradoxes *of* the semantical or set theoretical notions involved; (ii) The classification of notions as given is a natural one.

Not many philosophers today would accept Ramsey's own arguments for his distinction, but that does not mean that the issue is settled. The distinction is partly confirmed by the development of Zermelo-Fraenkel Set Theory. Russell's paradox is in this context clearly displayed as a paradox involving the notions of set and element. The paradoxical proof turns out to employ a too strong principle of set formation. That Russell's Paradox could be thus solved, where the Semantical Paradoxes are left untouched, seems to corroborate the idea that set in the sense of the Cumulative Hierarchy can be viewed as independent of notions as truth, proposition and the like.

Note however that the development of the notion of set can be seen as uncovering an ambiguity: on the one hand we have set as set of 'previously' existing objects, on the other hand set as property, attribute or concept. Set in the first sense seems to be implicit in the work of Cantor and Dedekind, set in the second sense in the work of Frege. Set in the second sense certainly is most naturally classified as a semantical concept, as is witnessed by the strong similarities between modern work on type free set theories and on type free truth. For sets in this sense Russell's Paradox is closely related or even identical to Grelling's Paradox.

I will not try to settle the issue of Ramsey's distinction here, but rather discuss the question what interest the study of the paradoxes has for both kinds of paradoxes together. In the rest of the chapter I concentrate however mainly on the liar.

Let us list some interests of the study of the paradoxes.

(i) *Foundational motivations.* Foundational questions are of this form: to articulate and justify principles on which to base a given practice.

The most important examples of foundational work connected with paradox are the development of Zermelo-Fraenkel Set Theory and the current development of systems for the foundations of Constructivistic Mathematics. Work in this area typically involves isolation and clarification of fundamental categories of objects and articulation of principles valid for these categories.

It must be stressed that Russell's paradox is just one foundational issue, in classical mathematics e.g. the Axiom of Choice, in constructivistic mathematics e.g. certain Continuity Principles another.

(ii) *Practical motivations.* A practical motivation is to extend an already accepted mathematical theory with e.g. a type free notion of truth or set for technical convenience. Typically but not necessarily such extensions will be conservative over the original language. For a forceful statement of such a practical program see Feferman's appendix to Aczel and Feferman [1980].

(iii) *Linguistical, semantical, philosophical motivations.* The primary aim from a philosophical or linguistical point of view is to understand by analysing the paradoxes the semantical notions involved better. Typically analyses of the paradoxes touch both on matters of detail like the proper treatment of implication or how to handle truth value gaps as on broad issues in the philosophy of language, like for example why do we have a notion of truth at all. For some examples of 'broad views' see e.g. Tarski [1944: 349] and Chihara [1979].

In the chapter I will mainly be guided by linguistical, semantical, and philosophical interest. A point where this will be particularly evident is in Part II, where I treat the idea that the problem of meaning for selfreferential sentences involving truth is wider than the problem of giving a theory of correct inference and judgement for these sentences.

2. VERSIONS OF THE LIAR

In this section versions of and variations on the Liar are presented. A closer discussion is given in Section 3.

2.1. *Sentential versions*

The simplest way to get a liar sentence is to write down:

This sentence is not true.

To arrive at Paradox, or Absurdity on no assumptions, reason as follows:

Suppose that sentence is true, i.e. it is not true. So it is true and it is not true, a contradiction. Hence the sentence is not true. But that is precisely what it says, so it is true. Conclude that the sentence is both true and not true, a contradiction on no assumptions. □

Clearly this version raises problems about the interaction of demonstratives with the truth predicate. Some of these will be treated in Section 3. For those who doubt that the context in which the sentence is presented is sufficient to disambiguate 'this' there is the following variant:

☞ This sentence is not true.☜

There are several ways to avoid the use of demonstratives. One is the use of an 'empirical' predicate.

The sentence printed on page 620, lines 27–29, of Part IV of the *Handbook of Philosophical Logic, Vol. IV*, edited by Gabbay and Guenthner is not true.

Another way employs baptism.

Consider:

L is not true.

Let us call above sentence "L". For later reference I give the paradoxical reasoning for the case of L in natural deduction style:

$$\neg E\,\dfrac{S\,\dfrac{L = \text{“}L\text{ is not true”} \quad T\,\dfrac{\neg I\,\dfrac{\neg E\,\dfrac{T\,\dfrac{S\,\dfrac{[L\text{ is true}]^{1} \quad L = \text{“}L\text{ is not true”}}{\text{“}L\text{ is not true” is true}}}{L\text{ is not true}} \quad [L\text{ is true}]^{1}}{\bot}}{L\text{ is not true}}\,1}{\text{“}L\text{ is not true” is true}}}{L\text{ is true}} \qquad \dfrac{\dfrac{\dfrac{\dfrac{[L\text{ is true}]^{2} \quad L = \text{“}L\text{ is not true”}}{\text{“}L\text{ is not true” is true}}}{L\text{ is not true}} \quad [L\text{ is true}]^{2}}{\bot}}{L\text{ is not true}}\,2}{\bot}$$

Here S stands for 'Substitutivity of Identity' and T for Tarski's Convention T.

Of the following 'versions' it is disputable whether they are "really" versions of the Liar rather than closely related paradoxes. I'm going to present them anyway.

The next 'version' is known as Grelling's Paradox. The German mathematician Kurt Grelling found it in 1908. It dispenses with the use of a selfreferential sentence altogether. Instead it employs self-application of words.

Let's call a word 'heterological' when it does not apply to itself, e.g. "long" is heterological. Is 'heterological' heterological? If it is, it applies to itself, so it is not heterological. Contradiction. So 'heterological' is not heterological, hence it is heterological. Contradiction on no assumptions.

It is worthwhile to observe that Grelling's Paradox can be converted in a 'metamathematical' version for languages containing their own satisfaction predicate. First note that the relevant 'words' for Grelling's definition of 'heterological' are predicate words. These may be replaced or mimicked by formulae having just x_0 free. Now define a formula standing for 'heterological(x_0)' by:

$$\neg\text{Sat}(x_0, \langle x_0\rangle).$$

Then Grelling's sentence '"heterological" is heterological' becomes:

$$\neg\text{Sat}(\text{“}(\neg\text{Sat}(x_0, \langle x_0\rangle))\text{”}, \langle\text{“}(\neg\text{Sat}(x_0, \langle x_0\rangle))\text{”}\rangle).$$

By the usual properties of Sat this quickly leads to Paradox. Above rephrasing of Grelling's sentence is closely related to Gödel's construction of his celebrated sentence: what corresponds to '$\neg\mathrm{Sat}(x_0, \langle x_0 \rangle)$' there, is '$\neg\mathrm{Prov}(\mathrm{sub}(x_0, x_0))$', where $\mathrm{sub}(a, b)$ is the function computing the Gödel number of the formula obtained by substituting the numeral of b for x_0 in the formula with Gödel number a.

The following sentence is Quine's. It is very similar to Grelling's, but for the fact that it does not use selfapplication but the operation of putting a string of symbols behind its quotation. This operation is metamathematically very simple. (For an elaboration see Smullyan [1957].) Quine's sentence is:

> "Yields a falsehood when appended to its own quotation"
>
> yields a falsehood when appended to its own quotation.

A final example, a bit more artificial maybe than the others in that it does not rest on previously accepted principles, employs a proposed extension of our stipulatory practices. It will be useful because it allows us to present the paradoxical reasoning purely in a propositional logical form.

First some explanation.

The usual stipulative definition of the Riemann Integral looks somewhat like this:

$$\int_a^b f(x)\mathrm{d}x = c\colon - a - b - c -$$

where $- a - b - c -$ is a sentence in the language of analysis not containing $\int$. Let us ignore the occurence of '$=$' on the left, which introduces extra problems.

The reasons why one does not admit $\int$ on the right is presumably that otherwise the definition would be circular. But consider:

$$x + y = z\colon (y = 0 \wedge x = z) \vee$$
$$\vee (\exists u \exists v \; y = u + 1 \wedge z = v + 1 \wedge x + u = v).$$

Here the domain of the definition is the natural numbers and '$+1$' is supposed to be given.

Clearly there is a way of reading the definition, familiar to those who know the Recursion Theorem, under which it is a perfectly sensible definition of $+$. For the moment let us vaguely say that we stipulate the left hand side to have the same meaning as the right hand side *whatever that will turn out to be*.

If we accept provisionally a liberalization of stipulative definitions in the spirit of above definition of $+$, we can have a Liar as follows:

$$\ell\colon \neg\ell.$$

Note the difference with L of the baptism way of constructing a Liar sentence: "L" is a *name* of the sentence under consideration, "ℓ" is a new atomic sentence, introduced in our language, that is stipulated to have the same meaning as "$\neg\ell$".

$$\dfrac{\dfrac{\dfrac{\dfrac{\dfrac{[\ell]^{1}}{\neg\ell}{*}\quad[\ell]^{1}}{\bot}}{\neg\ell}{1}}{\ell}{*} \qquad \dfrac{\dfrac{\dfrac{[\ell]^{2}}{\neg\ell}{*}\quad[\ell]^{2}}{\bot}}{\neg\ell}{2}}{\bot}$$

The *-ed steps are consequences of the stipulation – note that the stipulatory way involves neither selfreference nor selfapplication nor truth or satisfaction.

2.2. *Propositional Versions*

A good case can be made that truth, if a predicate at all, is a predicate of propositions not of sentences. Acceptance of this view does not disqualify the sentential versions, but at least makes it necessary to rephrase them: e.g.

> This sentence is not true

can be replaced by:

> This sentence does not express a true proposition

or alternatively by:

> This sentence expresses a false proposition.

Note that these paraphrases are really different: when we would try to deny that they express a proposition at all this would make the first seem true and the second seem false.

Let us look at directly propositional versions. The boldest is simply:

> This is false.

Hence the 'this' points via the sentence token at what is expressed: that that is false. The paradoxical reasoning then runs as follows:

> Suppose that that is true, i.e. it is false. Then it is both true and false, contradiction. Hence it is false, so true. Contradiction.

Variants are:

> What is expressed here is false

and

What this sentence expresses is false.

In this chapter I will not go into propositional versions, even if I believe that a full solution should include an account of propositions. Especially in connection with Kripke's paper I will discuss in what way this theory is compatible with a possible propositional solution.

3. TOPICS CONNECTED WITH THE LIAR

Consider the way L yields $\bot$. A fully explicit proof is given. Does that mean that we can simply write down a full list of 'assumptions' involved in the derivation of Paradox? No, assumptions here are assumptions of a semantic nature, e.g. the assumption that 'true' does not change its meaning during the proof. For any step in the proof one could conjure up any number of hidden assumptions involved in that step. (And that is more or less what happened in the literature.)

What the real assumptions involved in the proof are, will depend on what the true solution of the paradox is. But perhaps there is no true solution, maybe we should be content with a number of ways to block the paradox, the choice among which is to be governed by local considerations of utility and simplicity. Maybe language is like the skeleton of Aulonia Hexagona: there is no way to make it what it should be, a regular polygon, without giving up the original plan.

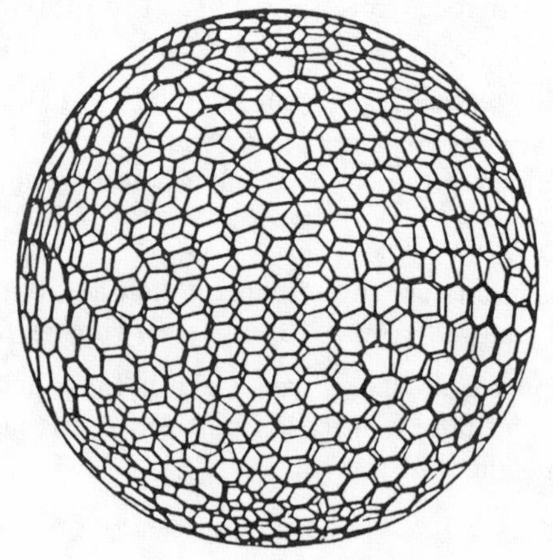

In this section I will present and discuss a list of 'assummptions', topics and problems in the construction of the Liar and in the derivation of Paradox. This list is surely far from exhaustive.

In the case of L the following ingredients are certainly present:

(a) There is a construction of selfreference.
(b) Truth is treated as a predicate of sentences including those involving Truth.
(c) Classical logic is used. (Actually less than classical logic.)
(d) Tarksi's convention T is applied.

Ingredients (a), (c), (d) are roughly Tarski's list of conditions sufficient for the derivation of paradox (see Tarski [1944], [1956]).

3.1. *Selfreference*

Many have felt that selfreference is the cause of the trouble with the Liar. They find fault with the Paradox in the very first stage of its construction.

This intuition immediately meets two formidable objections. The first is that selfreference is not necessary to generate the paradox; the second that selfreference is generally unproblematic.

Let us take a closer look at the first. E.g. Grelling's Paradox shows clearly that Liar-like phenomena occur even in the absence of anything that can be described as selfreference of a sentence, here selfapplication takes the place of selfreference. It might be argued that Grelling's Paradox is different from the Liar and should be treated differently, but nobody would want to argue thus against the following example: Consider

L_2 is not true

L_1 is true.

Call the first sentence "L_1", the second "L_2". As is easily seen L_1, L_2 lead to Paradox. (The example is in essence Buridan's 9th sophism (see e.g. Hughes [1982]). In a more playful form it is known as Jourdain's Card Paradox (discovered by the French mathematician P. E. B. Jourdain in 1913).

Neither L_1 nor L_2 refers to itself. There is of course a kind of circularity, but the problem is to give a precise description of what is supposed to be wrong. Thus the first objection takes the form of a challenge to elaborate the anti-selfreference view to cover a sufficiently broad class of cases.

I will not try to meet that challenge here. Rather I will try to get more grip on what is behind the feeling that something is amiss with self-reference. To focus on that question let us turn to the second objection.

The second objection points reaonably at the many unproblematic and even pedestrian occurrances of selfreference, like:

This sentence contains the letter "b" only quoted.

and the fully explicit metamathematical constructions of selfreference. The objection is eloquently voiced in Popper [1963] and Kripke [1975]. Kripke argues the claim that selfreference by baptism is unproblematic as follows (see Kripke [1975], p. 693):

> Let 'Jack' be a name of the sentence 'Jack is short' and we have a sentence that says of itself that it is short. I can see nothing wrong with "direct" selfreference of this type. If 'Jack' is not already a name in the language, why can we not introduce it as a name of any entity we please? In particular, why can it not be a name of the (uninterpreted) finite sequence of marks 'Jack is short'. (Would it be permissible to call this sequence of marks "Harry", but not "Jack"? Surely prohibitions on naming are arbitrary here.) There is no vicious circle in our procedure, since we need not *interpret* the sequence of marks 'Jack is short' before we name it. Yet if we name it "Jack", it at once becomes meaningful and true. (Note that I am speaking of self-referential sentences, not self-referential propositions.)

Surely Kripke is right that we can call the uninterpreted string of symbols "Jack is short" "Jack". What Semantical God could stop us?

But doesn't Kripke's example evade the problem? Why not argue thus: it is the *name* "Jack" and with it the meaningful sentence "Jack is short" that refers to Jack, the uninterpreted string of symbols. Moreover Jack, indeed, is short. But we want to predicate truth, not of a mere syntactical object, but of a meaningful sentence. Neither Jack nor L can sensibly be said to be true. (The sense of meaning involved here is not just linguistic meaning but the kind of meaning needed to make the question whether a sentence is true or false sensible at all.)

Note that above objection to Kripke is one possible elaboration of the anti selfreference intuitions: Liar-like sentences in order to be meaningful presuppose their own meaningfulness. This is the vicious circle in the very first stage of Paradox construction. Note that the objection in this form applies not just to L but also to all other sentential examples we treated.

The Kripke quote, as I read it, contains the germ of an answer to the objection: the passage about Jack becoming at once meaningful and true. The objection treats 'meaningful sentence' as an inseparable whole. Kripke on the other hand talks of the string of symbols becoming meaningful as soon as a certain word is interpreted. So maybe a meaningful sentence is simply a sentence as syntactical object that is meaningful. Here we have to move with some care. Let us assume a sentence as syntactical object is a *type*. (Not a token, we want to allow various occurrences of the same sentence.) It seems a bit strange to take a type to be meaningful *per se*. A type is only meaningful *at a context*. (Kripke writes things like: "If 'Jack' is not already a name in the language . . ." This suggests that he has an unambiguous formal language in mind. I want to discuss the construction of the Liar in English here.) Kaplan gives a theory of interpretation of sentence types at a given context, independent of utterances or tokens as links between context and type (see Kaplan [1978]). Whether we accept this or not, at least we want to keep the idea that a meaningful sentence can be represented by context and sentence type. (Maybe what one wants is a bit less specific: e.g. if contexts contain a sharp time and place it seems natural to 'divide out' aspects of the context not directly relevant for the interpretation. For example 'I am here' as said by me at the same place twice within 20 minutes, could be the same meaningful sentence.)

When we attribute truth to a sentence we do this at a context. If the context is not the context of our utterance, the relevant context is often *mentioned*, like in:

"I am small" as said by Napoleon is true.

Often when no relevant context is stated the appropriate intended context is the context of the truth claim itself. Let us adopt this as a simplistic convention.

The story of L on the present theory is as follows: first L is introduced as a string of symbols. Then by baptism a context is created at which "L" denotes and L can be interpreted: L becomes at once meaningful.

The story of "This sentence is not true" runs thus: the appropriate context c is created by the very act of writing down the sentence. The chain of interpretations is schematically:

(This sentence is not true) (c)

↓

(not true(this sentence, c)) (c)

↓

(not true(this sentence (c), c))

↓

(not true ("This sentence is not true", c)).

Note that we may want to see the first stage simply as an ellipsis for the second. In this case in writing "This sentence is not true" the context does not only function in interpreting the sentence but is also implicitly mentioned. Or we may want to view the first step already as part of the interpretation. For a further discussion of this and related points see Thomason [1975].

(Quine in his popular scientific American article (Quine [1962]) presents a remarkable argument:

> In an effort to clear up this antinomy it has been protested that the phrase "This sentence", so used, refers to nothing. This is claimed on the ground that you cannot get rid of the phrase by supplying a sentence that is referred to. For what sentence does the phrase refer to? "This sentence is false." If accordingly, we supplant the phrase "This sentence" by a quotation of the sentence referred to, we get: ""This sentence is false" is false". But the whole outside sentence here attributes falsity no longer to itself but merely to something other than itself, thereby engendering no paradox.

The fallacy here is clearly failure to note that "This sentence is false" is to be interpreted at a different context than ""This sentence is false" is false", so that in 'disquoting' the latter must shift contexts.)

I am not at all sure that the answer to the objection as sketched is successful. Certainly both the objection and the answer involve substantial philosophical assumptions. The fact that there are metamathematical reconstructions of the Liar does not add much to the argument. If meaningful sentences are not like Gödel-numbers at all, the appropriate reaction seems to be: so much the worse for metamathematics.

3.1.1. *Excurs: Non-standard Gödel numberings.* Kripke after discussing Jack mentions: the possibility to construct non-standard Gödel numberings as a way to obtain Gödel sentences directly. I give an example of how this can be done.

First the most naive construction. Let $\mathscr{L}_{\mathbf{PA}}$ be the language of Peano Arithmetic. Add a new constant c. Call the language so obtained $\mathscr{L}_{\mathbf{PA}}(c)$. Let $\varphi_0, \varphi_1, \ldots$ be an effective enumeration of the *sentences* of $\mathscr{L}_{\mathbf{PA}}(c)$. Let n be the non-standard Gödel number of $\varphi_n[\mathbf{n}/c]$, the result of substituting the numeral of n for c in φ_n. $\varphi_n[\mathbf{n}/c] \in \mathscr{L}_{\mathbf{PA}}$, moreover every sentence of $\mathscr{L}_{\mathbf{PA}}$ is so numbered. If we have chosen our enumeration wisely, there is a primitive recursive function $SG: \omega \to \omega$, such that $SG(n) = \#\varphi_n[\mathbf{n}/c]$. Here $\#\varphi$ is the standard Gödel number of φ (in a given standard Gödel numbering). SG can be represented in **PA**. Let $\mathrm{Prov}_{\mathbf{PA}}(x)$ be a standard provability predicate for our standard Gödel numbering, then $\mathrm{Prov}_{\mathbf{PA}}(SG(c)) = \varphi_N$. Then N is a non-standard Gödel number of $(\neg \mathrm{Prov}_{\mathbf{PA}}(SG(\mathbf{N})))$.

An inelegance is that non-standard Gödel numbers so constructed admit synonymity; we cannot exclude that for $n \neq k: \varphi_n[\mathbf{n}/c] = \varphi_k[\mathbf{k}/c]$.

To avoid this the construction must be slighly refined. First make the alphabet of $\mathscr{L}_{\mathbf{PA}}$ finite by taking as variables: $x, xS, xSSS, \ldots$, i.e. we view the variables as built of "x" and "S". Measure the length $l(\alpha)$ of a string of symbols α of $\mathscr{L}_{\mathbf{PA}}(c)$, by simply counting the number of symbols. Enumerate all *strings* of symbols of $\mathscr{L}_{\mathbf{PA}}(c)$, by first enumerating 'alphabetically' all strings of length 1, then all strings of length 2, etc. Let the ordering obtained be $\alpha_0, \alpha_1, \ldots$.

Define for $\alpha \in \mathscr{L}_{\mathbf{PA}}$:

$$⁂\varphi := (\text{the smallest } m \text{ such that } \varphi = \alpha_m[\mathbf{m}/c]).$$

⁂ is total and injective and for some primitive recursive $SG: SG(⁂\varphi) = \#\varphi$.

What we have to check is whether the selfreferential effect we are seeking is not lost. Clearly the effect is only of interest when there is at least one occurrence of c in a ψ for which we want to find a φ such that

$$\varphi = \psi[⁂\varphi/c].$$

Consider any $\psi \in \mathscr{L}_{\mathbf{PA}}(c)$ containing at least one occurrence of c. Say $\psi = \alpha_N$. It is sufficient to show:

$⁂\psi[\mathbf{N}/c] = N$. Surely N is *an* m such that

$\psi[\mathbf{N}/c] = \alpha_m[\mathbf{m}/c]$. Suppose for some $m < N$

$\psi[\mathbf{N}/c] = \alpha_m[\mathbf{m}/c]$. Define

$\chi := \psi[\mathbf{N}/c]$

$O_\alpha(\psi) :=$ the number of occurrences of string α in ψ.

$K :=$ the number of symbols in the alphabet of $\mathscr{L}_{\mathbf{PA}}(c)$.

We have:

(a) $\ell(\mathbf{p}) = 3p + 1$ for any number p ($\mathbf{p} = \underbrace{S(S \ldots S}_{p\ S's}(\mathbf{o}) \ldots))$.

$$\text{So } \ell(\chi) = \ell(\alpha_m) + 3O_c(\alpha_m) \cdot m$$
$$= \ell(\alpha_N) + 3O_c(\alpha_N) \cdot N.$$

$N > m$, hence $\ell(\alpha_N) \geqslant \ell(\alpha_m)$, so $O_c(\alpha_m) > O_c(\alpha_N)$.

(b) $O_{\mathbf{m}}(\chi) = O_c(\alpha_N) + O_{\mathbf{m}}(\alpha_N)$, for every occurrence of **N** in χ contains precisely one suboccurrence of **m**.
Moreover $O_{\mathbf{m}}(\chi) \geqslant O_c(\alpha_m)$ and by (a) $O_c(\alpha_m) > O_c(\alpha_N)$, so **m** occurs in α_N. Conclude $\ell(\alpha_N) \geqslant 3m + 1$.

(c) $O_c(\alpha_m) \leqslant \ell(\alpha_m)$ so

$$\begin{aligned} \ell(\chi) &\leqslant (3m + 1)\ell(\alpha_m) \\ &\leqslant (3m + 1)\ell(\alpha_N) \\ &\leqslant (\ell(\alpha_N))^2. \end{aligned}$$

(The last inequality is by (b).)

(d) $N > K^{\ell(\alpha_N)-1}$, for $K^{\ell(\alpha_N)-1}$ is the number of sequences of length $\ell(\alpha_N) - 1$. These precede α_N.

(e) $\ell(\chi) \geqslant 3N + 1$ for **N** occurs in χ.
By (d): $\ell(\chi) \geqslant 3K^{\ell(\alpha_N)-1} + 1$.
We may safely assume $K \geqslant 3$, hence $\ell(\chi) > 3^{\ell(\alpha_N)}$.

(f) By (c), (e): $3^{\ell(\alpha_N)} < \ell(\chi) \leqslant (\ell(\alpha_N))^2$.
Quod impossibile.

3.2. *Truth as a predicate*

Phenomena like ambiguity and indexicality block the possibility of taking truth (in natural language) as a simple predicate of sentences as types. If we have our language fragment specified in advance, free from indexical words and ambiguities, then we can successfully ignore the difference between sentences at a context and grammatical types. Note however that in natural language, due to its democratic character, no sentence is free from possible ambiguity. No semantical God can stop a fellow citizen to give a word new meaning.

We will however for the moment ignore such complications and consider only specified 'eternal' fragments.

First we may want to deny that truth is a predicate at all. For example for Frege truth is the denotation of sentences (mediated via their sense) (see Frege [1975]). Or for Grover truth is an inheritor (Grover [1977]). For reasons of space and presentation I do not want to go into the issues raised by these views and refer the reader to the relevant papers.

Secondly truth can be viewed as a predicate not of sentences but of propositions. On such a view truth can really be a simple predicate: all problems of indexicality and vagueness find their place in the question: What proposition does this sentence express? Moreover such a view seems to square well with common ways of speaking, like:

He said many striking and true things.
Especially that snow is white.

Clearly this could be true even if 'he' spoke French. Admittedly there are readings for our example on the philosophical market that do not involve commitment to propositions (see e.g. Davidson [1969]). Still the propositional reading has considerable plausibility.

Even if I do not plan to give a full discussion of the propositional view, at least I will pose the question of the compatibility of certain solutions with a propositional account. So it is appropriate to say a few things about such an account.

The basic picture is somewhat like this:

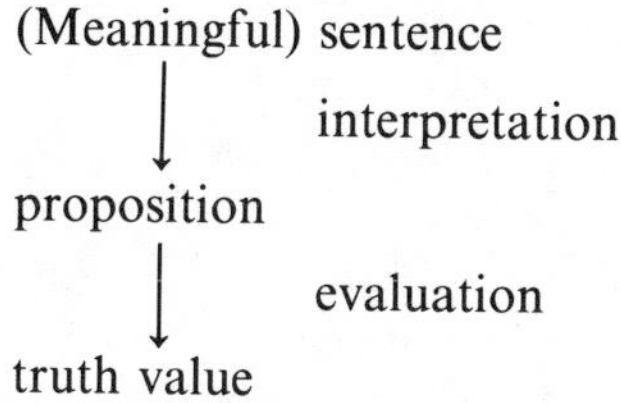

Any of the two arrows might fail (or at least I want to leave the possibility open): a meaningful sentence may fail to express a proposition, a proposition may lack a definite truth value. Propositions are precisely those things for which the question of truth or falsity may arise, not precisely those things that are definitely true or false. (The latter idea would prejudge by the very definition of proposition the debate on classical versus constructive logic.)

Kripke in his footnote 1, p. 691 or Kripke [1975] proposes an evasive strategy for the case that truth and falsity primarily apply to propositions: for a sentence to be true is to express a true proposition.

This translates:

This sentence is not true

to:

This sentence does not express a true proposition.

That translation seems to be a strengthened liar. Maybe we should rather look at:

This sentence is false

and translate:

This sentence expresses a false proposition.

However that may be the Liar variant studied in Kripke's paper has the form "This sentence is not true", so prima facie it is not clear whether Kripke's evasive strategy works at all. Moreover: if truth is primarily of propositions, then the sentential account is ultimately unsatisfactory: one wants an account in terms of the more basic notion.

A third possibility is: truth is a predicate of sentences, but not of all. Or maybe better: there are many different truth predicates but none for the whole language.

Such a view is of course Tarski's.

I want to give a brief presentation of Tarski's ideas and contrast these with what later became of them in the literature.

I follow the presentation of Tarski's popular Tarski [1944] rather than the basic Tarski [1956].

Tarski's problem is to give a 'materially adequate and formally correct' definition of Truth. In this context the Liar appears as an obstacle. Tarski analyses the way the Paradox arises as follows:

> (I) We have implicitly assumed the language in which the antinomy is constructed contains, in addition to its expressions, also the names of these expressions, as well as semantic terms such as the term "true" referring to sentences of this language; we have also assumed that all sentences which determine the adequate usage of this term can be asserted in the language. A language with these properties will be called "*semantically closed*".
>
> (II) We have assumed that in this language the ordinary laws of logic hold.
>
> (III) We have assumed that we can formulate and assert in our language an empirical premise such as the statement (2) which has occurred in our argument.
>
> (Tarski [1944], p. 348.)

Tarski's version of the Liar is of the "The sentence printed on page . . ." variety. The empirical premise involved is of the form: "The sentence printed on page . . . is "The sentence printed . . ."". Tarski rightly remarks that condition III can be circumvented. Other constructions have the same effect. (However if we have truth rather than satisfaction some replacement of III is necessary, see 3.2.1.) Rejection II would, he tells us, have grave consequences indeed, supposing it were possible at all. This leads him to consider I. Tarski decides not to try to solve the Paradox for natural language, for natural language seems to satisfy I and II. What is more: the problem of consistency for natural language has no clear meaning, natural language having no exactly specified structure. His solution is this: *only to use languages that are not semantically closed.* These include fragments of natural language and formalized languages. A formalized language is in Tarski's conception a language which has an exactly specified structure, where in specifying the structure we only refer to the form of the expressions involved.

Notoriously Tarski conflates language and theory. In the case of formalized languages axioms and inference rules belong to the language. He thinks of formalized languages and fragments of natural language as not different in kind.

Note that Tarski's "solution" of the paradox does not involve the object meta-language distinction at all. That distinction enters only when we try to solve his problem: to give a correct definition of truth.

> Since we have agreed not to employ semantically closed languages, we have to use two different languages in discussing the problem of the definition of truth and, more generally, any problems in the field of semantics. The first of these languages is the language which is "talked about" and which is the subject matter of the whole discussion; the definition of truth which we are seeking applies to the sentences of this language. The second is the language in which we "talk about" the first language, and in terms of which we wish, in particular, to construct the definition of truth for the first language. We shall refer to the first language as "*the object-language*" and to the second as "*the meta-language*".
>
> (Tarski [1944], pp. 349, 350.)

The important thing here to notice is that Tarski does not define some general distinction between language which is "talked about" and language in which we "talk about". Object and meta-language are not globalized versions of mention and use. He introduces these terms to be employed in a very specific context: the problem of giving a truth definition for one language, the subject matter of the discussion, in another language, the language we use to discuss the first language.

Tarski lists some features the meta-language should have to define truth for the object-language in it: it must contain the object-language as part, or at least the object-language must be translatable into it; the possibility must be present to construct names for the sentences of the object-language. Tarski does not stipulate these properties to be part of the definition of the meta-object-language pair. E.g. about the possibility of defining truth for the object-language in the meta-language he says:

> The solution turns out to be sometimes positive, sometimes negative. This depends upon some formal relations between the object-language and its meta-language; or, more specifically, upon the fact whether the meta-language in its logical part is "*essentially richer*" than the object-language or not. (Tarski [1944], p. 351.)

So "essential richness" is not part of the notion of meta-language.

Till now nothing has been said about the notion of a Hierarchy of Languages. The Hierarchy is introduced as follows:

> It should be noted that these terms "object-language" and "meta-language" have only a relative sense. If, for instance, we become interested in the notion of truth applying to sentences, not of our original object-language, but of its meta-language, the latter becomes automatically the object-language of our discussion; and in order to define truth for this language, we have to go to a new meta-language – so to speak; to a meta-language of higher level. In this way we arrive at a whole hierarchy of languages. (Tarski [1944], p. 350.)

To make sense of what Tarski says, we must, I think, assume that he means that we arrive at a Hierarchy of Languages if the project of defining truth is for each step carried out successfully. In this case the meta-languages will be essentially richer than their object-languages.

A hierarchy, built just on the idea of object and meta-language without adding the idea of success in constructing truth or satisfaction seems rather strange. "Object-language" and "meta-language" are terms, not reflecting some ontological distinction but referring to the roles certain languages play in a certain specific context. If for example we try on Monday to define truth for L_1 in L_2 and, having failed on Tuesday we have a go at defining truth for L_2 in L_1, then surely L_1 will not be two levels higher than itself (on Tuesday). L_1, L_2 and L_2, L_1 are only object-meta-pairs respectively on Monday and Tuesday. To call L_1 its own meta-meta-language is simply to think out of context. (Similarly we could call a barber shaving a barber a meta-barber, the barber who shaved the object-barber. Now suppose afterwards the latter shaves the former. Would that make him his own meta-meta-barber?)

Object- and meta-language may be fully interpreted but Tarski also considers the possibility that the meta-language *is* the object-language extended with undefined terms. These terms are then explained by axiomatizing their fundamental properties. This contrasts with for example an explicit definition of truth for number theory in the language of set theory.

In the literature and in conversation I have found any number of the following confused and conflated:

- object and meta-language as discussed above
- the language whose *grammar* is discussed versus the language in which this grammar is discussed
- mention and use
- a type theory for truth[1].

In particular 'making the object-language, meta-language distinction' has seemed the obvious solution to the Liar to some, because they *introduce* it as the, indeed obvious, distinction between mention and use (or more globally 'language mentioned' versus 'language used', without reference to the specific context of the semantic project) and subsequently *employ* the 'distinction' as a type theory for truth.

3.2.1. *Excurs: Classical models containing their own truth predicate.* As we have seen in the discussion of Grelling's paradox in Section 2.1, no classical model with quotation names for the formulae of its language in that language can contain its own satisfaction-relation. As a counterpoint: classical models with quotation names for the sentences of their languages in those languages can contain their own truth predicate.

This nice point is due to Gupta (see Gupta [1982]). To simplify matters a bit we put his construction on its head: Gupta starts with interpreted quotation names and an uninterpreted truth predicate and proceeds to interpret the truth predicate; we start with uninterpreted quotation names and an interpreted truth predicate and interpret the quotation names.

Let $\mathcal{M}$ be a model with language L such that:

(i) L is countable.

(ii) There is a formula $\varphi(x)$, with just x free, such that both the set X defined by $\varphi(x)$ in $\mathcal{M}$ and its complement Y in D, the domain of $\mathcal{M}$ are infinite.

Define a sequence of languages $(\mathrm{L}_i)_{i\in\omega}$ as follows:

$\mathrm{L}_0 := \mathrm{L}$

$\mathrm{L}_{n+1} :=$ the result of adding quotation names for the sentences of L_n to L_n and closing off in the usual manner.

Further: $\mathrm{L}_\omega := \cup_{n\in\omega}\mathrm{L}_n$. Clearly L_ω has quotation names for all sentences of L_ω.

We now proceed to extend $\mathscr{M}$ in stages $\mathscr{M}_0, \mathscr{M}_1, \ldots$ with interpretations for the quotation names of L_ω to $\mathscr{M}_\omega$, in such a way as to make $\varphi(x)$ the truth predicate for $\mathscr{M}_\omega$.

Let $X_1, X_2, \ldots$ be a sequence of pairwise disjoint countable subsets of X and $Y_1, Y_2, \ldots$ similarly of Y. We assume fixed enumerations of the X_i and Y_j.

Take $\mathscr{M}_0 := \mathscr{M}$. Suppose we have already defined $\mathscr{M}_k$ in such a way that $\mathscr{M}_k$ extends $\mathscr{M}$ by having interpretations for the quotation names occurring in L_k – so $\mathscr{M}_k$ is a model for L_k. Assume further that these interpretations are in $(X_1 \cup \cdots X_k) \cup (Y_1 \cup \cdots Y_k)$.

We construct $\mathscr{M}_{k+1}$ by interpreting the quotation names occurring in L_{k+1} not occurring in L_k. Let $\ulcorner\varphi_1\urcorner, \ulcorner\varphi_2\urcorner, \ldots$ enumerate these names. $\varphi_i \in L_k$, so the truth or falsity of φ_i is already fixed in $\mathscr{M}_k$.

Suppose $\ulcorner\varphi_1\urcorner, \ldots, \ulcorner\varphi_s\urcorner$ are interpreted already, define:

$$[\![\ulcorner\varphi_{s+1}\urcorner]\!] := \begin{cases} \text{the first } x \text{ in } X_{k+1} \text{ such that} \\ x \notin \{[\![\ulcorner\varphi_i\urcorner]\!] \mid i < s+1\} \text{ in case } \mathscr{M}_k \vDash \varphi_{s+1} \\ \text{the first } y \text{ in } Y_{k+1} \text{ such that} \\ y \notin \{[\![\ulcorner\varphi_i\urcorner]\!] \mid i < s+1\} \text{ in case } \mathscr{M}_k \nvDash \varphi_{s+1} . \end{cases}$$

Take $\mathscr{M}_\omega$ the union of the $\mathscr{M}_k$'s. As is easily seen:

$$\mathscr{M}_\omega \vDash (\psi \leftrightarrow \varphi(\ulcorner\psi\urcorner)) \text{ for all sentences } \psi \in L_\omega .$$

Note that we could as well have added a new predicate Tr satisfying the conditions above to $\mathscr{M}$ and then interpret quotation names as before for the extended language. Thus the above construction shows that if we extend a first order theory Γ, having only infinite models, in a language L to $\Gamma + \{(\varphi \leftrightarrow \mathrm{Tr}(\ulcorner\varphi\urcorner)) \mid \varphi \text{ a sentence of } L_\omega\}$, where L_ω is as before, then the new theory is conservative over Γ with respect to L.

3.2.2. *Excurs: Descending Hierarchies.* In his discussion of "Tarski's hierarchy of languages" Kripke states:

> One surprise to me was the fact that the orthodox approach by no means obviously guarantees groundedness in the intuitive sense mentioned above . . . Even if unrestricted truth-definitions are in question, standard theorems easily allow us to construct a *descending* chain of first order languages $L_0, L_1, L_2, \ldots$ such that L_i contains a truth predicate for L_{i+1}. I don't know whether such a chain can engender ungrounded sentences, or even quite how to state the problem here: some substantial questions in this area are yet to be solved.
>
> (Kripke [1975], pp. 697–698.)

To me "the orthodox approach" is simply too indefinite to be able to say whether it does or does not exclude a descending hierarchy. Here I just want to point out that *paradox like* phenomena may appear in descending hierarchies.

Let L be the language of arithmetic and let $L_i := L + Tr_i, Tr_{i+1}, \ldots$.

Take:

$T^+ :=$ Th($\mathbb{N}$) + Tarski style axioms for Tr_0 as truth predicate for L_1, for Tr_1 as truth predicate for L_2, etc.

So, e.g., if '$L_1(x)$' represents the arithmetical predicate 'to be a Gödel number of a sentence of L_1', then:

$$T^+ \vdash \forall x \forall y(L_1(x) \wedge L_1(y)) \to (Tr_0(\text{conj}(x, y)) \leftrightarrow$$

$$\leftrightarrow Tr_0(x) \wedge Tr_0(y)).$$

Note that we do not need the satisfaction relation to define truth here, arithmetic having the substitution function.

By a simple compactness argument we see that T^+ is consistent, for any finite set of axioms of T^+ can be interpreted in a finite *ascending* hierarchy (taking for example the empty set as interpretations of the truth predicates not occurring in our set of axioms.)

$$\text{Let } T^- := PA + \{Tr_n(\ulcorner\varphi\urcorner) \leftrightarrow \varphi \mid \varphi \in L_{n+1}, n \in \omega\}.$$

We claim T^- has no standard model, i.e. T^- has no model of the form:

$$\mathcal{M} = \mathbb{N}(\tilde{T}r_0, \tilde{T}r_1, \tilde{T}r_2, \ldots).$$

The moral here obviously is: even if strong T^+ is consistent in the usual sense, already weak T^- is 'semantically inconsistent' in the sense that it excludes the standard model.

Let $G_\Box, G_\neg, G_Q$ be (groups of) primitive recursive functions such that: for a standard Gödel numbering $\ulcorner\ \urcorner$:

$$G_\Box(\ulcorner\varphi\urcorner, \ulcorner\psi\urcorner) = \ulcorner(\varphi\Box\psi)\urcorner$$

$$G_\neg(\ulcorner\varphi\urcorner) = \ulcorner(\neg\varphi)\urcorner$$

$$G_Q(i, \ulcorner\varphi\urcorner) = \ulcorner(Qv_i\varphi)\urcorner$$

(where $\Box \in \{\wedge, \vee, \to, \leftrightarrow\}$, $Q \in \{\forall, \exists\}$).

Define by the recursion theorem for primitive recursive functions a primitive recursive f such that:

$$f(n) := \begin{cases} n \text{ if } n \text{ is not a Gödel number of a formula of} \\ \mathrm{L}_0 \text{ or } n = \ulcorner s = t \urcorner \\ G_{\Box}(f(\ulcorner \varphi \urcorner), f(\ulcorner \psi \urcorner)) \text{ if } n = \ulcorner (\varphi \Box \psi) \urcorner \\ G_{\neg}(f(\ulcorner \varphi \urcorner)) \text{ if } n = \ulcorner (\neg \varphi) \urcorner \\ G_{Q}(i, f(\ulcorner \varphi \urcorner)) \text{ if } n = \ulcorner (Qv_i \varphi) \urcorner \\ \ulcorner \mathrm{Tr}_{i+1}(f(t)) \urcorner \text{ if } n = \ulcorner \mathrm{Tr}_i(t) \urcorner. \end{cases}$$

Define further:

$$g(0, n) := n$$

$$g(m + 1, n) := f(g(m, n))$$

and:

$$\varphi^{(m)} \text{ is the sentence with Gödel number } g(m, \ulcorner \varphi \urcorner).$$

We can find a primitive recursive term t such that $t = \ulcorner \chi \urcorner$, where $\chi = (\forall x > 0 \neg \mathrm{Tr}_0(g(x, t)))$.

Suppose T^{-} has standard model $\mathscr{M}$. We reason 'in the model': the main point is that we conclude from $(\exists x \psi(x))$ to $\psi(\mathbf{n})$ for some numeral $\mathbf{n}$.

First we show: $\neg \chi^{(m)}$.

Right branch:

$$\frac{\frac{\frac{\frac{\frac{[\chi^{(m)}]^{①}}{\forall x > \mathbf{0} \neg \mathrm{Tr}_m (\overbrace{f \ldots f}^{m} (g(x, \ulcorner \boldsymbol{\chi} \urcorner)) \ldots)}}{\neg \mathrm{Tr}_m(\overbrace{f \ldots f}^{m} (g(\mathbf{1}, \ulcorner \boldsymbol{\chi} \urcorner)) \ldots)}}{\neg \mathrm{Tr}_m(\overbrace{f \ldots f}^{m+1} (\ulcorner \boldsymbol{\chi} \urcorner) \ldots)}}{\neg \mathrm{Tr}_m(\ulcorner \underline{\forall x > \mathbf{0} \neg \mathrm{Tr}_{m+1} (\overbrace{f \ldots f}^{m+1} (g(x, t)) \ldots)} \urcorner)}}{\exists x > \mathbf{0}\, \mathrm{Tr}_{m+1} (\overbrace{f \ldots f}^{m+1} (g(x, \ulcorner \boldsymbol{\chi} \urcorner)) \ldots)}$$

$$\frac{\exists x > \mathbf{0}\, \mathrm{Tr}_{m+1} (\overbrace{f \ldots f}^{m+1} (g(x, \ulcorner \boldsymbol{\chi} \urcorner)) \ldots)}{\frac{\frac{\mathrm{Tr}_{m+1}(\overbrace{f \ldots f}^{m+1} (g(\mathbf{n}, \ulcorner \boldsymbol{\chi} \urcorner)) \ldots)}{\chi^{(n+m+1)}}}{\mathrm{Tr}_m (g(\underline{n + m + 1}, \ulcorner \boldsymbol{\chi} \urcorner)}} \text{ some } n > 0$$

Left branch:

$$\frac{\frac{\frac{[\chi^{(m)}]^{①}}{\forall x > \mathbf{0} \neg \mathrm{Tr}_m (\overbrace{f \ldots f}^{m} (g(x, \ulcorner \boldsymbol{\chi} \urcorner) \ldots)}}{\neg \mathrm{Tr}_m (\overbrace{f \ldots f}^{m} (g(\underline{n + 1}, \ulcorner \boldsymbol{\chi} \urcorner) \ldots)}}{\neg \mathrm{Tr}_m(g (\underline{n + m + 1}, \ulcorner \boldsymbol{\chi} \urcorner))}$$

Conclusion:

$$\frac{\frac{\neg \mathrm{Tr}_m(g (\underline{n + m + 1}, \ulcorner \boldsymbol{\chi} \urcorner)) \qquad \mathrm{Tr}_m (g(\underline{n + m + 1}, \ulcorner \boldsymbol{\chi} \urcorner)}{\perp}}{\neg\, \chi^{(m)}}\ 1$$

Again reasoning in the model we derive absurdity:

$$\dfrac{\dfrac{\dfrac{\dfrac{\neg\chi \quad (\chi = \chi^{(0)})}{\exists x > \mathbf{0}\ \mathrm{Tr}_0(g(x, \ulcorner\boldsymbol{\chi}\urcorner))} \ \text{some } n > 0}{\mathrm{Tr}_0(g(\mathbf{n}, \ulcorner\boldsymbol{\chi}\urcorner))}}{\chi^{(n)}} \qquad \neg\chi^{(n)}}{\bot}$$

Hence no such $\mathscr{M}$ exists. □

3.3. *Logic*

Could it be that something is wrong with our logic or with our understanding of it?

We have to move with some care here. Logic as a deductive system is grounded in the semantics of the language. To discuss a change in logic is to discuss new or further insight in semantics. The problem of the Liar may interact with different views of semantics in different ways.

For convenience consider the derivation of paradox involving ℓ:

$$\dfrac{\dfrac{\neg I\ \dfrac{\neg E\ \dfrac{\dfrac{[\ell]^{①}}{\neg\ell} \qquad [\ell]^{①}}{\bot}}{\neg\ell}\ 1}{\ell} \qquad \dfrac{\dfrac{\dfrac{[\ell]^{②}}{\neg\ell} \qquad [\ell]^{②}}{\bot}}{\neg\ell}\ 2}{\bot}$$

(i) Naively it would seem that to block the derivation of paradox one must deny some step in the proof. This involves the idea that the global correctness depends only on the local correctness of the steps.

For example in a Prawitz/Martin–Löf style semantics where the semantical explanations are themselves in terms of proofs this need not be so. What makes a proof into a proof in this view is that we can see that it is reducible to a normal proof. What is wrong with above derivation (on a *variant* of such a semantics) is that it is not normalizable. See Prawitz [1965].

(ii) The step in the proof moving from ℓ to $\neg\ell$ or the one from $\neg\ell$ to ℓ. We could in fact represent the steps more explicitly, say:

$$\dfrac{\ell \equiv \neg\ell \quad \ell}{\neg\ell} \qquad \dfrac{\ell \equiv \neg\ell \quad \neg\ell}{\ell}$$

Where '$\equiv$' expresses the primary tie. Note that this tie is *prima facie* asymmetrical: the meaning of the right hand side is in some sense prior to the meaning of the left hand side.

This suggests that the relevant change in logic could very well be an *extension.* This point is made in Feferman [1983]. A solution along these lines is given in Aczel-Feferman [1980] and in Feferman [1983]. In fact these authors treat not truth or stipulative definitions but comprehension. But it seems their approach can be easily adapted.

(iii) $\neg I$ could fail.

The most popular approach under which $\neg I$ fails is clearly the Way of truth value gaps. The Liar fails to be definitely true or definitely false. Hence, even if we derive a contradiction from ℓ and so ℓ *cannot* be true, we need not conclude it is false. The idea to treat the paradoxes thus goes at least back to Bochvar (see Bochvar [1981]). (In fact it may go back to the Old Greeks, but Bochvar is certainly the first to give a worked out logical system, that is up to modern standards. On the other hand he does not construct a model.)

Just to claim that the Liar lacks a definite truth value, is next to saying nothing. Much is to be explained about the way undefinedness comes about. Sometimes this lack of a definite truth value is seen as a consequence of the outright meaninglessness of the Liar. But there are varieties of meaning and also varieties of lack of meaning. Certainly the Liar is not meaningless the way "Flubs blubs glorps" is. The meaninglessness approach seems to have things on its head: *that we in some sense perfectly understand the Liar is precisely the problem.*

Kripke is his footnote 18 (Kripke [1975], p. 700) takes a really surprising line: admission of truth value gaps does not invalidate classical logic. Why?

"Undefined" is not an extra truth value, anymore than – in Kleene's book – $\mathscr{U}$ is an extra *number* in Section 63. Nor should it be said that "classical logic" does not generally hold any more than (in Kleene) the use of partially defined functions invalidates the commutative law of addition. (Kripke [1975], footnote 18, p. 700.)

As I take it this argument compares $(\forall x \forall y\ x + y = y + x)$ with e.g. $(\forall p\, p \vee \neg p)$. Both of these sentences are considered to be true; in contrast not all sentences of the form $(\tau + \varrho) = (\varrho + \tau)$ and $(\phi \vee \neg \phi)$ need to be true: τ and ϱ are potentially partial numerical terms and ϕ a sentence not necessarily definitely true or false. Note that Kripke's discussion suggests that the range of '$\forall p$' is truth values rather than propositions. I did not yet fully quote footnote 18, it goes on thus:

If certain sentences express propositions, any tautological truth function of them expresses a true proposition. Of course formulas, even with the forms of tautologies, which have components that do not express propositions may have truth functions that do not express propositions either. (This happens under the Kleene valuation, but not under the van Fraassen.) Mere conventions for handling terms that do not designate numbers should not be called changes in arithmetic: conventions for handling sentences that do not express propositions are not in any philosophically significant sense "changes in logic". The term 'three-valued logic', occasionally used here, should not mislead. All our considerations can be formulated in a classical metalanguage.

(Kripke [1975], footnote 18, pp. 700–701.)

First a few side comments:

(1) The remark on 'a classical metalanguage' seems to be irrelevant: much of classical logic can be discussed in a, say, constructivistic metalanguage. That doesn't imply the term classical logic is inappropriately or misleadingly used.

(2) Kripke seems to view the valuation schemes both as conventions for handling sentences that do not express propositions and as conventions *creating* propositions. $(\phi \vee \neg \phi)$ expresses a proposition on the van Fraassen scheme, even if ϕ does not. Such a view seems to equate 'expressing a proposition' fully with 'having a definite truth value'. I think this is wrong. The 'expresses' aspect and the 'evaluation' aspect should be separated.

A tentative way to reconstruct Kripke's view is as follows:

(a) The proper locus of logic is at the level of propositions. Logic is about the evaluation of propositions.
(b) Propositions are all definitely true or false.
(c) Hence logic is classical, even if at the sentential level this need not show.

On this interpretation $\forall p(p \vee \neg p)$ quantifies over propositions.

Probably (a), (b), (c) are not what Kripke 'really' had in mind, but it is an interesting view anyway.

Proceeding from (a), (b), (c) there are two ways to go. We may view the evaluation schemes

I. As conventions for evaluating sentences that do not express propositions.
II. As the manifestations on the sentential level of the way sentences express propositions. E.g., in the van Fraassen valuation $(\phi \vee \neg \phi)$ is true, even in cases where ϕ does not express a proposition, just because $(\phi \vee \neg \phi)$ happens to express a proposition.

The first possibility looks philosophically suspect. For some sentences we abandon the idea that primarily the propositions they express are evaluated. The role of propositions is getting a bit obscure here. Moreover, consider the example of Nixon saying: "Everything Dean says about Watergate is false". Kripke points out that under unfavorable contingent circumstances Nixon's sentence may turn out to be paradoxical. One of the main advantages of his account is that under 'normal' conditions Nixon's sentence will be definitely true or false. But to get this advantage one needs a scheme like Strong Kleene or van Fraassen Supervaluations. I have the strange consequence that even if Nixon pronounces a true or false sentence, he does not say anything or: he does not make a statement, if Dean said e.g. "Nixon says something true about Watergate".

The second choice makes the *descriptive* problem for propositions very pressing. For example, suppose ϕ expresses a false proposition and χ does not express a proposition. On Strong Kleene $(\phi \wedge \chi)$ is false. But how is that proposition going to look, there being no constituent corresponding to χ?

I think it is more attractive to keep (a), but to deny (b).[2] Not all propositions are definitely true or false. Propositions are the *intended* domain of evluation, not the domain. This option will make Partial Logic a true logic.

A final point. Fixed points of the truth construction play a central role in Kripke's solution. As we will see these Fixed Points are intimately connected with the chosen valuation scheme. They are not like basic possible worlds over which one chooses this scheme or that. The point is that if a is a Fixed Point under scheme 1 one may go in a standard way from a to a fixed point under scheme 2, say b. But if we go from b to a fixed point c under scheme 1 this will not generally be a again. This suggests: *if* valuation schemes are more or less arbitrary, then so are Fixed Points.

(iv) $\neg E$ could fail. To opt for this possibility is to propose that logic is paraconsistent. Semantically this means that we take the Liar to be both true and false. I think this kind of view has much going for it, especially when thinking of various strengthened Liars. Note that a paraconsistent *semantics* need not necessarily licence judgment of $(\phi \wedge \neg \phi)$ in case ϕ is both true and false: for example we may only admit judgment of things that are uniquely true. Another possibility is that we employ an extension of van Fraassen Supervaluations. For further information see Priest [1979[3], 1984], Dowden [1979, 1984], Woodruff [1984], Visser [1984].

3.3.1 *Excurs: Negation and Implication, Partial and Constructive Logic.* Our reasoning involving ℓ is constructively valid. Constructively $\neg$ is often defined as $((\ ) \to \bot)$. $\neg I$ becomes a case of $\to I$ and $\neg E$ a case of Modus Ponens. $\bot E$ is not used in the proof, so in a paraphrased version just properties of $\to$ are employed. Thus we could obtain a 'local' version of the paradox as follows: let

t: The cat is on the mat

c: $(c \to t)$.

Reason as follows:

$$\dfrac{\dfrac{\dfrac{\dfrac{\dfrac{[c]^{①}}{c \to t} \quad [c]^{①}}{t}\,1}{(c \to t)}}{c} \qquad \dfrac{\dfrac{\dfrac{[c]^{②}}{c \to t} \quad [c]^{②}}{t}\,2}{(c \to t)}}{t}$$

Even if we do not have absurdity, we can use paradoxical reasoning to derive anything, so to speak locally. Note that above reasoning can be executed in minimal logic even. This version of the paradox is called Curry's Paradox (see Curry, Feys, Craig [1958]). It is to Löb's Theorem as the Liar is to Gödel's first incompleteness theorem.

Often people have blamed Excluded Third for the Paradox. But the above derivation shows that they are wrong.

Yet here is something puzzling. In the truth value gap approach a sentence ϕ has a definite truth value precisely if $(\phi \vee \neg \phi)$ is true. The lack of truth value of the Liar is supposed to be what blocks Paradox. So somehow it seems Excluded Third *is* intimately connected with the Paradox.

One source of the puzzlement is that we tend to infer from the fact that the paradoxical reasoning can be carried out in the deductive fragment of classical logic that is constructively valid, that the semantical *ground* of the paradox must be common to classical and constructive semantics.

To dissolve the puzzlement we can go two ways:

3.3.1.1. *Yes there is a common ground.* The point is simply that the Liar is blocked because it fails to express a proposition. Classically failure to express a proposition, assuming (a), (b) of 3.3 (iii), coincides with failure of Excluded Third for the relevant sentence. Constructively a proposition need not be definitely true or false. A sentence ϕ fails to express a proposition precisely if $(\phi \to \phi)$ fails.

Peter Aczel in his subtle paper on Frege Structures (Aczel [1980]) gives a solution that works both for constructive and for classical logic. His work employs the notion of proposition. However in my opinion his notion of proposition is closer to 'meaningful sentence' than to 'what a sentence expresses'. Interpreted classically Aczel's connectives turn out to be Weak Kleene, except his interpretation of $\to$ which, if we read $\to$ as $(\neg\, \chi \vee \varrho)$ in the metalanguage, has the table:

$\to$	T	F	$*$
T	T	F	$*$
F	T	T	T
$*$	$*$	$*$	$*$

It thus differs from Strong and from Weak Kleene $(\neg\, \chi \vee \varrho)$. Aczel's connectives taken together still do not make the Strong Kleene connectives definable. His clauses for implication are:

$(\phi \to \psi)$ is a proposition if (ϕ is a proposition and ψ is a proposition provided that ϕ is true)

$(\phi \to \psi)$ is true if (ψ is true if ϕ is true).

Aczel's approach could be adapted so that the classical reading will give Strong Kleene but then the clauses for 'proposition' will be quite awkward. Moreover for the constructive way of reading the definition they would make little sense (it seems to me).

Aczel's proposal is intended for the foundations of mathematics. If we would consider it as a proposal for natural language it is open to the objection that it does not handle cases of contingent paradoxicality well, as for example Nixon's sentence "Everything Dean says about Watergate is false". In case Dean said many true things about Watergate, but also said e.g. "Everything Nixon says about Watergate is true" then under a Weak Kleene style interpretation Nixon's sentence would not be false but undefined.

3.3.1.2. *No, there is no common ground.* The identification of $\neg$ with $((\ ) \to \bot)$ is a *change of meaning*. Classically if we take implication to be truth functional and consider only sentences that are definitely true or false, of course $\neg$ and $((\ ) \to \bot)$ coincide. Even assuming it is correct to take implication truth functionally on a classical fragment, i.e. a fragment where every sentence is definitely true or false, it does not follow that $\neg$ and $((\ ) \to \bot)$ are rightly identified if we consider a 'partial' fragment, a fragment in which some sentences do not have a definite truth value. Anyway: it seems to me that implication is not even truth functional on a classical fragment.

An example of a semantics for implication for partial valued logic is Richard Thomason's in Thomason [1969]. His implication is non-truth functional and monotonic; it satisfies both $\to I$ and $\to E$. His treatment is very similar to the Kripke Model semantics for Constructivistic Logic.

Assume for the moment negation and $(\ ) \to \bot$ are different connectives. It follows that the Liar and Curry's Paradox are really separate paradoxes. The fact that the classical derivation of Paradox from the Liar is similar to the constructive one is here more or less an accident.

If this is correct, it raises an important problem: can the usual solutions treat implicational paradoxes as Curry's? More specifically *can we extend for example Kripke's solution with, say, Thomasonian implication*?

A quick look shows that there is no obvious way to do this. The problem is that one would like to have e.g. $(\ell \leftrightarrow \neg\, \ell)$, which is clearly impossible.

I leave it to the reader to ponder on the choice between 3.3.1.1 and 3.3.1.2. The problem of extending Kripke's solution somehow with Thomasonian implication seems to me an important open problem. Evidently the philosophy of the *constraints* to be put on the solution is part of the problem!

3.4. *Tarski's convention T*

What is Tarski's convention T or, as some say, criterion T? Something like this: Let $\mathscr{L}$ be a language not involving "True" (or better "True-in-$\mathscr{L}$"), then for all φ in $\mathscr{L}$:

$$\ulcorner(\varphi \leftrightarrow \text{True (``}\varphi\text{'')})\urcorner \text{ is True.}$$

In stating T we employed Quine quotes; the expression between $\ulcorner$and$\urcorner$ including the quotes means:

The result of concatenating: "(", φ, "$\leftrightarrow$ True(" ", φ, " "))".

To disambiguate above explanation I should inform the reader that no comma occurs between quotes.

In our statement of the convention we *used* the notion of truth. For our present purposes that is not so bad, unlike Tarski we are not trying to define truth. Part of Tarski's problem was precisely to eliminate the (often suppressed) employment of truth above. He showed for example that for a language $\mathscr{L}$ with finitely many function and relation symbols and constants there is a finitely axiomatizable theory T_0 in $\mathscr{L}$ extended with "True" such that for all φ in $\mathscr{L}$:

$$T_0 \vdash (\varphi \leftrightarrow \text{True}(``\varphi")).$$

(By convention we leave the Quine quotes away after "$\vdash$".)

Moreover for any theory T_1 in $\mathscr{L}$ $T_1 + T_0$ is conservative over T_1 in $\mathscr{L}$. (Of course this need not be the case if certain *schemas* of T_1 are extended to $\mathscr{L}$ plus "True".)

Clearly when we are *told* the finitely many axioms of T_0 (supposing they are meaningful in *our* language we may – if we *accept* them – go on to judge true each instance (in $\mathscr{L}$) of $\ulcorner(\varphi \leftrightarrow \text{True}(``\varphi"))\urcorner$, without ever recurring to the general insight that for all $\varphi$ in $\mathscr{L}$ $\ulcorner(\varphi \leftrightarrow \text{True}(``\varphi"))\urcorner$ is True.

Problems about convention T both of detail and principle span the whole breadth of the Philosophy of Language. Certainly this is not the place to develop a perspective on all the issues. One point is that if we allow for example indexicals in $\mathscr{L}$ we cannot have above 'fixed point like' form of the convention but must bring in contexts.

The question that seems to be most urgent in the context of solutions to the Paradoxes is: what is the primary tie between φ and $\ulcorner\text{True}(``\varphi")\urcorner$. Tarski's convention gives a salient consequence of this tie for sentences expressing propositions using the classical connective $\leftrightarrow$. But is it very plausible that this tie is expressed adequately by $\leftrightarrow$? Moreover is it plausible to consider Tarski's first order theory T_0 as a definition? On a stipulatory account of the primary tie it would seem that this tie is asymmetric: $\ulcorner\text{True}(``\varphi")\urcorner$ is stipulated to have (in some sense) the same meaning as φ, not *vice versa*. On such a view breaches in the symmetry of Tarski's convention – even in absence of phenomena of indexicality and the like – could be admitted and perhaps even explained. The asymmetric view squares well with the iterative approaches to the Liar: the iterations reflect the basic stipulative direction: 'True("φ")$\urcorner$ being evaluated at each stage according to the evaluation of φ at earlier stages.

In the presence of truth value gaps one may doubt that φ and ⌜True("φ")⌝ have the same truth value even in unproblematic cases. On the "strong" view of truth ⌜True("φ")⌝ is 'intended' to be false, whenever φ is undefined. For a penetrating discussion of the strong versus the weak conception of truth, see Yablo [1985].

II: TWO MODERN APPROACHES TO THE LIAR

Part II treats work of Kripke, Herzberger and Gupta; it has structure 3 + 3 + 1. Roughly: the first '3' treats ideas mostly centered round fixed points; the second '3' treats ideas concerned with iterations; the '1' is a set of notes on Kripke fixed points and inductive definitions. Sections 1 and 4 are introductions to the ideas of Kripke respectively Herzberger and Gupta accompanied by some philosophical discussion. The reader not interested in the technological side could very well just read these. Sections 2 and 5 present some basic theory of fixed points and iterations in partial orders. Sections 3 and 6 give basic technical facts on Kripke fixed points respectively Herzberger/Gupta/Belnap iterations. I adopted the sub-division 3 + 3 to stress a philosophical point: not every aspect of the mathematical apparatus is intended to model a philosophical idea. For example iterations are a pleasant way to obtain fixed points, but there are other ways. If fixed points are the things one is after, one could dispense with iterations, if iterations are connected with the primary philosophical intuitions fixed points are (perhaps) just a nice bonus.

1. AN INFORMAL INTRODUCTION TO GROUNDEDNESS AND FIXED POINTS

In his Outline of a Theory of Truth Kripke opts for considering truth as a predicate. We follow him in this assumption here.

We start with an exposition of the concept of groundedness. We will try to see first whether the concept of groundedness all by itself can be philosophical foundation for a (partial) solution of the Liar. I will suggest that it is hard to see how the apparatus needed to explain certain intuitive distinctions can be motivated from the idea of groundedness alone. (The thesis that groundedness is primary should not be ascribed to Kripke: he holds that groundedness is an important notion. But he does not think that the use of fixed points should be motivated via groundedness.)

The idea of groundedness in semantics is originally due to Herzberger (see Herzberger [1970a]). Kripke explains this concept in terms of *the order of understanding*. The explanation runs somewhat like this:

Suppose someone, say Fred, is a competent speaker of language, as far as the part of language that is 'directly about the world' is concerned. So Fred has no problems at all with "The cat is on the mat." On the other hand Fred does not have the concept of truth yet.We offer Fred the following explanation:

(a) Truth is a predicate of sentences.

(b) You may assert ⌜"ϕ" is true⌝ precisely if you are entitled to assert ϕ; you may assert ⌜"ϕ" is not true⌝ precisely if you are entitled to assert not ϕ.

Fred starts puzzling out what we may have meant. The following is a 'picture' of his coming to grips with truth.

Stage 0: This is the initial stage, where Fred masters "The cat is on the mat", but not yet ""The cat is on the mat" is true". Fred does not yet know the truth conditions for the latter sentence.

There is a problem here, what about:

The cat is on the mat or "The cat is on the mat" is true.

In this stage the first disjunct has a definite truth value, the second has not. We may tell Fred to handle these cases by adopting some scheme of partial logic as Weak Kleene, Strong Kleene or Van Fraassen Supervaluations (see the chapter on Partial Valued Logic in this Handbook). When for example Fred adopts Strong Kleene the sentence receives a definite truth value just in cast the cat, indeed, is on the mat, and will not yet be evaluated in case the cat fails to be on the appropriate mat.

Stage 1: Fred applies (b) to see that he may assert ""The cat is on the mat" is true" if he may assert "The cat is on the mat". By (a) he may conclude things like: Some sentence is true. In stage 1 as in stage 0 Fred employs the chosen scheme for partial logic.

Stage 2: This stage is to stage 1, as stage 1 is to stage 0. here """The cat is on the mat" is true" is true" comes into the picture.

. . .

. . .

Stage ω: In stage ω Fred accumulates his previous insights. For example if $\phi_0 :=$ "The cat is on the mat", $\phi_{n+1} := \ulcorner$"ϕ_n" is true$\urcorner$, then for the first 'time' at ω *each* of these sentences is definitely evaluated. If the cat indeed is on the mat, Fred will be prepared to assert: for all n φ_n is true.

Stage $\omega + 1$: This stage is to stage ω as stage 1 is to stage 0. Example: if the cat is on the mat the sentence ""For all n φ_n is true" is true", gets evaluated true at $\omega + 1$.

. . .

. . .

Stage λ: This stage is to all stages α for $\alpha < \lambda$ as stage ω is to the finite stages.

. . .

Each stage extends (in the weak sense) the preceding stages. This fact depends on the proper choice of a scheme for partial logic: the schemes considered are monotonic. For example if "The cat is on the mat or "Bonzo is on the mat" is true" is evaluated true at stage 0, whatever ""Bonzo is on the mat" is true" is going to be at 1, the truth value remains fixed. It is easy to show that for some α stage α will be equal to stage α plus 1 (at least if the language can be treated as a set). Stage α will be a *fixed point* of the transition function that takes us from one stage to the next. Moreover as we will see stage α will be the *minimal* fixed point of the transition function.

Let us briefly look at what it means to be a fixed point of the transition function. Consider any sentence ϕ. First suppose that $\ulcorner$"ϕ" is true$\urcorner$ is true (false) at some stage β. Then certainly ϕ itself must be true (false) at β: how else could $\ulcorner$"ϕ" is true$\urcorner$ receive a truth value at β then by ϕ being true (false) at some earlier stage. Moreover truth (falsity) is preserved from earlier to later stages. Now consider the fixed point stage α. Suppose ϕ is true (false) at α, then $\ulcorner$"ϕ" is true$\urcorner$ will be true (false) at $\alpha + 1$. But stage $\alpha =$ stage $\alpha + 1$, so $\ulcorner$"ϕ" is true$\urcorner$ is true (false) at α. Combining we find:

ϕ is true (false) at α iff $\ulcorner$"ϕ" is true$\urcorner$ is true (false) at α.

A sentence is *grounded* if it gets evaluated at some stage. Clearly a sentence will be grounded just in case it has a truth value in the minimal fixed point of the transition function. Some illustrative examples. First consider:

This sentence is true.

Call this sentence "*S*", or "the Samesayer". *S* will not be evaluated at any stage. Certainly it is not evaluated at 0. Now suppose we are at α and that *S* is not evaluated before α. If α is a limit *S* will certainly not receive a truth value (being atomic). If α is a successor Fred must first inquire: What is denoted by "This sentence"? Why, *S*! Then Fred has to check the truth value of *S* at the predecessor stage. But there is none. So *S* is not evaluated at α.

Next consider:

No sentence spoken by a Cretan is true.

as said by Epimenides. Call this sentence "*E*".

Clearly if some Cretan once truly said "The cat is on the mat", *E* will be evaluated false at stage 1. On the other hand if all sentences, except *E*, ever spoken by Cretans are false, *E* will not get evaluated at any stage. Note that *E* could not become true. This example demonstrates an important aspect of the ungroundedness approach: whether sentences will be evaluated and, if so, at what stage is in general a matter of *contingent* fact. In a type theory we would have to provide a type in uttering the sentence. Types are added at the level of *syntax*. Stages are much like types, only they are at the level of *semantics*. In slogan: stages take care of themselves.

Remarks:

1.1. The first question that comes to mind is of course whether the metaphor of Fred is just a heuristic metaphor for the mathematical fixed point construction or an attempt to give a philosophical explanation and justification. I want to treat it as the latter. Maybe the metaphor should sometime be replaced by a 'straight' philosophical explanation, but it is not clear that such a thing can and should always be done.

One major problem with the story is the implicit move from assertability to evaluation. We want Fred to grasp the truth conditions of ""The cat is on the mat" is true" in general, not just when he knows whether or not a contextually determined cat is on a ditto mat. It seems that two moves are open to us. We could try to restate our explanation to Fred. He knows implicitly what it is for a sentence to be true or not, our explanation should just induce him to see the use of "is true" as articulation of his prior understanding. The second possibility is to provide an explanation of Fred's move from assertability to evaluation (for example by explaining evaluation in terms of assertability by an omniscient being). However, this problem is a problem for any philosophical attempt to explain truth. It occurs already in the first step in the stage picture. So I propose to leave it open here.

1.2. One definitely metaphorical aspect of Fred's story is the transfinite stages. Going through the transfinite stands for his insight in the generality of the process.

1.3. The picture of Fred and his stages is 'bottom to top'. There is a dual, 'top to bottom' picture of omniscient Paul, the genie of Semantics, who given a sentence involving truth in iterated ways starts evaluating the sentence, by finding the sentences on which the original sentence *depends* and proceeding to evaluate these, etcetera. For an eloquent mathematical commentary on Paul, see Yablo [1982].

1.4. Partiality enters the picture through the front door; as the partiality of *not yet*. At stage 0 sentences are given the 'value' undefined in case they are *not yet* evaluated. Similarly for the later stages. How to view partiality at the fixed point stage? We need not imagine Fred to share our insights in the special character of this stage, maybe he simply doggedly goes on with futile iterations. Here partiality has always the not yet character. More interestingly: we need not have told Fred that our explanation gave all there is to truth. Maybe truth is essentially open ended, say like the universe of ZFC. Then, even if he recognizes reaching a fixed point, for him partiality will keep the not yet character. The other possibility is to have Fred stop at the fixed point stage. We have told him: nothing more to truth than we have said. Now partiality is the partiality of *not ever*.

1.5. We also have instructed Fred how to handle falsity. At stage 0 the sentence ""The cat is on the mat" is true" does not receive a truth value. That does not mean that Fred at 0 may infer the sentence: ""The cat is on the mat" is not true". The sentence ""The cat is on the mat" is true" has just *not yet* received a truth value. Suppose, in fact, the cat is not on the mat. Then at 1 ""The cat is on the mat" is true" will be definitely evaluated as false, by our instruction. By monotonicity this is preserved to the later stages. What if a sentence is ungrounded? For example the Liar will have received no definite truth value at the fixed point stage. On the 'not ever' way of viewing the fixed point, it seems natural to say that hence the Liar is not true. Period. Here we can proceed distinguishing the *internal* negation of the stages picture and *external* negation, Fred arrives at by reflecting over all the stages up to the fixed point (see also Kripke's remarks and footnote on page 714). If we opt for the 'not yet' way of viewing the minimal fixed point, we are precluded from saying that the Liar is not true. Maybe some further explanation would make it true. (Homework problem: maybe Fred can *see* that whatever further explanation will be given none of them will make the Liar true. Elaborate.) Concluding: the 'not ever' picture raises the spectre of the strengthened Liar for us. It will then force us to go on with a further hierarchy. The 'not yet' picture seems to impose on us least to leave truth openended.

A second theme in Kripke's paper is the use of other fixed points of the transition function to analyze intuitive distinctions between sentences. Consider L, the Liar, and S, the Samesayer. There is an intuitive difference between these sentences. If L is true, then it is false, if false then true. S, in contrast, is true if true, false if false. We might want to say: L is paradoxical: it *cannot* be true, it *cannot* be false. S is what we might call 'biconsistent': it *could* be true, it *could* be false. Kripke's proposal is to analyze these intuitions in terms of all fixed points of the transition function. As we will show later: L has no definite truth value in any fixed point, but there are fixed points in which S is true, there are fixed points in which S is false. "S is true or S is not true" is true in some fixed points, false in none; but, if we employ Strong Kleene, not grounded.

Above ideas along generate the following notions:

True in – fixed points ↓ / False in – fixed points →	all	some	none
all	–	–	grounded true
some	–	biconsistent	only true
none	grounded false	only false	paradoxical

(Here: some = some, not all.)

Note that not all notions of this classification can be taken as 'meaning values' interacting in appropriate ways with the logical constants. For example both "S is true" and "S is not true" are biconsistent. However "S is true and S is not true" is only false, while "S is true and S is true" is biconsistent.

Above classification certainly does not exhaust the intuitive distinctions one might wish to make. Kripke gives the following example (for the Strong Kleene scheme):

This sentence is true or it isn't.

Call it "*I*". As is easily seen *I* is ungrounded, only true. So it cannot be given a truth value *arbitrarily*. In this it does not differ from "*S* is true or *S* is not true". To make "*S* is true or *S* is not true" true, however, we must make an arbitrary choice: either to make *S* true or to make *S* false. *I*, on the other hand, depends only on itself, so *I* itself only depends on sentences that cannot be given a truth value arbitrarily, and so on. *I* is so to speak heriditarily non-arbitrary, Kripke analyses this idea in terms of fixed points: a fixed point is non-arbitrary, Kripke calls it intrinsic, if it is compatible with any other fixed point, in other words: if it assigns to no sentence (formula) a truth value that conflicts with its truth value in any other fixed point. (Here, of course, we take truth and falsity to conflict, but undefined and truth, or undefined and falsity not.) Note that by monotonicity we need only demand compatibility on atomic formulas. Call a sentence *intrinsically true* if it is true in some intrinsic fixed point, *intrinsically false* if it is false in *some* intrinsic fixed point; *intrinsically undefined* if it is undefined in *all* intrinsic fixed points. We will see that there is a maximal intrinsic fixed point, an intrinsic fixed point extending all the others. So a sentence is intrinsically true (false, undefined) iff it is true (false, undefined) in the maximal intrinsic fixed point. (Note that this shows intrinsically true false and undefined *can* be treated as 'meaning values'.)

The maximal intrinsic fixed point is the largest non arbitrary fixed point that is in consonance with the intuitions behind Tarski's convention *T*. It might be considered as another proposal for a 'solution of the Liar'. Of course we would have to think of some philosophical motivation . . .

There is one point that stares one in the face from the pages of Kripke's paper. Nevertheless it took me quite some time to appreciate it. In the story of Fred the stages picture is supposed to be the ur-explanation. Arrival at the minimal fixed point is more or less just a happy consequence of the abundance of stages. Of course considered as a mathematical proof of the existence of the minimal fixed point the stages construction is just one way to do this. But philosophically speaking the fixed point is not the *terminus ad quem* but the stages construction is the basic thing. But if this is correct: *how came it about that other fixed points than the minimal one serve to analyze certain intuitive distinctions*?

It is not going to help to say that they evidently do their job nicely. Formal success is not at stake here. The question is what fixed points have at all to do with our understanding of sentences like L or S.

A closely related problem arises when we try to define validity.

Define first for a groundmodel $\mathscr{M}$ and partial predicate T:

$\Gamma \vDash_{\mathscr{M}(T)} \Delta$ iff either
if for all $\phi \in \Gamma$ $\mathscr{M}(T) \vDash \phi$ then for some $\psi \in \Delta$ $\mathscr{M}(T) \vDash \psi$
or
if for all $\psi \in \Delta$ $\mathscr{M}(T) \vDash \neg\omega$ then for some $\phi \in \Gamma$ $\mathscr{M}(T) \vDash \neg\phi$.

If the iteration process is the fundamental thing the following definition seems reasonable (assuming that we only consider models whose language is part of their domain):

$\Gamma \vDash \Delta$ if for all $\mathscr{M}(T)$, where T is the *minimal fixed point* of the transition function over $\mathscr{M}$ $\Gamma \vDash_{\mathscr{M}(T)} \Delta$.

But this definition gives us: $S \vDash L$, or the Samesayer logically implies the Liar. That seems counterintuitive. If we change the definition to:

$\Gamma \vDash \Delta$ if for all $\mathscr{M}(T)$, where T is *a* fixed point of the transition function over $\mathscr{M}$ $\Gamma \vDash_{\mathscr{M}(T)} \Delta$,

then clearly we will have:

$$L \vDash L,\ S \vDash S, \text{ but } S \nvDash L.$$

So in the most plausible definition of validity all fixed points play a role. This point seems to back up the idea that 'self referential meaning' is to be analysed in terms of all fixed points.

The idea that a good solution of the Liar should enable us to explicate intuitive distinctions like those above, strikes me as so convincing that I would like to make it an *adequacy condition* on solutions.

There seem to be several ways to proceed to answer the question.

A. One could argue that the fixed points come into the picture by the *next* reflection step of Fred. Fred takes stock of what has been accomplished by going through all those stages and recognizes he has arrived at a fixed point. Suppose he is told that the explanation given to him tells all there is to truth. One can imagine that Fred would then make the fixed point character of truth a principle of his reasoning. (Compare this to a

Martin-Löf style explanation of the meaning of 'or'. The introduction rule is fundamental; one arrives at the elimination rule by reflecting on the introduction rule and the fact that the introduction rule gives you all there is to disjunction.)

So the intuitive distinctions arise as a consequence of a secondary insight.

B. One may try to change the stages picture and give the intuitive distinctions in terms of the stages picture itself. Something like this is done in the Herzberger/Gupta solution. See 4.

C. One may dispute the claim that the stages picture is the fundamental thing. The minimal fixed point was the *terminus ad quem* after all. The basic insight in truth is nothing but that:

> ϕ is evaluated true (false) iff ⌜"ϕ" is true⌝ is evaluated true (false).

This seems to leave us without a rationale for choice among fixed points. But given our supposed basic insight why should one choose a designated fixed point? Presumably because we want a theory of judgment for sentences involving truth. If that is all we want, however, we do not need to designate a special fixed point: we can define validity outright in terms of all fixed points.

One way to do this for a given ground model $\mathcal{M}$ is in the style of partial valued logic (where also there is no designated value):

Define for a given ground model $\mathcal{M}$, assuming that $\mathcal{M}$'s language is part of the domain of $\mathcal{M}$:

$$\Gamma \vDash_{\mathcal{M}} \Delta \text{ iff for all fixed points } T \text{ of the transition function } \Gamma \vDash_{\mathcal{M}(T)} \Delta.$$

Clearly if T^* is the minimal fixed point:

$$\vDash_{\mathcal{M}} \phi \text{ iff } \mathcal{M}(T^*) \vDash \phi.$$

So the minimal fixed point has a role after all, but not by being designated. On the other hand:

$$S \vDash_{\mathcal{M}(T^*)} L, \text{ but } S \nvDash_{\mathcal{M}} L.$$

As we will see above definition can be reformulated in a neat way: first associate with ϕ a 'meaning value' over the ground model: $[\phi]_{\mathcal{M}} : [\phi]_{\mathcal{M}}$ is a partial function from fixed points to the truth values of ϕ in those fixed points. Then define $\vDash_{\mathcal{M}}$ in terms of meaning values.

2 FIXED POINTS AND STRUCTURES

2.1. DEFINITION. Consider any po (partial order) $\mathscr{D} = \langle D, \leqslant\rangle$. Let X be a subset of D.

(i) $\sqcup X(\mathscr{D})$, or when no confusion is possible $\sqcup X$, stands for the supremum of X in $\mathscr{D}$, if it exists.

(ii) $\sqcap X(\mathscr{D})$, or when no confusion is possible $\sqcap X$, stands for the infimum of X in $\mathscr{D}$, if it exists.

(iii) $\langle X, \leqslant\rangle$ stands for the po with domain X, where $\leqslant$ is the $\leqslant$ of $\mathscr{D}$ restricted to X.

(iv) for $d \in D$, define: $D(d) := \{d' \in D | d \leqslant d'\}$ and $\mathscr{D}(d) := \langle D(d), \leqslant\rangle$.

2.2. FACT. Consider a po $\mathscr{D} = \langle D, \leqslant\rangle$ and sets D' and X with $D' \subseteq D$ and $X \subseteq D'$.Let $\mathscr{D}' := \langle D', \leqslant\rangle$. Suppose $\sqcup X(\mathscr{D})$ exists and is in D', then $\sqcup X(\mathscr{D}')$ exists and $\sqcup X(\mathscr{D}) = \sqcup X(\mathscr{D}')$. Similarly for $\sqcap$.

Proof. trivial. □

2.3. DEFINITION. Let $\mathscr{D} = \langle D, \leqslant\rangle$ be a po.

(i) We call $X \subseteq D$ *consistent* if for every $x, y \in X$ there is a $z \in D$ such that $x \leqslant z$ and $y \leqslant z$.

(ii) We call $\mathscr{D}$ a *ccpo* or *coherent complete partial order* if every consistent $x \subseteq D$ has a supremum in $\mathscr{D}$.

2.4. FACT. Let $\mathscr{D} = \langle D, \leqslant\rangle$ be a ccpo, then:

(i) $\mathscr{D}$ has a bottom $*(\mathscr{D})$ or simply $*$.

(ii) Every *non empty* $X \subseteq D$ has an infimum.

(iii) For every $d \in D$ there is a maximal d' in D such that $d \leqslant d'$.

Proof. (i) $\varnothing$ is consistent and $* = \sqcup\varnothing$. (ii) Consider $X \subseteq D$ non empty; say $x_0 \in X$. Define:

$$\tilde{X} := \{y \in D | \text{ for every } x \in X\ y \leqslant x\}.$$

Every y in $\tilde{X}$ is smaller or equal to x_0, hence $\tilde{X}$ is consistent, so $\sqcup\tilde{X}$ exists. Every x in X is an upperbound for $\tilde{X}$, hence $x \geqslant \sqcup\tilde{X}$. So $\sqcup\tilde{X}$ is a lower bound for X. On the other hand any lower bound z of X is in $\tilde{X}$, hence

$z \leqslant \sqcup \tilde{X}$. Conclude $\sqcap X$ exists and $\sqcap X = \sqcup \tilde{X}$. (iii) By a routine application of Zorn's Lemma. □

2.5. REMARK. As the following examples show ccpo is a more restrictive notion than either complete lower semilattice or 'every directed $X \subseteq D$ has a supremum'. (A subset X of D is *directed* if for every x, y in X there is a z in X such that $x \leqslant z$ and $y \leqslant z$.)

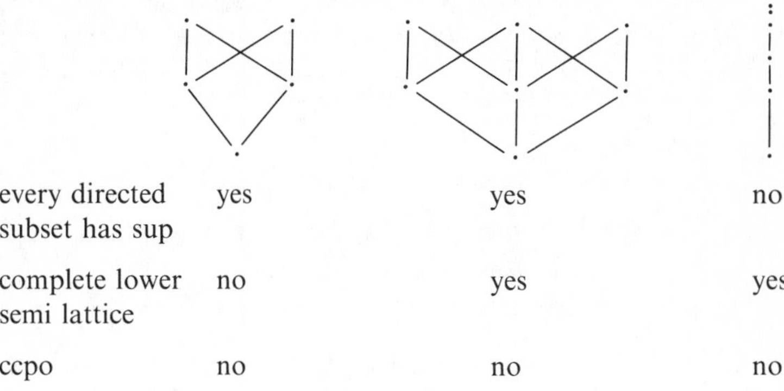

every directed subset has sup	yes	yes	no
complete lower semi lattice	no	yes	yes
ccpo	no	no	no

2.6. DEFINITION.

(i) $\mathbb{T} := \langle \{*, T, F\}, \leqslant \rangle$ where $\leqslant$ is given by:

(ii) Let $\mathscr{D} = \langle D, \leqslant \rangle$ be a po, A a set, then $\mathscr{D}^A := \langle D^A, \leqslant \rangle$, where $f \leqslant g$ if for all a in A $f(a) \leqslant g(a)$.

(iii) Let $\mathscr{D}_1 = \langle D_1, \leqslant \rangle, \ldots, \mathscr{D}_n = \langle D_n, \leqslant \rangle$ be po's then $\mathscr{D}_1 \times \mathscr{D}_2 \times \ldots \mathscr{D}_n := \langle D_1 \times D_2 \times \ldots D_n, \leqslant \rangle$, where $\langle d_1, \ldots, d_n \rangle \leqslant \langle d_1', \ldots, d_n' \rangle$ if $d_1 \leqslant d_1', \ldots, d_n \leqslant d_n'$.

2.7. FIRST GENERATION LEMMA.

(i) $\mathbb{T}$ *is a ccpo.*

(ii) *If* $\mathscr{D} = \langle D, \leqslant \rangle$ *is a ccpo,* $d \in D$, *then* $\mathscr{D}(d)$ *is a ccpo.*

(iii) *If* $\mathscr{D}$ *is a ccpo, A a set, then* $\mathscr{D}^A$ *is a ccpo.*

(iv) *If* $\mathscr{D}_1, \ldots, \mathscr{D}_n$ *are ccpo's, then* $\mathscr{D}_1 \times \mathscr{D}_2 \times \ldots \mathscr{D}_n$ *is a ccpo.*

Proof. (i), (ii) are trivial, (iv) is like (iii). (iii) Suppose $F \subseteq D^A$ is consistent in $\mathscr{D}^A$. For $a \in A$ define $F(a) := \{f(a) | f \in F\}$. Consider x, y in $F(a)$; there are f, g in F such that $x = f(a)$, $y = g(a)$. F is consistent, so for some

h in D^A: $h \geqslant f$, $h \geqslant g$. Hence $h(a) \geqslant f(a) = x$, $h(a) \geqslant g(a) = y$. Conclude $F(a)$ is consistent in $\mathscr{D}$.

Define $f_F : A \to D$ by: $f_F(a) := \sqcup F(a)(\mathscr{D})$.

For all $f \in F$ $f_F \geqslant f$. Consider any upperbound h of F. For any a in A, f in F: $h(a) \geqslant f(a)$, so $h(a) \geqslant \sqcup F(a)(\mathscr{D})$, i.e. $h \geqslant f_F$. Conclude: $f_F = \sqcup F(\mathscr{D}^A)$. □

2.8. DEFINITION. Let $\mathscr{D} = \langle D, \leqslant \rangle$ be a po, let f be monotonic on $\mathscr{D}$, i.e. $f: D \to D$ and for all d, d' in D if $d \leqslant d'$ then $f(d) \leqslant f(d')$. Define:

(i) $\mathrm{Up}\,(D, f) := \{d \in D \mid d \leqslant f(d)\}$
$\mathrm{Up}\,(\mathscr{D}, f) := \langle \mathrm{Up}(D, f), \leqslant \rangle$

(ii) $\mathrm{Down}\,(D, f) := \{d \in D \mid f(d) \leqslant d\}$
$\mathrm{Down}\,(\mathscr{D}, f) := \langle \mathrm{Down}\,(D, f), \leqslant \rangle$

(iii) $\mathrm{Fix}(D, f) := \{d \in D \mid f(d) = d\}$
$\mathrm{Fix}(\mathscr{D}, f) := \langle \mathrm{Fix}(D, f), \leqslant \rangle$.

2.9. LEMMA. *Let $\mathscr{D} = \langle D, \leqslant \rangle$ be a po, f monotonic on $\mathscr{D}$.*

(i) *Suppose $X \subseteq Up(D, f)$ and $\sqcup X(\mathscr{D})$ exists, then $\sqcup X(\mathscr{D})$ is in $Up(D, f)$ and hence $\sqcup X(Up(\mathscr{D}, f))$ exists and $\sqcup X(Up(\mathscr{D}, f)) = \sqcup X(\mathscr{D})$.*

(ii) *Suppose $X \subseteq Down(D, f)$ and $\sqcap X(\mathscr{D})$ exists. Then $\sqcap X(\mathscr{D})$ is in $Down(D, f)$ and hence $\sqcap X(Down(\mathscr{D}, f))$ exists and is equal to $\sqcap X(\mathscr{D})$.*

(iii) *Suppose $X \subseteq Fix(D, f)$ and $\sqcap X(Up(\mathscr{D}, f))$ exists, then $\sqcap X(Fix(\mathscr{D}, f))$ exists and is equal to $\sqcap X(Up(\mathscr{D}, f))$.*

Proof. (i) For x in $X: x \leqslant \sqcup X(\mathscr{D})$, hence $x \leqslant f(x) \leqslant f(\sqcup X(\mathscr{D}))$. So $\sqcup X(\mathscr{D}) \leqslant f(\sqcup X(\mathscr{D}))$, i.e. $\sqcup X(\mathscr{D})$ is in $\mathrm{Up}(D, f)$. The rest follows by 2.2. (ii) is the dual of (i). (iii) follows from (ii): $\mathrm{Fix}(\mathscr{D}, f) = \mathrm{Down}(\mathrm{Up}(\mathscr{D}, f), f \restriction \mathrm{Up}(D, f))$; $f \restriction \mathrm{Up}(D, f)$ is monotonic on $\mathrm{Up}(\mathscr{D}, f)$ and $\mathrm{Up}(D, f)$ is closed under f (if $d \leqslant f(d)$, then $f(d) \leqslant f(f(d))$). □

2.10. SECOND GENERATION LEMMA. If $\mathscr{D} = \langle D, \leqslant \rangle$ *is a ccpo and f monotonic on $\mathscr{D}$, then $Up(\mathscr{D}, f)$ is a ccpo.*

Proof. Suppose $X \subseteq \mathrm{Up}(D, f)$ is consistent in $\mathrm{Up}(\mathscr{D}, f)$ then X is *a fortiori* consistent in $\mathscr{D}$. hence $\sqcup X(\mathscr{D})$ exist, so by 2.9 (i) $\sqcup X(\mathrm{Up}(\mathscr{D}, f))$ exists. □

2.11. FACT. Suppose $\mathscr{D} = \langle D, \leqslant \rangle$ is a po, f monotonic on $\mathscr{D}$ and suppose m is maximal in $\mathrm{Up}(\mathscr{D}, f)$ then $m \in \mathrm{Fix}(D, f)$.

Proof. $m \leqslant f(m)$ hence $f(m) \leqslant f(f(m))$, so $f(m) \in \mathrm{Up}(D, f)$; m is maximal in $\mathrm{Up}(\mathscr{D}, f)$, so $m = f(m)$. □

2.12. DEFINITION. Let $\mathscr{D} = \langle D, \leqslant \rangle$ be a ccpo, f monotonic on $\mathscr{D}$. Define $(\ )_f^+ : \mathrm{Up}(D, f) \to \mathrm{Fix}(D, f)$ as follows:

$$(d)_f^+ := \sqcap \{d' \in \mathrm{Fix}(D, f) | d' \geqslant d\} \ (\mathrm{Up}(\mathscr{D}, f)).$$

Note. (1) There is a maximal element in $\mathrm{Up}(\mathscr{D}, f)$ above any d in $\mathrm{Up}(D, f)$, hence by 2.11 there is a fixed point above d. It follows that $\{d' \in \mathrm{Fix}(D, f) | d \leqslant d'\}$ is not empty. By 2.10 $\mathrm{Up}(\mathscr{D}, f)$ is a ccpo, hence $(d)_f^+$ exists.[4] (2) By 2.9 (iii): $(d)_f^+ \in \mathrm{Fix}(\mathscr{D}, f)$. (3) $(\ )_f^+$ is monotonic, inductive (i.e. for every $d \in \mathrm{Up}(D, f)$ $d \leqslant (d)_f^+$) and idempotent (i.e. for every $d \in \mathrm{Up}(D, f)$ $((d)_f^+)_f^+ = (d)_f^+$). In short: $(\ )_f^+$ is a *closure operation* on $\mathrm{Up}(\mathscr{D}, f)$.

2.13. THIRD GENERATION LEMMA. *Let $\mathscr{D} = \langle D, \leqslant \rangle$ be a ccpo, f monotonic on $\mathscr{D}$. Then $\mathrm{Fix}(\mathscr{D}, f)$ is a ccpo.*

Proof. Let X be consistent in $\mathrm{Fix}(\mathscr{D}, f)$, then X is *a fortiori* consistent in $\mathrm{Up}(\mathscr{D}, f)$. Hence $\sqcup X(\mathrm{Up}(\mathscr{D}, f))$ exists. Let $d_0 := (\sqcup X(\mathrm{Up}(\mathscr{D}, f)))_f^+$. First: d_0 is an upperbound for X in $\mathrm{Fix}(\mathscr{D}, f)$. Secondly any upperbound e of X in $\mathrm{Fix}(\mathscr{D}, f)$ is an upperbound for X in $\mathrm{Up}(\mathscr{D}, f)$, hence $\sqcup X(\mathrm{Up}(\mathscr{D}, f)) \leqslant e$. Conclude: $d_0 = (\sqcup X(\mathrm{Up}(\mathscr{D}, f)))_f^+ \leqslant (e)_f^+ = e$. Hence $d_0 = \sqcup X(\mathrm{Fix}(\mathscr{D}, f))$. □

2.14. REMARK. Note that $\sqcup$ in $\mathscr{D}$ and $\mathrm{Up}(\mathscr{D}, f)$ coincide and $\sqcap$ in $\mathrm{Up}(\mathscr{D}, f)$ and $\mathrm{Fix}(\mathscr{D}, f)$. The following generally do not coincide:

(i) $\sqcap$ on $\mathscr{D}$ and $\mathrm{Up}(\mathscr{D}, f)$

(ii) $\sqcup$ on $\mathrm{Up}(\mathscr{D}, f)$ and $\mathrm{Fix}(\mathscr{D}, f)$

(iii) $\sqcap$ on $\mathscr{D}$ and $\mathrm{Fix}(\mathscr{D}, f)$

(iv) $\sqcup$ on $\mathscr{D}$ and $\mathrm{Fix}(\mathscr{D}, f)$.

2.15. REMARK. Let $\mathscr{D} = \langle D, \leqslant \rangle$ be a ccpo, f monotonic on $\mathscr{D}$. $(\ )_f^+$ is defined on $\mathrm{Up}(\mathscr{D}, f)$. A natural question is whether the definition of $(\ )_f^+$ can be extended in a natural way to $\mathscr{D}$. There are several ways to do this. Here we treat one. Later we will meet another one.

2.15.1. *The First Stage.* Define for d in $\mathrm{Down}(D, f)$:

$$(d)_f^- := \sqcup \{x \leqslant d | x \leqslant f(x)\} \ (\mathscr{D}).$$

Clearly $\{x \leqslant d | x \leqslant f(x)\}$ is consistent in $\mathscr{D}$, hence $(d)_f^-$ is well defined. Set $z := (d)_f^-$. $z \leqslant d$, so $f(z) \leqslant f(d) \leqslant d$, so $f(z) \leqslant d$. By 2.9: $z \leqslant f(z)$, hence $f(z) \leqslant f(f(z))$. Conclude $f(z) \in \{x \leqslant d | x \leqslant f(x)\}$, so $f(z) \leqslant z$ and thus $f(z) = z$.

Evidently $(d)_f^-$ is the maximal fixed point below or equal to d. $(\ )_f^-$ is monotonic and idempotent on Down($\mathscr{D}$, f) and $(d)_f^- \leqslant d$ for $d \in$ Down(D, f).

2.15.2. *The Second Stage*. Consider $\sqcap\{f, id\}$ $(\mathscr{D}^D)$ or $f \sqcap id$ for short. As is easily seen:

$$\mathscr{D} \quad = \quad \text{Down}(\mathscr{D}, f \sqcap id).$$

Consider $(d)_{f\sqcap id}^-$, $(d)_{f\sqcap id}^- = f\sqcap id((d)_{f\sqcap id}^-) = f((d)_{f\sqcap id}^-) \sqcap (d)_{f\sqcap id}^-$, so $(d)_{f\sqcap id}^- \leqslant f((d)_{f\sqcap id}^-)$ or $(d)_{f\sqcap id}^- \in$ Up($\mathscr{D}, f$).

Define: $(\ )_f := (\ )_f^+ \circ (\ )_{f\sqcap id}^-$.

We have:

- if $d \in$ Up(D, f): $f\sqcap id(d) = d$ and hence $(d)_{f\sqcap id}^- = d$, so $(d)_f = (d)_f^+$
- if $d \in$ Down(D, f): every fixed point of f is a fixed point of $f \sqcap id$ hence $(d)_f^- \leqslant (d)_{f\sqcap id}^-$ on the other hand $(d)_{f\sqcap id}^-$ is in Up(D, f) hence by the proof of the first stage: $(d)_{f\sqcap id}^- \leqslant (d)_f^-$. Conclude $(d)_{f\sqcap id}^- = (d)_f^-$. Hence: $(d)_f = ((d)_{f\sqcap id}^-)_f^+ = ((d)_f^-)_f^+ = (d)_f^-$ ($(d)_f^-$ being a fixed point)
- $(\ )_f$ is monotonic and idempotent on $\mathscr{D}$.

2.16. SYSTEMATICAL REMARK. The choice of ccpo as the suitable notion for studying fixed points is motivated on the one hand by the fact that all relevant arguments work for this notion (in contrast to e.g. complete lower semilattices). Moreover if we want our class to contain $\mathbb{T}$ and to be closed under $(\mathscr{D})^A$ and Fix($\mathscr{D}, f$) (for f monotonic on $\mathscr{D}$) we find we need at least the ccpo's: one can show that for every ccpo $\mathscr{D} = \langle D, \leqslant \rangle$ there is an f, monotonic on $\mathbb{T}^D$ such that $\mathscr{D}$ is isomorphic with Fix($\mathbb{T}^D, f$).

2.17. HISTORICAL REMARK. The use of Up($\mathscr{D}, f$) as a means to show that Fix($\mathscr{D}, f$) is a ccpo is 'in essence' due to Martin and Woodruff (Martin and Woodruff [1975]) and independently to Manna and Shamir (Manna and Shamir [1978]). The construction of $(\ )_f$ in 2.15 is due to Manna and Shamir (Manna and Shamir [1978]).

2.18. *Intrinsicity*

2.18.1. DEFINITION. Let $\mathscr{D} = \langle D, \leqslant \rangle$ be a ccpo. $d \in D$ is called *intrinsic* if for all e in D $\{d, e\}$ is consistent, or: $d \sqcup e$ exists. Note that *prima facie* every ccpo has an intrinsic point, viz. $*(\mathscr{D})$.

2.18.2. THEOREM. *Let $\mathscr{D} = \langle D, \leqslant \rangle$ be a ccpo. Then the set I of intrinsic points in $\mathscr{D}$ has a maximum, the maximal intrinsic point i_0. In other words: $\langle I, \leqslant \rangle$ is a complete lattice.*

Moreover: $i_0 = \sqcap\{m \in D \mid m \text{ maximal in } \mathscr{D}\}$ $(\mathscr{D})$.

Proof. Down-up-Version: Clearly I is consistent, any point of it being consistent with any other point, even in D. Hence $\sqcup I$ exists. We claim: $i_0 = \sqcup I$. Consider $I \cup \{d\}$, for $d \in D$. I is consistent and d is consistent with every element of I, hence $I \cup \{d\}$ is consistent. So $\sqcup(I \cup \{d\})$ exists. $\sqcup I \leqslant \sqcup(I \cup \{d\})$, $d \leqslant \sqcup(I \cup \{d\})$, so $\{\sqcup I, d\}$ is consistent. Conclude $\sqcup I$ is intrinsic. Evidently $\sqcup I$ is maximal in I.

Up-Down Version: Let M be the set of maximal points of $\mathscr{D}$. We claim $i_0 = \sqcap M$. Consider an intrinsic d and an $m \in M$. $d \sqcup m$ exists, m is maximal hence $d \sqcup m = m$ or $d \leqslant m$. So $d \leqslant \sqcap M$.

Now consider any e in D. There is an m in M such that $e \leqslant m$, surely $\sqcap M \leqslant m$, hence $\{e, \sqcap M\}$ is consistent. Conclude $\sqcap M$ is intrinsic. □

2.18.3. HISTORICAL REMARK. The maximal intrinsic point was independently discovered by Kripke (Kripke [1975]) and by Manna and Shamir (see e.g. Manna and Shamir [1976]). Manna and Shamir call it the Optimal point. The observation $i_0 = \sqcap M$ is in their paper.

2.18.4. DEFINITION AND FACTS. let $\mathscr{D} = \langle D, \leqslant \rangle$ be a ccpo. For d in D $\mathscr{D}(d)$ is again a ccpo $\mathscr{D}(d)$ has its own maximal intrinsic point. In $\mathscr{D}$ this is, as it were, the maximal non arbitrary extension of d. We will call such extensions of elements of D: *relatively maximal intrinsic*. This suggests the following definition:

Define $i: D \to D$ as follows: $i(d) := \sqcap\{m \mid m \text{ maximal in } \mathscr{D}, m \geqslant d\}$ $(\mathscr{D})$.

Define further: $\text{Int}(\mathscr{D}) := \langle i(D), \leqslant \rangle$.

Note. (1) $i(d)$ is maximal intrinsic in $\mathscr{D}(d)$. (2) $i_0 = i(*(\mathscr{D}))$. (3) i is a *closure operation* on $\mathscr{D}$, i.e. i is *monotonic*, i is *inductive* (or for all $d \in D$ $d \leqslant i(d)$), i is *idempotent* (or for all $d \in D$ $i(d) = i(i(d))$). (4) $\text{Int}(\mathscr{D})$ is a ccpo, for $\text{Int}(\mathscr{D}) = \text{Fix}(\mathscr{D}, i)$ and i is monotonic on

$\mathscr{D}$. (5) $\sqcap(\text{Int}(\mathscr{D}))$ coincides with $\sqcap(\mathscr{D})$ for $\mathscr{D} = \text{Up}(\mathscr{D}, i)$ and $\text{Int}(\mathscr{D}) = \text{Fix}(\mathscr{D}, i)$, by remark 2.14.

2.18.5. FACT. Consider a ccpo $\mathscr{D} = \langle D, \leqslant \rangle$, f monotonic on $\mathscr{D}$. We have: $\text{Int}(\text{Up}(\mathscr{D}, f)) = \text{Int}(\text{Fix}(\mathscr{D}, f))$.

Proof. It is sufficient to show $\text{Up}(\mathscr{D}, f)$ and $\text{Fix}(\mathscr{D}, f)$ have the same relatively maximal intrinsic points. First note: m is maximal in $\text{Up}(\mathscr{D}, f)$ iff m is maximal in $\text{Fix}(\mathscr{D}, f)$ (see 2.11). So:

d is relatively maximal intrinsic in $\text{Up}(\mathscr{D}, f)$ iff

$d = \sqcap\{m | m \text{ maximal in } \text{Up}(\mathscr{D}, f), m \geqslant d\}\ (\text{Up}\ (\mathscr{D}, f))$ iff

$d = \sqcap\{m | m \text{ maximal in } \text{Fix}(\mathscr{D}, f), m \geqslant d\}\ (\text{Fix}(\mathscr{D}, f))$ iff

d is relatively maximal intrinsic in $\text{Fix}(\mathscr{D}, f)$.

(The second step uses 2.14.) □

2.18.6. OBSERVATION. The above shows that we can go directly from $\text{Up}(\mathscr{D}, f)$ to $\text{Int}(\text{Fix}(\mathscr{D}, f))$. Note the similarity of the proof in 2.18.5 with the construction of Fix $(\mathscr{D}, f)$ from $\text{Up}(\mathscr{D}, f)$.

2.18.7. COROLLARY. *Let $\mathscr{D} = \langle D, \leqslant \rangle$ be a ccpo, then $Int(Int(\mathscr{D})) = Int(\mathscr{D})$.*

By 2.18.5 $\text{Int}(\mathscr{D}) = \text{Int}(\text{Up}(\mathscr{D}, i))$
$= \text{Int}(\text{Fix}(\mathscr{D}, i))$. □

2.18.8. REMARK. Yet another, still fancier way of introducing intrinsicity and related concepts: Let $\mathscr{D} = \langle D, \leqslant \rangle$ be a ccpo. Define $\text{IndMon}(\mathscr{D})$ as the po of the inductive monotonic functions on $\mathscr{D}$ with the usual pointwise induced ordering. One can show:

(a) $\text{IndMon}(\mathscr{D})$ is a complete lattice.

(b) d is intrinsic in $\mathscr{D}$ iff for some f in $\text{IndMon}(D)$ $d = f(*(\mathscr{D}))$.

(c) For f, g in $\text{IndMon}(D)$: $f \circ g$ is in $\text{IndMon}(\mathscr{D})$ and $f \sqcup g \leqslant f \circ g$.

(d) The top i of $\text{IndMon}(\mathscr{D})$ satisfies $i \circ h = h \circ i = i$ for h in $\text{IndMon}(D)$, and hence $i \circ i = i$, or i is idempotent.

(e) $\langle i(D), \leqslant\rangle = \mathrm{Int}(\mathscr{D})$ (moreover for every indempotent h in IndMon(D): $\langle h(D), \leqslant\rangle$ is a ccpo).

2.19. *Preliminaries for comparing valuation schemes*

The following pair of theorems is useful for comparing the fixed points of different valuation schemes.

2.19.1. THEOREM. *Suppose $\mathscr{D}_1 = \langle D_1, \leqslant\rangle$ and $\mathscr{D}_2 = \langle D_2, \leqslant\rangle$ are ccpo's. Let g be monotonic from $\mathscr{D}_1$ to $\mathscr{D}_2$, f be monotonic from $\mathscr{D}_2$ to $\mathscr{D}_1$. Moreover suppose for x in D_1: $f(g(x)) \geqslant x$ and for $y \in D_2$: $g(f(y)) \leqslant y$. (In other words the pair $\langle g, f\rangle$ is an adjunction.)*
Then:

(i) *$f \circ g$ and $g \circ f$ are monotonic and idempotent.*

(ii) *$Fix(\mathscr{D}_1, f \circ g)$ is isomorphic with $Fix(\mathscr{D}_2, g \circ f)$ (by g and f).*

(iii) *$Int(Fix(\mathscr{D}_1, f \circ g)) = Int(\mathscr{D}_1)$.*

(iv) *$*(\mathscr{D}_2) = *(Fix(\mathscr{D}_2, g \circ f))$.*

Proof. (i) The only non trivial point is idempotency. We have: $x \leqslant f \circ g(x)$, so $g(x) \leqslant g \circ f \circ g(x)$, and $g \circ f \circ g(x) = g \circ f(g(x)) \leqslant g(x)$. Hence $g \circ f \circ g = g$, so $g \circ f \circ g \circ f = g \circ f$ and $f \circ g \circ f \circ g = f \circ g$. (ii) Consider d in $\mathrm{Fix}(D_1, f \circ g)$. Then $g(d)$ is in $\mathrm{Fix}(D_2, g \circ f)$, for: $g \circ f(g(d)) = g \circ f \circ g(d) = g(d)$. Moreover $f(g(d)) = f \circ g(d) = d$. Similarly for d' in $\mathrm{Fix}(D_2, g \circ f)$. Finally: f, g are monotonic. (iii) $f \circ g$ is monotonic and inductive, by 2.18.5:

$$\begin{aligned}\mathrm{Int}(\mathrm{Fix}(\mathscr{D}_1, f \circ g)) &= \mathrm{Int}(\mathrm{Up}(\mathscr{D}_1, f \circ g)) \\ &= \mathrm{Int}(\mathscr{D}_1).\end{aligned}$$

(iv) Easy. □

2.19.2. THEOREM. *Let $\mathscr{D} = \langle D, \leqslant\rangle$ be a ccpo. Suppose f, g monotonic on $\mathscr{D}$ and $f \leqslant g$ in $\mathscr{D}^D$. We have*: (i) *For $x \in Fix(D, f)$: $x \leqslant g(x)$. For $y \in Fix(D, g)$: $f(y) \leqslant y$.* (ii) *$(\ )^+_g$ and $(\ )^-_f$ are monotonic respectively from $Fix(\mathscr{D}, f)$ to $Fix(\mathscr{D}, g)$ and from $Fix(\mathscr{D}, g)$ to $Fix(\mathscr{D}, f)$. Moreover: for $x \in Fix(\mathscr{D}, f)$: $x \leqslant ((x)^+_g)^-_f$, for $y \in Fix(\mathscr{D}, g)$: $((y)^-_f)^+_g \leqslant y$.*

Proof. (i) $x = f(x) \leqslant g(x)$, $y = g(y) \geqslant f(y)$. (ii) E.g.: $x = (x)^-_f \leqslant ((x)^+_g)^-_f$. □

3. FIXED POINTS, STIPULATIONS AND SATISFACTION

After the technical preliminaries of Section 2, we just have to give the transition functions to get the appropriate fixed points for stipulations and satisfaction. This section can be very well read skipping the materials on Sat. Where the analogy of stipulations and satisfaction is very close I often just presented the material on stipulations.

3.1. DEFINITION. (i) Let A be a set. Define: $\mathscr{L}(A)$ is the smallest set such that $A \subseteq \mathscr{L}(A)$, $\perp \in \mathscr{L}(A)$ and if ϕ is in $\mathscr{L}(A)$ then $(\neg\phi)$ is in $\mathscr{L}(A)$ and if ϕ, ψ are in $\mathscr{L}(A)$ then $(\phi \wedge \psi)$, $(\phi \vee \psi)$ are in $\mathscr{L}(A)$.

A stipulation list for A is a function $\mathscr{S} : A \to \mathscr{L}(A)$. For a given stipulation list $\mathscr{S}$, $F_{\mathscr{S}}$ is a function $\{T, F, *\}^A \to \{T, F, *\}^A$. $F_{\mathscr{S}}$ is defined by: $F_{\mathscr{S}}(f)(a) = [\mathscr{S}(a)]f$. Here $[\phi]f$ is elliptic for either $[\phi]^{\text{SK}}f$ or $[\phi]^{\text{VF}}f$, i.e. respectively the Strong Kleene valuation of ϕ at f or the Van Fraassen valuation of ϕ at f. For convenience when treating points that hold equally for both schemes we will suppress the superscripts 'SK' and 'VF' for [ ] and *derived notions*.

Remarks. (ia) $F_{\mathscr{S}}$ is monotonic on $\mathbb{T}^A$. (ib) It would have been a bit more realistic to include the usual propositional variables in $\mathscr{L}(A)$ and relativize $F_{\mathscr{S}}$ to assignments on those variables. We prefer to leave this as an exercise to the reader: the only point below where the added generality would be interesting is 'axiomatization'. Even there the necessary adaptations are more or less trivial. (ic) *Notation*: we write for example

$$a : a \vee \neg b$$

$$b : a \wedge \neg b$$

$$c : c$$

for the list $\mathscr{S}$: $\{a, b, c\} \to \mathscr{L}(\{a, b, c\})$ with $\mathscr{S}(a) = (a \vee \neg b)$, $\mathscr{S}(b) = (a \wedge \neg b)$, $\mathscr{S}(c) = c$.

(ii) Let $\mathscr{M}$ be an *acceptable structure* (see Moschovakis [1974]). The reader not familiar with acceptable structures can simply think of $\mathbb{N} = \langle \omega; (\ )', +, \cdot, \uparrow; \text{o}\rangle$, the standard model of the natural numbers with successor, addition, multiplication, exponentiation and zero. Suppose the domain of $\mathscr{M}$ is M. Some properties of acceptable structures we will use:

- The relation '$y = \langle x_1, \ldots, x_n\rangle$' is representable in $\mathscr{M}$. For notational convenience we will proceed *as if* '$\langle t_1, \ldots, t_n\rangle$' is a term of the language of $\mathscr{M}$ in case $t_1, \ldots, t_n$ are terms of that language.

- Let $\mathscr{L}$ be the language of $\mathscr{M}$ extended with a two place relation symbol Sat. We assume – to simplify notations – that there are constants **m** in $\mathscr{L}$ for each m in M.
- $\mathscr{L}$ is embedded in some standard way in M (see Moschovakis [1974] pp. 67, 68). Say ϕ is represented by $\ulcorner\phi\urcorner$.

Let S be a *partial* two place relation on M, i.e. $S: M \times M \to \{T, F, *\}$. We write $[\![\phi]\!]\ (\mathscr{M}(S), f)$ for the evaluation of formula ϕ at $\mathscr{M}(S)$ and assignment $f (f: \{x_1, x_2, \ldots\} \to M)$. As in (i) we adopt the policy of being neutral between SK and VF, unless we want to make a point specific to one of the two. Then we write $[\![\ ]\!]^{SK}$ etc.

Suppose $m = \langle m_1, \ldots, m_k\rangle$. Write h^m for the assignment with:

$$h^m(x_i) = \begin{cases} m_i \text{ iff } i \leqslant k \\ 0 \text{ otherwise.} \end{cases}$$

Here 0 is some fixed element of M. Define:

$$\mathrm{Sat}_{\mathscr{M}(S)}(n, m) := \begin{cases} [\![\phi]\!]\ (\mathscr{M}(S), h^m) \text{ if} \\ \quad n = \ulcorner\phi\urcorner \text{ and } m = \langle m_1, \ldots, m_k\rangle \\ * \text{ otherwise.} \end{cases}$$

Define the transition function as follows: $\Phi_{\mathscr{M}}(S) := \mathrm{Sat}_{\mathscr{M}(S)}$.

Remarks. (iia) $\Phi_{\mathscr{M}}$ is monotonic on $\mathbb{T}^{M \times M}$. (iib) The restriction to acceptable structures is for nearly all purposes too heavy and definitely wrong for axiomatization. We adopt it to simplify the presentation.

3.2. FACT. (i) Let A be a set and $\mathscr{S}$ be a stipulation list on A. By 2.13:

$$\mathrm{Fix}(\mathbb{T}^A, F_{\mathscr{S}}) \text{ is a ccpo.}$$

Write: $\mathrm{Fix}(\mathscr{S}) := \mathrm{Fix}(\mathbb{T}^A, F_{\mathscr{S}})$.

Let f be in the domain of $\mathrm{Fix}(\mathscr{S})$, let a be in A, say $\mathscr{S}(a) = \phi$; we have: $[a]f = f(a) = F_{\mathscr{S}}(f)(a) = [\mathscr{S}(a)]f = [\phi]f$.

(ii) Let $\mathscr{M}$ be acceptable, M its domain. $\mathbb{T}^{M \times M}$ is a ccpo. Hence by 2.13 so is $\mathrm{Fix}(\mathbb{T}^{M \times M}, \Phi_{\mathscr{M}})$.

Write: $\mathrm{Fix}(\mathscr{M}) := \mathrm{Fix}(\mathbb{T}^{M \times M}, \Phi_{\mathscr{M}})$.

Consider S in the domain of $\mathrm{Fix}(\mathscr{M})$, we have:

$$S(n, m) = \Phi_{\mathscr{M}}(S)(n, m) = \mathrm{Sat}_{\mathscr{M}(S)}(n, m).$$

Hence:

$$S(\ulcorner\phi\urcorner, \langle m_1, \ldots, m_k\rangle) \;=\; [\![\phi]\!](\mathcal{M}(S), h^{\langle m_1, \ldots, m_k\rangle}).$$

Write '$\chi \leftrightarrow \varrho$' (or if necessary '$\chi \leftrightarrow_S \varrho$' or even $\chi \leftrightarrow_{\mathcal{M}(S)} \varrho$) for:

for all assignments g: $[\![\chi]\!]\ (\mathcal{M}(S), g) = [\![\varrho]\!]\ (\mathcal{M}(S), g)$.

Suppose the free variables of ϕ are among $x_1, \ldots, x_k$, we have:

$$[\![\mathrm{Sat}(\ulcorner\phi\urcorner, \langle x_1, \ldots, x_k\rangle)]\!]\ (\mathcal{M}(S), g) \;=\; S(\ulcorner\phi\urcorner, \langle g(x_1), \ldots, g(x_k)\rangle) = [\![\phi]\!]\ (\mathcal{M}(S), h^{\langle g(x_1), \ldots, g(x_k)\rangle}) = $$
$$= [\![\phi]\!]\ (\mathcal{M}(S), g).$$

(The last step is by a standard lemma.)
Hence:

$$\mathrm{Sat}(\ulcorner\phi\urcorner, \langle x_1, \ldots, x_k\rangle) \leftrightarrow \phi.$$

More generally one easily shows:
$\mathrm{Sat}(\ulcorner\phi\urcorner, \langle t_1, \ldots, t_k\rangle) \leftrightarrow \phi[t_1/x_1] \ldots [t_k/x_k]$ provided that all free variables of ϕ are among $x_1, \ldots, x_k$ and the t_i are substitutable for x_i in ϕ. Note that this essentially depends on the fact that we have no undefined terms in our model.

3.3. REMARK. Using 2.16 one can show that for every *finite* ccpo $\mathcal{D} = \langle D, \leqslant\rangle$ there is a finite set A (in fact A can be chosen to be $D \cup \{\ell\}$ for $\ell \notin D$) and a stipulation list $\mathcal{S}$ for A such that $\mathcal{D}$ is isomorphic to $\mathrm{Fix}^{\mathrm{SK}}(\mathcal{S})$.

3.4. THE PREDIAGONAL LEMMA FOR SAT. *Let $\mathcal{M}$ be as in* 3.1. *Extend $\mathcal{L}$ with $\in$ and a set variable X (occurring only free) to, say, $\mathcal{L}(X)$. Define a mapping $(\ )[\{x_i|\psi\}/X]$ from $\mathcal{L}(X)$ to $\mathcal{L}$ as follows: $(\phi)[\{x_i|\psi\}/X]$ is the result of substituting $\psi[t/x_i]$ for $(t \in X)$ in ϕ. Here we assume that none of the variables of t is a bound variable of ψ and that none of the free variables of ψ except possibly x_i is a bound variable of ϕ. If this is not the case rename bound variables.*
Write

'$\phi(\{x_i|\psi\})$' *for* '$(\phi(X))[\{x_i|\psi\}/X]$'.

Let $\phi(x_1, x_2, \ldots, x_s, X)$ *be a formula of* $\mathscr{L}(X)$ *with all free variables shown, then there is a* $\psi(x_1, \ldots, x_s)$ *in* $\mathscr{L}$ *such that for all assignments* g:

$$[\![\psi(x_1, \ldots, x_s]\!](\mathscr{M}(\Phi_{\mathscr{M}}(S)), g) =$$
$$= [\![\phi(x_1, \ldots, x_s, \{x_1 | \psi\})]\!](\mathscr{M}(S), g).$$

Proof. Consider:

$$\chi(x_1, \ldots, x_s, x_{s+1}) := \phi(x_1, \ldots, x_s,$$
$$\{x_1 | \mathrm{Sat}(x_{s+1}, \langle x_1, \ldots, x_s, x_{s+1}\rangle)\}).$$

Let $m := \ulcorner\chi\urcorner$, define:

$$\psi(x_1, \ldots, x_s) := \mathrm{Sat}(\mathbf{m}, \langle x_1, \ldots, x_s, \mathbf{m}\rangle).$$

We have:

$$[\![\psi(x_1, \ldots, x_s)]\!]\ (\mathscr{M}(\Phi_{\mathscr{M}}(S)), g) =$$
$$= [\![\mathrm{Sat}(\mathbf{m}, \langle x_1, \ldots, x_s, \mathbf{m}\rangle)]\!]\ (\mathscr{M}(\Phi_{\mathscr{M}}(S)), g) =$$
$$= \Phi_{\mathscr{M}}(S)(m, \langle g(x_1), \ldots, g(x_s), m\rangle) =$$
$$= [\![\chi(x_1, \ldots, x_s, x_{s+1})]\!]\ (\mathscr{M}(S), h^{\langle g(x_1), \ldots g(x_s), m\rangle}) =$$
$$= [\![\chi(x_1, \ldots, x_s, \mathbf{m})]\!]\ (\mathscr{M}(S), g) =$$
$$= [\![\phi(x_1, \ldots, x_s, \{x_1 | \mathrm{Sat}(\mathbf{m}, \langle x_1, \ldots, x_s, \mathbf{m}\rangle)\})]\!]$$
$$(\mathscr{M}(S), g) = [\![\phi(x_1, \ldots, x_s, \{x_1 | \psi\})]\!]\ (\mathscr{M}(S), g). \quad \square$$

3.5. THE DIAGONAL LEMMA FOR SAT. *For any formula* $\phi(x_1, \ldots, x_s, X)$ *of* $\mathscr{L}(X)$ *there is a formula* $\psi(x_1, \ldots, x_s)$ *of* $\mathscr{L}$ *such that for any fixed point S of* $\Phi_{\mathscr{M}}$:

$$\psi(x_1, \ldots, x_s) \leftrightarrow_S \phi(x_1, \ldots, x_s, \{x_1 | \psi\}).$$

Proof. Immediate from 3.4. □

3.6. REMARK. We leave it to the reader to prove the *simultaneous* prediagonal and diagonal lemma. E.g. the simultaneous diagonal lemma is stated as follows: let S be a fixed point of $\Phi_{\mathscr{M}}$, then:
for every $\phi_1(x_1, \ldots x_s, X_1, \ldots, X_k), \ldots, \phi_k(x_1, \ldots, x_s, X_1, \ldots, X_k)$ there are $\psi_1(x_1, \ldots, x_s), \ldots, \psi_k(x_1, \ldots, x_s)$ such that for $i = 1, \ldots, k$:

$$\psi_i(x_1, \ldots, x_s) \leftrightarrow_S \phi_i(x_1, \ldots, x_s, \{x_1 | \psi_1\}, \ldots, \{x_1 | \psi_k\}).$$

3.7. SOME EXAMPLES. (i) *For stipulations*:

List	$\mathrm{Fix}^{\mathrm{SK}}(\mathscr{L})$	$\mathrm{Fix}^{\mathrm{VF}}(\mathscr{L})$	Example of . . .
$s:s$	s $\neg s$	*idem*	s is biconsistent
$\ell:\neg\ell$	$\cdot$	*idem*	ℓ is paradoxical
$i:i \vee \neg i$	i	$\cdot i$	i is SK only true, SK maximally intrinsically true, not SK minimally true.
$\ell:\neg\ell$ $a:\ell \wedge \neg\ell$ $b:a \vee b$	b	$\neg a, b$ $\neg a, \neg b$ $\neg a$	The SK maximal intrinsic fixed point and the VF maximal intrinsic fixed point are incomparable (one can show: they are alwasy consistent).
$a:a \vee b$ $b:\neg a$	$a, \neg b$	*idem*	The VF maximal intrinsic fixed point is not the VF minimal one.

(ii) *For Satisfaction*. Let $\mathscr{M}$ be as usual, let P be any partial valued subset of M, i.e. $P: M \to \{T, F, *\}$. Then there is a fixed point S of $\Phi_{\mathscr{M}}$ and a formula $\psi(x_1)$ having just x_1 free, such that ψ represents P in $\mathscr{M}(S)$, i.e. for every n in M and any g:

$$P(n) = [\![\psi(\mathbf{n})]\!]\,(\mathscr{M}(X), g).$$

Proof. Consider the formula $(x_1 \in X)$ of $\mathscr{L}(X)$. Let ψ be the corresponding diagonal sentence constructed in 3.5, clearly:

$$\begin{aligned} \chi(x_1, x_2) &= (x_1 \in \{x_1 \mid \mathrm{Sat}(x_2, \langle x_1, x_2\rangle)\}) \\ &= \mathrm{Sat}(x_2, \langle x_1, x_2\rangle), \\ m &= \ulcorner\chi\urcorner, \\ \psi(x_1) &= \mathrm{Sat}(\mathbf{m}, \langle x_1, \mathbf{m}\rangle). \end{aligned}$$

Define:

$$S_P(q, r) := \begin{cases} P(n) \text{ if } q = m,\ r = \langle n, m\rangle \\ * \text{ otherwise.} \end{cases}$$

(Here n is arbitrary and $m = \ulcorner\chi\urcorner$ as above.)

We have:

$$\Phi_{\mathscr{M}}(S_P)(m, \langle n, m\rangle) = [\![\chi]\!](\mathscr{M}(S_P), h^{\langle n,m\rangle}) = S_P(m, \langle n, m\rangle).$$

Hence: $S_P \leqslant \Phi_{\mathscr{M}}(S_P)$, by 2.15.2 there is a minimal fixed point S of $\Phi_{\mathscr{M}}$ extending S_P. We find:

$$\begin{aligned} P(n) &= S_P(m, \langle n, m\rangle) \leqslant S(m, \langle n, m\rangle) \\ &= [\![\mathrm{Sat}(\mathbf{m}, \langle \mathbf{n}, \mathbf{m}\rangle)]\!]\ (\mathscr{M}(S), g) = [\![\psi(\mathbf{n})]\!]\ (\mathscr{M}(S), g). \end{aligned}$$

To see that identity holds, consider:

$$P^+ \text{ with } P^+(n) := \begin{cases} P(n) \text{ if } P(n) \text{ is } T \text{ or } F \\ T \text{ otherwise} \end{cases}$$

and

$$P^- \text{ with } P^-(n) := \begin{cases} P(n) \text{ if } P(n) \text{ is } T \text{ or } F \\ F \text{ otherwise.} \end{cases}$$

Define S_{P+}, $S_{P_{+}^{-}}$ similar to S_P above; there are just as for S_P minimal fixed points S^+ and S^- extending S_{P+} and S_{P-} respectively. We have: $S \leqslant S^+$, $S \leqslant S^-$. So if $P(n)$ is undefined we have:

$$S(m, \langle n, m\rangle) \leqslant S^+(m, \langle n, m\rangle) = T$$

$$S(m, \langle n, m\rangle) \leqslant S^-(m, \langle n, m\rangle) = F.$$

Hence $S(m, \langle n, m\rangle) = *$.
Conclude $\psi(x_1)$ represents P in $\mathscr{M}(S)$. □

Remark. Above theorem shows that $\mathbb{T}^M$ can be embedded in Fix($\mathscr{M}$). As a consequence $\mathscr{P}(M)$ and $\mathbb{R} = \langle R, \leqslant\rangle$ can be embedded in Fix($\mathscr{M}$).

3.8. *Comparing Strong Kleene and Van Fraassen*

We just consider stipulations, Sat being analogous. How to compare Strong Kleene to Van Fraassen fixed points. One idea is to look at natural ways to go from one kind to the other. Consider:

$$\mathrm{SK} := (\ )_{F^{\mathrm{SK}}_{\mathscr{S}}}\ \mathrm{VF} := (\ )_{F^{\mathrm{VF}}_{\mathscr{S}}}.$$

Clearly $F^{\mathrm{SK}}_{\mathscr{S}} \leqslant F^{\mathrm{VF}}_{\mathscr{S}}$. Let f be a strong Kleene and g be a Van Fraassen fixed point. By 2.19.2 (i): $f \leqslant F^{\mathrm{VF}}_{\mathscr{S}}(f)$, $g \geqslant F^{\mathrm{SK}}_{\mathscr{S}}(g)$. Hence by 2.15: VF($f$) is the *minimal* Van Fraassen fixed point extending f, SK(g) is the *maximal* Strong Kleene fixed point below g. So SK and VF suggest themselves as natural.

By 2.19.2 (ii) SK and VF are monotonic respectively from $\mathrm{Fix}^{\mathrm{VF}}(\mathscr{S})$ to $\mathrm{Fix}^{\mathrm{SK}}(\mathscr{S})$ and from $\mathrm{Fix}^{\mathrm{SK}}(\mathscr{S})$ to $\mathrm{Fix}^{\mathrm{VF}}(\mathscr{S})$. Moreover as is easily seen: if f is a Strong Kleene fixed point: $f \leqslant \mathrm{SK} \circ \mathrm{VF}(f)$. So SK, VF, $\mathrm{Fix}^{\mathrm{SK}}(\mathscr{S})$, $\mathrm{Fix}^{\mathrm{VF}}(\mathscr{S})$ satisfy the conditions of 2.19.1.

Call a Strong Kleene fixed point f stable if $f = \mathrm{SK} \circ \mathrm{VF}(f)$, call a Van Fraassen fixed point g stable if $g = \mathrm{VF} \circ \mathrm{SK}(g)$. The structures of these stable fixed points are:

$$\widetilde{\mathrm{Fix}}^{\mathrm{SK}}(\mathscr{S}) := \mathrm{Fix}(\mathrm{Fix}^{\mathrm{SK}}(\mathscr{S}), \mathrm{SK} \circ \mathrm{VF})$$

$$\widetilde{\mathrm{Fix}}^{\mathrm{VF}}(\mathscr{S}) := \mathrm{Fix}(\mathrm{Fix}^{\mathrm{VF}}(\mathscr{S}), \mathrm{VF} \circ \mathrm{SK}).$$

By 2.19.1 (ii): $\widetilde{\mathrm{Fix}}^{\mathrm{SK}}(\mathscr{S})$ and $\widetilde{\mathrm{Fix}}^{\mathrm{VF}}(\mathscr{S})$ are isomorphic.
By 2.19.1 (iii): Int $(\widetilde{\mathrm{Fix}}^{\mathrm{SK}}(\mathscr{S})) = \mathrm{Int}(\mathrm{Fix}^{\mathrm{SK}}(\mathscr{S}))$. Or: the intrinsic Strong Kleene fixed points are precisely the intrinsic stable fixed points. By 2.19.1 (iv): the minimal Van Fraassen fixed point is stable.

Example: Let $\mathscr{S}$ be:

$\ell : \neg\ell$

$a : a \vee (\ell \vee \neg\ell)$

$b : c \wedge (\ell \vee \neg\ell)$

$c : b \vee \neg c.$

We have:

$\mathrm{Fix}^{\mathrm{SK}}(\mathscr{S})$	$\mathrm{Fix}^{\mathrm{VF}}(\mathscr{S})$	$\widetilde{\mathrm{Fix}}^{\mathrm{SK}}(\mathscr{S})$ $\widetilde{\mathrm{Fix}}^{\mathrm{VF}}(\mathscr{S})$
• a	• a, b, c	
|	|	
•	• a	• a

So: the maximal intrinsic VF fixed point is *not* stable, nor is the minimal SK fixed point.

3.9. *Fixed Point Valued Semantics*

We develop Fixed Point Valued Semantics here for the case of Stipulations. The case of Sat is fully analogous (except for the problem of axiomatization). We treat just the SK case. Technically there is no obstacle to do the VF case, philosophically however it is very awkward: doubling an idea as it were. As will be illustrated Fixed Point Valued Semantics does not evade the problem of *choice*: one can always consider interesting substructures of fixed points to build the meaning values.

Let $\mathbb{T}^+ := \langle\{T, F, *\}, \sqsubseteq\rangle$ where $\sqsubseteq$ is given by:

T — $*$ — F (vertical chain: T top, $*$ middle, F bottom)

Clearly $\mathbb{T}^+$ is a complete lattice.

Let A be a set, $\mathscr{S}$ a stipulation list on A. Consider the following possible sets of meaningvalues:

$$V_F := \{\Psi \mid \Psi \text{ monotonic from } \mathrm{Fix}^{\mathrm{SK}}(\mathscr{S}) \text{ to } \mathbb{T}\}$$

$$V_I^- := \{\Psi \mid \Psi \text{ monotonic from } \mathrm{Int}(\mathrm{Fix}^{\mathrm{SK}}(\mathscr{S})) \text{ to } \mathbb{T}\}$$

$$V_M := \{\Psi \mid \Psi \text{ from } \{\mathrm{m} \mid \mathrm{m} \text{ maximal in } \mathrm{Fix}^{\mathrm{SK}}(\mathscr{S})\} \text{ to } \{\mathrm{T}, \mathrm{F}, *\}\}.$$

(Here: 'F' stands for 'Full', 'I' for 'Intrinsic' 'M' for 'Maximal', 'V' for 'Values'.)

Let τ be in $\{F, I, M\}$, let $\sqsubseteq$ be the order on V_τ induced by $\sqsubseteq$ of $\mathbb{T}^+$: i.e. $\Psi \sqsubseteq \Psi'$ iff for all f in the appropriate domain $\Psi(f) \sqsubseteq \Psi'(f)$.
Define $\tau\text{-val}(\mathscr{S}) := \langle V_\tau, \sqsubseteq\rangle$.
As is easily seen $\tau\text{-val}(\mathscr{S})$ is a complete lattice.
We write $[\phi]_\tau$ for the element of V_τ such that for all f of the appropriate domain $[\phi]_\tau(f) = [\phi]^{\mathrm{SK}}f$.

3.9.1. *Example* (*coloured samesayers*). Samesayers receive different meaningvalues by having different 'colours'. E.g. let $\mathscr{S}$ be:

$$a : a \quad b : b.$$

Then $[a]_\tau \neq [b]_\tau$.

This illustrates that meaningvalues are not 'propositional' but capture a notion 'closer to syntax'.

Let $\Gamma, \Delta \subseteq \mathscr{L}(A)$, define:

$$\Gamma \models_\tau \Delta \text{ iff } \sqcap\{[\phi]_\tau \mid \phi \in \Gamma\}\ (\tau\text{-val}(\mathscr{S})) \sqsubseteq \sqcup\{[\psi]_\tau \mid \psi \in \Delta\}\ (\tau\text{-val}(\mathscr{S})).$$

Call the domain of $\phi \in V_\tau : D_\tau$.

We have:

$$\Gamma \models_\tau \Delta \text{ iff for all } f \text{ in } D_\tau$$
$$\sqcap\{[\phi]_\tau f \mid \phi \in \Gamma\}\ (\mathbb{T}^+) \sqsubseteq \sqcup\{[\psi]_\tau f \mid \psi \in \Delta\}\ (\mathbb{T}^+)$$
$$\text{iff for all } f \text{ in } D_\tau$$
$$\begin{cases} \text{if for all } \phi \in \Gamma\ [\phi]^{\text{SK}} f = T \text{ then for some } \psi \in \Delta\ [\psi]^{\text{SK}} f = T \\ \text{and} \\ \text{if for all } \psi \in \Delta\ [\psi]^{\text{SK}} f = F \text{ then for some } \phi \in \Gamma\ [\phi]^{\text{SK}} f = F. \end{cases}$$

As is easily seen:

$$\Gamma \models_F \Delta \Rightarrow \Gamma \models_I \Delta$$
$$\Gamma \models_I \Delta \Rightarrow \Gamma \models_M \Delta.$$

3.9.2. *Example*. Let $\mathscr{S}$ be:

$a:\neg b$

$b:\neg a$

$c:d \wedge (c \vee \neg c)$

$d:d.$

Then:

$d \models_I c$, but $d \not\models_F c$

$\models_M \neg(a \wedge b)$, but $\not\models_I \neg(a \wedge b)$.

Possibly $\models_M$ captures our tendency to say of a, b: they *can't* both be true.

3.9.3. AXIOMATIZATION. Given $\mathscr{S}$, can we axiomatize $\models_F$? Consider the relation $\vdash_F$ generated by: (i) the rules of propositional partial valued logic, (ii) $a \dashv\vdash_F \phi$, for $a \in A$, where $\phi = \mathscr{S}(a)$. (Here $\chi \dashv\vdash_F \varrho$, stands for $\chi \vdash_F \varrho$ and $\varrho \vdash_F \chi$). A partial valuation f is in $\text{Fix}(\{T, F, *\}^A, F^{\text{SK}}_{\mathscr{S}})$ precisely if for all $a \in A$ and $\phi = \mathscr{S}(a)$:
$a \models_F \phi$ and $\phi \models_F a$, i.e. $[a]^{\text{SK}}_f \sqsubseteq [\phi]^{\text{SK}}_f$ and $[\phi]^{\text{SK}}_f \sqsubseteq [a]^{\text{SK}} f$, in other words $[a]^{\text{SK}} f = [\phi]^{\text{SK}} f$. Hence a partial valuation f is in $\text{Fix}(\{T, F, *\}^A, F^{\text{SK}}_{\mathscr{S}})$

precisely if it satisfies $a \dashv\vdash_F \phi$ for all $a \in A$, $\phi = \mathcal{S}(a)$. Applying the completeness theorem for propositional partial logic we find $\vdash_F = \models_F$.

3.9.3.1. *Open Question.* Axiomatize $\models_I$, $\models_M$.

To get a similar result for Sat we must adapt our approach by also considering 'non-standard acceptable structures'. The result will then be that the proper axiomatization (for the F-case) of Sat extending a given theory having only non-standard acceptable structures as models, is of the form:

$$\mathrm{Sat}(\ulcorner\phi\urcorner, \langle x_1, \ldots, x_n\rangle) \dashv\vdash \phi(x_1, \ldots, x_n, \mathbf{0}, \ldots, \mathbf{0}).$$

4. THE REVISION PICTURE OF TRUTH

Gupta and Herzberger independently discovered the idea of treating the paradoxes using iterations of classical structures.

Suppose Socrates says:

Aristotle speaks falsely

and suppose Aristotle says:

Socrates speaks truly.

Suppose further these are the only things they say. Call above sentences respectively S and A. Clearly if S is true, A is true and then S must be false. S's falsity will make A false. So S becomes true. Etc.

Of course we can formalize above reasoning in predicate logic and arrive at a contradiction. But let us view what is going on more semantically: as the unwinding of our semantical intuitions. The reasoning can be represented as follows:

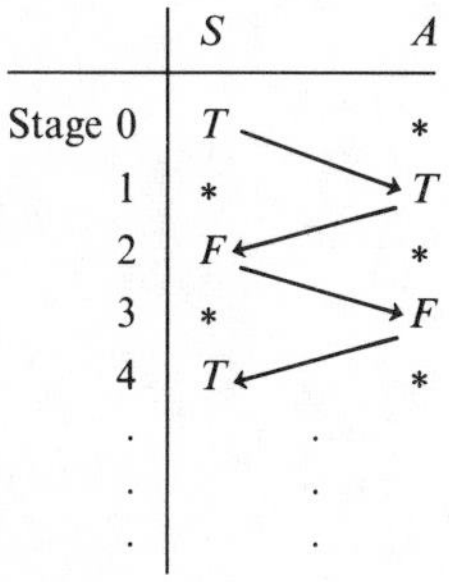

	S	A
Stage 0	T	*
1	*	T
2	F	*
3	*	F
4	T	*
.	.	
.	.	
.	.	

This representation is not yet the kind of process Gupta and Herzberger consider. Their idea is that a stage is like a classical world, a world in which the truth predicate is already fully evaluated – either by stipulation or by inheritance from earlier stages. So we should start by taking for example S true and A false. We get:

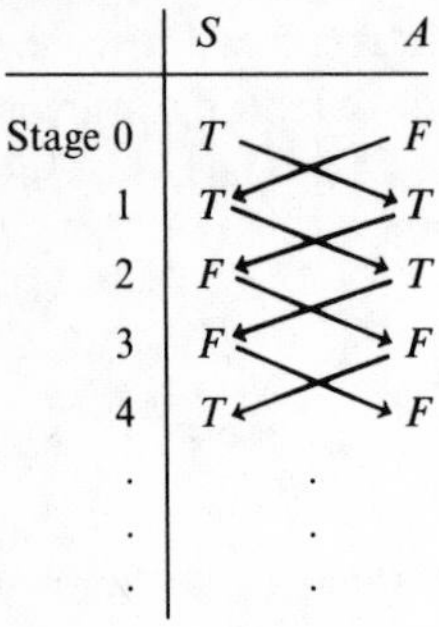

	S	A
Stage 0	T	F
1	T	T
2	F	T
3	F	F
4	T	F
⋮	⋮	⋮

Note that the reevaluations occur with the ticks of an inexorable clock for all sentences at the same time. Let L_0 be:

L_0 is false

and let L_1 be:

L_1 is false.

Gupta/Herzberger unwindings are e.g.:

	L_0	L_1
Stage 0	T	F
	↓	↓
1	F	T
	↓	↓
2	T	F
	↓	↓
3	F	T
⋮	⋮	⋮

and

	L_0	L_1
Stage 0	T	T
	↓	↓
1	F	F
	↓	↓
2	T	T
	↓	↓
3	F	F
⋮	⋮	⋮

But not the following "Einsteinean" process

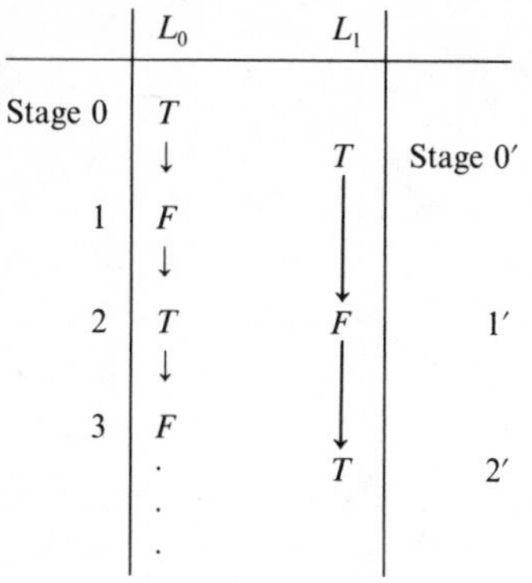

	L_0	L_1	
Stage 0	T		
	↓	T	Stage 0′
1	F	│	
	↓	↓	
2	T	F	1′
	↓	│	
3	F	↓	
	·	T	2′
	·		
	·		

A consequence is that in Gupta/Herzberger process the sentence $C := $ "L_0 is false $\leftrightarrow L_1$ is false" has a stable value: either 'always' true or 'always' false.

A first pressing question is: what to take as initial stages? Herzberger seems to have some preference to start with taking everything false – or at least this choice seems most coherent with his limit rule. See below. Such a choice could be defended by claiming some fundamental asymmetry between truth and falsity, e.g. because falsity is just lack of truth, i.e. falsity is really the gap. Gupta takes a different track: all choices have equal status. We should consider all initial stages and thus 'quantify out' the arbitrariness of choice. One could view the initial stages as fully classical worlds one puts 'experimentally' for one's mind's eye – but probably Gupta would not approve of such a psychologistic way of looking at them.

To get anything like a realistic picture we should admit transfinite stages. Consider for example the sequence:

$$\varphi_0 := \text{"Snow is white"}$$

$$\varphi_{n+1} := \ulcorner \text{"}\varphi_n\text{" is true} \urcorner$$

$$\varphi_\omega := \text{"For all } n \ \varphi_n \text{ is true"} .$$

A Gupta/Herzberger process for the φ_i looks as follows:

	φ_0	φ_1	φ_2	φ_3	...	φ_ω
Stage 0	F	F	F	F	...	F
1	T	F	F	F	...	F
2	T	T	F	F	...	F
3	T	T	T	F	...	F
4	T	T	T	T	...	F
.	.	.	.	.	...	.
.	.	.	.	.		.
.	.	.	.	.		.

Note that in our scheme we give the extension of the truth predicate in each stage: the so called declared truths and falsities, not what is true or false at that stage considered as a model. This is the reason that φ_0 is false in stage 0: we start with nothing in the extension of the truth predicate.

At all finite stages φ_ω is going to be false, but φ_ω is intuitively true. So a second pressing question is: what to do at the limit stages?

A partial solution suggests itself immediately. Let us say a sentence ψ is stably true towards limit λ if there is an $\alpha < \lambda$ such that in all stages β with $\alpha \leqslant \beta < \lambda$ the sentence is true. 'Stably false' is defined similarly. Now take ψ true at λ if ψ is stably true towards λ, false at λ if stably false towards λ. The idea here is something like this: the later stages are in a sense better then the earlier ones. Each stage can be seen as fitting our intuitive idea of truth better than the earlier stages. Stability of a given sentence is a sign that it has reached a correct value. (Note however that stability towards λ may be lost after λ.)

Above point is well illustrated in the example of the φ_i and φ_ω. The first few transfinite stages are:

	φ_0	φ_1	φ_2	φ_3	...	φ_ω
Stage ω	T	T	T	T	...	F
$\omega + 1$	T	T	T	T	...	T
$\omega + 2$	T	T	T	T	...	T
.	.	.	.	.	...	.
.	.	.	.	.		.
.	.	.	.	.		.

The φ_i and φ_ω converge to the intuitively correct value in this way.

What to do with the unstable sentences? There are many possibilities. The literature provides three outstanding ways:

(i) Herzberger's: make everything that is unstable towards λ false at λ. To understand this decision one could see falsity as the gap. Because the process told us nothing definite about the unstable sentence it is simply *not* true at the limit, i.e. false.

(ii) Gupta's: adapt the limit stage to the initial stage; if the unstable sentence was initially true, make it true at the limit, if false: false.

The idea here must be something like: we have already made an arbitrary decision, so why not stick to it?

(iii) Belnap's; give the unstable sentence an arbitrary value at the limit and quantify out by considering all processes corresponding to all possible choices.

This elegant solution is clearly in line with Gupta's solution of the problem of the initial stages.

Given a process Π we call a sentence φ *stably true* in Π if there is an α such that φ is true in all stages $\beta \geqslant \alpha$.

Similarly we may define: *stably false* in Π.

φ is *stable* in Π if φ is stably true or stably false in Π.

φ is *unstable* in Π if φ is not stable in Π.

Because there is only one Herzberger process (given the model) above notions give us intuitive distinctions for sentences *à la* Herzberger. In case we consider Gupta/Belnap processes there are more distinctions. We can represent these by having three coordinates standing for:

stably true in _ processes Π

stably false in _ processes Π

unstable in _ processes Π.

The possible values are: all, some (in the sense of some, not all), none. These are to be substituted for the "_". A sentence is for example (some, some, none) if it is stably true in some processes, stably false in some processes, unstable in no processes. We find (leaving out nonsensical vectors):

(all, none, none) : fully stable true
(some, some, some) : mixed

(some, some, none) : processwise stable
(some, none, some) : only stably true
(none, all, none) : fully stably false
(none, some, some) : only stably false
(none, none, all) : fully unstable, or: paradoxical.

For example the Samesayer *S* is processwise stable, the Liar fully unstable. In the next section we give examples of the other possibilities.

Processes with the Gupta/Herzberger limit rules eventually enter in a Grand Loop, as Herzberger calls it. This fact was first shown by Herzberger. In this loop there is a special stage – Herzberger coined it the alignment point with the following remarkable property: the stably true sentences are true, the stably false are false and the unstable sentences are false if we have the Herzberger limit rule or have the values they had in the initial stage if we have the Gupta limit rule. Pretty as this insight may be, its philosophical meaning remains obscure.

We briefly dwelled upon seeing the Gupta/Herzberger processes as unwinding certain intuitions surfacing in the reasoning about the paradoxes. It is very hard to extract a more definite explanation or justification from Herzberger's or Gupta's work. I concentrate here on what Gupta says in Gupta [1982]. There seem to be two main strands in Gupta's ideas. The first strand is that a theory of truth should separate unproblematic from problematic selfreference in such a way that if only unproblematic selfreference is present the relevant Tarski biconditionals become true (in Gupta's theory: fully stably true). A large part of Gupta [1982] is devoted to spelling out this constraint and analyzing the notion of unproblematic selfreference. (One of Gupta's main arguments against Kripke is that Kripke's construction doesn't satisfy the constraint: Kripke gets certain sentences ungrounded even when no problematic selfreference is present.) Now perhaps the following line can be defended. The proper task of a theory of truth is to analyze the meaning of sentences involving the truth predicate and to provide their assertability conditions. Any solution should obey Gupta's constraint: it should separate problematic from unproblematic selfreference in the appropriate way. This is *all* such a theory has to do. It is not necessary that the way of generating the solution has a further explanatory value. So the processes are to be viewed as *mathematical artifacts* to generate the appropriate classes of sentences and to provide assertability conditions for the unproblematic (here: fully stably true or fully stably false) sentences.

The second strand in Gupta's ideas makes clear that this cannot be quite the line he takes. He writes:

> I propose . . . that we view the concept of truth as characterized by a *revision* procedure. I suggest that truth, unlike ordinary concepts such as red, blue, and sum, does not in general have an *application* procedure associated with it . . . When we learn the meaning of 'true' what we learn is a rule that enables us to *improve* on a proposed candidate for the extension of truth. (Gupta [1982], p. 37.)

This surely is an interesting idea. The question is: can it be expanded to an explanation of the processes *including* initial stages and limit rules? Perhaps we can dream up the story of Marc who upon hearing the definition of truth experimentally puts a classical world including a fully evaluated truth predicate before his mind's eye and starts improving it using the revision rule. Whether such a metaphor can be made plausible I am not sure and I leave it to the reader to ponder.

4.1. *Excurs: Gupta's criticisms of Kripke*

We restrict ourselves to Gupta's criticisms that are directed against the minimal Strong Kleene fixed point or against all Strong Kleene fixed points at once.

The criticisms are – in slightly abbreviated form:

(1) The Law of Non Contradiction ($\forall x \neg(x$ is true $\wedge \neg x$ is true$)$) is *paradoxical* if there are Liar type sentences in the language.
(2) Even if there is no vicious selfreference in the language the Law of Non Contradiction is still *ungrounded.*
(3) Call the following sentences "A_1", "A_2", "B" respectively:

    B is true

    B is false

    At most one of A_1, A_2 is true.

    Clearly A_1, A_2 contradict each other, so B must be true, hence A_1 is true and A_2 is false. But A_1, A_2, B are *ungrounded.*
(4) The minimal fixed point *never* makes all Tarski biconditionals true *even* if no unproblematic selfreference is present.

The point I want to make is not that these criticisms are *wrong* but that they are misdirected. First consider the first negation of $\forall x \neg(x$ is true $\wedge \neg x$ is true$)$: intuitively it is not the usual classical negation. If it were the sentence would be equivalent to $\forall x$ ($\neg x$ is true $\vee$ x is true). But surely we could have nothing both true and not true (no *over*definedness) without having everything either not true or true (no *un*definedness)? So the problem for Kripke is more that he does not offer a theory for this negation. Similar remarks hold for the "at most" of (3): "at most one of A_1, A_2 is true" can even be paraphrased as: $\neg(A_1$ is true $\wedge$ A_2 is true), where $\neg$ has to be taken as the non classical negation we discussed before. Secondly consider (4): here the point should be that Kripke's structures do not have a reasonable implication and biimplication. What we *do* have in Kripke's fixed points is: $\ulcorner Q \simeq$ "Q" is true$\urcorner$. Only $\simeq$ is not part of the language for which the fixed points are constructed. One can argue that above non classical negation can be paraphrased by $(\ ) \to \bot$. If this is true Gupta's whole criticism comes to: Kripke structures lack a good implication.

5. ITERATIONS AND STRUCTURES

5.1. DEFINITION. Let $\mathscr{D} = \langle D, \leqslant \rangle$ be a po, define for $X, Y \subseteq D$:
$X \leqslant_1 Y$ iff for all x in X there is a y in Y such that $x \leqslant y$
$X \leqslant_2 Y$ iff for all y in Y there is an x in X such that $x \leqslant y$
$X =_i Y$ iff $X \leqslant_i Y$ and $Y \leqslant_i X$ for $i = 1, 2$.

5.2. FACT. Let $\mathscr{D} = \langle D, \leqslant \rangle$ be a ccpo; $X, Y \subseteq D$. (i) If $X \leqslant_1 Y$ and $\sqcup Y(\mathscr{D})$ exists, then $\sqcup X(\mathscr{D})$ exists and $\sqcup X(\mathscr{D}) \leqslant \sqcup Y(\mathscr{D})$. (ii) If $X \leqslant_2 Y$ and $\sqcap Y(\mathscr{D})$ exists, i.e. $Y \neq \varnothing$, then $\sqcap X(\mathscr{D})$ exists and $\sqcap X(\mathscr{D}) \leqslant \sqcap Y(\mathscr{D})$.

Proof. trivial. □

5.3. DEFINITION. Let $\mathscr{D} = \langle D, \leqslant \rangle$ be a ccpo. Suppose $(x_\alpha)_{\alpha \in ON}$ is a sequence of elements of D indexed by ordinals. Define for limit λ:

$$\liminf_{\alpha \to \lambda} x_\alpha := \bigsqcup_{\alpha < \lambda} \bigsqcap_{\alpha \leqslant \beta < \lambda} x_\beta .$$

(Or more precisely:

$$\liminf_{\alpha \to \lambda} x_\alpha(\mathscr{D}) := \bigsqcup_{\alpha < \lambda} \Big(\bigsqcap_{\alpha \leqslant \beta < \lambda} x_\beta(\mathscr{D}))(\mathscr{D})\Big).$$

5.4. FACT. Under the conditions of 5.3:

(i) $\liminf_{\alpha \to \lambda} x_\alpha$ exists.

(ii) $\liminf_{\alpha \to \lambda} x_{\beta+\alpha} = \liminf_{\alpha \to \beta+\lambda} x_\alpha .$

(iii) If $(x_\alpha)_{\alpha \in On}$ is weakly monotonically increasing for $\alpha < \lambda$, then $\liminf_{\alpha \to \lambda} x_\alpha = \sqcup_{\alpha < \lambda} x_\alpha$.

(iv) If $(x_\alpha)_{\alpha \in ON}$ is weakly monotonically decreasing for $\alpha < \lambda$, then $\liminf_{\alpha \to \lambda} x = \sqcap_{\alpha < \lambda} x_\alpha$.

Proof. (i) $\{\sqcap_{\alpha \leqslant \beta < \lambda} x_\beta | \alpha < \lambda\}$ is consistent (even a chain) as is easily seen. (ii) Let $X := \{\sqcap_{\alpha \leqslant \gamma < \beta + \lambda} x_\gamma | \alpha < \beta + \lambda\}$ and $Y := \{\sqcap_{\alpha \leqslant \gamma < \lambda} x_{\beta+\gamma} | \alpha < \lambda\}$. Clearly: $Y = \{\sqcap_{\alpha \leqslant \gamma < \beta + \lambda} x_\gamma | \beta \leqslant \alpha < \beta + \lambda\}$. So: $X =_1 Y$, and thus $\sqcup X = \sqcup Y$. (iii) If our sequence increases weakly monotonically for $\alpha < \lambda$, then $\sqcap_{\alpha \leqslant \beta < \lambda} x_\beta = x_\alpha$. (iv) If the sequence decreases weakly monotonically for $\alpha < \lambda$, then: $\{x_\beta | \alpha \leqslant \beta < \lambda\} =_2 \{x_\beta | \beta < \lambda\}$. So $\sqcap_{\alpha \leqslant \beta < \lambda} x_\beta = \sqcap_{\beta < \lambda} x_\beta$. Hence clearly:

$$\liminf_{\alpha \to \lambda} x_\alpha = \bigsqcap_{\alpha < \lambda} x_\alpha. \qquad \square$$

5.5. DEFINITION. let $\mathscr{D} = \langle D, \leqslant \rangle$ be a ccpo, $f: D \to D$, x in D. Define the *fundamental sequence* generated by f from x as follows.

$$(x)_0 := x$$

$$(x)_{\alpha+1} := f((x)_\alpha)$$

$$(x)_\lambda := \liminf_{\alpha \to \lambda} (x)_\alpha.$$

(If necessary we use more explicit notations like $(x)_\alpha(f)$ or even $(x)_\alpha(\mathscr{D}, f)$.)

Further define:

$$(x)^f := (x)_\infty := (x)_{\aleph(D)},$$

where $\aleph(D)$ is the smallest cardinal bigger than $\max(\mathrm{card}(D), \aleph_0)$. (A more natural definition of $(x)_\infty$ would be $\liminf_{\alpha \in ON} (x)_\alpha$, but it will turn out that for the fundamental sequence this definition and ours coincide.)

5.6. FACT. Under the conditions of 5.5:

(i) $(x)_{\alpha+\beta} = ((x)_\alpha)_\beta$ (and so $(x)_\infty = ((x)_\alpha)_\infty$).

(ii) If $(x)_\alpha = (x)_\beta$, then for all γ $(x)_{\alpha+\gamma} = (x)_{\beta+\gamma}$

(iii) For some $\alpha, \beta < \aleph(D)$, with $\alpha \neq \beta$: $(x)_\alpha = (x)_\beta$.

Proof. (i) By an easy induction on β, using 5.4 (ii). (ii) is immediate from (i). (iii) $((x)_\alpha)_{\alpha < \aleph(D)}$ maps $\aleph(D)$ in D. The cardinality of D is smaller than $\aleph(D)$, hence the mapping cannot be injective. $\square$

5.7. THEOREM. *Let* $\mathscr{D} = \langle D, \leqslant \rangle$ *be a ccpo, let* f *be monotonic on* $\mathscr{D}$.

(i) *Let* x *be in* $Up(\mathscr{D}, f)$, *we have*: (ia) $((x)_\alpha)_{\alpha \in ON}$ *is weakly monotonically increasing*. (ib) *For some* $\alpha < \aleph(D)$: $(x)_\alpha = (x)_{\alpha+1} = (x)_\infty$. (ic) $(x)_\infty$ *is the minimal fixed point of* f *above* x, *or*: $(x)^f = (x)_f$.

(ii) *Let* x *be in* $Down(\mathscr{D}, f)$, *we have*: (iia) $((x)_\alpha)_{\alpha \in ON}$ *is weakly monotonically decreasing*. (iib) *For some* $\alpha < \aleph(D)$: $(x)_\alpha = (x)_{\alpha+1} = (x)_\infty$. (iic) $(x)_\infty$ *is the maximal fixed point below* x, *or*: $(x)^f = (x)_f$.

Proof. (ia) We show by transfinite induction: for all α: $(x)_\alpha \leqslant (x)_{\alpha+1}$. (ia) then follows by a single second induction. The only interesting case is where α is a limit: Assume for all $\beta < \alpha$ $(x)_\beta \leqslant (x)_{\beta+1}$. By a simple induction we can show that $((x)_\beta)_{\beta<\alpha}$ is weakly monotonically increasing, hence $(x)_\alpha = \sqcup_{\beta<\alpha} (x)_\beta$. For $\beta < \alpha$ $(x)_\beta \leqslant (x)_\alpha$, hence $(x)_\beta \leqslant (x)_{\beta+1} \leqslant (x)_{\alpha+1}$. Conclude $\sqcup_{\beta<\alpha}(x)_\beta \leqslant (x)_{\alpha+1}$ i.e. $(x)_\alpha \leqslant (x)_{\alpha+1}$. (ib) For some $\alpha < \beta < \aleph(D)$ we have $(x)_\alpha = (x)_\beta$. Our sequence is weakly monotonically increasing, hence $(x)_\alpha = (x)_{\alpha+1}$. By a trivial induction $(x)_\alpha = (x)_{\alpha+\gamma}$ for all γ, so $(x)_\alpha = (x)_\infty$. (ic) By an easy induction $(x)_\alpha \leqslant (x)_f$, so $(x)_\infty \leqslant (x)_f$. $(x)_\infty$ is a fixed point above x and $(x)_f$ is the minimal fixed point above x. Hence $(x)_\infty = (x)_f$.

(ii) By dual reasoning. □

5.8. LEMMA. *Let* $\mathscr{D} = \langle D, \leqslant \rangle$ *be a ccpo*, $f: D \to D$. *Suppose there are* γ, β, λ *with* $\beta \neq 0$, λ *a limit and for all* $\sigma < \lambda$ $(x)_\gamma = (x)_{\gamma+\beta\sigma}$, *then*: $(x)_{\gamma+\beta\lambda} = \sqcap_{\nu<\beta}(x)_{\gamma+\nu}$.

Proof. $(x)_{\gamma+\beta\lambda} = \sqcup_{\alpha<\beta\lambda}\sqcap_{\alpha\leqslant\mu<\beta\lambda}(x)_{\gamma+\mu}$, so it is sufficient to show:

$$\{(x)_{\gamma+\mu} | \alpha \leqslant \mu < \beta\lambda\} = \{(x)_{\gamma+\nu} | \nu < \beta\}.$$

By a standard result of set theory for a given $\mu < \beta\lambda$ there are unique σ, ν with $\sigma < \lambda$ and $\nu < \beta$ and $\mu = \beta\sigma + \nu$. So $(x)_{\gamma+\mu} = (x)_{\gamma+\beta\sigma+\nu} = (x)_{\gamma+\nu}$. Conversely if $\nu < \beta$ for any $\alpha < \beta\lambda$, there is a σ such that $\alpha \leqslant \beta\sigma + \nu < \beta\lambda$. We have: $(x)_{\gamma+\nu} = (x)_{\gamma+\beta\sigma+\nu}$. □

5.9. THEOREM. *Let* $\mathscr{D} = \langle D, \leqslant \rangle$ *be a ccpo*, f *monotonic on* $\mathscr{D}$, x *in* D. *Then there is a* $\sigma < \aleph(D)$ *such that* $(x)_\sigma$ *is a fixed point of* f *and hence* $(x)_\sigma = (x)_\infty$.

Proof. There are $\alpha < \beta < \aleph(D)$ with $(x)_\alpha = (x)_\beta$. Suppose $\beta = \alpha + \gamma$. By 5.6: $(x)_\alpha = (x)_{\alpha+\gamma n}$. Hence by 5.8 $(x)_{\alpha+\gamma\omega} = \sqcap_{\nu<\gamma}(x)_{\alpha+\nu}$.

Further for $\nu < \gamma$:

$$(x)_{\alpha+\gamma\omega+1} = f((x)_{\alpha+\gamma\omega}) \leqslant f((x)_{\alpha+\nu}) = (x)_{\alpha+\nu+1}.$$

By an easy induction we show that for $1 \leqslant \nu \leqslant \gamma$: $(x)_{\alpha+\gamma\omega+1} \leqslant (x)_{\alpha+\nu}$. Moreover $(x)_\alpha = (x)_{\alpha+\gamma}$, so the same holds for $0 \leqslant \nu < \gamma$. Hence:

$$(x)_{\alpha+\gamma\omega+1} \leqslant \prod_{\nu<\gamma} (x)_{\alpha+\nu} = (x)_{\alpha+\gamma\omega}.$$

By 5.7 (ii) there is a $\delta < \aleph(D)$ such that:

$$(x)_{\alpha+\gamma\omega+\delta} = ((x)_{\alpha+\gamma\omega})_\delta = ((x)_{\alpha+\gamma\omega})_\infty = (x)_\infty.$$

Here $(x)_\infty$ is a fixed point. Take $\sigma = \alpha + \gamma\omega + \delta$. As is easily seen: $\sigma < \aleph(D)$. □

5.10. ELABORATION. Let $\mathscr{D} = \langle D, \leqslant\rangle$ be a ccpo, f monotonic on $\mathscr{D}$. 5.9 shows us that:

$$(\ )^f: D \to \mathrm{Fix}(D, f)$$

By an easy induction we see that for $x \leqslant y$: $(x)_\alpha \leqslant (y)_\alpha$ and hence $(x)^f \leqslant (y)^f$. Also: $((x)^f)^f = (x)^f$. 5.7 shows that for x in $\mathrm{Up}(D, f) \cup \mathrm{Down}(D, f)$: $(x)^f = (x)_f$. In other words $(\ )^f$ is an extension of $(\ )_f^+$ and $(\ )_f^-$ too. By a simple induction using $id \sqcap f \leqslant f$, we see that $(x)_f \leqslant (x)^f$.

The following example:

F T
*

$$f(*) = *,\ f(F) = f(T) = T$$

shows that *not* in general $(\ )_f = (\ )^f$.

Warning. neither $(\ )^f$ nor $(\ )_f$ preserve $\sqcup$.

Consider:

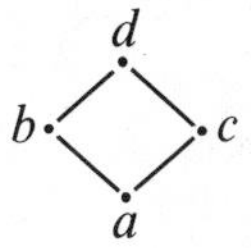

d
b c
a

$$f(a) = f(b) = f(c) = a$$
$$f(d) = d.$$

Here: $(\sqcup\{b, c\}\ (\mathscr{D}))^f = d \neq a = \sqcup\{(b)^f, (c)^f\}(\mathrm{Fix}(\mathscr{D}, f))$. Similarly for $(\ )_f$.

Let M be the set of maximal elements of $\mathscr{D}$, N be the set of maximal elements of $\mathrm{Fix}(\mathscr{D}, f)$. $N \subseteq \{(m)^f | m \in M\}$: consider $n \in N$: there is an $m \in M$ with $n \leqslant m$, hence $(m)^f \leqslant (n)^f = n$. n is maximal in $\mathrm{Fix}(\mathscr{D}, f)$, hence $(m)^f = n$.

Define: $j := \sqcap\{(m)^f | m \in M\}\ (\mathrm{Fix}(\mathscr{D}, f))$.

j is another salient fixed point connected both to Kripke's and to Herzberger and Gupta's ideas. Clearly $j \leqslant i_0$, where i_0 is the maximal intrinsic point of $\mathrm{Fix}(\mathscr{D}, f)$.

As the following example shows

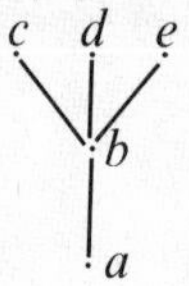

$$f(a) = a, f(b) = b, f(c) = d,$$
$$f(d) = c, f(e) = e$$

it is possible that $*(\mathrm{Fix}(\mathscr{D}, f)) < j < i_0$, for $*(\mathrm{Fix}(\mathscr{D}, f)) = a$, $j = b$, $i_0 = e$.

5.11. *The non-monotonic case*

The material in this section is due to Herzberger.
What happens in the non-monotonic case? Simple: the process will enter into a possibly gigantic cycle or, as Herzberger aptly coined it: The Grand Loop. The general picture is like this:

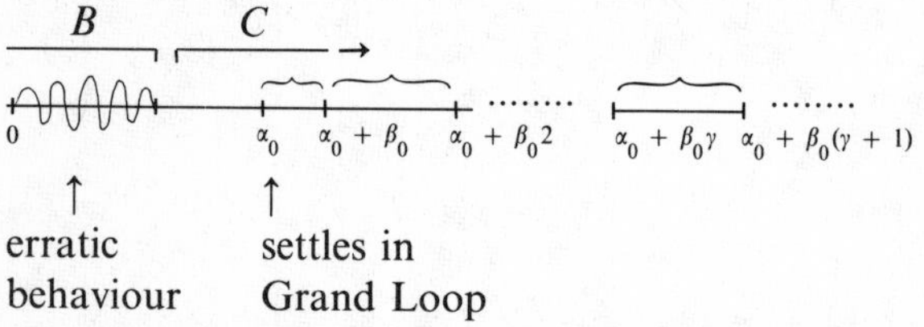

Here B are the bounded elements that only occur at some initial stages of the process. C are the cofinal ones that occur again and again. As we shall see the process enters the loop before $\aleph(D)$ and $(x)_{\alpha_0} = (x)_\infty = \sqcap C$. To work this out:
Let $\mathscr{D} = \langle D, \leqslant \rangle$ be a ccpo, $f\colon D \to D$, x in D. Define:

$$A := \{(x)_\alpha \mid \alpha < \aleph(D)\}$$
$$B := \{y \in A \mid \exists\beta < \aleph(D) \forall\gamma < \aleph(D)\ \beta \leqslant \gamma \Rightarrow x_\gamma \neq y\}$$
$$C := \{y \in A \mid \forall\beta < \aleph(D) \exists\gamma < \aleph(D)\ \beta \leqslant \gamma \text{ and } (x)_\gamma = y\}.$$

ini: $A \to \aleph(D)$ is defined by

$$\mathrm{ini}(y) := (\text{the smallest } \alpha \text{ such that } y = (x)_\alpha),$$

bo: $B \to \aleph(D)$ is defined by:

$$\mathrm{bo}(y) := (\text{the smallest } \alpha \text{ such that for all}$$
$$\beta \leqslant \aleph(D)\ \beta > \alpha \Rightarrow (x)_\beta \neq y).$$

(The restrictions to $\aleph(D)$ will later be shown to be spurious.)

5.11.1. FACT.

(i) For $\alpha < \aleph(D)$: if $(x)_\alpha \in B$ and $\beta \leqslant \alpha$ then $(x)_\beta \in B$.

(ii) For $\beta < \aleph(D)$: if $(x)_\alpha \in C$, $\alpha \leqslant \beta$ then $(x)_\beta \in C$.

(iii) $C \neq \emptyset$.

(iv) For some $\alpha_0 < \aleph(D)$ $(x)_{\alpha_0} = (x)_\infty = \sqcap C$.

Proof. (i) and (ii): by 5.6 (ii). (iii) Let $X := \{\text{bo}(y) \mid y \in B\}$; the cardinality of X is smaller than $\aleph(D)$. $\aleph(D)$ is a successor cardinal, hence $\aleph(D)$ is regular, so X is not cofinal in $\aleph(D)$, i.e. for some $\sigma < \aleph(D)$, for all β in X $\beta < \sigma$. Clearly $(x)_\sigma \in C$. (iv) Let $Y := \{\text{ini}(y) \mid y \in C\}$, as X in (iii) Y is bounded in $\aleph(D)$. Let σ_0 be the smallest element of y and let σ_1 be the smallest ordinal such that $(x)_{\sigma_1} = (x)_{\sigma_0}$ and for all $v \in Y$ $\sigma_1 \geqslant v$. Clearly $\sigma_1 < \aleph(D)$. Say $\sigma_1 = \sigma_0 + \varepsilon$. Consider $(x)_{\sigma_0 + \varepsilon\omega}$. By 5.8:

$$(x)_{\sigma_0 + \varepsilon\omega} = \prod_{\sigma_0 \leqslant \alpha < \varepsilon} (x)_\alpha = \sqcap C.$$

So as is easily seen $(x)_\infty = \sqcap C$. □

5.11.2. THEOREM. *There are α_0, β_0 below $\aleph(D)$ such that for all γ in ON $(x)_{\alpha_0 + \beta_0\gamma} = (x)_{\alpha_0}$.*

Proof. Take

$$\alpha_0 := (\text{the smallest } \alpha \text{ such that } (x)_\alpha = \sqcap C)$$

$$\beta_0 := (\text{the smallest } \beta \text{ such that } (x)_{\alpha_0 + \beta} = \sqcap C \text{ and } \beta \neq 0).$$

By 5.11.1 α_0 and β_0 are below $\aleph(D)$.
We show by induction that $(x)_{\alpha_0 + \beta_0\gamma} = \sqcap C$.
The cases where $\gamma = 0$ or γ is a successor are trivial. If γ is a limit then by IH and 5.8: $(x)_{\alpha_0 + \beta_0\gamma} = \sqcap_{\sigma < \beta_0} (x)_{\alpha_0 + \sigma}$. Clearly the $(x)_{\alpha_0 + \sigma}$ are in C so $\sqcap_{\sigma < \beta_0} (x)_{\alpha_0 + \sigma} \geqslant \sqcap C$. On the other hand $(x)_{\alpha_0} = \sqcap C$. Hence $(x)_{\alpha_0 + \beta_0\gamma} = \sqcap C$. □

5.11.3. *Remarks and some further information*

(1) Let α_0, β_0 be as in 5.11.2, then the following can be shown: Suppose for some α and β for all γ: $(x)_\alpha = (x)_{\alpha + \beta\gamma}$ then $\alpha = \alpha_0 + \beta_0\sigma$ for some σ and $\beta = \beta_0\varepsilon$ for some ε. Hence $(x)_\alpha = \sqcap C$.

(2) For every δ there are unique σ and ϱ with $\varrho < \beta_0$ and $\delta = \beta_0\sigma + \varrho$. Write: $\varrho = \delta \bmod \beta_0$. By 5.11.2: for all δ: $(x)_{\alpha_0+\delta} = (x)_{\alpha_0+\delta \bmod \beta_0}$.

(3) The process settles as we have seen in the Grand Loop before $\aleph(D)$, so clearly the restriction to $\aleph(D)$ was inessential:

$$\liminf_{\alpha \in ON} (x)_\alpha = (x)_{\alpha_0} = (x)_\infty.$$

(4) In case $\mathscr{D} = \mathscr{E}^A$, $\mathscr{E} = \langle E, \leqslant \rangle$ a ccpo, A a set, then one can take $\aleph(D) := (\max(\text{card}(E), \text{card}(A), \aleph_0))^+$. This point is essentially due to John Burgess and Vann McGee.

(5) If D is finite we can restrict the process to ω^ω.

6. ITERATIONS, STIPULATIONS AND SATISFACTION

After the technical preliminaries of Section 5, we just have to introduce the appropriate structures and transition functions to study the processes of Kripke, Gupta and Herzberger.

First we have a brief look at iterations of partial valued structures.

6.1. *The Partial Valued Case*

We use the conventions and notations of Definition 3.1.

(i) Let A be a non empty set and $\mathscr{S}$ a stipulation list in A. Let $*$ be the bottom element of $\mathbb{T}^A$. (Clearly $((*)_\alpha)_{\alpha \in ON}$, the fundamental sequence generated by $F_{\mathscr{S}}$ from $*$, is Kripke's ascending stages process.) $(*)^{F_{\mathscr{S}}}$ is Kripke's minimal fixed point. One easily sees: if A is finite, then the fixed point is reached at some finite stage, if A is infinite the fixed point is reached at some stage $\leqslant \omega$.

An example where the fixed point is reached only at stage ω is:

$$a_0 \colon T$$

$$a_{n+1} \colon a_n.$$

(ii) Let $\mathbb{N}$ be the standard model of Arithmetic. Let $*$ be the bottom element of $\mathbb{N}^{\omega \times \omega}$. $((*)_\alpha)_{\alpha \in ON}$, the fundamental sequence generated by $\Phi_{\mathbb{N}}$ from $*$, is Kripke's ascending stages process and $(*)^{\Phi_{\mathbb{N}}}$ is Kripke's minimal fixed point. In Section 7 we will see that the fixed point is reached precisely at stage ω_1^{CK}.

(iii) Consider the stipulation list $\mathscr{S}$ with:

$$a\colon a \vee \neg a$$

$$b\colon \neg b \vee \neg c$$

$$c\colon \neg b \wedge \neg c.$$

For the transition function $F^{\mathrm{SK}}_{\mathscr{S}}$, we have:

- the minimal fixed point is $\begin{pmatrix} * \\ * \\ * \end{pmatrix}$
- the maximal intrinsic fixed point is $\begin{pmatrix} T \\ T \\ F \end{pmatrix}$
- the fixed point j of 5.10 is $\begin{pmatrix} T \\ * \\ * \end{pmatrix}$.

(iv) Consider the stipulation list $\mathscr{S}$ with:

$$\ell : \neg \ell$$

$$a_0 : \ell \vee \neg \ell$$

$$a_1 : a_0 .$$

The fixed point $\begin{pmatrix} T \\ T \\ T \end{pmatrix}^{F^{\mathrm{SK}}_{\mathscr{S}}}$ is $\begin{pmatrix} * \\ * \\ * \end{pmatrix}$.

It is reached at stage $\omega + 2$. This shows that for arbitrary starting points even for finite stipulation lists the process need not reach a fixed point before or at ω. One easily shows that for finite stipulation lists the fixed point will be reached strictly before $\omega.2$.

6.2. *The Total Valued Case*

6.2.1. DEFINITION. We introduce here the Herzberger/Gupta/Belnap processes. To fit the Herzberger/Gupta processes into the framework of Section 5 we must specify structures in which their limits are liminf's.

Define:

$$\mathbb{T}_0 := \langle \{T, F\}, \leqslant \rangle, \text{ where } \leqslant \text{ is given by: } F \leqslant T \quad (T \text{ above } F)$$

In a sense truth is here the definite element and falsity the gap. The Herzberger style iterations operate on objects from $\mathbb{T}_0^A$ for certain non empty sets A.

The Herzberger sequence for a given function $H\colon \{T, F\}^A \to \{T, F\}^A$ is the fundamental sequence $((f_0)_\alpha)_{\alpha \in ON}$ in $\mathbb{T}_0^A$ generated by H from f_0. Here $f_0(a) = F$ for all $a \in A$.

To get the Gupta limits we can best view the Gupta processes as Herzberger processes with *varying* gap. We stipulate the gap to be at some places T, at some places F. The starting point f encodes these decisions: if $f(a) = T$ then T is the gap *at* a, if $f(a) = F$ then F is the gap *at* a. This induces an ordering on $\{T, F\}^A$ of which f is the bottom element. The easiest way to implement this idea is to map the Gupta ordering on the Herzberger ordering.

Define $(g)^f : A \to \{T, F\}$, by:

$$(g)^f(a) := \begin{cases} g(a) \text{ if } f(a) = F \\ \neg g(a) \text{ if } f(a) = T. \end{cases}$$

(Or: $(g)^f(a) := (g(a) \leftrightarrow \neg f(a))$.)
For g, h in $\{T, F\}^A$ we define: $g \leqslant^f h :\leftrightarrow (g)^f \leqslant (h)^f$.
(Here $\leqslant$ is the ordering on $\mathbb{T}_0^A$.)

$$\mathbb{T}_0^{f,A} := \langle \{T, F\}^A, \leqslant^f \rangle.$$

We leave it to the reader to verify that the limits of a Gupta style process in $\{T, F\}^A$ with starting point f are precisely the liminf's of $\mathbb{T}_0^{f,A}$. The Gupta sequence for a given function $H: \{T, F\}^A \to \{T, F\}^A$ from a given starting point f is the fundamental sequence $((f)_\alpha)_{\alpha \in ON}$ in $\mathbb{T}_0^{f,A}$ generated by H from f.

Finally a sequence $([f]_\alpha)_{\alpha \in ON}$ is a Belnap sequence for a given transition function H: $\{T, F\}^A \to \{T, F\}^A$ and a given starting point $f \in \{T, F\}^A$ if:
- $[f]_0 = f$
- $[f]_{\alpha+1} = H([f]_\alpha)$
- if $([f]_\alpha(a))_{\alpha<\lambda}$ is stably $T(F)$ towards λ, then $[f]_\lambda(a)$ is $T(F)$ for $a \in A$.

The transition functions for all three kinds of processes that we are interested in are the following.

Let $\mathscr{S}$ be a stipulation list on A. We define the transition function $F_{\mathscr{S}}$ on $\{T, F\}^A$ as $F_{\mathscr{S}}(g)(a) := [\mathscr{S}(a)]^c g$, where $[\varphi]^c g$ is the classical valuation of φ at g.

Let $\mathscr{M}$ be an acceptable structure with domain M. First define $\mathrm{Sat}_{\mathscr{M}(S)}$ for S: $M \times M \to \{T, F\}$ by:

$$\mathrm{Sat}_{\mathscr{M}(S)}(n, m) := \begin{cases} [\![\varphi]\!]^c(\mathscr{M}(S), h^m) \text{ if } n = \ulcorner\varphi\urcorner \text{ and} \\ \qquad m = \langle m_1, \ldots, m_k \rangle \\ F \text{ otherwise.} \end{cases}$$

Define $\Phi_{\mathscr{M}}(S) := \mathrm{Sat}_{\mathscr{M}(S)}$.
Here h^m is as in 3.1 (ii); $[\![\ ]\!]^c$ is the classical evaluation function.

$F_{\mathscr{S}}$ and $\Phi_{\mathscr{M}}$ are non-monotonic transition functions on the ccpo's $\mathbb{T}_0^A$, $\mathbb{T}_0^{f,A}$ respectively $\mathbb{T}_0^{M \times M}$, $\mathbb{T}_0^{S,M \times M}$. So the results of Section 5 on the behaviour of non-monotonic processes are applicable.

6.2.2. *Kinds of sentences*

6.2.2.1. *Examples of the intuitive distinctions for Gupta/Belnap processes.*

e is an example of	Stipulation list
fully stable true	$e: T$
mixed	$a: a$
	$b: b$
	$e: a \wedge (b \vee \neg e)$
processwise stable	$e: e$
only stably true	$a: a$
	$e: a \vee \neg e$
fully stably false	$e: \bot$
only stably false	$a: a$
	$e: a \wedge \neg e$
paradoxical	$e: \neg e$

6.2.2.2. *Differences between the kinds of processes.* The kinds of processes differ in the classifications of sentences they make. We consider for example paradoxicality. Clearly we have:

$$\begin{aligned}\text{Belnap paradoxical} &\Rightarrow \text{Gupta paradoxical}\\ &\Rightarrow \text{Herzberger paradoxical.}\end{aligned}$$

(Vann McGee in McGee [1983] shows the stronger fact that above statement holds even when we allow Herzberger processes with arbitrary starting points.) The arrows cannot be reversed.

(i) Consider

$$a: a$$

$$b: a \vee \neg b.$$

Clearly b is Herzberger paradoxical, but Gupta only stably true.

(ii) Examples of a Gupta paradoxical, not Belnap paradoxical sentence were independently given by Anil Gupta and Vann McGee. These examples involve predicate logic in an essential way. One may show that for finite stipulation lists Gupta and Belnap paradoxicality coincide (see Visser [1984]). I conjecture that this holds for any stipulation list.

6.2.3. *Alignment*

Consider a Gupta process generated by H: $\{T, F\}^A \to \{T, F\}^A$ from a given starting point g. Let C be the set of cofinal elements of the sequence. We have seen that for certain α_0 $(g)_{\alpha_0} = \sqcap C(\mathbb{T}_0^{g,A})$.

Define for given H:

- the alignment ordinal $O(g)$ by:

$$O(g) := \text{the smallest } \alpha_0 \text{ such that } (g)_{\alpha_0} = \sqcap C(\mathbb{T}_0^{g,A}) \text{ in the Gupta process from } g.$$

- the alignment value $(g)^G$ by:

$$(g)^G := \sqcap C(\mathbb{T}_0^{g,A}).$$

Clearly $(g)^G$ is a distinguished stage in the Gupta process. One may ask: is it stable in the sense that $(g)^G = ((g)^G)^G$?

6.2.3.1. THEOREM. $(g)^G = ((g)^G)^G$.

Proof. Let $\alpha_0 := O(g)$. Clearly: $g \leqslant^g (g)^G \leqslant^g (g)_{\alpha_0+\alpha}$. It is sufficient to show: for all α: $(g)_{\alpha_0+\alpha} = ((g)^G)_\alpha$. (Note that the liminf's of these processes are different.) We prove this last fact by induction on α. The cases $\alpha = 0$ and $\alpha = \beta + 1$ are trivial. Suppose α is a limit. We show:

$$\liminf_{\beta\to\alpha_0+\alpha} (g)_\beta(\mathbb{T}_0^{g,A}) = \liminf_{\beta\to\alpha} ((g)^G)_\beta(\mathbb{T}_0^{(g)^G,A}):$$

By the Induction Hypothesis we have that for all $\beta < \alpha$ $(g)_{\alpha_0+\beta} = ((g)^G)_\beta$. Moreover:

$$\liminf_{\beta\to\alpha_0+\alpha} (g)_\beta(\mathbb{T}_0^{g,A}) = \liminf_{\beta\to\alpha} (g)_{\alpha_0+\beta}(\mathbb{T}_0^{g,A}).$$

So it is sufficient to prove:

$$\ell_0 := \liminf_{\beta\to\alpha} (g)_{\alpha_0+\beta}(\mathbb{T}_0^{g,A}) = \liminf_{\beta\to\alpha} (g)_{\alpha_0+\beta}(\mathbb{T}_0^{(g)^G,A}) := \ell_1.$$

Let $a \in A$ and consider $((g)_{\alpha_0+\beta}(a))_{\beta<\alpha}$. If this sequence is stable towards α clearly $\ell_0(a) = \ell_1(a)$. Suppose it is unstable. We have: $\ell_0(a) = g(a)$, $\ell_1(a) = ((g)^G)(a)$. Now $(g)^G \leqslant^g \ell_0$ - ℓ_0 being a cofinal stage - so $((g)^G)^g(a) \leqslant (\ell_0)^g(a)$.

But

$$(\ell_0)^g(a) = \begin{cases} g(a) \text{ if } g(a) = F \\ \neg g(a) \text{ if } g(a) = T. \end{cases}$$

Ergo

$$(\ell_0)^g(a) \;=\; F, \text{ so } ((g)^G)^g(a) \;=\; F.$$

Conclude

$$(g)^G(a) \;=\; g(a). \qquad \square$$

Above fact was discovered independently by Vann McGee and the author (see McGee [1983]).

Which alignment ordinals are possible? For the case of (say) Arithmetic and satisfaction Vann McGee shows that all and only countable ordinals can be alignment ordinals (see McGee [1983]). We treat the case of finite stipulation lists here.

6.2.3.2. THEOREM. *The alignment ordinals of Gupta (Herzberger) processes for finite stipulation lists are all and only ordinals strictly below* ω^ω.

Proof. (i) For any Gupta (Herzberger) process given by a finite stipulation list the alignment ordinal is strictly smaller than ω^ω. This can be easily seen by sharpening the proof of 5.11 a bit for the finite case.

(ii) Every ordinal strictly below ω^ω is alignment ordinal for some Herzberger (and hence also: Gupta) process given by a finite stipulation list.

Call the set of alignment ordinals under consideration O. It is clearly sufficient to show:

(a) $0 \in O$ and $1 \in O$

(b) if $\alpha \in O$ and $\beta \in O$ then $\alpha + \beta \in O$

(c) if $\alpha \in O$, then $\alpha\omega \in O$.

To show this first note that for every $H{:}\{T, F\}^A \to \{T, F\}^A$ – where A is finite, non empty – there is a stipulation list $\mathcal{S}$ such that $H = F_{\mathcal{S}}$. This follows by functional completeness of the standard classical connectives. So we only have to look for the appropriate transition functions.

Case a: trivial.

Case b: suppose transition functions H_0, H_1 are given on respectively $\{T, F\}^{A_0}$, $\{T, F\}^{A_1}$, such that the alignment stages of the Herzberger processes based on H_0, H_1 are respectively α and β. The case $\beta = 0$ being trivial, we assume $\beta \neq 0$.

Without loss of generality we may take $A_0 = \{0, \ldots, n_0\}$, $A_1 = \{0, \ldots, n_1\}$. Let g_0 be the alignment value of the process based on H_0. Let f_0, f_1 be the constantly F functions respectively on A_0 and A_1.

Define $n := \max(n_0, n_1)$, $A := \{0, \ldots, n+1\}$.

We specify H on $\{T, F\}^A$ such that the Herzberger process based on H has alignment ordinal $\alpha + \beta$. Consider f in $\{T, F\}^A$.

In case $f(n+1) = T$ or $f \upharpoonright A_0 \neq g_0$: let $t := 0$ if $f(n+1) = F$ and $t := 1$ if $f(n+1) = T$ and define:

$$H(f)(k) := \begin{cases} H_t(f \upharpoonright A_t)(k) \text{ if } k \leqslant n_t \\ f(n+1) \text{ if } n_t < k \leqslant n+1. \end{cases}$$

In case $f(n+1) = F$ and $f \upharpoonright A_0 = g_0$:

$$H(f)(k) := \begin{cases} H_1(f_1)(k) \text{ if } k \leqslant n_1 \\ T \text{ if } n_1 < k \leqslant n+1. \end{cases}$$

Assuming, e.g., $n_0 \leqslant n_1$ we may represent the process so obtained as follows:

Stage	0	1	2	$\cdots$	α	$\alpha+1$	$\alpha+2$	$\cdots$	$\alpha+\beta$
	f_0	$H_0(f_0)$	$H_0^2(f_0)$		g_0				
	F	F	F		F				
	.	.	.	$\cdots$	.	$H_1(f_1)$	$H_1^2(f_1)$	$\cdots$	g_1
	.	.	.		.				
	.	.	.		.				
	F	F	F		F	T	T		T

Clearly g_1/T is the alignment value of the process.

Case c: we make the same assumptions on H_0 and A_0 as under Case b. Let $A_1 = \{0, \ldots, n_0, n_0+1, n_0+2, n_0+3\}$. Instead of specifying the desired transition function H or $\{T, F\}^A$ we immediately give the process, leaving it to the reader to read off H. We assume $\alpha \neq 0$.

Stage	0	1	2	$\cdots$	α	$\alpha+1$	$\alpha+2$	$\cdots$	$a+1+\alpha$	$\alpha+1+\alpha+1$	$\cdots$	$\alpha.\omega$	$\alpha.\omega+1$
	f_0	$H_0(f_0)$	$H_0^2(f_0)$		g_0	f_0	$H_0(f_0)$		g_0	f_0		f_0	f_0
	F	F	F	$\ldots$	F	T	T	$\ldots$	T	F	$\ldots$	F	F
	F	T	T		T	F	F		F	T		F	F
	F	T	T		T	T	T		T	T		T	T

□

6.2.4. *A result by Vann McGee and John Burgess.* Consider the Belnap processes for $\mathbb{N}$ with Satisfaction. Vann McGee and John Burgess independently showed that we can get the same distinctions between sentences when we restrict ourselves to initial Belnap processes of length ω_1 as for the full processes. See McGee [1983], Burgess [1985].

7. NOTES ON INDUCTIVE DEFINITIONS AND SATISFACTION

The present section treats the connection between inductive definitions and Kripke fixed points. (Kripke fixed points are in this section: Strong Kleene Kripke fixed points for Sat over an acceptable structure.) As I see it this connection is two sided: the theory of Inductive Definitions can be used to study Kripke fixed points on the one hand, Kripke fixed points yield a natural formulation of many elementary facts about Inductive Definitions on the other.

This section is not a self contained introduction to Inductive Definitions. The reader is advised to consult Moschovakis [1974], when necessary.

7.1. *Preliminaries on partial sets*

First consider a partial model $\mathcal{N}$ with domain N and language L_0. Let $L_0(X)$ be L_0 extended with $\in$ and a free set variable X. A partial set X_0 of elements of N is simply a function from N to $\{*, T, F\}$. We write:

$$\text{‘}n \in X_0\text{’ for: } X_0(n) = T$$

$$\text{‘}n \notin X_0\text{’ for: } X_0(n) = F$$

$$\text{‘not } n \in X_0\text{’ for: } X_0(n) \neq T.$$

Note that a set in the classical sense can be considered as a total partial set.

For $\varphi \in L_0(X)$: $[\![\varphi]\!](\mathcal{N}(X_0), f)$ is defined in the obvious way using the Strong Kleene interpretation of the connectives.

C^+, C^- and C^* are functions on $\{*, T, F\}$ such that:

$$C^+(*) := F \quad C^-(*) := T \quad C^*(*) := *$$

$$C^+(T) := T \quad C^-(T) := T \quad C^*(T) := T$$

$$C^+(F) := F \quad C^-(F) := F \quad C^*(F) := *.$$

Define: $X_0^+ := C^+ \circ X_0$, $X_0^- := C^- \circ X_0$, $X_0^* := C^* \circ X_0$.

We write:

$$X_0 = {}^{+}X_1 := \text{for all } n \in N\ X_0(n) = T \Leftrightarrow X_1(n) = T$$

$$X_0 = {}^{-}X_1 := \text{for all } n \in N\ X_0(n) = F \Leftrightarrow X_1(n) = F.$$

Note: $X_0 = {}^{+}X_1 \Leftrightarrow X_0^+ = X_1^+$ and $X_0 = {}^{-}X_1 \Leftrightarrow X_0^- = X_1^-$.

Finally define for φ in L_0:

$\{x_1 | \varphi(x_1)\}(\mathcal{N}, f)$ is the partial set X_0 with

$X_0(n) := [\![\varphi(x_1)]\!](\mathcal{N}, f[n/x_1])$.

We collect some elementary facts in a lemma.

7.1.1. LEMMA. (i) *Let $\varphi(x_1)$ be a formula of L_0, let $X_0 := \{x_1 | \varphi(x_1)\}$ $(\mathcal{N}, f)$. We have for any $\Psi(X)$ of $L_0(X)$: $[\![\Psi(\{x_1 | \varphi(x_1)\})]\!]\ (\mathcal{N}, f) = [\![\Psi(X)]\!]\ (\mathcal{N}(X_0), f)$.*

(ii) *Consider a formula $\Psi(X)$ of $L_0(X)$ such that X occurs only on positive places in $\Psi(X)$, then:*

– if $X_0 = {}^{+}X_1$: $\mathcal{N}(X_0), f \vDash \Psi(X) \Leftrightarrow \mathcal{N}(X_1), f \vDash \Psi(X)$

– if $X_0 = {}^{-}X_1$: $\mathcal{N}(X_0), f \vDash \neg\Psi(X) \Leftrightarrow \mathcal{N}(X_1), f \vDash \neg\Psi(X)$.

Proof. By honest toil. □

Let $\varphi(x_1, X)$ be a formula of $L_0(X)$. Define $F(\varphi, \mathcal{N}, f)$ as an operator from partial subsets of N to partial subsets of N by: $F(\varphi, \mathcal{N}, f)(Y_0) := \{x_1 | \varphi(x_1, X)\}(\mathcal{N}(Y_0), f)$. We have:

7.1.2. LEMMA. *Suppose X occurs only positively in $\varphi(x_1, X)$. Let $F := F(\varphi, \mathcal{N}, f)$. Suppose X_0 is a fixed point of F.*

(i) $$X_0 = {}^{+}X_1 \Rightarrow F(X_1) = {}^{+}X_1,$$

$$X_0 = {}^{-}X_1 \Rightarrow F(X_1) = {}^{-}X_1.$$

(ii) *Suppose $\mathcal{N}$ is classical, then:*

$$F(X_0^+) = X_0^+,$$

$$F(X_0^-) = X_0^-.$$

If X_0 is the minimal fixed point in the partial sense then X_0^+ is the minimal fixed point in the classical sense and X_0 the maximal fixed point in the classical sense.

Proof. (i) $X_1 = {}^{+}X_0 = F(X_0) = {}^{+}F(X_1)$. (The last identity is by 7.1.1 (ii).)

(ii) By (i) and the fact that $\mathcal{N}(X^+)$ and $\mathcal{N}(X^-)$ are classical models. (The second point is more or less trivial.) □

7.1.3. LEMMA. *Under the conditions of* 7.1.2: *assume* U *is a sentence of* L_0 *such that* $[\![U]\!](\mathcal{N}, f) = *$. *Define* $G := F((\varphi \vee U), \mathcal{N}, f)$. *We have*: (i) *If* X_0 *is a fixed point of* F, *then* X_0^* *is a fixed point of* G. (ii) *For every fixed point* Y_0 *of* G *there is a fixed point* X_0 *of* F *such that* $Y_0 = X_0^*$.

Proof. (i) Clearly $X_0 = {}^+X_0^*$, hence

$$\begin{aligned}\mathcal{N}(X_0^*), f[n/x_1] \vDash \varphi(x_1, X) \vee U &\Leftrightarrow \mathcal{N}(X_0^*), f[n/x_1] \vDash \varphi(x_1, X)\\ &\Leftrightarrow \mathcal{N}(X_0), f[n/x_1] \vDash \varphi(x_1, X)\\ &\Leftrightarrow \mathcal{N}(X_0), f[n/x_1] \vDash x_1 \in X\\ &\Leftrightarrow \mathcal{N}(X_0^*), f[n/x_1] \vDash x_1 \in X.\end{aligned}$$

Moreover for all n $X_0^*(n) \neq F$ and $\{x_1 | \varphi(x_1, X) \vee U\}(\mathcal{N}, f)(n) \neq F$.

(ii) Let Y_0 be a fixed point of G. It can be shown that $F(Y_0) = {}^+Y_0$ and hence $Y \leqslant F(Y_0)$. Let X_0 be the minimal fixed point of F above Y_0. It is easily seen that $X_0 = {}^+Y_0$, or else X_0' with:

$$X_0'(n) := \begin{cases} T \text{ if } Y_0(n) = T\\ F \text{ if } X_0(n) = F\\ * \text{ otherwise}\end{cases}$$

would be a smaller fixed point. □

7.2. *Partial Sets and Satisfaction*

Consider a (classical!) acceptable structure $\mathcal{M}$ with domain M and language L_0. L is L_0 extended with 'Sat'.

We say that Y_0 is *representable* in $\mathcal{M}(S)$ if there is an assignment g and a formula $\varphi(x_1)$ in L such that: $Y_0 = \{x_1 | \varphi(x_1)\}(\mathcal{M}(S), g)$. We say that Y_0 is *semirepresentable* in $\mathcal{M}(S)$ if there is an assignment g and a formula $\varphi(x_1)$ in L such that: $Y_0 = {}^+\{x_1 | \varphi(x_1)\}(\mathcal{M}(S), g)$.

7.2.1. THEOREM. *Let* $\varphi(x_1, X)$ *be a formula of* $L(X)$. *Let* $\Psi(x_1)$ *be the fixed point formula provided by the Prediagonal Lemma* 3.4 *for* φ. $F := F(\varphi, \mathcal{M}, f)$. (i) *For every fixed point* S *of* $\Phi_{\mathcal{M}}$ $\{x_1 | \Psi(x_1)\}(\mathcal{M}(S), f)$ *is a fixed point of* F. (ii) *For every fixed point* X_0 *of* F *there is a minimal fixed point* S *of* $\Phi_{\mathcal{M}}$ *such that* $X_0 = \{x_1 | \Psi(x_1)\}(\mathcal{M}(S), f)$.

Proof. (i) By 3.5 and 7.1.1(i). (ii) Let m be the code of χ in the construction of Ψ (see the proof of 3.4). Define:

$$S_0(k, \ell) := \begin{cases} X_0(n) \text{ if } k = m \text{ and} \\ \quad \ell = \langle n, f(x_2), \ldots, f(x_s), m\rangle \\ * \text{ otherwise.} \end{cases}$$

Note that this gives: $X_0 = \{x_1 | \Psi(x_1)\}\ (\mathcal{M}(S_0), f)$.

We have:

$$\begin{aligned} \Phi_{\mathcal{M}}(S_0)(m, \langle n, f(x_2), \ldots, f(x_s), m\rangle) &= [\![\Psi(x_1)]\!]\ (\mathcal{M}(\Phi_{\mathcal{M}}(S_0)), \\ &\qquad f[n/x_1]) \\ &= [\![\varphi(x_1, \{x_1 | \Psi\})]\!] \\ &\qquad (\mathcal{M}(S_0), f[n/x_1]) \quad (3.4) \\ &= [\![\varphi(x_1, X)]\!]\ (\mathcal{M}(X_0), \\ &\qquad f[n/x_1]) \qquad (7.1.1(i)) \\ &= [\![x_1 \in X]\!]\ (\mathcal{M}(X_0), f[n/x_1]) \\ &= X_0(n). \end{aligned}$$

Conclude that $S_0 \leqslant \Phi_{\mathcal{M}}(S_0)$. Let S be the minimal fixed point above S_0. One easily shows (by an argument like the above) that S' with:

$$S'(k, \ell) := \begin{cases} X_0(n) \text{ if } k = m \text{ and} \\ \quad \ell = \langle n, f(x_2), \ldots, f(x_s), m\rangle \\ S(k, \ell) \text{ otherwise} \end{cases}$$

satisfies $\Phi_{\mathcal{M}}(S') \leqslant S'$. It follows that there is a fixed point between S_0 and S' and hence that $S' = S$.

Conclude $X_0 = \{x_1 | \Psi(x_1)\}\ (\mathcal{M}(S), f)$. □

7.2.2. THEOREM. *Under the conditions of* 7.2.1: *suppose that X occurs only positively in φ; then $\Psi(x_1)$ represents the minimal fixed point X_0 of F in S_*, the minimal fixed point of $\Phi_{\mathcal{M}}$, and in S_{int}, the maximal intrinsic fixed point of $\Phi_{\mathcal{M}}$.*

Proof. The first fact is immediate by 7.2.1. The second follows, because X_0^+ and X_0 are both fixed points of F, by 7.1.2 (ii), and hence represented by Ψ in certain fixed points S^+, S^- of $\Phi_{\mathcal{M}}$, by 7.2.1 (ii). □

7.2.3. COROLLARY. *Every (classical) inductive set over* $\mathcal{M}$ *is semi-representable in* S_* *and* S_{int}.

Proof. By 7.2.2 and 7.1.2 (ii). □

7.2.4. FACT. Under the conditions of 7.2.2. Let L be a Liar sentence in L.

$$G := F((\varphi \vee \mathrm{L}), \mathcal{M}(S_*), f).$$

(Note that for any fixed points S of $\Phi_{\mathcal{M}}$: $G = \mathrm{F}((\varphi \vee \mathrm{L}), \mathcal{M}(S), f)$.) Let $\Psi_{\mathrm{int}}(x_1)$ be the diagonal formula for $(\varphi \vee \mathrm{L})$ provided by the Prediagonal Lemma 3.4.

As is easily seen the full analogue of 7.2.1 holds for G and Ψ_{int}.

Let X_1 be any fixed point of G. Consider:

$$S_1(\mathrm{k}, \ell) := \begin{cases} X_1(n) \text{ if } k = m' \text{ and} \\ \quad \ell = \langle n, f(x_2), \ldots, f(x_s), m' \rangle \\ * \text{ otherwise.} \end{cases}$$

Here m' is the code of χ provided by the Prediagonal Lemma for $\varphi \vee \mathrm{L}$. Let S be the minimal fixed point above S_1.

Consider any fixed point T of $\Phi_{\mathcal{M}}$, by the analogue of 7.2.1 we see that S_1 and T are consistent. It is easily seen that $S_1 \sqcup T \leqslant \Phi_{\mathcal{M}}(S_1 \sqcup T)$. Let V be the minimal fixed point above $S_1 \sqcup T$. Clearly $S \leqslant V$, $T \leqslant V$. Hence S and T are consistent. Conclude: S, the minimal fixed point in which Ψ_{int} represents X_1, is intrinsic.

Consider any fixed point X_1 of G. Let X_0 be the minimal fixed point of $F = F(\varphi, \mathcal{M}, f)$. Let Z_0 be a fixed point of F such that $Z_0^* = X_1$. Clearly $X_0 \leqslant Z_0$ and hence $X_0^* \leqslant Z_0^* = X_1$. So X_0^* is the minimal fixed point of G. X_0^- is a fixed point of F, hence $(X_0^-)^*$ is a fixed point of G. Moreover if $X_1(n) = T$, then $Z_0(n) = T$, and so $X_0(n) \neq F$, hence $X_0^-(n) = T$ and $(X_0^-)^*(n) = T$. So $X_1(n) \leqslant (X_0^-)^*(n)$. Ergo $(X_0^-)^*$ is the maximal fixed point of G.

By the analogue of 7.2.1 we see: $\Psi_{\mathrm{int}}(x_1)$ represents $(X_0)^*$ and hence semirepresents X_0^+ in S_*; $\Psi_{\mathrm{int}}(x_1)$ represents $(X_0^-)^*$ and hence semi-represents X_0^- in S_{int}.

7.2.5. *Set and Antiset.* Let $\varrho(X)$ be in $L_0(X)$. Define $\varrho(\neg X)$ as the result of replacing in $\varrho(X)$ every occurrence of the form '$(t \in X)$' by '$\neg(t \in X)$'. Define $\breve{\varrho}(X)$ as $\neg\varrho(\neg X)$.

Consider a formula $\varphi(x_1, X)$ with X occurring only at positive places. Clearly in $\check{\varphi}(x_1, X)$, X occurs also only positively.

Let $F := F(\varphi, \mathcal{M}, f)$, $\check{F} := F(\check{\varphi}, \mathcal{M}, f)$. Let X_0 be the minimal fixed point of F; let $\check{X}_0$ be the minimal fixed point of $\check{F}$.

It may be shown that: $\check{X}_0(n) = \neg X_0(n)$. Let $\Psi(x_1)$ be the formula provided by the Prediagonal Lemma for $\varphi(x_1, X)$. It follows that $\neg\Psi(x_1)$ represents $\check{X}_0$ in S_*. Clearly $(\check{X}_0)^+(n) = \neg(X_0^-(n))$. We call $(\check{X}_0)^+$ the *antiset* of $(X_0)^+$. The antiset contains the negative information positively given by the inductive definition of $(X_0)^+$. $(\check{X}_0)^+$ is a classical inductive fixed point, hence X_0^- is coinductive. Conversely every complement of an inductive fixed point is $(X_0)^-$ for the minimal fixed point X_0 of some $F(\varphi, \mathcal{M}, f)$. This follows easily from the fact that $\check{\check{\varrho}}$ is equivalent in partial logic to ϱ.

7.2.6. COROLLARY. *The coinductive sets are semirepresentable in* $\mathcal{M}(S_{int})$.

Proof. Immediate from 7.2.4, 7.2.5. □

7.2.7. REMARKS. (i) In 7.2.5: X_0^- is the *kernel* of the operator defined by $\varphi(x_1, X)$, see Aczel [1977], pp. 749, 750 for examples.

(ii) A proof of the fact that the coinductive sets are semirepresentable in the maximal intrinsic fixed point of $\Phi_{\mathcal{M}}$ was shown to me by Vann McGee in correspondence. A stronger fact was proved by John Burgess and Saul Kripke, see 7.2.11.

(iii) In Manna and Shamir [1976] it is shown that the arithmetical sets are representable in the Maximal Intrinsic Fixed Point for the case of Recursion Theory.

7.2.8. DEFINITION. (i) Let R be a partial n-place relation in M. Define the classical subset $\mathrm{Clas}_n(R)$ as follows:

$$\mathrm{Clas}_n(R) := \{\langle x, y\rangle \mid x \in M \wedge ((y = 0 \wedge R((x)_1, \ldots, (x)_n) = T) \vee (y = 1 \wedge R((x)_1, \ldots, (x)_n) = F)\}.$$

Here 0 and 1 are certain designated elements of M, with $0 \neq 1$.

(ii) Let Y_0 be a classical subset of M. Define a partial n-place relation $\mathrm{Pa}_n(Y)$ by:

$$\mathrm{Pa}_n(Y_0)(\vec{m}) := \begin{cases} T \text{ if } \langle\langle m_1, \ldots, m_n\rangle, 0\rangle \in Y_0 \\ F \text{ if } \langle\langle m_1, \ldots, m_n\rangle, 1\rangle \in Y_0 \\ * \text{ otherwise.} \end{cases}$$

(iii) Let $\mathscr{A}_n := \langle$the partial n-place relations in M, $\leqslant\rangle$ and $\mathscr{B} := \langle$the classical subsets of M, $\subseteq\rangle$.

Clearly Clas_n and Pa_n are monotone respectively from $\mathscr{A}_n$ to $\mathscr{B}$ and from $\mathscr{B}$ to $\mathscr{A}_n$. Moreover $\text{Clas}_n(\text{Pa}_n(X_0)) \subseteq X_0$ and $\text{Pa}_n(\text{Clas}_n(R)) = R$. (In categorical terminology: $\mathscr{B}_n$ is a retract of $\mathscr{A}$; also: Clas_n is the left adjoint of Pa_n.)

7.2.9. THEOREM. (i) *$\text{Clas}_2(S_*)$ is inductive over $\mathscr{M}$.* (ii) *Every classical set that is semirepresentable over $\mathscr{M}(S_*)$ is inductive over $\mathscr{M}$.* (iii) *Every classical set representable over $\mathscr{M}(S_*)$ is hyperelementary over $\mathscr{M}$.*

Proof. (i) One can prove this using the Positive Induction Completeness Theorem (see Moschovakis [1974], p. 87) or directly as Kripke does in his footnote 24, see Kripke [1975], p. 706. We treat the latter proof, which is a minor adaptation of the construction in section 5B of Moschovakis [1974], pp. 67, 68. The reader is referred to Moschovakis' book for the details!

We assume PrVal of Moschovakis is defined on codes for the atomic formulas of L_0. Moreover we assume we have defined a term evaluating function $\text{Ev}(m, n)$, that evaluates the term *coded* by m relative to the assignment sequence n to its *uncoded* value. We code '$n \in X$' by $\langle 7, \ulcorner n \urcorner\rangle$.

Put: $\chi\,(x_1, X) :=$

$$\exists a \exists x \exists t[x_1 = \langle\langle a, x\rangle, t\rangle \wedge \text{Fmla}\,(a) \wedge \{\text{PrVal}\,(a, x, t) \vee$$

$$\vee\ (a = \langle 4, (a)_2\rangle \wedge \langle\langle (a)_2, x\rangle, 1\rangle \in X \wedge t = 0) \vee$$

$$\ldots$$

$$(a = \langle 7, (a)_2\rangle \wedge \exists z(\text{Ev}((a)_2, x) = z \wedge z \in X \wedge (z)_2 =$$

$$= 0 \wedge t = 0) \vee (a = \langle 7, (a)_2\rangle \wedge \exists z(\text{Ev}((a)_2, x) =$$

$$= z \wedge z \in X \wedge (z)_2 = 1 \wedge t = 1)\}].$$

Let X_0 be the classical minimal fixed point of $F(\chi, \mathscr{M}, f)$. One may show: $\text{Pa}_2(X_0)$ is a fixed point of $\Phi_{\mathscr{M}}$.

Thus $S_* \leqslant \text{Pa}_2(X_0)$. Clearly X_0 is in the range of Clas_2. Hence $X_0 = \text{Clas}_2(\text{Pa}_2(X_0))$. Moreover: $\text{Clas}_2(S_*) \subseteq X_0$.

On the other hand one may show: $\text{Clas}_2(S_*)$ is a fixed point of $F(\chi, \mathscr{M}, f)$. So $X_0 \subseteq \text{Clas}_2(S_*)$.

Conclude $X_0 = \text{Clas}_2(S_*)$. Hence $\text{Clas}_2(S_*)$ is inductive over $\mathscr{M}$.

(ii) Define functions $(\ )^0$ and $(\ )^1$ from L to L_0 as follows:

$$(P)^0 := P \text{ if } P \text{ is atomic in } L_0$$
$$(\mathrm{Sat}(t, n))^0 := (\langle\langle t, n\rangle, 0\rangle \in X)$$
$$(.)^0 \text{ commutes with } \vee, \wedge, \forall, \exists$$
$$(v \to \varrho)^0 := ((v)^1 \vee (\varrho)^0)$$
$$(\neg v)^0 := (v)^1$$
$$(P)^1 := \neg P \text{ if } P \text{ is atomic in } L_0$$
$$(\mathrm{Sat}(t, n))^1 := (\langle\langle t, n\rangle, 1\rangle \in X)$$
$$(v \wedge \varrho)^1 := ((v)^1 \vee (\varrho)^1)$$
$$(v \vee \varrho)^1 := ((v)^1 \wedge (\varrho)^1)$$
$$(v \to \varrho)^1 := ((v)^0 \wedge (\varrho)^1)$$
$$(\neg v)^1 := (v)^0$$
$$(\forall x\varrho)^1 := (\exists x(\varrho)^1)$$
$$(\exists x\varrho)^1 := (\forall x(\varrho)^1).$$

One shows by induction on ϱ, using (i): for X_0 as in (i):

$$\mathscr{M}(S_*), f \models \varrho \Leftrightarrow \mathscr{M}(X_0), f \models (\varrho)^0.$$

Suppose Y_0 is semirepresentable in $\mathscr{M}(S_*)$ by $\varrho(x_1)$ then:

$$m \in Y_0 \Leftrightarrow \mathscr{M}(S_*), f[m/x_1] \models \varrho(x_1)$$
$$\Leftrightarrow \mathscr{M}(X_0), f[m/x_1] \models (\varrho(x_1))^0.$$

X occurs only positively in $(\varrho(x_1))^0$, so it follows from Theorem 1C.A of Moschovakis [1974], p. 17, that Y_0 is inductive over $\mathscr{M}$.

(iii) Immediate from (ii). □

7.2.10. FACT. Let S_0 be the nowhere defined binary relation on M. The sequence $((S_0)_\alpha)_{\alpha \in ON}$ based on $\Phi_{\mathscr{M}}$ goes up to S_*. Let $\alpha_* =$ (the minimal α such that $(S_0)_\alpha = S_*$). Then $\alpha_* = k^{\mathscr{M}}$, as defined in Moschovakis [1974], p. 31. It follows that when $\mathscr{M}$ is the usual structure of the Natural Numbers, the minimal fixed point is reached at stage ω_1^{CK}.

Sketch of the proof. Consider $\varphi(x_1, X)$ in $L_0(X)$, where X occurs only positively in φ. Let $\Psi(x_1)$ be the formula provided by the Prediagonal

Lemma for φ. Let $X^\alpha := I_\varphi^{<\alpha}$. (See Moschovakis [1974], p. 7.) One shows by induction on α:

$$X^\alpha = \{x_1 | \Psi(x_1)\} \, (\mathcal{M}((S_0)_\alpha, f)).$$

It follows that $k^{\mathcal{M}} \leqslant \alpha_*$. On the other hand consider χ as defined in the proof of 7.2.9 (i). Let $X_0^\alpha := I_\chi^{<\alpha}$. One shows: $\mathrm{Pa}_2(X_0^\alpha) \leqslant (S_0)_\alpha$, ergo $S_* = \mathrm{Pa}_2(X_0) = \mathrm{Pa}_2(X_0{}^{k^{\mathcal{M}}}) \leqslant (S_0)_{k^{\mathcal{M}}} \leqslant S_*$. Hence $S_* = (S_0)_{k^{\mathcal{M}}}$, so $\alpha_* \leqslant k^{\mathcal{M}}$; combining: $\alpha_* = k^{\mathcal{M}}$.

7.2.11. *Perspectives.* For further information on the complexity of fixed points and classes of sentences connected with the H/G/B processes the reader is referred to Burgess [1985]. I quote some of the results in this paper:

- Truth in the SK maximal intrinsic fixed point over arithmetic is complete Σ_1^1-in-a-Π_1^1-parameter, i.e. complete Σ_1^1-in-Kleene's-$\mathcal{O}$ (The upperbound side of this theorem is due to Kripke. The lower bound side to Burgess. McGee independently found the somewhat weaker upper-bound Δ_2^1.)
- Truth in the VF maximal intrinsic fixed point over arithmetic is at least Π_2^1, at most Σ_2^1-in-a-Π_2^1-parameter.
- Truth in some VF fixed point over arithmetic is complete Σ_2^1.
- Eventually stable truth in all *H* or *G* or *B* sequences over arithmetic is complete Π_2^1. (Here *H* is taken with varying starting points.)

Moreover John Burgess announces the following result (to be presented in a sequel to The Truth is Never Simple on Quasi Inductive Definitions):

- Eventually stable truth in the sequence starting with nothing true for *H* or *G* over arithmetic is complete $\Sigma_1(\mathrm{L}_\alpha)$, where α is the least admissible such that there is a $\beta > \alpha$ admissible, with $\mathrm{L}_\alpha \prec_{\Sigma_2} \mathrm{L}_\beta$.

Rijksuniversiteit Utrecht

NOTES

[1] A type theory for truth or satisfaction coincides with a hierarchy of languages, where each step just adds semantic terms and where each language is really part of, not just translatable in, the subsequent language.

[2] This is not to claim that sentences *always* express propositions.

[3] I think Priest's paper is seriously flawed by a number of unsubstantiated and uninformed claims about Gödel's incompleteness theorems.

[4] We could have used relatively maximal intrinsic elements instead of maximal elements in this argument, thereby avoiding the use of Zorn's Lemma, see 2.18.

REFERENCES

Aczel, P.: 1977, 'An introduction to inductive definitions', in J. Barwise (ed.), 1977, pp. 739–782.

Aczel, P.: 1980, 'Frege structures and the notions of proposition, truth and set', in J. Barwise *et al.* (eds.), 1980, pp. 31–59.

Aczel, P. and Feferman, S.: 1980, 'Consistency of the unrestricted abstraction principle using an intensional equivalence operator', in J. P. Seldin and J. R. Hindley (eds.) 1980, pp. 67–98.

Barwise, J. (ed.): 1977, *Handbook of Mathematical Logic*, North Holland, 1977.

Barwise, J., Keisler, H. J., and Kunen, K. (eds.): 1910, *The Kleene Symposium*, North Holland, Amsterdam.

Behmann, H.: 1931, 'Zu den Widersprüchen der Logik und der Mengenlehre', *Jahresbericht der Deutschen Mathematiker-Vereinigung* **40**, 37–48.

Belnap, N. D., Jr.: 1982, 'Gupta's rule of revision theory of truth', *Journal of Philosophical Logic* **11**, 103–116.

Blau, U.: 1983, 'Vom Henker, vom Lügner und von ihrem Ende', *Erkenntnis* **19**, 27–44.

Blau, U.: 1984, 'Wahrheit von innen und aussen', unpublished manuscript, Seminar für Philosophie, Logik und Wissenschaftstheorie, Univ. München.

Bochvar, D. A.: 1981, 'On a three-valued logical calculus and its application to the analysis of the classical extended functional calculus', English translation in: *History and Philosophy of Logic* **2**, 87–112.

Bunder, M. W.: 1980, 'The naturalness of illative combinatory logic as a basis for mathematics', in J. P. Seldin and J. R. Hindley (eds.), 1980, pp. 55–64.

Burge, T.: 1979, 'Semantical paradox', *Journal of Philosophy* **76**, 169–198. Reprinted in Martin [1984], pp. 83–118.

Burge, T.: 1982, 'The liar paradox: Tangles and chains', *Philosophical Studies* **41**, 353–366.

Burgess, J. P.: to appear, 'The truth is never simple', Department of Philosophy, Princeton University, 1879 Hall, Princeton, New Jersey 08544.

Cantini, A.: 1979, '"Tarski extensions" of theories', *preprint*, Mathematisches Institut, München.

Cantini, A.: 1980, 'A note on three-valued logic and Tarski theorem on truth definitions', *Studia Logica* **39**, 405–414.

Chihara, C. S.: 1973, *Ontology and the Vicious Circle Principle*, Cornell University Press, Ithaca & London.

Chihara, C. S.: 1979, 'The semantic paradoxes; A diagnostic investigation', *Philosophical Review* **88**, 590–618.

Chihara, C. S.: 1984, 'Priest, the Liar, and Gödel', *Journal of Philosophical Logic* **13**, 117–124.

Cowan, D. A.: 1980, *Language & Negation*, Joseph Publishing Company, San Mateo (Ca.).

Curry, H. B., Feys, R., Craig, W.: 1958, *Combinatory Logic, Vol. I*, North Holland, Amsterdam.

Davidson, D.: 1969, 'On saying that', in D. Davidson and J. Hintikka, (eds.), *Words and Objections, Essays on the Work of W. V. Quine*, Reidel, Dordrecht, pp. 158–174.

Dowden, B. H.: 1979, '*The Liar Paradox and Tarski's Undefinability theorem*', dissertation, Stanford University.

Dowden, B. H.: 1984, 'Accepting inconsistencies from the paradoxes', *Journal of Philosophical Logic* **13**, 125–130.

Feferman, S.: 1967, 'Set-theoretical foundations of category theory', *Reports of the Midwest Category Seminar. III*, Lecture Notes in Mathematics, vol. 106, Springer, Berlin, pp. 207–246.

Feferman, S.: 1975a, 'A language and axioms for explicit mathematics', *Algebra and Logic*, Lecture Notes in Mathematics, vol. 450, Springer, Berlin, pp. 87–139.

Feferman, S.: 1975b, 'Investigative logic for theories of partial functions and relations. I and II', unpublished notes, Stanford University, Stanford, California.

Feferman, S.: 1975c, 'Non-extensional type free theories of partial operations and classifications. I', ⊧ *ISILC Proof Theory Symposion, Kiel, 1974*, Lecture Notes in Mathematics, vol. 500, Springer, Berlin, pp. 73–118.

Feferman, S.: 1976, 'Comparison of some type-free semantic and mathematical theories', unpublished notes, Stanford University, Stanford, California.

Feferman, S.: 1977, 'Categorical foundations and foundations of category theory', in R. Butts and J. Hintikka (eds.), 1977, *Logic, Foundations of Mathematics and Computability Theory*, D. Reidel, Dordrecht, pp. 149–169.

Feferman, S.: 1979, 'Constructive theories of functions and classes', in M. Boffa, D. van Dalen and K. McAloon (eds.), *Logic Colloquium '78*, North Holland, Amsterdam, pp. 159–224.

Feferman, S.: 1984, 'Toward useful type-free theories. I', *Journal of Symbolic Logic* **49**, 75–111. Reprinted in Martin [1984], pp. 237–288.

Finsler, P.: 1975, *Aufsätze zur Mengenlehre*, Wissenschaftliche Buchgesellschaft, Darmstadt.

Fitch, F. B.: 1948, 'An extension of basic logic', *Journal of Symbolic Logic* **13**, 95–106.

Fitch, F. B.: 1963, 'The system Cd of combinatory logic', *Journal of Symbolic Logic* **28**, 87–97.

Fitch, F. B.: 1966, 'A consistent modal set theory', abstract, *Journal of Symbolic Logic* **31**, 701.

Fitch, F. B.: 1974, *Elements of Combinatory Logic*, Yale U. P., New Haven & London.

Fitch, F. B.: 1980, 'A consistent combinatory logic with an inverse to equality', *Journal of Symbolic Logic* **45**, 529–543.

Frege, G.: 1975, *Funktion, Begriff, Bedeutung*, Vandenhoeck & Ruprecht, Göttingen.

Gilmore, P. C.: 1974, 'The consistency of partial set theory without extensionality', in *Axiomatic Set Theory*, Proc. of Symposia in Pure Mathematics, vol. 13, Part II, AMS, Providence R.I., pp. 147–153.

Gilmore, P. C.: 1980, 'Combining unrestricted abstraction with universal quantification', in J. P. Seldin and J. R. Hindley (eds.), 1980, pp. 99–123.

Gödel, K.: 1944, 'Russell's mathematical logic', in P. A. Schilpp (ed.), *The Philosophy of Bertrand Russell*, Tudor, New York, pp. 123–153. Reprinted in P. Benacerraf and H. Putnam, (eds.), 1964, *Philosophy of Mathematics*, Prentice Hall, Englewood Cliffs, New Jersey, pp. 211–232.

Grover, D.: 1977, 'Inheritors and paradox', *The Journal of Philosophy* **74**, 590–604.

Gupta, A.: 1982, 'Truth and paradox', *Journal of Philosophical Logic* **11**, 1–60. Reprinted in Martin [1984], pp. 175–236.

Gupta, A. and Martin, R. L.: 1984, 'A fixed point theorem for the weak Kleene valuation scheme', *Journal of Philosophical Logic* **13**, 131–135.

Gupta, A.: 1987, 'The meaning of truth', in E. LePore, *New Directions in Semantics*, Academic Press, London, pp. 453–480.

Haack, S.: 1978, *Philosophy of Logic*, Cambridge University Press, Cambridge.

Hansson, B.: 1978, 'Paradoxes in a semantic perspective', in J. Hintikka, I. Niiniluoto, and E. Saarinen, (ed.), *Essays on Mathematical and Philosophical Logic*, D. Reidel, Dordrecht, pp. 371–385.

Herzberger, H. G.: 1970, 'Paradoxes of grounding in semantics', *The Journal of Philosophy* **67**, 145–167.

Herzberger, H. G.: 1980, 'Notes on periodicity', unpublished manuscript.

Herzberger, H. G.: 1982a, 'Naive semantics and the Liar paradox', *The Journal of Philosophy* **79**, 479–497.

Herzberger, H. G.: 1982b, 'Notes on naive semantics', *Journal of Philosophical Logic* **11**, 61–102.

Hughes, G. E.: 1982, *John Buridan on Self-Reference*, Cambridge U.P., Cambridge.

Kaplan, D.: 1978, 'On the logic of demonstratives', *Journal of Philosophical Logic* **8**, 81–98.

Kindt, W.: 1976, 'Über Sprachen mit Wahrheitsprädikat', in C. Habel and S. Kanngiesser (eds.), 1978, *Sprachdynamik und Sprachstruktur*, Niemeyer, Tübingen.

Kindt, W.: 1978, 'The introduction of truth predicates into first-order languages', in F. Guenthner, and S. J. Schmidt, (eds.), *Formal Semantics and Pragmatics for Natural Languages*, D. Reidel, Dordrecht, pp. 359–371.

Koyré, A.: 1947, *Épiménide le Menteur*, Hermann et Cie, Paris.

Kripke, S.: 1975, 'Outline of a theory of truth', *The Journal of Philosophy* **72**, 690–716.

Lopez-Escobar, E. G. K.: 1979, 'A formal logic for the study of paradoxes', Technical Report TR 79–11, University of Maryland, Department of Mathematics, College Park, Maryland.

Manna, Z., and Shamir, A.: 1976, 'The theoretical aspects of the optimal fixed-point', *Siam Journal of Computing* **5**, 414–426.

Manna, Z. and Shamir, A.: 1978, 'The convergence of functions to fixed-points of recursive definitions', *Theoretical Computer Science* **6**, 109–141.

Martin, R. L. (ed.): 1970, *The Paradox of the Liar*, Yale University Press, New Haven (Second edition: 1978, Ridgeview Pub., Atascadero (Ca.)).

Martin, R. L. and Woodruff, P. W.: 1975, 'On representing "true-in-*L*" in *L*', *Philosophia* **5**, 217–221. Reprinted in Martin [1984], pp. 47–52.

Martin, R. L. (ed.): 1984. *Recent Essays on Truth and the Liar paradox*, Oxford University Press, Oxford.

Martin, R. M.: 1979, 'The truth about Kripke's "Truth"', in R. M. Martin: *Pragmatics, Truth and Language*, D. Reidel, Dordrecht, pp. 173–180.

McGee, V.: to appear, 'Technical notes on three systems of naive semantics', Group in Logic and the Methodology of Science U. C. Berkeley, CA 94720.

Moschovakis, Y. N.: 1974, *Elementary Induction on Abstract Structures*, North Holland, Amsterdam.

Moschovakis, Y. N.: 1977, 'On the basic notions in the theory of induction', in R. E. Butts and J. Hintikka (eds.): *Logic, Foundations of Mathematics and Computability Theory*, D. Reidel, Dordrecht, pp. 207–236.

Myhill, J.: 1984, 'Paradoxes', *Synthese* **60**, 129–143.

Nepeĭvoda, N. N.: 1973, 'A new notion of predicative truth and definability', *Mathematičeskie Zumetki* **13**, 735–745; English translation: *Mathematical Notes of the Academy of Sciences of the USSR* **13**, pp. 439–445.

Parsons, C.: 1974, 'The liar paradox', *Journal of Philosophical Logic* **3**, 381–42. Reprinted in Martin [1984], pp. 9–46.

Parsons, C.: 1982, 'Postscript to "The liar paradox"', to appear in *Mathematics in Philosophy*, Cornell University Press, Ithaca, N.Y.

Parsons, T.: 1984, 'Assertion, denial, and the Liar paradox', *Journal of Philosophical Logic* **13**, 137–152.

Pollock, J. L.: 1977, 'The liar strikes back', *The Journal of Philosophy* **74**, 604–606.

Popper, Sir K. R.: 1954, 'Self reference and meaning in ordinary language', *Mind* **63**, 162–169. Reprinted in Sir K. R. Popper: 1972, *Conjectures and Refutations*, Routledge and Kegan Paul, London, pp. 304–311.

Prawitz, D.: 1965, *Natural Deduction: A Prooftheoretical Study*, Almqvist and Wiksell, Stockholm.

Priest, G.: 1979, 'The logic of paradox', *Journal of Philosophical Logic* **8**, 219–241.

Priest, G.: 1984, 'Logic of paradox revisited', *Journal of Philosophical Logic* **13**, 153–179.

Quine, W. V. O.: 1962, 'Paradox', *Scientific American of April 1962*, 84–96.

Ramsey, F. P.: 1925, 'The foundations of mathematics', *Proceedings of the London Mathematical Society, Nr. 2, Vol. 25*, pp. 338–384, also in F. P. Ramsey: 1978, *Foundations*, Routledge & Kegan Paul, London, pp. 152–212.

Scott, D. S.: 1960, 'Combinators and classes', in *λ-Calculus and Computer Science Theory*, Lecture Notes in Computer Science, vol. 37, Springer Verlag, Berlin, pp. 1–26.

Seldin, J. P. and Hindley, J. R. (eds.): 1980, '*To H. B. Curry: Essays on Combinatory Logic, Lambda Calculus and Formalism*, Academic Press, New York.

Skyrms, B.: 1970, 'Return of the Liar, three valued logic and the concept of truth', *American Philosophical Quarterly* **7**, 153–161.

Skyrms, B: 1984, 'Intensional aspects of semantical self-reference', in Martin [1984], pp. 119–132.

Smoryński, C.: 1977, 'The incompleteness theorems', in Barwise [1977], pp. 821–865.

Smullyan, R. M.: 1957, 'Languages in which self-reference is possible', *Journal of Symbolic Logic* **22**, 55–67. Reprinted in J. Hintikka (ed.), 1969, *The Philosophy of Mathematics*, Oxford University Press, Oxford, pp. 64–77.

Smullyan, R. M.: 1981, *What Is the Name of This Book?*, Penguin Books, Harmondsworth.

Smullyan, R. M.: 1984, 'Chameleonic Languages', *Synthese* **60**, 201–224.

Tarski, A.: 1944, 'The semantic conception of truth', *Philosophy and Phenomenological Research* **4**, 341–375.

Tarski, A.: 1956, 'The concept of truth in formalized languages', in A. Tarski: *Logic, Semantics, Metamathematics*, Oxford University Press, Oxford, 152–278.

Tarski, A.: 1969, 'Truth and proof', *Scientific American of June 1969*, pp. 63–70, 75–77.

Thomason, R. H.: 1969, 'A semantical study of constructive falsity', *Zeitschrift für mathematische Logik und Grundlagen der Mathematik* **15**, 247–257.

Thomason, R. H.: 1975, 'Necessity, quotation and truth: An indexical theory', *Philosophia* **5**, 219–241.

Van Fraassen, B. C.: 1968, 'Presupposition, implication and self-reference', *Journal of Philosophy* **65**, 135–152.

Van Fraassen, B. C.: 1972, 'Inference and selfreference', in D. Davidson and G. Harman (eds.), *Semantics of Natural Language*, D. Reidel, Dordrecht, pp. 695–708.

Visser, A.: 1984, 'Four valued semantics and the Liar', *Journal of Philosophical Logic* **13**, 181–212.

Woodruff, P. W.: 1984, 'Paradox, truth and logic, Part 1', *Journal of Philosophical Logic* **13**, 213–232.

Yablo, S.: 1982, 'Grounding, dependence and paradox', *Journal of Philosophical Logic* **11**, 117–137.

Yablo, S.: 1985, 'Truth and reflection', *Journal of Philosophical Logic* **14**, 297–349.

NAME INDEX

SUBJECT INDEX

CONTENTS TO VOLUME I

CONTENTS TO VOLUME II

CONTENTS TO VOLUME III